SCOTT, FORESMAN

Discover SCIENCE

Teacher's Annotated Edition Authors

Robert G. Guy
Classroom Teacher
Big Lake Elementary School
Sedro-Wooley, Washington

Dr. Robert J. Miller
Professor of Science Education
Eastern Kentucky University
Richmond, Kentucky

Mary Jane Roscoe
Teacher and Team Coordinator
Fairwood Alternative Elementary
School of Individually
Guided Education
Columbus, Ohio

Anita Snell
Elementary Coordinator for Early
Childhood Education
Spring Branch Independent
School District
Houston, Texas

Sheri L. Thomas
Classroom Teacher
McLouth Unified School
District #342
McLouth, Kansas

Pupil's Edition Authors

Dr. Michael R. Cohen
Professor of Science and
Environmental Education
School of Education
Indiana University
Indianapolis, Indiana

Dr. Timothy M. Cooney
Chairperson K-12 Science Program
Malcolm Price Laboratory School
University of Northern Iowa
Cedar Falls, Iowa

Cheryl M. Hawthorne
Science Curriculum Specialist
Mathematics, Engineering, Science
Achievement Program (MESA)
Stanford University
Stanford, California

Dr. Alan J. McCormack
Professor of Science Education
San Diego State University
San Diego, California

Dr. Jay M. Pasachoff
Director, Hopkins Observatory
Williams College
Williamstown, Massachusetts

Dr. Naomi Pasachoff
Research Associate
Williams College
Williamstown, Massachusetts

Karin L. Rhines
Science/Educational Consultant
Valhalla, New York

Dr. Irwin L. Slesnick
Professor of Biology
Western Washington University
Bellingham, Washington

Scott Foresman and Company

Sunnyvale, California
Tucker, Georgia
Glenview, Illinois
Oakland, New Jersey
Carrollton, Texas

Cooperative Learning Consultant

Dr. Robert E. Slavin
Director, Elementary School Program
Center for Research on Elementary and Middle Schools
Johns Hopkins University
Baltimore, Maryland

Gifted Education Consultants

Hilda P. Hobson
Teacher of the Gifted
W.B. Wicker School
Sanford, North Carolina

Christine Kuehn
Assistant Professor of Education
University of South Carolina
Columbia, South Carolina

Nancy Linkel York
Teacher of the Gifted
W.B. Wicker School
Sanford, North Carolina

Special Education Consultants

Susan E. Affleck
Classroom Teacher
Salt Creek Elementary School
Elk Grove Village, Illinois

Dr. Dale R. Jordan
Director
Jordan Diagnostic Center
Oklahoma City, Oklahoma

Dr. Shirley T. King
Learning Disabilities Teacher
Helfrich Park Middle School
Evansville, Indiana

Jeannie Rae McCoun
Learning Disabilities Teacher
Mary M. McClelland Elementary School
Indianapolis, Indiana

Thinking Skills Consultant

Dr. Joseph P. Riley II
Professor of Science Education
University of Georgia
Athens, Georgia

Reading Consultants

Patricia T. Hinske
Reading Specialist
Cardinal Stritch College
Milwaukee, Wisconsin

Dr. Robert A. Pavlik
Professor and Chairperson of Reading/Language Arts Department
Cardinal Stritch College
Milwaukee, Wisconsin

Dr. Alfredo Schifini
Reading Consultant
Downey, California

Special Content Consultant

Dr. Abraham S. Flexor
Science Education Consultant
Boulder, Colorado

Health Consultant

Dr. Julius B. Richmond
John D. MacArthur Professor of Health Policy
Director, Division of Health Policy Research and Education
Harvard University
Advisor on Child Health Policy
Children's Hospital of Boston
Boston, Massachusetts

Safety Consultant

Dr. Jack A. Gerlovich
Science Education Safety Consultant/Author
Des Moines, Iowa

Process Skills Consultant

Dr. Alfred Devito
Professor Emeritus Science Education
Purdue University
West Lafayette, Indiana

Activity Consultants

Edward Al Pankow
Teacher
Petaluma City Schools
Petaluma, California

Valerie Pankow
Teacher and Writer
Petaluma City Schools
Petaluma, California

Science and Technology Consultant

Dr. David E. Newton
Adjunct Professor—Science and Social Issues
University of San Francisco
College of Professional Studies
San Francisco, California

ISBN: 0-673-35693-0

123456789-GBC-009998979695949392

Reviewers and Content Specialists

Dr. Ramona J. Anshutz
Science Specialist
Kansas State Department of Education
Topeka, Kansas

Teresa M. Auldridge
Science Education Consultant
Amelia, Virginia

Annette M. Barzal
Classroom Teacher
Willetts Middle School
Brunswick, Ohio

James Haggard Brannon
Classroom Teacher
Ames Community Schools
Ames, Iowa

Priscilla L. Callison
Science Teacher
Topeka Adventure Center
Topeka, Kansas

Rochelle F. Cohen
Education Coordinator
Indianapolis Head Start
Indianapolis, Indiana

Linda Lewis Cundiff
Classroom Teacher
R. F. Bayless Elem. School
Lubbock, Texas

Dr. Patricia Dahl
Classroom Teacher
Bloomington Oak Grove Intermediate School
Bloomington, Minnesota

Audrey J. Dick
Supervisor, Elementary Education
Cincinnati Public Schools
Cincinnati, Ohio

Nancy B. Drabik
Reading Specialist
George Washington School
Wyckoff, New Jersey

Bennie Y. Fleming
Science Supervisor
Providence School District
Providence, Rhode Island

Mike Graf
Classroom Teacher
Branch Elementary School
Arroyo Grande, California

Thelma Robinson Graham
Classroom Teacher
Pearl Spann Elem. School
Jackson, Mississippi

Dr. Claude A. Hanson
Science Supervisor
Boise Public Schools
Boise, Idaho

Dr. Jean D. Harlan
Psychologist, Early Childhood Consultant
Lighthouse Counseling Associates
Racine, Wisconsin

Dr. Rebecca P. Harlin
Assistant Prof. of Reading
State University of New York—Geneseo
Geneseo, New York

Richard L. Ingraham
Professor of Biology
San José State University
San José, California

Ron Jones
Science Coordinator
Salem Keizer Public Schools
Salem, Oregon

Sara A. Jones
Classroom Teacher
Burroughs-Molette Elementary School
Brunswick, Georgia

Dr. Judy LaCavera
Director of Curriculum and Instruction
Learning Alternatives
Vienna, Ohio

Jack Laubisch
Science, Health, Outdoor Education Coordinator
West Clermont Local School District
Amelia, Ohio

Douglas M. McPhee
Classroom Teacher/Consultant
Del Mar Hills Elem. School
Del Mar, California

Larry Miller
Classroom Teacher
Caldwell Elementary School
Caldwell, Kansas

Jan Murphy
Classroom Teacher
Rosemeade Elem. School
Carrollton, Texas

Sam Murr
Teacher—Gifted Science
Mid Del Schools
Midwest City—Del City, Oklahoma

Janet Nakai
Classroom Teacher
Community Consolidated School District #65
Evanston, Illinois

Patricia Osborne
Classroom Teacher
Valley Heights Elementary School
Waterville, Kansas

Elisa Pinzón-Umaña
Classroom Teacher
Coronado Academy
Albuquerque, New Mexico

Dr. Jeanne Phillips
Director of Curriculum and Instruction
Meridian Municipal School District
Meridian, Mississippi

Maria Guadalupe Ramos
Classroom Teacher
Metz Elementary School
Austin, Texas

Elissa Richards
Math/Science Teacher Leader
Granite School District
Salt Lake City, Utah

Sister Mary Christelle Sawicki, C. S. S. F.
Science Curriculum Coordinator
Department of Catholic Education
Buffalo, New York

Linda Shepard
Classroom Teacher
Oscar Hinger School
Canyon, Texas

Ray E. Smalley
Classroom Teacher/Science Specialist
Cleveland School of Science
Cleveland, Ohio

Norman Sperling
Chabot Observatory
Oakland, California

Lisa D. Torres
Science Coordinator
Lebanon School District
Lebanon, New Hampshire

Alice C. Webb
Early Childhood Resource Teacher
Primary Education Office
Rockledge, Florida

Tina Ziegler
Classroom Teacher
Evanston, Illinois

Contents

Discover Science . . . for a lifetime

Complete science content coverage . . . PLUS opportunities to develop the skills needed to

- *understand how science works*
- *learn new concepts*
- *solve problems*
- *and make decisions in today's technological society*

Scott, Foresman
Discover SCIENCE
Scott, Foresman
Discover SCIENCE
Scott, Foresman
Discover SCIENCE
Scott, Foresman
Discover SCIENCE

Series Scope and Sequence

Content in Discover SCIENCE *is organized into four major units and several content strands. Each unit and strand is developed continuously throughout the program.*

Unit	Strand	Kindergarten	Grade 1	Grade 2
Life Science	*Plants*	**Chapter 4** Seeds and Plants	**Chapter 4** Learning About Plants	**Chapter 1** How Plants Are Different
	Animals	**Chapter 5** Animals (growth, habitats)	**Chapter 5** Learning About Animals (differences, pets)	**Chapter 2** How Animals Are Different
	Ecology	**Chapter 3** Living and Nonliving (characteristics and uses of)	**Chapter 3** Living and Nonliving (needs of living things)	**Chapter 3** Life on Earth Long Ago
Physical Science	*Matter*	**Chapter 6** Comparing and Grouping Matter	**Chapter 6** Grouping Things (solids, liquids, gases)	**Chapter 4** Matter Around You (grouping, changing)
	Energy	**Chapter 7** Heat and Cold **Chapter 8** Sound **Chapter 9** Movement	**Chapter 7** Light, Sound, and Heat **Chapter 8** Moving and Working	**Chapter 5** Heat, Light, and Sound **Chapter 6** Machines and Electricity
Earth Science	*Earth*	**Chapter 12** Seasons **Chapter 13** Helping Our World (conservation)	**Chapter 9** The Earth (landforms, water, air)	**Chapter 7** Water and Air
	Weather	**Chapter 11** Weather (air, wind, temperature)	**Chapter 10** Weather and Seasons (kinds, changes)	**Chapter 8** Changes in Weather
	Space	**Chapter 10** Earth and Sky	**Chapter 11** The Sky (sun, moon, stars)	**Chapter 9** The Sun and Other Stars
Human Body		**Chapter 1** Your Senses **Chapter 2** Growing and Changing (body parts, teeth, eating, exercising)	**Chapter 1** Your Senses (learning, using senses) **Chapter 2** Growing and Changing	**Chapter 10** How Your Body Works **Chapter 11** Keeping Healthy

Grade 3	Grade 4	Grade 5	Grade 6
Chapter 1 Plant Growth	**Chapter 1** Flowering Plants	**Chapter 2** Plant Processes	**Chapter 4** Plant Response
Chapter 2 How Animals Grow and Change	**Chapter 2** Animal Behavior	**Chapter 3** Invertebrates and Vertebrates	**Chapter 1** Cells and Heredity
Chapter 3 Living Things Need Each Other **Chapter 4** How People Affect Plants and Animals	**Chapter 3** Food Chains and Food Webs **Chapter 4** Animals and Plant Adaptations	**Chapter 1** Classifying Living Things **Chapter 4** Populations and Communities	**Chapter 2** The Fossil Record **Chapter 3** Change Through Time **Chapter 5** Ecosystems and Biomes
Chapter 5 Properties of Matter	**Chapter 5** Measuring Matter (length, volume, mass, density)	**Chapter 5** Investigating Matter (properties, composition) **Chapter 6** Heat and Matter	**Chapter 6** Structure of Matter **Chapter 7** Investigating Compounds
Chapter 6 Work and Machines **Chapter 7** Forms of Energy **Chapter 8** Sound	**Chapter 6** Work and Energy **Chapter 7** Electricity and Magnetism **Chapter 8** Light and Sound	**Chapter 7** Changing Forms of Energy **Chapter 8** Energy Resources	**Chapter 8** Gravity and Motion **Chapter 9** Electrical Energy **Chapter 10** Investigating Light and Sound
Chapter 9 Rocks and Soil **Chapter 10** Changes in the Earth	**Chapter 10** Changes in Landforms **Chapter 11** Oceans	**Chapter 9** Earth's Changing Crust **Chapter 10** Protecting the Environment	**Chapter 12** Earth's Moving Plates **Chapter 13** Resources and Conservation
Chapter 11 Clouds and Storms	**Chapter 9** Measuring Weather Conditions	**Chapter 11** Climate	**Chapter 11** Forecasting Weather
Chapter 12 The Sun, Moon, and Planets	**Chapter 12** Movement in the Solar System	**Chapter 12** Mapping the Stars	**Chapter 14** Exploring Space
Chapter 13 The Body's Support **Chapter 14** Your Body's Health Needs	**Chapter 13** Digestion and Circulation **Chapter 14** Your Brain and Your Sense Organs	**Chapter 13** Body Support, Movement, and Growth **Chapter 14** Respiration and Excretion	**Chapter 15** The Body's Control Systems **Chapter 16** Growing Up Healthy

Grade 3 Components

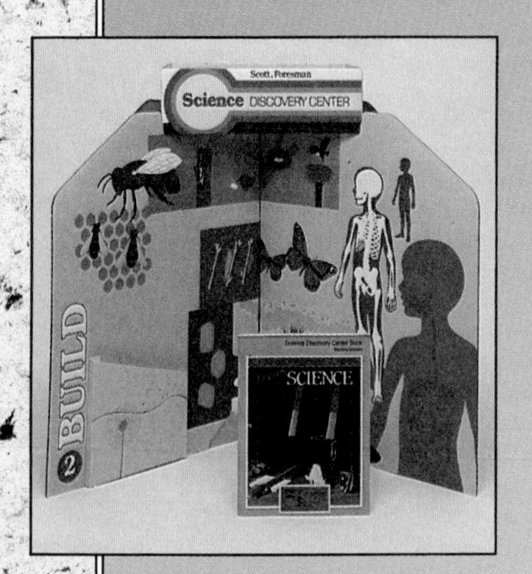

Science Discovery Center

Equipment Kit

Grade 3 Scope and Sequence

Life Science

Plants

Chapter 1
Plant Growth

- ★Lesson 1 Importance of roots, stems, and leaves
- ★Lesson 2 Importance of flowers and cones
- Lesson 3 How plants grow from seeds

Animals

Chapter 2
How Animals Grow

- Lesson 1 How animals can be grouped
- ★Lesson 2 How some animals with backbones grow and change
- ★Lesson 3 How some animals without backbones grow and change

Ecology

Chapter 3
Living Things Need Each Other

- Lesson 1 The five groups of living things
- ★Lesson 2 How organisms live together
- ★Lesson 3 How organisms get food

Chapter 4
How People Affect Plants and Animals

- ★Lesson 1 How people change the lives of plants and animals
- Lesson 2 How people protect plants and animals
- Lesson 3 How people use plants and animals

Physical Science

Matter

Chapter 5
Properties of Matter

- Lesson 1 What matter is
- ★Lesson 2 The make-up of matter
- ★Lesson 3 How matter can change

Energy

Chapter 6
Work and Machines

- Lesson 1 What work is
- ★Lesson 2 What simple machines are
- ★Lesson 3 What compound machines are

Chapter 7
Forms of Energy

- ★Lesson 1 Some kinds of energy
- ★Lesson 2 What electricity is
- ★Lesson 3 How matter is heated
- ★Lesson 4 How light travels

Chapter 8
Sound

- ★Lesson 1 What sound is
- ★Lesson 2 How sound travels
- Lesson 3 How people make sound

Earth Science

Earth

Chapter 9
Rocks and Soil

- Lesson 1 How rocks are formed
- ★Lesson 2 What soil is made of
- ★Lesson 3 How people use rocks and soil

Chapter 10
Changes in the Earth

- Lesson 1 What the inside of the earth is like
- ★Lesson 2 How water and wind change the earth's crust
- ★Lesson 3 How earthquakes and volcanoes change the earth's crust
- Lesson 4 How living things change the earth's crust

Weather

Chapter 11
Clouds and Storms

- ★Lesson 1 How clouds form
- ★Lesson 2 Water in clouds
- ★Lesson 3 Causes of storms

Space

Chapter 12
The Sun, Moon, and Planets

- ★Lesson 1 How the earth and the moon move
- Lesson 2 What the sun and the planets are like
- ★Lesson 3 What scientists learn from space travel

Human Body

Chapter 13
The Body's Support

- ★Lesson 1 The make-up of the body
- ★Lesson 2 The importance of bones
- ★Lesson 3 The importance of muscles

Chapter 14
Your Body's Health Needs

- ★Lesson 1 How to stay healthy
- Lesson 2 Causes of disease
- ★Lesson 3 How alcohol, tobacco, and drugs affect the body

Note: If time does not permit you to cover all of the material in this book, it is recommended that the lessons with a star be presented as a core program. This core program represents a balanced treatment of topics throughout the series.

Content

Organized

One of the most important features of Scott, Foresman's *Discover SCIENCE* centers on its highly organized lesson structure. The sample on these pages illustrates the one-to-one correlation between objective, content, and review. This tight structure improves student comprehension.

The authors of *Discover SCIENCE* encourage you to check for this benefit in any science series you review.

4 How Do Living Things Change the Earth's Crust?

LESSON GOALS

You will learn
- how people change the earth's crust.
- how plants and animals change the earth's crust.

Whenever you dig a hole in the ground or split a rock in half, you are changing the earth's crust. People, plants, and animals change the earth's crust every day.

How Animals and Plants Change the Earth's Crust

Some animals, such as earthworms, live in the earth's crust. These animals change the earth's crust as they dig through the soil. Other animals look for food in the soil. How is the animal in the picture changing the land?[1]

Plants also help change the earth's crust. Plants can protect the land from erosion by holding the soil down. How can plants help change rocks into soil?[2]

[1]digging holes
[2]growing roots crack rocks

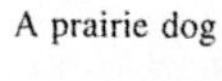
A prairie dog

Tight lesson structure improves science comprehension

1 **Each goal at the beginning of the lesson . . .**

2 **relates to one specific subhead within the lesson . . .**

How People Change the Earth's Crust

When people build things, they change the land. People dig rocks and soil out of the ground to build roads and buildings. People cut down plants. Then wind and water can easily erode the soil. You learned in Chapter 9 that people dig mines to reach ores. Digging mines changes the earth's crust. Look at the picture. How are these workers changing the land?

SCIENCE IN YOUR LIFE

You might have seen cracked sidewalks near trees. As the trees' roots grew, they pushed up on the sidewalks. The pushing caused the sidewalks to break apart.

Building a road

[1]moving soil to build a road

Lesson Review

1. How can plants and animals change the earth's crust?
2. How do people change the land when they build things?
3. **Challenge!** Why do plants grow better in soil in which animals live or dig?

Study on your own, pages 334–335.

EARTH SCIENCE

FIND OUT ON YOUR OWN

Long ago, there were no living things on the earth. Try to imagine what the earth might have looked like then. How was the earth's surface different from the surface today? Write a short paragraph that describes what you think the earth's crust used to be like. Then, use library books to find out what scientists think the earth was like at that time in history.

223

Up to Date

A highly qualified authorship and review team incorporated the most recent scientific data available. References to new technology, new information, and recent global events help prepare you and your students for science in the 1990s.

Relevant

Most examples and illustrations are presented in the context of the students' experience. Research shows that students are better able to learn new concepts when they are permitted to draw on their own background experience.

3 **which in turn relates to one specific review question at the end of the lesson.**

Process Skills

A flexible activity program allows teachers to customize hands-on experience to their needs.

An easy-to-do *DISCOVER* activity begins each chapter and provides a common experience that students can later draw on to assimilate new lesson concepts.

An *Investigate* feature in each chapter challenges students to apply scientific methods as they develop hypotheses and test their hypotheses with experiments.

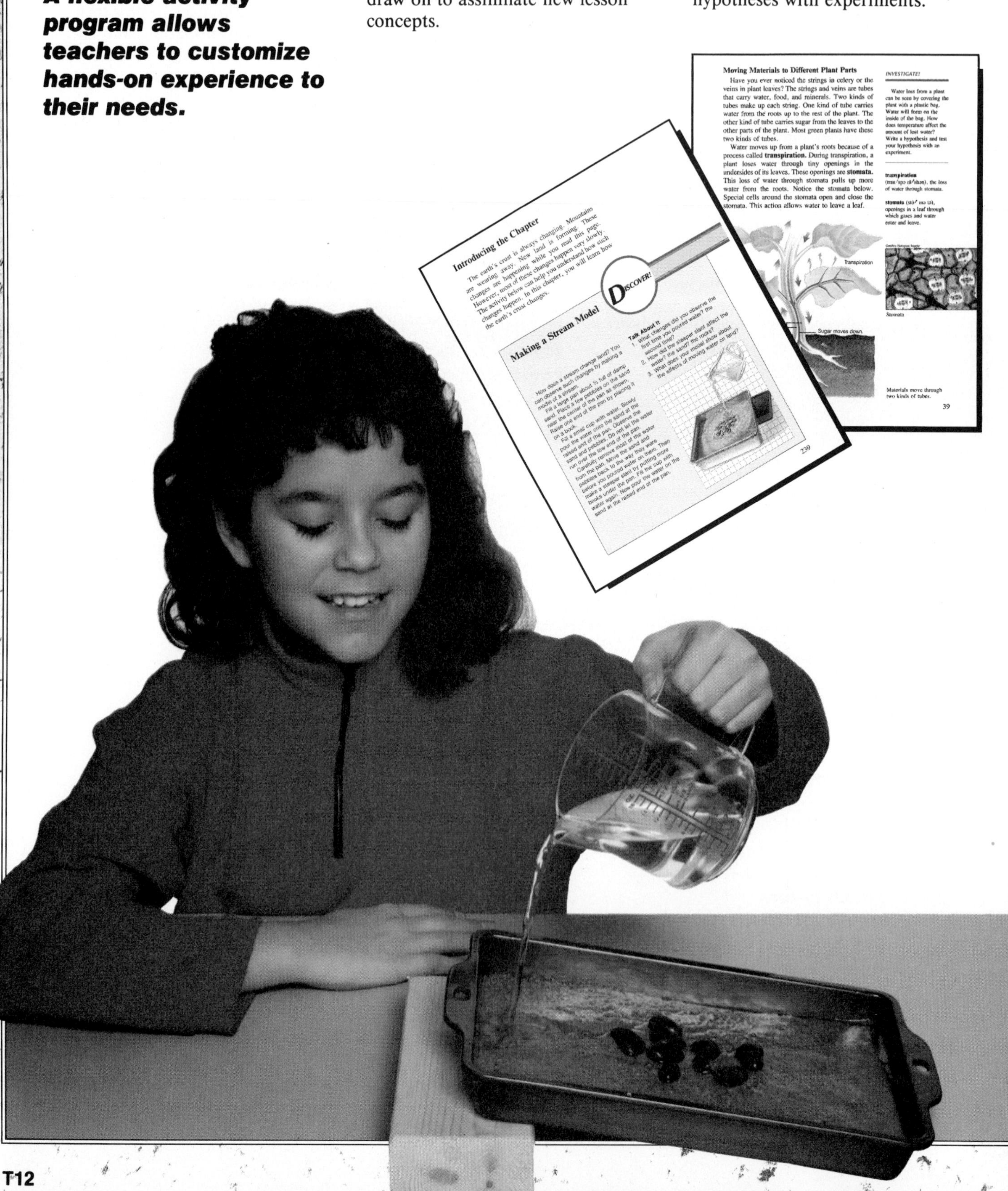

Introducing the Chapter

The earth's crust is always changing. Mountains are wearing away. New land is forming. These changes are happening while you read this page. However, most of these changes happen very slowly. The activity below can help you understand how such changes happen. In this chapter, you will learn how the earth's crust changes.

DISCOVER!

Making a Stream Model

How does a stream change land? You can observe such changes by making a model of a stream.

Fill a large pan about ½ full of damp sand. Place a few pebbles on the sand near the center of the pan as shown. Raise one end of the pan by placing it on a book.

Fill a small cup with water. Slowly pour the water onto the sand at the raised end of the pan. Observe the sand and pebbles. Do not let the water run over the low end of the pan.

Carefully remove most of the water from the pan. Move the sand and pebbles back to the way they were before you poured water on them. Then make a steeper slant by putting more books under the pan. Fill the cup with water again. Now pour the water on the sand at the raised end of the pan.

Talk About It

1. What changes did you observe the first time you poured water? the second time?
2. How did the steeper slant affect the water? the sand? the rocks?
3. What does your model show about the effects of moving water on land?

239

Moving Materials to Different Plant Parts

Have you ever noticed the strings in celery or the veins in plant leaves? The strings and veins are tubes that carry water, food, and minerals. Two kinds of tubes make up each string. One kind of tube carries water from the roots up to the rest of the plant. The other kind of tube carries sugar from the leaves to the other parts of the plant. Most green plants have these two kinds of tubes.

Water moves up from a plant's roots because of a process called **transpiration.** During transpiration, a plant loses water through tiny openings in the undersides of its leaves. These openings are **stomata.** This loss of water through stomata pulls up more water from the roots. Notice the stomata below. Special cells around the stomata open and close the stomata. This action allows water to leave a leaf.

INVESTIGATE!

Water loss from a plant can be seen by covering the plant with a plastic bag. Water will form on the inside of the bag. How does temperature affect the amount of lost water? Write a hypothesis and test your hypothesis with an experiment.

transpiration (tran′spə rā′shən), the loss of water through stomata.

stomata (stō′ mə tə), openings in a leaf through which gases and water enter and leave.

Stomata

Materials move through two kinds of tubes.

39

Two full-page activities in each chapter help students learn how to follow scientific procedure and interpret data. An additional activity per chapter in the *Workbook* reinforces the use of these skills.

An *Experiment Skills* feature for each chapter teaches students basic skills necessary to do experiments. Each feature develops these skills through an experiment, questions, and practice.

ACTIVITY

Wear cover goggles for this activity.

Observing Changes in Milk

Suggested grouping: pairs

Purpose
Observe how an acid and an enzyme change milk.

Gather These Materials
• 4 clear plastic cups • masking tape • milk • plastic wrap • 4 rubber bands • 4 plastic spoons • water • enzyme • weak acid

Follow This Procedure
1. Use a chart like the one shown to record your observations.
2. Label 4 cups 1 through 4 by putting the numbers on masking tape and putting the tape on the side of each cup.
3. Put 3 spoonfuls of milk in each cup.
4. Use a clean spoon to add 2 spoonfuls of water to cup number 2 and stir.
5. Use a clean spoon to add 2 spoonfuls of acid to cup number 3 and stir.
6. Use a clean spoon to add 2 spoonfuls of enzyme to cup number 4 and stir.
7. Cover each cup with plastic wrap. Use a rubber band to hold the plastic wrap in place.
8. Wait for 1 minute and observe the milk. Record what the milk looks like.
9. Wait for 1-2 hours and observe the milk. Record any changes.

Record Your Results

Cup number	Changes after 1 minute	Changes after 1-2 hours
1. Milk		
2. Milk and water	See Activity Results below.	
3. Milk and acid		
4. Milk and enzyme		

State Your Conclusion
1. Describe how water, an acid, and an enzyme change milk.
2. What difference did you notice in the way the acid and the enzyme acted?

Use What You Learned
The enzyme in this activity is from a meat tenderizer. People sometimes buy meat tenderizer in the grocery store. They use it on meat to make the meat more tender. Why do you think meat tenderizer can make meat more tender?

329

Chapter 6 Experiment Skills

Setting Up an Experiment

Julie's science class made salt solutions on Friday. When the bell rang, the students put their jars of solution in a cupboard. Julie left her jar on a sunny windowsill in the classroom.

On Monday, Julie found that only salt crystals were left in her jar. The water had evaporated from her solution. She noticed that the jars in the cupboard still contained solutions. Julie decided that heat from the sunlight caused her solution to evaporate quickly.

She set up an experiment to see how temperature affects the evaporation of salt water. She left jars of salt water open to air at different temperatures.

Read Julie's experiment on the next page. Then answer the questions on this page.

Thinking About the Experiment

Julie knew that water slowly disappears if it is left in an open container.
1. What happened to the water in Julie's solution?
2. How could she have kept her solution from drying up?
3. Could Julie have set up her experiment using tap water instead of salt water? Explain.

Julie set up her experiment so that only one thing is different for the jars. The part of the experiment that is different is the variable being tested.
4. What is the same for each jar?
5. What is the variable begin tested in Julie's experiment?

Julie has a control in her experiment. In this experiment, the control is the part of the setup that shows how fast water evaporates in Julie's classroom.
6. Which part of the setup is the control?

See pages 434-437 to review scientific methods and safety.

Experimenting with Evaporation

Problem
Does the temperature of the air have an effect on how fast water evaporates from an open container?

Hypothesis
Water evaporates faster in warm air than in cold air.

Materials
3 jars of same size
salt
spoon
graduated cylinder
water
marker
200-watt lamp and stand

3. Pour 20 mL of the salt solution into each of the 3 jars. Mark the water level on the side of each jar.
4. Set jar A under the lighted lamp. Place jar B somewhere in the room where it will not be disturbed. Set jar C in a cold place. Make sure none of the jars is in a drafty place.
5. Check the jars at the end of each day for 3 days. Record your observations in a chart like the one below.

Data and Observations

Day	Water level A	Water level B	Water level C
1			
2			
3			

Conclusions
You might do Julie's experiment to find your own data and make observations. Draw conclusions based on your data.

Practice Setting Up an Experiment
Suppose you want to find out how air temperature affects the evaporation of a liquid other than water.
1. What liquid would you use?
2. What might be your hypothesis?
3. How would you set up an experiment to test your hypothesis?

449

Process Skills

A separate *Science Process Skills Book* instructs and assesses whether students understand the basic process skills they use in doing activities. This booklet also teaches the manipulative skills students need to use basic science equipment.

Fun activities at the *Science Discovery Center* provide independent opportunity for enrichment of each chapter's content.

Name

Process Skill Activity

Predicting Temperatures

Lisa heated water to 74° Celsius. She wanted to find out which container, a plastic foam cup or a tin can, would keep the water hotter. She measured the temperature every minute. The results are shown on the graph. Study the graph and use it to answer the questions.

1. At 4 minutes, how hot was the water in the tin can?

   How hot was the water in the plastic foam cup?

2. Use the graph to predict the water temperature in the tin can at:

   6 minutes
   7 minutes
   8 minutes

3. Predict the temperature of the plastic foam cup at:

   6 minutes
   7 minutes
   8 minutes

4. Which container would keep an ice cube colder for the longer time, the tin can or the plastic foam cup? Explain.

5. Lisa also tested a glass cup. Do you predict that it kept the water as hot as the plastic foam cup? Explain.

Hot Water Chart

Temperature (°C)

Minutes

KEY
Tin Can
Plastic Cup

39

Manipulative Skill

Dropper
Water drop

63

A *Teacher Demonstration* in the *Teacher's Edition* provides an active way to introduce every lesson.

NAME SCIENCE DISCOVERY CENTER CHAPTER 6

EXPLORE 3

To Move or Not To Move
or
Why does a moving object usually keep moving?

Have you ever seen a magician pull a tablecloth off a table, leaving all the dishes standing on the table? This trick is possible because of inertia. Inertia is the resistance of any object to a change in its state of motion. Thus a moving object tends to keep moving. An object that is not moving tends to stay still. This activity will help you understand inertia. You will need a cup, a penny, and a slip of paper. Place the slip of paper over the top of the cup. Then put the penny on top of the paper. Slowly slide the paper off the cup.

1. What happens to the penny?

2. Set up the cup, paper, and penny again. Now pull the slip of paper straight out from under the penny as fast as you can. What happens to the penny?

3. Which object, the paper or the penny, shows that an object that is not moving tends to stay still?

4. What force acts on the penny *after* you pull away the slip of paper?

5. What is the result of this force?

Next place a sugar cube on top of a toy car. Put a book at one end of a table. Roll the car across the table, toward the book.

6. What happens to the toy car and the sugar cube when they hit the book?

7. Which object, the toy car or the sugar cube, shows that an object that is moving tends to keep moving?

8. Think about the toy car and the sugar cube. Why do you think it is important to wear a seat belt in an automobile?

69

Thinking Skills

Each chapter ends with a skills feature that teaches students to collect, organize, and interpret data to solve problems.

Skills for Solving Problems

Using Pictographs and Time Lines

Problem: How has the population of trumpeter swans changed since 1850?

Part A. Using Pictographs to Collect Information

1. The key shows that each picture in the pictograph stands for a number of swans. What do the large pictures stand for? medium-sized pictures? small pictures?
2. The pictograph shows how many swans were alive in certain years. To find out how many were alive in 1850, add the numbers the pictures for that year stand for. Many hunters shot trumpeter swans. Habitats were changed. Later, people protected the swans. How many swans were alive in 1900? in 1950? in 1980?

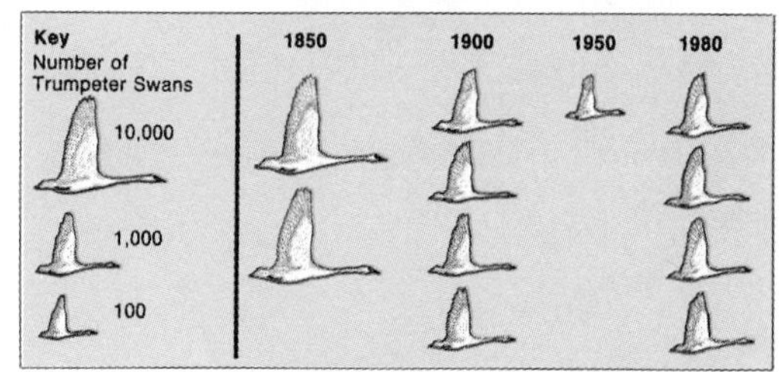

Part B. Using a Time Line to Organize and Interpret Information

3. The time line contains the information you collected. The spaces on the line stand for lengths of time. What is the length of time between the marks on the time line?

82

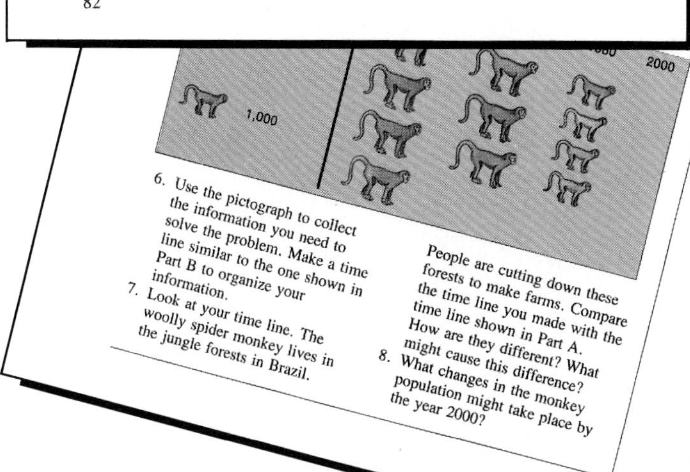

6. Use the pictograph to collect the information you need to solve the problem. Make a time line similar to the one shown in Part B to organize your information.
7. Look at your time line. The woolly spider monkey lives in the jungle forests in Brazil.

People are cutting down these forests to make farms. Compare the time line you made with the time line shown in Part A. How are they different? What might cause this difference?

8. What changes in the monkey population might take place by the year 2000?

83

Other features designed to develop thinking skills

Pupil's Edition

- *DISCOVER* activities at the beginning of each chapter
- In-text questions
- *Challenge!* question in each *Lesson Review*
- *Find Out On Your Own* at the end of each lesson
- *Investigate* in each chapter
- Two formal activities per chapter
- *What Do You Think?* questions in the *Science and Technology* or *Science and People* feature in each chapter
- Interpretation and application questions in the chapter and unit reviews
- A *Science and Society* feature at the end of each unit

Teacher's Edition

- Suggested questions, discussion topics, and activities for every page
- Highlighting of skills used in all *Pupil's Edition* and *Teacher's Edition* material

Ancillary Components

- *Science Process Skills Book* for each grade
- *Science and Reading* sheets in the *Resource Book*
- *Science Skills* and *Science Activity* sheets in the *Workbook*

Evaluation

Check students' progress at every level for continuous feedback.

Chapter Test

- Two forms for grades 3–6
- Keyed to lesson objectives

Independent Study Guide

- Additional questions for review

Lesson Review

- Provides frequent assessment
- Correlates with lesson goals

Unit Review

- Integrates chapter concepts
- Includes essay questions for application

Chapter Review

- Students recall, interpret, and apply information

Unit Test

- Includes picture interpretation questions
- Keyed to lesson objectives

Application

Numerous special features relate science to the students' real world.

Careers

"To become a dietician, I need to go to college."

Science in the News

"The paper shows a cold front moving in today."

Science and Technology

"Fiber optics is replacing copper wiring in telephone lines."

Science and People

"We know a lot about animals because of people like Jane Goodall."

Science and Society

"Should oil drilling be allowed off the coast of my state?"

How It Works

"My doorbell works on a circuit."

Science in Your Life

"My supermarket uses laser beams to checkout groceries."

Teacher's Annotated Edition

A clearly defined two-part format lets you more easily plan your day.

A 3-step **TEACHING PLAN** makes the lesson easy to teach.

1. Motivate

Demonstration and follow-up discussion to begin each lesson effectively

2. Teach

Time-saving tips to reinforce and extend the content

Includes questions to ask students—with answers

Includes possible misconceptions—information to help you correct commonly misunderstood science ideas

Includes answers to *Investigate* activities

3. Assess

Answers to lesson review questions and *Find Out On Your Own*

A **Teaching Options** section provides added flexibility for your planning.

Background information, extra activities, projects, discussion topics, and strategies found under these headings:

Science Background
Science Anecdote
Reading Strategies
Cooperative Learning
Applying Whole Language
Reinforcement
Enrichment
Special Education
Science and (other subject)
Game Suggestion
Reteaching Suggestion

LESSON 4 pages 222–223

TEACHING PLAN

1. Focus

Lesson Objectives

- *Explain* how animals and plants change the earth's crust. **2D**
- *Explain* how people change the earth's crust. **2D**

Demonstration Activity ♦ 4D

Fill a clear plastic tray with soil and several earthworms, or bring an ant farm to class. Point out that the soil is part of the earth's crust. Ask students to *observe* and comment on how the earthworms (or ants) can change the crust.

2. Instruct

- **Possible Misconception:** Students may think that the crust refers only to the top layer of the earth underground. The crust also, however, includes the surface and near surface.

3. Guided Practice

Draw a chart with three columns labeled *people, plants,* and *animals.* Ask students for examples of how each of these living things changes the earth's crust. *List* their examples in the appropriate columns.

4 How Do Living Things Change the Earth's Crust?

LESSON GOALS

You will learn
- how people change the earth's crust.
- how plants and animals change the earth's crust.

Whenever you dig a hole in the ground or split a rock in half, you are changing the earth's crust. People, plants, and animals change the earth's crust every day.

How Animals and Plants Change the Earth's Crust

Some animals, such as earthworms, live in the earth's crust. These animals change the earth's crust as they dig through the soil. Other animals look for food in the soil. How is the animal in the picture changing the land?[1]

Plants also help change the earth's crust. Plants can protect the land from erosion by holding the soil down. How can plants help change rocks into soil?[2]

[1]digging holes
[2]growing roots crack rocks

A prairie dog

Teaching Options

Science Background

The Surface Mining Control and Reclamation Act of 1977 was passed to control the environmental impact of coal strip-mining activities. Mining companies must prove they can reclaim—restore—the land before receiving a mining permit. After mining, the land must be restored so that it can be used for the purpose it served before mining. Restoration includes filling holes, contouring the land, removing wastes, and replanting vegetation. Mining companies must use the best available technology to prevent the pollution of streams and groundwater from toxic and acidic materials that wash off the mining wastes. Most experts agree that this law can help protect the environment if it is properly enforced.

Reading Strategies ♦

1. Guide students' pre-reading by asking: Which parts of the lesson are familiar and which parts are new?
2. Assign these strategies: Visualizing Information and Writing a Memory Sentence. (See pages T26-T29.)
3. Pair students to share what information is clear and unclear and initiate discussion using students' unanswered questions.

♦ *Suitable as a language development activity*

How People Change the Earth's Crust

When people build things, they change the land. People dig rocks and soil out of the ground to build roads and buildings. People cut down plants. Then wind and water can easily erode the soil. You learned in Chapter 9 that people dig mines to reach ores. Digging mines changes the earth's crust. Look at the picture. How are these workers changing the land?[1]

Lesson Review

1. How can plants and animals change the earth's crust?
2. How do people change the land when they build things?
3. **Challenge!** Why do plants grow better in soil in which animals live or dig?

Study on your own, pages 334–335.

SCIENCE IN YOUR LIFE

You might have seen cracked sidewalks near trees. As the trees' roots grew, they pushed up on the sidewalks. The pushing caused the sidewalks to break apart.

Building a road

[1]moving soil to build a road

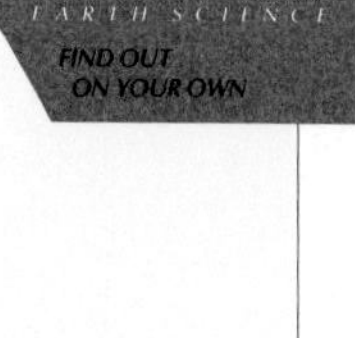

Long ago, there were no living things on the earth. Try to imagine what the earth might have looked like then. How was the earth's surface different from the surface today? Write a short paragraph that describes what you think the earth's crust used to be like. Then, use library books to find out what scientists think the earth was like at that time in history.

223

4. Reteach

Have students discuss photographs of ways that people, animals, and plants have changed the earth's crust.

5. Independent Practice

Have students answer the Lesson Review questions, then complete the Study Guide exercise on page 335.

Answers *Lesson Review*
1. Animals dig in the soil and plant roots help break rocks apart. **2D**
2. by digging rocks and soil and cutting down trees **2D**
3. Challenge! Plant roots grow more easily in loose soil. **Thinking Skill:** *Recognizing cause and effect* **2D**

6. Closure

Answer *Find Out On Your Own*
Students might say the surface looked rocky because there were no plants or animals. **Thinking Skill:** *Visualizing* **4A, 6D**

Review/Summary
Have students draw before and after pictures of how plants, animals, and people change the earth's crust.

Larger Context
Point out that changing the earth's crust is one way people endanger plants and animals.

Workbook page 59 *

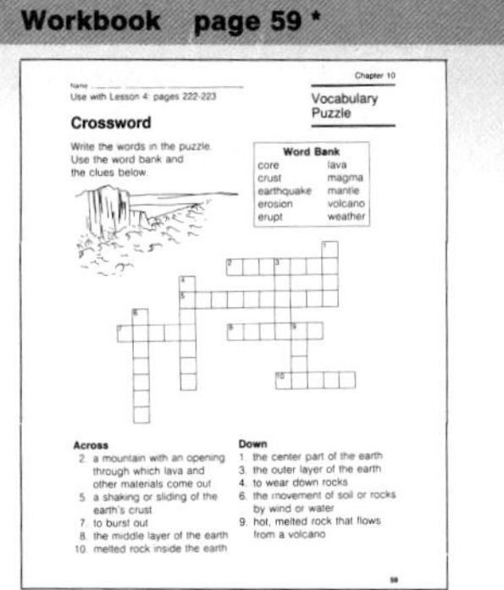
Use with Lesson 4: pages 222-223 — Chapter 10 — Vocabulary Puzzle

Crossword

Write the words in the puzzle. Use the word bank and the clues below.

Word Bank: core, lava, crust, magma, earthquake, mantle, erosion, volcano, erupt, weather

Across
2. a mountain with an opening through which lava and other materials come out
5. a shaking or sliding of the earth's crust
7. to burst out
8. the middle layer of the earth
10. melted rock inside the earth

Down
1. the center part of the earth
3. the outer layer of the earth
4. to wear down rocks
6. the movement of soil or rocks by wind or water
9. hot, melted rock that flows from a volcano

Special Education 4B

Ask students with learning disabilities to *describe* ways that the earth's crust has been changed near their homes or near the school.

Resource Book page 119 *

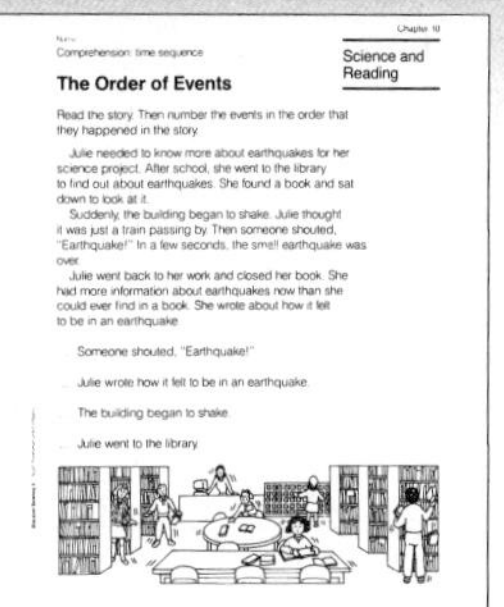
Comprehension: time sequence — Chapter 10 — Science and Reading

The Order of Events

Read the story. Then number the events in the order that they happened in the story.

Julie needed to know more about earthquakes for her science project. After school, she went to the library to find out about earthquakes. She found a book and sat down to look at it.

Suddenly, the building began to shake. Julie thought it was just a train passing by. Then someone shouted, "Earthquake!" In a few seconds, the small earthquake was over.

Julie went back to her work and closed her book. She had more information about earthquakes now than she could ever find in a book. She wrote about how it felt to be in an earthquake.

Someone shouted, "Earthquake!"
Julie wrote how it felt to be in an earthquake.
The building began to shake.
Julie went to the library.

*** Answers to masters on pages 206E–206H**

Unit 3 Chapter 10 Lesson 4 **223**

Other features designed for convenient and successful teaching:

- Two-page insert before each unit includes a Planning Guide and resources (books, community resources, audio visuals, and software).
- Four-page insert before each chapter includes a Planning Guide, a list of preteaching suggestions, and bulletin-board ideas.
- Reduced copies of resource masters appear on the pages where they are best used and before each chapter with answers.
- Clearly defined objectives for each lesson and activity aid in planning.
- Safety tips for all activities let you teach with greater confidence.
- Activity Results photographs, where appropriate, help you anticipate outcomes.
- Thinking skills are highlighted wherever they appear in the margin notes so that you can more easily plan to meet specific goals.

Thinking and Process Skills in Science

Joseph P. Riley II
University of Georgia

Developing Thinking Skills

A major goal of an elementary science program should be the development of thinking skills that will serve students in becoming better problem solvers and intelligent decision makers. Thinking skills include skills people use to collect, understand, interpret, apply, and evaluate information and ideas. The thinking skills emphasized in *Discover SCIENCE* are listed in the chart on page T21. Their classification is based on Bloom's Taxonomy of Educational Objectives, which forms a continuum from basic skills used in information gathering to higher order skills used in synthesizing and evaluating information.

Both basic and higher order thinking skills can be and should be introduced at all elementary grade levels. Cognitive research into the nature of thinking indicates that the activities commonly associated with higher order thinking are not necessarily limited to advanced levels of development. For example, young children engage in complex problem-solving behaviors when they decode a word in reading or imagine beyond the information contained in a story. These same skills can be called upon in science to encourage students to go beyond the data—to provide inferences, make predictions, and analyze information.

Such skills can be successfully employed if the students have a concrete experience or solid knowledge base from which to start. For example, from the common observation that wet pavement dries, students can make inferences about where the water goes. They can make predictions about how long large and small puddles will take to dry.

Teachers can encourage the development of all levels of thinking skills by:

- allowing students to draw upon their familiar, concrete experiences to extend their knowledge
- involving students in learning activities that include hands-on experiences
- developing a science knowledge base of specific facts and organizing principles
- providing extended practice in the use of thinking skills and
- challenging students with questions and problems that require the use of various thinking skills to solve.

Discover SCIENCE incorporates all these points. Program features that develop thinking skills are listed and described on page T21.

Developing Process Skills

Thinking skills most closely related to scientific inquiry are identified as science process skills. These are the tools of scientific investigation. They are applied daily by scientists in the field and in the laboratory as part of a procedure that is commonly called the scientific method. Although no one specific method is strictly followed, scientific procedures employ the following process skills.

Science Process Skills
- Observing
- Communicating
- Measuring
- Classifying
- Making and using models
- Recognizing space or time relationships
- Collecting and interpreting data
- Inferring
- Predicting
- Identifying and controlling variables
- Experimenting
- Formulating questions and hypotheses
- Making operational definitions

As with other thinking skills, the development of science process skills should be an integral part of an elementary science program. By their very nature, process skills are best utilized when students are engaged in hands-on experiences—when they are actively involved in scientific investigation. The authors and editors of *Discover SCIENCE* have recognized this maxim. The development of process skills in *Discover SCIENCE* by "doing science" is shown on page T21.

Elementary students bring to class the most basic and important ingredient for scientific inquiry—curiosity about the natural world. With it, students will develop, to a degree, certain thinking and process skills on their own. A program that incorporates thinking and process skills will direct and amplify that development, allowing students to fulfill their potentials as problem solvers and decision makers.

Skills Development

Pupil's Edition
- *DISCOVER* activities begin each chapter with a focus on process skills development.
- Two additional activities per chapter focus on process skills as students apply newly acquired knowledge.

- In-text questions focus students' thinking.
- General text integrates thinking skills by explaining how we came to know what we know.
- Lesson review questions, including a *Challenge!* question, extend students' thinking. *FIND OUT ON YOUR OWN* at the end of each lesson emphasizes information gathering and communicating skills.
- *Skills for Solving Problems* focuses on the skills of *making decisions/identifying and solving problems* and *interpreting charts, maps, and graphs.*
- Chapter review and unit review questions require the use of a continuum of thinking skills.
- A *Science and Society* feature at the end of each unit focuses on evaluative and decision-making skills.

Teacher's Edition

- Suggested questions, discussion topics, and activities reinforce and extend thinking and process skills.
- Skills used in all Pupil's Edition and Teacher's Edition material are listed in heavy italics in the sidenotes.

Ancillary Components

- A *Science Process Skills Book* for each grade contains worksheets that teach and assess student mastery of science process skills. Other worksheets teach manipulative skills for using basic science equipment.
- *Science and Reading* sheets in the *Resource Book* focus on skills associated with reading science content.
- *Science Skills* sheets in the *Workbook* concentrate on higher order thinking skills.
- *Science Activity* sheets in the *Workbook* focus on process skills.

Thinking Skills Developed in *Discover SCIENCE*

1 Gathering and Recalling Information

Collecting Information/Data
- Measuring
- Observing

Recalling Facts

Communicating
- Naming
- Listing
- Describing
- Recording

2 Understanding and Interpreting Information/Data

Recognizing the Main Idea and Supporting Details
Recognizing and Using Space or Time Relationships
Sequencing (Time, Steps in a Process, Following Directions, Place)
Recognizing Cause and Effect
Classifying
Organizing Information
- Recognizing Patterns and Relationships
- Comparing (Identifying Similarities)
- Contrasting (Identifying Differences)
- Summarizing

Restating or Explaining Ideas
Interpreting Charts, Maps, and Graphs
Comprehending Meaning
Visualizing
Making Physical Models
Identifying Variables
Interpreting Data

3 Applying, Analyzing, Synthesizing, and Evaluating

Making Generalizations
Making Inferences/Inferring
Making Analogies
Drawing Conclusions
Formulating Questions and Hypotheses
Identifying and Suggesting Alternatives
Controlling Variables
Predicting
Judging and Evaluating
- Distinguishing Fact from Opinion, Bias, and Propaganda
- Ranking Ideas and Information According to Criteria
- Reconciling Inconsistent Criteria
- Recognizing Relevant Information and Data
- Recognizing Factual and Logical Inconsistencies

Making Decisions/Identifying and Solving Problems
Applying Information to New Situations
Experimenting
Making Operational Definitions

High-Potential and Mainstreamed Students

Christine Kuehn
University of South Carolina

High-Potential Students

Perhaps more so than other students, high-potential students benefit from the opportunity to make individual study choices and to structure an activity in a manner that is personally challenging and interesting. Therefore, the *Classroom Management*page in the front of each chapter in the Teacher's Edition includes an experiment or other project designed for high-potential students.

Many other items throughout the Pupil's Edition and Teacher's Edition of *Discover SCIENCE* provide individuals the opportunities to investigate science concepts beyond the basic material presented in the chapters. Such items include science fair projects, a variety of extensions in the *Teaching Option* section, and questions at the end of each lesson and chapter in the Pupil's Edition that stimulate higher-level thinking skills.

Mainstreamed Students

Discover SCIENCE offers teachers extensive opportunity to make the study of science more rewarding and relevant to mainstreamed students—those who have emotional handicaps; orthopedic, visual, or hearing impairments; learning disabilities; or are mentally retarded. A suggestion for helping students with learning disabilities is included in the *Teaching Options* for each lesson under the heading *Special Education.* These ideas incorporate alternative learning modes.

The *Classroom Management page* in front of each chapter in the Teacher's Edition includes specific ideas to help students with handicaps or impairments experience the fullness of learning. These ideas are chapter-specific, but the following general guidelines can also be applied.

Emotionally Handicapped

- Define class rules clearly and be consistent in their enforcement.
- Provide a highly structured environment for activities.
- Encourage interaction in discussions and activities.
- Provide responsibilities such as helping to clean the room or care for a classroom plant or pet.
- Reinforce appropriate behaviors.

Orthopedically Handicapped

- Consult with the school nurse or student's physical therapist to identify the student's limitations.
- Adapt the physical environment of the classroom accordingly.
- Regarding field trips and outdoor activities: choose environments and activities that accommodate the capabilities of all the students.

Visually Impaired

- Use white chalk on a black chalkboard.
- Say out loud what you are writing on the board.
- Seat the student as close to the front as necessary.
- Stress the use of touching and hearing.
- Investigate the use of special learning aids such as sheets of magnifying plastic for reading materials.

Hearing Impaired

- Seat the student near the front of the room.
- Speak clearly.
- Use visual aids whenever possible.

Mentally Retarded

- Provide concrete examples of explanations.
- Use repetition and reinforcing activities.
- Provide hands-on experiences.
- Reinforce appropriate behaviors.

Get to know your mainstreamed students. If they fear being singled out, avoid highlighting them in class. Special needs may be taken care of outside of class time.

Language Skills for LEP Students

Alfredo Schifini, Ph.D.
Reading Consultant

Although students who are native speakers of English may experience some difficulty with new science terminology and concepts, students who do not speak English natively are particularly at risk. Limited English proficient (LEP) students do not have the English language skills of listening comprehension, speaking, reading, and writing necessary to sufficiently engage in instruction in English. While some students appear to handle English reasonably well in everyday situations, and are academically quite capable and motivated, they lack the language skills to reason, hypothesize, make predictions, and form and defend logical arguments in English.

For both native- and second-language speakers of English, language is acquired through meaningful communication—by engaging in a genuine exchange of information in a variety of low-anxiety settings. In classrooms this communication involves surrounding learners with language they can understand and devising tasks to use newly acquired language.

Language Development Activities
Discover SCIENCE has integrated language development activities designed to build academic language proficiency and to enhance science instruction for all students. These activities are denoted by a ♦ in the Teacher's Edition for each chapter and include the following:

- *Cooperative Learning* (on the first page of each chapter and the first page of each Chapter Review)
- *Applying Whole Language* (on the second page of each chapter)
- *DISCOVER* (on the second page of each chapter in the Pupil's Edition)
- *Demonstration Activity* (on the first page of each lesson)
- *Reading Strategies* (on the first page of each lesson)
- *Reteaching Suggestion* (on the last page of each lesson)
- *Game Suggestion* (at least one per chapter)

The cooperative learning activities give students the opportunity to engage in group interaction. Such interactions are particularly appropriate for LEP students. They provide a low-anxiety setting in which students with more advanced skills may help their peers who are newer to the language.

The other activities listed above are suitable for developing language because they involve the use of contextual clues and capitalize on prior knowledge. Simply put, contextual clues may be anything that helps the learner follow what is being said. The demonstration of new science concepts, the use of real objects, and games all assist in getting the main points of a lesson across to students who do not possess fully developed language skills in English.

Success in learning new science information depends heavily on the learner's prior knowledge of the world. The *DISCOVER* activities and *Reading Strategies* particularly utilize and focus students' prior knowledge to increase comprehension of new material.

Useful Teaching Techniques
Language development activities can be used successfully with all students. However, teachers of students who are at low levels of linguistic competence in English are encouraged to use a slower but natural speech rate, shorter sentences, and ample repetitions. Visual aids, audio-visual assistance, and the use of gestures and mime will also serve to make English more comprehensible to LEP students.

Cooperative Learning in Science

Robert E. Slavin
Center for Research on Elementary and Middle Schools
Johns Hopkins University

In *cooperative learning,* students work in four- to five-member learning teams to help one another master academic knowledge and skills. Research on cooperative learning has found that if students work on teams that are rewarded based on the learning of all team members, they learn more than do students in traditional settings.

To use cooperative learning in *Discover SCIENCE,* first assign students to four-member learning teams. If the class does not divide evenly by four, a few five-member teams may be assigned. Each team should have one relatively high achiever, one low achiever, and two or three average achievers, and should be mixed in sex and ethnicity. Let students choose team names and sit together during science periods. You might change team assignments every four to six weeks.

Two principal cooperative learning formats that can be used in *Discover SCIENCE* are *Jigsaw* and *STAD* (Student Teams—Achievement Divisions).

Jigsaw
In *Jigsaw* each team member becomes an "expert" on a unique topic for a chapter. Four topics are suggested in the *Teaching Options* at the beginning of each chapter. Assign topics at random. If a team has five members, two may share a topic.

All students should read the entire chapter, but they should especially be looking for information on their own topics. At the end of the chapter, students from different teams who had the same topic meet in "expert groups" to discuss their topics. You may wish to circulate among the groups to guide their discussion and to see that they are focusing on the right information.

After twenty to thirty minutes, have students return to their teams and take turns presenting what they have learned to their teammates. Finally, all students take a test covering all topics. Teams that average 90% or more on the test may earn attractive Superteam certificates; those that average 80-89% may earn smaller Greatteam certificates. These criteria and rewards may be adjusted to your preferences and circumstances.

STAD
In *STAD* students study together to help one another prepare for individual tests. At the end of each chapter, give students about one class period to work in their learning teams to master the material presented in the Chapter Review. After the teams have had enough time to complete the Chapter Review and study the material, give students individual tests covering the chapter content. Teams may earn certificates or other rewards based on average team scores, as described above.

You may wish to combine *Jigsaw* and *STAD* by allowing students to study the Chapter Review after they have finished reporting to the team on their topics.

Teaching Safety in the Classroom

Dr. Jack A. Gerlovich
Science Education Safety Consultant/Author
Des Moines, Iowa

Activities throughout *Discover SCIENCE* reinforce and extend science concepts using materials and procedures that are inherently safe. *Discover SCIENCE* teaches that safe procedure is part of sound scientific inquiry. Students who use this program learn not only how to safely investigate the topics at hand; they also develop safety habits that will serve them well in future scientific endeavors.

How does *Discover SCIENCE* accomplish this task? First and foremost, by performing the activities in the text, students learn that simple, safe materials can be used extensively to investigate science concepts. Second, safety reminders regarding procedure are given in the Pupil's Edition wherever appropriate. These include *CAUTION* statements and a cover goggle symbol and statement on appropriate pages. Third, the Teacher's Edition includes safety tips for the various student activities and teacher demonstrations that appear throughout the program. Following is a list of the most general of these tips for the elementary science classroom. If followed from the start, these guidelines should be easily assimilated into classroom procedures by teachers and students alike.

- The proper use of cover goggles that meet American National Standards Institute (ANSI z87.1, 1979) standards should be demonstrated to students. Cover goggles should be worn whenever the potential for eye injury exists; for example, when heating any substance, when using any chemicals including "ordinary" substances such as vinegar, and when using glassware. Even relatively safe items such as rubber bands and balloons can cause eye injury and warrant the use of goggles.
- To prevent student interference with each other and to assist the safe exit of students from the room in case of an emergency, teachers should try to assure that rooms are not overcrowded, that students understand exit procedures, and that aisles are kept uncluttered.
- Teachers should periodically conduct simulations with students for dealing with foreseeable emergencies. Examples might include exiting the room due to an emergency, coping with a fire, aiding someone who has been splashed by a chemical, and helping a fall victim.
- Prior to using any equipment or chemicals, teachers should be certain they understand the proper function and hazards associated with the use of those items. This information should be communicated to the students.
- Unless you know the outcome is safe, you should never mix substances "just to see what happens." No hazardous substances are used in *Discover SCIENCE.* However, the *combining* of certain substances might pose safety problems. For example, mixing ammonia with bleach produces particularly dangerous fumes. Notes about the dangers of mixing chemicals are included on the appropriate pages throughout the program.
- All equipment should be properly stored. The more dangerous items should be kept under lock and key.
- Whenever possible, plastic items should replace glass. If glass containers are essential, temperature- and break-resistant glassware should be selected.
- To prevent slipping and falls, any liquids spilled on tile or hardwood floors should be wiped up immediately.
- If the teacher cannot satisfy himself or herself that all foreseeable dangers have been reduced to an acceptable level, the activity should be altered or eliminated.

Teachers should be aware of all applicable federal, state, and local regulations and relevant guidelines from professional organizations which apply to the activites being performed. Examples would include Occupational Safety and Health Administration (OSHA) standards for workplace safety; state laws relating to cover goggles; local fire department requirements regarding the use of open flame, fire extinguishers, and fire blankets; and National Science Teachers Association (NSTA) suggestions regarding overcrowding. Refer to the following materials for other information about classroom safety.

Downs, G. et al. *Science Safety for Elementary School Teachers,* 1983. Iowa State University Press, Ames, Iowa 50010.

Gerlovich, J., Gerard, T., and Downs, G. *School Science Safety: Elementary,* 1984. Flinn Scientific Co., P.O. Box 219, Batavia, IL 60510.

Reading Science

Robert A. Pavlik
Cardinal Stritch College

Discover SCIENCE incorporates reading as a teaching-learning tool in science:

- to build and refine students' knowledge background.
- to help students make connections in their reading.
- to challenge students to think critically.

Reading Comprehension

Monitoring Readability

The authors and editors assessed each lesson using either the Dale-Chall or the Spache Readability Formula. The vocabulary and sentence lengths are appropriate for your students working either independently or under your guidance. The use of reading formulae, however, is only one factor in achieving maximum reading comprehension.

Clarifying Organization

Each chapter is divided into short lessons. Each lesson has a question-title and two or more goals. The lesson content contains only those main ideas and supporting details that pertain to the question-title and lesson goals.

Motivating Purposeful Reading

The second page of each chapter contains an action-oriented feature entitled *TRY THIS* in grades 1 and 2 and *DISCOVER* in grades 3–6. This feature invites students to become involved in scientific problem solving *before* reading new information.

The *Science and Technology, Science and People,* and *Careers* features motivate students' reading by focusing on the imagination, courage, hard work, frustration, and excitement involved in scientific endeavors. The features of *Science in Your Life, Find Out On Your Own,* and *How It Works* help students connect the concepts and processes of science with their lives.

Correlating Illustrations to Text

The text contains phrases, statements, or questions that direct students to the purpose, features, or importance of the illustrations. Our goal is to have students use the illustrations to reinforce and expand their understanding of the content. The tight correlation between text and illustration helps the reader get the most out of both elements.

Developing Vocabulary

New science words appear in boldface type when they are first used in the text. These words are defined in context and in the margins, along with pronunciation aids. The *Glossary* provides definitions of all vocabulary and other key terms.

Using an Engaging Writing Style

A distinguishing feature of *Discover SCIENCE* is the degree to which an active, engaging writing style has been incorporated.

- The regular use of the active voice yields more forceful passages and helps students clarify cause-effect relationships.
- In-text questions help students activate their backgrounds, summarize, interact with the illustrations, and predict outcomes.
- A judicious use of *you* appears throughout the text to promote students' identification with information and processes.

Assessing and Evaluating

Each lesson contains *Lesson Review* questions to help students verify their initial comprehension of the lesson's goals and key words. The *Chapter Review* contains a variety of formats to assess, reinforce, and expand student's comprehension and understanding of the material.

Reading Strategies

A New Teaching Option

Even with a highly readable science text, students may encounter situations in which their reading simply does not make sense to them. Strategic readers know how to activate their backgrounds; describe the structure of the material; set their own purposes for reading; apply specific strategies; and reflect on their reading. To help students become strategic readers, every lesson in the Teacher's Edition includes a feature called *Reading Strategies.* This feature appears in the *Teaching Options* on the first page of each lesson as follows.

> **Reading Strategies**
>
> 1. Guide students' prereading by asking: (suggestions vary)
> 2. Assign these strategies: (suggestions vary) See pages T26–T29.
> 3. Pair students to share what information is clear and unclear and initiate discussion using students' unanswered questions.

Each sentence in this feature addresses one or more of the strategic reader traits described above. Sentence 1 recommends a whole-class, prereading strategy. Sentence 2 recommends two during-reading strategies. These are chosen from the eight strategies described in detail on the next four pages. We suggest each of the two strategies be assigned to different halves of the class. Sentence 3 recommends post-reading strategies. The student pairs should consist of students who have been assigned different during-reading strategies.

When introducing each during-reading strategy, provide a model, using either the example given or a topic familiar to the students. Such modeling will not be necessary for every lesson.

Visualizing Information

Purpose

This strategy lets students use their interest and ability in drawing to visualize information in a way that makes sense to them.

Directions for the Students

1. Make a drawing that could be used to explain part or all of the lesson's information to someone else. The drawing may be similar to one in the textbook.
2. Title the drawing.
3. Label the parts of the drawing, if necessary.
4. Beneath the drawing, write some questions that you can answer using the drawing.
5. Write some questions about the lesson topics that you cannot answer even with the use of the drawing. These questions may or may not be answered in the lesson but can all be discussed.

Example

Book 3 Chapter 10 Changes in the Earth
Lesson 1 What Is the Inside of the Earth Like?

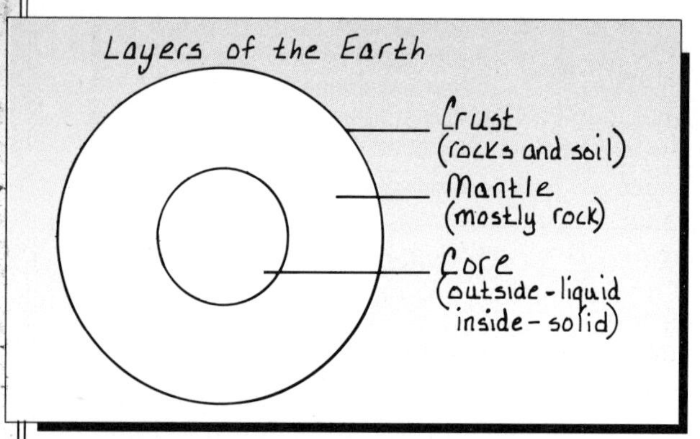

Questions answered from drawing:

- How many layers does the earth have?
- What layer do we live on?
- Is the earth hollow?
- Is the earth all solid rock?

Questions for discussion:

- What kinds of rock is the crust, mantle, and core each mostly made of ?
- How thick is each layer of the earth?

The first three strategies are recommended for use in grade 3. You may want to use the other strategies too.

Writing a Memory Sentence

Purpose

This strategy encourages students to set a purpose for their reading by writing down what they want to remember and why they want to remember it.

Directions for the Students

1. Write the beginning of a memory sentence with: "I want to remember . . . "
2. Add what you want to remember to the sentence.
3. Add why you might want to remember this information outside of schoolwork.
4. Under the memory sentence, write some questions your sentence answers.
5. Write some questions your sentence does not answer. These questions may not be answered in the lesson but can be discussed.

Example

Book 3 Chapter 2 How Animals Grow and Change
Lesson 1 How Can Animals Be Grouped?

Memory sentences:

- I want to remember the two main groups of animals—animals with and without backbones—so that I can recognize which group the animals I see belong to, and so that I can know something about each animal I see.

Questions answered by sentence:

- What two main groups do animals belong to?
- Do all animals have backbones?

Questions for discussion:

- What is a backbone?
- How does having a backbone help an animal?
- What are some animals with backbones?
- What are some animals without backbones?

Mapping Examples

Purpose

This strategy helps students graphically organize various examples of a key term.

Directions for the Students

1. Write the key term the teacher assigns you and draw a box around it.
2. As you read the lesson, search for examples of the key term.
3. Write the examples under the key term with a line connecting the key term to each example.
4. Under the example map, write some questions you can answer using your map.
5. Write some questions about the lesson that you cannot answer just by studying your map. These questions may or may not be answered in the lesson but can all be discussed.

Example

Book 3 Chapter 6 Work and Machines
Lesson 2 What are Simple Machines?

Key term: simple machines

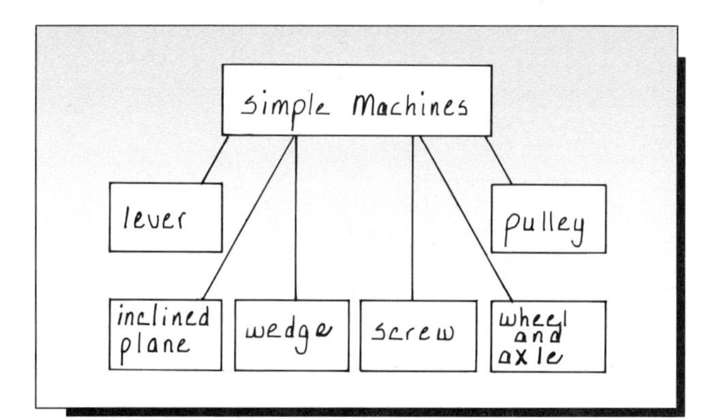

Questions answered from map:

- What are six simple machines?
- Is a seesaw an example of a simple machine?
- What kind of simple machine is a ramp?

Questions for discussion:

- What is a simple machine?
- What is an example of each kind of simple machine?
- How do simple machines help you?

Writing Sentences for Vocabulary Words

Purpose

This strategy helps students develop a personal and in-depth understanding of science vocabulary words.

Directions for the Students

1. Find two sentences in the lesson that contain the same vocabulary word. The definitions in the margins and glossary do not count.
2. Copy the sentences exactly as they appear in the textbook.
3. Read the sentences thinking about how they answer who/what/where/when/why/how questions.
4. Write a new sentence using the vocabulary word.
5. Write down any questions you might wish to discuss from the sentences you have read.
6. Follow the five steps above for other vocabulary words.

Example

Book 4 Chapter 9 Measuring Weather Conditions
Lesson 2 How Does Temperature Affect Air Pressure and Wind?

Vocabulary word: air pressure

Text sentences:

- This pressing down of air is called air pressure.
- Knowing what the air pressure is in different places helps people predict the weather.

New sentence:

- Air pressure gets lower as you go higher in the atmosphere.

Questions for discussion:

- How is air pressure measured?
- How can air pressure be used to predict weather?

Charting Like and Unlike Properties

Purpose

This strategy helps students clarify similar and different properties for two or more objects.

Directions for the Students

1. Make a chart like that shown in the example below.
2. In the left column, list objects that are being compared in the lesson.
3. In each box across the top, write a phrase that describes one or more of the objects.
4. For each object, place an "X" under the phrase or phrases that apply.
5. Below the chart, write some questions you can answer by using the chart.
6. Write some questions about the lesson topics that you cannot answer by using the chart. These questions may or may not be answered in the lesson but can all be discussed.

Example

Book 4 Chapter 1 Flowering Plants
Lesson 1 How Are Plants Classified?

	Makes seeds	Makes spores	Has flowers	Has cones
Conifer	X			X
Apple tree	X		X	
Fern		X		
Moss		X		
Pine tree	X			X

Questions answered by chart:

- Do conifers make seeds?
- Do ferns have flowers?
- What plants make spores?
- What plants have cones?

Questions for discussion:

- What do flowers do for a plant?
- What are spores?
- What do conifers look like?

Mapping Causes and Effects

Purpose

This strategy helps students describe, explain, and evaluate cause-effect relationships.

Directions for the Students

1. Using the diagram shown in the example, write any causes and effects identified in the lesson goals.
2. As you read the lesson, diagram any other cause-effect connections, or relationships, you come across. Keep in mind that different causes can have the same effect, and that the same cause can have different effects.
3. Under the diagram, write some questions you can answer by looking at the relationships you have written.
4. Write some questions about the lesson topics that you cannot answer by using the cause-effect relationships alone. These questions may or may not be answered in the lesson but can all be discussed.

Example

Book 4 Chapter 12 Movement in the Solar System
Lesson 2 What Does the Moon's Movement Cause?

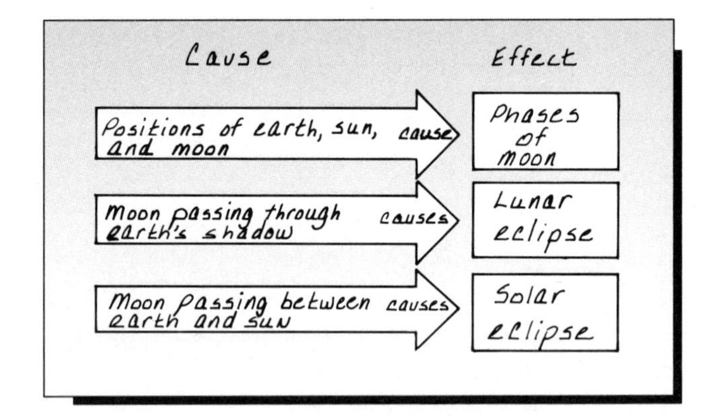

Questions answered from diagram:

- What causes the phases of the moon?
- How does a lunar eclipse differ from a solar eclipse?

Questions for discussion:

- How long do the phases of the moon last?
- Why does only part of the earth experience a solar or lunar eclipse at one time?

Circling Stages

Purpose

This strategy helps students visualize how living things change as they pass through natural stages in their life cycles.

Directions for the Students

1. As you read the lesson, pick out separate stages that are being described. Write the name of the first stage in a circle. Also write in the circle a sentence, phrase, or a few key words that will remind you what happens during that stage.
2. Continue drawing and labeling circles for all the stages in the life cycle, as shown in the example. If a stage is not named, choose labels that make sense to you.
3. Number the stages from first to last.
4. Under the diagram, write some questions that you can answer just by using the diagram.
5. Write some questions about the lesson topics that you cannot answer just by using the diagram. These questions may or may not be answered in the lesson but can all be discussed.

Example

Book 5 Chapter 2 Plant Processes
Lesson 4 How Do Plants Produce Seeds?

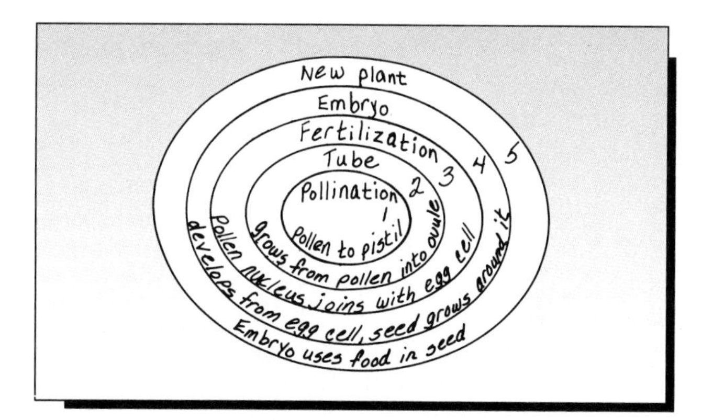

Questions answered from diagram:

- What does the embryo of a plant grow from?
- What must occur before fertilization can take place?

Questions for discussion:

- How does a pollen grain get from a stamen to a pistil?
- What do seeds need to grow?

Charting a Process

Purpose

This strategy helps students improve their understanding of the steps within a scientific process.

Directions for the Students

1. Construct a chart with three columns labeled as in the example.
2. As you read, find the important actions that are involved in the process you are assigned. List them in the "Actions" column.
3. List the important objects, organisms, or conditions for each action in the "Things" column.
4. List the results of each action in the "Results" column.
5. Check the accuracy of your chart.
6. Under the chart, write some questions that can be answered from the chart.
7. Write some questions about the lesson topics that cannot be answered from the chart. These questions may or may not be answered in the lesson but can all be discussed.

Example

Book 6 Chapter 11 Forecasting Weather
Lesson 4 How Is Severe Weather Predicted?

Process: thunderstorm formation
(Only part of this process is charted in this example.)

Things	Actions	Results
droplets in cumulus clouds	collide	large drops
large drops	fall from cloud	rain
electrons	flow between positive and negative	lightning

Questions answered by chart:

- What happens when tiny droplets collide in a cloud?
- What causes the sound of thunder?

Questions for discussion:

- Do thunderstorms form only along cold fronts?
- Why do we hear the thunder after we see the lightning?

Whole Language Instruction in Science

by Robert A. Pavlik and Patricia T. Hinske

Discover *Discover Science* invites you and your students to advance beyond merely teaching and learning about new scientific knowledge. Through whole language instruction, your students develop a personal science literacy that carries beyond the classroom. This literacy involves the abilities to produce one's own understandings and applications of science. Whole language instruction helps build and refine science literacy by using a wide variety of activities that provide a full spectrum of language opportunities.

A good whole language activity involves students in:

- reading, writing, speaking, and listening opportunities.
- raising their own questions and hypotheses.
- taking risks by sharing authority and responsibility for planning, implementing, and evaluating the activity.
- cooperating in the management of specific activity components.
- investigating beyond typical classroom resources.
- making connections within and among content areas.

To help you implement whole language instruction in your science class, a feature called *Applying Whole Language* appears in the *Teaching Options* on the second page of each chapter. This feature serves as a model to show how a whole language framework can be applied to any activity you select in the *Teaching Options.*

As you can see in the example in the box, the framework includes four components: purpose, context, decisions, and evaluation. The ideas contained in this framework are only brief samples. The descriptions of the components that follow will show some of the many ways you can stretch and fill out this framework.

Applying Whole Language

Discuss the whole language framework with each Teaching Option you select. Here is an example applied to the Science and Language Arts option on p. 33.

1. **Purpose:** To make a class poetry book about animals
2. **Context:** The roles of poets and publishers
3. **Decisions:** What steps are involved in writing poems? What information can we get from science? from other subjects? How can technology help? (computer)
4. **Evaluation:** What would our readers and listeners learn about animals? (See p. T30.)

1. Purpose: Whole language instruction in science begins with teacher-student discussions to focus on the direction a particular activity might take. Your discussions will clarify options and establish realistic goals.

2. Context: The second phase of whole language is open-ended. It provides the opportunity to clarify the reasons for doing the activity and to identify the procedures and variables of the activity. Some of these variables might include setting, time, roles of participants, group size, warnings, and exceptions.

3. Decisions: Once the whole language activity has a clear purpose and context, students develop additional voice through creative problem solving. All participate in making decisions and carrying out appropriate actions. Decisions might involve materials, resources, permission, alternatives, travel, time, sequence, and presentation.

4. Evaluation: Discussions during and after the activity give students a chance to reflect upon their actions; how the activity has helped them, others, and their science awareness. You could initiate discussion using affective questions as well as process- and result-oriented questions. Here are some ideas:

- Which parts of the activity would we keep or change? Why?
- What science information did we need to know in order to be successful?
- What information from other subjects did we need to know in order to be successful?
- What questions did the activity help us raise?
- What did we learn about improving our reading, writing, speaking, and listening?
- What did we learn about organization?
- What new information did we discover? by ourselves? from others?
- What did we learn that we did not expect?
- Who would be interested in our activity?
- What advice would we give to others who do the activity?

A whole language approach allows whole concept learning. Some students will see only pieces and parts. Others will make connections. Science literacy builds as meaningful language interactions occur in the classroom. In a whole language environment, students learn from their own involvement and from the curiosities of others. These interactions mirror the interactions of people in "real world" projects. Success will be assured if children are allowed to think about and investigate questions of their own wonder. We are confident that, through whole language instruction, your students will lift their understandings of science concepts to new levels of awareness and purpose.

Materials List

For Activities and Discover

Quantities are based on a class size of 30 students.

A

★ Aluminum foil . . . . 1 roll

B

★ Baking soda . . . . 1 box

★ Balls
- large 10
- small 10
- dark-colored 15

Batteries, flashlight (size D) 15

Bones
- chicken leg 15
- beef 15

Books, hardcover 10

★ Bowls, large 15

Boxes, small with lids 15

★ Bulbs, flashlight 15

★ Buttons, assorted 150

C

Cabbage, red . . . . 1 head

★ Cans, shiny metal 15

★ Cardboard
- large lightweight . . . . 30 sheets
- small heavy . . . . 15 pieces

Celery, with leaves . . . . 2 bunches

★ Chalk . . . . 25 pieces

★ Clay, modeling . . . . 30 sticks

Clock, large with second hand 1

★ Cloth
- cotton (10 cm x 10 cm) . . . . 15 pieces
- linen (10 cm x 10 cm) . . . . 15 pieces

★ Coat hangers 15

★ Coffee cans, with lids 12

Compasses, math (for drawing circles) 15

Construction paper
- assorted colors . . . . 105 sheets
- white . . . . 15 sheets
- black . . . . 15 sheets

★ Cover slips 15

Crayons . . . . 30 sets

★ Cups
- clear-plastic 90
- tall clear-plastic 95
- paper 60

F

★ Feathers, bird 15

File cards 120

★ Flashlights, with batteries 15

Food, assorted samples 60

★ Food coloring, red . . . . 5 small bottles

G

Glue, white . . . . 35 bottles

H

★ Hand lenses 15

J

★ Jars, large 15

L

Leaves, assorted tree 60

M

★ Magnets 15

Marking pens, colored 25

★ Medicine droppers 15

★ Meter sticks 10

★ Microscope slides 15

O

★ Objects, assorted small 45

Onion, large 1

P

★ Paint brushes 30

Paper
- drawing . . . . 30 sheets
- white, unlined . . . . 15 sheets
- white, heavy . . . . 55 sheets
- notebook . . . . 15 sheets

Paper clips, large 30

★ Paper fasteners 15

★ Paper plates
- small 375
- large 15

Paper punches. 15

Paper towels . . . . .3 rolls

Pencils, with erasers 30

Pennies 15

Photographs, large
- spider 1
- insect 1

★ Pipe cleaners. 240

★ Plastic bags
- zip-type 30
- regular. 30

★ Plastic wrap. . . . . .1 roll

★ Pliers. . . . . .15 pairs

Popcorn, kernels . . . . .5 bags

★ Prisms. 10

R

★ Rocks, assorted. 45

Rubber bands, large. 30

★ Rulers, metric (plastic) 30

S

★ Salt . . . . .1 box

★ Sand . . . . .2 bags

Scissors . . . . .30 pairs

★ Seeds
- radish (60 seeds). . . . . .1 package
- sunflower . . . . .1 bag

Shoeboxes .30

★ Soil
- potting. . . . . .3 bags
- garden. . . . . .1 bucket
- clay . . . . .1 bucket

★ Spoons, plastic 90

★ Stirring sticks. 15

★ Stones, small rough 180

★ Straws. 30

★ String . . . . .5 balls

★ Sugar . . . . .2 boxes

★ Sugar cubes . . . . .1 box

★ Sugar test paper . . . . .5 bottles

T

Tape
- masking. . . . . .6 rolls
- transparent . . . . .15 rolls

★ Thermometer, Celsius (alcohol only). .15

★ Tongue depressors 30

★ Toothpicks. . . . . .1 box

Towels, small terry. 15

V

★ Vinegar . . . . .1 bottle

W

★ Washers, large metal 150

★ Waxed paper . . . . .2 rolls

★ Wire, thin insulated . . . . .900 cm

★ Wood blocks, small 10

★ *Items included in Scott, Foresman* Discover Science *Master Lab Kit Quantities in the kits may vary from quantities listed here.*

Scott, Foresman

Discover SCIENCE

Authors

Dr. Michael R. Cohen
Professor of Science and Environmental Education
School of Education
Indiana University
Indianapolis, Indiana

Dr. Timothy M. Cooney
Professor of Earth Science and Science Education
Earth Science Department
University of Northern Iowa
Cedar Falls, Iowa

Cheryl M. Hawthorne
Science Curriculum Specialist
Mathematics, Engineering, Science Achievement Program (MESA)
Stanford University
Stanford, California

Dr. Alan J. McCormack
Professor of Science Education
San Diego State University
San Diego, California

Dr. Jay M. Pasachoff
Director, Hopkins Observatory
Williams College
Williamstown, Massachusetts

Dr. Naomi Pasachoff
Research Associate
Williams College
Williamstown, Massachusetts

Karin L. Rhines
Science/Educational Consultant
Valhalla, New York

Dr. Irwin L. Slesnick
Professor of Biology
Western Washington University
Bellingham, Washington

Scott, Foresman and Company
Editorial Offices: Glenview, Illinois

Regional Offices: Sunnyvale, California • Tucker, Georgia • Glenview, Illinois • Oakland, New Jersey • Dallas, Texas

Consultants

Special Content Consultant

Dr. Abraham S. Flexer
Science Education Consultant
Boulder, Colorado

Health Consultant

Dr. Julius B. Richmond
John D. MacArthur Professor of Health Policy
Director, Division of Health Policy Research and Education
Harvard University
Advisor on Child Health Policy
Children's Hospital of Boston
Boston, Massachusetts

Safety Consultant

Dr. Jack A. Gerlovich
Science Education Safety Consultant/Author
Des Moines, Iowa

Process Skills Consultant

Dr. Alfred DeVito
Professor Emeritus Science Education
Purdue University
West Lafayette, Indiana

Activity Consultants

Edward Al Pankow
Teacher
Petaluma City Schools
Petaluma, California

Valerie Pankow
Teacher and Writer
Petaluma City Schools
Petaluma, California

Science and Technology Consultant

Dr. David E. Newton
Adjunct Professor—Science and Social Issues
University of San Francisco
College of Professional Studies
San Francisco, California

Cooperative Learning Consultant

Dr. Robert E. Slavin
Director, Elementary School Program
Center for Research on Elementary and Middle Schools
Johns Hopkins University
Baltimore, Maryland

Gifted Education Consultants

Hilda P. Hobson
Teacher of the Gifted
W.B. Wicker School
Sanford, North Carolina

Christine Kuehn
Assistant Professor of Education
University of South Carolina
Columbia, South Carolina

Nancy Linkel York
Teacher of the Gifted
W.B. Wicker School
Sanford, North Carolina

Special Education Consultants

Susan E. Affleck
Classroom Teacher
Salt Creek Elementary School
Elk Grove Village, Illinois

Dr. Dale R. Jordan
Director
Jordan Diagnostic Center
Oklahoma City, Oklahoma

Dr. Shirley T. King
Learning Disabilities Teacher
Helfrich Park Middle School
Evansville, Indiana

Jeannie Rae McCoun
Learning Disabilities Teacher
Mary M. McClelland Elementary School
Indianapolis, Indiana

Thinking Skills Consultant

Dr. Joseph P. Riley II
Professor of Science Education
University of Georgia
Athens, Georgia

Reading Consultants

Patricia T. Hinske
Reading Specialist
Cardinal Stritch College
Milwaukee, Wisconsin

Dr. Robert A. Pavlik
Professor and Chairperson of Reading/Language Arts Department
Cardinal Stritch College

Dr. Alfredo Schifini
Reading Consultant
Downey, California

Cover painting commissioned by Scott, Foresman
Artist: Ralph Giguere

ISBN 0-673-35687-6

12345678910RRW9998979695949392

Reviewers and Content Specialists

Dr. Ramona J. Anshutz
Science Specialist
Kansas State Department of Education
Topeka, Kansas

Teresa M. Auldridge
Science Education Consultant
Amelia, Virginia

Annette M. Barzal
Classroom Teacher
Willetts Middle School
Brunswick, Ohio

James Haggard Brannon
Classroom Teacher
Ames Community Schools
Ames, Iowa

Priscilla L. Callison
Science Teacher
Topeka Adventure Center
Topeka, Kansas

Rochelle F. Cohen
Education Coordinator
Indianapolis Head Start
Indianapolis, Indiana

Linda Lewis Cundiff
Classroom Teacher
R. F. Bayless Elementary School
Lubbock, Texas

Dr. Patricia Dahl
Classroom Teacher
Bloomington Oak Grove Intermediate School
Bloomington, Minnesota

Audrey J. Dick
Supervisor, Elementary Education
Cincinnati Public Schools
Cincinnati, Ohio

Nancy B. Drabik
Reading Specialist
George Washington School
Wyckoff, New Jersey

Bennie Y. Fleming
Science Supervisor
Providence School District
Providence, Rhode Island

Mike Graf
Classroom Teacher
Branch Elementary School
Arroyo Grande, California

Thelma Robinson Graham
Classroom Teacher
Pearl Spann Elementary School
Jackson, Mississippi

Robert G. Guy
Classroom Teacher
Big Lake Elementary School
Sedro-Woolley, Washington

Dr. Claude A. Hanson
Science Supervisor
Boise Public Schools
Boise, Idaho

Dr. Jean D. Harlan
Psychologist, Early Childhood Consultant
Lighthouse Counseling Associates
Racine, Wisconsin

Dr. Rebecca P. Harlin
Assistant Professor of Reading
State University of New York—Geneseo
Geneseo, New York

Richard L. Ingraham
Professor of Biology
San José State University
San José, California

Ron Jones
Science Coordinator
Salem Keizer Public Schools
Salem, Oregon

Sara A. Jones
Classroom Teacher
Burroughs-Molette Elementary School
Brunswick, Georgia

Dr. Judy LaCavera
Director of Curriculum and Instruction
Learning Alternatives
Vienna, Ohio

Jack Laubisch
K-12 Science, Health, and Outdoor Education Coordinator
West Clermont Local School District
Amelia, Ohio

Douglas M. McPhee
Classroom Teacher/Consultant
Del Mar Hills Elementary School
Del Mar, California

Larry Miller
Classroom Teacher
Caldwell Elementary School
Caldwell, Kansas

Dr. Robert J. Miller
Professor of Science Education
Eastern Kentucky University
Richmond, Kentucky

Jan Murphy
Classroom Teacher
Rosemeade Elementary School
Carrollton, Texas

Sam Murr
Teacher—Elementary Gifted Science
Academic Center for Enrichment—Mid Del Schools
Midwest City—Del City, Oklahoma

Janet Nakai
Classroom Teacher
Community Consolidated School District #65
Evanston, Illinois

Patricia Osborne
Classroom Teacher
Valley Heights Elementary School
Waterville, Kansas

Elisa Pinzón-Umaña
Classroom Teacher
Coronado Academy
Albuquerque, New Mexico

Dr. Jeanne Phillips
Director of Curriculum and Instruction
Meridian Municipal School District
Meridian, Mississippi

Maria Guadalupe Ramos
Classroom Teacher
Metz Elementary School
Austin, Texas

Elissa Richards
Math/Science Teacher Leader
Granite School District
Salt Lake City, Utah

Mary Jane Roscoe
Teacher and Team Coordinator
Fairwood Alternative Elementary School of Individually Guided Education
Columbus, Ohio

Sister Mary Christelle Sawicki, C. S. S. F.
Science Curriculum Coordinator
Department of Catholic Education Diocese of Buffalo
Buffalo, New York

Linda Shepard
Classroom Teacher
Oscar Hinger School
Canyon, Texas

Ray E. Smalley
Classroom Teacher/Science Specialist
Cleveland School of Science
Cleveland, Ohio

Anita Snell
Elementary Coordinator for Early Childhood Education
Spring Branch Independent School District
Houston, Texas

Norman Sperling
Chabot Observatory
Oakland, California

Sheri L. Thomas
Classroom Teacher
McLouth Unified School District #342
McLouth, Kansas

Lisa D. Torres
Science Coordinator
Lebanon School District
Lebanon, New Hampshire

Alice C. Webb
Early Childhood Resource Teacher
Primary Education Office
Rockledge, Florida

Unit 2 Physical Science 90

Unit 4

Human Body 272

x

TEACHING PLAN

Overview
This introductory section provides descriptions of the steps involved in scientific methods. An ongoing example helps students relate these steps to questions they may have about nature.

Discussion
Ask students what they think science is and what scientists do. Write their responses on the board. Then explain that science is the search for answers to questions about nature. Science also includes the knowledge that has been gained by searching for those answers. Therefore, when studying science, students will make discoveries themselves as well as learn what others have discovered. Relate these ideas to those students gave.

Science Background
A musical sound, or tone, is made when air vibrates a certain number of times each second. These vibrations are sound waves. A musical instrument has features that produce, contain, control, and amplify sound waves to produce musical sounds.

A flute is a woodwind instrument. Its sounds are produced by blowing across, not into, a hole in the mouthpiece.

Discovering Science

Scientific Methods

The instruments in an orchestra make many different sounds. Some instruments make loud, sharp sounds. Some make soft, low sounds. Others make sounds in between. How do the instruments make the sounds? You can use a **scientific method** to find the answer to this question.

Scientists use scientific methods to study problems. These methods have certain steps. The order of the steps might change depending on the problem. Read on to see how you can use these steps to answer some of your own questions.

Identify Problem

The problem is the question you want to answer.

Nicole had a question about the flute. She wondered how the many different sounds of the flute were made. Her problem was: How does the flute make different sounds?

Make Observations

Observations are anything you notice about the problem.

The first step in finding the answer to the problem is to make observations. Nicole noticed that a flute has several holes. A flute player blows across the mouthpiece of the flute. Each time the player opens a hole the flute makes a different sound. Nicole wrote down what she observed.

1

Teaching Tips

- Students might need a clarification of the use of the plural versus the singular form of the term *scientific method*. The plural *scientific methods* refers to the many combinations of steps scientists use. For example, one method might include all the steps in the order given in this section. Other methods include different combinations of these steps.
- The concepts related to Nicole's experiment are discussed in Chapter 8. Refer to this introductory section when discussing Chapter 8.
- Point out to students that *data* is the plural form of *datum*.
- Emphasize that a hypothesis is based not only on *observations* about the problem but also on *knowledge* related to the problem.
- It may be helpful to distinguish between hypothesis and theory. The difference is not always clear, but a theory is usually more general than a hypothesis and is backed by more evidence. A hypothesis sometimes develops into a theory but more often a single theory incorporates several hypotheses.
- Point out to students that *hypotheses* is the plural form of *hypothesis*.

TEACHING PLAN

Teaching Tips

- Emphasize that an experiment may have many variables, but only one should change at a time between setups within the experiment or between different trials of the experiment.
- Point out that a control is part of the experiment that is not changed by the variable being tested. With a control, you can tell if the tested variable caused the experiment's results or if the results would have happened anyway. In Nicole's experiment, she could not conclude that the amount of air inside a flute affects its sound if the uncut straw made the same sounds as the cut ones.
- Emphasize that data can be more than just numbers. Written descriptions, drawings, diagrams, photographs, sounds, videos, and other information that can be recorded can all qualify as data.
- For concrete examples and applications of charts and graphs, have students skim through a newspaper or newsmagazine, pointing out the charts and graphs. Discuss how each organizes and summarizes data and how it provides a lot of information at a glance.
- Refer students to some of the *Skills for Solving Problems* features that appear before each *Chapter Review* to compare ways of organizing data in charts and graphs.

State Hypothesis

A hypothesis is a possible answer to the problem.

Nicole thought about her problem and her observations. She knows that there is air in the flute. You change the amount of air inside the flute when you open different holes. Nicole thought of a hypothesis. Her hypothesis was that changing the amount of air inside a flute causes different sounds.

Test Hypothesis

You can test a hypothesis by doing an experiment.

Nicole thought about how she could do an experiment to test her hypothesis. She decided to use four straws. Nicole cut three of the straws. One straw she made 14 centimeters long. Another straw she made 10 centimeters long. Nicole made the third straw 6 centimeters long. So each straw had a different amount of air inside. Then Nicole blew across the top of each straw to find out if the sound changed. She started with the longest straw.

Anything in an experiment that can be changed is a **variable.** The variable Nicole changed was the length of the straws. Would it matter if she blew harder across each straw? Yes, because then she would be changing two variables. She would not know which variable caused the different sounds.

If possible, an experiment should have a **control.** The control is the part of the experiment that does not change. In Nicole's experiment, the uncut straw is the control.

Collect Data

Data are observations from the experiment.

To collect her data, Nicole blew across the top of each straw. She noticed the sound each straw made.

Study Data

Decide what the information means.

Nicole looked at her data and thought about what it meant. To help study the data, she put some of it into a chart. Then she could quickly see what the experiment showed.

Length of Straw	Sound
whole	Very Low
14 cm	Low
10 cm	Higher
6 cm	Highest

Make Conclusions

Decide if your hypothesis is correct.

Nicole decided that changing the amount of air inside a flute causes different sounds. Notice how her conclusion is like her hypothesis.

Teaching Tips

- Ask students why it is important that scientists share the results of their work with others. (Scientists use the knowledge gained by others in doing their own work. Also, scientists try to duplicate the experiments to compare and verify the results.)
- Students will have many opportunities to use scientific methods in *Discover Science*. These include *INVESTIGATE!* margin features in each chapter and the *Experiment Skills* section on pages 348–375.

TEACHING PLAN

Science Background

Thaddeus Cahill made the first electronic instrument in 1906. It weighed about 180 metric tons and transmitted sounds over telephone wires. A variety of electronic instruments have been invented since that time. In 1966, the first commercially available synthesizers were designed by Robert Moog and by Donald Buchla. The Moog synthesizer included a keyboard attachment. The Buchla synthesizer produced sounds by pressing a touch-sensitive plate that had no individual keys. Both types of synthesizers had controls for regulating the volume, pitch, and tone of the sounds produced. Today, synthesizers are used in various types of music, including classical, rock, jazz, and rhythm and blues.

Applying Science

Synthesizers

Scientists discovered new ways to make musical sounds.

Some sounds from musical instruments are made by plucking or rubbing strings. Other sounds are made by blowing into a tube or hitting the instrument. Today, some musicians use a new kind of instrument called a synthesizer (sin′ thə sī′ zər).

A scientist named Robert Moog invented the synthesizer in 1964. This instrument uses electric current to make sounds.

A synthesizer has a keyboard and many switches and dials. Some musicians use synthesizers because they can sound like different instruments. By flipping the right switch or turning the right dial, a synthesizer can sound like a trumpet, a drum, a piano, or a violin. It can also make sounds like wind, rainfall, and thunder. Most musicians use synthesizers because they can make sounds that no other instrument can make.

Today, synthesizers use computers to put together the sounds of many instruments. From this one machine, you can get the sounds of a whole orchestra!

Teaching Tips

- Have students listen to musical recordings produced by synthesizers. You might play a recording of *Switched on Bach,* which was played on a Moog synthesizer. This recording, which was released in 1968, introduced the synthesizer to the general public. Ask students to describe the types of sounds they hear on the recordings.
- Directs students' attention to the photograph at the bottom of page 5. Point out the computer that is used to put together the different sounds made by the synthesizer.
- Students who play electronic instruments might like to bring their instruments to class and demonstrate the sounds the instruments make.

1

UNIT 1/Life Science

Planning Guide

Unit Components			Pages
Unit Overview/*SCIENCE IN THE NEWS*			6–7
Chapter 1 Plant Growth	**Lesson 1** How Are Roots, Stems, and Leaves Important? **Lesson 2** How Are Flowers and Cones Important?	**Lesson 3** How Do Plants Grow From Seeds?	8–27
Chapter 2 How Animals Grow and Change	**Lesson 1** How Can Animals Be Grouped? **Lesson 2** How Do Some Animals With Backbones Grow and Change?	**Lesson 3** How Do Some Animals Without Backbones Live and Grow?	28–47
Chapter 3 Living Things Need Each Other	**Lesson 1** What Are the Five Groups of Living Things? **Lesson 2** How Do Organisms Live Together?	**Lesson 3** How Do Organisms Get Food?	48–65
Chapter 4 How People Affect Plants and Animals	**Lesson 1** How Do People Change the Lives of Plants and Animals? **Lesson 2** How Do People Protect Plants and Animals?	**Lesson 3** How Do People Use Plants and Animals?	66–85
Careers	forester, gardener, fish-culture technician, microbiologist, meat inspector		86
How It Works	An Aquarium Filter		87
Unit Review			88
Unit Projects and Books	Science Projects Science and Society Books About Science		89

Science Process Skills Book

The Science Process Skills Book contains worksheets that can be used to teach and then assess student mastery of the basic science process skills. In addition, other worksheets in this book teach students the manipulative skills they will need to use basic science equipment. You might use some or all of these worksheets

- at the beginning of the year to teach and assess skills.
- directly before students encounter a particular skill in an activity.
- as a remedial tool when students begin to have trouble with specific skills.

:ience Resources for the Unit

:esource Books

·ton, Robert. *Beginnings of Life.* ·acts on File, 1986. Discusses the ife cycles of plants and animals.

·ton, Maurice. *Warm-blooded ·nimals.* Facts on File, 1986. :xplores the differences between varm-blooded mammals and ther animal species.

·ton, Maurice. *Cold-blooded 1nimals.* Facts on File, 1986. ;urveys the entire animal ingdom.

lliday, Tim R., and Adler, Kraig, ·ds. *The Encyclopedia of Reptiles ınd Amphibians.* Facts on File, 986. Describes the range, ıabitats, and behaviors of reptiles ınd amphibians.

nbert, David. *Planet Earth 2000.* ·acts on File, 1985. Discusses the ·roblems of diminishing natural esources and growing ›opulations.

cFarlane, Ruth B. Alford. *:ollecting and Preserving Plants ·or Science and Pleasure.* Arco, 985. Offers guidance on ıandling and displaying plants.

Γoole, Christopher, ed. *The :ncyclopedia of Insects.* Facts on ·ile, 1986. Describes the range, ıabitat, characteristics and ›ehavior of insects.

ver, Donald M. *The Animal Vorld: From Single-Cell Creatures o Giants of the Land and Sea.* Random House, 1987. Describes ıll the major animal groups.

Community Resources

Check with local greenhouse operators, foresters, or garden clubs about speaking to the class or about having the class visit them to learn more about plants. You might contact the following to inquire about grants for young gardeners: The National Gardening Association, 180 Flynn Avenue, Burlington, VT 05401.

Take a field trip to a local zoo, aquarium, or natural history museum.

Audio-Visual Resources

Animals and How They Grow. National Geographic. Filmstrips with cassettes, 13–14 minutes each.

Flowers to Seed. Churchill. Film, 8 1/2 minutes. Time-lapse photography of seed production.

Get Ready, Get Set, Grow!. Bullfrog Children's Films. Film or video, 15 minutes. Combines live action with animation for a science lesson on how plants reproduce and how plants and insects interact.

How Living Things Depend on Each Other. National Geographic. Filmstrip with cassette, 13 minutes. Introduces the concept of interdependence.

Learning About Leaves. Britannica. Film, 11 minutes. Viewers learn about pine needles, vegetable leaves, blades of grass, and deciduous leaves.

Turn Off Pollution. Britannica. Film, 11 minutes. Children take steps to eliminate pollution from their community and in the process they identify pollution's major sources.

The World of Plants. National Geographic. Filmstrips with cassettes, 12–13 minutes each. *The Parts of a Plant, How Plants Grow, Kinds of Plants, Where Plants Grow,* and *Plants and People.*

Computer Software

Animal Facts. Aquarius. Apple II Series. 48K, TRS80 III 32K, two cassettes, or TRS80 I, III, 32K. Introduction to animals.

Animal Photo Fun. DLM. Apple IIe+, IIe, IIc, 48K. 5–10 minutes, 6 activities. Six progressively more difficult games challenge the student to classify animals by habitat.

Mammals. Right-On. Apple II Series, 48K or Commodore PET 64, 32K. Discusses animals in general and mammals in particular.

Micro Gardener. Educational Activities. Three simulations of growing plants.

Plants and How They Grow. Right-On. Apple II Series, 48K or Commodore 64, 64K. Shows the parts of plants and trees, describes what each part does, and explains how seeds are carried.

TEACHING PLAN

Unit Overview
This unit explores plants and animals. It describes how plants are alike and different, plant propagation, and plant growth. It also describes how animals live and grow in different ways. The unit describes how plants and animals interact, and the effects of human actions on other living things.

About the Photograph
The herd of zebras in the photograph represents one population in the grassland community. Zebras depend on the grasses in their environment for food. Herds of zebras band together to flee from grassland predators, such as lions and cheetahs.

6

Teaching Options

Chapter 1 Preview

Plant Growth
This chapter focuses on basic plant structure and growth. Students will observe the movement of fluids through a vascular plant, learn about the differences between various types of plants, and find out about the life cycle of a plant, from seed dispersal to pollination.

Chapter 2 Preview

How Animals Grow and Change
In this chapter students will learn about vertebrates and invertebrates. Characteristics of birds, reptiles, fish, amphibians, and mammals are described. Students will study the characteristics of the main groups of invertebrates and the life cycle of an insect that undergoes metamorphosis.

Life Science

Unit 1

These zebras stay together most of the time. Together, they look for food and water. Zebras usually live in places where tall grass grows.

Many other living things live in groups. In this unit, you will learn about living things. You will discover ways living things affect each other.

SCIENCE IN THE NEWS During the next few weeks, look in newspapers or magazines for stories about plants and animals. Also look for news about places, such as parks or forests, where plants and animals live. Share the news with your class.

Teaching Tips

- Before starting this unit, lead a discussion about plants and animals. Ask students to *name* different plants and animals. List the examples on the chalkboard.
- Direct students' attention to the photograph on page 6. Questions: **What kind of living thing is shown?** (zebras, which are animals) **What are the zebras in the background doing?** (eating grass) **What kinds of animals might kill and eat the zebras?** (lions, cheetahs, and hyenas)
- Explain that in this unit students will learn more about different kinds of plants and animals and how they depend on each other.

Chapter 3 Preview

Living Things Need Each Other

In this chapter the students will be introduced to the concept of cells as the basis for all organisms. The dependence of organisms on each other and the balance of organisms in a community will be explored. Basic five kingdom classification is presented, along with the concepts of predator/prey, producer/consumer, and food chains.

Chapter 4 Preview

How People Affect Plants and Animals

In this chapter humans' place in, and their effect on, the environment and other organisms' habitats are explored. Ways of preserving habitats and protecting organisms are also presented. The place of animals and plants in people's everyday lives is described, as well as some hazardous plants and animals.

Chapter 1 Plant Growth Planning Guide

TEACHING PLAN

Chapter Components	Skills	Materials
Chapter Opener/*DISCOVER:* Observing Leaves pp. 8–9	*DISCOVER* p. 9 Science Process Skills *Classifying, Observing*	*DISCOVER* p. 9 (individual) 30 large sheets heavy paper, 60 assorted tree leaves, 30 sets of crayons
Lesson 1 How Are Root, Stems, and Leaves Important? pp. 10–13	Thinking Skills Challenge!: *Inferring* Find Out On Your Own: *Restating or explaining ideas*	Demonstration p. 10 carrot root, celery, lettuce leaf, rice grains, apple
Activity Observing the Movement of Water in Plants p. 14	Science Process Skills *Observing, Communicating, Measuring, Collecting and interpreting data, Predicting*	(groups of 2) 15 tall plastic cups, 5 bottles red food coloring, 2 large bunches of celery with leaves, 15 pairs of scissors
Science and People Helping Roots Do Their Jobs p. 15	Thinking Skills *Recognizing the main idea and supporting details, Inferring*	
Lesson 2 How Are Flowers and Cones Important? pp. 16–19	Thinking Skills Challenge!: *Drawing conclusions* Find Out On Your Own: *Collecting information, Comparing, Contrasting*	Demonstration p. 16 variety of flowers (tulips, daisies, and potted flowers)
Lesson 3 How Do Plants Grow from Seeds? pp. 20–22	Thinking Skills Challenge!: *Making inferences* Find Out On Your Own: *Collecting information, Restating and explaining ideas, Making physical models*	Demonstration p. 20 soaked kidney bean, corn kernel
Activity Observing How Seeds Germinate and Grow Without Light p. 23	Science Process Skills *Observing, Communicating, Inferring, Identifying variables*	(groups of 2) 30 plastic cups, 2 large bags of potting soil, 60 radish seeds, 15 plastic spoons
Skills for Solving Problems Using Thermometers and Bar Graphs pp. 24–25	Problem Solving Skills *Making decisions/Identifying and solving problems, Interpreting charts, maps, and graphs*	
Chapter Review pp. 26–27	Thinking Skills *Restating or explaining ideas, Recognizing the main idea and supporting details, Contrasting, Making inferences, Drawing conclusions*	

Teaching Options

Strategies	Extensions		Resource Masters
Cooperative Learning p. 8 (Also see p. T23.) Applying Whole Language p. 9 (Also see p. T30.)			Family Letter: *Resource Book* p. 3
Reading Strategies p. 10 (Also see pp. T26–T29.)	Science and Social Studies p. 11 Special Education p. 11 Reinforcement p. 12	Enrichment p. 12 Game Suggestion p. 13 Reteaching Suggestion p. 13	Vocabulary Preview: *Workbook* p. 1 Science and Reading: *Workbook* p. 2
			Activity Worksheet: *Resource Book* p. 7
Reading Strategies p. 16 (Also see pp. T26–T29.)	Science and Math p. 17 Special Education p. 17 Reinforcement pp. 17, 18	Enrichment p. 18 Game Suggestion p. 19 Reteaching Suggestion p. 19	Science Skills: *Workbook* p. 3
Reading Strategies p. 20 (Also see pp. T26–T29.)	Enrichment p. 21 Special Education p. 21	Reteaching Suggestion p. 22	Science Activity: *Workbook* p. 4 Vocabulary Puzzle: *Workbook* p. 5
			Activity Worksheet: *Resource Book* p. 9
	Reteaching Suggestion p. 25		Science and Reading: *Resource Book* p. 11
Cooperative Learning p. 26 (Also see p. T23.)			Chapter Tests: Forms A and B *Test Book* pp. 9–12

Classroom Management

Advance Preparation

Demonstration, page 10
Bring a carrot, a celery stalk, lettuce leaves, rice grains, and an apple to class.

Activity, page 14
Give students celery stalks about 10 centimeters long. The section of celery with leaves works best. Use about 1/2 teaspoon of food coloring in each cup of water.

Demonstration, page 16
Collect different kinds of cut flowers or wildflowers for this demonstration.

Demonstration, page 20
Soak a kidney bean for at least eight hours before doing this demonstration.

Activity, page 23
Provide four radish seeds and two cups almost full of potting soil for each group. Moisten the soil before the students plant their seeds. Students will see results sooner if you presoak seeds for about 8 hours.

Vocabulary Review

Use the following sentences with your students to review the meanings of the italicized words.

1. The *roots* of plants are usually underground and take in water for the plants to grow.
2. A plant's *stem* is usually long and thin and helps hold the plant upright.
3. The seeds of a pine tree are in its *cones.*
4. When the *temperature* of the air is low, it feels cold outside.
5. Some plants grow from *bulbs*—large, round underground plant parts that store food for the young plant.

High-Potential Students

Ask students to locate and read the story, *Jack and the Beanstalk.* Encourage them to pretend they are television reporters who have been called to a school where the giant beanstalk has grown out of sight into the sky. Direct students to write an account of the event. The account should include a description of the root system, stem, leaves, flowers, and seeds of the plant. It should also contain a description of how the seeds might have germinated and how pollination of the flowers could occur. Ask students to consider how this plant could be beneficial to the town. They should draw a picture of the beanstalk during the news report.

Mainstreamed Students

Emotionally Handicapped
Grow plants in the classroom as you study this chapter. Encourage emotionally handicapped students to care for the plants.

Visually Impaired
Uproot a plant and allow visually impaired students to handle it. Encourage students to note the texture of the leaves, the firmness of the stem, and the length and fineness of the plant's roots as they feel these plant parts.

Science Fair Projects

The Experiment Skills on p. 348 and the Investigate feature in the chapter can be used for science fair projects. You also might encourage interested students to do one of the following projects:

1. Identify and press as many specimens of wildflowers as you can. Try to include leaves and seeds with the pressed flower specimens.
2. Collect and compare conifer needles. Sort into groups, such as needles that are in bunches, and needles that are not in bunches. Place cones with the needles and clearly identify the type of the tree that the needles came from.
3. If a marine environment is nearby, collect different varieties of algae (seaweed). Dry the specimens and display them.

Classroom Resources

Bulletin Board

Encourage students to find pictures of other ways plants are important, such as for lumber, paper, furniture, and landscaping.

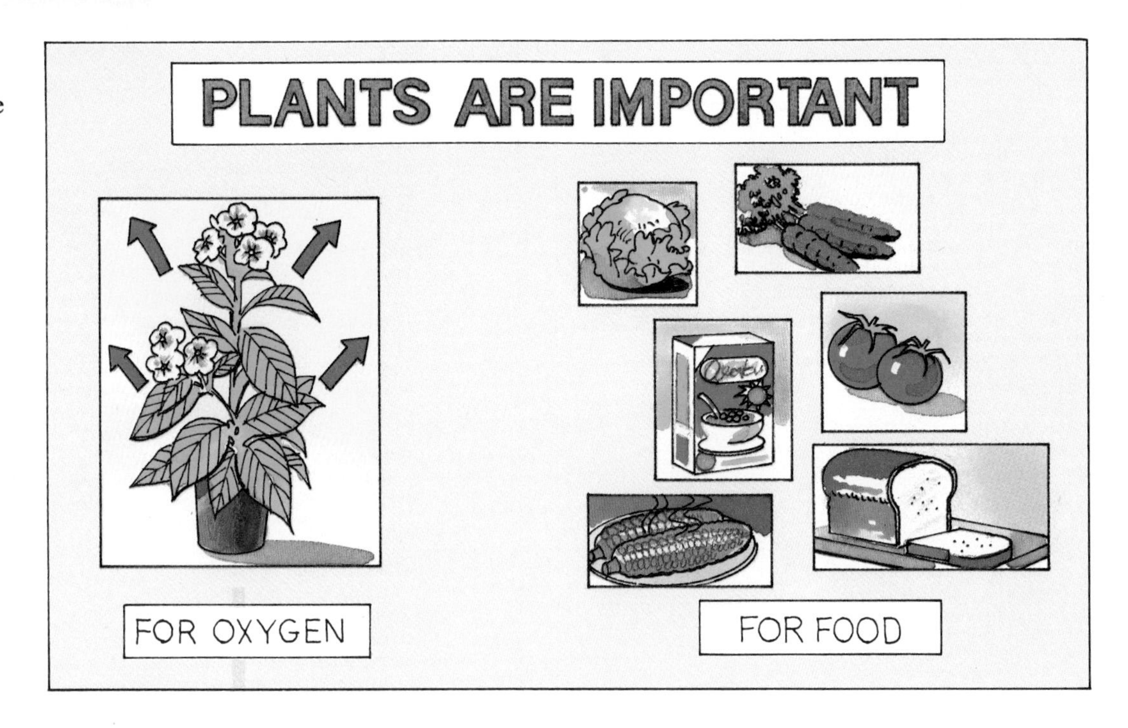

Chapter 1 Poster

Science Discovery Center

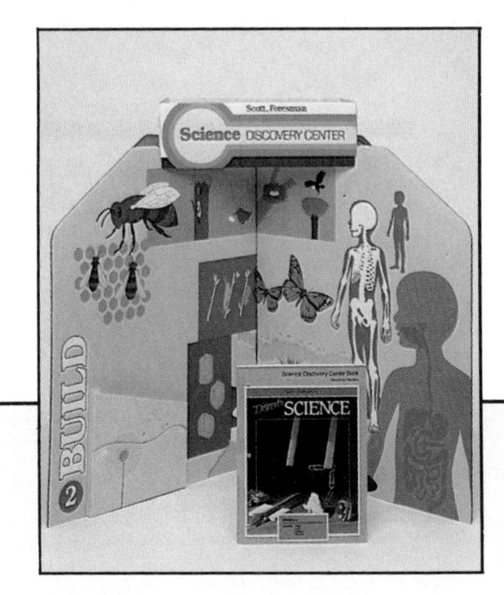

Use pages 15–20 from the *Science Discovery Center Book.* Place these worksheets in the appropriate pockets in the Science Discovery Center.

Overhead Transparencies

Use Transparencies 1, 2, and 3 from the package of color overhead transparencies.

CHAPTER 1 COPY MASTERS

Teacher's Resource Book

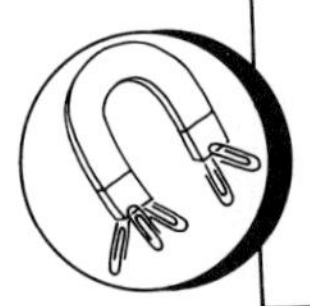

Dear Family,
Your student will be reading **Chapter 1: Plant Growth** in *Discover Science*, published by Scott, Foresman. In this chapter, we learn about how roots, stems, and leaves help a plant grow. Roots absorb water and minerals. Tubes in the stem carry water and food. In the leaves, water and carbon dioxide (a gas from the air) are turned into sugar (the basic plant food) and oxygen.
You and your student can do this activity together to learn more about plants.

Trip to a Florist

1. Take a trip to a florist shop.
2. Look at all the plants and flowers. Choose your favorites.
3. Talk with the florist and find out about the plants and flowers.
   - Where do the plants and flowers come from?
   - What do you do to keep the plants alive?
   - Do all the plants need the same care?
   - Which plants are the hardest to keep alive? Why?
   - How do you keep cut flowers fresh?
   - How long do cut flowers last?
   - Why do people buy plants and flowers?
4. Draw a picture of your favorite plant and cut flower. Write down how to care for each.

Workbook

Name ______

Use with Lesson 1: pages 10-13

Chapter 1

Vocabulary Preview

Plant Growth

Chapter Vocabulary		
carbon dioxide	oxygen	seed coat
germinate	petal	seed leaf
life cycle	pollen	seedling
mineral	pollinate	

Vocabulary Cards

1. Write each word on a card.
2. Find each word in the glossary. Copy the pronunciation under the word on the card.
3. Practice saying the words with a partner.

Pronunciation Match-Up

1. Work with a partner. Put two sets of vocabulary cards together. Shuffle the cards.
2. Place the cards face down in rows of three. The last row will have only two cards.
3. Take turns. The first player will turn two cards over. If the two words match, the player must try to pronounce them correctly.
4. Correctly pronounced words are kept by the player. If the words do not match or are not pronounced correctly, the cards are turned face down again.
5. Play continues until all words are matched. The player with the most cards wins.
6. After the game, give a complete set of cards back to each player.

At Home

1. Practice saying the words. Learn their meanings.
2. Look for the words and pictures of the words in newspapers and magazines. You can make a poster.

Teacher's Notes: Encourage the students to write the definitions on the word cards as they come across the words in the chapter.

Teacher's Resource Book

Name ______

Use with Lesson 1: pages 10-13

Chapter 1

Science and Reading

A Plant That Eats Insects

Read the story. Then answer the questions.

Most plants get all the minerals they need from the soil. But the Venus's-flytrap grows in soil that does not have enough minerals. Where does the Venus's-flytrap get the minerals it needs? It "traps flies" and "eats" them. It gets its minerals from the insects it catches.

Each leaf of the flytrap has six hairs. When an insect touches these hairs, the sides of the leaf close. The insect gets caught in the leaf. The insect cannot escape.

The Venus's-flytrap has special juices that break down the insect so that the minerals in the insect can be used by the plant. The plant uses the minerals from the insect to help the plant grow. It takes the plant 5 to 10 days to use up the insect. Then the leaf opens up and is ready to catch another insect.

The Venus's-flytrap grows only in one small part of the United States. This part is a strip of wet, swampy ground in North Carolina. This area is only about 1,000 square kilometers.

Circle the best answer.

1. The Venus's-flytrap is unusual because it
   a. grows in soil.
   (b.) catches insects.
   c. does not need water.
2. The Venus's-flytrap needs insects
   (a.) for minerals.
   b. for water.
   c. to move.
3. The Venus's-flytrap catches insects with
   a. its flowers.
   (b.) its leaves.
   c. its stem.
4. The Venus's-flytrap grows in
   a. dry, sandy deserts.
   b. snowy mountains.
   (c.) wet, swampy ground.

Teacher's Notes: Encourage students to read about other carnivorous plants, such as the sundew and pitcher plant.

Teacher's Resource Book

Name ______

Use with Lesson 1: page 14

Chapter 1

Activity Worksheet

Observing the Movement of Water in Plants

Record Your Results

What you think might happen	What happened	Distance the color moved

State Your Conclusion

1. Explain the color changes you saw in the celery stem.

2. Suppose you left the celery in the food coloring for two more hours. How would you expect the color to change in the celery? Explain your answer.

Use What You Learned

Suppose the stem of a green plant was broken. What might happen to the plant?

Name ______________________

Use with Lesson 2: pages 16-19

Chapter 1

Science Skills

Looking at Parts of Plants

Study the pictures of plant parts that have seeds.
1. Match two parts that look alike by drawing a line from one to the other.
2. Circle the plant parts that are fruits.
3. Put an X on the plant parts that are cones.

Teacher's Notes: Extend the lesson by pointing out that nuts are a kind of fruit. Ask students to make lists of kinds of nuts.

Name ______________________

Use with Lesson 3: pages 20-22

Chapter 1

Science Activity

Do Seeds Need Water?

Gather These Materials

• 4 paper cups • soil • 16 bean seeds • spoon • measuring cup

Follow This Procedure

1. Label the first cup: No Water. Label the second cup: One Spoon. Label the third cup: 50 mL. Label the fourth cup: 100 mL.
2. Half fill each cup with soil. Plant four beans in each cup.
3. Add water to the cups. The labels tell how much water.
4. Check the bean seeds every day. Record what happens.

Record Your Results **Answers may vary.**

	No Water	One Spoon	50 mL	100 mL
Day 1	**No Growth**			
Day 2	**No Growth**	**Water**		
Day 3	**No Growth**	**should not be**	**Some seeds**	**Seeds will**
Day 4	**No Growth**	**enough to let**	**may grow.**	**grow best.**
Day 5	**No Growth**	**seeds grow.**		

State Your Conclusions **Answers may vary.**

1. In which cup did the bean seeds grow best?
   Example: The bean seeds grew best in 100 mL of water.
2. How much water do bean seeds need to grow?
   Example: Seeds need to be very wet to sprout.

Teacher's Notes: For best results, use white bean seeds. Do *not* soak the beans before planting.

Name ______________________

Use with Lesson 3: pages 20-22

Chapter 1

Vocabulary Puzzle

Key Words

Use the clues to fill in the missing letters.

1. The outside covering of a seed
2. To begin to grow and develop
3. To carry pollen to the center part of a flower
4. Roots take in these nonliving materials
5. The part of a seed that looks like a leaf
6. The stages in the life of a plant
7. The fine yellowish powder in a flower
8. A gas that most living things need to stay alive
9. A young plant that grows from a seed
10. The outside parts of a flower that are often colored

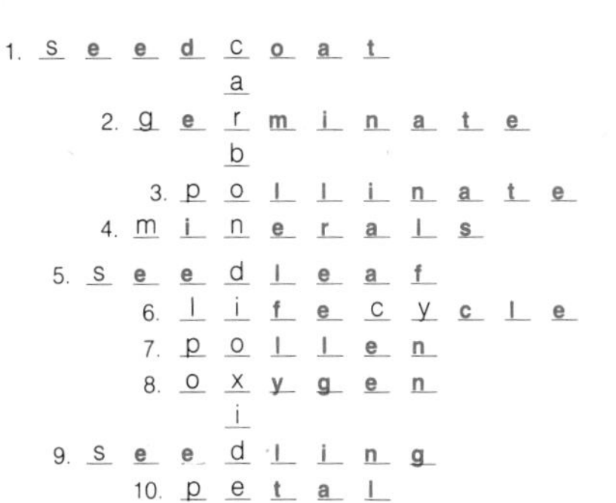
1. s e e d c o a t
a
2. g e r m i n a t e
b
3. p o l l i n a t e
4. m i n e r a l s
5. s e e d l e a f
6. l i f e c y c l e
7. p o l l e n
8. o x y g e n
i
9. s e e d l i n g
10. p e t a l

Teacher's Notes: Ask students to draw pictures showing the life cycle stages of a bean plant.

Name ______________________

Use with Lesson 3: page 23

Observing How Seeds Germinate and Grow Without Light

Chapter 1

Activity Worksheet

Record Your Results

	Drawing	Description
Plant in light		
Plant in dark		

State Your Conclusion

1. How do the plants grown in the dark look different from the plants grown in the light?
2. Suppose you kept your plants in the dark for several weeks. What would happen to the plants? Explain your answer.

Use What You Learned

Sometimes people use special lights to shine on plants all the time. How would you expect such plants to appear?

Name ______

Comprehension: time sequence

Chapter 1

Science and Reading

Dandelions

Read the story. Then answer the questions.

Samantha likes to play in the field next to her house. She loves when the dandelions are in bloom and the ground looks like it is covered with a bright yellow blanket.

She likes it even better, though, after the flowers dry up. Soft white fuzz forms on the tip of the stem. The fuzz is so light that the wind picks it from the plant and carries it away. The white fuzz carries the dandelion seeds, and wherever it falls, new yellow flowers will grow.

1. What do dandelions look like when they are in bloom?

   They are bright yellow.

2. What happens after the flowers dry up?

   The seeds have soft white fuzz.

3. How are the seeds carried to different places?

   The wind picks the fuzz from the plant and carries it away.

4. What happens after the seeds fall to the ground?

   New yellow flowers grow.

Name ______

Chapter 1
Test A

Multiple Choice Choose the best answer.

1. The parts of a plant that hold it tightly in the soil are the (1-1)
   a. leaves.
   b. stem.
   (c.) roots.
2. The plant part that carries water and minerals from the roots to the leaves is the (1-2)
   (a.) stem.
   b. bud.
   c. bark.
3. Food needed for plant growth is made in the (1-3)
   a. stem.
   (b.) leaf.
   c. bud.
4. The center of a flower contains parts that make (2-1)
   (a.) seeds.
   b. carbon dioxide.
   c. petals.
5. Pollination takes place when pollen reaches (2-2)
   a. part of an animal's body.
   b. the tips of flower petals.
   (c.) the center of a flower.
6. Trees such as spruce and fir have leaves that are shaped like (2-3)
   (a.) needles.
   b. cones.
   c. flower petals.
7. The outside covering of a seed is the (3-1)
   a. seed leaf.
   (b.) seed coat.
   c. seedling.
8. With enough air, water, and the proper temperature, plants usually will (3-3)
   (a.) germinate.
   b. pollinate.
   c. produce cones.
9. A young plant that has just sprouted is called a (3-3)
   a. bud.
   b. seed.
   (c.) seedling.
10. The life cycle of a plant includes all the (3-3)
    (a.) stages of its life.
    b. parts of its flower.
    c. seeds in its flowers.

Numbers in parentheses after each question refer to the lesson number and the objective of that lesson.

Name ______

Chapter 1
Test A

Short Answer Label the parts of the plant.

1. Flower
2. Stem
3. Leaf
4. Roots

Short Essay Use complete sentences to answer each question.

1. Name three ways that pollen is scattered. Explain how each could happen. (2-2)

   Animals can scatter pollen because pollen can become attached to their bodies.
   The wind can scatter pollen because it can blow pollen from place to place.
   Water can carry pollen from place to place as it flows.

2. How are bean seeds different from corn seeds? (3-3)

   Bean seeds have food stored in the seed leaves. A corn seed has food stored outside the seed leaf.

Numbers in parentheses after each question refer to the lesson number and the objective of that lesson.

Name ______

Chapter 1
Test B

Multiple Choice Choose the best answer.

1. The materials used by a plant that are taken in by its roots are called (1-1)
   (a.) minerals.
   b. weeds.
   c. sugar.
2. The part of a plant that holds the plant above the ground is the (1-2)
   a. leaf.
   (b.) stem.
   c. bud.
3. A plant uses some of its stored food to (1-3)
   a. share with other plants.
   b. put minerals into the soil.
   (c.) make new leaves.
4. The yellow powder inside a flower that is needed to form seeds is (2-1)
   a. talc.
   (b.) pollen.
   c. food.
5. After a flower is pollinated, part of the flower swells and becomes a (2-2)
   a. petal.
   b. new flower.
   (c.) fruit.
6. Most trees that have cones also have leaves shaped like (2-3)
   a. hearts.
   (b.) needles.
   c. fans.
7. The seed coat of a seed helps to (3-1)
   (a.) protect the seed.
   b. attract insects.
   c. pollinate the seed.
8. A plant germinates when the small plant inside the seed begins to (3-3)
   a. produce fruits.
   b. dry up and die.
   c. grow and develop.
9. Seed, seedling, and a grown plant are three stages of every plant's (3-3)
   (a.) life cycle.
   b. germination.
   c. seed leaf.
10. One way pollen can be scattered is by (3-3)
    a. sunlight.
    b. leaves.
    (c.) insects.

Numbers in parentheses after each question refer to the lesson number and the objective of that lesson.

Name ____________________

Chapter 1
Test B

Short Answer Tell the job that each plant part does.

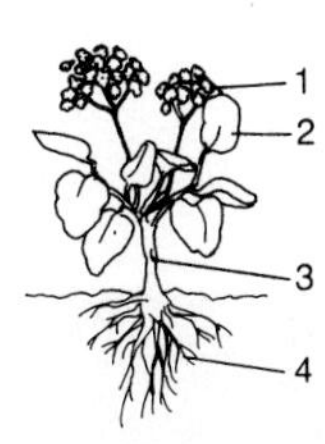

1. Produces seeds that grow into new plants
2. Holds plant parts above the ground
3. Makes the food a plant needs
4. Takes in water and minerals

Short Essay Use complete sentences to answer each question.

1. How do animals help scatter pollen? (2-2)

Pollen sticks to parts of animals such as bees and butterflies. As these animals move from flower to flower, the pollen comes off and gets to the center of the flower.

2. How do leaves make the food that a plant needs? (1-2)

Water goes from the roots through the stems to the leavès. Carbon dioxide goes into the leaves from the air. The green leaves change water and carbon dioxide into sugar and oxygen. The plants uses the sugar for its food.

Numbers in parentheses after each question refer to the lesson number and the objective of that lesson.

TEACHING PLAN

Major Concepts

Lesson 1 Roots, stems, and leaves help plants live and grow.
Lesson 2 Flowers and cones form seeds that can produce new plants.
Lesson 3 Germination, growth, and formation of new seeds are stages in the life cycle of a plant.

Chapter Vocabulary

carbon dioxide, germinate, life cycle, mineral, oxygen, petal, pollen, pollinate, seed coat, seed leaf, seedling

Getting Started

Ask students to point to different parts of the plants they see in the picture. Have students discuss why they think some plants have brightly colored flowers. Ask them to describe their favorite flower, or the most unusual flower, the largest flower, or the smallest flower they have seen.

Chapter 1

Plant Growth

Have you seen plants like some of the ones in this garden? Notice how the plants look different from each other. Yet all these plants are alike in some ways.

Teaching Options

Cooperative Learning ♦

Jigsaw Format (See page T23.)
Assign the following topics at random to your cooperative learning teams.
Topic A: What are the main parts of a plant and what do they do?
Topic B: How do plants make new plants like themselves?
Topic C: What do plants need to live, and how do they get what they need?
Topic D: How do plants help animals, and how do some animals help plants?
Have students search for information on their topic as they read the chapter. Then let all students with the same topic meet in an expert group to discuss the information. When students return to their teams, they may take turns presenting their topics to the team. Then give students a test covering all topics to complete individually (Chapter 1 Test A or B in the *Test Book*). Award Superteam certificates to teams whose average test scores exceed 90%, and Greatteam certificates to teams whose average test scores exceed 80%.

♦ ***Suitable as a language development activity***

Introducing the Chapter

In this chapter, you will learn about the parts of a plant. You also will learn how plants live and grow. The activity below will help you learn how leaves of some plants look different.

Observing Leaves

You probably have seen leaves with many shapes and sizes. A maple leaf is about the size and shape of your hand. An oak leaf can be long and thin. A horse chestnut leaf has more than one part. You can make a leaf collection to compare the sizes and shapes of leaves.

Make leaf rubbings to record the sizes and shapes of your leaves. First, cover your desk with heavy paper. Place a leaf on the paper with the rough side facing up. Next, place a sheet of white paper on top of the leaf. Then rub the paper lightly with a pencil or crayon.

Compare your leaf rubbings with the leaves in the picture. Notice how some of your leaves look alike. Group your leaves in two different ways, such as by color and size.

Talk About It

1. Describe the sizes and shapes of the leaves you found.
2. How did you group your leaves?

DISCOVER

Objective

This optional *DISCOVER* activity will help students explore and build background information about the concept of classifying leaves. Later in the chapter, students will be able to draw on this experience to help them assimilate the new content.

Science Process Skills

Classifying, Observing

Materials

For each student: heavy paper (such as newspaper), white paper, assorted leaves

Teaching Tips

- Provide students with a reference book that shows the different types of patterns and shapes leaves can have.
- Have students ***describe*** the texture of the leaves. (Some leaves have a smooth, waxy texture; some have clear leaf divisions; and some leaves have a bumpy surface or many veins.)

Answers

Talk About It

1. Answers will vary. Accept answers that describe the sizes and shapes of the leaves students are working with.

2. Answers might include: by size, shape, edge shape, or color.

Applying Whole Language ♦

Discuss the whole language framework with each Teaching Option you select. Here is an example applied to the Science and Social Studies option on p. 11.

1. Purpose: To make a brochure about food plants

2. Context: Assuming the role of a health aide

3. Decisions: What information is needed from science and other subjects? other people? places? Who should have our brochure?

4. Evaluation: How did listening help this activity? (See p. T30.)

Resource Book page 3

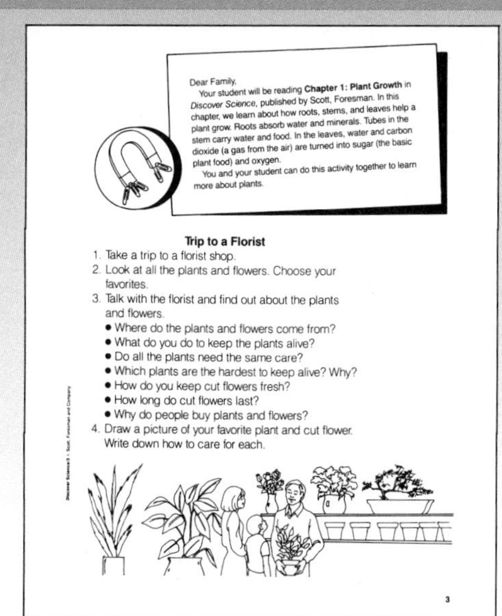

Dear Family,

Your student will be reading **Chapter 1: Plant Growth** in *Discover Science*, published by Scott, Foresman. In this chapter, we learn about how roots, stems, and leaves help a plant grow. Roots absorb water and minerals. Tubes in the stem carry water and food. In the leaves, water and carbon dioxide (a gas from the air) are turned into sugar (the basic plant food) and oxygen.

You and your student can do this activity together to learn more about plants.

Trip to a Florist

1. Take a trip to a florist shop.
2. Look at all the plants and flowers. Choose your favorites.
3. Talk with the florist and find out about the plants and flowers.
   - Where do the plants and flowers come from?
   - What do you do to keep the plants alive?
   - Do all the plants need the same care?
   - Which plants are the hardest to keep alive? Why?
   - How do you keep cut flowers fresh?
   - How long do cut flowers last?
   - Why do people buy plants and flowers?
4. Draw a picture of your favorite plant and cut flower. Write down how to care for each.

3

Science Background

Simple leaves, such as those of oaks and maples, consist of one main blade with a small stalk connected to the plant stem. Compound leaves, such as those of honey locusts and buckeyes, consist of several leaflets, each of which has a small stalk. Veins of leaves are small tubes that carry food and water through the leaf.

TEACHING PLAN

Lesson Objectives

- *Describe* plant roots, and *identify* roots as plant parts that anchor plants and absorb and conduct water and minerals from the ground.
- *Explain* how stems support plants, and transport water and minerals.
- *Explain* the process by which green leaves make the food a plant needs.

Lesson Vocabulary

carbon dioxide, mineral, oxygen

1. Motivate

Demonstration Activity ♦

Draw an "anyplant" on the chalkboard. Include a root, stem, leaf, seed, and fruit. Obtain a carrot root, an asparagus stem, lettuce leaf, rice grains, and an apple. Have the students try to match each food item with the appropriate part of the plant.

Discussion

Write the following words on a chalkboard: *root, leaf, stem, fruit,* and *seed.* Ask students to name other foods from plant parts (examples: beet—root; spinach—leaf; broccoli—stem; orange—fruit; peanut—seed)

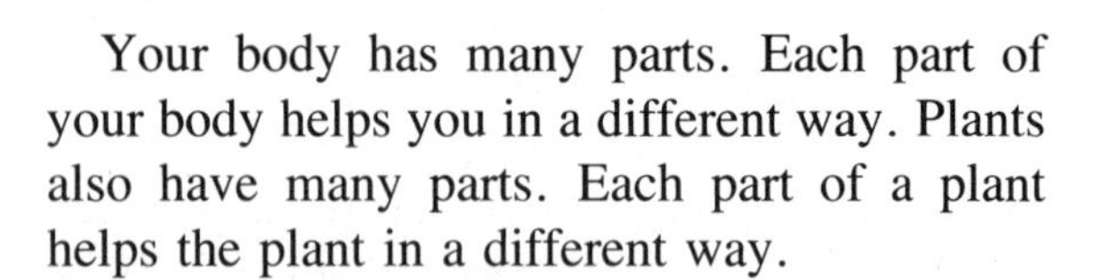

1 How Are Roots, Stems, and Leaves Important?

LESSON GOALS

You will learn
- how plant roots help plants live and grow.
- how plant stems are important to plants.
- how green leaves make food for plants.

mineral (min′ər əl), a material that was never alive and that can be found in soil.

Your body has many parts. Each part of your body helps you in a different way. Plants also have many parts. Each part of a plant helps the plant in a different way.

How Roots Are Important

Imagine trying to pull a weed like this one out of the ground. You might find that the weed does not come out easily. Compare the pictures of plant roots. Notice that the weed has a long, thick root that grows deep into the soil. The grass plant has thin roots that spread out under the plant. Both kinds of roots hold plants tightly in soil.

Roots take in water and **minerals**—materials in the soil that were never alive. Plants need water and minerals to live.

Teaching Options

Science Background

Roots anchor plants in the soil and help hold soil around plants. Root hairs grow from the outside layer of a plant's root. The root hairs help the root take in water and minerals. Xylem and phloem make up the tubelike structures that go through the roots, stems, and leaves of vascular plants. Xylem tissues carry water and minerals through a plant. Phloem cells carry food from the leaves through the plant. In leaves, green plants make food through photosynthesis. Chlorophyll and other pigments in leaves absorb light energy from the sun, and change light into chemical energy. The plant uses the stored chemical energy to change carbon dioxide and water into glucose (a sugar) and oxygen.

Reading Strategies ♦

1. Guide students' pre-reading by asking: Which parts of the lesson are known to you and which parts are new?
2. Assign these strategies: Visualizing Information and Writing a Memory Sentence. (See pages T26–T29.)
3. Pair students to share what information is clear and unclear and initiate discussion using students' unanswered questions.

♦ *Suitable as a language development activity*

How Stems Are Important

The stems of most plants hold up the leaves and other plant parts that grow above the ground. Look at the different kinds of stems in the picture. Find the plants with thin stems growing along the ground. Which plant has thick, woody stems that hold up many leaves?[1]

Stems have tiny tubes. These tubes carry water and minerals from the roots to the other parts of a plant. The tubes also carry food from the leaves to the roots.

[1]the bush

11

2. Teach

Teaching Tips

- Help students learn about how plants adapt to different climates. Describe an arid climate where ground water lies several meters below the surface. Questions: **How might the roots of a plant look in an arid climate?** (They might be very long and extend deep into the ground.) Next, describe a moderate climate where water is near the ground's surface. Have students *infer* the possible differences in root structure between the two environments. **What might the roots look like where water is close to the surface?** (Roots might be shallower and spread out more.)
- **Possible Misconception:** Some students might think that a plant absorbs water and minerals separately. Explain that minerals must be dissolved in water before a plant can absorb them.

Workbook page 1 *

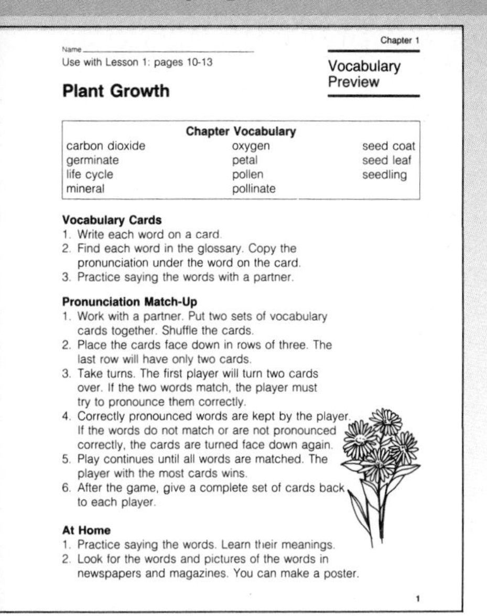

Name ______

Use with Lesson 1: pages 10-13

Chapter 1

Vocabulary Preview

Plant Growth

Chapter Vocabulary		
carbon dioxide	oxygen	seed coat
germinate	petal	seed leaf
life cycle	pollen	seedling
mineral	pollinate	

Vocabulary Cards
1. Write each word on a card.
2. Find each word in the glossary. Copy the pronunciation under the word on the card.
3. Practice saying the words with a partner.

Pronunciation Match-Up
1. Work with a partner. Put two sets of vocabulary cards together. Shuffle the cards.
2. Place the cards face down in rows of three. The last row will have only two cards.
3. Take turns. The first player will turn two cards over. If the two words match, the player must try to pronounce them correctly.
4. Correctly pronounced words are kept by the player. If the words do not match or are not pronounced correctly, the cards are turned face down again.
5. Play continues until all words are matched. The player with the most cards wins.
6. After the game, give a complete set of cards back to each player.

At Home
1. Practice saying the words. Learn their meanings.
2. Look for the words and pictures of the words in newspapers and magazines. You can make a poster.

1

* *Answers to masters on pages 8E–8H*

Science and Social Studies

Encourage students to find out which plants are considered "cash crops" in your state. You might want to discuss what types of plants can be grown in the different geographic regions of the country.

Special Education

Have students with learning disabilities draw pictures of plants and label the roots, stems, and leaves. Display the pictures and discuss the function of each of the plant parts labeled. You might want to bring examples of different types of plant roots to show the students.

TEACHING PLAN

Teaching Tips

- Refer students to the *DISCOVER* on page 9 in which they made rubbings of leaves. Have students examine their rubbings and *observe* the lines left by veins. Question: **What is the purpose of the lines in leaves?** (Lines represent veins, which are tubes that carry sugar and water between the leaves and the other parts of the plant.)
- Have students *compare* and *contrast* the patterns of the vein lines in different leaves. In some leaves, veins are parallel. In others, veins form a netted pattern. Question: **Do the lines in leaves form special patterns?** (Yes, each plant has a characteristic vein pattern.)

carbon dioxide (kär′bən dī ok′sīd), a gas in the air that plants use to make food.

oxygen (ok′sə jən), a gas in the air that living things need to stay alive.

Green Plants Make Food

How Leaves Are Important

Think about different kinds of leaves you have seen. The leaves of green plants make most of the food a plant needs. This food is sugar.

Look at the picture as you read about the way a plant makes sugar. Find the arrow that goes from the soil to the leaves. This arrow shows that water goes from the soil through the roots and stems to the leaves. Now find the arrows that point to the leaf. A gas from the air, called **carbon dioxide,** goes into the plant through tiny openings in the leaves. Green leaves use sunlight to change water and carbon dioxide to sugar and **oxygen.** Oxygen is a gas in the air that living things need to stay alive. The oxygen from the plant goes into the air. Plants use the sugar to live and grow.

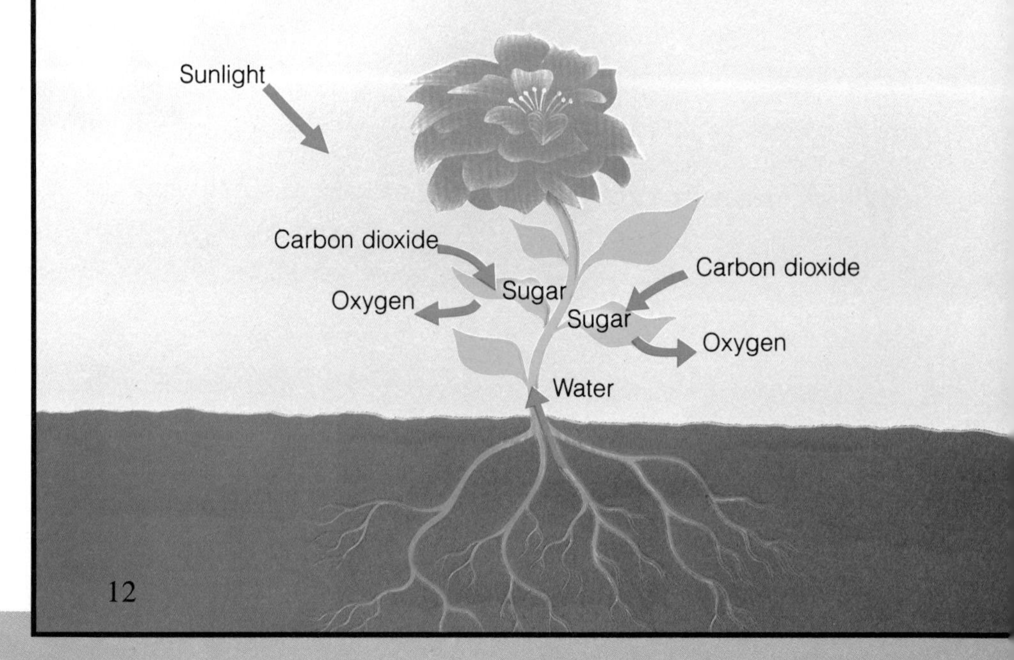

12

Teaching Options

Science Anecdote

A winter rye plant can produce 611 km (380 miles) of roots in .05 m³ (1.8 ft³) of soil.

Reinforcement

Explain that roots have hairs that help take water and minerals into the plant. If possible, bring in an entire root system of a dandelion plant or a grass plant. Carefully wash the soil from the root and let students examine the root hairs with a hand lens. *CAUTION:* Instruct students not to use lenses to concentrate the sun's rays onto skin or paper. Have students wash their hands thoroughly after this activity.

Enrichment

Measure the rate of imbibition (rise of fluids in a plant) in paper plants. Cut paper "plants" from blotter paper. Place the paper plants in colored water and measure (in cm per minute) the rate at which the paper plant imbibes the water. *CAUTION:* Immediately wipe up any water that spills on the floor. Try the experiment again using a different type of paper. Have the students compare the results. Question: **Do you think that all plants take in water at the same rate?** (no)

♦ ***Suitable as a language development activity***

Spring

Summer

Autumn

Winter

Most plants store some of the sugar they make in their stems and roots. Plants use stored food when the leaves cannot make enough food for the plant.

Notice how this tree changes from season to season. How do leaves help the tree in the summer?[1] What happens to the leaves in the autumn?[2] You can see that the tree has no leaves in the winter. Then the tree uses stored food. The tree also uses stored food to make new leaves in the spring.

SCIENCE IN YOUR LIFE

People eat some plant roots and stems that store food. You eat roots when you eat carrots, radishes, or beets. You eat stems when you eat potatoes or asparagus.

[1]they make food
[2]they fall off

Lesson Review

1. How do roots help plants?
2. How are stems important to plants?
3. How do green leaves make food for plants?
4. **Challenge!** What makes this tree need more water in the summer than in the winter?

Study on your own, pages 316–317.

Look in a book about plants to find out what chemical makes some plants green. Write a few sentences explaining how this chemical helps plants make food.

PHYSICAL SCIENCE
FIND OUT ON YOUR OWN
CONNECTION

3. Assess

Lesson Review

1. Roots hold plants in the ground and carry in water and minerals.
2. Stems hold up leaves and other plant parts. They carry water and minerals from roots, and food from leaves.
3. Green plants use sunlight to change carbon dioxide and water into sugar and oxygen.
4. Challenge! The plant needs more water in summer to help make food. **Thinking Skill:** *Inferring*

Find Out On Your Own

Green indicates the presence of chlorophyll. Chlorophyll captures energy from the sun. The leaf uses this energy to change water and carbon dioxide to sugar and oxygen. **Thinking Skill:** *Restating or explaining Ideas*

Workbook page 2 *

Name

Use with Lesson 1: pages 10-13

Chapter 1

Science and Reading

A Plant That Eats Insects

Read the story. Then answer the questions.

Most plants get all the minerals they need from the soil. But the Venus's-flytrap grows in soil that does not have enough minerals. Where does the Venus's-flytrap get the minerals it needs? It "traps flies" and "eats" them. It gets its minerals from the insects it catches.

Each leaf of the flytrap has six hairs. When an insect touches these hairs, the sides of the leaf close. The insect gets caught in the leaf. The insect cannot escape.

The Venus's-flytrap has special juices that break down the insect so that the minerals in the insect can be used by the plant. The plant uses the minerals from the insect to help the plant grow. It takes the plant 5 to 10 days to use up the insect. Then the leaf opens up and is ready to catch another insect.

The Venus's-flytrap grows only in one small part of the United States. This part is a strip of wet, swampy ground in North Carolina. This area is only about 1,000 square kilometers.

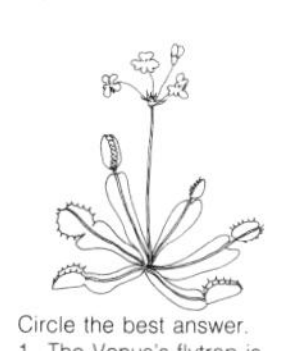

Circle the best answer.

1. The Venus's-flytrap is unusual because it
   a. grows in soil.
   b. catches insects.
   c. does not need water.
2. The Venus's-flytrap needs insects
   a. for minerals.
   b. for water.
   c. to move.
3. The Venus's-flytrap catches insects with
   a. its flowers.
   b. its leaves.
   c. its stem.
4. The Venus's-flytrap grows in
   a. dry, sandy deserts.
   b. snowy mountains.
   c. wet, swampy ground.

2

* ***Answers to masters on pages 8E–8H***

Game Suggestion ♦

Make up a Bingo-like game board for each student. Label the rows with the letters P L A N T. At random, write the names or draw pictures of plant parts (roots, stems, leaves, and so on) in the squares on the boards. Then, hold up large cards with the names or pictures of plant parts. Direct students to cover the appropriate squares on their boards until someone has a *plant bingo*.

Reteaching Suggestion ♦

Bring to class several roots, stems, and leaves. Provide hand lenses and allow students to *observe* and *compare* varieties of each plant part. *CAUTION:* Direct students not to use lenses to concentrate the sun's rays onto skin or paper. Questions: **Are the roots thick or thin, long or short? Are the stems thin and flexible or thick and woody? What are the patterns, edges, and textures of the leaves?** Discuss how each plant part contributes to a plant's growth.

Concept
Tubes in celery stalks conduct water through the stalk to the leaves.

Objectives/Process Skills
- *Predict* how water will move through a celery stalk.
- *Observe* how water moves through a piece of celery.
- *Measure* the distance the colored water moved up the celery stalk.
- *Record* observations.

Time Allotment
Allow two 15-minute periods about 30 minutes apart.

Safety Tips (See page T24.)
- Make certain that young students use only round-tipped (not pointed) scissors.
- Insist that any spilled water is wiped up immediately to prevent falls.

Teaching Tips
- The distance the color migrates up the stalk will vary. Evaporation from leaves pulls water up the stalk faster on a warm day.
- Place a white flower, such as a carnation, in a glass of green food coloring for a few days. The petals of the flower will turn green.

Answers
State Your Conclusions

1. The water and food coloring moved through the tubes in the stalk.

2. The tubes would carry the water and food coloring farther up the stalk. The rest of the stalk and the leaves would become red.

Use What You Learned

The plant might die. Food and water could not move through the stalk to the leaves. Then, the leaves could not make the food the plant needed to live and grow. **Thinking Skills:** *Applying information to new situations, Inferring*

Observing the Movement of Water in Plants

Suggested grouping: pairs

Purpose
Observe the movement of water up through celery.

Gather These Materials
• cup containing red food coloring in water • celery stalk • scissors

Follow This Procedure
1. Use a chart like the one shown to record your observations.
2. Using scissors, cut a small piece off the bottom of the celery. *CAUTION: Use scissors carefully.*
3. Put the cut end of the celery into the cup of food coloring like you see in the picture.
4. Record what you think will happen to the celery.
5. After 30 minutes, remove the celery from the cup.
6. Use a pair of scissors to cut the celery in half.

7. Notice any color changes in the cut ends of the celery. If you see color in both ends, cut the celery higher to see where the color stops. If you do not see color in the ends, cut lower to find the color.
8. Put the pieces of celery back in order. Measure how far the food coloring went up the celery.

Record Your Results

What you think might happen	What happened	Distance the color moved
Results will vary. Color moves about 2 cm every 10 minutes.		

State Your Conclusion
1. Explain the color changes you saw in the celery stalk.
2. Suppose you left the celery in the food coloring for two more hours. How would you expect the color to change in the celery? Explain your answer.

Use What You Learned
Suppose the stem of a green plant was broken. What might happen to the plant?

14

Activity Results
The color moves up the stalk toward the leaves.

Resource Book page 7

Chapter 1

Name ____________

Use with Lesson 1: page 14

Activity Worksheet

Observing the Movement of Water in Plants

Record Your Results

What you think might happen	What happened	Distance the color moved

State Your Conclusion

1. Explain the color changes you saw in the celery stem.

2. Suppose you left the celery in the food coloring for two more hours. How would you expect the color to change in the celery? Explain your answer.

Use What You Learned

Suppose the stem of a green plant was broken. What might happen to the plant?

7

Helping Roots Do Their Jobs

Roots have many jobs. One of these jobs is to hold a plant in the soil. Another job is to hold the soil in its place. Roots help keep soil from being washed away. When a forest is cut down, the tree roots no longer hold the soil. The soil washes away. New plants have trouble growing on the bare land. Without plants, animals leave the area to find food or hiding places.

The same problem can happen in oceans. The water can become polluted—filled with harmful materials. This can kill sea grasses that grow on the ocean floor. Then the sea grass roots do not hold down the sand. The ocean floor can get stirred up, making the water muddy. Fish and other animals leave the area when the plants are gone.

Dr. Anitra Thorhaug is working to help protect the oceans. She finds places where polluted water has killed the sea grasses. Then, she plants new grasses. They spread out and cover the bare area. The grasses then hold down the sand.

If the polluted water has been cleaned up, Dr. Thorhaug plants the same kind of grasses that were there before. If the water is still polluted, she plants other grasses. These other grasses are tough. They can live even if the area is still polluted.

Dr. Anitra Thorhaug

To do her job, Dr. Thorhaug has to swim in rough water. Sometimes the water is muddy and dark. In the picture, you can see her working at her difficult job. She also teaches other people how to plant sea grasses. Dr. Thorhaug goes all over the world to help protect the oceans and their plants.

What Do You Think?

1. How do sea grasses help protect the ocean floor?
2. Why do you think Dr. Thorhaug does not plant the same kind of grasses in areas that still have a pollution problem?

15

Science and People

TEACHING PLAN

Discussion

Ask students to think of areas where they have seen soil being washed away because plant roots no longer hold the soil down. Possible answers might include the area under a swing where grass is worn down, the land around a house under construction, strips of land along roadways, and so on.

Teaching Tips

- Ask students to imagine the roots of a plant to be like fingers reaching down into the soil, holding it tightly in place.
- Point out that sand or soil can be washed away in the ocean just as on land. The ocean can move loose sand and deposit it on a beach or move it to another part of the ocean floor.

Answers

What Do You Think?

1. Their roots hold down the soil and keep it from being washed away. **Thinking Skill:** ***Recognizing the main idea and supporting details***

2. Those plants probably would die because of the pollution. **Thinking Skill:** *Inferring*

Teaching Options

Science Background

Dr. Anitra Thorhaug is an oceanographer. She studied at the University of Miami, Florida. Her work in replanting sea grasses and in teaching others to do this type of work helps to ensure that damage caused by pollution will be repaired. The conservation of the ocean is important to people, since people rely on fish and other organisms from the oceans as food sources. Keeping the ocean floor stable also helps prevent beaches from eroding, and helps keep debris from the ocean floor from washing onto beaches.

TEACHING PLAN

Lesson Objectives
- *Describe* pollination and explain how plants with flowers form seeds.
- *Describe* two ways pollen can be scattered.
- *Explain* how plants with cones form seeds.

Lesson Vocabulary
petal, pollen, pollinate

1. Motivate

Demonstration Activity ♦
Show students a variety of flowers, including tulips and daisies, if possible. Potted flowering plants also may be used. *CAUTION:* Use only nontoxic plants. Ask students to ***observe*** the different kinds of flowers and their parts.

Discussion
Question: **In what ways are these flowers alike?** (Flowers have similar parts, even though they look different.) Describe and point out each part of one flower: stem, petal, and pollen-containing organ (stamen). Point to the same parts in another flower. Mention that a plant also has roots and leaves.

2 How Are Flowers and Cones Important?

LESSON GOALS

You will learn
- how plants with flowers form seeds.
- two ways pollen is scattered.
- how plants with cones form seeds.

petal (pet′l), the outside parts of a flower that often are colored.

You might have seen flowers like the ones in the picture. Flowers look different from each other. Some plants have only one flower on a stem. Other plants have many flowers on one stem. Seeds grow inside each flower. These seeds can grow into new plants.

How Flowers Form Seeds

Look at the different colors of these flowers. **Petals** are the outside parts of the flower that often are colored. Some flowers have more petals than other flowers. Notice the sizes and shapes of these petals. What are the colors of these flower petals?[1]

[1]red and yellow

Seeds grow inside flowers.

Teaching Options

Science Background

The stamen is the male reproductive organ in a flower. The pistil is the female reproductive organ. The ovary, which contains one or more ovules, is at the base of the pistil. After pollination occurs, the pollen grain produces a tube that grows through the pistil, into the ovary. Fertilization occurs when a sperm cell nucleus joins an egg cell in the ovule. Pollination is necessary before a plant can form seeds. Seeds are mature ovules.

Reading Strategies ♦

1. Guide students' pre-reading by asking: Which parts of the lesson are known to you and which parts are new?
2. Assign these strategies: Visualizing Information and Writing a Memory Sentence. (See pages T26–T29.)
3. Pair students to share what information is clear and unclear and initiate discussion using students' unanswered questions.

♦ ***Suitable as a language development activity***

Look inside the flower petals in the picture below. The center part of the flower makes seeds. A yellow powder called **pollen** must move to the center part of the flower before seeds can form.

The picture in the margin shows what happens after pollen reaches the center part of the flower. This part of the flower swells and changes into a fruit. Then the flower petals dry up and fall off the plant.

You can see seeds inside fruit. Find the seeds in the picture. The fruit helps protect the seeds as they grow. Many foods people eat are fruits. You probably have seen seeds inside apples and oranges. What other kinds of fruit can you name?[1]

SCIENCE IN YOUR LIFE

You might have seen peas growing in a long, thin cover called a pod. The peas people eat are seeds. The pod is the fruit of the pea plant.

pollen (pol′ən), a fine yellowish powder in a flower.

Fruit with seeds

[1]Answers can include any type of fruit.

The parts of a flower

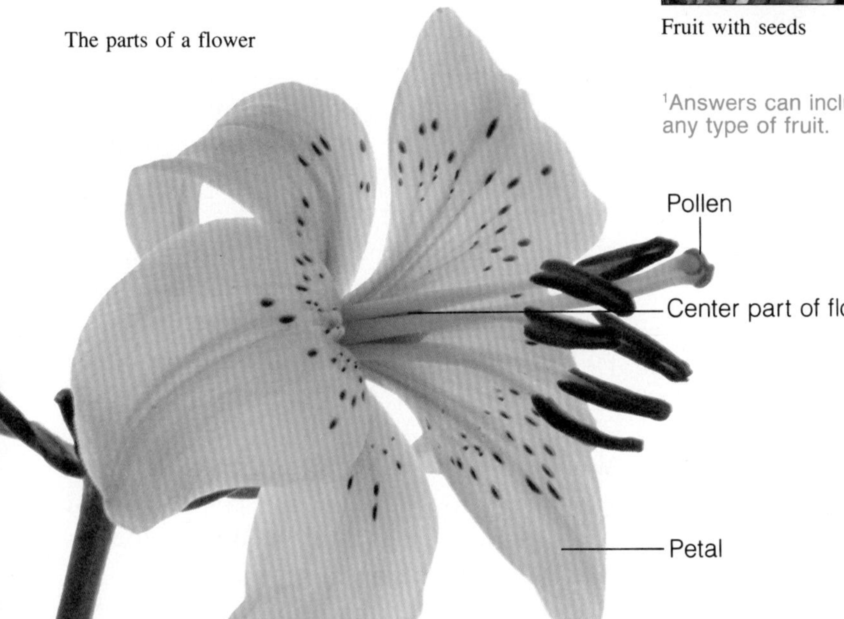

2. Teach

Teaching Tips

- Question: **How many petals do you see on each flower?** (Answers will vary.) Explain that some flowers have three petals or multiples of three, while other flowers have petals in fours and fives.
- Have volunteers *describe* their favorite flowers. Students might include characteristics such as color, number of petals, and so on. Other students could *identify* the flower by its characteristics.
- **Possible Misconception:** Some students might have trouble thinking of trees as flowering plants. You might show them pictures of flowering trees.

Science and Math

Bring in several fresh beans or pea pods. (These are usually available in the fresh food section of a grocery store.) Give several beans or pods to each student. *CAUTION:* Instruct students not to place plant parts in their mouths. On the chalkboard (as a group), or on paper (individually), have students graph the number of beans or peas per pod. Determine which amounts of seeds occur the most and which occur the least.

Special Education

Discuss the concept of pollination by having students with learning disabilities tell about the adventures of a bit of pollen. Encourage students to use their imaginations as they tell about where and how the pollen traveled.

Reinforcement

Construct a large cutaway of a flower stem with only a few petals. Tape the flower to a chalkboard or pin it to a bulletin board. Construct large petals, stamen, pistil, pollen grains, and seeds. Have the students pin or tape the parts of the flower to the appropriate part of the plant. Help students understand that the pollen grains attach to the pistil, and the seeds grow inside the center part of the flower (in the ovary).

TEACHING PLAN

Teaching Tips

- After students have read page 18, discuss why many farmers have their crop dusting done after dark. (If necessary, explain that crop dusting is a way that pesticides—chemicals that kill unwanted bugs—are applied to crops.) Lead students to ***infer*** that crop dusting is done after dark because bees do most of their pollinating during daylight. Spraying at night helps minimize the effect of pesticides on bees.
- Point out that some farmers set up bee hives near their crops and bring in bees to help pollinate crops.
- You might want to bring some pine cones to class and show students where the seeds are located.

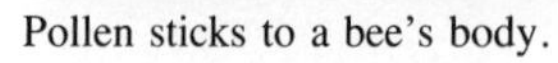

Pollen sticks to a bee's body.

A moth pollinates a flower.

pollinate (pol′ə nāt), to carry pollen to the center part of a flower.

How Pollen Is Scattered

You might have seen bees buzzing near flowers. Bees help flowers form seeds by moving pollen from one flower to another.

A bee lands on a flower and brushes against pollen. The picture shows that pollen sticks to a bee's body. The bee carries the pollen to another flower. Some of the pollen falls off the bee and sticks to the second flower. When a bee moves pollen to the center part of a flower, it **pollinates** the flower. Other animals, such as butterflies and hummingbirds, also help pollinate flowers. Look at the picture of the moth pollinating a flower.

Wind scatters the pollen of some flowers. Wind can blow pollen off a flower and carry the pollen through the air. The pollen can land on the flower of another plant. Wind pollinates corn and many other grasses and trees.

Teaching Options

Science Anecdote

Sea grasses are plants that grow in the ocean. The ocean water moves the pollen from the stamen to the pistils of the plants' flowers. The water also scatters the seeds that form.

Reinforcement

Ask students to create cartoon frames that tell the story of a flower that is waiting to be pollinated. The cartoon should tell how the flower gets pollinated and what happens after pollination.

Enrichment

Depending on the time of year, obtain pollen from several different types of flowers. Place each type of pollen in a separate plastic bag. Let students use a hand lens to observe the pollen, or set up a microscope for students' use. *CAUTION:* Do not permit students with plant allergies to do this activity. Instruct students not to use hand lenses to concentrate the sun's rays onto skin or paper.

♦ ***Suitable as a language development activity***

How Cones Form Seeds

You might have seen small objects called cones growing on a tree. Look at the cones on this pine branch. Some cones have pollen. Wind blows the pollen into the air. Some of the pollen reaches other cones. Then new seeds grow inside these cones.

Most trees with cones have leaves shaped like needles. These trees usually keep their leaves all year. Pine trees, spruce trees, and fir trees are trees with cones.

Cones on a pine tree

Lesson Review

1. How do plants with flowers form seeds?
2. What are two ways pollen is scattered?
3. How do cones form seeds?
4. **Challenge!** What might happen to a plant if you cut off all the flowers?

Study on your own, pages 316–317.

EARTH SCIENCE — *FIND OUT ON YOUR OWN* — *CONNECTION*

Ferns and mosses live mainly in wet places on earth. Use library books to find out how these plants are different from other kinds of plants. Also find out why water is especially important to them.

Fern

Moss

3. Assess

Lesson Review

1. A flower forms seeds when pollen reaches the center part of the flower. The center swells to form a fruit. Seeds grow inside the fruit.
2. Animals and wind can help scatter pollen.
3. Seeds form when pollen reaches the cone.
4. Challenge! The plant would not be able to form seeds. **Thinking Skill:** *Drawing conclusions*

Find Out On Your Own

Ferns and mosses do not produce seeds. They produce spores. Water is needed for the spores to grow into new plants. **Thinking Skills:** *Collecting information, Comparing, Contrasting*

Workbook page 3 *

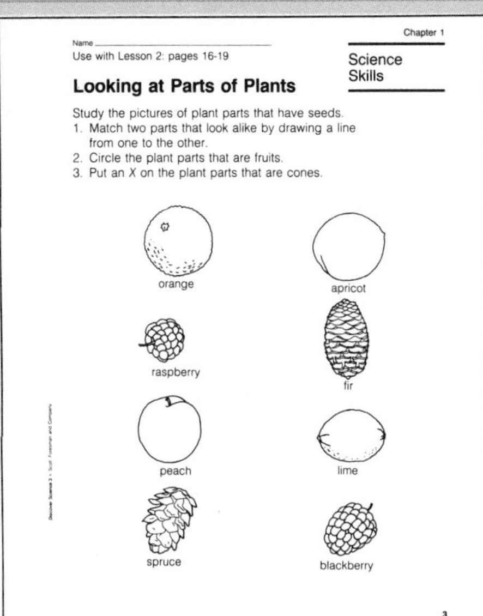
Chapter 1

Name ______________________

Use with Lesson 2: pages 16-19

Science Skills

Looking at Parts of Plants

Study the pictures of plant parts that have seeds.
1. Match two parts that look alike by drawing a line from one to the other.
2. Circle the plant parts that are fruits.
3. Put an *X* on the plant parts that are cones.

orange
apricot
raspberry
fir
peach
lime
spruce
blackberry

3

*** Answers to masters on pages 8E–8H**

Game Suggestion ♦

Mount several different types of seeds on cards. Place pictures or drawings around the room of the parent plants that the seeds come from. On the back of each picture (in the lower right corner), make an answer key by taping or gluing one of the seeds that matches that plant. Then let students try to match the seed cards to the correct parent plants. Students can check their accuracy by looking at the back of the plant pictures.

Reteaching Suggestion ♦

Pass out a paper cup to each student. Have them cut the cups into petal shapes while leaving the lower 1/4 of each cup intact. Let them use straws or pipe cleaners to form the inside parts (stamens and pistils) of their flowers. Add pollen made of finely shredded yellow tissue. Challenge the students to find as many ways as possible to spread their pollen to other flowers. (Blowing, tipping, and applying glue stick to "insect fingers" are possible ways of spreading the tissue pollen.) *CAUTION*: Students should use round-tipped scissors.

TEACHING PLAN

Lesson Objectives
- *Describe* the parts of a seed.
- *Explain* how seeds are scattered.
- *Explain* how a seed germinates and produces a new plant.

Lesson Vocabulary
germinate, life cycle, seed coat, seed leaf, seedling

1. Motivate

Demonstration Activity ♦
Bring a soaked kidney bean to class. Separate the two parts, or seed leaves, of the bean and pass them around for students to examine. Explain that some seeds have two parts and other seeds have only one part. Show students some seeds with only one part, such as corn.

Discussion
Have the students name some types of seeds. Draw an outline or simple sketch of each type of seed as it is mentioned. Question: **If a seed has two parts, can each part grow into a plant?** (No, both parts make up just one seed. The seed leaves become parts of the growing plant.)

3 How Do Plants Grow from Seeds?

LESSON GOALS

You will learn
- the parts of a seed.
- different ways seeds are scattered.
- how a seed produces a new plant.

seed leaf (sēd lēf), a part that looks like a leaf and is inside each seed.

seed coat (sēd kōt), the outside covering of a seed.

Think about different seeds you have seen. The seeds in the picture are different sizes and shapes, but they are alike in one way. These seeds can grow into new plants.

Parts of a Seed
The drawings show that a tiny new plant grows inside each seed. This new plant uses stored food to grow.

Find the **seed leaf** in each seed. Some seeds, such as the bean seed, have two seed leaves. Food is stored in these seed leaves. Other seeds, such as corn, have only one seed leaf. Notice that food is stored outside of this seed leaf. A hard covering, called the **seed coat,** protects the seed. Find the seed coat of each of these seeds.

Parts of a seed

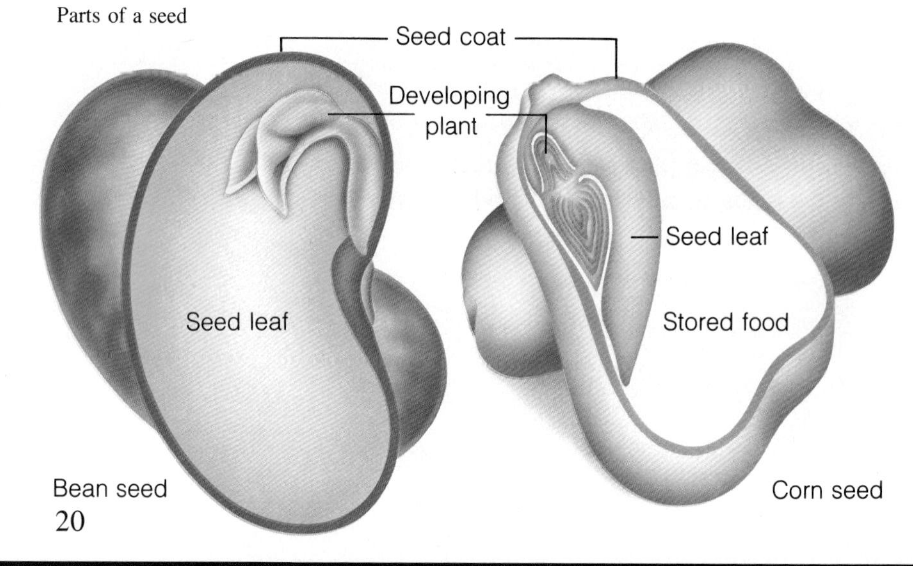

Teaching Options

Science Background
The fertilized egg inside an ovule forms an embryo, or young plant, inside a seed. Cotyledons, or seed leaves, contain food for the embryo. Dicot seeds, such as those found in bean plants, contain two cotyledons. Monocot seeds, such as those found in corn, contain one cotyledon. The walls of the ovule become the seed coat.

Reading Strategies ♦
1. Guide students' pre-reading by asking: Which parts of the lesson are known to you and which parts are new?
2. Assign these strategies: Visualizing Information and Mapping Examples of how seeds are scattered. (See pages T26–T29.)
3. Pair students to share what information is clear and unclear and initiate discussion using students' unanswered questions.

♦ ***Suitable as a language development activity***

Water lily

Milkweed plant

How Seeds Are Scattered

Suppose you planted seeds in a garden. You probably would spread out the seeds in the soil. Scattering seeds helps each plant get enough water and sunlight to grow.

Animals can help scatter seeds. Some fruits can stick to an animal's fur or a bird's feathers. Some of the fruits fall to the ground as the animals move from place to place.

Water and wind also help scatter seeds. The fruits of this water lily can float long distances on water. How can wind help scatter the seeds of this milkweed plant?[1]

How a Seed Grows into a New Plant

Many seeds grow into new plants. A seed **germinates** when the small plant inside begins to grow. A seed germinates only when it gets enough air and water. A seed also needs the proper temperature to germinate.

INVESTIGATE!

Find out if plant stems grow toward light. Write a hypothesis and test your hypothesis with an experiment. You might observe a house plant sitting on a window ledge for several days and a similar plant elsewhere in the room.

germinate (jėr′mə nāt), begin to grow and develop.

[1]can blow seeds to different places

2. Teach

Teaching Tips

- Discuss some of the ways that seeds are dispersed. Question: **What would happen if a person ate a watermelon outdoors and dropped the seeds on the ground?** (If conditions were right, the seeds would grow new plants.)
- Germinate some bean seeds in the classroom. Allow students to *observe* the life cycle of the plant from seed to mature plant. Ask students to make a point of noting when the seed leaves fall off of the plant.
- **Possible Misconception:** Some students might think that seeds must be covered by soil in order to grow. Explain that seeds can grow where they have been scattered if they have the correct conditions such as enough water, air, and a suitable temperature.

Investigate!

Accept any testable hypothesis. Most students' data will support this hypothesis: *Plant stems will turn toward the direction where they get the most light.* Students should find that if a plant grows toward the light, its stem will become curved rather than straight.

Workbook page 4 *

Name ______

Use with Lesson 3: pages 20-22

Chapter 1

Science Activity

Do Seeds Need Water?

Gather These Materials

• 4 paper cups • soil • 16 bean seeds • spoon • measuring cup

Follow This Procedure

1. Label the first cup: No Water. Label the second cup: One Spoon. Label the third cup: 50 mL. Label the fourth cup: 100 mL.
2. Half fill each cup with soil. Plant four beans in each cup.
3. Add water to the cups. The labels tell how much water.
4. Check the bean seeds every day. Record what happens.

NO WATER
ONE SPOON
50 mL
100 mL
300 mL
200 mL
100 mL

Record Your Results

	No Water	One Spoon	50 mL	100 mL
Day 1				
Day 2				
Day 3				
Day 4				
Day 5				

State Your Conclusions

1. In which cup did the bean seeds grow best?

______

2. How much water do bean seeds need to grow?

______

4

* *Answers to masters on pages 8E–8H*

Enrichment

Place a wet cotton ball into a plastic bag. Put a bean or corn seed (soaked for 24 hours ahead of time) next to the cotton. Attach the bag to a window sill that receives sunlight. Observe the young plant as it grows. The leaf grows toward the light and the root grows away from the light (phototropism). Lay the seed horizontally, so that the root end is not at the bottom. Place the bag out of direct light. Observe that the leaves grow up and the root grows down. Turn the seed over and observe the changes (geotropism).

Special Education

Allow students with learning disabilities to examine seeds. *CAUTION:* Remind students not to place seeds in their noses, ears, or mouths. Have them find the tiny young plant (embryo) inside, the seed coat, and the seed leaves.

TEACHING PLAN

3. Assess

Lesson Review

1. The parts of a seed include the seed coat, seed leaf, and the tiny plant inside.
2. Animals, wind, and water help scatter seeds.
3. As a seed germinates, the new root pushes through the seed coat and into the soil. The young seedling uses stored food in order to grow.
4. Challenge! The amount of air or water, or the temperature, might not have been what the seeds needed for germinating. **Thinking Skill:** *Making inferences*

Find Out On Your Own

A bulb is a tiny stem with hairy roots and tightly packed leaves that store food. When the bulb is planted, the inner leaves grow up through the soil and begin to make food. Tulips, lilies, and daffodils are common bulb plants. **Thinking Skills:** *Collecting information, Restating and explaining ideas, Making physical models*

seedling (sēd′ling), a young plant that grows from a seed.

life cycle (sī′kəl), the stages in the life of a plant or animal.

The pictures show how a bean plant grows from a seed. The root pushes through the seed coat and into the soil. Then the young plant, or **seedling,** grows out of the ground. The seedling uses the stored food in the seed leaves. Finally, the seed leaves fall off and the new plant begins to make its own food.

All the stages in the life of a plant make up the **life cycle** of a plant. A seed grows into a new plant that forms seeds. Then the new seeds repeat the life cycle.

Lesson Review

1. What are the parts of a seed?
2. What are three ways seeds are scattered?
3. How does a seed grow into a new plant?
4. **Challenge!** Explain why some seeds people plant might not germinate.

Study on your own, pages 316–317.

LIFE SCIENCE

FIND OUT ON YOUR OWN

Look in a book about flowers to find out how plants grow from bulbs. Draw a picture of a bulb. Under your picture, write a few sentences telling about bulbs.

Teaching Options

Science Anecdote

In the right temperature and moisture conditions, a seed coat can protect a seed for a long time. Indian lotus seeds have been known to lie dormant for 400 years. One species of water lily was grown from seed after 800 years.

Reteaching Suggestion ♦

Ask students to write a few paragraphs that tell the story of the life of a plant, from seed to maturity. Students can make their stories creative or humorous, but they should include information about what type of seed was planted, how many seed leaves it had, how the plant germinated and grew, what happened to the seed leaves, and when the plant began to make its own food.

Workbook page 5 *

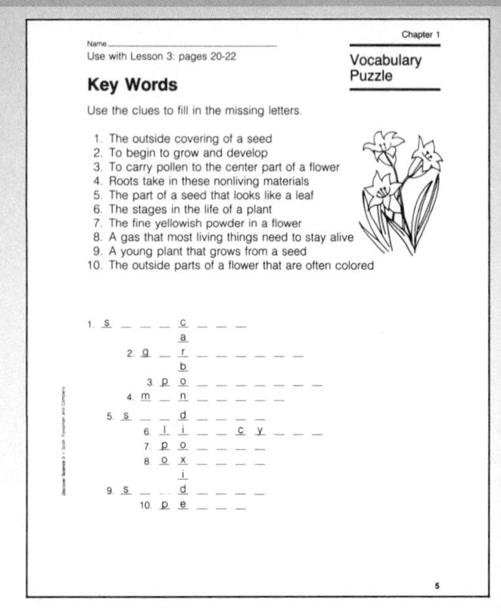

Name ____________

Use with Lesson 3: pages 20-22

Chapter 1

Vocabulary Puzzle

Key Words

Use the clues to fill in the missing letters.

1. The outside covering of a seed
2. To begin to grow and develop
3. To carry pollen to the center part of a flower
4. Roots take in these nonliving materials
5. The part of a seed that looks like a leaf
6. The stages in the life of a plant
7. The fine yellowish powder in a flower
8. A gas that most living things need to stay alive
9. A young plant that grows from a seed
10. The outside parts of a flower that are often colored

1. s _ _ _ c _ _ _
a
2. g _ r _ _ _ _ _ _
b
3. p o _ _ _ _ _ _ _
4. m _ n _ _ _ _ _
5. s _ _ d _ _ _ _
6. l i _ _ c y _ _ _
7. p o _ _ _ _
8. o x _ _ _ _
i
9. s _ _ d _ _ _ _
10. p e _ _ _

5

♦ *Suitable as a language development activity*

Observing How Seeds Germinate and Grow Without Light

Suggested grouping: pairs

Purpose

Observe how seeds germinate and grow without light.

Gather These Materials

• 2 cups • potting soil • pencil
• 4 radish seeds • spoon

Follow This Procedure

1. Use a chart like the one shown to record your observations.
2. Fill your cups almost full with potting soil as shown.
3. Poke about 4 centimeters of the pencil into the soil in each cup.
4. Plant 2 seeds in each cup.
5. Cover the seeds with soil. Pour 2 spoonfuls of water over the seeds in each cup.
6. Put your name on each cup. Label one cup *light*. Put this cup in a well-lighted place.

Label the second cup *no light*. Place this cup in a dark place.
7. Give both of your plants a little water every few days.
8. When the plants in one cup are about 6–8 centimeters tall, put both cups on your desk.
9. Draw a picture of each plant. Describe the way each plant appears.

Record Your Results

	Drawing	Description
Plant in light	Bright green leaves	
Plant in dark	Pale or white leaves	

State Your Conclusion

1. How do the plants grown in the dark look different from the plants grown in the light?
2. If you kept your plants in the dark for several weeks, what would happen to the plants? Explain your answer.

Use What You Learned

Sometimes people use special lights to shine on plants all the time. How would you expect such plants to appear?

Concept

Green plants need light to grow.

Objectives/Process Skills

- *Observe* how radish seeds germinate and grow in light and darkness.
- *Compare* the growth of the plants.
- *Record* observations.
- *Infer* what would happen to plants kept in darkness for several weeks.

Time Allotment

Allow two 20-minute periods.

Safety Tips (See page T24.)

- Tell students not to taste the seeds.
- Be certain to wipe up spilled water immediately to prevent falls.
- Instruct students on the safe handling of sharpened pencils.

Teaching Tips

- A cardboard carton can be a convenient place to keep plants in the dark.
- While students water the plants in the dark, they should try to not let light reach the plants.
- Try putting some of the plants from the dark into low light to see if these plants can still produce chlorophyll.
- **Helpful hint:** Radish, mung, or bean seeds work well.

Answers

State Your Conclusion

1. The plant grown in the light has green leaves. The plant grown in the dark has pale, almost white, leaves.

2. The plants would not be able to make enough food to live and grow. The plants would die.

Use What You Learned

The plants would use the light to make food to live and grow. They would be able to make a lot of food because they would receive a lot of light. The plants would be green and grow quickly. **Thinking Skill:** *Applying information to new situations*

Resource Book page 9

Name

Use with Lesson 3: page 23

Chapter 1

Activity Worksheet

Observing How Seeds Germinate and Grow Without Light

Record Your Results

	Drawing	Description
Plant in light		
Plant in dark		

State Your Conclusion

1. How do the plants grown in the dark look different from the plants grown in the light?
2. Suppose you kept your plants in the dark for several weeks. What would happen to the plants? Explain your answer.

Use What You Learned

Sometimes people use special lights to shine on plants all the time. How would you expect such plants to appear?

9

*** *Answers to masters on pages 8E–8H***

Activity Results

TEACHING PLAN

Purpose
To develop the skills of collecting and organizing information using thermometers and bar graphs to solve problems.

1. Motivate

Discussion
Display different kinds of thermometers. Assess the students' understanding of how to read a thermometer. Question: **What does a thermometer measure?** (temperature) If necessary, show students how to find the temperature using a thermometer.

2. Teach

Teaching Tips
- Explain that scientists often need precise measurements of temperature in order to understand many processes.
- **Helpful Hint:** Describe how to construct and interpret a bar graph. Discuss the concepts of parallel and horizontal lines in terms of carrying values across the graph to establish the tops of the bars.

Skills for Solving Problems

Using Thermometers and Bar Graphs

Problem: How does temperature affect the number of seeds that germinate?

Part A. Using Thermometers to Collect Information

1. The thermometers in the picture measure temperature in degrees Celsius. Notice that the heating lamp is 20 cm away from Tray A and 100 cm away from Tray B. Tray C is outdoors in February. Twenty seeds were planted in each tray. What is the temperature of Tray A? How many seeds have germinated in the tray?
2. What is the temperature of Tray B? of Tray C? How many plants have germinated in Tray B? in Tray C?

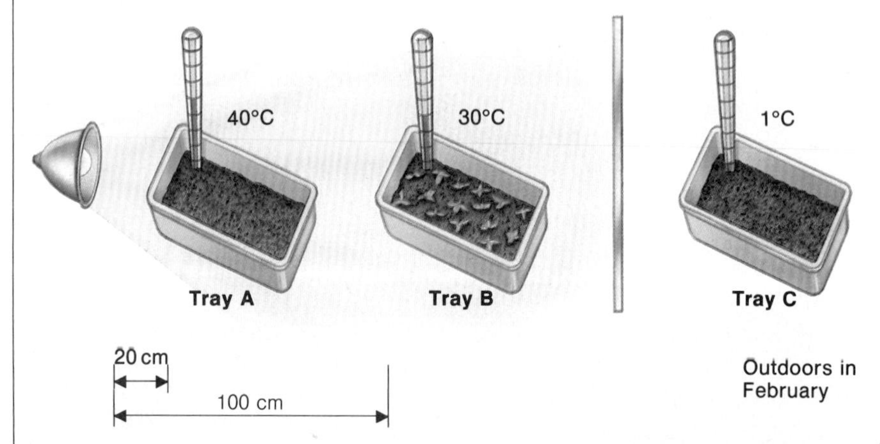

Part B. Using Bar Graphs to Organize and Interpret Information

3. The bar graph contains the information you collected about thermometer readings for each tray in Part A. Look at the scale on the left side of the graph. What does each line on the scale stand for?
4. Look at the first bar. It shows the temperature of Tray A. With your finger, follow across from the top of this bar to the left side of the graph. What is the temperature of Tray A?

24

Teaching Options

Sample Bar Graph for Part C

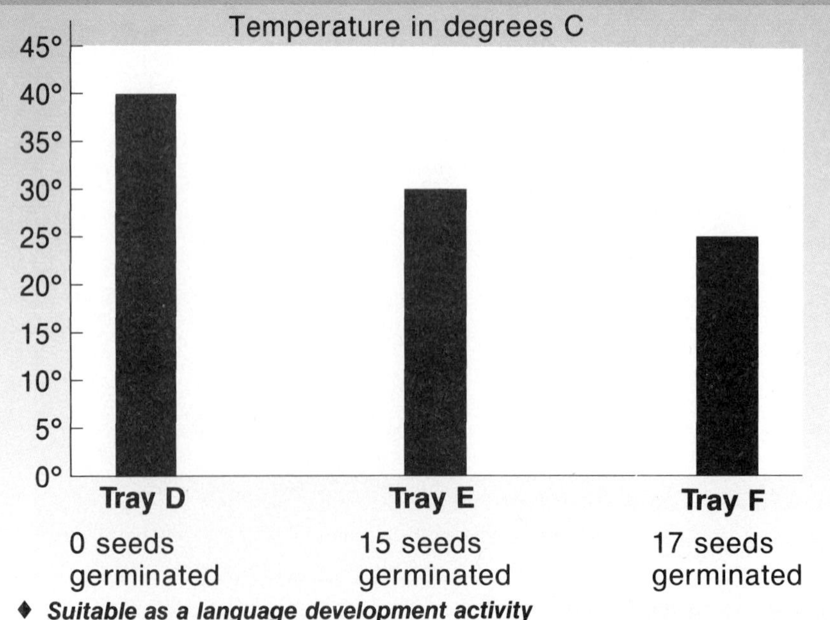

♦ ***Suitable as a language development activity***

How many seeds germinated in Tray A?
5. What is the temperature of Tray B? Tray C? How many seeds germinated in Tray B? in Tray C?
6. Which temperature seemed best for seeds to germinate?

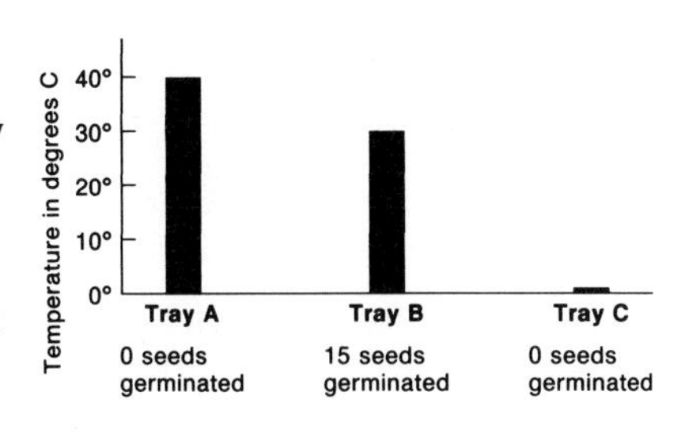

Part C. Using Thermometers and Bar Graphs to Solve a Problem

Problem: How do temperatures in a different season affect the number of seeds that germinate?

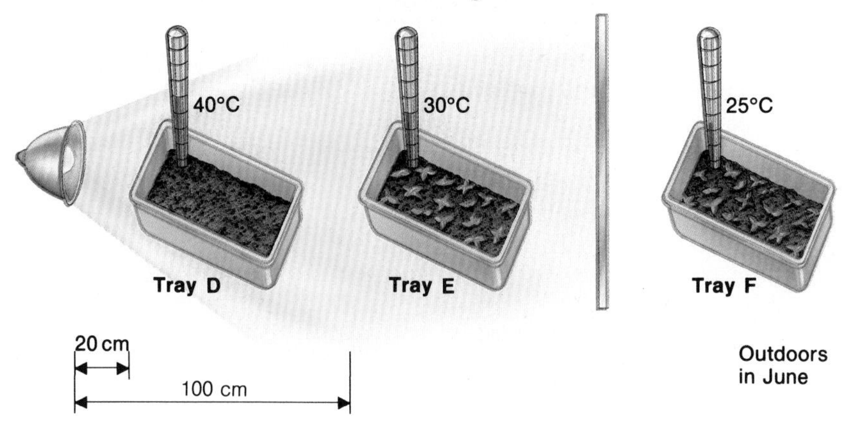

7. Use the thermometers to collect the information you need to solve the problem. Make a bar graph similar to the one shown in Part B to organize your information.
8. Look at your bar graph. Compare it to the bar graph in Part B. How does the temperature in a different season affect how many seeds germinate?
9. You might want to do this experiment in another season and use your results to make a bar graph.

25

3. Assess

Part A

1. 40 ° C; none
2. 30 ° C; 1 ° C; 15 plants; none

Part B

3. 10 ° C
4. 40 ° C; no seeds
5. 30 ° C; 1° C; 15 seeds; none
6. 30 ° C

Part C

7. See Sample Bar Graph for Part C in Teaching Options.
8. Weather that is too cold or too warm at different seasons can cause poor seed germination.
9. Be sure students understand that all conditions except temperature must be kept the same among the trays.

Reteaching Suggestion ♦

Obtain three bowls and fill them with cold, lukewarm, and hot water. *CAUTION:* Do not use water so hot as to cause discomfort. Have students use thermometers to find out the water temperature in each bowl. Then draw the vertical column of a bar graph, showing ten degree increments from 0 to 50 degrees Celsius. Have students draw bars to show the result of each temperature reading of water.

Resource Book page 11 *

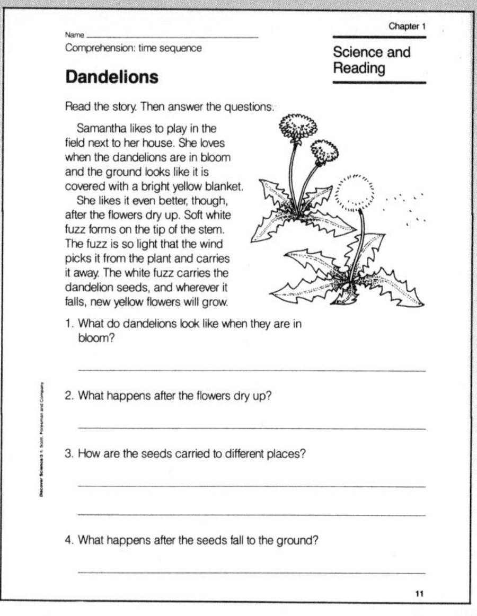
Name

Comprehension: time sequence

Chapter 1

Science and Reading

Dandelions

Read the story. Then answer the questions.

Samantha likes to play in the field next to her house. She loves when the dandelions are in bloom and the ground looks like it is covered with a bright yellow blanket.

She likes it even better, though, after the flowers dry up. Soft white fuzz forms on the tip of the stem. The fuzz is so light that the wind picks it from the plant and carries it away. The white fuzz carries the dandelion seeds, and wherever it falls, new yellow flowers will grow.

1. What do dandelions look like when they are in bloom?
2. What happens after the flowers dry up?
3. How are the seeds carried to different places?
4. What happens after the seeds fall to the ground?

11

REVIEW PLAN

Reviewing Science Words

1. seed coat
2. life cycle
3. petal
4. pollinate
5. seed leaf
6. oxygen
7. minerals
8. pollen
9. carbon dioxide
10. germinates
11. seedling

Reviewing What You Learned

1. d **3.** a **5.** d **7.** a
2. b **4.** b **6.** c **8.** c

Chapter 1 Review

Chapter Main Ideas

Lesson 1 • Roots hold plants in soil and take in water and minerals the plants need. • Stems hold up plant parts that grow above the ground and carry water and minerals through the plants. • Leaves make the food a plant needs.

Lesson 2 • Plants with flowers form seeds when pollen reaches the center part of the flower. • Animals and wind help pollinate flowers. • Seeds form inside a cone when pollen reaches the cone.

Lesson 3 • A seed has a small plant, at least one seed leaf, and stored food. • Animals, wind, and water help scatter seeds. • A seed can germinate and grow into a new plant that forms seeds.

Reviewing Science Words

carbon dioxide	oxygen	seed coat
germinates	petal	seed leaf
life cycle	pollen	seedling
minerals	pollinate	

Copy each sentence. Fill in the blank with the correct word from the list.

1. The ___ protects a seed.
2. All the stages in the life of a plant make up the ___ of the plant.
3. A ___ is an outside part of a flower that often is colored.
4. A bee can help ___ a flower.
5. The ___ of a bean seed has stored food.
6. Green leaves make sugar and ___.
7. Roots take in water and ___ from the soil.
8. Seeds often form when ___ reaches the center part of a flower.
9. Plants need ___ from the air to make food.
10. A seed ___ when the small plant inside starts to grow.
11. A young plant is a ___.

26

Review Options

Cooperative Learning ♦

STAD Format (See page T23.) Assign students to work in four- to five-member teams to study Chapter 1 Review. Students should work together to make sure that they and their teammates know the material in the chapter. After students have had enough time to study together, give them a test to complete individually (Chapter 1 Test A or B in the *Test Book*). Award Superteam certificates to teams whose average test scores exceed 90%, and Greatteam certificates to teams whose average test scores exceed 80%.

Test Book page 9 *

Name ____________

Chapter 1
Test A

Multiple Choice Choose the best answer.

1. The parts of a plant that hold it tightly in the soil are the
   a. leaves.
   b. stem.
   c. roots.
2. The plant part that carries water and minerals from the roots to the leaves is the
   a. stem.
   b. bud.
   c. bark.
3. Food needed for plant growth is made in the
   a. stem.
   b. leaf.
   c. bud.
4. The center of a flower contains parts that make
   a. seeds.
   b. carbon dioxide.
   c. petals.
5. Pollination takes place when pollen reaches
   a. part of an animal's body.
   b. the tips of flower petals.
   c. the center of a flower.
6. Trees such as spruce and fir have leaves that are shaped like
   a. needles.
   b. cones.
   c. flower petals.
7. The outside covering of a seed is the
   a. seed leaf.
   b. seed coat.
   c. seedling.
8. With enough air, water, and the proper temperature, plants usually will
   a. germinate.
   b. pollinate.
   c. produce cones.
9. A young plant that has just sprouted is called a
   a. bud.
   b. seed.
   c. seedling.
10. The life cycle of a plant includes all the
    a. stages of its life.
    b. parts of its flower.
    c. seeds in its flowers.

9

♦ ***Suitable as a language development activity***

Reviewing What You Learned

Write the letter of the best answer.

1. The part of a plant that forms seeds is the
   (a) stem. (b) root. (c) leaf. (d) flower.
2. A seedling uses stored food from the
   (a) petals. (b) seed leaf. (c) pollen. (d) seed coat.
3. What helps protect seeds?
   (a) fruit (b) seed leaf (c) bees (d) wind
4. What part holds a plant tightly in the soil?
   (a) stem (b) root (c) leaf (d) flower
5. Leaves use sunlight to make
   (a) carbon dioxide. (b) cones. (c) water. (d) sugar.
6. The seeds of a pine tree form in the
   (a) flowers. (b) leaves. (c) cones. (d) trunk.
7. What is the number of seed leaves in a bean seed?
   (a) two (b) three (c) one (d) zero
8. When a bee lands on a flower, it brushes against
   (a) seeds. (b) cones. (c) pollen. (d) fruits.

Interpreting What You Learned

Write a short answer for each question or statement.

1. How can animals help scatter seeds?
2. How do trees live without leaves in winter?
3. Describe two kinds of plant roots.
4. How do a bean seed and a corn seed differ?
5. How can you tell that a tomato is a fruit?

Extending Your Thinking

Write a paragraph to answer each question or statement.

1. Can a seed germinate without sunlight? Explain your answer.
2. Suppose you took off all the petals of a flower. Could the flower form seeds? Explain your answer.

To explore scientific methods, see Experiment Skills on pages 348–349.

Interpreting What You Learned

1. Fruits can stick to an animal's fur or a bird's feathers. The fruits can fall off as the animal moves around. **Thinking Skill:** *Restating or explaining ideas*
2. They live on stored food. **Thinking Skill:** *Recognizing the main idea and supporting details*
3. Two kinds of roots are thick roots that grow deep into the soil and thin roots that spread out under the plant. **Thinking Skill:** *Restating or explaining ideas*
4. Bean seed has two seed leaves; food is stored in seed leaves. Corn seed has one seed leaf; food is stored outside the seed leaf. **Thinking Skill:** *Contrasting*
5. Tomato is a fruit because it has seeds. **Thinking Skill:** *Recognizing the main idea and supporting details*

Extending Your Thinking

1. Yes. The plant inside the seed uses stored food to grow; it does not need sunlight because it is not making its own food. **Thinking Skill:** *Making inferences*
2. Seeds could form. The plant can form seed if pollen reaches the center part of the flower. **Thinking Skill:** *Drawing conclusions*

Test Book page 10 *

Name ____________

Chapter 1
Test A

Short Answer Label the parts of the plant.

1. ________
2. ________
3. ________
4. ________

Short Essay Use complete sentences to answer each question.

1. Name three ways that pollen is scattered. Explain how each could happen.

2. How are bean seeds different from corn seeds?

10

Test Book page 11 *

Name ____________

Chapter 1
Test B

Multiple Choice Choose the best answer.

1. The materials used by a plant that are taken in by its roots are called
   a. minerals.
   b. weeds.
   c. sugar.
2. The part of a plant that holds the plant above the ground is the
   a. leaf.
   b. stem.
   c. bud.
3. A plant uses some of its stored food to
   a. share with other plants.
   b. put minerals into the soil.
   c. make new leaves.
4. The yellow powder inside a flower that is needed to form seeds is
   a. talc.
   b. pollen.
   c. food.
5. After a flower is pollinated, part of the flower swells and becomes a
   a. petal.
   b. new flower.
   c. fruit.
6. Most trees that have cones also have leaves shaped like
   a. hearts.
   b. needles.
   c. fans.
7. The seed coat of a seed helps to
   a. protect the seed.
   b. attract insects.
   c. pollinate the seed.
8. A plant germinates when the small plant inside the seed begins to
   a. produce fruits.
   b. dry up and die.
   c. grow and develop.
9. Seed, seedling, and a grown plant are three stages of every plant's
   a. life cycle.
   b. germination.
   c. seed leaf.
10. One way pollen can be scattered is by
    a. sunlight.
    b. leaves.
    c. insects.

11

Test Book page 12 *

Name ____________

Chapter 1
Test B

Short Answer Tell the job that each plant part does.

1. ________
2. ________
3. ________
4. ________

Short Essay Use complete sentences to answer each question.

1. How do animals help scatter pollen?

2. How do leaves make the food that a plant needs?

12

* *Answers to masters on pages 8E–8H*

Chapter 2 How Animals Grow and Change Planning Guide

TEACHING PLAN

Chapter Components	Skills	Materials
Chapter Opener/*DISCOVER:* Classifying Animals pp. 28–29	*DISCOVER* p. 29 Science Process Skills *Observing, Classifying*	*DISCOVER* p. 29 (groups of 2) 30 sheets of drawing paper, 30 sets of crayons
Lesson 1 How Can Animals Be Grouped? pp. 30–31	Thinking Skills Challenge!: *Inferring* Find Out On Your Own: *Judging and evaluating*	Demonstration p. 30 pictures of invertebrates and vertebrates
Lesson 2 How Do Some Animals With Backbones Grow and Change? pp. 32–36	Thinking Skills Challenge!: *Drawing Conclusions* Find Out On Your Own: *Restating or explaining ideas*	Demonstration p. 32 pictures of birds, reptiles, fish, and mammals
Activity Observing Bones and Feathers of a Bird p. 37	Science Process Skills *Observing, Communicating, Inferring*	(groups of 2) 15 bird feathers, 15 hand lenses, 15 chicken leg bones, 15 pairs of pliers, 15 beef bones
Lesson 3 How Do Some Animals Without Backbones Live and Grow? pp. 38–41	Thinking Skills Challenge!: *Drawing conclusions* Find Out On Your Own: *Drawing conclusions*	Demonstration p. 38 pictures/specimens of earthworms, spiders, snails, and insects, clear container with soil
Activity Making Models of Spiders and Insects p. 42	Science Process Skills *Observing, Classifying, Making models*	(groups of 2) 1 large photograph of a spider, 1 large photograph of an insect, 30 sticks of modeling clay, 90 toothpicks, 240 pipe cleaners
Science and People Seeing How Insects Behave p. 43	Thinking Skills *Drawing conclusions, Formulating questions and hypotheses*	
Skills for Solving Problems Using Calendars and Pictographs pp. 44–45	Problem Solving Skills *Making decisions/Identifying and solving problems, Interpreting charts, maps, and graphs*	
Chapter Review pp. 46–47	Thinking Skills *Recognizing main idea and supporting details, Restating and explaining ideas, Comprehending meaning, Inferring*	

Teaching Options

Strategies	Extensions		Resource Masters
Cooperative Learning p. 28 (Also see p. T23.) Applying Whole Language p. 29 (Also see p. T30.)			Family Letter: *Resource Book* p. 15
Reading Strategies p. 30 (Also see pp. T26–T29.)	Reteaching Suggestion p. 31		Vocabulary Preview: *Workbook* p. 7 Science Skills: *Workbook* p. 8
Reading Strategies p. 32 (Also see pp. T26–T29.)	Reinforcement pp. 33, 34, 35 Special Education p. 33 Science and Language Arts p. 33	Enrichment pp. 34, 35 Game Suggestion p. 35 Reteaching Suggestion p. 36	Science and Math: *Workbook* p. 9
			Activity Worksheet: *Resource Book* p. 19
Reading Strategies p. 38 (Also see pp. T26–T29.)	Reinforcement pp. 39, 40 Special Education p. 39 Science and Art p. 39	Enrichment p. 40 Reteaching Suggestion p. 41	Science Activity: *Workbook* p. 10 Vocabulary Puzzle: *Workbook* p. 11
			Activity Worksheet: *Resource Book* p. 21
	Reteaching Suggestion p. 45		Science and Reading: *Resource Book* p. 23
Cooperative Learning p. 46 (Also see p. T23.)			Chapter Tests: Forms A and B *Test Book* pp. 17–20

Classroom Management

Advance Preparation

Activity, page 37
Obtain a chicken bone, a beef bone, and a pair of pliers for each group of students. You can get chicken feathers and bones from a grocery store, a butcher, a poultry store, or a biological supply company. Cook chicken bones, remove meat, and thoroughly clean the bones. Obtain beef bones from a grocery store or a butcher. Ask the butcher to cut a long beef bone into sections about 10 cm long. Boil the bones for about 15 minutes and clean them. Store all the bones in the refrigerator or freezer until you are ready to use them. If you cannot get feathers from the same source as the chicken bones, obtain them from a biological supply company or from a craft store.

Demonstration, page 38
Collect earthworms, spiders, snails, and insects, and place them separately in clear containers with some soil.

Vocabulary Review

Use the following sentences with your students to review the meanings of the italicized words.

1. A person's *backbone* helps support the body.
2. *Oxygen* is a gas in the air.
3. The way an animal *behaves* is the way it acts.
4. A sign showing children crossing the street is a type of *pictograph.*
5. Flies, bees, and grasshoppers are kinds of *insects.*

High-Potential Students

The majority of animals in the world are invertebrates. Ask students to collect pictures of different kinds of invertebrates from magazines and use them to make a collage. Ask the students who made collages to identify some of the invertebrates pictured for the other students.

Mainstreamed Students

Visually Impaired
Have the students feel their own backbone and the backbone of another student. Then ask them to feel the body of a worm to observe the difference between animals with and without backbones.

Orthopedically Handicapped
Encourage orthopedically handicapped students to take a tour of the schoolyard or neighborhood with another student to observe insects, worms, and other animals studied in the chapter.

Science Fair Projects

The Experiment Skills on p. 350 and the Investigate feature in the chapter can be used for science fair projects. You also might encourage interested students to do one of the following projects:

1. Begin an ant farm with a kit available from a pet store, hobby store, or a biological supply company. Record daily observations of the ants and prepare a report of the findings.
2. Refer to encyclopedias and other reference sources to investigate how different animals eat. Design a chart to record information about animals with and without backbones, and how they obtain their food.
3. Raise tadpoles in an aquarium. Make drawings or take photographs to show how the tadpoles develop.

lassroom Resources

Bulletin Board

Encourage students to write a short story or describe what it might be like going through the stages of the butterfly's life cycle.

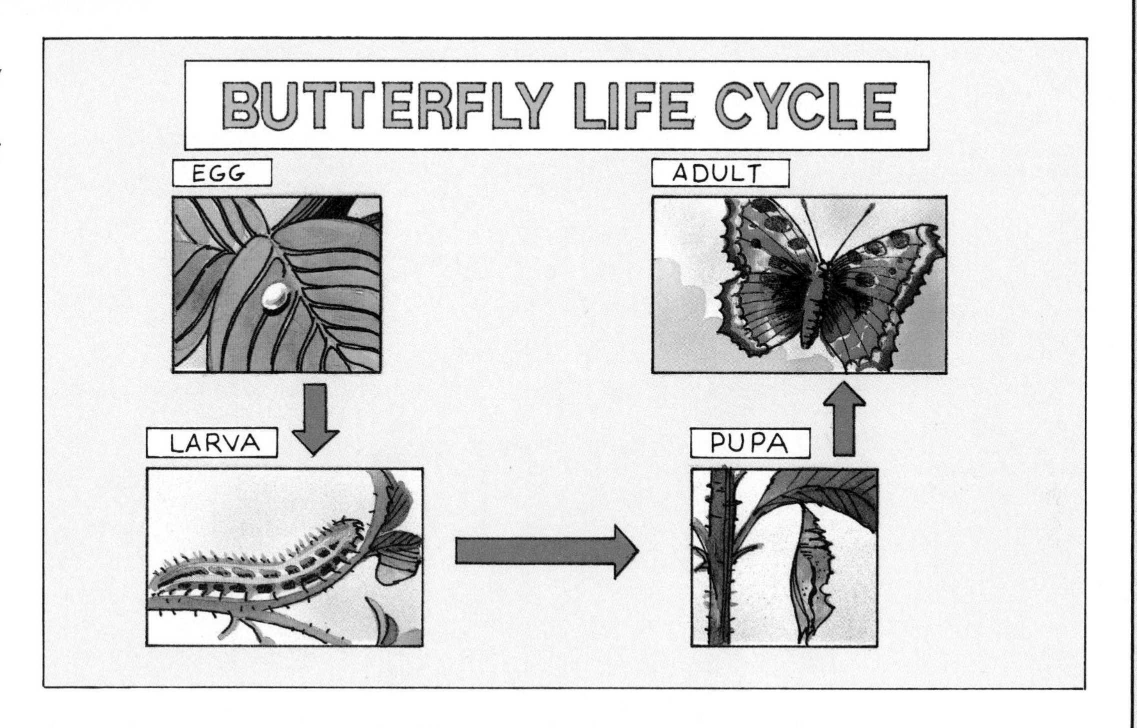

Chapter 2 Poster

Science Discovery Center

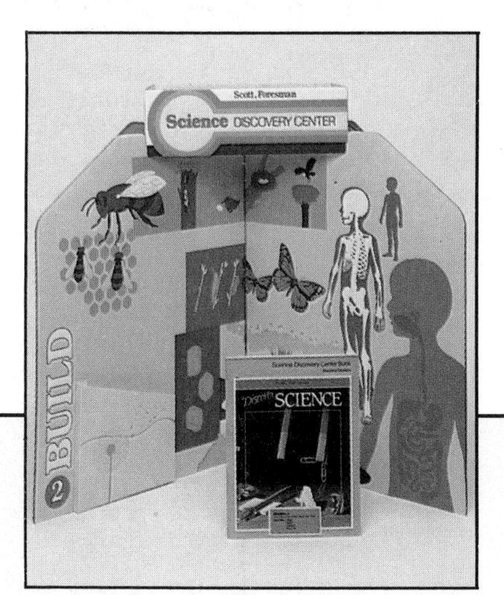

Use pages 25–30 from the *Science Discovery Center Book.* Place these worksheets in the appropriate pockets in the Science Discovery Center.

Overhead Transparencies

Use Transparency 4 from the package of color overhead transparencies.

CHAPTER 2 COPY MASTERS

Teacher's Resource Book

Dear Family,
Your student will be reading **Chapter 2: How Animals Grow and Change** in *Discover Science*, published by Scott, Foresman. We will learn that animals can be grouped into two groups—those with backbones and those without. Birds, reptiles, and fish have backbones. Earthworms, snails, and butterflies do not have backbones.
You and your student can do this activity together to learn more about animals with backbones.

Chicken Skeleton

You will need a whole, uncut chicken, liquid bleach, and modeling clay.

1. Ask your student (after washing hands) to feel the bones through the skin and meat.
2. Boil the chicken in water until the meat comes off the bones. (Save soup and meat to eat later.)
3. Soak the bones in bleach for 5 minutes. Drain and dry them.
4. Find the bones that look alike. There are many pairs of bones in the wings and legs.
5. There are many neck bones. Ask, "Are they exactly alike?" (No, each has a slightly different shape.)
6. Let your student try to figure out which bones go together.
7. The place where the bones come together is called a "joint." Use the modeling clay to hold the joints together.

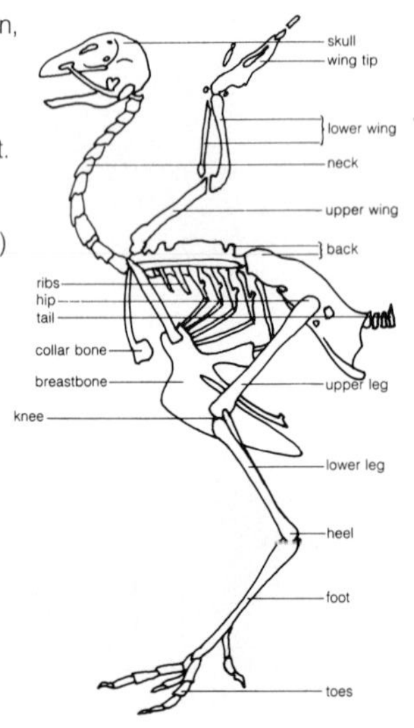

Workbook

Name ______

Chapter 2

Use with Lesson 1: pages 30-31

Vocabulary Preview

How Animals Grow and Change

Chapter Vocabulary		
cold-blooded	mammal	tadpole
gill	pupa	warm-blooded
larva	reptile	

Vocabulary Cards

1. Write each word on a card.
2. Find each word in the glossary. Copy the pronunciation under the word on the card.
3. Practice saying the words with a partner.

Word Meanings

1. Draw pictures for two of the words in the list. Use the backs of their cards for your pictures.
2. Find the words from the list in sentences in Chapter 2 of your book. Copy one sentence for each word on a separate sheet of paper.
3. Now write your own definitions of these words on the backs of their cards. Compare your definitions with the ones in the glossary.

At Home

1. Practice saying the words. Learn their meanings.
2. Look for the words and pictures of the words in newspapers and magazines. You can make a poster.

Teacher's Notes: Have students look at the pictures on the page. Ask them to tell how each picture is different.

Workbook

Name ______

Chapter 2

Use with Lesson 1: pages 30-31

Science Skills

Finding Backbones in Animals

Study the four skeletons.
Write the name of each on the line.
Find and color the backbone in each skeleton.

bird

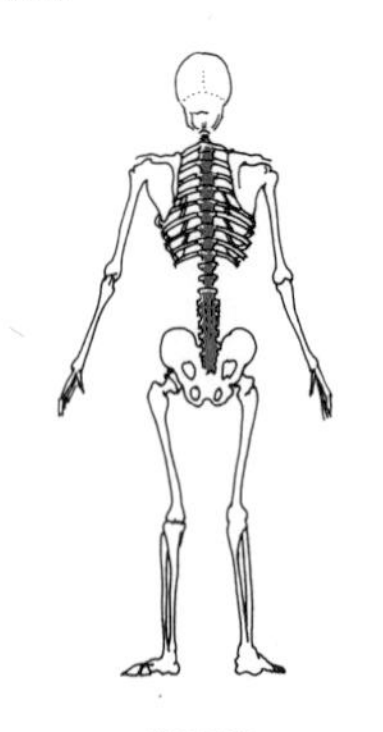

person

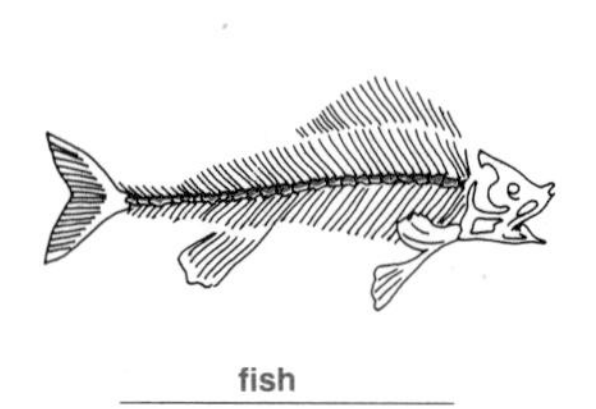

fish

snake

Teacher's Notes: Ask students to name or list as many animals with backbones as they can.

Workbook

Name ______

Chapter 2

Use with Lesson 2: pages 32-36

Science and Mathematics

Graphing Animal Lives

Some animals can live for many years. Many box turtles live to be 123 years old. Other animals live short lives. Most blue jays only live to be 4 years old.

Study the chart below. Use the chart to complete the graph at the bottom of the page.

Length of Life in Years			
Buffalo	10	Robin	12
Cat	16	Squirrel	9
Dog	13	Tiger	11
Monkey	15	Wolf	12

Length of Life

Years: 0 1 2 3 4 5 6 7 8 9 10 11 12 13 14 15 16 17

Buffalo Cat Dog Monkey Robin Squirrel Tiger Wolf

Animals

Teacher's Notes: Ask "Which animal in the graph has the longest life? The shortest? Which animals live the same amount of time?"

Chapter 2

Name ______________________

Use with Lesson 2: page 37

Activity Worksheet

Observing Bones and Feathers of a Bird

Record Your Results

Bird feather	
Chicken bone	
Beef bone	

State Your Conclusion

1. Describe the bird feather and the chicken bone.

2. What makes a chicken bone more suitable for flying than a beef bone?

Use What You Learned

Suppose you wanted to make a kite. Which material would you use, a thin tube of paper or a thick block of wood? Explain your answer.

Chapter 2

Name ______________________

Use with Lesson 3: pages 38-41

Science Activity

What Are Snails Like?

Gather These Materials

• small potted plant • large clear jar • garden snail • spoon

Follow This Procedure

1. Place the snail on the potted plant. Cover the plant and snail with a large jar. Watch what happens for several days.
2. Lightly tap the side of the jar with a spoon. What does the snail do?

Record Your Results

1. What does the snail eat?
   The snail eats plants.
2. How does the snail move?
   The snail moves slowly on one foot.
3. What does the snail do when you tap the jar?
   The snail pulls into its shell.

State Your Conclusions

1. How does a snail protect itself?
   Snails protect themselves by withdrawing into their shells.
2. What do snails do to plants?
   Snails eat and destroy plants.
3. What might happen if birds did not eat snails?
   If birds did not eat snails, there would be too many snails.
   The snails would destroy too many plants.

Teacher's Notes: There are a variety of land snails that students could observe. With adjustments, aquatic snails can be observed.

Chapter 2

Name ______________________

Use with Lesson 3: pages 38-41

Vocabulary Puzzle

Wordsearch

Match each word with its meaning.
Then find and circle each word in the puzzle below.

1. cold-blooded
2. gills
3. larva
4. mammal
5. pupa
6. reptile
7. tadpole
8. warm-blooded

a. the young of an animal that is different from the adult
b. the stage in the insect life cycle between larva and adult
c. a cold-blooded animal that has a backbone, scales, and lungs
d. animal with a body temperature that does not change
e. animals with body temperatures that change with the air or water around them
f. the part of a fish used to take in oxygen
g. a young frog
h. an animal that has a backbone and has hair

G T L C O L D B L O O D E D V
I M A M M A L C S P F M R E P
L B R E P T I L E K E S G K U
L C V W A R M B L O O D E D P
S W A R T A D P O L E L K E A

Teacher's Notes: Invite the students to make illustrations to define the following words: *larva*, *pupa*, *reptile*, and *tadpole*.

Chapter 2

Name ______________________

Use with Lesson 3: page 42

Activity Worksheet

Making Models of Spiders and Insects

Record Your Results

Spider	Insect

State Your Conclusion

1. How many body parts and legs does a spider have?

2. How does an insect differ from a spider?

Use What You Learned

Look carefully at the daddy long-legs shown in picture C. Is a daddy long-legs an insect? Explain your answer.

Name ______________________

Chapter 2

Science and Reading

Comprehension: main idea and supporting details

The Ant Colony

Read the paragraphs. Then answer the questions.

Every member of an ant colony has a job to do. The queen lays eggs that hatch into larvae. The larvae live in cocoons until they become fully-grown ants and break through the cocoons.

There are many jobs for the worker ants to do. Some take care of the larvae, and some help the ants break through their cocoons. Others find and bring back food for the colony or build and take care of the colony's living quarters.

1. What is the topic of the paragraphs?

   The topic is the ant colony.

2. Which sentence states the main idea?

   Every member of an ant colony has a job to do.

3. What does the queen do?

   The queen lays eggs.

4. What are three jobs that workers ants do?

   Accept any three: take care of larvae; help ants break through cocoons; find and bring back food; build and take care of the living quarters.

Name ______________________

Chapter 2 Test A

Multiple Choice Choose the best answer.

1. Animals are divided into two main groups based on whether or not they (1-1)
   (a.) have backbones.
   b. live in water.
   c. have fur.

2. An animal with feathers, feet, and lungs is a (2-1)
   a. reptile.
   (b.) bird.
   c. mammal.

3. Animals that change temperature with the air around them are (2-1)
   a. warm-blooded.
   (b.) cold-blooded.
   c. mammals.

4. Cold-blooded animals that live in water and have gills and backbones are (2-1)
   a. turtles.
   (b.) fish.
   c. alligators.

5. An example of a mammal is a (2-3)
   (a.) human.
   b. turtle.
   c. lizard.

6. An example of an animal that looks different from its parents when it is young is a (2-2)
   (a.) frog.
   b. bird.
   c. dog.

7. Animals that breathe through lungs, have hair, and feed milk to their young are (2-3)
   a. birds.
   (b.) mammals.
   c. reptiles.

8. An example of an animal with a soft body, shell, and no backbone is (3-1)
   a. a caterpillar.
   (b.) a snail.
   c. an ant.

9. An animal whose young look much like the adults is
   a. the butterfly.
   (b.) the treehopper.
   c. the ant. (3-1)

10. The caterpillar stage of a butterfly is also known as the (3-2)
    a. adult stage.
    b. pupa stage.
    (c.) larva stage.

Numbers in parentheses after each question refer to the lesson number and the objective of that lesson.

Name ______________________

Chapter 2 Test A

Short Answer Label the stages of the butterfly's life cycle. Write each word on the line under the correct picture.

adult larva egg pupa

1	2	3	4
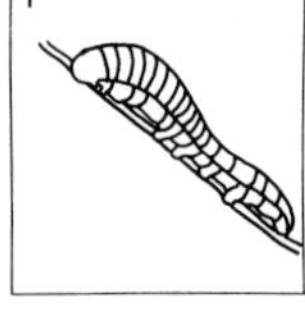		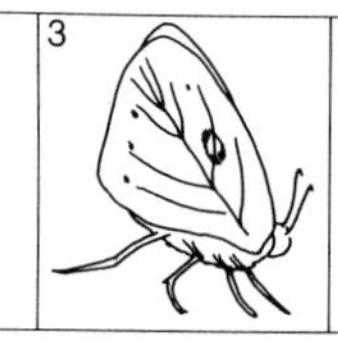	

1. **egg** 2. **larva** 3. **pupa** 4. **adult**

Short Essay Use complete sentences to answer each question.

1. Name two ways that mammals and reptiles are different. (2-2, 2-5)

   Mammals are warm-blooded and their young grow inside the mothers' bodies until they are born. Reptiles are cold-blooded and their young hatch from eggs.

2. Name four different groups of animals with backbones. Describe what the animals are like in each group. (2-1, 2-2, 2-3, 2-5)

   Mammals are warm-blooded, have hair, are fed their mothers' milk, and grow inside their mothers until they are born. Birds are warm-blooded, have feathers, and hatch from eggs. Fish have fins and scales, breathe through gills, live in water, and their young hatch from eggs. Reptiles are cold-blooded and their young hatch from eggs.

Numbers in parentheses after each question refer to the lesson number and the objective of that lesson.

Name ______________________

Chapter 2 Test B

Multiple Choice Choose the best answer.

1. The two main groups of animals are (1-1)
   (a.) animals with backbones and those without backbones.
   b. animals that live on land and those living in water.
   c. animals with hair and those without hair.

2. Animals that keep nearly the same body temperature at all times are (2-1)
   (a.) warm-blooded.
   b. cold-blooded.
   c. hot-blooded.

3. Lizards, turtles, and alligators are (2-2)
   a. mammals.
   (b.) reptiles.
   c. fish.

4. A young frog is called a (2-4)
   a. reptile.
   (b.) tadpole.
   c. larva.

5. Animals that have backbones and fins, lay eggs, and breath through gills are called (2-5)
   a. birds.
   (b.) fish.
   c. reptiles.

6. An example of an animal with a thin, soft body and no backbone is (3-1)
   a. a spider.
   b. a turtle.
   (c.) a worm.

7. A snail's hard shell helps the snail by (3-2)
   (a.) protecting its soft body.
   b. keeping it warm.
   c. storing food and water.

8. An example of an animal with two main body parts and eight legs is the (3-3)
   a. beetle.
   (b.) spider.
   c. butterfly.

9. An animal with three body parts, six legs, and no backbone is (3-4)
   a. a worm.
   b. a spider.
   (c.) an insect.

10. The first stage of a butterfly's life cycle is the (3-5)
    (a.) egg stage.
    b. pupa stage.
    c. larva stage.

Numbers in parentheses after each question refer to the lesson number and the objective of that lesson.

Name ____________________

Chapter 2
Test B

Short Answer Match each word on the left with the correct drawing of the butterfly's life cycle. Write the correct letter in the blank.

1. egg D
2. larva A
3. adult C
4. pupa B

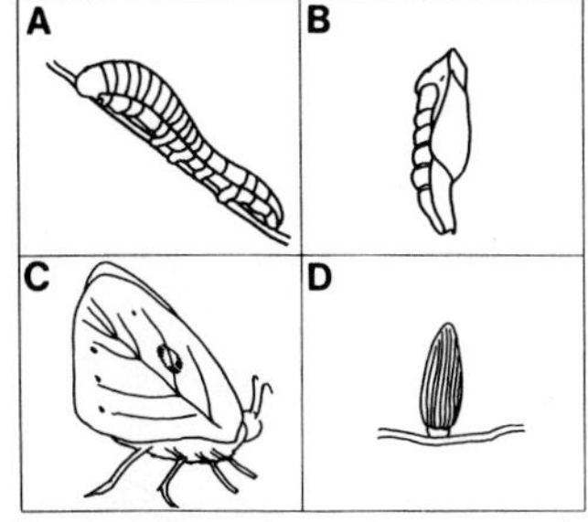

Short Essay Use complete sentences to answer each question.

1. Explain the difference between warm-blooded and cold-blooded animals. (2-2)

   A cold-blooded animal's body temperature changes with the air or water around it. A warm-blooded animal's body temperature stays about the same regardless of the air or water temperature around it.

2. How are mammals different from other animals with backbones? (2-3)

   All mammals have fur or hair. Young mammals grow inside their mothers' bodies until they are born. Baby mammals are fed their mothers' milk and need care for a long time.

Numbers in parentheses after each question refer to the lesson number and the objective of that lesson.

TEACHING PLAN

Major Concepts

Lesson 1 Animals can be grouped as animals with backbones and animals without backbones.
Lesson 2 Different kinds of animals with backbones live and grow in different ways.
Lesson 3 Different kinds of animals without backbones live and grow in different ways.

Chapter Vocabulary

amphibian, cold-blooded, gills, larva, mammal, pupa, reptile, tadpole, warm-blooded

Getting Started

Ask students how these cubs are different from their mother. Encourage students to think of other kinds of baby animals they have seen and how the babies might change as they grow.

Chapter 2

How Animals Grow and Change

A mother cheetah feeds her babies and keeps them safe. This mother takes care of her cubs. The cubs will stay with their mother until they can take care of themselves.

28

Teaching Options

Cooperative Learning ♦

Jigsaw Format (See page T23.)
Assign the following topics at random to your cooperative learning teams.

Topic A: Explain how scientists group animals. What are some other ways animals could be grouped?
Topic B: Describe the characteristics of vertebrates. What animal groups are vertebrates, and how are they different from invertebrates?
Topic C: Identify the largest group of animals on earth, and describe the characteristics common to its members.
Topic D: Describe the characteristics of spiders and insects. What animal groups do they belong in? How are they different from each other?

Have students search for information on their topic as they read the chapter. Then let all students with the same topic meet in an expert group to discuss the information. When students return to their teams, they may take turns presenting their topics to the team. Then give students a test covering all topics to complete individually (Chapter 2 Test A or B in the *Test Book*). Award Superteam certificates to teams whose average test scores exceed 90%, and Greatteam certificates to teams whose average test scores exceed 80%.

♦ ***Suitable as a language development activity***

Introducing the Chapter

You probably have seen many animals that look different from each other. Different animals live and grow in different ways. In this chapter, you will learn how scientists group animals. The activity below will help you learn about some ways to group animals.

Classifying Animals

Think about how you might describe the animals in the picture. You might tell about the sizes or colors of the animals. You also might tell how the animals move from one place to another.

On a piece of paper, draw a picture of an imaginary animal. Your animal can be any size, shape, or color. It can walk on land, swim in the water, or fly in the air. Now exchange papers with one of your classmates.

Talk About It

1. How does each animal in this picture move from one place to another?
2. How can you tell whether the animal in your classmate's drawing walked, swam, or flew?

29

DISCOVER

Objective ♦
This optional *DISCOVER* activity will help students explore and build background information about the concept of classification. Later in the chapter, students will be able to draw on this experience to help them assimilate the new content.

Science Process Skills
Observing, Classifying

Materials
For each pair of students: 2 sheets drawing paper, 2 boxes crayons

Teaching Tips

- Encourage students to ***describe*** any experiences they have had with the animals pictured.
- Have the students ***compare*** and ***contrast*** their animals.
- Ask the students to ***describe*** some important features of animals that walk, swim, and fly.

Answers
Talk About It
1. The rabbit walks on land, the seal swims, and the bird flies.
2. Answers will vary with students' pictures. Body parts that show how animals move from one place to another should be identified.

Applying Whole Language ♦

Discuss the whole language framework with each Teaching Option you select. Here is an example applied to the Science and Language Arts option on p. 33.
1. Purpose: To make a class poetry book about animals
2. Context: The roles of poets and publishers
3. Decisions: What steps are involved in writing poems? What information can we get from science? from other subjects? How can technology help? (computer)
4. Evaluation: What would our readers and listeners learn about animals? (See p. T30.)

Resource Book page 15

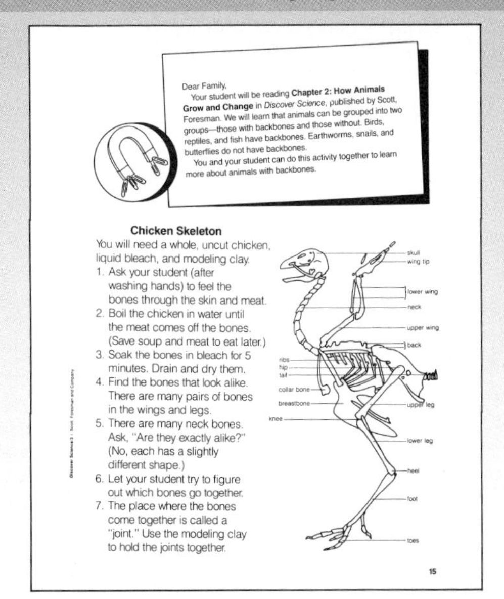

Dear Family,
Your student will be reading **Chapter 2: How Animals Grow and Change** in *Discover Science*, published by Scott, Foresman. We will learn that animals can be grouped into two groups—those with backbones and those without. Birds, reptiles, and fish have backbones. Earthworms, snails, and butterflies do not have backbones.
You and your student can do this activity together to learn more about animals with backbones.

Chicken Skeleton
You will need a whole, uncut chicken, liquid bleach, and modeling clay.
1. Ask your student (after washing hands) to feel the bones through the skin and meat.
2. Boil the chicken in water until the meat comes off the bones. (Save soup and meat to eat later.)
3. Soak the bones in bleach for 5 minutes. Drain and dry them.
4. Find the bones that look alike. There are many pairs of bones in the wings and legs.
5. There are many neck bones. Ask, "Are they exactly alike?" (No, each has a slightly different shape.)
6. Let your student try to figure out which bones go together.
7. The place where the bones come together is called a "joint." Use the modeling clay to hold the joints together.

15

Science Background

In the fourth century B.C., Aristotle began classifying organisms. He realized that one has to look at many traits in order to classify. For example, the ability to fly could be used to group birds, bats, and insects. Other traits, though, show how dissimilar these animals are.

TEACHING PLAN

Lesson Objectives

• *Explain* that animals with backbones make up one main group of animals and *list* examples of animals with backbones.
• *Explain* that animals without backbones make up the other main group of animals and *list* examples of animals without backbones.

1. Motivate

Demonstration Activity ♦

Obtain several pictures of vertebrates and invertebrates. Allow the students to examine the pictures.

Discussion

Ask students to *classify* the animals into groups based on similar characteristics, *listing* groups on paper. Question: **How did you *classify* the animals?** (Answers may include by color, size, similarities of appearance, and so on.) Explain that scientists group animals into two main groups: animals with backbones, and animals without backbones. Question: **Which animals do you think have backbones?** (mammals, fish, amphibians, reptiles, and birds)

1 How Can Animals Be Grouped?

LESSON GOALS

You will learn
• that animals with backbones make up one main group of animals.
• that animals without backbones make up the other main group of animals.

Think about groups of animals at the zoo. You might find all the bears in one part of the zoo and the sea lions in another part of the zoo.

Scientists group animals in a different way. The main groups of animals are animals with backbones and animals without backbones.

Animals with Backbones

Feel the backbone in the middle of your back. Your backbone helps you stand straight. Your backbone also helps you bend and move.

You probably have seen many animals with backbones. Bears and sea lions are animals with backbones. Birds, fish, and turtles also have backbones. Look at the animals with backbones in the pictures.

Tree frog

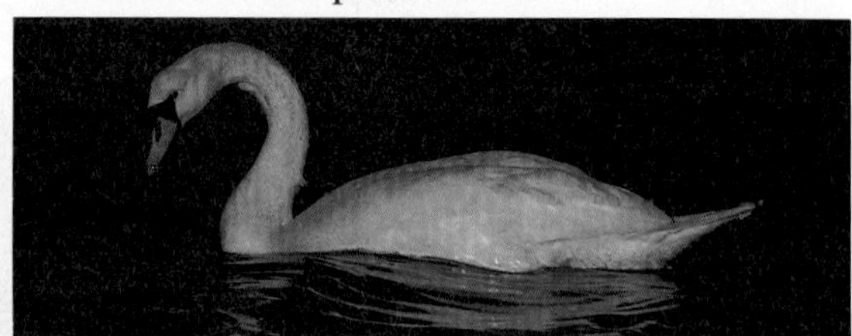

Swan

Whitetail buck

Teaching Options

Science Background

Animals are a kingdom of many-celled organisms that cannot make their own food. Most of the species of animals in the world are invertebrates, animals without backbones. These invertebrates include arthropods, mollusks, echinoderms, worms, sponges, and coelenterates. Vertebrates, animals with backbones, include mammals, birds, fish, reptiles, and amphibians.

Reading Strategies ♦

1. Guide students' pre-reading by asking: Which parts of the lesson look easy, which parts look hard, and why?
2. Assign these strategies: Visualizing Information and Writing a Memory Sentence (See pages T26–T29.)
3. Pair students to share what information is clear and unclear and initiate discussion using students' unanswered questions.

♦ ***Suitable as a language development activity***

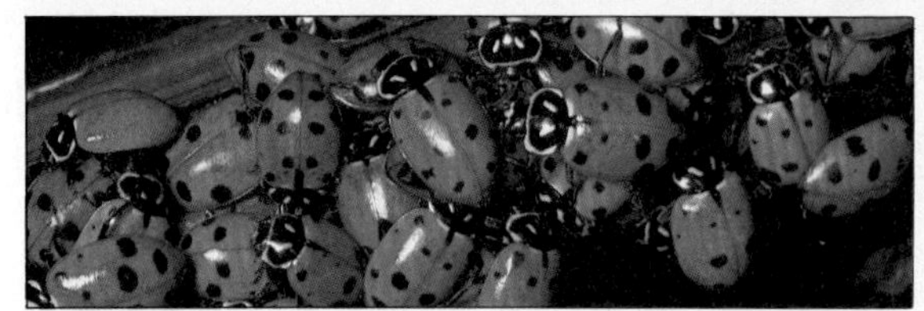

Ladybird beetles

Praying mantis

Starfish on purple coral

Animals Without Backbones

Suppose you were having a picnic. You might notice bees or ants. A spider might crawl near you. You might see a worm in the dirt. Bees, ants, spiders, and worms are animals without backbones. Other animals without backbones are in the pictures.

SCIENCE IN YOUR LIFE

You might have seen different kinds of seashells on a beach. Seashells come from animals without backbones, such as clams. Most of the time, these shells are empty by the time you find them. The animals have been eaten by other sea animals or birds.

Lesson Review

1. What are three animals with backbones?
2. What are three animals without backbones?
3. **Challenge!** Could a dog sit if it did not have a backbone? Explain your answer.

Study on your own, pages 318–319.

LIFE SCIENCE

FIND OUT ON YOUR OWN

Use pictures from magazines to make a poster that shows three animals with backbones and three animals without backbones. List the animals you chose from each group.

2. Teach

Teaching Tips

- **Possible Misconception:** Some students might have difficulty understanding that, while animals may have different physical characteristics, they can belong to the same group.
- Remind students of the *DISCOVER* on page 29 in which they ***classified*** animals based on characteristics.

3. Assess

Lesson Review

1. Answers can include animals with backbones that are shown or listed in the text.

2. Answers can include animals without backbones that are shown or listed in the text.

3. Challenge! No. The dog needs a backbone to support it in the sitting position. **Thinking Skill:** *Inferring*

Find Out On Your Own

Student posters should include pictures of three animals with backbones and three animals without backbones. **Thinking Skill:** *Judging and evaluating*

Workbook page 7 *

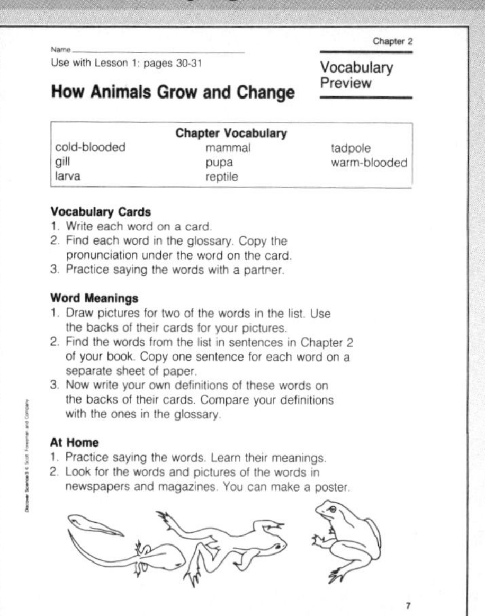

Name ______ Chapter 2

Use with Lesson 1: pages 30-31

How Animals Grow and Change — Vocabulary Preview

Chapter Vocabulary		
cold-blooded	mammal	tadpole
gill	pupa	warm-blooded
larva	reptile	

Vocabulary Cards
1. Write each word on a card.
2. Find each word in the glossary. Copy the pronunciation under the word on the card.
3. Practice saying the words with a partner.

Word Meanings
1. Draw pictures for two of the words in the list. Use the backs of their cards for your pictures.
2. Find the words from the list in sentences in Chapter 2 of your book. Copy one sentence for each word on a separate sheet of paper.
3. Now write your own definitions of these words on the backs of their cards. Compare your definitions with the ones in the glossary.

At Home
1. Practice saying the words. Learn their meanings.
2. Look for the words and pictures of the words in newspapers and magazines. You can make a poster.

7

Workbook page 8 *

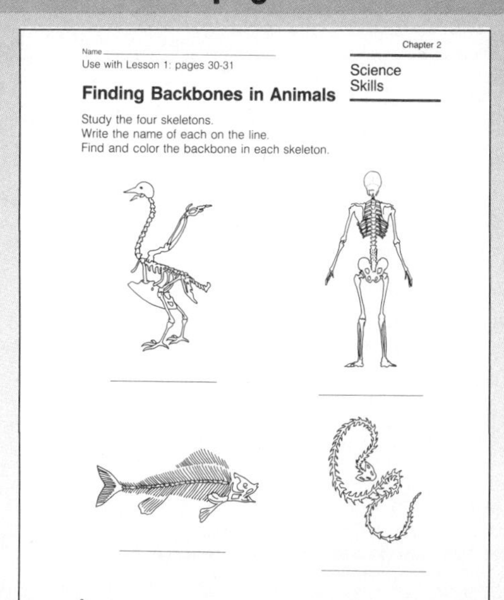

Name ______ Chapter 2

Use with Lesson 1: pages 30-31

Finding Backbones in Animals — Science Skills

Study the four skeletons.
Write the name of each on the line.
Find and color the backbone in each skeleton.

8

Reteaching Suggestion ♦

Obtain two file folders and label them *Animals with backbones* and *Animals without backbones*. Obtain enough pictures of animals from each group to supply one per student. Place the pictures in a box. Ask each student to choose a picture and file it in the correct folder. When all pictures are filed, review the pictures in each folder and discuss the similarities and differences among the animals.

* ***Answers to masters on pages 28E-28H***

TEACHING PLAN

Lesson Objectives

- *Explain* how birds, reptiles, and fish live and take care of their young.
- *Explain* how a frog changes as it grows.
- *Describe* some characteristics of mammals.

Lesson Vocabulary

amphibian, cold-blooded, gills, mammal, reptile, tadpole, warm-blooded

1. Motivate

Demonstration Activity ♦

Allow students to examine several pictures of birds, reptiles, fish, and mammals.

Discussion

Ask students to *describe* how these animals are alike and different. (Answers may include that some have fur, some have feathers, some can swim, and so on.) Questions: **How would you classify these animals?** (They can be grouped as mammals, reptiles, birds, and fish.) **Do these animals have backbones?** (Yes.)

2 How Do Some Animals With Backbones Grow and Change?

LESSON GOALS

You will learn
- how birds, reptiles, and fish live and take care of their young.
- how a frog grows and changes.
- how mammals are different from other animals.

warm-blooded (wôrm′-blud′id), animals that have about the same body temperature even when the air or water temperature around them changes.

Think about sizes and shapes of some animals with backbones. These animals often look very different from each other. Animals with backbones live and grow in different ways.

How Birds, Reptiles, and Fish Live

Imagine a bird flying high in the air. Most birds use their wings and feathers to fly. Birds also have feet and breathe with lungs.

Baby birds hatch from eggs. The parents usually protect and care for young birds. Notice how this bird feeds its young. The nest keeps the young birds warm.

Feathers also help keep birds warm. Birds are **warm-blooded** animals because they keep about the same body temperature when the temperature of the air around them changes.

A robin takes care of its young.

32

Teaching Options

Science Background

Vertebrates are a subphylum of the phylum Chordata. Vertebrates live in almost every type of habitat. In vertebrates, the nerve cord is differentiated into a complex nervous system. All vertebrates have a vertebral column, or backbone. Vertebrates also have an internal skeleton made of bone and/or cartilage. The internal skeleton supports the body and, along with the muscles, enables the animal to move. Every vertebrate also has a closed circulatory system with a heart that has two or more chambers. This chambered heart provides efficient circulation of the blood.

Reading Strategies ♦

1. Guide students' pre-reading by asking: Which parts of the lesson look easy, which parts look hard, and why?
2. Assign these strategies: Visualizing Information and Writing a Memory Sentence (See pages T26–T29.)
3. Pair students to share what information is clear and unclear and initiate discussion using students' unanswered questions.

♦ *Suitable as a language development activity*

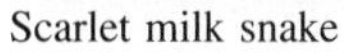

Scarlet milk snake

Angelfish

You might have seen animals like those in the pictures. Snakes, turtles, lizards, and alligators are **reptiles.** These animals are **cold-blooded** because their body temperatures change with the temperature of the air around them. How does heat from sunlight change the body temperature of this snake?[1]

Reptiles have scales and breathe with lungs. Many reptiles live on land, but some reptiles spend most of their lives in water. Young reptiles hatch from eggs. Most young reptiles can take care of themselves right away.

Which animal in the picture lives in water all its life?[2] Find the fins a fish uses for swimming. Fish also have parts called **gills** that take in oxygen from the air in water. Fish are cold-blooded animals with scales. Fish lay eggs in the water. Young fish can care for themselves as soon as they hatch.

reptile (rep′təl), a cold-blooded animal that has a backbone, has scales, and breathes with lungs.

cold-blooded (kōld′-blud′id), an animal with a body temperature that changes with the temperature of the air or water around it.

gills (gils), the parts of fish that are used to take in oxygen from the water.

[1]makes it warmer
[2]fish

2. Teach

Teaching Tips

- **Possible Misconception:** Some students might think that cold-blooded animals have cold blood, while warm-blooded animals have warm blood. Make sure students understand the correct meaning of these terms.
- Point out that humans are warm-blooded, with a body temperature of 37° C (98.6° F). Questions: **Does your body temperature change on a very cold day? a very hot day?** (No. Students should *infer* that it stays the same.)
- Discuss some characteristics of birds, fish, and reptiles. Questions: **What are some birds that can also swim?** (Answers may include ducks, geese, and penguins.) **How are young birds, fish, and reptiles born?** (Birds, reptiles, and fish all lay eggs.) **In what way are they different?** (Fish lay eggs in the water while birds and reptiles lay eggs on land.)

Reinforcement

Ask students to write the terms *warm-blooded* and *cold-blooded* on a sheet of paper. Have students *list* examples of each type of animal. (Examples of warm-blooded animals include birds, cats, dogs, and so on; examples of cold-blooded animals include fish and reptiles.)

Special Education

Have students with learning disabilities draw pictures of a bird, a reptile, and a fish. Instruct students to show the animal's backbone in their drawings.

Science and Language Arts

Encourage students to write poems about the various animals discussed in the lesson. Allow volunteers to read their poems to the class. Duplicate the poems and make a class poetry booklet for all to enjoy.

LESSON 2 pages 32–36

TEACHING PLAN

Teaching Tips

- Direct students' attention to the illustrations of the frog and tadpole on this page. Question: **What is the difference between the way a tadpole breathes and the way a frog breathes?** (A tadpole breathes with gills and a frog breathes with lungs.) Ask students to *describe* any experiences they might have had with tadpoles and frogs.
- Point out to students some other examples of young animals that look very different from their parents. (butterfly, moth, mealworm beetle)

Investigate!

Accept any testable hypothesis. Most students' data will support this hypothesis: ***Cold-blooded animals are more active in warm temperatures than in cold.*** Students should find that the activity levels of most cold-blooded animals change according to the outside temperature. For example, some fishes become sluggish when the water cools.

INVESTIGATE!

Find out how temperature affects cold-blooded animals. Write a hypothesis and test it with an experiment. You might observe how fishes in an aquarium act when the water cools.

tadpole (tad′pōl′), a young frog.

How a Frog Changes as It Grows

Some young animals look very different from their parents. Look at the pictures of the frog. A young frog is a **tadpole.** How does the tadpole look different from its parents?[1]

A frog is an **amphibian**—a cold-blooded animal with a backbone that lives part of its life in water and part on land. Young frogs hatch from eggs. A tadpole has gills and swims in the water. A fully grown frog breathes with lungs. An adult frog can live on land or swim in water.

Frogs have smooth, wet skin. They eat such animals as flies, mosquitoes, and moths.

Tadpole

Adult frog

[1]tadpole has a tail and no legs; frog has legs

amphibian (am fib′ē ən), a cold-blooded animal with a backbone that lives part of its life in water and part on land.

34

Teaching Options

Science Anecdote

The largest frog is the Goliath frog of West Africa. One such frog that was caught measured more than 80 centimeters (about 32 inches) in length. The smallest frog in the world could fit inside a thimble.

Reinforcement

Collect pictures of young and adult animals. Divide the students into small groups. Provide each group with a random assortment of pictures, making sure that pictures of both the young and the adult animal from each pair are included in each assortment. Ask the students to match each young animal with the adult animal. Allow the groups to trade the pictures until the students have matched all pairs.

Enrichment

Visit a nearby aquarium, pond, or lake to *observe* reptiles, fish, tadpoles, and frogs. Encourage interested students to *compare* and *contrast* the animals' characteristics.

♦ ***Suitable as a language development activity***

Bat

Whale

Jaguar

How Mammals Live

Imagine touching a puppy's fur. What other animals have fur?[1] Animals that have backbones and have hair or fur are **mammals**. This covering helps keep mammals warm. Dogs have a great deal of hair. Other mammals, such as dolphins, have very little hair.

The pictures show different kinds of mammals. All mammals breathe with lungs. Whales and dolphins are mammals that live in water. Notice how the whale comes out of the water for air. Many mammals live on land. Look at the pictures. Find a mammal that climbs trees. Name a mammal that can fly.[2]

mammal (mam′əl), an animal that has a backbone and has hair.

[1]answers might include cat, lion, rabbit, horse

[2]bat

35

Teaching Tips

- Show the students a variety of pictures of mammals. Have students ***compare*** and ***contrast*** the mammals in terms of habitat and amount of fur.
- Point out that the blue whale—the largest mammal—can grow as long as 32 meters (105 feet). Question: **How are mammals that live in water different from fish?** (Fish breathe with gills, have scales, and are cold-blooded; mammals breathe with lungs, have fur or hair, and are warm-blooded.)

Reinforcement

Ask students to *list* some mammals that people keep as pets. (Answers may include dogs, cats, rabbits, gerbils, and hamsters.) Use students' lists to *compare* and *contrast* the habits of mammals to the habits of other animals with backbones.

Enrichment

Visit a nearby farm or zoo to *observe* mammals and their young. Have students *record* observations about the habitat, type of hair or fur, and feeding habits of each mammal observed.

Game Suggestion ♦

Prepare labels listing the names of various reptiles, fish, birds, and mammals. Divide the students into two teams. Alternate between teams as students choose a label and act out the animal printed on it . The other team must try to *identify* the animal.

TEACHING PLAN

3. Assess

Lesson Review

1. Birds feed and protect their young. Young reptiles and fish take care of themselves as soon as they hatch.
2. A tadpole has gills, swims in water, has a tail, and has no legs. A frog breathes with lungs, can live on land or swim in water, has legs, and has no tail.
3. A young mammal grows inside its mother's body and, after birth, gets milk from its mother. A mammal needs care from its parents for a long time. A mammal has hair or fur.
4. Challenge! Reptiles get their body heat from the air around them.
Thinking Skill: *Drawing conclusions*

Find Out On Your Own

A frog's coloring is a camouflage that can help it hide. Turtles hide in their shells. Porcupine quills get stuck in an animal when the porcupine touches it.
Thinking Skill: *Restating or explaining ideas*

A mother lion feeds her cubs.

Most young mammals grow inside their mothers' bodies until they are born. Young mammals get milk from their mothers. Notice how this lion feeds her cubs. Baby mammals need care for a long time from their parents. People are mammals. What kinds of care did you need when you were a baby?

Lesson Review

1. How do birds, reptiles, and fish take care of their young?
2. Explain how a frog changes as it grows.
3. How is a mammal different from other animals?
4. **Challenge!** Why do most reptiles live in places that have warm weather?

Study on your own, pages 318–319.

LIFE SCIENCE

FIND OUT ON YOUR OWN

Animals with backbones protect themselves in different ways. Use library books to find out about frogs, turtles, and porcupines. Write sentences telling how each animal can protect itself.

36

Teaching Options

Science Anecdote

When born, lion cubs are helpless. They stay hidden in long grass for about seven months. Although they usually nurse for about seven months, they might start eating meat at three months of age.

Reteaching Suggestion ♦

Write the following words on the board: *Birds, Reptiles, Fish,* and *Mammals.* Ask students to *list* the characteristics of each group of animals. (Student descriptions might include that mammals are warm-blooded, fish have scales and gills, mammals grow inside mother's body and have hair or fur, and so on.)

Workbook page 9 *

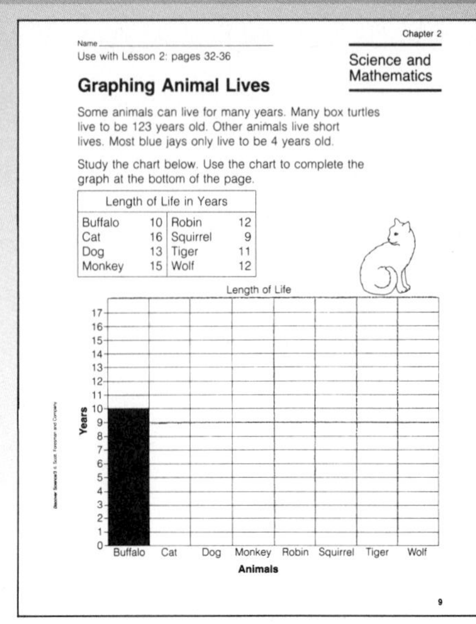
Name ____________

Use with Lesson 2: pages 32-36

Chapter 2

Science and Mathematics

Graphing Animal Lives

Some animals can live for many years. Many box turtles live to be 123 years old. Other animals live short lives. Most blue jays only live to be 4 years old.

Study the chart below. Use the chart to complete the graph at the bottom of the page.

Length of Life in Years			
Buffalo	10	Robin	12
Cat	16	Squirrel	9
Dog	13	Tiger	11
Monkey	15	Wolf	12

9

♦ *Suitable as a language development activity*

Observing Bones and Feathers of a Bird

Suggested grouping: pairs

Purpose

Observe that a bird's bones and feathers are light, have hollow parts, and enable a bird to fly.

Gather These Materials

• bird feather • scissors • hand lens • chicken bone • pliers • beef bone

Follow This Procedure

1. Use a chart like the one shown to record your observations.
2. Use scissors to cut the feather in half as shown in the picture. *CAUTION: Be careful with the scissors.*
3. Use a hand lens to look at the cut ends of the feather. On the chart, record what the inside of the feather looks like.
4. Use pliers to break the chicken bone into two pieces. *Caution: Be careful with the pliers.*
5. Use a hand lens to look at the cut ends of the chicken bone. On the chart, record what the bone looks like.
6. Look at the surface and the ends of the beef bone. On the chart, describe what you see.

Record Your Results

Bird feather	hollow
Chicken bone	hollow thin walls
Beef bone	thick walls

State Your Conclusion

1. Describe the bird feather and the chicken bone.
2. What makes a chicken bone more suitable for flying than a beef bone?

Use What You Learned

Suppose you wanted to make a kite. Which material would you use, a thin tube of paper or a thick block of wood? Explain your answer.

37

Activity

Concept

A bird's light weight and hollow bones and feathers enable it to fly.

Objectives/Process Skills

- *Observe* a feather and a bone from a bird.
- *Record* observations.
- *Compare* the chicken bone and beef bone.
- *Infer* that feathers and hollow bones enable a bird to fly.

Time Allotment

Allow 30 minutes.

Safety Tips (See page T24.)

- Tell students with allergies to feathers to avoid handling the feathers used in this activity.
- Boil all bones thoroughly before students handle them. All meat should be removed by the boiling process.
- Remind students of the hazards of using a hand lens to concentrate sunlight on skin or paper.
- Assist students in proper use of pliers to prevent injuries to the fingers.
- If bones are brittle, and shatter when broken with pliers, require students to wear cover goggles. Make certain to check the bones before students begin the activity.
- Use only round-tip safety scissors.
- Instruct students not to put their fingers into their mouths during the activity, and direct them to wash their hands.

Answers

State Your Conclusion

1. The central part of the feather is hollow; the bone has hollow parts.

2. The walls of the chicken bone are thin. The walls of the beef bone are thick. The beef bone is heavier than the chicken bone. The chicken bone is light enough to stay in the air, but the beef bone is not.

Use What You Learned

Thin tubes of paper would be light enough to stay in the air. A thick block of wood would be too heavy to stay in the air. **Thinking Skill:** *Applying information to new situations*

Resource Book page 19

Name
Use with Lesson 2: page 37
Observing Bones and Feathers of a Bird

Chapter 2
Activity Worksheet

Record Your Results

Bird feather	
Chicken bone	
Beef bone	

State Your Conclusion

1. Describe the bird feather and the chicken bone.
2. What makes a chicken bone more suitable for flying than a beef bone?

Use What You Learned

Suppose you wanted to make a kite. Which material would you use, a thin tube of paper or a thick block of wood? Explain your answer.

19

Activity Results

*** Answers to masters on pages 28E-28H**

TEACHING PLAN

Lesson Objectives

- *Describe* some characteristics of worms, snails, spiders, and insects.
- *Explain* how a butterfly changes as it grows.

Lesson Vocabulary

larva, pupa

1. Motivate

Demonstration Activity ♦

Show pictures or collect specimens of earthworms, spiders, snails, and insects from the school playground or your yard, and display them in a clear container with some soil. Allow the students to *observe* the animals.

Discussion

Ask students to *describe* some characteristics of these animals. (Student descriptions might include that earthworms have long, narrow, round bodies; spiders have brown bodies and eight legs; and ants have tiny black bodies and six legs, and so on.) Have students *compare* and *contrast* the physical features of these animals. Explain that these animals are examples of the many types of animals without backbones.

3 How Do Some Animals Without Backbones Live and Grow?

LESSON GOALS

You will learn
- how worms, snails, spiders, and insects are different from each other and from other animals.
- how a butterfly changes as it grows.

You probably have seen many animals without backbones. Most of these animals can take care of themselves as soon as they hatch from eggs. Most of the animals in the world have no backbones. Different animals without backbones live and grow in different ways.

How Some Animals Without Backbones Live

A worm has a thin, soft body and no legs. Some worms live in water. Others live on land. You might have seen an earthworm like this one. Earthworms make tunnels in the soil. Air and water can move through the tunnels. Plants need this air and water to live and grow.

An earthworm digs through the soil.

Teaching Options

Science Background

There are more species of arthropods than all the other species of invertebrates and vertebrates combined. Insects make up a large part of the arthropod phylum. Arthropods adapt to many different environments. Members of this phylum live in oceans, in fresh water, on land, in the air, as well as in and on other organisms.

Reading Strategies ♦

1. Guide students' pre-reading by asking: Which parts of the lesson look easy, which parts look hard, and why?
2. Assign these strategies: Visualizing Information to answer the question-title and Writing a Memory Sentence to answer the question-title (See pages T26–T29.)
3. Pair students to share what information is clear and unclear and initiate discussion using students' unanswered questions.

♦ ***Suitable as a language development activity***

Snail

Treehopper

The snail in the picture lives on land. Other snails live in water. Look at the snail's soft body and hard shell. How does a shell help protect a snail?[1]

Think about how a spider looks. A spider has two main body parts and eight legs. Most spiders live on land. A spider spins a web out of silk that it makes inside its body. Flies and other insects get caught in spider webs. The spider uses these insects for food.

Insects are the largest group of animals without backbones. Insects have three main body parts and six legs. Many insects also have wings. Most insects lay eggs. Some young insects, like this treehopper, look like their parents. Other young insects, such as caterpillars, look very different from their parents.

SCIENCE IN YOUR LIFE

Many insects are useful to people. For example, bees carry pollen from plant to plant. Then the plants can form seeds. People eat many of these plants. Some bees also make honey that people use for food.

[1]covers its soft body and protects it from injury

Teaching Tips

- Discuss with students how earthworms are helpful in gardens. Guide students to ***infer*** that the tunnels earthworms create help plants receive air and water.
- Draw a diagram of a spider and an insect on the board and have the students do the same on a sheet of paper. Question: **How are spiders and insects different?** (Spiders have two main body parts and eight legs and insects have three main body parts and six legs.) Instruct the students to label their drawings as a spider or an insect. Have a volunteer label the drawing on the board. Explain that spiders and insects are from two different groups of arthropods.

Reinforcement

Take the class on a nature walk around the schoolyard and neighborhood to ***observe*** earthworms, spiders, ants, and other insects. Have students ***record*** the physical features, method of movement, and habitat of each animal observed. *CAUTION:* If students handle soil or organisms, have them thoroughly wash their hands after this acitivity.

Special Education

Have students with learning disabilities ***observe*** a worm. Demonstrate that the worm does not have a backbone by gently touching the worm. Allow the students to do the same if they wish. *CAUTION:* Direct students to thoroughly wash their hands after completing this activity.

Science and Art

Provide students with a square piece of polystyrene foam, toothpicks, and colored yarn. Instruct students to randomly insert the toothpicks into the foam and then create a spider's web by winding the yarn around the toothpicks. *CAUTION:* Instruct students to handle toothpicks with care. Display the webs in the classroom.

TEACHING PLAN

Teaching Tips

- **Possible Misconception:** Some students may not recognize a caterpillar as the larval stage of a butterfly. Help students understand that a caterpillar hatches from the egg of a butterfly and eventually develops into an adult butterfly.
- Have the students use their fingers to trace the stages of growth of a butterfly.
- Question: **What are the four stages of a butterfly's development?** (egg, larva, pupa, adult butterfly)

larva (lär′və), the young of an animal that is different from the adult. [Plural: **larvae** (lär′vē)]

pupa (pyü′pə), stage in the insect life cycle between larva and adult. [Plural: **pupae** (pyü′pē)]

How a Butterfly Grows and Changes

All the stages in the life of an animal make up the animal's life cycle. The pictures show the stages in the life cycle of the butterfly.

The egg is the first stage of the life cycle. Butterflies often lay eggs on leaves the insects can eat after they hatch.

Find the picture of the caterpillar that hatches from the egg. The caterpillar is the second stage—or **larva**—of the butterfly. The larva looks different from the adult. A larva eats all the time and grows very quickly.

The insect is called a **pupa** when the larva makes a hard covering for itself. Inside this covering, the pupa changes.

Finally, the hard case around the pupa splits open. A butterfly like the one in the picture comes out. An adult is the fourth stage of the butterfly's life cycle.

The Life Cycle of a Butterfly

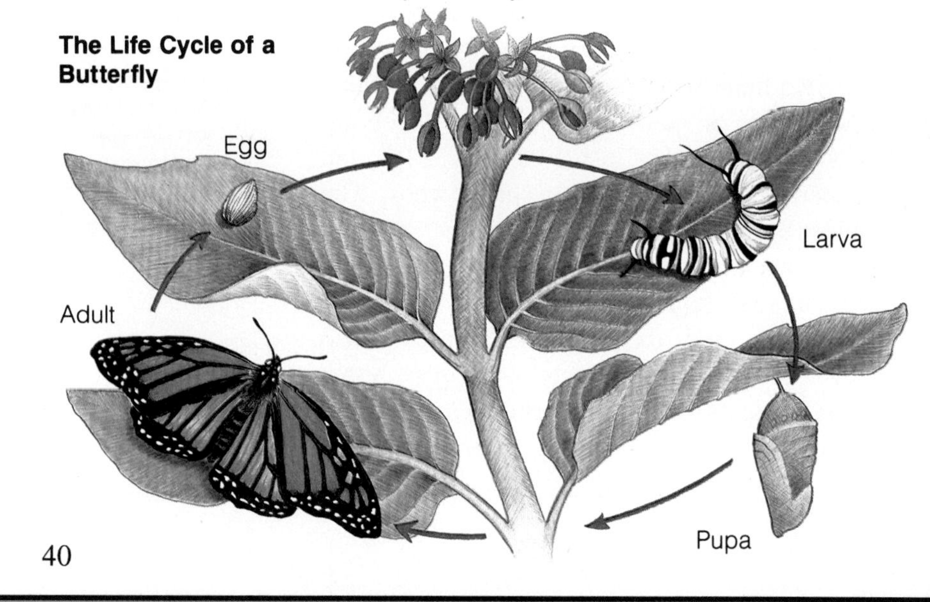

40

Teaching Options

Science Anecdote

Some insects make a buzzing sound when they fly. This is the sound of the insects' wings moving back and forth. The wings of mosquitoes can move as fast as 300 times a second.

Reinforcement

Have students draw the life cycles of a frog and a butterfly. Use students' drawings to *compare* the two animals studied that change as they grow.

Enrichment

Encourage interested students to use encyclopedias and other reference sources to investigate which animals without backbones are useful to humans. For example, the earthworm aerates soil, the bee pollinates flowers to make honey, and the ladybird beetle eats pests in gardens.

◆ ***Suitable as a language development activity***

An adult butterfly

A new adult butterfly has damp wings when it comes out of its case. When the butterfly's wings dry out, the insect will fly away.

Lesson Review

1. How are worms, snails, spiders, and insects different from each other?
2. What are the four stages of a butterfly's life cycle?
3. **Challenge!** How can the larva of the butterfly be harmful to plants?

Study on your own, pages 318–319.

PHYSICAL SCIENCE
FIND OUT ON YOUR OWN
CONNECTION

Ants live in nests that have many tunnels. Some ants bring food back to the nest. These ants collect seeds and carry them underground. Look in a book about insects to find out what other work ants do. Write a few sentences describing the work.

Teaching Tip

- Ask the students if they think a young butterfly cares for itself after it comes out of the pupa. (Guide students to *infer* that it does because when the damp wings dry, it is able to fly away.)

3. Assess

Lesson Review

1. Worms have thin, soft bodies with no legs. Snails have soft bodies and shells. Spiders have two main body parts and eight legs. Insects have three main body parts and six legs.
2. egg, larva, pupa, adult
3. Challenge! It eats the leaves. **Thinking Skill:** *Drawing conclusions*

Find Out On Your Own

Some ants in the colony have the job of carrying eggs from the queen to the nursery. Some ants cut down leaves and carry them into underground gardens. **Thinking Skill:** *Drawing conclusions*

Workbook page 10 *

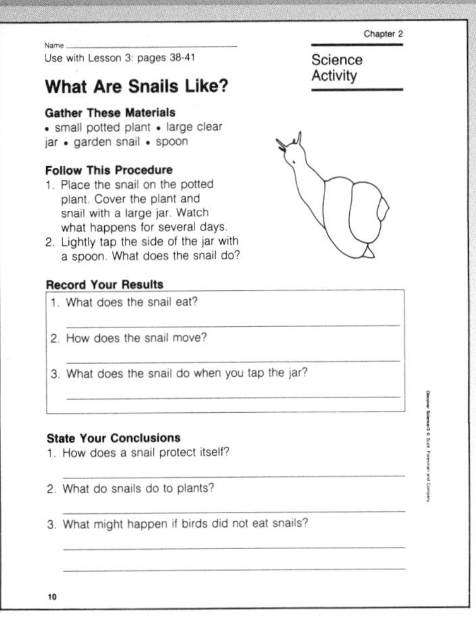

Name ____
Use with Lesson 3: pages 38-41

Chapter 2
Science Activity

What Are Snails Like?

Gather These Materials
• small potted plant • large clear jar • garden snail • spoon

Follow This Procedure
1. Place the snail on the potted plant. Cover the plant and snail with a large jar. Watch what happens for several days.
2. Lightly tap the side of the jar with a spoon. What does the snail do?

Record Your Results
1. What does the snail eat?
2. How does the snail move?
3. What does the snail do when you tap the jar?

State Your Conclusions
1. How does a snail protect itself?
2. What do snails do to plants?
3. What might happen if birds did not eat snails?

10

Workbook page 11 *

Name ____
Use with Lesson 3: pages 38-41

Chapter 2
Vocabulary Puzzle

Wordsearch

Match each word with its meaning.
Then find and circle each word in the puzzle below.

1. cold-blooded — a. the young of an animal that is different from the adult
2. gills — b. the stage in the insect life cycle between larva and adult
3. larva — c. a cold-blooded animal that has a backbone, scales, and lungs
4. mammal — d. animal with a body temperature that does not change
5. pupa — e. animals with body temperatures that change with the air or water around them
6. reptile — f. the part of a fish used to take in oxygen
7. tadpole — g. a young frog
8. warm-blooded — h. an animal that has a backbone and has hair

G T L C O L D B L O O D E D V
I M A M M A L C S P F M R E P
L B R E P T I L E K E S G K U
L C V W A R M B L O O D E D P
S W A R T A D P O L E L K E A

11

Reteaching Suggestion ♦

Instruct students to make four columns on a sheet of paper. Have them label the columns as follows: *Worms*, *Snails*, *Spiders*, and *Insects*. Ask students to *list* characteristics of each animal in the correct column. Use the students' lists to *compare* and *contrast* animals without backbones.

* **Answers to masters on pages 28E-28H**

Concept

Spiders and insects are different kinds of invertebrates.

Objectives/Process Skills

- *Make models* of a spider and an insect.
- *Observe* the differences between spiders and insects.
- *Record* data.

Time Allotment

Allow 30 minutes.

Safety Tip (See page T24.)

- Warn students to carefully use the toothpicks to avoid puncturing their skin.

Teaching Tips

- **Helpful Hint:** A model of an insect with distinct body parts, such as an ant or bee, would be easy for the students to build. A large spider, such as the tarantula, would also be easy to build a model of.
- Have the students build models of butterflies, using unruled paper for wings. Paper can be colored with felt tip pens.

Answers

State Your Conclusion

1. A spider has two body parts and eight legs.

2. An insect has three body parts and six legs.

Use What You Learned

1. A Daddy Long-Legs is not an insect because it has eight legs. An insect has six legs. **Thinking Skill:** *Drawing conclusions*

Making Models of Spiders and Insects

Suggested grouping: 1-2 students

Purpose

Make and use models of spiders and insects.

Gather These Materials

- photograph of a spider
- photograph of an insect • clay
- tooth picks • pipe cleaners
- photograph of daddy long-legs

Follow This Procedure

1. Use a chart like the one shown to record your observations.
2. Look at the photograph of the spider.
3. Make a spider body out of clay. Use toothpicks to join the body parts as in picture A.
4. Use pipe cleaners to make legs for your spider. Attach the legs to the correct body part.
5. Look at the photograph of the insect.
6. Make an insect body using clay. Use toothpicks to join the body parts.
7. Use pipe cleaners to make legs for your insect. Attach the legs to the correct body part as shown in picture B.
8. Look again at the photographs. Are there any other body parts you can add to your models?
9. In the chart, draw pictures of your spider model and your insect model. Label the parts of each animal.

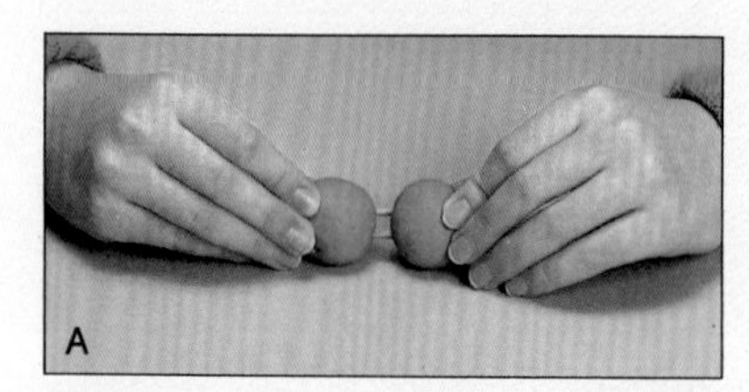

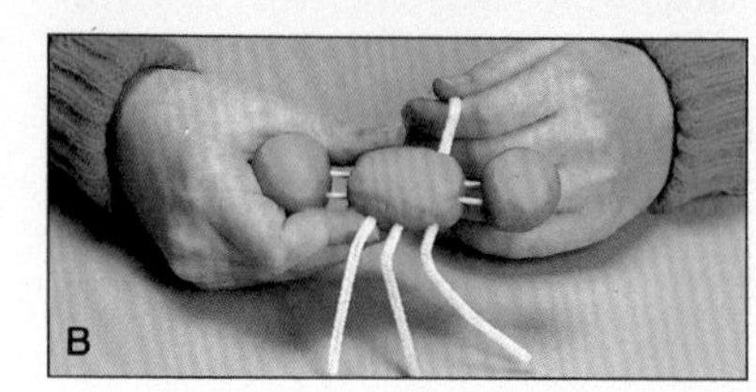

Record Your Results

Spider	Insect
2 body parts 8 legs	3 body parts 6 legs

State Your Conclusion

1. How many body parts and legs does a spider have?
2. How does an insect differ from a spider?

Use What You Learned

Look carefully at the daddy long-legs shown in the picture. Is a daddy long-legs an insect? Explain your answer.

42

Activity Results

Resource Book page 21

Name ______

Use with Lesson 3: page 42

Chapter 2

Activity Worksheet

Making Models of Spiders and Insects

Record Your Results

Spider	Insect

State Your Conclusion

1. How many body parts and legs does a spider have?

______

2. How does an insect differ from a spider?

______

Use What You Learned

Look carefully at the daddy long-legs shown in picture C. Is a daddy long-legs an insect? Explain your answer.

______

21

Seeing How Insects Behave

The bee in the picture flies back and forth between its hive and the yellow flower. How does it find its way? Can it tell apart the colors of flowers? More than a hundred years ago, Dr. Charles Henry Turner wondered about questions like these. As a boy, he watched bees, wasps, ants, and other insects. Later, as a scientist, he made up experiments to find out how insects behave.

Dr. Turner discovered that bees and wasps find their way by remembering how their surroundings look. When Dr. Turner changed the surroundings, the insects became confused. In another experiment, Dr. Turner studied bees and color. He used colored circles of paper and colored boxes filled with honey. He discovered that bees use color as well as smell to help them find flowers. Dr. Turner found that bees can even tell colors apart when they are far away.

In other experiments, Dr. Turner studied how wasps get food for their young. He also found out how insects act when light is shined on them. His work has helped people understand how insects behave. Also, his experiments showed other scientists ways to make discoveries of their own.

Bee pollinating flower

Dr. Charles Henry Turner

What Do You Think?

1. If you moved a beehive, what do you think would happen to the bees? Why?
2. Pretend you are Dr. Turner. Then, choose an insect. Think of a question you might ask about how that insect behaves.

TEACHING PLAN

Discussion

Discuss with students the insect behaviors they have ***observed*** and the behaviors they find interesting. (Answers might include bees visiting flowers, ants carrying large loads, and so on.)

Teaching Tips

- Tell students that different insects are able to see different colors. Bees cannot see red, which appears black to them. As a result, bees are attracted mainly to flowers of other colors, especially yellow and blue. Butterflies can see red, and are attracted to red and orange flowers.
- Ask students which sense they think is most important for insects. Encourage interested students to find out more about the senses of insects.

Answers

What Do You Think?

1. The bees might become confused and unable to find their way home, because bees rely on how their surroundings look to find their way. **Thinking Skill:** *Drawing conclusions*

2. Answers will vary. Students might ask questions such as: How do butterflies find their way when they migrate? How can ants carry such heavy loads? **Thinking Skill:** *Formulating questions and hypotheses*

Teaching Options

Science Background

Charles Henry Turner (1861–1923) studied science at the University of Cincinnati and received a Ph.D. magna cum laude from the University of Chicago. He spend his life teaching and researching. He devised many types of experiments with insects, other invertebrates, and small animals, and during his career he published over 50 articles in leading scientific journals. He was recognized as a major contributor in his field.

TEACHING PLAN

Purpose
To develop the skill of collecting and organizing information using calendars and pictographs to solve problems.

1. Motivate

Discussion
Display a calendar that shows the entire year on one sheet to illustrate the order of days and weeks, and the extension of these into months. Question: **How do you and your family use a calendar?** (to record appointments, plan future activities, and so on)

2. Teach

Teaching Tips
- Explain that calendars record past, present, and future time. Point out that calendars may have different time frames. For example, geological calendars measure time in millions of years.
- **Helpful Hint:** Explain that in the pictographs shown, each individual item stands for a single day. A fraction of a day can be represented by a part of the picture used in the pictograph.

Skills for Solving Problems

Using Calendars and Pictographs

Problem: Does how long a bird stays in its egg affect whether the bird is active as soon as it hatches?

Part A. Using Calendars to Collect Information

1. A calendar is divided into squares. Each month is divided into days. How many days are in a week? a month?
2. Look at the first calendar. How long does a chicken stay in its egg? Is a newly hatched chicken active?
3. How long does the killdeer stay in its egg? the meadowlark? Is a newly hatched killdeer active? a newly hatched meadowlark?

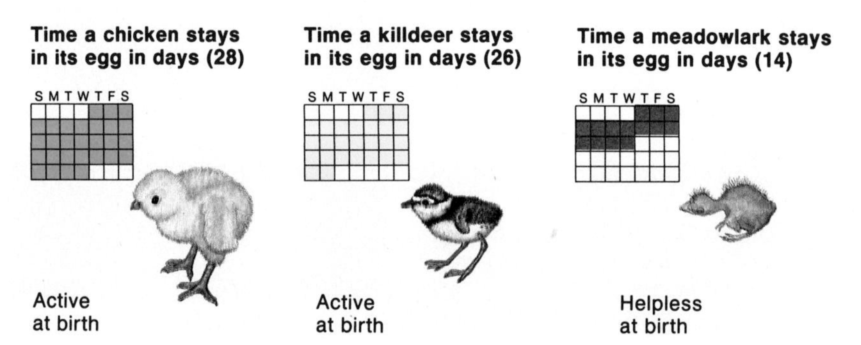

Part B. Using a Pictograph to Organize and Interpret Information.

4. This pictograph contains the information you collected about how long each chick stays in its egg. How many days was the chicken in its egg? the killdeer? the meadowlark?

44

Teaching Options

Sample Pictograph for Part C

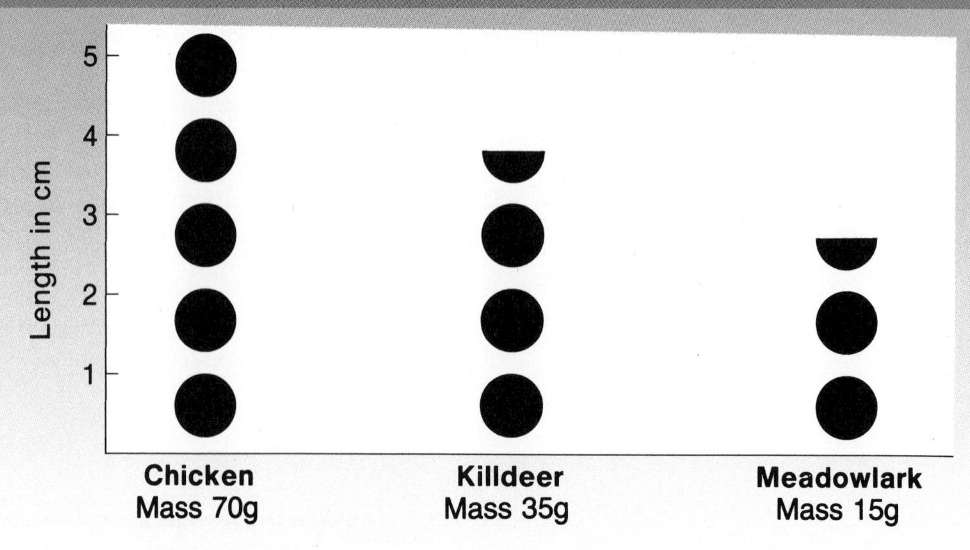

♦ *Suitable as a language development activity*

5. A bird that is active and can look for food as soon as it hatches is more likely to stay alive. Compare how active these three birds are when they hatch. How do you think the amount of time spent in the egg affects how active the bird is as soon as it hatches?

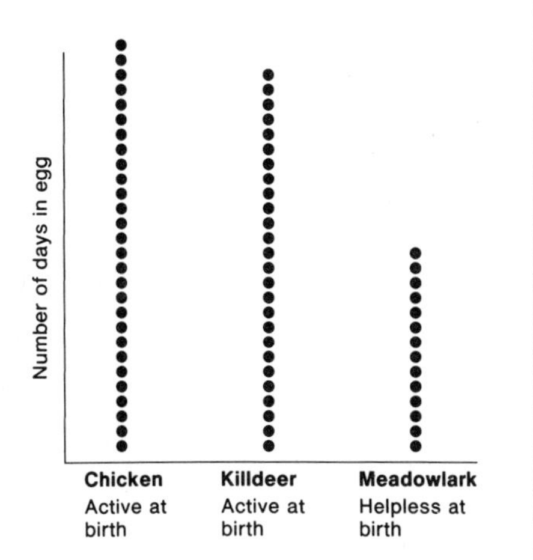

Part C. Using Calendars and Pictographs to Solve a Problem

Problem: Does the size of the egg affect how long the bird stays in the egg?

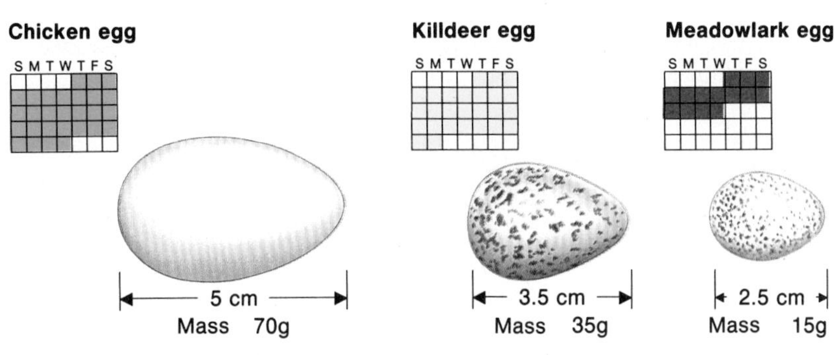

6. Use the calendars to collect the information you need to solve the problem. The calendars show how many days each bird stays in its egg. Make a pictograph like the one shown in Part B to organize your information.

7. Look at your pictograph. How does the size of an egg affect how long the bird stays in the egg? What might your pictograph look like if you used the mass of an egg instead of its size?

45

3. Assess

Part A

1. 7; 28–31
2. 28; yes
3. 26 days; 14 days; yes; no

Part B

4. 28; 26; 14
5. The chicken and the killdeer can look for food as soon as they hatched. The meadowlark cannot. From the evidence given, the longer the incubation period, the more active and independent the bird is upon hatching.

Part C

6. See Sample Pictograph for Part C in Teaching Options.
7. From the evidence given, the larger the egg, the longer the bird stays in the egg. The pictograph would probably look about the same if the mass of the eggs were used. Larger eggs probably have more mass.

Reteaching Suggestion ♦

Explain to students that gestation means the amount of time it takes for a young mammal to develop within its mother, until the time it is born. Write the following approximate gestation times on the board: *wolf: 60 days; rabbit: 30 days; mouse: 21 days.* Then draw blank calendars, labeling them with the names of each animal. Have volunteers shade in the appropriate gestation time for each animal. Then help students develop a pictograph, using squares. Each square should represent one day.

Resource Book page 23 *

Name
Comprehension: main idea and supporting details
Chapter 2
Science and Reading

The Ant Colony

Read the paragraphs. Then answer the questions.

Every member of an ant colony has a job to do. The queen lays eggs that hatch into larvae. The larvae live in cocoons until they become fully-grown ants and break through the cocoons.

There are many jobs for the worker ants to do. Some take care of the larvae, and some help the ants break through their cocoons. Others find and bring back food for the colony or build and take care of the colony's living quarters.

1. What is the topic of the paragraphs?

2. Which sentence states the main idea?

3. What does the queen do?

4. What are three jobs that workers ants do?

23

REVIEW PLAN

Reviewing Science Words

1. gills
2. pupa
3. tadpole
4. cold-blooded
5. larva
6. warm-blooded
7. reptile
8. mammal
9. amphibian

Reviewing What You Learned

1. b **3.** a **5.** c **7.** d
2. d **4.** b **6.** b **8.** d

Chapter 2 Review

Chapter Main Ideas

Lesson 1 • Animals with backbones make up one main group of animals. • Animals without backbones make up the other main group of animals.

Lesson 2 • Birds are warm-blooded animals with backbones, feathers, and wings. Reptiles are cold-blooded and have backbones. Fish are cold-blooded, have backbones, and live in water. • Amphibians live part of their lives on land and part in water. • Mammals have backbones and hair and feed their young with mother's milk.

Lesson 3 • Worms have thin, soft bodies and no backbones or legs. Snails have soft bodies, shells, and no backbones. Spiders have two main body parts, eight legs, and no backbones. Insects have three main body parts, six legs, and no backbones. • Egg, larva, pupa, and adult are the four stages in the life cycle of a butterfly.

Reviewing Science Words

amphibian	larva	reptile
cold-blooded	mammal	tadpole
gills	pupa	warm-blooded

Copy each sentence. Fill in the blank with the correct word from the list.

1. Fish use ___ to get oxygen from the water.
2. A hard case forms around the ___ of the butterfly.
3. A ___ is a young frog.
4. A ___ animal has a body temperature that changes with the temperature of the air around it.
5. The ___ of a butterfly eats all the time.
6. A ___ animal keeps about the same body temperature even when the air temperature around it changes.
7. A ___ is a cold-blooded animal that has scales and breathes with lungs.
8. An animal that has a backbone and has hair is a ___.
9. A frog is an ___.

Review Options

Cooperative Learning ♦

STAD Format (See page T23.)
Assign students to work in four- to five-member teams to study Chapter 2 Review. Students should work together to make sure that they and their teammates know the material in the chapter. After students have had enough time to study together, give them a test to complete individually (Chapter 2 Test A or B in the *Test Book*). Award Superteam certificates to teams whose average test scores exceed 90%, and Greatteam certificates to teams whose average test scores exceed 80%.

Test Book page 17 *

Name ____________________

Chapter 2
Test A

Multiple Choice Choose the best answer.

1. Animals are divided into two main groups based on whether or not they
   a. have backbones.
   b. live in water.
   c. have fur.
2. An animal with feathers, feet, and lungs is a
   a. reptile.
   b. bird.
   c. mammal.
3. Animals that change temperature with the air around them are
   a. warm-blooded.
   b. cold-blooded.
   c. mammals.
4. Cold-blooded animals that live in water and have gills and backbones are
   a. turtles.
   b. fish.
   c. alligators.
5. An example of a mammal is a
   a. human.
   b. turtle.
   c. lizard.
6. An example of an animal that looks different from its parents when it is young is a
   a. frog.
   b. bird.
   c. dog.
7. Animals that breathe through lungs, have hair, and feed milk to their young are
   a. birds.
   b. mammals.
   c. reptiles.
8. An example of an animal with a soft body, shell, and no backbone is
   a. a caterpillar.
   b. a snail.
   c. an ant.
9. An animal whose young look much like the adults is
   a. the butterfly.
   b. the grasshopper.
   c. the ant.
10. The caterpillar stage of a butterfly is also known as the
    a. adult stage.
    b. pupa stage.
    c. larva stage.

17

♦ ***Suitable as a language development activity***

Reviewing What You Learned

Write the letter of the best answer.

1. Young animals that can care for themselves as soon as they hatch are
   (a) birds. (b) fish. (c) dogs. (d) lions.
2. What is the third stage of a butterfly's life cycle?
   (a) larva (b) tadpole (c) egg (d) pupa
3. One animal with a backbone is a
   (a) fish. (b) grasshopper. (c) spider. (d) worm.
4. An animal with two main body parts and eight legs is a
   (a) insect. (b) spider. (c) larva. (d) reptile.
5. What part of a snail helps protect its body?
   (a) gills (b) scales (c) shell (d) pupa
6. One young animal that gets milk from its mother is a
   (a) turtle. (b) whale. (c) bee. (d) frog.
7. How many legs do insects have?
   (a) four (b) eight (c) ten (d) six
8. One animal that is warm-blooded is a
   (a) snake. (b) fish. (c) frog. (d) bird.

Interpreting What You Learned

Write a short answer for each question or statement.

1. How does a tadpole get oxygen from the water?
2. How do birds keep warm?
3. List two young animals that need care from their parents.
4. What are the two main groups of animals?
5. List three kinds of reptiles.

Extending Your Thinking

Write a paragraph to answer each question or statement.

1. How might some kinds of worms be useful to farmers?
2. Mammals have fewer young at one time than fish. Explain how this helps young mammals live and grow.

To explore scientific methods, see Experiment Skills on pages 350–351.

Interpreting What You Learned

1. with gills **Thinking Skill:** *Recognizing main idea and supporting details*
2. with feathers and in nests **Thinking Skill:** *Restating and explaining ideas*
3. birds, any mammal **Thinking Skill:** *Comprehending meaning*
4. animals with backbones; animals without backbones **Thinking Skill:** *Comprehending meaning*
5. accept any three: snakes, turtles, lizards, alligators **Thinking Skill:** *Restating or explaining ideas*

Extending Your Thinking

1. Some worms make space in the soil for air and water. The air and water are important for plants the farmers might be growing. **Thinking Skill:** *Inferring*
2. Young fish take care of themselves. Young mammals need care. An adult mammal is better able to care for fewer numbers of young. **Thinking Skill:** *Inferring*

Test Book page 18 *

Name ____________

Chapter 2
Test A

Short Answer Label the stages of the butterfly's life cycle. Write each word on the line under the correct picture.
adult larva egg pupa

1. ____ 2. ____ 3. ____ 4. ____

Short Essay Use complete sentences to answer each question.
1. Name two ways that mammals and reptiles are different.

2. Name four different groups of animals with backbones. Describe what the animals are like in each group.

18

Test Book page 19 *

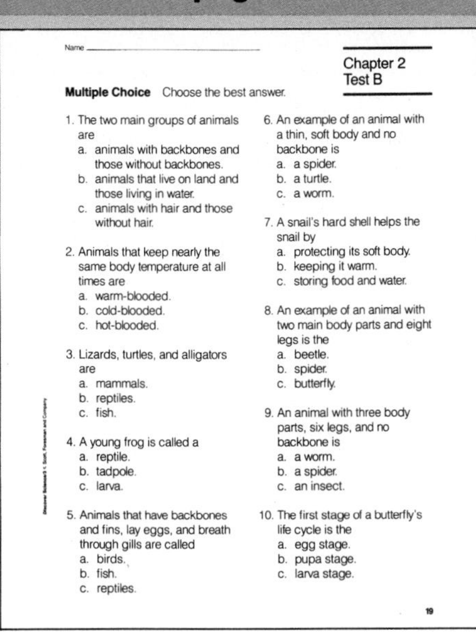

Name ____________

Chapter 2
Test B

Multiple Choice Choose the best answer.

1. The two main groups of animals are
   a. animals with backbones and those without backbones.
   b. animals that live on land and those living in water.
   c. animals with hair and those without hair.
2. Animals that keep nearly the same body temperature at all times are
   a. warm-blooded.
   b. cold-blooded.
   c. hot-blooded.
3. Lizards, turtles, and alligators are
   a. mammals.
   b. reptiles.
   c. fish.
4. A young frog is called a
   a. reptile.
   b. tadpole.
   c. larva.
5. Animals that have backbones and fins, lay eggs, and breathe through gills are called
   a. birds.
   b. fish.
   c. reptiles.
6. An example of an animal with a thin, soft body and no backbone is
   a. a spider.
   b. a turtle.
   c. a worm.
7. A snail's hard shell helps the snail by
   a. protecting its soft body.
   b. keeping it warm.
   c. storing food and water.
8. An example of an animal with two main body parts and eight legs is the
   a. beetle.
   b. spider.
   c. butterfly.
9. An animal with three body parts, six legs, and no backbone is
   a. a worm.
   b. a spider.
   c. an insect.
10. The first stage of a butterfly's life cycle is the
    a. egg stage.
    b. pupa stage.
    c. larva stage.

19

Test Book page 20 *

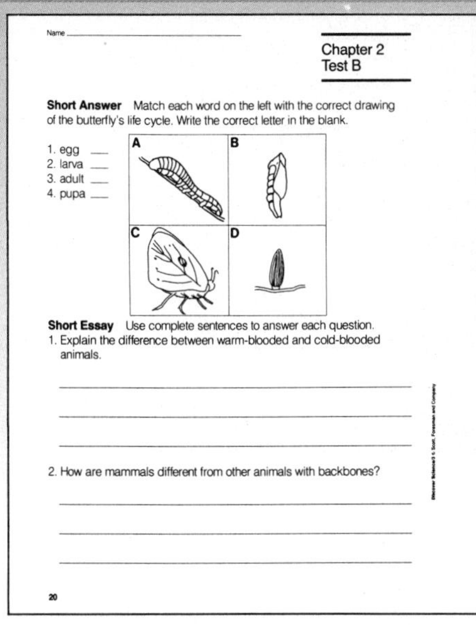

Name ____________

Chapter 2
Test B

Short Answer Match each word on the left with the correct drawing of the butterfly's life cycle. Write the correct letter in the blank.

1. egg ___
2. larva ___
3. adult ___
4. pupa ___

Short Essay Use complete sentences to answer each question.
1. Explain the difference between warm-blooded and cold-blooded animals.

2. How are mammals different from other animals with backbones?

20

*** Answers to masters on pages 28E-28H**

Chapter 3 Living Things Need Each Other Planning Guide

TEACHING PLAN

Chapter Components	Skills	Materials
Chapter Opener/*DISCOVER*: Classifying Objects pp. 48–49	*DISCOVER* p. 49 Science Process Skills *Classifying*	*DISCOVER* p. 49 (groups of 2) 15 small plastic bags, 150 assorted buttons
Lesson 1 What Are the Five Groups of Living Things? pp. 50–52	Thinking Skills Challenge!: *Comparing* Find Out On Your Own: *Collecting information, Communicating*	Demonstration p. 50 small toy car or truck, plant
Activity Feeding the Jays p. 53	Science Process Skills *Communicating, Measuring, Collecting and interpreting data, Making models*	(groups of 2) 375 small paper plates, 15 colored markers, 1 bag sunflower seeds (300 seeds)
Lesson 2 How Do Organisms Live Together? pp. 54–56	Thinking Skills Challenge!: *Recognizing patterns and relationships* Find Out On Your Own: *Organizing information*	Demonstration p. 54 pictures of an ocean community and a town
Activity Making a Model of a Food Chain p. 57	Science Process Skills *Communicating, Classifying, Making models*	(groups of 2) 15 sheets construction paper, 15 rolls transparent tape, 15 crayons or colored pencils
Lesson 3 How Do Organisms Get Food? pp. 58–60	Thinking Skills Challenge!: *Drawing conclusions* Find Out On Your Own: *Collecting information, Summarizing*	Demonstration p. 58 no materials needed
Science and Technology Helping Pandas Survive p. 61	Thinking Skills *Restating or explaining ideas, Inferring*	
Skills for Solving Problems Using Diagrams and Pie Graphs pp. 62–63	Problem Solving Skills *Making decisions/Identifying and solving problems, Interpreting charts, maps, and graphs*	
Chapter Review pp. 64–65	Thinking Skills *Predict, Contrasting, Sequencing, Drawing conclusions*	

Teaching Options

Strategies	Extensions		Resource Masters
Cooperative Learning p. 49 (Also see p. T23.) Applying Whole Language p. 49 (Also see p. T30.)			Family Letter: *Resource Book* p. 27
Reading Strategies p. 50 (Also see pp. T26–T29.)	Special Education p. 51 Enrichment p. 51	Reteaching Suggestion p. 52	Vocabulary Preview: *Workbook* p. 13 Science and Language Arts: *Workbook* p. 14
			Activity Worksheet: *Resource Book* p. 31
Reading Strategies p. 54 (Also see pp. T26–T29.)	Reinforcement p. 55 Enrichment p. 55	Special Education p. 55 Reteaching Suggestion p. 56	Science Activity: *Workbook* p. 15
			Activity Worksheet: *Resource Book* p. 33
Reading Strategies p. 58 (Also see pp. T26–T29.)	Reinforcement p. 59 Special Education p. 59	Reteaching Suggestion p. 60	Science Skills: *Workbook* p. 16 Vocabulary Puzzle: *Workbook* p. 17
	Reteaching Suggestion p. 63		Science and Reading: *Resource Book* p. 35
Cooperative Learning p. 64 (Also see p. T23.)			Chapter Tests: Forms A and B *Test Book* pp. 25–28

Classroom Management

Advance Preparation

Discover, page 49
Provide each group of students with ten assorted buttons.

Demonstration, page 50
Obtain a toy car or truck and a live plant for this demonstration.

Activity, page 53
Each group will need 25 paper plates and 20 sunflower seeds.

Demonstration, page 54
You will need pictures of an ocean community and pictures of a town for this demonstration.

Vocabulary Review

Use the following sentences with your students to review the meanings of the italicized words.

1. *Mold* is a kind of fungus that can spoil food.
2. After a flower is *pollinated*, seeds begin to form.
3. *Yeast* is a kind of fungus that is used in bread dough.
4. Green plants make—or *produce*—their own food.
5. Animals must eat—or *consume*—food, because they cannot make it themselves.

High-Potential Students

Let interested students observe the behavior of ants in an ant farm. Ant farm kits are available from pet shops, hobby stores, and biological supply companies. Encourage students to place different foods in the ant colony. Ask students to observe and record any changes in behavior in terms of how the ants locate and react to food. Encourage students to draw conclusions about ant behavior and communication based on their observations.

Mainstreamed Students

Visually Impaired
When studying communities, take visually impaired students to various nearby natural environments such as a pond or forest. If possible, allow students to touch examples of various nontoxic plants in each environment.

Hearing Impaired
When taking a class trip to a pond or forest environment, make sure to pair hearing-impaired students with hearing students who can indicate when and what kinds of sounds are heard in the environment.

Science Fair Projects

The Experiment Skills on p. 352 and the Investigate feature in the chapter can be used for science fair projects. You also might encourage interested students to do one of the following projects:

1. Build a habitat for one of the following organisms: fish, reptiles, or insects. Prepare a chart with information about how the habitat provides the things the organisms need.
2. Design a food web for an ocean, desert, or arctic habitat. Illustrate the food web with pictures of the organisms in the web.

assroom Resources

Bulletin Board

Ask students to draw or bring in pictures of other living things and assemble other natural communities.

Chapter 3 Poster

How do these living things need each other?

Science Discovery Center

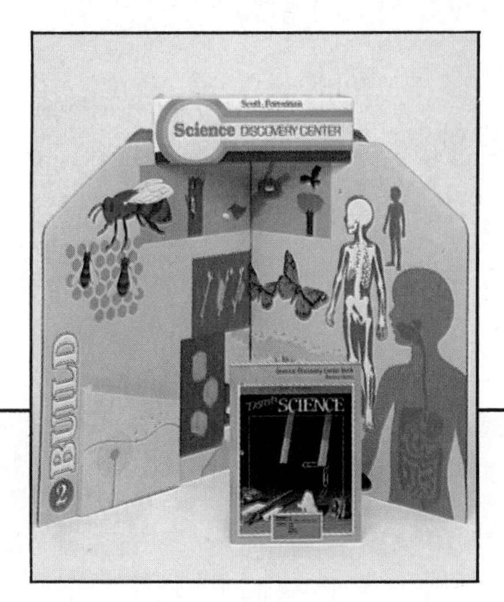

Use pages 35–40 from the *Science Discovery Center Book.* Place these worksheets in the appropriate pockets in the Science Discovery Center.

Overhead Transparencies

Use Transparencies 5 and 6 from the package of color overhead transparencies.

CHAPTER 3 COPY MASTERS

Teacher's Resource Book

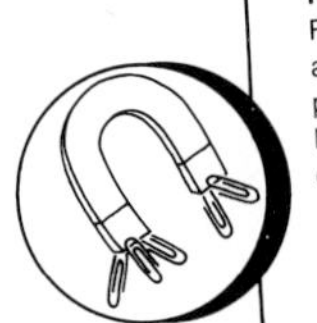

Dear Family,
Your student will be reading **Chapter 3: Living Things Need Each Other** in *Discover Science*, published by Scott, Foresman. We will learn about bacteria, protists, fungi, plants, and animals. Living things (organisms) live in habitats that provide food, water, shelter, and space. The organisms that live together in a habitat form a community, and they depend on each other.
You and your student can do this activity together to learn more about organisms and communities.

The Community in a Lawn

You will need a lawn or weedy area that you have permission to dig up, ruler, shovel or spoon, newspaper, magnifying lens, notebook and pencil. This activity works best if done in the warm weather.

1. Measure a square of lawn 30 cm (12 inches) on each side.
2. Look for living things in this area. Do you see plants, animals, fungi (mushrooms)? Write down what you see. If you do not know an organism's name, draw its picture. Later, you can look for its name in a book.
3. Dig around the edge of your square down to 8 cm (3 inches). Lift the square onto the newspaper.
4. Break your piece of lawn into smaller pieces. Look for plants that are not grass plants.
5. Use the magnifying lens to look for tiny animals.
6. Write the name or draw each organism. How many of each kind did you find?

27

Workbook

Name ____________

Use with Lesson 1: pages 50-52

Chapter 3

Vocabulary Preview

Living Things Need Each Other

Chapter Vocabulary

bacteria	consumer	habitat	predators
cell	food chain	organisms	prey
community	fungi	population	producer
			protist

Vocabulary Cards

1. Write each word on a card.
2. Find each word in the glossary. Copy the pronunciation under the word on the card.
3. Practice saying the words with a partner.

Fill in the Blanks

Use words from the vocabulary list to complete the sentences. The first letter of the word is given.

Pam and I walked quietly through the forest. Soon we saw some deer. Pam told me that a large group, or Population, of deer live in this forest. The forest is the deer's habitat. Deer, squirrels, birds, trees, and other Organisms make up this forest Community. We met a forest ranger. The ranger told us that people are not allowed to hunt the deer. Very few other animals hunt deer for food in this forest. Deer have few Predators here.

At Home

1. Practice saying the words. Learn their meanings.
2. Look for the words and pictures of the words in newspapers and magazines. You can make a poster.

Teacher's Notes: Ask students to write sentences, using the vocabulary words. Suggest they find they words in the glossary.

13

Workbook

Name ____________

Use with Lesson 1: pages 50-52

Chapter 3

Science and Language Arts

Organism Riddles

Each riddle describes one of the five groups of organisms. Read each riddle. Then write the organism group on the line.

1. We have only one cell.
   Some of us can make medicine.
   Some of us can make you sick.
   What are we?
   Bacteria
2. We may have one cell or many cells.
   We live in wet places.
   Some of us make our own food.
   What are we?
   Protists
3. Most of us have more than one cell.
   We cannot make our own food.
   Some grow on food or dead material.
   What are we?
   Fungi
4. We have many cells.
   We make our own food.
   We cannot move around on our own.
   What are we?
   Plants
5. We have many cells.
   We must eat food.
   We can move from place to place.
   What are we?
   Animals

Teacher's Notes: Invite students to write their own "What Am I?" riddles about specific living things in the five groups.

14

Teacher's Resource Book

Name ____________

Use with Lesson 1: page 53

Chapter 3

Activity Worksheet

Feeding the Jays

Record Your Results

Number of Jays	Seeds each Jay had to eat
4	
5	
10	
20	
25	

State Your Conclusion

1. How did changing the number of Jays change the number of seeds that each Jay could eat?
2. How many Jays can get food in your habitat?

Use What Your Learned

How might the population of Jays change when the population of sunflowers changes?

31

Workbook

Chapter 3

Science Activity

Name ____________________

Use with Lesson 2: pages 54-56

How Do Snails Help Plants?

Gather These Materials
- aquatic snail • *Elodea* plant
- pond or aquarium water • test tube
- tight fitting cork

Follow This Procedure
1. Fill the test tube three-quarters full of water. Place the *Elodea* and a snail in the tube. Tightly cork the tube.
2. Observe what happens. Make a daily record of what you see.

Record Your Results

Day 1	
Day 2	
Day 3	
Day 4	
Day 5	

State Your Conclusions
1. Where does the snail get the food and oxygen it needs?
   from the plant
2. Where does the plant get the carbon dioxide it needs?
   from the snail
3. Why do most aquariums and ponds have both plants and animals?
   The plants give the animals food and oxygen. The animals give the plants carbon dioxide.

Teacher's Notes: *Elodea* and aquatic snails can be obtained at pet and fish supply stores.

Teacher's Resource Book

Chapter 3

Activity Worksheet

Name ____________________

Use with Lesson 2: page 57

Making a Model of a Food Chain

Record Your Results

________ ← ________ ← ________ ← ________

State Your Conclusion
1. Explain how the organisms you chose make up a food chain.
2. What happens to a food chain if one of the links breaks?

Use What You Learned
Explain why two producers cannot be in the same food chain.

Workbook

Chapter 3

Science Skills

Name ____________________

Use with Lesson 3: pages 58-60

Investigating a Food Chain

Study the pond community below.
1. Color the producers *green* and the consumers *yellow*.
2. Circle the predators in *red* and the prey in *blue*.
3. Draw a *black line* to show a food chain.

Color green: plants
Color yellow: fish and birds
Circle in red: birds
Circle in blue: fish
Black line from plants to fish to birds

Teacher's Notes: Discuss another example of a food chain, such as plants, cows, humans.

Workbook

Chapter 3

Vocabulary Puzzle

Name ____________________

Use with Lesson 3: pages 58-60

Hidden Message

Write the missing letters in the sentences.

1. C E L L S are tiny parts of organisms. (C = 12)
2. A living thing is an O R G A N I S M. (O = 13, A = 1, I = 3)
3. B A C T E R I A are made of one cell. (A = 1, T = 7, R = 14, A = 1)
4. P R O T I S T S live in wet places. (R = 14, S = 9)
5. Mold and mushrooms are F U N G I. (G = 6)
6. A P O P U L A T I O N of swans L I V E S in the pond. (L = 2, N = 5, V = 4)
7. The pond C O M M U N I T Y has a variety of animals. (C = 12, N = 5)
8. The rabbit gets what it needs from its H A B I T A T. (H = 8, T = 7, T = 7)
9. A plant is a P R O D U C E R, but an animal is a C O N S U M E R. (D = 11, E = 10, N = 5, R = 14)
10. Food passes through a F O O D C H A I N in a community. (O = 13, D = 11, A = 1, N = 5)
11. A hawk is a flying P R E D A T O R. (R = 14, D = 11, R = 14)
12. A mouse is P R E Y to a hawk. (E = 10)

Below, write the letters that match the numbers.

A L L   L I V I N G   T H I N G S
1 2 2   2 3 4 3 5 6   7 8 3 5 6 9

N E E D   E A C H   O T H E R.
5 10 10 11   10 1 12 8   13 7 8 10 14

Teacher's Notes: Encourage students to enrich their vocabularies by learning names of organisms that are consumers or producers.

Teacher's Resource Book

Name

Comprehension: classifying and comparing

Chapter 3

Science and Reading

Living Things

Read the story and follow the directions.

When Randy prepares food, he is very careful to keep his knives and cutting board clean. He does not want any germs on them that could cause sickness.

Today Randy is making a salad of lettuce, sliced mushrooms, grated carrots, and dried seaweed. He is going to give some of it to his pet rabbit.

Complete the chart. Write each underlined word from the story in the correct column.

Bacteria	Protists	Fungi	Plants	Animals/ People
germs	seaweed	mushrooms	lettuce carrots	Randy rabbit

Complete the chart. Place an X in the proper column to tell how many cells each group has.

	Have One Cell	Have More Than One Cell
bacteria	X	
protists	X	X
fungi	X	X
plants		X
animals/people		X

Test Book

Name

Chapter 3 Test A

Multiple Choice Choose the best answer.

1. All living things on earth are made up of (1-1)
   a. bacteria.
   (b.) cells.
   c. energy.
2. Organisms of one or more cells that live in wet places are (1-2)
   a. bacteria.
   (b.) protists.
   c. fungi.
3. An example of organisms that often are used to make medicines are (1-2)
   a. bacteria.
   b. protists.
   (c.) fungi.
4. A common fungi is (1-2)
   a. bacteria.
   b. protists.
   (c.) mold.
5. A group of organisms of the same kind that live in the same place is called a (2-1)
   a. habitat.
   (b.) population.
   c. collection.
6. The place where each organism in a community lives is its (2-3)
   (a.) habitat.
   b. population.
   c. community.
7. Organisms that make their own food are (3-1)
   a. consumers.
   (b.) producers.
   c. predators.
8. Animals caught and eaten by other animals are (3-2)
   (a.) prey.
   b. producers.
   c. fungi.
9. Food passes from one organism in a community to another in a (3-3)
   a. food community.
   (b.) food chain.
   c. food habitat.
10. Organisms that cannot make their own food are (3-1)
   a. prey.
   b. workers.
   (c.) consumers.

Numbers in parentheses after each question refer to the lesson number and the objective of that lesson.

Test Book

Name

Chapter 3 Test A

Short Answer Label each drawing correctly. Use these words:
organism population community habitat

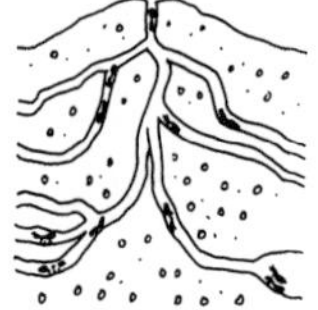

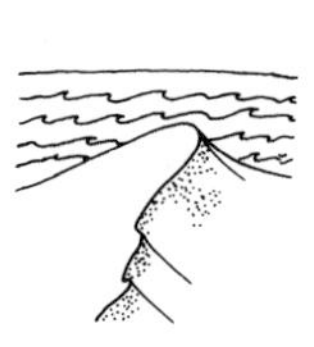

1. population 2. organism 3. habitat 4. community

Short Essay Use complete sentences to answer each question.

1. What are the five groups of organisms? (1-2)

The five groups of organisms are plants, animals, protists, fungi, and bacteria.

2. Explain the difference between producers and consumers. Give an example of each. (3-1)

Producers are able to make their own food. A plant is an example of a producer. Consumers are not able to make their own food. An animal is an example of a consumer.

Numbers in parentheses after each question refer to the lesson number and the objective of that lesson.

Test Book

Name

Chapter 3 Test B

Multiple Choice Choose the best answer.

1. The smallest living parts of an organism are the (1-1)
   (a.) cells.
   b. bacteria.
   c. blood.
2. One-celled organisms are (1-2)
   (a.) bacteria.
   b. plants.
   c. fungi.
3. Most seaweeds are (1-2)
   a. bacteria.
   (b.) protists.
   c. fungi.
4. Organisms that can make their own food are (1-2)
   a. animals.
   (b.) plants.
   c. fungi.
5. Enough food, water, and sun will make plants (2-1)
   (a.) grow larger.
   b. stay the same.
   c. die out.
6. Different organisms living in the same place are a (2-2)
   a. population.
   (b.) community.
   c. habitat.
7. Organisms get food, shelter, water, and space from their
   (a.) habitat.
   b. population.
   c. community. (2-3)
8. Organisms that do not make their own food are (3-2)
   a. plants.
   (b.) consumers.
   c. producers.
9. Animals that hunt are (3-2)
   (a.) predators.
   b. prey.
   c. producers.
10. A food chain begins with (3-3)
   a. bacteria.
   (b.) plants.
   c. animals.
11. A group of the same organisms living in one place form a (2-2)
   a. community.
   b. habitat.
   (c.) population.

Numbers in parentheses after each question refer to the lesson number and the objective of that lesson.

Name ____________________

Chapter 3
Test B

Short Answer Correctly label the picture of each habitat.
Use these words:

forest meadow pond swamp

1. forest 2. pond 3. swamp 4. meadow

Short Essay Use complete sentences to answer each question.

1. What is necessary for plant populations to become larger? **(2-1)**

   To become larger, plant populations need plenty of food, water, and sunlight. This will help them grow and reproduce.

2. In what ways do people depend on other organisms for food and shelter? **(3-3)**

   People depend on plants and animals for food. Trees and other plants provide shelter, and plants and some animals provide clothing.

Numbers in parentheses after each question refer to the lesson number and the objective of that lesson.

TEACHING PLAN

Major Concepts

Lesson 1 All living things have certain characteristics in common, but they also have differences. By studying the differences, scientists can divide organisms into five groups: bacteria, protists, fungi, plants, and animals.
Lesson 2 Organisms live together in populations and communities.
Lesson 3 Living things depend on each other for food.

Chapter Vocabulary

bacteria, cell, community, consumer, food chain, fungus, habitat, organism, population, predator, prey, producer, protist

Getting Started

Ask students to describe the hummingbird's features, such as its size, its coloring, or how it flies. Have them describe other birds they are familiar with and how they interact with plants.

Chapter 3

Living Things Need Each Other

The hummingbird is having a meal. It is using its long, thin beak to suck sweet-tasting liquid from the flower. The bird also eats small insects that are on the flower.

48

Teaching Options

Cooperative Learning ♦

Jigsaw Format (See page T23.)
Assign the following topics at random to your cooperative learning teams.

Topic A: What would happen to a community if winter came three months early?
Topic B: How do scientists group organisms?
Topic C: What characteristics do all plants have in common?
Topic D: How are consumers and producers alike? How are they different?

Have students search for information on their topic as they read the chapter. Then let all students with the same topic meet in an expert group to discuss the information. When students return to their teams, they may take turns presenting their topics to the team. Then give students a test covering all topics to complete individually (Chapter 3 Test A or B in the *Test Book*). Award Superteam certificates to teams whose average test scores exceed 90%, and Greatteam certificates to teams whose average test scores exceed 80%.

♦ ***Suitable as a language development activity***

Introducing the Chapter

The activity below will help you learn how to group objects. In this chapter, you will learn how scientists divide living things into groups. You also will learn how living things need other living things to stay alive.

Classifying Objects

Look at the buttons you get from your teacher. Think about how you would describe the different buttons.

Divide the buttons into two groups. You might group the buttons by their sizes, shapes, or colors. What other ways can you divide your buttons into groups? Ask a classmate to figure out how you grouped the buttons. Now think of another way to divide the buttons into groups. Ask your classmate to try to figure out how you grouped the buttons the second time.

Talk About It

1. What two ways did you group the buttons?
2. How could you divide one of your groups into two smaller groups?

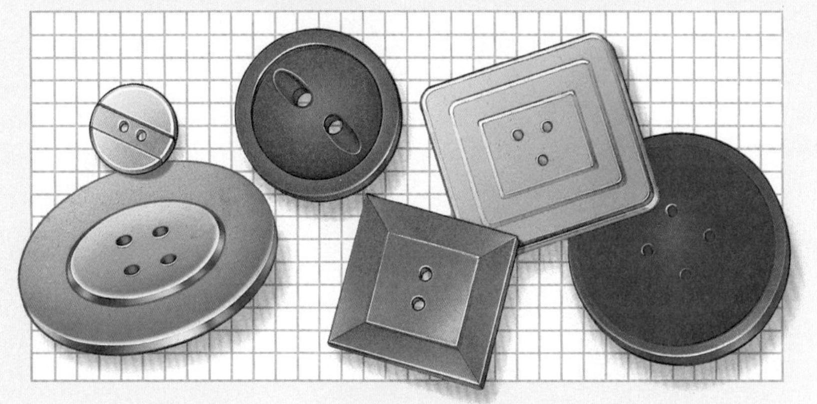

DISCOVER

Objective ♦

This optional *DISCOVER* activity will help students explore and build background information about the concept of classfying organisms. Later in the chapter students will be able to draw on this experience to help assimilate the new content.

Science Process Skills

Organizing information, Comparing, Contrasting

Materials

For each pair of students: one small plastic bag, 10 assorted buttons

Safety Tip (See page T24.)

- Remind students not to put buttons in mouth (ears, nose, etc.)

Teaching Tip

- Allow students to ***group*** the buttons in different ways and give reasons for their choices. (Students may refer to size, color, number of holes, shank type, etc.)

Answers

Talk About It

1. Answers might include grouping objects by size, shape, color, texture, or type or number of holes for thread.
2. Answers might include any of those mentioned in answer 1.

Applying Whole Language ♦

Discuss the whole language framework with each Teaching Option you select. Here is an example applied to the Enrichment option on p. 55.

1. Purpose: To write news articles about how people affect a plant or animal population
2. Context: News reporter
3. Decisions: What resource material and people might help? What warnings and exceptions might be part of the story? What visits might be needed?
4. Evaluation: How did we get reliable information? What questions did this activity raise? How did we use language? (See p. T30.)

Resource Book page 27

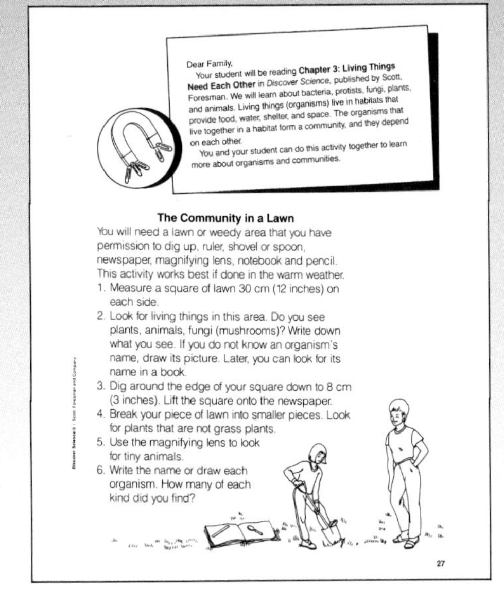

Dear Family,

Your student will be reading **Chapter 3: Living Things Need Each Other** in *Discover Science*, published by Scott, Foresman. We will learn about bacteria, protists, fungi, plants, and animals. Living things (organisms) live in habitats that provide food, water, shelter, and space. The organisms that live together in a habitat form a community, and they depend on each other.

You and your student can do this activity together to learn more about organisms and communities.

The Community in a Lawn

You will need a lawn or weedy area that you have permission to dig up, ruler, shovel or spoon, newspaper, magnifying lens, notebook and pencil. This activity works best if done in the warm weather.

1. Measure a square of lawn 30 cm (12 inches) on each side.
2. Look for living things in this area. Do you see plants, animals, fungi (mushrooms)? Write down what you see. If you do not know an organism's name, draw its picture. Later, you can look for its name in a book.
3. Dig around the edge of your square down to 8 cm (3 inches). Lift the square onto the newspaper.
4. Break your piece of lawn into smaller pieces. Look for plants that are not grass plants.
5. Use the magnifying lens to look for tiny animals.
6. Write the name or draw each organism. How many of each kind did you find?

27

Science Background

In classifying living things, biologists have studied the makeup and behavior of more than 1,500,000 different kinds of organisms. Biologists use these studies to group organisms by their similarities and differences.

TEACHING PLAN

Lesson Objectives
- *Describe* how organisms are alike.
- *List* the five kingdoms of organisms.

Lesson Vocabulary
bacteria, cell, fungus, organism, protists

1. Motivate

Demonstration Activity ♦
Obtain a small car or truck, and a plant. Demonstrate how the toy moves by pushing it.

Discussion
Ask students to ***name*** the object that is living. Allow students to explain their answers. (Some students may think that the toy is living because it moves.) Questions: **Which one needs food?** (plant) **Which one grows?** (plant) **Which one can make more organisms like itself?** (plant) Explain that living things are made of one or more cells. Discuss how the car does not grow, need food, or make more cars; therefore, it is a non-living thing.

1 What Are the Five Groups of Living Things?

LESSON GOALS

You will learn
- how living things are alike.
- how scientists group living things.

organism (ôr′gə niz′əm), a living thing.

cell (sel), the basic unit of an organism.

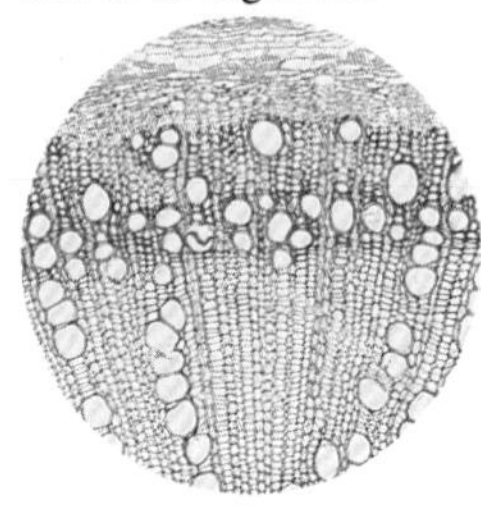

Plant cells

Look at the pictures at the bottom of these two pages. Notice that each picture looks different. All of these pictures show things that are alive.

How Living Things Are Alike

You cannot always easily tell what is alive and what is not. How can you tell if something is alive? Living things—or **organisms**—are alike in certain ways. All organisms grow. Organisms need food. They make more organisms like themselves. Organisms are made of one or more cells.

Cells are the basic units of an organism. Think about how bricks can make up a building. In much the same way, cells make up an organism. The picture on the left shows plant cells. Most cells are so tiny you need to use a microscope to see them.

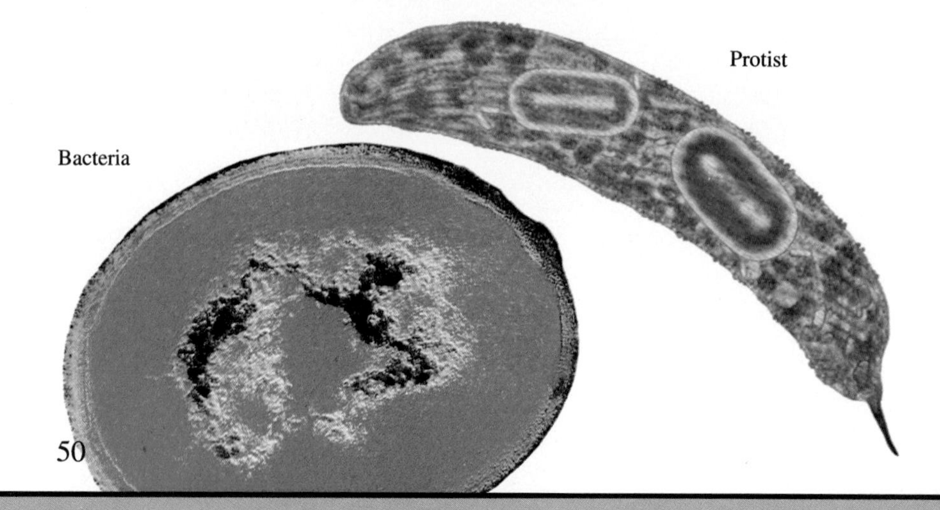

Protist

Bacteria

50

Teaching Options

Science Background

Today scientists divide living organisms into five kingdoms. Monerans are simple organisms that lack a nuclear membrane. All other living things have cells with a nucleus surrounded by a membrane. Protists are often one-celled, but they are more complex than monerans. Fungi range from one-celled yeasts to complex multicelled mushrooms. Their cells have a cell wall. Fungi get their food from outside sources. Plants are rooted in one place. Plants have a cell wall and produce their own food using a green pigment, chlorophyll. Animals get their food from other organisms. Their cells do not have a cell wall. Most animals have some type of locomotion.

Reading Strategies ♦

1. Guide students' pre-reading by asking: What two or three questions do you have for each subheading?
2. Assign these strategies: Mapping Examples and Writing a Memory Sentence (See pages T26–T29.)
3. Pair students to share what information is clear and unclear and initiate discussion using students' unanswered questions.

♦ ***Suitable as a language development activity***

How Scientists Group Living Things

Millions of different kinds of organisms live on the earth. Scientists have divided these organisms into five groups: bacteria, protists, fungi, plants, and animals. These pictures show an organism from each group.

Bacteria are made of one cell. Some bacteria use other living things for food. Some bacteria can make their own food. Many kinds of bacteria help people. For example, certain bacteria are used to make medicines. A few bacteria cause illnesses, such as sore throats.

Protists are made of one or more cells. They live in wet places. Many protists have parts that help them move. Some protists can make their own food and some cannot. The picture shows a protist that is made of one cell. This protist moves by whipping its tail back and forth. Most seaweeds are protists. Seaweeds have many cells and make their own food.

bacteria (bak tir/ē ə), organisms made of one cell that can be seen through a microscope.

protists (prō/tists), organisms that live in wet places and have one or more cells.

2. Teach

Teaching Tips

- Refer students to the *DISCOVER* on page 49 in which they grouped buttons by similarities. Point out that scientists also group living things by similarities and differences.
- **Possible Misconception:** Some students might think that anything that is green is a plant, and anything that moves is an animal. Point out that the protist shown on page 50 is green and moves but is not a plant or animal.
- If possible, prepare slides of bacteria and plant cells for students to examine. Have them ***compare*** the cells of each organism. Point out that bacteria are classified by their shape.
- Question: **How do plants make food?** (Guide students to ***describe*** how plants use sunlight, water, and air when they make food.)

Workbook page 13 *

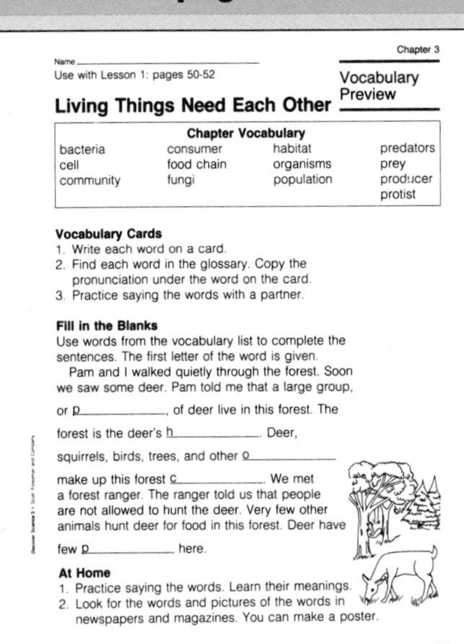

Chapter 3

Name ____________

Use with Lesson 1: pages 50-52

Vocabulary Preview

Living Things Need Each Other

Chapter Vocabulary			
bacteria	consumer	habitat	predators
cell	food chain	organisms	prey
community	fungi	population	producer
			protist

Vocabulary Cards
1. Write each word on a card.
2. Find each word in the glossary. Copy the pronunciation under the word on the card.
3. Practice saying the words with a partner.

Fill in the Blanks
Use words from the vocabulary list to complete the sentences. The first letter of the word is given.

Pam and I walked quietly through the forest. Soon we saw some deer. Pam told me that a large group, or p__________, of deer live in this forest. The forest is the deer's h__________. Deer, squirrels, birds, trees, and other o__________ make up this forest c__________. We met a forest ranger. The ranger told us that people are not allowed to hunt the deer. Very few other animals hunt deer for food in this forest. Deer have few p__________ here.

At Home
1. Practice saying the words. Learn their meanings.
2. Look for the words and pictures of the words in newspapers and magazines. You can make a poster.

13

**** Answers to masters on pages 48E–48H***

Enrichment

Demonstrate how plants reproduce by studying lima bean seeds. Soak the beans in water overnight. Slice some of the beans in half vertically and point out the structures inside each seed. Explain how these structures will become the root and the stem of the plant. Plant the remaining seeds in a pot filled with soil. Place the pot in a sunny window. Water the plants regularly and encourage students to ***observe*** their growth. *CAUTION:* Immediately wipe up any water that spills on the floor to prevent slipping accidents.

Special Education

Instruct students to divide a poster board in half vertically. Have students with learning disabilities make a collage of magazine pictures showing living things on one side and nonliving things on the other.

TEACHING PLAN

Teaching Tips

- Ask students if they have ever tasted or seen moldy bread or cheese. Explain that the mold is a kind of fungi.
- Explain that yeast used to bake bread is a kind of fungi. Yeast is a one-celled organism which needs sugar to grow and live.
- Question: **Where do you think yeast gets sugar to grow?** (from the flour used to make bread)

3. Assess

Lesson Review

1. Living things grow, need food, make more living things like themselves, and are made of one or more cells.
2. The five groups of organisms are bacteria, protists, fungi, plants, and animals.
3. Challenge! Both animals and fungi grow, they make more organisms like themselves, they are made of cells, and they depend on other organisms for food. **Thinking Skill:** *Comparing*

Find Out On Your Own

Yeasts used in bread dough ferment the sugar in the dough, releasing carbon dioxide. The carbon dioxide causes the bread to rise. **Thinking Skills:** *Collecting information, Communicating*

SCIENCE IN YOUR LIFE

Kelp are large protists that live in oceans. They are used as food and in making ice cream and some medicines.

fungus (fung′gəs), an organism, such as a mold or mushroom, that gets food from dead material or by growing on foods or living things. [Plural: **fungi** fun′jī]

A **fungus** can be only one cell, but most fungi are more than one cell. Fungi cannot make their own food. Some fungi grow on foods. The mold in the picture is a fungus that grows on bread. Other fungi grow on living things and can make them sick. Mushrooms are fungi that get food from dead material in the soil.

Plants and animals are made of many cells. Plants can make food. Animals must eat food. Plants do not move from place to place on their own. Their roots keep them attached to the soil. Animals can move around. What are some ways animals move from place to place?[1]

Mold

Lesson Review

1. How are living things alike?
2. What are the five groups of organisms?
3. **Challenge!** How are animals and fungi alike?

Study on your own, pages 320–321.

[1]fly, run, swim, crawl

PHYSICAL SCIENCE — FIND OUT ON YOUR OWN — CONNECTION

Yeast is a fungus used in baking bread. Find out what gas yeast gives off that causes bread to rise.

52

Teaching Options

Science Anecdote

Some kinds of bacteria are used to make yogurt and other dairy products. Yogurt is made by adding bacteria to milk. The bacteria change the sugar in milk (lactose) into lactic acid. The acid thickens the milk and changes it into yogurt.

Reteaching Suggestion ♦

Take the students on a walk around the school or neighborhood. Have students bring paper and pencil. Ask them to *list* the living and nonliving things they see as they walk. Use the students' lists to *compare* the living and nonliving things in terms of the four characteristics of living things. Question: **How do these living and nonliving things affect your lives?** (Accept reasonable answers.)

Workbook page 14 *

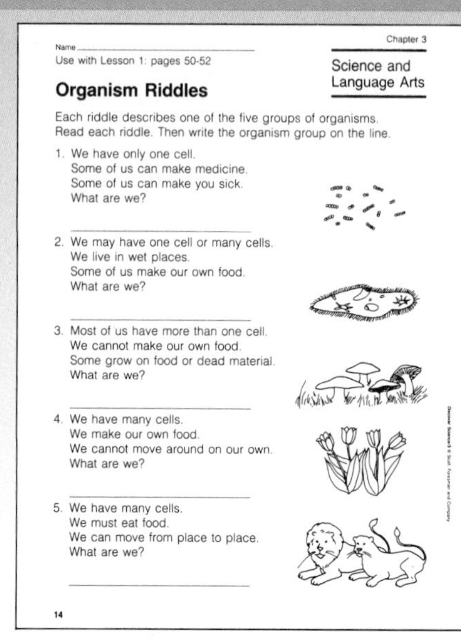

Name ______
Use with Lesson 1: pages 50-52

Chapter 3

Science and Language Arts

Organism Riddles

Each riddle describes one of the five groups of organisms. Read each riddle. Then write the organism group on the line.

1. We have only one cell.
   Some of us can make medicine.
   Some of us can make you sick.
   What are we? ______
2. We may have one cell or many cells.
   We live in wet places.
   Some of us make our own food.
   What are we? ______
3. Most of us have more than one cell.
   We cannot make our own food.
   Some grow on food or dead material.
   What are we? ______
4. We have many cells.
   We make our own food.
   We cannot move around on our own.
   What are we? ______
5. We have many cells.
   We must eat food.
   We can move from place to place.
   What are we? ______

14

♦ *Suitable as a language development activity*

Feeding the Jays

Suggested grouping: 1–2 students

Purpose
Observe how the food supply affects a population in a habitat.

Gather These Materials
• 25 small paper plates • colored markers • 20 sunflower seeds

Follow This Procedure
1. Use a chart like the one shown to record your observations.
2. Write the word *Jay* on 20 paper plates.
3. Place 4 of the plates in front of you and put 5 sunflower seeds on each plate, as shown in the picture. Each Jay gets to eat 5 seeds.
4. Record the number of Jays and how many seeds each Jay gets to eat.
5. Place 1 more plate in front of you. Divide the seeds equally among all the Jays.
6. Repeat step 4.
7. Place 5 more Jay plates in front of you. Divide the seeds equally among all the Jays.
8. Repeat step 4.
9. Lay all the Jay plates out and divide the seeds equally among all the Jays.
10. Repeat step 4.
11. Make 5 more Jay plates. Divide the seeds equally.

Record Your Results

Number of Jays	Seeds each Jay had to eat
4	5
5	4
10	2
20	1
25	5 Jays had no food.

State Your Conclusion
1. How did changing the number of Jays change the number of seeds that each Jay could eat?
2. How many Jays can get food in your habitat?

Use What You Learned
How might the population of Jays change when the population of sunflowers changes?

53

Concept
Food supply can limit the size of a population in a habitat.

Objectives/Process Skills
- *Make a model* to find how food supply affects population size.
- *Observe* how an increase in population affects food.
- *Measure* and *record* data.

Time Allotment
Allow 20 minutes.

Safety Tips (See page T24.)
- Be certain to use nontoxic colored markers.
- Have students wash their hands when finished with the activity.

Teaching Tips
- Instead of using paper plates for birds, have the students make their own birds from construction paper and a bird pattern.
- Be sure students understand that a real Jay needs many seeds, and not just one to survive.

Answers
State Your Conclusion
1. As the number of Jays got larger, the number of seeds each Jay could eat got smaller.
2. The habitat will support 20 or fewer Jays.

Use What You Learned
There would be more Jays living in the habitat when there were more sunflowers. Fewer Jays would live in the habitat when there were fewer sunflowers.
Thinking Skill: *Predicting*

Resource Book page 31

Name ______ Chapter 3
Use with Lesson 1: page 53
Activity Worksheet
Feeding the Jays

Record Your Results

Number of Jays	Seeds each Jay had to eat
4	
5	
10	
20	
25	

State Your Conclusion
1. How did changing the number of Jays change the number of seeds that each Jay could eat?
2. How many Jays can get food in your habitat?

Use What Your Learned
How might the population of Jays change when the population of sunflowers changes?

31

Activity Results
4 Jays could eat 5 seeds apiece. 5 Jays could eat 4 seeds apiece. 10 Jays could eat 2 seeds apiece. 20 Jays could eat 1 seed apiece. In a habitat containing 25 Jays there would not be enough seeds for each Jay to have one.

* **Answers to masters on pages 48E–48H**

TEACHING PLAN

Lesson Objectives

- *Define* population and *explain* how the size of a population can change.
- *Explain* how populations make up a community.
- *Define* an organism's habitat.

Lesson Vocabulary

community, habitat, population

1. Motivate

Demonstration Activity ♦

Display pictures of an ocean community and of a town. Write the words *population* and *community* on the board.

Discussion

Questions: **What organisms live in an ocean?** (seaweed, fish, crabs, coral) List students' responses under population. Explain that these are all different populations that live in an ocean. **What different populations live in the town?** (people, trees, animals, etc.) List students' responses under *populations*. Explain that all the populations that live in the same place make up a community. Write the words *ocean* and *town* under *community*.

2 How Do Organisms Live Together?

LESSON GOALS

You will learn

- what a population is and what can cause the size of a population to change.
- how groups of different kinds of organisms live in a community.
- how each organism in a community lives in its own place.

population (pop′yə lā′shən), organisms of the same kind that live in the same place.

[1]plants (cattails)

The swans in the picture stay together most of the time. They look for food together, drink water together, and rest together. Many animals and other organisms live in groups.

Populations

A **population** is a group of organisms of the same kind that live in the same place. These swans are part of a population. What other population do you see in the picture?[1]

The size of a population can change from time to time. Every year, some members of the population die and new members are born. Populations often grow larger when there is plenty of food. Populations often become smaller when there is not enough food. How might the amount of sunlight and water cause the size of a plant population to change?[2]

Population of swans

[2]Population might grow larger if there is plenty of sunlight and water; population might grow smaller if there is not enough sunlight or water.

Teaching Options

Science Background

A population is a group of one kind of organism living in a specific place. A community is formed by the interacting populations of an area.

Populations within a community have a specific habitat and niche. An organism's habitat includes everything the organism needs to stay alive and reproduce. An organism's niche is what the organism does within the community. Often the niche is described by the organism's position within a food chain. This is a useful description, but it is not the entire definition. A niche also includes non-feeding relationships such as symbiotic relationships.

Reading Strategies ♦

1. Guide students' pre-reading by asking: What two or three questions do you have for each subheading?
2. Assign these strategies: Visualizing Information and Writing a Memory Sentence (See pages T26–T29.)
3. Pair students to share what information is clear and unclear and initiate discussion using students' unanswered questions.

♦ *Suitable as a language development activity*

Communities

Think of your favorite outdoor place. How many different populations live there?[1] Most places have more than one kind of population. For example, a meadow has populations of birds, deer, and trees. An ocean has populations of fish, crabs, and seaweeds.

All the populations that live together in the same place make up a **community**. The organisms in a community depend on each other for food and shelter. The picture shows a forest community. Think about how the organisms in a forest depend on each other. This squirrel makes its nest in the tree. The rabbit eats plants. The birds feed on worms and insects that live in the forest. Insects, such as this butterfly, pollinate the plants.

Forest community

INVESTIGATE!

The populations in a garden need space, sun, soil, and water. Find out how pulling the weeds out of a garden affects the growth of other populations. Write a hypothesis and test it with an experiment.

community (kə myü′nə tē), all plants, animals, and other organisms that live in the same place.

[1]Answers might include populations of plants, animals, or other organisms.

55

2. Teach

Teaching Tips

- Question: **What are some populations in our school community?** (students, teachers, administrators, maintenance staff)
- Question: **What is a group of students called?** (class) **What is a group of zebras called?** (herd) Explain that animal groups are given special names. Have students *describe* and *compare* how different animal groups such as a school of fish, or a flock of birds live, move, rest, and look for food together.
- **Possible Misconception:** Some students may have difficulty comprehending that plants are part of communities. Explain how similar plants exist in the same area, such as desert plants, forest plants, and ocean plants.

Investigate!

Accept any testable hypothesis. Most students' data will support this hypothesis: *Pulling the weeds out of a garden helps other plants grow better.* If students have no access to a garden, you might suggest that they set up two "gardens" in paint trays or shoe boxes and use one as a control. Students should find out that removing the competitors in a garden ensures the best growth for garden plants.

Enrichment

Take the students on a walk around the school neighborhood. Have them bring a pencil and paper. Ask students to *list* the various populations of plants and animals living in the community. Use students' lists to discuss how the different populations interact in the community. (Animals feed on plants. Insects pollinate flowers. People play and work together.)

Special Education

Allow students with learning disabilities to choose a type of community such as a forest, ocean, or desert. Have students make a collage of magazine pictures showing the various populations found in each community.

Reinforcement

Show students pictures of a forest community and a desert community. Have students *list* the types of populations found in each community. Use students' lists to *describe* how the members of each population interact in the community.

TEACHING PLAN

Teaching Tips

- Question: **Where is your habitat?** (Students' homes are their habitats.)
- Encourage students to *identify* and *describe* the habitats of plants and animals in their neighborhood. (trees, soil, houses)

3. Assess

Lesson Review

1. New members are born, and old members die. Populations get larger when food is plentiful, and smaller when food is scarce.
2. A community is all the populations that live together in the same place.
3. An organism gets food, water, shelter, and space from its habitat.
4. Challenge! The food we eat comes from plants and animals. We use wood, which comes from plants, to build houses. **Thinking Skill:** *Recognizing patterns and relationships*

Find Out On Your Own

Students' maps will vary depending on the organisms chosen. **Thinking Skill:** *Organizing information*

habitat (hab′ə tat), the place where an organism lives.

Red-necked grebe

Habitats

Each organism in a community lives in its own special place—or **habitat.** An organism gets everything it needs from its habitat. It gets food, water, shelter, and space to live.

The bird in the picture lives in a pond community. The bird's habitat is its nest and the part of the pond where it swims and finds food.

Lesson Review

1. What might cause the size of a population to change?
2. What makes up a community?
3. What does an organism get from its habitat?
4. **Challenge!** How do you depend on other organisms for food and shelter?

Study on your own, pages 320–321.

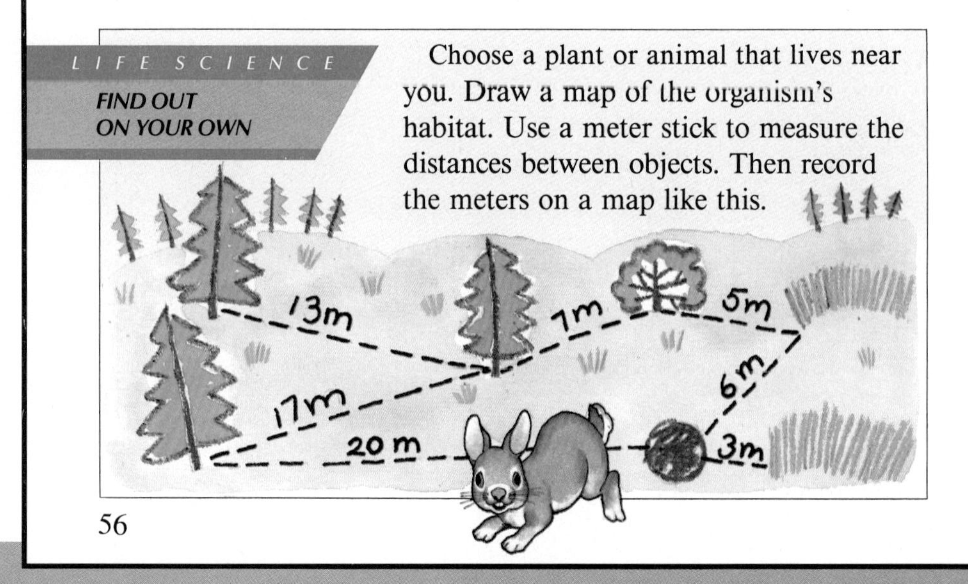

Choose a plant or animal that lives near you. Draw a map of the organism's habitat. Use a meter stick to measure the distances between objects. Then record the meters on a map like this.

56

Teaching Options

Science Anecdote

Earth's biosphere is the region around the earth that can support life. The biosphere extends 10 km (about 6 mi.) above and below the earth's surface.

Reteaching Suggestion ♦

Obtain a picture of a forest showing plants and animals. Have students *identify* and *describe* the populations and habitats in the picture of the forest community. (populations—birds, squirrels, ants, fox, insects, trees, shrubs, moss, and so on; habitats—trees, tree stumps, leaves, flowers, soil, etc.) Use the lists to discuss how the organisms in the community interact.

Workbook page 15 *

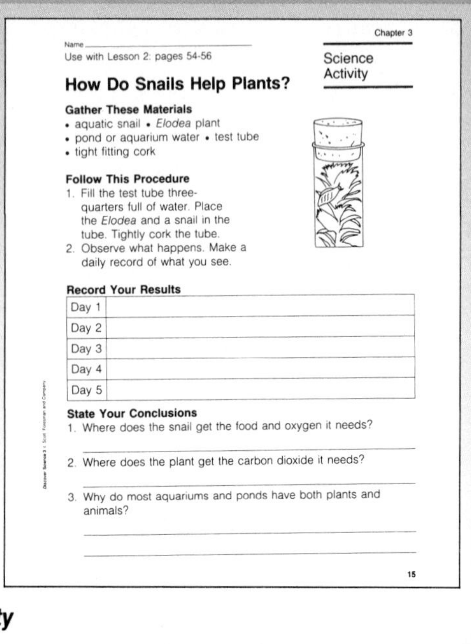

Name

Use with Lesson 2: pages 54-56

Chapter 3

Science Activity

How Do Snails Help Plants?

Gather These Materials
- aquatic snail • *Elodea* plant
- pond or aquarium water • test tube
- tight fitting cork

Follow This Procedure
1. Fill the test tube three-quarters full of water. Place the *Elodea* and a snail in the tube. Tightly cork the tube.
2. Observe what happens. Make a daily record of what you see.

Record Your Results

Day 1	
Day 2	
Day 3	
Day 4	
Day 5	

State Your Conclusions
1. Where does the snail get the food and oxygen it needs?
2. Where does the plant get the carbon dioxide it needs?
3. Why do most aquariums and ponds have both plants and animals?

15

♦ *Suitable as a language development activity*

Making a Model of a Food Chain

Suggested grouping: pairs

Purpose
Make a model of a food chain and *observe* that the links of a food chain depend on each other.

Gather These Materials
•strips of paper • tape • crayons or pencil

Follow This Procedure
1. Use a chart like the one shown to record your observations.

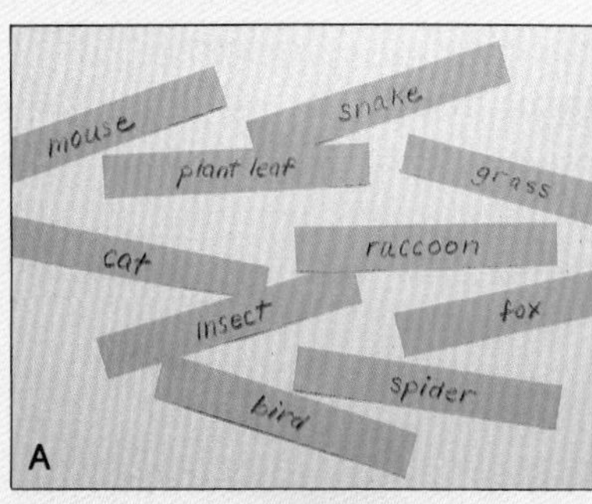

A

B

2. Picture A shows the names of plants and animals. Choose at least four of these names to make a food chain. Write the name of each of your organisms on a separate strip of paper.
3. Arrange the strips in the correct order for a food chain.
4. Tape the first paper strip so that it makes a circle.
5. Loop the second strip through the first strip and tape it to make a chain as shown in the picture.
6. Add the other links to your chain.
7. Record the names of the organisms you used.

Record Your Results

Food chain

grass ← insect ← bird ← cat
Answers will vary.

State Your Conclusion
1. Explain how the organisms you chose make up a food chain.
2. What happens to a food chain if one of the links breaks?

Use What You Learned
Explain why two producers cannot be in the same food chain.

57

Activity

Concept
All living things depend on other living things through food chains.

Objectives/Process Skills
- *Make a model* of a food chain.
- *Observe* how a food chain connects living things.
- *Record* data.
- *Describe* how a food chain is affected by a broken link.

Time Allotment
Allow about 30 minutes.

Safety Tips (See page T24.)
- If students use scissors remind them to be careful.
- Students should handle their pencils and paper carefully to avoid cutting and poking one another.

Teaching Tips
- Students can write the names of each organism in their food chain on an index card, and tie the index cards together in order with yarn.
- Food chains can be hung in the classroom.

Answers
State Your Conclusion
1. Answers will vary according to organisms chosen. However answers should indicate that plants are always the first link in the chain, and that each member of the chain depends on another member for food.
2. The organisms that depended on that link in the food chain will not get enough food.

Use What You Learned
Producers make their own food. Plants are producers. Plants cannot eat other plants. For this reason, two producers cannot be in the same food chain.
Thinking Skill: *Drawing conclusions*

Resource Book page 33

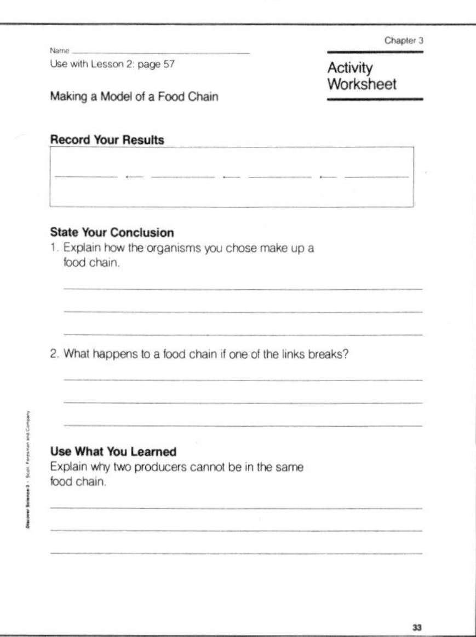
Name ____ Chapter 3
Use with Lesson 2: page 57
Activity Worksheet
Making a Model of a Food Chain

Record Your Results

State Your Conclusion
1. Explain how the organisms you chose make up a food chain.

2. What happens to a food chain if one of the links breaks?

Use What You Learned
Explain why two producers cannot be in the same food chain.

33

Activity Results

*** Answers to masters on pages 48E–48H**

TEACHING PLAN

Lesson Objectives

- *Define* the terms producer and consumer.
- *Explain* the difference between predators and prey.
- *Describe* a food chain in a community.

Lesson Vocabulary

consumer, food chain, predator, prey, producer

1. Motivate

Demonstration Activity ♦

Have students *list* the foods they eat for lunch or dinner. Ask volunteers to write their lists on the chalkboard.

Discussion

Questions: **Which foods listed come from plants?** (bread, grains, vegetables, fruits) **Which foods listed come from animals?** (meat, fish, dairy products) Explain that in this lesson, the students will learn how organisms get food in different ways.

3 How Do Organisms Get Food?

LESSON GOALS

You will learn
- that, to get food, some organisms make food and some organisms eat food.
- how some animals catch and eat other animals.
- how organisms depend on each other for food.

producer (prə dü′sər), an organism that makes its own food.

consumer (kən sü′mər), an organism that eats food.

Think about the different kinds of food you eat every day. All living things need food. Different organisms get food in different ways.

Producers and Consumers

You learned in Chapter 1 that green plants use sunlight to make—or produce—their own food. Plants and other organisms that make their own food are called **producers.**

Most organisms cannot make their own food. They must eat—or consume—food. These organisms are called **consumers.** Some consumers, such as mice, rabbits, and deer, eat plants. Other consumers, such as hawks, wolves, and tigers, eat other animals. Which organisms in the desert community in the picture are consumers?[1] Which are producers?[2]

[1]Iguana, bird, desert rat [2]cacti, grass

58

Teaching Options

Science Background

Organisms must have food to survive. Producers—plants and some monerans and protists—can make their own food. Consumers—all animals, fungi, and some protists—must get food by eating other organisms. Some consumers eat producers, some eat other consumers.

The major and most easily observed interactions among populations within a community are feeding relationships. These can be described in food chains, a sequential listing of who eats whom. In any community there are numerous, overlapping food chains that form the community's food web.

Reading Strategies ♦

1. Guide students' pre-reading by asking: What two or three questions do you have for each subheading?
2. Assign these strategies: Writing a Memory Sentence and Mapping Examples (See pages T26–T29.)
3. Pair students to share what information is clear and unclear and initiate discussion using students' unanswered questions.

♦ *Suitable as a language development activity*

Lion hunting for food

Predators and Prey

Many consumers hunt for their food. Hawks fly over fields and look for animals to catch and eat. Some snakes quietly wait for small animals to come near. Then they catch and eat the animals. Tigers follow herds of antelope. They wait for an antelope to wander away from the herd. Then the tigers chase and catch the antelope for food.

Animals that hunt for their food are **predators.** The animals they catch and eat are their **prey.** Which animal in the picture is the predator?[1] Which animals are the prey?[2]

predator (pred′ə tər) organism that captures and eats other organisms.

prey (prā), organism that is captured and eaten by another organism.

[1]lion
[2]zebras

2. Teach

Teaching Tips

- Questions: **Where do green plants get energy to make food?** (Students should *infer* that plants use air, sunlight, and water to make food.) **Are humans producers or consumers? Why?** (Consumers—humans cannot make their own food.)

Direct students' attention to the illustration on page 59. Questions: **What would happen to the number of zebras in an area if all the lions died?** (Guide students to *infer* that the number of zebras would gradually increase.) **How could too many zebras in an area hurt the zebras?** (Guide students to *infer* that there would be a shortage of food and the zebras would begin to die.) Emphasize how predator/prey relationships are important to the balance of a community.

- Explain that an animal can be both a predator and prey. Guide students to *infer* that some animals, such as a frog, are both. A frog is a predator when it catches an insect; and it is prey when it is eaten by a bird.
- **Possible Misconception:** Some students may think that predatory animals are bad. Help students to understand that predation is not evil. It is one way that organisms get food.

Workbook page 16 *

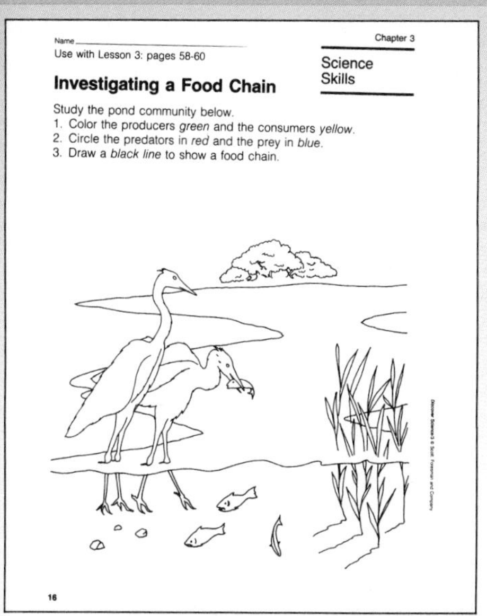

Name ______ Chapter 3

Use with Lesson 3: pages 58-60

Science Skills

Investigating a Food Chain

Study the pond community below.

1. Color the producers *green* and the consumers *yellow*.
2. Circle the predators in *red* and the prey in *blue*.
3. Draw a *black line* to show a food chain.

16

*** Answers to masters on pages 48E–48H**

Special Education

Glue pictures of plants and animals to index cards. Have students with learning disabilities sort the cards into producers and consumers. Then ask the students to make predator/prey pairs from the stack of consumer pictures.

Reinforcement

Direct the students to label a large sheet of paper as follows: *Producers; Consumers; Predator/Prey*. Have them find pictures in magazines of producers, consumers, and predators and prey. Instruct students to glue the pictures under the correct heading. Ask students to glue the pictures of predators and prey together and have them label the predator and prey in each pair. Have students *compare* the plants and animals found and discuss how some animals can be both predators and prey.

LESSON 3 pages 58–60

TEACHING PLAN

Teaching Tip

• Question: **Can the order of the food chain be rearranged?** (Guide students to *infer* that plants must come first; however, insects might also feed on grass and plants, hawks might eat frogs, or a snake might eat a baby hawk.) Discuss how the order of food chains can change in soil, on land, and in water.

3. Assess

Lesson Review

1. A producer is an organism that makes its own food. A consumer is an organism that eats food.
2. Predators hunt for and eat other animals. Prey are animals that are hunted for and eaten by other animals.
3. Grass Mice Snakes Hawks
4. Challenge! Food chains begin with producers because producers make their own food. **Thinking Skill:** *Drawing conclusions*

Find Out On Your Own

Answers will vary but should show correct feeding relationships among marsh organisms. One possible food chain is: plants ⇨ insects ⇨ frogs ⇨ ducks. **Thinking Skills:** *Collecting information, Summarizing*

Food chain

SCIENCE IN YOUR LIFE

You are part of a food chain when you eat a hamburger. The food chain begins with plants, which make their own food. Cattle eat the plants. Then you eat hamburger meat which comes from cattle.

food chain, the way food passes from one organism to another organism in a community.

Food Chains

Organisms in a community depend on each other for food. A **food chain** is the way food passes from one organism to another in a community. The picture shows a food chain in a meadow. The grass makes its own food. Mice feed on the grass. Snakes eat the mice. Coyotes eat the snakes.

All communities have food chains. Food chains are found in the soil, on land, and in water.

Lesson Review

1. What is a producer? What is a consumer?
2. How are predators and prey different?
3. What is an example of a food chain in a meadow?
4. **Challenge!** Why do food chains begin with plants or some other kinds of producers?

Study on your own, pages 320–321.

LIFE SCIENCE
FIND OUT ON YOUR OWN

Look in library books to find out about marshes and the organisms that live in them. Write a few sentences explaining how these organisms depend on each other for food.

60

Teaching Options

Science Anecdote

Some organisms are both producers and consumers. Venus's flytrap is a green plant that undergoes photosynthesis. It also has specialized leaves that trap and digest insects.

Reteaching Suggestion ♦

On index cards, print the names or place pictures of various familiar plants and animals. Pass one card out to each student. Challenge students to build a food chain around the animal or plant listed on the card. Repeat the activity by having students exchange cards with a classmate.

Workbook page 17 *

Name ______
Use with Lesson 3: pages 58-60

Chapter 3
Vocabulary Puzzle

Hidden Message

Write the missing letters in the sentences.

1. __E L L S are tiny parts of organisms. (12)
2. A living thing is an __R G__N__S M. (13 1 3)
3. B__C__E__I__ are made of one cell. (1 7 14 1)
4. P__O T I__T S live in wet places. (14 9)
5. Mold and mushrooms are F U N__I. (6)
6. A P O P U__A T I O__ of swans L I__E S in the pond. (2 5 4)
7. The pond __O M M U__I T Y has a variety of animals. (12 5)
8. The rabbit gets what it needs from its __A B I__A__. (8 7 7)
9. A plant is a P R O__U C__R, but an animal is a C O__S U M E__. (11 10 5 14)
10. Food passes through a F__O__ C H__I__ in a community. (13 11 1 5)
11. A hawk is a flying P__E__A T O__. (14 11 14)
12. A mouse is P R__Y to a hawk. (10)

Below, write the letters that match the numbers.

1 2 2 2 3 4 3 5 6 7 8 3 5 6 9

5 10 10 11 10 1 12 8 13 7 8 10 14

17

♦ *Suitable as a language development activity*

Helping Pandas Survive

The Problem You might have trouble thinking of the animal in the picture as a picky eater. Giant pandas like this one eat a lot of mostly one food—bamboo. Wild giant pandas live in bamboo forests in the mountains of China. The problem is that populations of pandas are getting smaller. Only about 700 remain in the whole world. China is trying to protect pandas by not letting people hunt them. Even so, sometimes pandas fall into traps set for other animals. Scientists are trying to understand how pandas live to try to save them. Understanding the problem is the first step to solving it.

The Breakthrough During the last ten years, photographs of panda habitats were taken from space satellites. These photographs show how the clearing of forests has hurt panda populations. Now scientists know that when trees are cut down, the pandas move. Sometimes they cannot find bamboo in their new home. Cleared forest areas split pandas into groups of fewer than twenty animals. If they cannot find bamboo, small groups are easily wiped out.

New Technology Scientists have made a special kind of community

Giant panda

to help pandas live and grow. One community is in a large protected area in China. About one hundred pandas live there. Each panda in the community has its own house. A hungry panda can go to feed in a nearby bamboo thicket. When a female panda is ready to mate, she can walk around until she finds a male. Pandas do have babies, but many do not live to be adults in the wild. Scientists hope that the pandas born in their new community can be released into the wild.

What Do You Think?

1. How have people harmed pandas?
2. Why is it especially hard for a small group of pandas in the wild to produce a new generation?

Science and Technology

Discussion

Display a picture of a beautiful forest. Discuss resources that people get from a forest. (lumber, fuel, food, and so on) Question: **Why should people not cut down all the world's forests?** (The forests hold water and minerals; clearing forests can threaten many different types of living things; also cutting down all of the trees could cause erosion.)

Teaching Tip

- Point out that most of 700 or so wild pandas live in one of the 12 panda reserves set aside by the Chinese government to protect the declining natural habitats of the pandas.

Answers

What Do You Think?

1. By clearing forests, people have destroyed the bamboo thickets where pandas eat. By setting traps for other animals, people sometimes kill pandas by mistake. **Thinking Skill:** *Restating or explaining ideas*

2. Answers may vary. In the wild, small groups of pandas are wiped out if they cannot find bamboo; the group might be all females or all males; many young pandas do not survive to adulthood. **Thinking Skill:** *Inferring*

Science Background

In 1987, the World Wildlife Fund declared a "panda emergency"—a campaign to save the dwindling number of pandas—after a census showed that the panda population had dropped by about 200 in the preceding decade.

* **Answers to masters on pages 48E–48H**

TEACHING PLAN

Purpose
To develop the skills of collecting and organizing information using diagrams and pie graphs to solve problems.

1. Motivate

Discussion
Review with students that a community is a part of an ecosystem. Remind them that each community is made of populations. Question: **How do the students in this classroom make up a population?** (The students in the room are a group of like organisms that live together.)

2. Teach

Teaching Tip
- Discuss the kinds of information a pie graph can illustrate. (relative amounts, fractions, percentages, proportions, and so on) Explain that in this activity, the pie graph illustrates information about producers and consumers.

Skills for Solving Problems

Using Diagrams and Pie Graphs

Problem: How does the number of producers in a marsh community compare to the number of consumers?

Part A. Using a Diagram to Collect Information

1. Look at the diagram below. All the organisms in an area make up a community. How many organisms are shown in the diagram?
2. Producers are organisms that make their own food. Plants use sunlight to make their food. How many producers are in the community?
3. Organisms that cannot make their own food are consumers. All animals are consumers. How many consumers are in the community?

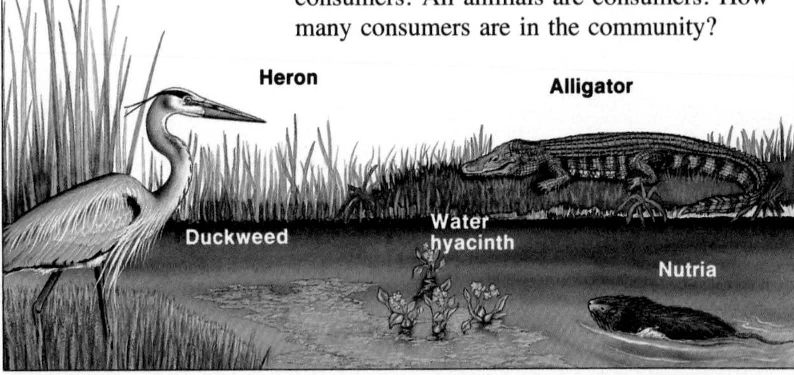

Marsh community

Part B. Using Pie Graphs to Organize and Interpret Information

4. This pie graph contains the information you collected in Part A about producers and consumers. A pie graph is a circle divided into pieces like slices. What does each slice of the graph stand for? How many organisms are shown in the pie graph?

62

Teaching Options

Sample Pie Graph for Part C

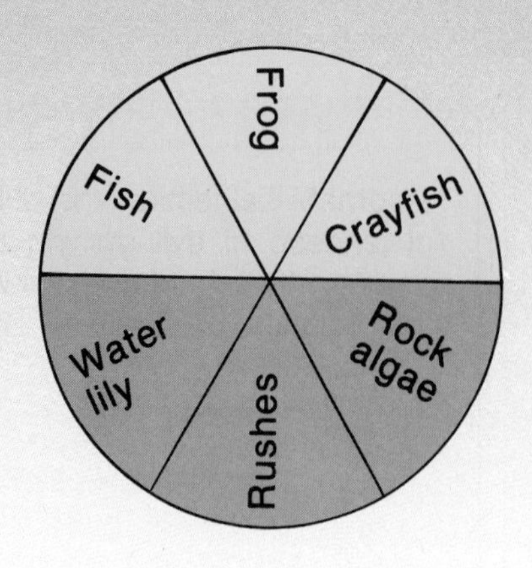

♦ *Suitable as a language development activity*

5. In the graph, producers are pink and consumers are green. How many producers are shown? How many consumers are shown?
6. Can the number of living things in a community change? How would this affect the number of slices in the pie graph?

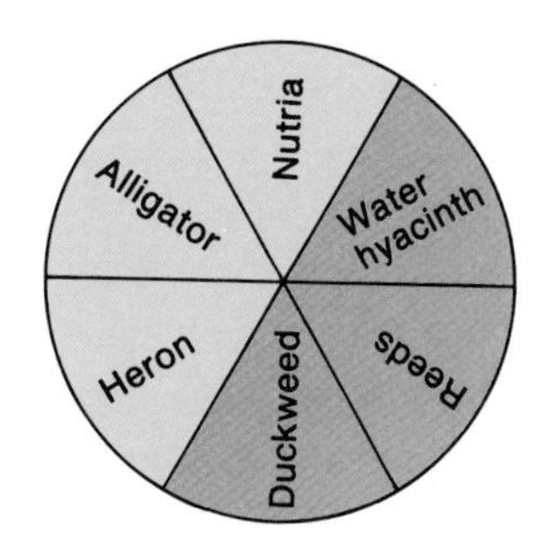

Part C. Using a Diagram and a Pie Graph to Solve a Problem

Problem: How does the number of producers in a pond community compare to the number of consumers in the community?

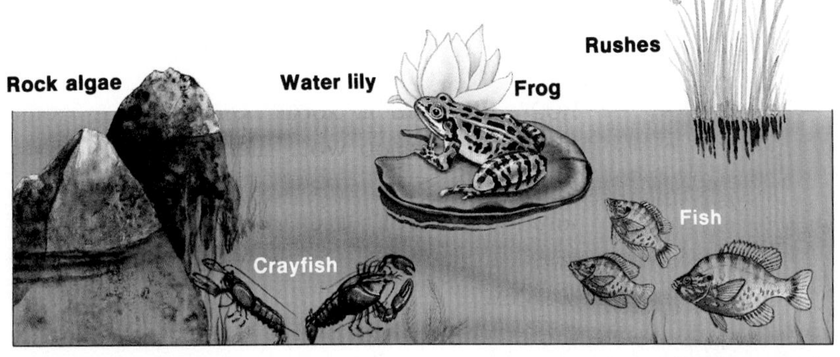

Pond community

7. Use the diagram to collect the information you need to solve the problem. Make a pie graph similar to the one shown in Part B to organize your information.
8. How many organisms are in the pond community shown here? How many of these organisms are producers? How many are consumers? How many slices does your pie graph have?
9. How is the pond community like the marsh community shown in Part A? How is it different?

63

3. Assess

Part A

1. six
2. three
3. three

Part B

4. one of the organisms in the community; six
5. three; three
6. Yes, the number can increase or decrease. The number of parts or pieces of the pie graph would increase or decrease in the same way.

Part C

7. See Sample Pie Graph for Part C in Teaching Options.
8. six; three; three; six
9. Both communities are made up of living organisms, including producers and consumers. They are different in the kinds of organisms.

Reteaching Suggestion ♦

Obtain or draw a diagram of an ocean habitat that includes sharks, whales, large fish, sea birds, small fish, and shrimp. Have students distinguish the individual populations from one another, as well as the predators from the prey. (predators: shark, whale, large fish, sea birds; prey: small fish, shrimp) Draw a circle on the board. Have students calculate the number of slices that would be needed in a pie graph of different populations in this habitat. Have them label each slice. Then ask students to shade the slices that represent predators.

Resource Book page 35 *

Name ______ Chapter 3

Comprehension: classifying and comparing

Science and Reading

Living Things

Read the story and follow the directions.

When Randy prepares food, he is very careful to keep his knives and cutting board clean. He does not want any germs on them that could cause sickness.
Today Randy is making a salad of lettuce, sliced mushrooms, grated carrots, and dried seaweed. He is going to give some of it to his pet rabbit.

Complete the chart. Write each underlined word from the story in the correct column.

Bacteria	Protists	Fungi	Plants	Animals/ People

Complete the chart. Place an X in the proper column to tell how many cells each group has.

	Have One Cell	Have More Than One Cell
bacteria		
protists		
fungi		
plants		
animals/people		

35

* ***Answers to masters on pages 48E–48H***

REVIEW PLAN

Reviewing Science Words

1. food chain
2. prey
3. habitat
4. bacteria
5. fungi
6. population
7. cell
8. producer
9. community
10. organisms
11. predator
12. protist
13. consumers

Reviewing What You Learned

1. b **2.** d **3.** a **4.** c **5.** b

Chapter 3 Review

Chapter Main Ideas

Lesson 1 • All living things grow, need food, make more organisms like themselves, and are made of one or more cells. • Scientists group living things into bacteria, protists, fungi, plants, and animals.

Lesson 2 • Organisms live in populations. • Different populations live in a community. • Each organism in a community has its own habitat.

Lesson 3 • Organisms can be producers or consumers. • Some animals catch and eat other animals. • Organisms depend on each other for food.

Reviewing Science Words

bacteria	fungi	predator
cell	habitat	prey
community	organisms	producer
consumers	population	protist
food chain		

Copy each sentence. Fill in the blank with the correct word from the list.

1. A ___ is the way food passes from one organism to another in a community.
2. An animal that is captured and eaten by another animal is called ___.
3. Each organism in a community lives in its own ___.
4. ___ are organisms that are made of one cell and sometimes cause illnesses.
5. ___ are organisms that get food from dead material.
6. A ___ is a group of organisms of the same kind that live in the same place.
7. The smallest living part of an organism is a ___.
8. An organism that makes its own food is called a ___.
9. All the organisms that live in one place make up a ___.
10. Scientists have divided ___ into five main groups.

64

Review Options

Cooperative Learning ♦

STAD Format (See page T23.) Assign students to work in four- to five-member teams to study Chapter 3 Review. Students should work together to make sure that they and their teammates know the material in the chapter. After students have had enough time to study together, give them a test to complete individually (Chapter 3 Test A or B in the *Test Book*). Award Superteam certificates to teams whose average test scores exceed 90%, and Greatteam certificates to teams whose average test scores exceed 80%.

Test Book page 25 *

Name ___

Chapter 3
Test A

Multiple Choice Choose the best answer.

1. All living things on earth are made up of
   a. bacteria.
   b. cells.
   c. energy.
2. Organisms of one or more cells that live in wet places are
   a. bacteria.
   b. protists.
   c. fungi.
3. An example of organisms that often are used to make medicines are
   a. bacteria.
   b. protists.
   c. fungi.
4. A common fungi is
   a. bacteria.
   b. protists.
   c. mold.
5. A group of organisms of the same kind that live in the same place is called a
   a. habitat.
   b. population.
   c. collection.
6. The place where each organism in a community lives is its
   a. habitat.
   b. population.
   c. community.
7. Organisms that make their own food are
   a. consumers.
   b. producers.
   c. predators.
8. Animals caught and eaten by other animals are
   a. prey.
   b. producers.
   c. fungi.
9. Food passes from one organism in a community to another in a
   a. food community.
   b. food chain.
   c. food habitat.
10. Organisms that cannot make their own food are
    a. prey.
    b. workers.
    c. consumers.

25

♦ ***Suitable as a language development activity***

11. A ▒▒ catches and eats another organism.
12. A ▒▒ is an organism that lives in a wet place and has one or more cells.
13. Living things that depend on other living things for food are ▒▒.

Reviewing What You Learned

Write the letter of the best answer.

1. A group of goldfish that live in a pond make up a
   (a) community. (b) population. food chain. (d) habitat.
2. All organisms are made of one or more
   (a) bacteria. (b) food chains. (c) prey. (d) cells.
3. A frog that catches and swallows a grasshopper is a
   (a) predator. (b) producer. (c) protist. (d) prey.
4. Most seaweeds are
   (a) fungi. (b) animals. (c) protists. (d) bacteria.
5. A pine tree is a
   (a) predator. (b) producer. (c) consumer. (d) protist.

Interpreting What You Learned

Write a short answer for each question or statement.

1. What might happen to a plant population if the weather is unusually dry for a long period of time?
2. What are two ways plants and animals differ?
3. List these steps in a food chain in order: a bird eats a grasshopper, the grasshopper eats a plant, a fox eats the bird, the plant makes its own food.

Extending Your Thinking

Write a paragraph to answer each question.

1. What would happen to the population of mice in a field if the population of snakes grew larger? Explain.
2. Suppose all the plants in a pond died. What would happen to the food chain in the pond? Explain your answer.

To explore scientific methods, see Experiment Skills on pages 352–353.

Interpreting What You Learned

1. The plants might not get enough water to make their own food; some of the plants might die and the plant population would get smaller. **Thinking Skill:** *Predicting*

2. Many plants can make their own food, animals must eat food; plants cannot move from place to place on their own, animals can move from place to place. **Thinking Skill:** *Contrasting*

3. plants make food grasshopper eats plants bird eats grasshopper fox eats bird **Thinking Skill:** *Sequencing*

Extending Your Thinking

1. The population of mice might get smaller because the snakes might eat the mice. **Thinking Skill:** *Predicting*

2. If the plants die, there would be no food for the other organisms in the food chain; the other organisms might move to another place to find food or they might die. **Thinking Skill:** *Drawing conclusions*

Test Book page 26 *

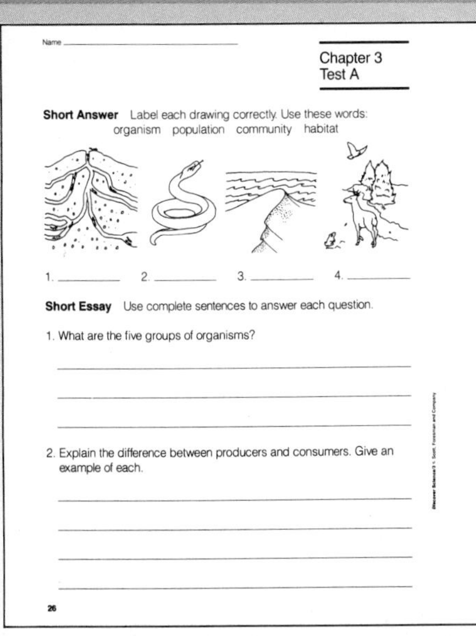

Name

Chapter 3
Test A

Short Answer Label each drawing correctly. Use these words:
organism population community habitat

1. ____ 2. ____ 3. ____ 4. ____

Short Essay Use complete sentences to answer each question.

1. What are the five groups of organisms?

2. Explain the difference between producers and consumers. Give an example of each.

26

Test Book page 27 *

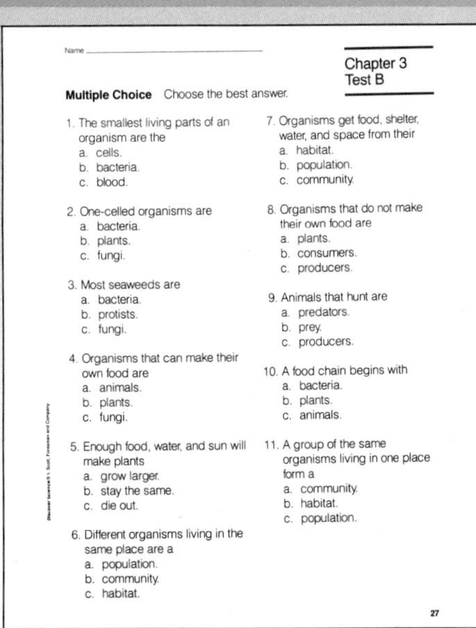

Name

Chapter 3
Test B

Multiple Choice Choose the best answer.

1. The smallest living parts of an organism are the
   a. cells.
   b. bacteria.
   c. blood.
2. One-celled organisms are
   a. bacteria.
   b. plants.
   c. fungi.
3. Most seaweeds are
   a. bacteria.
   b. protists.
   c. fungi.
4. Organisms that can make their own food are
   a. animals.
   b. plants.
   c. fungi.
5. Enough food, water, and sun will make plants
   a. grow larger.
   b. stay the same.
   c. die out.
6. Different organisms living in the same place are a
   a. population.
   b. community.
   c. habitat.
7. Organisms get food, shelter, water, and space from their
   a. habitat.
   b. population.
   c. community.
8. Organisms that do not make their own food are
   a. plants.
   b. consumers.
   c. producers.
9. Animals that hunt are
   a. predators.
   b. prey.
   c. producers.
10. A food chain begins with
    a. bacteria.
    b. plants.
    c. animals.
11. A group of the same organisms living in one place form a
    a. community.
    b. habitat.
    c. population.

27

Test Book page 28 *

Name

Chapter 3
Test B

Short Answer Correctly label the picture of each habitat. Use these words:
forest meadow pond swamp

1. ____ 2. ____ 3. ____ 4. ____

Short Essay Use complete sentences to answer each question.

1. What is necessary for plant populations to become larger?

2. In what ways do people depend on other organisms for food and shelter?

28

*** Answers to masters on pages 48E–48H**

How People Affect Plants and Animals Planning Guide

TEACHING PLAN

Chapter Components	Skills	Materials
Chapter Opener/*DISCOVER*: Making a Plant Dye pp. 66–67	*DISCOVER* p. 67 Science Process Skills *Observing*	*DISCOVER* p. 67 (groups of 2) 1 large head of red cabbage, 15 plastic cups, 15 plastic spoons, 30 sheets drawing paper, 30 paint brushes, water
Lesson 1 How Do People Change the Lives of Plants and Animals? pp. 68–71	Thinking Skills Challenge!: *Applying information to new situations* Find Out On Your Own: *Collecting information, Communicating*	Demonstration p. 68 no materials needed
Lesson 2 How Do People Protect Plants and Animals? pp. 72–74	Thinking Skills Challenge!: *Applying information to new situations* Find Out On Your Own: *Collecting information, Communicating*	Demonstration p. 72 pictures of endangered species, things made from the endangered species (fur coat, ivory jewelry, shoes, belts, purses)
Activity Making a Model of an Animal Habitat p. 75	Science Process Skills *Making models, Predicting, Collecting and interpreting data*	(individual) 30 sets crayons or colored markers, 30 scissors, 90 sheets assorted construction paper, 10 bottles white glue, 30 shoeboxes
Lesson 3 How Do People Use Plants and Animals? pp. 76–79	Thinking Skills Challenge!: *Applying information to new situations* Find Out On Your Own: *Collecting information*	Demonstration p. 76 pictures of cheese, butter, milk, leather shoes, hamburger or steak
Activity Looking at Fibers p. 80	Science Process Skills *Observing, Collecting and interpreting data*	(groups of 2) 15 sheets white paper, 15 squares cotton cloth, 15 squares linen cloth, 15 hand lenses, 15 large paper plates, colored water
Science and People Protecting the Wilderness p. 81	Thinking Skills *Recognizing the main idea and supporting details, Predicting, Recognizing cause and effect*	
Skills for Solving Problems Using Pictographs and Time Lines pp. 82–83	Problem Solving Skills *Making decisions/Identifying and solving problems, Interpreting charts, maps, and graphs*	
Chapter Review pp. 84–85	Thinking Skills *Restating or explaining ideas, Recognizing cause and effect, Comprehending meaning, Drawing conclusions*	

Teaching Options

Strategies	Extensions		Resource Masters
Cooperative Learning p. 66 (Also see p. T23.) Applying Whole Language p. 67 (Also see p. T30.)			Family Letter: *Resource Book* p. 39
Reading Strategies p. 68 (Also see pp. T26–T29.)	Reinforcement pp. 69, 70 Special Education p. 69 Enrichment p. 70	Game Suggestion p. 71 Reteaching Suggestion p. 71	Vocabulary Preview: *Workbook* p. 19 Science Skills: *Workbook* p. 20
Reading Strategies p. 72 (Also see pp. T26–T29.)	Reinforcement p. 73 Special Education p. 73	Science and Social Studies p. 73 Reteaching Suggestion p. 74	Science and Language Arts: *Workbook* p. 21
			Activity Worksheet: *Resource Book* p. 43
Reading Strategies p. 76 (Also see pp. T26–T29.)	Reinforcement pp. 77, 78 Special Education p. 77 Enrichment p. 78	Game Suggestion p. 79 Reteaching Suggestion p. 79	Science Activity: *Workbook* p. 22 Vocabulary Puzzle: *Workbook* p. 23
			Activity Worksheet: *Resource Book* p. 45
	Reteaching Suggestion p. 83		Science and Reading: *Resource Book* p. 47
Cooperative Learning p. 84 (Also see p. T23.)			Chapter Tests: Forms A and B *Test Book* pp. 33–36

Classroom Management

Advance Preparation

DISCOVER, page 67
Obtain one large, red cabbage for this activity.

Demonstration, page 72
Collect pictures of endangered animals that are hunted by humans and articles made of the fur, skin, or other parts of those animals.

Activity, page 75
Provide a shoebox for each student. Bring in pictures, books, or magazines students can use for reference when selecting and drawing animals. You may wish to ask students to bring in shoeboxes and magazines for this activity.

Demonstration, page 76
You will need pictures of cheese, butter, milk, leather shoes, and hamburger or steak.

Activity, page 80
Obtain solid-colored remnant samples of cotton and linen cloth from a fabric store. Cut the cloth to make squares about 10 cm wide. Make the colored water using food coloring.

Vocabulary Review

Use the following sentences with your students to review the meanings of the italicized words.

1. The area where an organism lives is its *habitat.*
2. To *pollute* is to dirty the air, water, or land by adding unsafe chemicals or things that cause disease.
3. Wheat, oats, and corn are types of *grain.*
4. *Flax* is a material that comes from plants and is used to make linen.
5. *Leather* is a material made from animal skins.
6. *Wool* is a material made from sheep's hair.

High-Potential Students

Have students use an encyclopedia or other reference materials to find out about the Environmental Protection Agency (EPA). Ask them to make a list of manufacturing industries and transportation vehicles that might contribute to pollution in your area. Then have students write several statements involving these sources of pollution, and involvement by the EPA. For example, one statement could be "The EPA should not interfere in local decisions about what industries in our community can dump into nearby rivers or lakes." Have each student survey 8–12 adults and record their responses under one of the following headings: *Strongly Agree, Agree, Disagree, Strongly Disagree.*

Mainstreamed Students

Emotionally Handicapped
Allow students to do a skit acting out what it would be like to be an animal in today's world, where animal habitats are constantly being disrupted by people. Some students might want to act out how a particular type of animal became extinct.

Visually Impaired
Bring in objects made from plant and animal materials that students can touch. Help students identify and list characteristics of these materials. Discuss reasons why conserving habitats of these and other plants and animals is important.

Science Fair Projects

The Experiment Skills on p. 354 and the Investigate feature in the chapter can be used for science fair projects. You also might encourage interested students to do one of the following projects:

1. Demonstrate some of the ways that humans depend on plants by displaying some items or pictures of items that are made of plants, such as food, medicine, and clothing.
2. Make a diorama that shows different types of animal habitats in the forest, desert, grasslands, and so on. Make a second diorama that shows possible changes that humans could make that would endanger animal habitats, such as building roads, buildings, dams, airports, and so on.
3. Make an exhibit that shows some of the different ways that people are attempting to conserve natural resources, and some of the ways these resources are still being abused. Present plans for ways people could better conserve resources.

Bulletin Board

Encourage students to find out about these animals and reasons for them becoming extinct.

EXTINCTION IS FOREVER

Chapter 4 Poster

Science Discovery Center

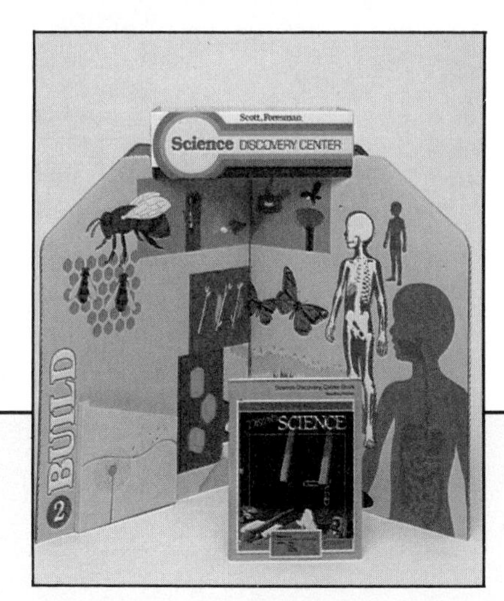

Use pages 45–50 from the *Science Discovery Center Book*. Place these worksheets in the appropriate pockets in the Science Discovery Center.

Overhead Transparencies

Use Transparency 7 from the package of color overhead transparencies.

Teacher's Resource Book

Dear Family,
Your student will be reading **Chapter 4: How People Affect Plants and Animals,** in *Discover Science,* published by Scott, Foresman. We will learn how people can change the lives of plants and animals (organisms). When people change habitats, many organisms in the habitat might die. The organisms may become endangered. If an organism disappears entirely, it becomes extinct.
You and your student can do this activity together to learn more about how much we depend on plants and animals.

What If This Organism Became Extinct?

You will need paper, pencil, and imagination.

1. With your student, think of an organism, for example, the cotton plant. Then list all the ways that you and your family use it, for example, towels, sheets, socks.
2. Then play the "What if . . ." game. Say, "What if this organism became extinct?" Then tell all the ways that this fact would change your life. For example, what if the cotton plant became extinct? You would not be able to get cotton sheets, towels or socks.
3. Then think of substitutes that are from organisms (for example, wool). Ask, "Would the substitute be as good?" Then ask, "What if that organism (sheep) became extinct?"
4. Continue going for as long as you can. When you run out of substitute organisms, think of an artificial substitute.
5. Then go on to another organism (for example, a cow). Play for as long as you can.

Workbook

Name ____________

Use with Lesson 1: pages 68-71

Chapter 4

Vocabulary Preview

People, Plants, and Animals

Chapter Vocabulary			
endangered organisms	extinct organisms	fibers	pollution

Vocabulary Cards

1. Write each word on a card.
2. Find each word in the glossary. Copy the pronunciation under the word on the card.
3. Practice saying the words with a partner.

Vocabulary Riddles

Use the words from the vocabulary list to answer the riddles. Write the answer on the line. You can use the glossary to help you.

1. It is anything harmful in the air, water, or land. It kills organisms and destroys habitats.
   What is it? pollution
2. These are strong thin threads in plants. Cotton plants have these. What are they? fibers
3. Plants and animals become this when they are no longer found on the earth. The dodo bird has become this. What is this? extinct organisms

Find the Picture

Look at the pictures and follow the directions.

1. Circle the endangered organism.
2. Put a box around the extinct organism.
3. Put an *X* on the picture of pollution.

At Home

1. Practice saying the words. Learn their meanings.
2. Look for the words and pictures of the words in newspapers and magazines. You can make a poster.

Teacher's Notes: Ask students to find the hidden words in the term *endangered organisms*. Examples: danger, end, organ, red.

Workbook

Name ____________

Use with Lesson 1: pages 68-71

Chapter 4

Science Skills

Endangered Species

Read about endangered species. Decide why the species has become endangered. Then circle the reason.

1. The California condor is one of the world's largest birds. It once lived throughout the West. But when settlers moved in, there was little room left for the condor. It now survives only in zoos. These zoos want to save the condor. One day, they hope to return it to the wild.
   hunting   pollution   (habitat destruction)
2. The Mexican grizzly bear is known for its size, power, and courage. Because of this, many people want its fur as a trophy. Tracked down for sport, few of these bears have survived. Grizzly bears once lived throughout the American Southwest.
   (hunting)   pollution   habitat destruction
3. The fur of the giant otter is among the most beautiful in the world. The otter is easy to catch because it is curious and unafraid. At one time giant otters were often seen swimming in rivers throughout South America. Now, few otters are seen.
   (hunting)   pollution   habitat destruction
4. When chemicals are dumped into a river, many fish die. The fish that live, carry the poison in their bodies. The pelican pictured below has become endangered because it eats fish that have been poisoned with chemicals.
   hunting   (pollution)   habitat destruction

Dalmatian Pelican

Teacher's Notes: Other endangered species include the whooping crane, peacock, wolf, leopard, Asian elephant, panda, and koala.

Workbook

Name ____________

Use with Lesson 2: pages 72-74

Chapter 4

Science and Language Arts

Making Bumper Stickers

Study the bumper stickers designed by students. One bumper sticker contains words. Another contains pictures. The third contains both words and pictures. The bumper stickers show that the students care about what happens to the earth and to living things.

Think about an "earth care" message for a bumper sticker. In the space below, design a colorful bumper sticker.

Teacher's Notes: Students can transfer their designs to white contact paper, and color them with markers.

Teacher's Resource Book

Name ______________________

Use with Lesson 2: page 75

Chapter 4

Activity Worksheet

Making a Model of an Animal Habitat

Record Your Results

Part of Habitat Destroyed	What Might Happen to Animal
1.	
2.	
3.	

State Your Conclusion

1. What kind of a model habitat did you make?

2. Why is the habitat you made a good place for your animal to live in?

Use What You Learned

What might happen to your animal if a road were built through the animal's habitat?

Workbook

Name ______________________

Use with Lesson 3: pages 76-79

Chapter 4

Science Activity

How Can Plant Dyes Be Used?

Gather These Materials

• beets • strawberries • onion skins • purple grapes • grater • scissors • 4 cups • masking tape • felt marker • water • pieces of white cloth • glue • construction paper

Follow This Procedure

1. Write the name of each plant on masking tape. Use each name to label one cup.
2. Cut or shred each type of plant. Place it in its own cup. Add 25 mL of water to each cup.
3. Stir the plant pieces and water. Then add a few small pieces of cloth to each cup. Stir until the cloth becomes colored. Take out the cloth. Let it dry.
4. Use the pieces of cloth to make a colorful picture. In the squares on the right, glue a small piece of cloth from each cup.

BEETS · STRAWBERRIES · ONION SKINS · PURPLE GRAPES

Record Your Results

Students will glue one sample of dyed cloth in each square. Beet Dye	Strawberry Dye
Onion Skin Dye	Purple Grape Dye

State Your Conclusions

1. What happened to the cloth when it was put into the water with the plant pieces?

The cloth became colored. It was dyed by the plant parts.

2. What could be made with large pieces of cloth dyed with plant parts?

Answers will vary. Students might mention clothes, drapes, artwork.

Teacher's Notes: Other plant parts may be used, such as flowers, leaves, roots, or bark. Do not use wild berries; some are toxic.

Workbook

Name ______________________

Use with Lesson 3: pages 76-79

Chapter 4

Vocabulary Puzzle

Word Scramble

Unscramble the letters in each pie to write a word. Write on the line below the pie. One is done for you.

1. A strong thin thread found in plants that can be used to make cloth

R B F E I

FIBER

2. A kind of organism that is no longer found on the earth (two words)

E I T T C X N · M O G N S R A I

EXTINCT ORGANISM

3. Anything harmful added to the air, water, or land

L T O N I L O U P

POLLUTION

4. A kind of organism that is very few in number and might someday no longer be found on the earth (two words)

E D G A D N E R N · I R M G N O S A

ENDANGERED ORGANISM

Teacher's Notes: Suggest that the students use the vocabulary words to make their own puzzles for others to solve.

Teacher's Resource Book

Name ______________________

Use with Lesson 3: page 78

Chapter 4

Activity Worksheet

Looking at Fibers

Record Your Results

	How Fibers Looked	Order Samples Soaked Water
Paper		
Cotton		
Linen		

State Your Conclusions

1. What do plant fibers look like?

2. Which of the three samples would you use to wipe up a water spill? Why?

Use What You Learned

How do you think the fibers in wool cloth would compare with the cotton and linen fibers?

Teacher's Resource Book

Name ______________________

Chapter 4

Comprehension: main idea

Science and Reading

Elephants Help Each Other

Read the paragraph. Underline the main idea. Then follow the directions.

Elephants work together to help each other. Sometimes an elephant falls into a game trap, which is a very large, deep hole dug by people who want to kill elephants so they can sell their ivory tusks. When an elephant is trapped, elephants who hear its cries come to the rescue. They trample the ground to make it slant by the hole, so that it is easier for the elephant to climb out. Some of them push the elephant from behind, and others wrap their trunks around the elephant and pull from the front. The process is very difficult and takes a long time, but the elephants work together until the trapped elephant is free.

Circle the letter of each sentence that gives details from the story.

(a.) Elephants hear the cries of the trapped elephant.
b. The people who set the trap are frightened away.
(c.) The elephants trample the ground to make it slant.
d. The elephants push and pull the trapped elephant.
(e.) The elephants quickly free the trapped elephant.

Test Book

Name ______________________

Chapter 4 Test A

Multiple Choice Choose the best answer.

1. People sometimes harm an organism's habitat when they **(1-1)**
   a. make large parks.
   (b.) pollute.
   c. make new habitats.
2. Organisms that are very few in number are **(1-2)**
   (a.) endangered.
   b. dangerous.
   c. extinct.
3. The dodo bird is an animal that is **(1-3)**
   a. dangerous.
   b. endangered.
   (c.) extinct.
4. People can help keep plants and animals from dying out by saving their **(2-1)**
   a. fibers.
   b. furs.
   (c.) habitats.
5. Areas where people cannot build houses, hunt animals, or collect plants are **(2-1)**
   (a.) parks.
   b. habitats.
   c. forests.
6. People can help protect the habitats of organisms by **(2-1)**
   (a.) making parks.
   b. closing zoos.
   c. spreading pollution.
7. Bread, macaroni, and cereal are made from **(3-1)**
   (a.) grains.
   b. vegetables.
   c. leaves.
8. People make cheese, butter, yogurt, and ice cream from **(3-1)**
   a. grains.
   b. eggs.
   (c.) milk.
9. The strong, thin threads of cotton and flax plants are called **(3-2)**
   (a.) fibers.
   b. silks.
   c. linen.
10. Buttercups and mistletoe can be dangerous to people when these plants are **(3-3)**
    a. touched.
    b. worn.
    (c.) eaten.

Numbers in parentheses after each question refer to the lesson number and the objective of that lesson.

Test Book

Name ______________________

Chapter 3 Test A

Matching Match each plant or animal with the product that comes from it.

a.

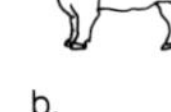
b.

c.

d.

1. __b__ milk
2. __a__ wool
3. __d__ cotton
4. __c__ wood

Short Essay Use complete sentences to answer each question.

1. Name two ways people can change an animal's habitat. (1-1)

An animal's habitat can be changed by cutting down trees to build road and houses. When people add harmful things to the air or water, this can change a habitat by polluting it.

2. Name three plant parts people eat. Give an example of each. (3-1)

Three plant parts people eat are seeds, leaves, and stems. Grains, such as wheat, oats, and corn are seeds. Lettuce is leaves, and potatoes are stems.

Numbers in parentheses after each question refer to the lesson number and the objective of that lesson.

Test Book

Name ______________________

Chapter 4 Test B

Multiple Choice Choose the best answer.

1. An organism gets its food, shelter, and water from **(1-1)**
   a. other organisms.
   b. pollution.
   (c.) its habitat.
2. Anything harmful added to the air, water, or land is **(1-2)**
   (a.) pollution.
   b. a fiber.
   c. a chemical.
3. Organisms that are no longer found on earth are **(1-3)**
   a. endangered.
   (b.) extinct.
   c. dangerous.
4. People help protect plants, animals, and their habitats by making **(2-1)**
   a. forests.
   (b.) parks.
   c. swamps.
5. For protection, some endangered animals are placed in **(2-2)**
   (a.) zoos.
   b. forests.
   c. kennels.
6. Grains, such as wheat, oats, and corn are **(3-1)**
   a. leaves.
   b. stems.
   (c.) seeds.
7. Rattlesnakes and poison ivy are **(3-3)**
   a. endangered organisms.
   (b.) dangerous to people.
   c. extinct.
8. Strong thin threads from plants that can be used to make cloth are **(3-2)**
   (a.) fibers.
   b. wood.
   c. wool.
9. Leather used for making some shoes comes from **(3-2)**
   a. chemicals.
   b. plant parts.
   (c.) animal skins.
10. Scientists protect endangered plants by saving their **(2-2)**
    (a.) seeds.
    b. fibers.
    c. leaves.

Numbers in parentheses after each question refer to the lesson number and the objective of that lesson.

Name ______________________

Chapter 4 Test B

Matching Match each food in the pictures with the plant part that people eat.

a. b. c. d.

1. c stem
2. a leaves
3. b root
4. d seeds

Short Essay Use complete sentences to answer each question.

1. Name three ways people can protect organisms and their habitats. **(2-1)**

People can protect organisms by making laws that do not allow people to hunt certain kinds of animals. Scientists can protect endangered plants by saving their seeds. People can protect habitats by trying not to pollute them with litter or chemical wastes.

2. Name one endangered animal and one extinct animal. **(1-3)**

An animal that is endangered is the tiger. The dodo bird is an extinct animal.

Numbers in parentheses after each question refer to the lesson number and the objective of that lesson.

Name ______________________

Unit Test 1

Multiple Choice Choose the best answer.

1. Pollination takes place when pollen reaches **(1-2-2)**
   a. part of an animal's body.
   b. the tips of flower petals.
   (c.) the center of a flower.

2. The outside covering of a seed is the **(1-3-1)**
   a. seed leaf.
   (b.) seed coat.
   c. seedling.

3. Lizards, turtles, and alligators are **(2-2-1)**
   a. mammals.
   (b.) reptiles.
   c. fish.

4. A snail's hard shell helps the snail by **(2-3-1)**
   a. keeping it warm.
   (b.) protecting its soft body.
   c. storing food and water.

5. An example of a mammal is a **(2-2-3)**
   (a.) human.
   b. turtle.
   c. lizard.

Use the picture to answer questions 6 and 7.

1. 2. 3. 4.

6. Food needed for plant growth is made in which part? **(1-1-3)**
   (a.) 2 b. 3 c. 4

7. Which plant part helps give support above the ground? **(1-1-2)**
   a. 1 (b.) 3 c. 4

8. The materials used by a plant that are taken in by its roots are called **(2-1-1)**
   (a.) minerals.
   b. weeds.
   c. sugar.

9. A young frog is called a **(2-2-3)**
   a. reptile.
   (b.) tadpole.
   c. larva.

1. ⓐ ⓑ ●
2. ⓐ ● ⓒ
3. ⓐ ● ⓒ
4. ⓐ ● ⓒ
5. ● ⓑ ⓒ
6. ● ⓑ ⓒ
7. ⓐ ● ⓒ
8. ● ⓑ ⓒ
9. ⓐ ● ⓒ

Numbers in parentheses after each question refer to the chapter and lesson number and the objective of that lesson.

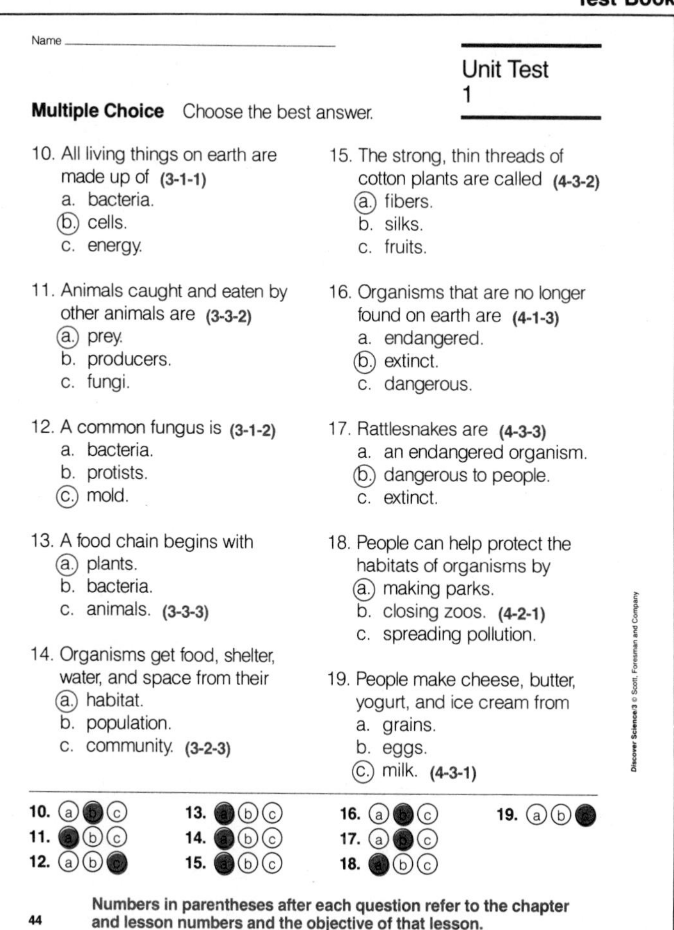

Name ______________________

Unit Test 1

Multiple Choice Choose the best answer.

10. All living things on earth are made up of **(3-1-1)**
    a. bacteria.
    (b.) cells.
    c. energy.

11. Animals caught and eaten by other animals are **(3-3-2)**
    (a.) prey.
    b. producers.
    c. fungi.

12. A common fungus is **(3-1-2)**
    a. bacteria.
    b. protists.
    (c.) mold.

13. A food chain begins with
    (a.) plants.
    b. bacteria.
    c. animals. **(3-3-3)**

14. Organisms get food, shelter, water, and space from their
    (a.) habitat.
    b. population.
    c. community. **(3-2-3)**

15. The strong, thin threads of cotton plants are called **(4-3-2)**
    (a.) fibers.
    b. silks.
    c. fruits.

16. Organisms that are no longer found on earth are **(4-1-3)**
    a. endangered.
    (b.) extinct.
    c. dangerous.

17. Rattlesnakes are **(4-3-3)**
    a. an endangered organism.
    (b.) dangerous to people.
    c. extinct.

18. People can help protect the habitats of organisms by
    (a.) making parks.
    b. closing zoos. **(4-2-1)**
    c. spreading pollution.

19. People make cheese, butter, yogurt, and ice cream from
    a. grains.
    b. eggs.
    (c.) milk. **(4-3-1)**

10. ⓐ ● ⓒ
11. ● ⓑ ⓒ
12. ⓐ ⓑ ●
13. ● ⓑ ⓒ
14. ● ⓑ ⓒ
15. ● ⓑ ⓒ
16. ⓐ ● ⓒ
17. ⓐ ● ⓒ
18. ● ⓑ ⓒ
19. ⓐ ⓑ ●

Numbers in parentheses after each question refer to the chapter and lesson numbers and the objective of that lesson.

TEACHING PLAN

Major Concepts

Lesson 1 People can make environmental changes that affect plants and animals.

Lesson 2 People protect plants and animals by protecting their habitats and by protecting endangered species.

Lesson 3 People use plants and animals for food, clothing, and other useful products.

Chapter Vocabulary

endangered organisms, extinct organisms, fibers, pollution

Getting Started

Ask students if they have ever seen raccoons near their homes. Have students *name* other "wild" animals that can be found in residential areas. (Answers will vary but might include deer, skunks, squirrels, birds, and crayfish.) Initiate a discussion about why some wild animals live near people while others do not.

Chapter 4

How People Affect Plants and Animals

These raccoons find food, water, shelter, and space to live in their wooded habitat.

66

Teaching Options

Cooperative Learning ♦

Jigsaw Format (See page T23.)
Assign the following topics at random to your cooperative learning teams.

Topic A: What are some ways in which people can change the lives of plants and animals?
Topic B: What are endangered plants and animals, and what are some ways that people protect them?
Topic C: How are plants and animals useful to people? How can they be harmful?
Topic D: Name two kinds of animals that have become extinct. How did they become extinct? How can extinction of animals be prevented?

Have students search for information on their topic as they read the chapter. Then let all students with the same topic meet in an expert group to discuss the information. When students return to their teams, they may take turns presenting their topics to the team. Then give students a test covering all topics to complete individually (Chapter 4 Test A or B in the *Test Book*). Award Superteam certificates to teams whose average test scores exceed 90%, and Greatteam certificates to teams whose average test scores exceed 80%.

♦ ***Suitable as a language development activity***

Introducing the Chapter

In this chapter you will learn how people affect plants and animals. You will also learn how plants and animals affect people. People make many things from plants and animals. In the activity below you will make a dye from a cabbage plant.

Making a Plant Dye

Plants have different colors. Some plants are green. Others are red, yellow, or purple. People make dyes from the colored parts of some plants. They use the dyes to color fabrics, paper, and other things.

You can use a red cabbage plant to make a dye. Then, you can use the dye to paint a picture.

Tear some cabbage into small pieces. Place the pieces into a cup. Add a few tablespoons of water to the cup. Stir the pieces of cabbage until the water becomes darkly colored. Remove the cabbage from the cup. Use the dye in the cup to paint a picture.

Talk About It

1. What part of the cabbage plant did the dye in your cup come from?
2. What are some other things people make from plants?

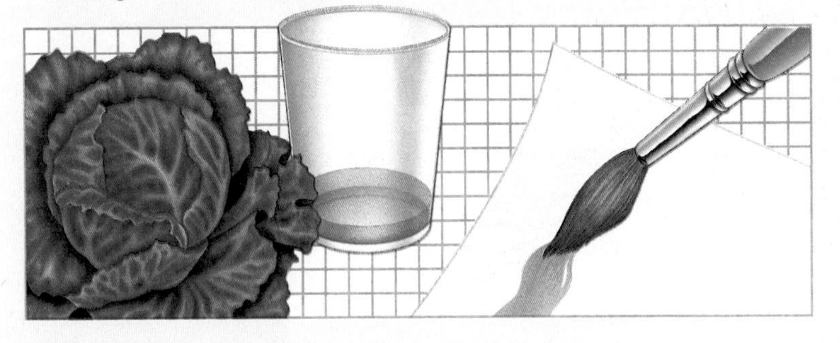

DISCOVER

Objective ♦
This optional *DISCOVER* activity will help students explore and build background information about the concept that people make useful things from plants. Later in the chapter, students will be able to draw on this experience to help them assimilate the new content.

Science Process Skills
Observing

Materials
For each pair of students: red cabbage, plastic cup, plastic spoon, 2 sheets drawing paper, 2 paintbrushes, water

Safety Tips (See page T24.)

- Remind students not to taste materials used in science activities, such as the dye used in this activity.
- Remind students to immediately wipe up any water that spills on the floor to prevent falls.

Teaching Tip

- Encourage students to find out what other plants are used to make dyes.

Answers
Talk About It
1. the leaf of the cabbage plant
2. Answers will vary, but might include paper, cloth, and wood items.

Applying Whole Language ♦

Discuss the whole language framework with each Teaching Option you select. Here is an example applied to the Reinforcement option on p. 70.

1. Purpose: To make a display showing populations of endangered species

2. Context: Small group contributions to whole group purpose. Include a phone call or interview to gather information.

3. Decisions: Choices of animals, references, sequence of events. How to display. Who should see it?

4. Evaluation: What questions did the activity raise? (See p. T30.)

Resource Book page 39

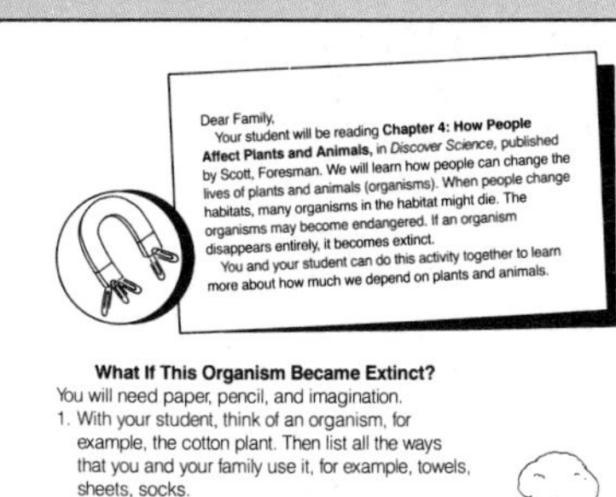
Dear Family,
Your student will be reading **Chapter 4: How People Affect Plants and Animals,** in *Discover Science*, published by Scott, Foresman. We will learn how people can change the lives of plants and animals (organisms). When people change habitats, many organisms in the habitat might die. The organisms may become endangered. If an organism disappears entirely, it becomes extinct.
You and your student can do this activity together to learn more about how much we depend on plants and animals.

What If This Organism Became Extinct?
You will need paper, pencil, and imagination.
1. With your student, think of an organism, for example, the cotton plant. Then list all the ways that you and your family use it, for example, towels, sheets, socks.
2. Then play the "What if . . ." game. Say, "What if this organism became extinct?" Then tell all the ways that this fact would change your life. For example, what if the cotton plant became extinct? You would not be able to get cotton sheets, towels or socks.
3. Then think of substitutes that are from organisms (for example, wool). Ask, "Would the substitute be as good?" Then ask, "What if that organism (sheep) became extinct?"
4. Continue going for as long as you can. When you run out of substitute organisms, think of an artificial substitute.
5. Then go on to another organism (for example, a cow). Play for as long as you can.

sheets
towels
socks

39

Science Background

In addition to dyes, many plants produce chemicals, called *alkaloids*. These alkaloids are helpful to plants because they discourage animals and insects from eating them. Humans find alkaloids useful, too. Many drugs are made from the alkaloids found in various plants.

TEACHING PLAN

Lesson Objectives

- *Explain* how people change the habitats of plants and animals.
- *Describe* how people have caused some plants and animals to become endangered.
- *Define* the term *extinct*.

Lesson Vocabulary

endangered organisms, extinct organisms, pollution

1. Motivate

Demonstration Activity ♦

With students' help, list things they require in order to live on the chalkboard.

Discussion

Ask students where the items they listed come from. (Answers will vary.) Select one general item, such as food, to discuss. Question: **What would happen to you if all the stores that sold this item disappeared?** (Answers will vary.)

1 How Do People Change the Lives of Plants and Animals?

LESSON GOALS

You will learn
- how people change habitats.
- about endangered organisms.
- about extinct organisms.

The owl in the picture lives in a forest. Suppose people cut down the forest so they could build new houses. Then the owl would have to look for a new home.

How People Change Habitats

When people build houses, roads, or factories, they change the habitats of many organisms. An organism gets everything it needs from its habitat, such as food, water, and shelter. If the habitat changes, the organism might not be able to live there anymore.

Owl in tree

Teaching Options

Science Background

When habitats are destroyed, organisms are forced to find new areas in which to live. Many organisms cannot make this transition. Something unique in their habitats—specific nesting sites, food sources, and the like—may not be found within migrating distance. When this happens, the local population of these organisms dies out. If other populations of the same species also die out, the number of individual organisms declines and the species becomes endangered. If the decline continues, the species may become extinct. Many species that have lived on earth are now extinct due to natural causes.

Reading Strategies ♦

1. Guide students' pre-reading by asking: How would you ask the question-title in your own words?
2. Assign these strategies: Writing a Memory Sentence and Visualizing Information (See pages T26–T29.)
3. Pair students to share what information is clear and unclear and initiate discussion using students' unanswered questions.

♦ ***Suitable as a language development activity***

Sometimes, people change habitats by adding harmful things to them. **Pollution** is anything harmful added to the air, water, or land. Pollution can kill organisms and can destroy their habitats.

People pollute the air when they burn fuels in their homes, cars, and factories. Gasoline, coal, and oil are some kinds of fuels. When a fuel burns, it gives off harmful chemicals.

Look at the water pollution in the picture. Some factories pollute water by dumping chemical wastes into lakes and streams. The chemicals can poison fish and other organisms in the water.

When people litter, they pollute the land. Litter can change the appearance of the land. It can also change the habitats of plants and animals.

pollution (pə lü′shən), anything harmful added to the air, water, or land.

INVESTIGATE!

Find out how people can affect the plants and animals in a habitat. Write a hypothesis and test your hypothesis with an experiment. You might set up two terrariums and then observe them for a few weeks, changing one variable at a time in one of the terrariums.

Polluted river

2. Teach

Teaching Tips

- Question: **How does man harm the environments of plants and animals?** (by polluting the environments or by changing the land so that animal and plant habitats are destroyed)
- Tell students that farmers in Washington state used poisons (pesticides) to kill insects that tried to eat their crops. Explain that the pesticides got into the drinking water in some areas, and the residents had to import drinking water from other places because they could not use the water in their wells. Question: **Whose habitats were changed?** (the habitats of plants, animals, and people) Make sure students understand that sometimes attempting to improve an area actually causes trouble for plants, animals, and humans. Stress that sometimes, the problems created cannot be reversed or fixed.

Investigate!

Accept any testable hypothesis. One possible hypothesis is: *Feeding the animals in a habitat can help them grow better.* Terrariums may be considered as a means of conveniently securing certain observations and interpretations. In most instances, students should find out that feeding the animals in a terrarium can have positive results.

Reinforcement

Have the students draw or write about some of the ways that man changes plant and animal habitats. Display students' pictures and essays on a classroom bulletin board with a slogan such as *Endangering Animals Can Endanger You!* or *Respect the Rights of Animals — Be Careful to Preserve Their Habitats!*

Special Education

Take students with learning disabilities on a nature walk around the school. Have the students look for animal habitats. Discuss what each animal gets from its habitat (such as food and shelter). Ask students to think about where the animal would go if its habitat were destroyed.

Workbook page 19 *

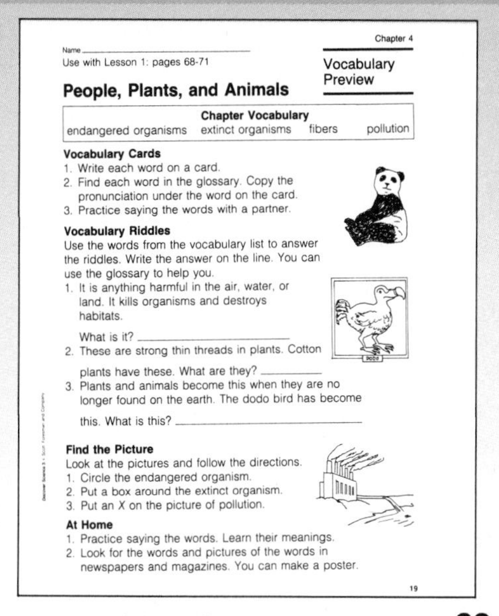

Name ______

Use with Lesson 1: pages 68-71

Chapter 4

Vocabulary Preview

People, Plants, and Animals

Chapter Vocabulary

endangered organisms   extinct organisms   fibers   pollution

Vocabulary Cards

1. Write each word on a card.
2. Find each word in the glossary. Copy the pronunciation under the word on the card.
3. Practice saying the words with a partner.

Vocabulary Riddles

Use the words from the vocabulary list to answer the riddles. Write the answer on the line. You can use the glossary to help you.

1. It is anything harmful in the air, water, or land. It kills organisms and destroys habitats.
   What is it? ______
2. These are strong thin threads in plants. Cotton plants have these. What are they? ______
3. Plants and animals become this when they are no longer found on the earth. The dodo bird has become this. What is this? ______

Find the Picture

Look at the pictures and follow the directions.

1. Circle the endangered organism.
2. Put a box around the extinct organism.
3. Put an *X* on the picture of pollution.

At Home

1. Practice saying the words. Learn their meanings.
2. Look for the words and pictures of the words in newspapers and magazines. You can make a poster.

19

*** Answers to masters on pages 66E–66H**

TEACHING PLAN

Teaching Tips

- Print the following list on the board: Blue Whale (humpback whale), Vicuna, Giant Panda, Polar Bear, Gorilla, Orangutan, California Condor, Hawaiian Silversword, and Whooping Crane. Question: **What do you think these organisms have in common?** (Guide students to *infer* that people have affected each of these organisms' habitats; the organisms are endangered and could become extinct.)
- **Possible Misconception:** Some students might think that organisms can always travel to other habitats if their original habitats are destroyed. Point out that, in many cases, organisms whose habitats have been destroyed are not able to find new homes and thus become extinct.
- Explain that sometimes people do not realize an organism is endangered until it is too late to prevent its extinction.

SCIENCE IN YOUR LIFE

Scientists thought the ivory-billed woodpecker was extinct. It lived in forests in the United States. When the forests were cut down, the bird's habitat was destroyed. To their surprise, scientists found a few of the woodpeckers in Cuba in 1986.

endangered (en dān′jərd) **organisms,** kinds of organisms that are very few in number and might someday no longer be found on the earth.

[1] Pitcher's thistle flower
Loggerhead turtle

Endangered Plants and Animals

Some organisms can find new homes when their habitats are changed. Other organisms cannot find new homes. Those organisms die.

When many organisms of the same kind die, that kind of organism might become endangered. **Endangered organisms** are kinds of organisms that are very few in number. Someday these organisms might no longer be found on the earth.

People have caused many kinds of organisms to become endangered by changing their habitats or by killing the organisms. Tigers are endangered animals. They used to live in many parts of Asia. Over the years, people killed many tigers for sport and for their fur. People also cut down the forests where the tigers lived. Endangered organisms live in many parts of the world. What endangered organisms do you see in the pictures?[1]

Pitcher's thistle flower

Loggerhead turtle

70

Teaching Options

Science Anecdote

Billions of passenger pigeons used to live in the forests of North America. Beginning in the 1850s, people destroyed much of the pigeons' habitat. They also killed a large number of pigeons for food and sport. Eventually, the passenger pigeon became extinct.

Reinforcement

Have students draw a cartoon strip that shows what happened to the population of dodo birds from 1400 to the present. Students can use one cartoon panel for each significant event or to show how the dodo population changed over time.

Enrichment

Encourage interested students to contact a local Sierra club chapter, a branch of the National Audubon Society, or a community library to request information about endangered species and extinction. You might want to invite a representative to speak to the class about preserving wildlife and caring for the land.

♦ ***Suitable as a language development activity***

Extinct Plants and Animals

Some kinds of plants and animals no longer are found on the earth. These organisms are **extinct.** The picture on the right shows a fossil of an extinct plant.

extinct (ek stingkt′) **organisms,** kinds of organisms that no longer are found on the earth.

The dodo bird in the picture used to live on an island that had no people. In the 1500s, sailors landed on the island. They killed many dodo birds for food. The sailors' dogs destroyed the eggs of the dodo birds. The number of dodos became smaller and smaller. Finally, dodo birds became extinct.

Leaves of Maidenhair tree

Dodo bird

Lesson Review

1. How do people change the habitats of plants and animals?
2. How have people caused some plants and animals to become endangered?
3. What are extinct plants and animals?
4. **Challenge!** How can people prevent tigers from becoming extinct?

Study on your own, pages 322–323.

EARTH SCIENCE
FIND OUT ON YOUR OWN
CONNECTION

Scientists think the weather might have caused dinosaurs to become extinct millions of years ago. Use an encyclopedia to find out how the weather might have changed the dinosaurs' habitat. Write a report about what you learned.

3. Assess

Lesson Review

1. People change the habitats of plants and animals when they build things and when they pollute.
2. People have endangered plants and animals by changing their habitats and by killing them in large numbers.
3. Extinct plants and animals are kinds of plants and animals that no longer live on the earth.
4. Challenge! People can protect tigers by not killing them, by not changing their habitats, and by placing some tigers in zoos. **Thinking Skill:** *Applying information to new situations*

Find Out On Your Own

Scientists think that during the time when dinosaurs lived, the climate might have become colder. It is thought that the dinosaurs could not survive in the cold weather and they eventually became extinct. **Thinking Skills:** *Collecting information, Communicating*

Workbook page 20 *

Name ________________
Use with Lesson 1: pages 68-71

Chapter 4
Science Skills

Endangered Species

Read about endangered species. Decide why the species has become endangered. Then circle the reason.

1. The California condor is one of the world's largest birds. It once lived throughout the West. But when settlers moved in, there was little room left for the condor. It now survives only in zoos. These zoos want to save the condor. One day, they hope to return it to the wild.

hunting pollution habitat destruction

2. The Mexican grizzly bear is known for its size, power, and courage. Because of this, many people want its fur as a trophy. Tracked down for sport, few of these bears have survived. Grizzly bears once lived throughout the American Southwest.

hunting pollution habitat destruction

3. The fur of the giant otter is among the most beautiful in the world. The otter is easy to catch because it is curious and unafraid. At one time giant otters were often seen swimming in rivers throughout South America. Now, few otters are seen.

hunting pollution habitat destruction

4. When chemicals are dumped into a river, many fish die. The fish that live, carry the poison in their bodies. The pelican pictured below has become endangered because it eats fish that have been poisoned with chemicals.

hunting pollution habitat destruction

Dalmatian Pelican

20

*** Answers to masters on pages 66E–66H**

Game Suggestion ♦

Print *Endangered Species* on the board. Instruct students to connect the following words, crossword-style, to the letters in the phrase: *extinct, habitat, pollution, coal, factories, dodo.*

Reteaching Suggestion ♦

Ask the students to make a list of things that they have lost and never found. Ask students to ***describe*** how they felt when they knew the items were really lost. Question: **Have you ever experienced the death of a pet?** (Answers will vary.) Allow students to discuss how they felt upon losing an irreplaceable animal. Then, compare the death of a pet with the extinction of a certain type of plant or animal. Tell students that the pet cannot be replaced, just as extinct animals and plants cannot be replaced.

TEACHING PLAN

Lesson Objectives
- *Describe* how people can protect the habitats of plants and animals.
- *Describe* ways people protect endangered plants and animals.

1. Motivate

Demonstration Activity ♦
Obtain pictures of endangered species that humans hunt (crocodiles, elephants, snow leopards, and so on), and place the pictures on the board. Next to each picture, place a picture of something that was made from the animal (fur coat; ivory jewelry; shoes, belts, and purses, and so on)

Discussion
Question: **What would happen if everyone wore a leopard coat or ivory jewelry?** (Allow students to respond.) Point out that fashions change frequently, but when the animals used for fashions become extinct, they are gone forever. Point out that it is illegal to hunt endangered animals or to bring products made from these animals into the United States.

2 How Do People Protect Plants and Animals?

LESSON GOALS

You will learn
- how people protect the habitats of plants and animals.
- how people protect endangered plants and animals.

The bird in the picture is a bald eagle. At one time only 15,000 bald eagles were alive. Then, people began to protect these birds. Today, about 37,000 bald eagles live on the earth. Without the help of people, these birds might have become extinct.

Mountain Goat in Glacier National Park

Protecting Habitats of Plants and Animals
People can protect plants and animals from becoming endangered or extinct. One way of protecting organisms is by protecting their habitats. Look at the park in the picture on the left. People have made large parks such as this one to protect plants and animals. No one is allowed to disturb the habitats in these parks. People cannot build houses there. They cannot hunt animals. No one can collect plants or animals in the parks.

72

Teaching Options

Science Background
Protecting species from extinction benefits humans. Studying other animals, especially nonhuman primates, can help us understand ourselves better. Understanding environmental factors that harm other living things can help us protect ourselves as well.

Reading Strategies ♦
1. Guide students' pre-reading by asking: How would you ask the question-title in your own words?
2. Assign these strategies: Visualizing Information and Writing a Memory Sentence (See pages T26–T29.)
3. Pair students to share what information is clear and unclear and initiate discussion using students' unanswered questions.

♦ *Suitable as a language development activity*

People have made laws that protect organisms and their habitats. Hunting laws do not allow people to hunt certain kinds of animals. Some laws limit the number of trees that can be cut down in forests. Other laws do not allow people to build houses, stores, or factories in certain places.

People can also protect habitats by trying not to pollute them. For example, some factories remove chemical wastes from water before returning the used water to lakes and streams. Factories also use certain kinds of coal that cause less air pollution when they are burned.

SCIENCE IN YOUR LIFE

Many bluebirds used to live in the eastern United States. Today, these birds are rare. Other birds took their nesting places. Now, people are trying to protect bluebirds. Some people have made boxes that are safe places for bluebirds to build nests.

2. Teach

Teaching Tips

- On the chalkboard, write a list of some of the parks in your area. Question: **Have you ever been to any of these parks? What kind of animals and plants did you see there?** (Allow volunteers to share their experiences.) Remind students that it is important to obey posted park rules regarding feeding animals and birds, picking plants, and so on, and that they should ask park personnel if they are in doubt about the rules.
- **Possible Misconception:** Some students might think that extinct organisms will eventually reappear on the earth. Reinforce that extinction is permanent.

Reinforcement

Ask students to draw two columns on a piece of paper and to label the columns *endanger* and *protect*. Have the students list (in the appropriate column) activities that help and hurt plants and animals. Have the students place an *X* by every item on the list that humans can improve by taking some action or by avoiding other actions. (Most items will have an *X* next to them.)

Special Education

Have students with learning disabilities draw pictures or write stories about places they have visited that provide a safe environment for plants and animals. (Examples might include zoos, botanical gardens, and so on.) Encourage students to select one of these environments to sponsor, and to volunteer their services.

Science and Social Studies

Display a map of the world or the western hemisphere. Point out that the salmon that are hatched in a river on the west coast will return to that same river to lay their eggs. Ask students to use an encyclopedia to locate the places salmon travel during their years in salt water. Ask students to ***infer*** what would happen to the salmon if a dam were built over the river and the salmon could not get over the dam.

TEACHING PLAN

3. Assess

Lesson Review

1. People protect habitats by making large parks where habitats cannot be disturbed, by making laws to protect habitats, and by trying not to pollute habitats.
2. People protect endangered plants by saving their seeds and growing them indoors. People protect endangered animals by taking care of them in zoos.
3. Challenge! Answers will vary, but might include not littering, riding a bicycle instead of riding in a car whenever possible, and using public transportation. **Thinking Skill:** *Applying information to new situations*

Find Out On Your Own

One population of black-footed ferrets remains alive. It lives in the state of Wyoming. **Thinking Skills:** *Collecting information, Communicating*

Field of wildflowers

Protecting Endangered Plants and Animals

Scientists protect endangered plants by saving their seeds, growing the seeds indoors, and later putting the plants in their habitats. These people protect wildflowers.

Some endangered animals are placed in zoos. Someday the animals might be returned to their habitats.

Lesson Review

1. How do people protect the habitats of plants and animals?
2. How do people protect endangered plants and animals?
3. **Challenge!** How can you help fight pollution?

Study on your own, pages 322–323.

LIFE SCIENCE
FIND OUT ON YOUR OWN

The black-footed ferret is an endangered animal. Use library books to find out how many populations of black-footed ferrets remain alive, and where these animals live. Write a few sentences telling what you learned.

Teaching Options

Science Anecdote

Yellowstone National Park was the world's first national park. It was established in 1872.

Reteaching Suggestion ♦

Have students write a public service announcement for television. Direct them to provide a description of what is to be shown, as well as a script telling students to tell Americans what can be done to protect endangered species and what can harm endangered species.

Workbook page 21 *

Name ____________________

Use with Lesson 2: pages 72-74

Chapter 4

Science and Language Arts

Making Bumper Stickers

Study the bumper stickers designed by students. One bumper sticker contains words. Another contains pictures. The third contains both words and pictures. The bumper stickers show that the students care about what happens to the earth and to living things.

Think about an "earth care" message for a bumper sticker. In the space below, design a colorful bumper sticker.

21

♦ *Suitable as a language development activity*

Making a Model of an Animal Habitat

Suggested grouping: individual

Purpose
Identify the kind of habitat an animal lives in and *make a model* of that habitat.

Gather These Materials
- crayons or colored markers
- scissors • construction paper
- glue • shoe box

Follow This Procedure
1. Use a chart like the one shown to record your observations.
2. Choose one wild animal, such as a tiger. Draw a picture of it. Color and cut out your picture.
3. Make a model of the animal's habitat. Use construction paper for different parts of the habitat. You might cut out the outline of a tree or a rock. Glue the parts of the habitat inside the shoe box.
4. Glue the picture of the animal in the habitat.
5. List all the parts of the habitat in the chart. For each part of the habitat, write what might happen to the animal if that part was destroyed.

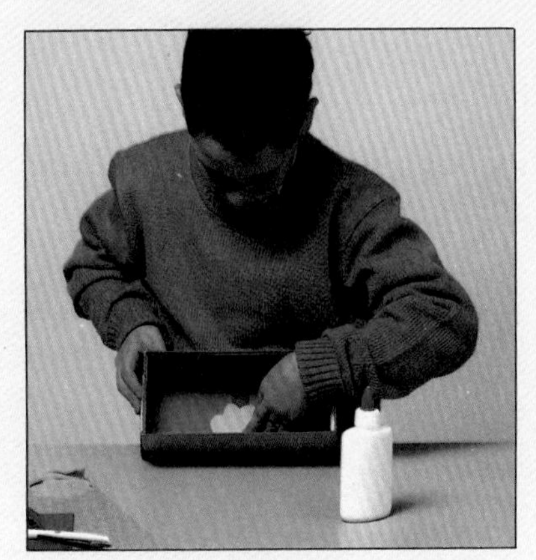

Record Your Results

Part of Habitat Destroyed	What Might Happen to Animal
1.	Answers will vary, but should include references to food, water, and shelter.
2.	
3.	

State Your Conclusion
1. What kind of a model habitat did you make?
2. Why is the habitat you made a good place for your animal to live in?

Use What You Learned
What might happen to your animal if a road were built through the animal's habitat?

75

Concept
Animals live in specific habitats.

Objectives/Process Skills
- *Identify* the habitat of an animal.
- *Make a model* of the animal's habitat.
- *Predict* what might happen to the animal if its habitat were disturbed.
- *Record* data.

Time Allotment
Allow 30 minutes.

Safety Tips (See page T24.)
- Be certain to use only nontoxic markers and glue.
- Warn students to be cautious when handling paper to avoid paper cuts.
- Use round tipped scissors.

Teaching Tips
- Discuss characteristics of different kinds of habitats so students can decide which features to include in their models.
- Suggest that students exchange habitats while keeping their animal drawings. Ask students to explain whether or not their animals could survive in the new habitats.

Answers
State Your Conclusion
1. Answers will vary depending on the animal chosen.
2. Answers will vary but should include adequate food and water supplies.

Use What You Learned
Answers will vary, but might include that the animal could move to another area, that the animal could die because it did not get enough food or water or lost its shelter, or that the animal could be run over by a vehicle on the road. **Thinking Skill:** *Predicting*

Resource Book page 43

Name ______ Chapter 4
Use with Lesson 2: page 75
Activity Worksheet
Making a Model of an Animal Habitat

Record Your Results

Part of Habitat Destroyed	What Might Happen to Animal
1.	
2.	
3.	

State Your Conclusion
1. What kind of a model habitat did you make?

2. Why is the habitat you made a good place for your animal to live in?

Use What You Learned
What might happen to your animal if a road were built through the animal's habitat?

43

*** Answers to masters on pages 66E–66H**

Activity Results

Models will vary depending on the animal chosen. Models may show habitats in forests, prairies, lakes, and so on.

TEACHING PLAN

Lesson Objectives
- *Identify* foods that come from plants and animals.
- *Identify* useful products that come from plants and animals.
- *Describe* how some plants and animals are harmful to people.

Lesson Vocabulary
fibers

1. Motivate

Demonstration Activity ♦
Place pictures of the following items on the board: cheese, butter, milk, leather shoes, and a hamburger or steak.

Discussion
Questions: **What do the items shown on the board have in common?** (They are all products made from something a cow gives.) **Can you think of other things we get from cows?** (Yogurt, ice cream, belts, purses, and coats are made from something obtained from a cow; sometimes items are made from the horns of cows.)

3 How Do People Use Plants and Animals?

LESSON GOALS

You will learn
- how people get foods from plants and animals.
- that some useful things are made from plants and animals.
- that some plants and animals are harmful.

Plants and animals are important to you. Many foods you eat come from plants and animals. Some of the clothes you wear also come from plants and animals. Plants and animals help keep you alive.

Food from Plants and Animals

Look at the foods in the picture. Where do these foods come from? Fruits and vegetables come from plants. Grains, such as wheat, oats, and corn, also come from plants. People use grains to make bread, macaroni, and cereal. You can see in the picture that people eat almost every kind of plant part. Grains are seeds. Lettuce is leaves. Potatoes are stems. Apples are fruits. What kind of plant part are carrots?[1]

[1] Carrots are roots.

Teaching Options

Science Background

Most of the food we eat has been altered or processed in some way. Processing includes pickling, canning, freezing, pasteurizing, and adding chemical additives. Chemicals are used to enhance the taste, color, or nutritional value of food and/or to preserve food. Some people think that additives can be dangerous to human health; the Food and Drug Administration, however, tests all additives and must approve them as safe to use. If later research finds an additive to be harmful, the FDA will revoke its approval or require the use of warnings on the labels of products containing the substance.

Reading Strategies ♦

1. Guide students' pre-reading by asking: How would you ask the question-title in your own words?
2. Assign these strategies: Visualizing Information and Writing a Memory Sentence (See pages T26–T29.)
3. Pair students to share what information is clear and unclear and initiate discussion using students' unanswered questions.

♦ *Suitable as a language development activity*

Meat comes from animals. A chicken leg, a hamburger, and a piece of fish are some kinds of meat. Milk and eggs also come from animals. People use milk in many ways. People drink milk. They also make cheese, butter, yogurt, and ice cream from it.

Many of the foods you eat look different from the plants or animals from which they came. Look at the foods below. What plant did the juice in the picture come from?[1] What animal did the eggs come from?[2]

[1] orange tree
[2] chicken

2. Teach

Teaching Tips

- Pass out a grape to each student. Question: **How are grapes useful to people?** (They can be eaten as is, dried to make raisins, or squeezed to make juice. The plants also provide oxygen.)
- Bring in several empty food packages that contain plant and/or animal ingredients. Have the students ***identify*** the plant and animal ingredients in each. Point out that many foods use a wide variety of products from plants and animals to make a certain food. Question: **What are the plant and animal ingredients in pizza? in tacos?** (Pizza contains cheese, and sausage, pepperoni, beef, or Canadian bacon from animals, and tomato sauce, spices, and green peppers and onions from plants. Tacos contain beef and cheese from animals, and a corn shell, tomatoes, lettuce, and taco sauce from plants.)
- When discussing things that humans get from plants and animals, remind students of the *DISCOVER* on page 67, where they made dye from plant parts.
- **Possible Misconception:** Students might not be aware that grains, such as oats, wheat, and rye, come from plants. Point out that grains are the seeds of plants.

Reinforcement

Divide students into groups. Give each group a different grocery list and have them ***identify*** the items on the list as originating from plants, animals, both, or neither. Have the students share their group's analysis of their list with the class.

Special Education

Have each student with learning disabilities name a food he or she has recently eaten. Direct another student with learning disabilities to ***identify*** the food as coming from plants, animals, or both. Then have each student draw the food he/she mentioned as well as its source.

Workbook page 22 *

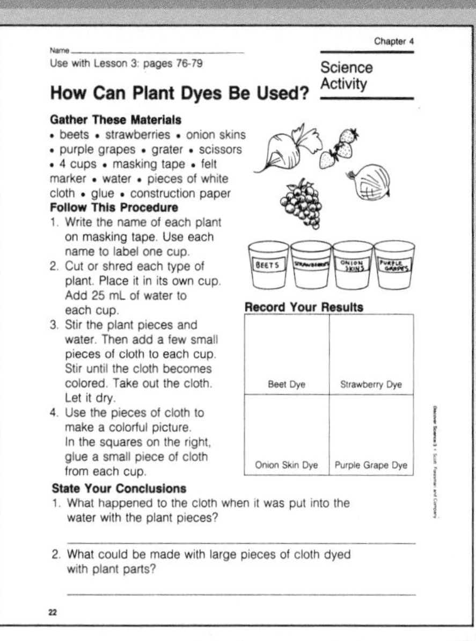
Name

Use with Lesson 3: pages 76-79

Chapter 4

Science Activity

How Can Plant Dyes Be Used?

Gather These Materials

• beets • strawberries • onion skins • purple grapes • grater • scissors • 4 cups • masking tape • felt marker • water • pieces of white cloth • glue • construction paper

Follow This Procedure

1. Write the name of each plant on masking tape. Use each name to label one cup.
2. Cut or shred each type of plant. Place it in its own cup. Add 25 mL of water to each cup.
3. Stir the plant pieces and water. Then add a few small pieces of cloth to each cup. Stir until the cloth becomes colored. Take out the cloth. Let it dry.
4. Use the pieces of cloth to make a colorful picture. In the squares on the right, glue a small piece of cloth from each cup.

Record Your Results

Beet Dye	Strawberry Dye
Onion Skin Dye	Purple Grape Dye

State Your Conclusions

1. What happened to the cloth when it was put into the water with the plant pieces?
2. What could be made with large pieces of cloth dyed with plant parts?

22

Answers to masters on pages 66E–66H

TEACHING PLAN

Teaching Tip

• Question: **How can plants be harmful to people?** (Allow students to respond.) Discuss the poisonous plants indigenous to your region, as well as poisonous houseplants, and caution students not to ingest *any* unfamiliar plant parts. Some examples of poisonous plants and plant parts include: acorn leaves, sprouts, and bulbs; amaryllis bulbs; angle trumpet; apricot pits; azalea; bittersweet leaves and fruit; black locust; bleeding heart; bloodroot; buckeye flowers, sprouts, and nuts; blackthorn; castor beans; cherry pits; Chinese lantern leaves and unripe fruit; Christmas berry leaves; delphinium; foxglove leaves and seeds; geraniums (*all* assumed poisonous although the California geranium is the only kind that has caused death); glory lilies; hemlock; horse chestnut flowers, sprouts, and seeds; ivy (any kind); Japanese yew; Jerusalem cherry leaves and unripe fruit; jonquils; lily of the valley; larkspur; mistletoe; mushrooms (*all* are assumed to be deadly poisonous); nightshade; oleander; philodendron; peach leaves, seeds, and pits; poison ivy; prinet; potato sprouts, vines, and unripe tubers; rhubarb leaves; snow drop bulbs; and tomato leaves and vines.

SCIENCE IN YOUR LIFE

Silk cloth comes from caterpillars called silkworms. Silkworms make long, thin threads of silk and then wrap themselves up in the silk. This silk covering is called a cocoon. People get silk by unwinding the silk threads of cocoons.

fibers (fī′bərz), strong thin threads in plants that can be used to make cloth.

Useful Things from Plants and Animals

You get more than food from plants and animals. You also get things you can use. For example, wood comes from trees. People use wood to build houses and to make furniture, paper, and toys. What are some other ways people use wood?[1]

Look at the clothes the girl is wearing. Her blouse is made of cotton cloth. Cotton comes from cotton plants. These plants have strong, thin threads—**fibers**—that are used to make cloth. The girl's skirt is linen. Flax plants have fibers that are used to make linen cloth. The girl is also wearing leather shoes. People make leather from animal skins. The girl's sweater is woolen. Wool is made from sheep's hair.

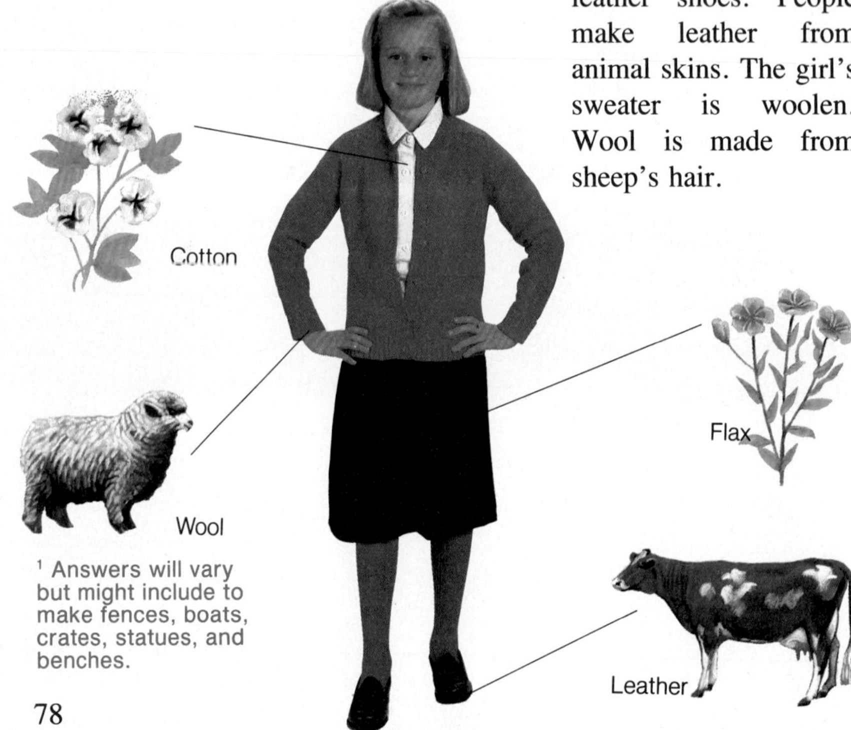

[1] Answers will vary but might include to make fences, boats, crates, statues, and benches.

78

Teaching Options

Science Anecdote

Digitalis is a medicine that is made from the leaves of a garden plant called purple foxglove. Digitalis is used to treat certain heart diseases. Foxglove leaves and seeds are, however, poisonous to healthy people.

Reinforcement

Collect mosses and lichen or purchase them from a science supply house. Place them in enough boiling water to cover them and bring the water to a slow boil. Place a white cloth in the water and have students observe the natural dye. *CAUTION:* Do not let students touch the dyed cloth or the pot until both have cooled.

Enrichment

Encourage interested students to investigate various poisonous plants and animals found in your area. Direct them to use reference books to find pictures of these plants and animals, and to find out the symptoms that an encounter with each of these organisms would cause. Have students draw pictures of each plant and animal investigated, listing the associated symptoms below each picture. Encourage them to find out appropriate first aid measures for each situation. Allow students to share their findings with the class.

♦ ***Suitable as a language development activity***

Black widow spider

Poison sumac

Harmful Animals and Plants

Some animals can hurt people. When the spider in the picture bites, it releases a poison that can make people very sick. What other animals are dangerous to people?[1]

Some plants also can hurt people. If you touched the poison sumac in the picture, you might get an itchy rash. Some plants, such as buttercups, are poisonous to eat.

[1] Answers will vary, but might include bears, lions, and wolves.

Lesson Review

1. What are three kinds of food people get from plants? from animals?
2. What are two useful things people get from plants? from animals?
3. How are some plants harmful to people?
4. **Challenge!** What things did you use today that came from plants or animals?

Study on your own, pages 322–323.

Use an encyclopedia to find out what chemicals are used in tanning leather to make clothes.

PHYSICAL SCIENCE
FIND OUT ON YOUR OWN
CONNECTION

3. Assess

Lesson Review

1. People get fruits, vegetables, and grains from plants. People get milk, meat, and eggs from animals.
2. Wood comes from plants. Linen and cotton cloth are made from plant fibers. Leather and wool come from animals.
3. Some plants cause a rash; others are poisonous to eat.
4. Challenge! Answers will vary. **Thinking Skill:** *Applying information to new situations*

Find Out On Your Own

Some chemicals used in tanning leather are salt, lime, sulfuric acid, tannin, and borax. **Thinking Skill:** *Collecting information*

Workbook page 23 *

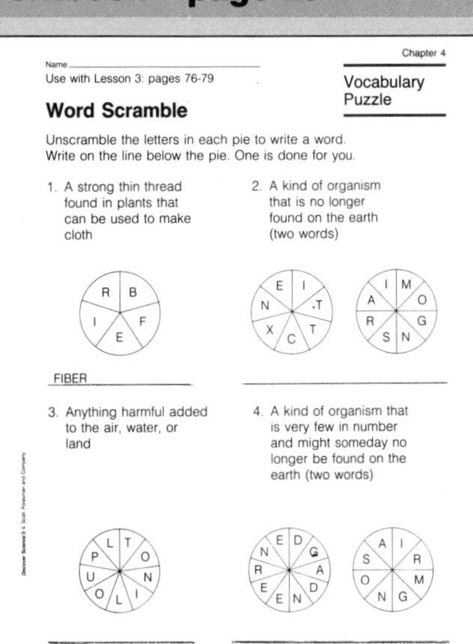

Chapter 4

Name ______

Use with Lesson 3: pages 76-79

Vocabulary Puzzle

Word Scramble

Unscramble the letters in each pie to write a word. Write on the line below the pie. One is done for you.

1. A strong thin thread found in plants that can be used to make cloth

R B F E I

FIBER

2. A kind of organism that is no longer found on the earth (two words)

E I T T C X N   I M O G N S R A

3. Anything harmful added to the air, water, or land

L T O N I L O U P

4. A kind of organism that is very few in number and might someday no longer be found on the earth (two words)

E D G A D N E E R N   A I R M G N O S

23

*** *Answers to masters on pages 66E–66H***

Game Suggestion ♦

Play a variation of "Twenty Questions," using questions that can be answered with "yes," "no," "I don't know," "neither," or "both." Choose an item from the room and direct students to ask you questions about it to determine what you are thinking of. Encourage them to find out whether the item is made of plant materials, animal materials, both, or neither. Allow the student who correctly identifies the object to choose the next object and answer the other students' questions. Assist when needed.

Reteaching Suggestion ♦

Have students plan what they will wear for a day and what they will eat and drink for a day. Tell students that what they wear and eat must come from either plants or animals. Have the students identify what each food or clothing item comes from, and let them share their lists with the class when everyone is finished planning.

Concept
Paper, linen, and cotton cloth are made from plant fibers.

Objectives/Process Skills
- *Observe* fibers in objects made from plants.
- *Record* observations.
- *Compare* the fibers in paper, linen, and cotton cloth.

Time Allotment
Allow 20 minutes.

Safety Tips (see page T24.)
- Use only nontoxic markers and food colorings.
- Instruct students to immediately wipe up any water spilled onto the floor to prevent slipping and falls.
- Warn students of the hazards of focusing the sun's rays through the hand lens onto skin or paper.

Teaching Tips
- Discuss the word *absorb* with students.
- Suggest that students bring different brands of paper towels to class and *compare* how well the towels absorb water.

Answers
State Your Conclusion

1. The fibers look like stiff threads.

2. Students should answer that they would use cotton cloth, because cotton cloth soaks up water faster than linen or paper.

Use What You Learned

The fibers look similar. **Thinking Skill:** ***Comparing***

Looking at Fibers

Suggested grouping: pairs

Purpose
Compare the fibers in objects made from plants.

Gather These Materials
• paper • cotton cloth • linen cloth • hand lens • paper plate • colored water

Follow This Procedure
1. Use a chart like the one shown to record your observations.
2. Tear the paper and the cloths two ways—up and down and across. Look at the edges with the hand lens. Record your observations.

3. Place a torn edge of each sample on the paper plate. Leave the opposite edge out of the plate, as shown in the picture.
4. Slowly pour some colored water into the plate until it reaches the edges of the samples. Compare how fast the samples soak up the water. Record the order of the samples from fastest to slowest.

Record Your Results

	How Fibers Looked	Order Samples Soaked Water
Paper	like	3
Cotton	stiff	1
Linen	threads	2

State Your Conclusions
1. What do plant fibers look like?
2. Which of the three samples would you use to wipe up a water spill? Why?

Use What You Learned
How do you think the fibers in wool cloth would compare with the cotton and linen fibers?

80

Activity Results
Cotton soaks up water faster than linen or paper. Linen soaks up water faster than paper.

Resource Book page 45

Name ______
Use with Lesson 3: page 78

Looking at Fibers

Chapter 4

Activity Worksheet

Record Your Results

	How Fibers Looked	Order Samples Soaked Water
Paper		
Cotton		
Linen		

State Your Conclusions
1. What do plant fibers look like?

2. Which of the three samples would you use to wipe up a water spill? Why?

Use What You Learned
How do you think the fibers in wool cloth would compare with the cotton and linen fibers?

45

Protecting the Wilderness

More than 100 years ago, a man named John Muir decided to take a trip. He had had an accident that almost left him blind. Now he could see again. He wanted to see the beautiful world with his own eyes. On his trip, he studied plants and animals. He did see quite a lot on his trip. He walked all the way from Kentucky to Florida!

John Muir

Waterfall in Yosemite Park

This long walk was not the only trip John Muir made. Throughout his life, he enjoyed living outdoors and traveling to new places. He visited the Yosemite Valley in California. There he saw beautiful mountains and giant trees. He could also see that cutting trees, mining, and grazing sheep might soon ruin the wilderness. This valley, like all habitats, was formed over a long period of time. If people destroyed it, they could not build it over again. Muir felt that the best idea was to save the land before it was harmed.

John Muir wrote about the valley in books and magazines. He told about trees that were thousands of years old. He told about winter snows and spring flowers and birds. He also talked to people who make laws. Many people agreed that saving the wilderness was important. Due to his hard work, the land he loved became Yosemite National Park.

Throughout his life, John Muir kept working to save wild lands and make them national parks. He wanted people like you to be able to enjoy the lands just as he did.

What Do You Think?

1. John Muir is sometimes called the father of the national parks. Why do you think people gave him that name?
2. If a forest was cut down, what do you think would happen to the plants and animals that lived there?

TEACHING PLAN

Discussion

Tell students that some people help conserve land by caring for their own land, by helping to pass laws to protect wilderness areas, and by controlling pollution that would harm wild habitats. Point out that there are many ways to practice conservation, and ask students to discuss their ideas.

Teaching Tip

- Tell students that one of John Muir's first conservation efforts was to ask his brother-in-law to fence in a woodland pond on his farm so that the plants would not be trampled by cattle. Ask students if they can think of similar ways to save land.

Answers

What Do You Think?

1. Because he worked hard to save lands as national parks. **Thinking Skill:** ***Recognizing the main idea and supporting detail***

2. Small plants would have no shade; birds and other small animals would not have places to nest; animals that use trees as food would lose a food source. **Thinking Skills:** ***Predicting, Recognizing cause and effect***

Books to Read

Fox, Stephen R. *John Muir and His Legacy: The American Conservation Movement.* Little, 1981. (for the teacher)

Teaching Options

Science Background

Yosemite Valley is named after the Native American tribe that once lived there. White people first discovered the area in 1851. By 1855, sightseers visited Yosemite, and in 1864, Congress made it a park. Because there was no national park system in effect then, the state of California was given control of the park, but the park was not well protected from loggers, miners, and farmers. Muir fought hard to ensure the protection of the park. In 1890, the area surrounding the existing park was made a national park, and in later years, the original park was also put under federal control. Despite Muir's efforts, one valley in the park was flooded when a dam was built to provide water for San Francisco residents.

TEACHING PLAN

Purpose
To develop the skills of collecting and organizing information using pictographs and time lines to solve problems.

1. Motivate

Discussion
Direct each student to draw a line 10–12 cm long, making a mark at the beginning of the line. Tell students that each centimeter on the time line represents one year in their lives. Question: **What does the mark at the beginning of the line represent?** (day of birth) Help students plot important events in their lives on their time lines.

2. Teach

Teaching Tip
- **Helpful Hint:** Make sure that students understand that using pictographs involves using symbols that represent numerical values, and provide concrete examples to help them develop this skill. For example, students could use circles of different sizes and practice adding up the total values of groups of circles.

Skills for Solving Problems

Using Pictographs and Time Lines

Problem: How has the population of trumpeter swans changed since 1850?

Part A. Using Pictographs to Collect Information

1. The key shows that each picture in the pictograph stands for a number of swans. What do the large pictures stand for? medium-sized pictures? small pictures?
2. The pictograph shows how many swans were alive in certain years. To find out how many were alive in 1850, add the numbers the pictures for that year stand for. Many hunters shot trumpeter swans. Habitats were changed. Later, people protected the swans. How many swans were alive in 1900? in 1950? in 1980?

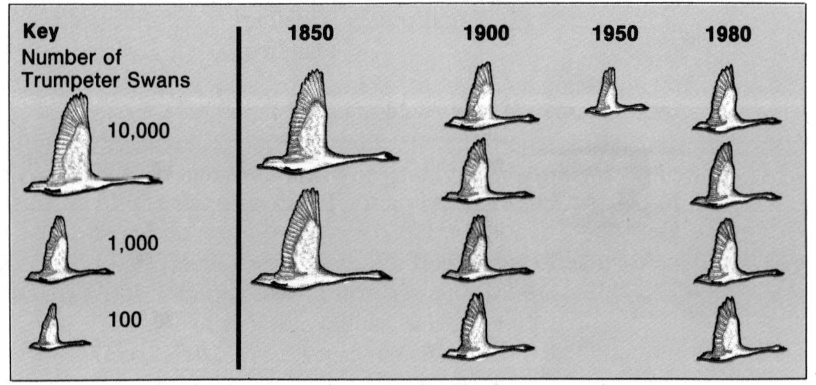

Part B. Using a Time Line to Organize and Interpret Information

3. The time line contains the information you collected. The spaces on the line stand for lengths of time. What is the length of time between the marks on the time line?

82

Teaching Options

Sample Time Line for Part C

1900
1950
1980
2000

400,000 Spider monkeys
300,000 Spider monkeys
4,000 Spider monkeys

◆ ***Suitable as a language development activity***

4. What happened to the swan population between the years 1850 and 1900? between 1900 and 1950? between 1950 and 1980?
5. What might have changed the population of swans between 1850 and 1980? What changes could you infer might take place by the year 2000?

1850 — 20,000 Trumpeter swans
1900 — 4,000 Trumpeter swans
1950 — 100 Trumpeter swans
1980 — 4,000 Trumpeter swans
2000

Part C. Using Pictographs and Time Lines to Solve a Problem

Problem: How has the woolly spider monkey population changed?

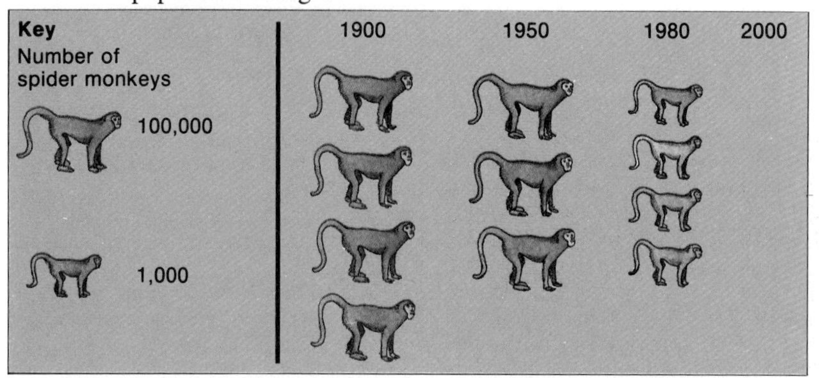

6. Use the pictograph to collect the information you need to solve the problem. Make a time line similar to the one shown in Part B to organize your information.
7. Look at your time line. The woolly spider monkey lives in the jungle forests in Brazil. People are cutting down these forests to make farms. Compare the time line you made with the time line shown in Part A. How are they different? What might cause this difference?
8. What changes in the monkey population might take place by the year 2000?

83

3. Assess

Part A

1. 10,000 swans; 1,000 swans; 100 swans

2. 20,000 in 1850; 4,000 in 1900; 100 in 1950; 4,000 in 1980

Part B

3. 50 years, 50 years, 30 years, 20 years

4. Between 1850 and 1900, the trumpeter swan population decreased. Between 1900 and 1950, the trumpeter swan population continued to decrease. Between 1950 and 1980, the swan population began to increase.

5. Changing habitats and too much hunting might have caused the population decrease between 1850 and 1950. Protection of the swans and their habitat might have enabled them to increase in numbers from 1950 to 1980. The swans might continue to increase in numbers through the year 2000.

Part C

6. See Sample Time Line for Part C.

7. The monkey population has changed from 400,000 to 4,000. The time line in Part A shows an increase in population between 1950 and 1980. The monkeys are not protected by laws against killing them and destroying their habitats.

8. Unless they are protected, they might be extinct by the year 2000.

Reteaching Suggestion ♦

Point out to students that musk oxen, large animals that were nearly extinct earlier in this century, were recently brought to Greenland to see if their population might grow. Write the following years and approximate numbers of Greenland musk oxen on the board: *1962: 25; 1977: 150; 1982: 500; 1989: 1500*. Have students use circles, each representing 50, to create a pictograph of Greenland musk oxen for each of those years. Have them use this pictograph to create a time line about the change in the population. Help students ***conclude*** that the population has steadily increased.

Resource Book page 47 *

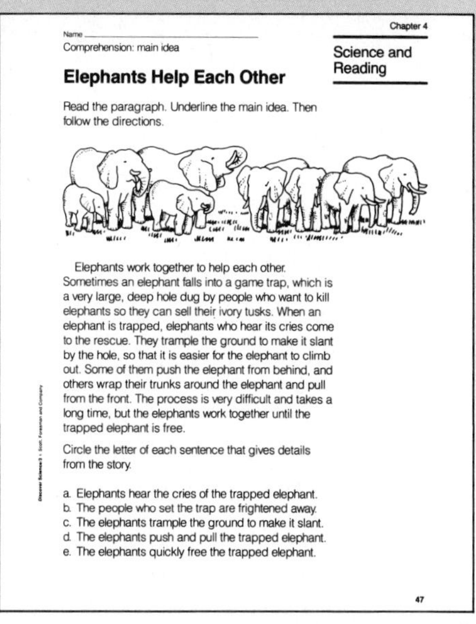
Name
Comprehension: main idea
Chapter 4
Science and Reading

Elephants Help Each Other

Read the paragraph. Underline the main idea. Then follow the directions.

Elephants work together to help each other. Sometimes an elephant falls into a game trap, which is a very large, deep hole dug by people who want to kill elephants so they can sell their ivory tusks. When an elephant is trapped, elephants who hear its cries come to the rescue. They trample the ground to make it slant by the hole, so that it is easier for the elephant to climb out. Some of them push the elephant from behind, and others wrap their trunks around the elephant and pull from the front. The process is very difficult and takes a long time, but the elephants work together until the trapped elephant is free.

Circle the letter of each sentence that gives details from the story.

a. Elephants hear the cries of the trapped elephant.
b. The people who set the trap are frightened away.
c. The elephants trample the ground to make it slant.
d. The elephants push and pull the trapped elephant.
e. The elephants quickly free the trapped elephant.

47

* Answers to masters on pages 66E–66H

REVIEW PLAN

Reviewing Science Words
1. pollution
2. fibers
3. extinct organisms
4. endangered organisms

Reviewing What You Learned
1. d **3.** a **5.** d **7.** b **9.** b
2. b **4.** c **6.** a **8.** d **10.** b

Interpreting What You Learned
1. The animal might not be able to find the things it needs to live, such as food, water, or shelter. **Thinking Skill:** *Restating or explaining ideas*

Chapter 4 Review

Chapter Main Ideas

Lesson 1 • People change the habitats of plants and animals when they build things and when they pollute. • Many kinds of organisms are endangered because people changed their habitats or killed many of them. • An extinct organism is a kind of organism that no longer is found on the earth.

Lesson 2 • People protect habitats by making large parks where habitats cannot be disturbed, by passing laws that protect habitats, and by trying not to pollute habitats. • Scientists protect endangered plants by saving their seeds and by growing the seeds indoors. Zoos protect endangered animals.

Lesson 3 • Fruits, vegetables, and grains come from plants. Meat, milk, and eggs come from animals. • Wood, linen, and cotton cloth are useful things that come from plants. Leather and wool are useful things that come from animals. • Some plants and animals are harmful to people.

Reviewing Science Words

endangered organisms   fibers
extinct organisms   pollution

Copy each sentence. Fill in the blank with the correct word from the list.

1. ___ can change the habitats of plants and animals.
2. Cotton cloth is made from the ___ of a cotton plant.
3. ___ are kinds of organisms that no longer are found on the earth.
4. ___ are so few in number they might someday no longer be found on the earth.

Reviewing What You Learned

Write the letter of the best answer.

1. An organism gets everything it needs from its
   (a) community. (b) population. (c) food. (d) habitat.

84

Review Options

Cooperative Learning ♦

STAD Format (See page T23.)
Assign students to work in four- to five-member teams to study Chapter 4 Review. Students should work together to make sure that they and their teammates know the material in the chapter. After students have had enough time to study together, give them a test to complete individually (Chapter 4 Test A or B in the *Test Book*). Award Superteam certificates to teams whose average test scores exceed 90%, and Greatteam certificates to teams whose average test scores exceed 80%.

Test Book page 33 *

Name ____________

Chapter 4
Test A

Multiple Choice Choose the best answer.

1. People sometimes harm an organism's habitat when they
   a. make large parks.
   b. pollute.
   c. make new habitats.
2. Organisms that are very few in number are
   a. endangered.
   b. dangerous.
   c. extinct.
3. The dodo bird is an animal that is
   a. dangerous.
   b. endangered.
   c. extinct.
4. People can help keep plants and animals from dying out by saving their
   a. fibers.
   b. furs.
   c. habitats.
5. Areas where people cannot build houses, hunt animals, or collect plants are
   a. parks.
   b. habitats.
   c. forests.
6. People can help protect the habitats of organisms by
   a. making parks.
   b. closing zoos.
   c. spreading pollution.
7. Bread, macaroni, and cereal are made from
   a. grains.
   b. vegetables.
   c. leaves.
8. People make cheese, butter, yogurt, and ice cream from
   a. grains.
   b. eggs.
   c. milk.
9. The strong, thin threads of cotton and flax plants are called
   a. fibers.
   b. silks.
   c. linen.
10. Buttercups and mistletoe can be dangerous to people when these plants are
   a. touched.
   b. worn.
   c. eaten.

33

♦ ***Suitable as a language development activity***

2. When a fuel is burned, it often releases harmful
(a) plants. (b) chemicals. (c) fibers. (d) light.
3. To protect the habitats of organisms, people have
(a) made parks. (b) stopped making chemicals.
(c) closed zoos. (d) opened museums.
4. Which of the following animals are extinct?
(a) tigers (b) lions (c) dodo birds (d) goldfish
5. Scientists protect endangered plants by saving their
(a) leaves. (b) names. (c) pictures. (d) seeds.
6. Which of these animals is endangered?
(a) tiger (b) bee (c) dog (d) dodo bird
7. Grains are
(a) roots. (b) seeds. (c) stems. (d) leaves.
8. Wool is made from
(a) cotton plants. (b) sheep's hair. (c) flax plants.
(d) animal skins.
9. When a black widow spider bites, it releases
(a) an odor. (b) a poison. (c) water. (d) food.

Interpreting What You Learned

Write a short answer for each question.

1. Why might an animal not be able to live in its habitat if the habitat is changed?
2. How did tigers become endangered organisms?
3. What is the difference between endangered organisms and extinct organisms?
4. How are plants and animals important to people?

Extending Your Thinking

Write a paragraph to answer each question.

1. How might an endangered organism become an extinct organism?
2. How do people help themselves when they protect plants and animals?

To explore scientific methods, see Experiment Skills on pages 354–355.

2. Tigers became endangered when people killed many tigers and cut down forests where they lived. **Thinking Skill:** *Recognizing cause and effect*
3. Endangered organisms are very few in number, but they still are found on the earth. Extinct organisms are no longer found on the earth. **Thinking Skill:** *Comprehending meaning*
4. People use plants and animals for food and to make many kinds of useful things. **Thinking Skill:** *Restating or explaining ideas*

Extending Your Thinking

1. Endangered organisms are very few in number. Human activities, such as hunting and building, and natural occurrences, such as bad weather or disease might kill the remaining organisms. Then the species organisms would be extinct. **Thinking Skill:** *Drawing conclusions*
2. People use plants and animals in many ways. By making sure that these organisms do not become endangered or extinct, people are making sure that they will be able to use them in the future. **Thinking Skill:** *Recognizing cause and effect*

Test Book page 34 *

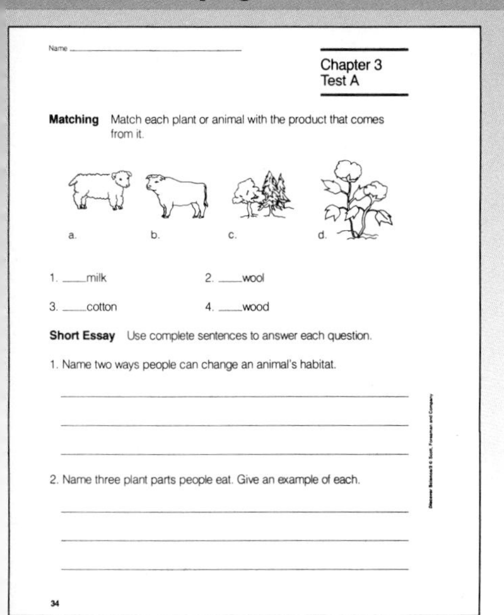

Name ____

Chapter 3
Test A

Matching Match each plant or animal with the product that comes from it.

a. b. c. d.

1. ___milk 2. ___wool
3. ___cotton 4. ___wood

Short Essay Use complete sentences to answer each question.

1. Name two ways people can change an animal's habitat.

2. Name three plant parts people eat. Give an example of each.

34

Test Book page 35 *

Name ____

Chapter 4
Test B

Multiple Choice Choose the best answer.

1. An organism gets its food, shelter, and water from
a. other organisms.
b. pollution.
c. its habitat.
2. Anything harmful added to the air, water, or land is
a. pollution.
b. a fiber.
c. a chemical.
3. Organisms that are no longer found on earth are
a. endangered.
b. extinct.
c. dangerous.
4. People help protect plants, animals, and their habitats by making
a. forests.
b. parks.
c. swamps.
5. For protection, some endangered animals are placed in
a. zoos.
b. forests.
c. kennels.
6. Grains, such as wheat, oats, and corn are
a. leaves.
b. stems.
c. seeds.
7. Rattlesnakes and poison ivy are
a. endangered organisms.
b. dangerous to people.
c. extinct.
8. Strong thin threads from plants that can be used to make cloth are
a. fibers.
b. wood.
c. wool.
9. Leather used for making some shoes comes from
a. chemicals.
b. plant parts.
c. animal skins.
10. Scientists protect endangered plants by saving their
a. seeds.
b. fibers.
c. leaves.

35

Test Book page 36 *

Name ____

Chapter 4
Test B

Matching Match each food in the pictures with the plant part that people eat.

a. b. c. d.

1. ___stem 2. ___leaves
3. ___root 4. ___seeds

Short Essay Use complete sentences to answer each question.

1. Name three ways people can protect organisms and their habitats.

2. Name one endangered animal and one extinct animal.

36

* *Answers to masters on pages 66E–66H*

TEACHING PLAN

Purpose

Several careers that involve working with living things are presented in this feature. Students might not be familiar with all the careers presented. Veterinary medicine, nursery work, and farming are other occupations related to life science.

Teaching Tips

- Ask students how they and their families might benefit from the work of the people in these occupations.
- Ask students who fish to ***describe*** the different kinds of fish they have caught.
- If possible, have students use microscopes to examine prepared slides of various bacteria.
- Ask students to discuss any vegetable or flower gardening they do at home. Ask them to tell what they like most about working in the garden.

Careers

You can probably think of many different kinds of living things. Just imagine how many different jobs there must be for people who like living things.

Fish-culture technician

Foresters like to work with trees. They help protect forests from fire, harmful insects, and disease. They might make maps of forest areas. Foresters spend a great deal of time outside. To become a forester, you need to go to college.

Like foresters, **gardeners** work outside a great deal. They plant flower or vegetable gardens. Then they take care of the plants by watering them and pulling weeds. Gardeners usually learn their skills by working with other gardeners.

A person who likes fish might enjoy becoming a **fish-culture technician.** This person raises young fish and then lets them go into lakes or rivers. Fish-culture technicians can learn their skills on the job, or they can go to college.

Microbiologists study the tiny living things that can be seen only under a microscope. Some microbiologists study bacteria that cause disease. To become a microbiologist, you must graduate from college. Then you need to take special classes.

Some people are interested in the food we get from animals. **Meat inspectors** check meat and poultry to make sure it is safe for people to eat. They also check the place where food is handled. They make sure the place and the equipment are clean. People who want to be food inspectors must pass a special test.

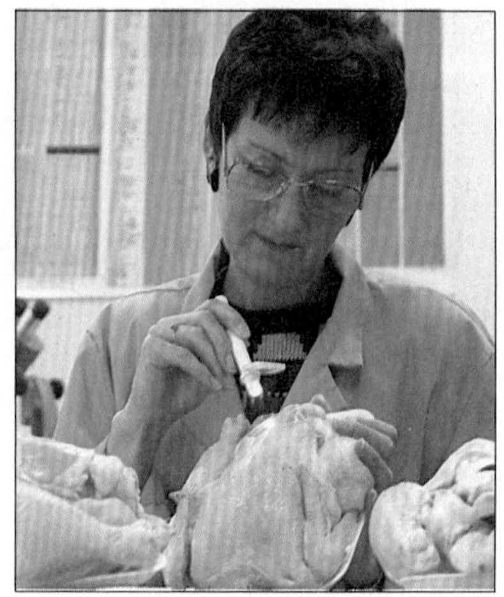
Meat inspector

Teaching Options

Enrichment

Invite a member of a local garden club to talk with students about caring for plants. You might see if someone working in the fisheries division of your state's conservation department could visit the class. Have students look on meat at home to see if it is marked "inspected."

Where To Write

Forester: American Forestry Association, 1319 Eighteenth Street NW, Washington DC 20036.
Gardener: American Society for Horticultural Science, 701 N. Saint Asaph Street, Alexandria, VA 22314.
Fish-culture Technician: American Fisheries Society, 5410 Grosvenor Lane, Bethesda, MD 20814.
Microbiologist: American Society for Microbiology, 1913 I Street NW, Washington, DC 20006.
Meat Inspector: United States Department of Agriculture, Administration Building, Fourteenth and Independence Avenue SW, Washington, DC 20250.

An Aquarium Filter

Air, water, and gravel are three things that pet fish need to stay healthy. In a good aquarium, water always moves. It travels in a circle from the top of the tank to the bottom and back up again. At the bottom of the tank, the water passes through a gravel filter.

The filter helps keep the fish healthy. It collects leftover food and wastes from the tank. The wastes and old food could make the fish sick if they stayed in the water. Here is how an aquarium filter works.

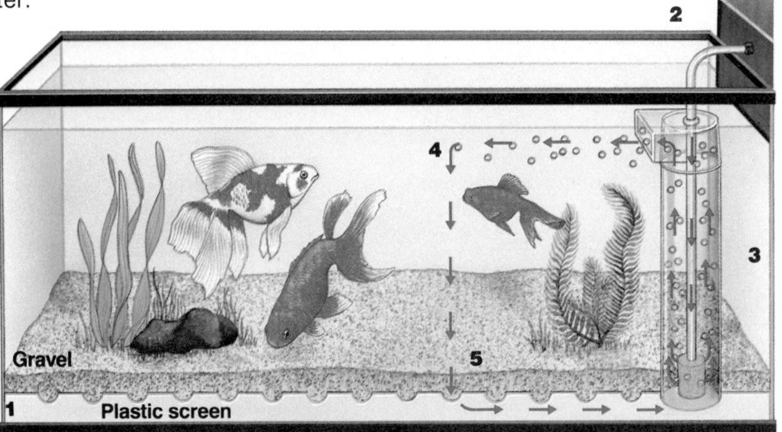

1 A plastic screen covers the bottom of the aquarium. The plate contains many small holes. A layer of gravel rests on top of the screen.

2 An air pump pushes air into the tank. The air goes down a small tube to the bottom of the tank.

3 Then, the air goes up the large tube that is around the small tube. The flow of air up the tube pulls water along with it. Here, oxygen in the air mixes with the water. Fish take in the oxygen that is mixed with the water.

4 When water is pulled up the tube, it makes water underneath the filter move toward the tubes. This movement makes the water in the tank move down through the filter. The water carries wastes and old food along with it.

5 Wastes and food are trapped in the gravel filter. Bacteria on the gravel eat the wastes and food. The water becomes clean as it moves through the filter.

How It Works

Teaching Tips

- If possible, allow students to make an aquarium as a class project. The owner of a local pet or fish store might be willing to talk with the class about the steps in setting up and operating a good aquarium.
- Review with students the way in which fish obtain oxygen from water. Use this review as the basis for explaining why fish bowls are not recommended for use as aquariums. Compare the surface area where water and air are in contact in a fish bowl with that of an aquarium.
- Help students ***compare*** the advantages and disadvantages of using various additions, such as live plants vs. plastic plants and metal sculptures vs. stone sculptures, to an aquarium.

Science Background

Devices for moving water (a pump) and for removing wastes (a filter) are essential in any properly functioning aquarium. Without movement, the upper layers of water tend to become too warm and the lower layers too cool. The decay of fish wastes and excess food tends to produce by-products that are harmful or toxic to fish. Live plants are a useful addition to any aquarium. They produce oxygen for fish and use waste materials trapped in the gravel filter as fertilizer. The mutual dependence of fish and plants in an aquarium is a good example of the kind of equilibrium that occurs in a natural setting.

Complete the Sentence

1. pollution
2. gills
3. fibers
4. oxygen
5. larva
6. cell
7. germinate
8. prey
9. reptiles
10. habitat
11. pupa
12. petals

Short Answer

1. Both carry water and other materials to the leaves of the plants, and both support the plants' weight. *(Identifying similarities)*

2. by the bodies of insects or other animals, or by the wind *(Summarizing)*

3. because they need warm surroundings to keep their body temperatures high *(Cause and effect)*

4. egg, larva, pupa, adult *(Sequencing)*

5. protists and fungi *(Identifying similarities)*

6. A larger habitat will usually support more lions than a smaller habitat. *(Generalizations)*

7. Animals cannot produce their own food within their body cells; they must consume other organisms. *(Explaining ideas)*

8. Wide-set eyes give rabbits a greater range of vision, which helps rabbits spot predators more quickly. *(Formulating hypotheses)*

9. Too many members of a species may be killed, or their habitats may be destroyed. *(Summarizing)*

10. for food, building materials, cloth, paper, etc. *(Summarizing)*

Essay

1. Both are vertebrates, have scales, and are cold-blooded. Their young hatch from eggs and can take care of themselves right away. They are different because fish breathe with gills while reptiles breathe with lungs, and fish live in the water all the time while reptiles can live out of the water. *(Identifying similarities and differences)*

2. The important parts of a plant's habitat are the soil it grows in, the gases in the air or water, the sun's rays, and other things that help it grow and reproduce. Other important parts of the habitat include the animals that might eat the plant or help pollinate the flowers. *(Identifying variables)*

Unit 1 Review

Complete the Sentence

Fill in the blank with the correct word or words from the list.

cell	oxygen
fibers	petals
germinate	pollution
gills	prey
habitat	pupa
larva	reptiles

1. ___ takes place when harmful things are added to the air, water, or land.
2. A fish uses its ___ to take in oxygen from the water.
3. The ___ of plants can be used to make cloth.
4. Plants give off the gas ___ into the air.
5. A caterpillar is the ___ of a butterfly.
6. The ___ is the basic unit of an organism.
7. A seed will ___ only when it gets enough air and water.
8. Mice, birds, and other small animals often are the ___ of snakes.
9. Snakes, turtles, and alligators are ___.
10. An organism finds everything it needs to live in its ___.
11. An insect is a ___ when it forms a hard shell around itself.
12. The ___ of a plant often are colored.

Short Answer

Write a short answer for each question or statement.

1. How are trunks of trees like stems of small plants?
2. Name two ways that pollen can be carried to the center part of a flower.
3. Why do most cold-blooded animals live in warm places?
4. Tell the order of these stages in an insect's life cycle: pupa, egg, adult, larva.
5. Name two groups of organisms that can have one cell or many cells.
6. How might a population of lions be affected by the size of their habitat?
7. Explain why all animals are consumers.
8. How do you think that eyes set toward the sides of their heads help rabbits survive?
9. What causes plants and animals to become endangered?
10. Name four ways that people use plants.

Essay

Write a paragraph for each question or statement.

1. How are reptiles and fish alike? How are they different?
2. Describe the important parts of a plant's habitat.

88

Test Book page 41 *

Name ____________

Unit Test 1

Multiple Choice Choose the best answer.

1. Pollination takes place when pollen reaches
   a. part of an animal's body.
   b. the tips of flower petals.
   c. the center of a flower.
2. The outside covering of a seed is the
   a. seed leaf.
   b. seed coat.
   c. seedling.
3. Lizards, turtles, and alligators are
   a. mammals.
   b. reptiles.
   c. fish.
4. A snail's hard shell helps the snail by
   a. keeping it warm.
   b. protecting its soft body.
   c. storing food and water.
5. An example of a mammal is a
   a. human.
   b. turtle.
   c. lizard.

Use the picture to answer questions 6 and 7.

6. Food needed for plant growth is made in which part?
   a. 2 b. 3 c. 4
7. Which plant part helps give support above the ground?
   a. 1 b. 3 c. 4
8. The materials used by a plant that are taken in by its roots are called
   a. minerals.
   b. weeds.
   c. sugar.
9. A young frog is called a
   a. reptile.
   b. tadpole.
   c. larva.

1. ⓐⓑⓒ 2. ⓐⓑⓒ 3. ⓐⓑⓒ 4. ⓐⓑⓒ 5. ⓐⓑⓒ 6. ⓐⓑⓒ 7. ⓐⓑⓒ 8. ⓐⓑⓒ 9. ⓐⓑⓒ

41

Test Book page 42 *

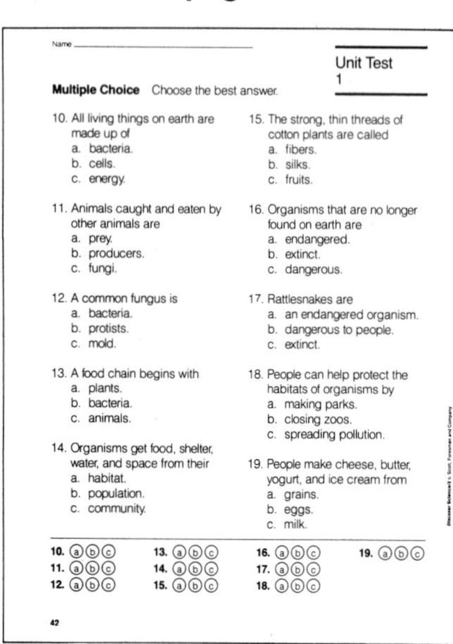

Name ____________

Unit Test 1

Multiple Choice Choose the best answer.

10. All living things on earth are made up of
    a. bacteria.
    b. cells.
    c. energy.
11. Animals caught and eaten by other animals are
    a. prey.
    b. producers.
    c. fungi.
12. A common fungus is
    a. bacteria.
    b. protists.
    c. mold.
13. A food chain begins with
    a. plants.
    b. bacteria.
    c. animals.
14. Organisms get food, shelter, water, and space from their
    a. habitat.
    b. population.
    c. community.
15. The strong, thin threads of cotton plants are called
    a. fibers.
    b. silks.
    c. fruits.
16. Organisms that are no longer found on earth are
    a. endangered.
    b. extinct.
    c. dangerous.
17. Rattlesnakes are
    a. an endangered organism.
    b. dangerous to people.
    c. extinct.
18. People can help protect the habitats of organisms by
    a. making parks.
    b. closing zoos.
    c. spreading pollution.
19. People make cheese, butter, yogurt, and ice cream from
    a. grains.
    b. eggs.
    c. milk.

10. ⓐⓑⓒ 11. ⓐⓑⓒ 12. ⓐⓑⓒ 13. ⓐⓑⓒ 14. ⓐⓑⓒ 15. ⓐⓑⓒ 16. ⓐⓑⓒ 17. ⓐⓑⓒ 18. ⓐⓑⓒ 19. ⓐⓑⓒ

42

Unit Projects and Books

Science Projects

1. Ask permission to cut off a small piece of a house plant, such as ivy, that has trailing vines. Use scissors to cut a piece that is about 12 centimeters long. Be careful not to cut yourself with the scissors. Put the cut end of the plant in a jar of water. Observe the cut end of the plant for several weeks. Draw a picture of the plant and its new roots.
2. Obtain several mealworms from a pet store. Put the mealworms in a container with a small amount of oatmeal. Cover the container with a lid that has several tiny holes. Feed the mealworms every few days with oatmeal. Observe the mealworms for several weeks. Describe the life cycle of a mealworm. Draw pictures to illustrate the life cycle.
3. A terrarium can be thought of as a small habitat for plants. Use library books to help you make a terrarium in a clear plastic container with a lid. Watch your project for several weeks. How is your terrarium like the real world?
4. Look through magazines and cut out pictures of plants and animals. Sort these pictures into two groups. One group should be helpful organisms; the other group should be harmful organisms. Using the pictures, make a poster of helpful and harmful organisms.

Books About Science

Plants and Flowers by Brian Holley. Penworthy, 1986. Learn about plants and their flowers. Gr. 3-7

The Puffins Are Back! by Gail Gibbons. HarperCollins, 1991. Find out how scientists helped this Atlantic seabird.

Science and Society

Protecting Habitats Some people in the community want to build a new baseball field. But other people object. The new field would be built in Henson's meadow. This meadow is the only place the Mariposa blue butterfly still lives. If the ball field is built, the butterfly might become extinct. Many members of the community do not see the problem. They wonder what difference it makes if one kind of butterfly becomes extinct. They know that playing baseball helps keep boys and girls out of trouble. They think baseball is more important than a butterfly. What are some good reasons to build the new ball field? What are some good reasons not to?

89

TEACHING PLAN

Science Projects

1. Other plants that will root in water are philodendron, coleus, Swedish ivy, and wandering Jew. If such plants are available in the classroom and some students do not have plants at home, you may wish to give them a cutting so they can carry out this project.
2. Provide a place in the classroom where students can keep their containers of mealworms.
3. Have a terrarium set up in the classroom so students will have a model to follow when making their own terrarium.
4. Have magazines available for students to use.

Science and Society

Encourage students to explore both sides of the issue in class or on their own. In class, you may wish to set up a debate on the issue. On their own, students might do further research and choose one side of the issue to support. Throughout this exercise, emphasize the importance of respecting the opinions of others.

Thinking Skill: *Judging and evaluating*

Books in Spanish

Boyer, Marie-France. *El libro de las flores.* Altea, 1986. Discusses the growth and life cycle of flowers. (grades 4–6)
Fernandez, Flora. *Una mirada a los arboles.* Fernandez, 1984. Describes the structure of trees. (grades K–3)
Sabugo Pintor, Angel. *El libro del medio ambiente.* Bilingual Publications, 1984. Discusses different types of ecosystems. (grades 3–6)
Selsam, Millicent E. *Como crecen los perritos.* Scholastic Books, 1986. Describes how puppies grow. (grades K–3)

Teaching Options

Science Background

Science and Society The greatest threat humans pose to wildlife is the development of wild habitats. While the issue of clearing a field for baseball versus saving one species of butterfly may not seem to be of great importance, the results can effect the surrounding ecosystem. A variety of organisms is needed to maintain different ecosystems. Different species of organisms are dependent on one another for their survival. If any organism becomes extinct, it could affect an entire ecosystem. Other people point out that growing human populations need more space and resources, and these human needs must be met.

*Answers to masters on pages 66E–66H

2 UNIT 2/Physical Science

Planning Guide

Unit Components			Pages
Unit Overview/*SCIENCE IN THE NEWS*			90–91
Chapter 5 Properties of Matter	**Lesson 1** What Is Matter? **Lesson 2** What Makes Up Matter?	**Lesson 3** How Can Matter Change?	92–113
Chapter 6 Work and Machines	**Lesson 1** What Is Work? **Lesson 2** What Are Simple Machines?	**Lesson 3** What Are Compound Machines?	114–135
Chapter 7 Forms of Energy	**Lesson 1** What Are Some Kinds of Energy? **Lesson 2** What Is Electricity?	**Lesson 3** How Is Matter Heated? **Lesson 4** How Does Light Travel?	136–161
Chapter 8 Sound	**Lesson 1** What Is Sound? **Lesson 2** How Does Sound Travel?	**Lesson 3** How Do People Make Sound?	162–179
Careers	chemist, electrician, carpenter, musician		180
How It Works	Violin		181
Unit Review			182
Unit Projects and Books	Science Projects Science and Society Books About Science		183

Science Process Skills Book

The Science Process Skills Book contains worksheets that can be used to teach and then assess student mastery of the basic science process skills. In addition, other worksheets in this book teach students the manipulative skills they will need to use basic science equipment. You might use some or all of these worksheets
- at the beginning of the year to teach and assess skills.
- directly before students encounter a particular skill in an activity.
- as a remedial tool when students begin to have trouble with specific skills.

cience Resources for the Unit

Resource Books

dley, Neil. *Making Things Move.* Watts, 1984. Introduces basic physical principles by way of demonstrations.

rger, Melvin. *Atoms, Molecules, and Quarks.* Putnam, 1986. Studies the tiny particles that make up the basic units of matter.

ttlekamp, Larry. *Magic of Sound.* Morrow, 1982. Simple experiments probe aspects of sound.

cCormack, Alan J. *Inventors Workshop.* Pitman Learning, 1982. Creativity and imagination are the keys for over 25 projects and activities.

cci, Salvatore. *Chemistry Around You: Experiments and Projects with Everyday Products.* Arco, 1985.

Community Resources

ke the class on a tour of the local e department to learn about the nple machines used there dders, pulleys, wheels).
Take the class to visit a local rdware store or marine shop to ew examples of simple and mpound machines and how they ork.
Visit a local telephone company learn about how electricity and agnetism are used in the lephone. The power company ight provide a person to give a fety talk.

Audio-Visual Resources

Electricity and How It's Made. Britannica. Film, 16 minutes. Illustrates how electricity is produced, what it does, and how it is used.

Electricity—How to Make A Circuit (Using the Dry Cell). Britannica. Film, 11 minutes. Introduces basic concepts about electric current.

Energy . . . Ability to Do Work. AIMS. Film or video, 9 minutes. A basic approach to understanding energy.

Energy from the Sun. National Geographic. Wonders of Learning Kit. Shows how the sun affects plants and weather.

Heat and How We Use It. Britannica. Film, 11 minutes. Explains the basic principles of heat, its sources, the ways in which it travels, and its many uses.

Learning About Liquids, Solids, and Gases. AIMS. Film or video, 11 minutes. Children explore the properties of matter.

Learning About Solar Energy. AIMS. Film or video, 12 minutes. Children find ways to heat water for an outdoor shower and to cook a hamburger in a miniature solar furnace.

Learning About Sounds. Britannica. Sound filmstrip series, average 48 frames/8 minutes each. Encourages students to observe, describe, and classify familiar and unfamilar sounds.

Let's Talk About Machines. RMI. Film, 9 minutes. Introduces simple machines: lever, gear, wheel, axle, pulley, and inclined plane.

Movement Everywhere. Britannica. Film, 11 minutes. Shows that when one thing moves another, work is being done.

Things Change: Solids, Liquids, Gases. Britannica. Film, 10 minutes. Helps young students identify solids, liquids, and gases and observe how matter can change from one state to another.

Water and What It Does. Britannica. Film, 11 minutes. Shows that water is composed of tiny, constantly moving particles.

Wondering About Air. Britannica. Film, 13 minutes. Through misadventures with a balloon, a kite, and a beach ball, three clowns discover that air can only be seen by observing its effects.

Computer Software

Exploring Science II. Queue. Apple II Series. 48K or IBM PC, 64K. Tutorial and drill on various science topics.

Mystery Objects. MECC. Apple, 128K. Enables students to identify objects by performing tests to define physical properties of the objects.

Sharing Science. Micrograms. Commodore PET, 16K. Provides a series of true/false statements in seven categories.

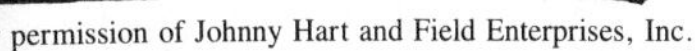

TEACHING PLAN

Unit Overview

This unit explores the students' physical world. Students will learn about the states and properties of matter, about work and machines, about forms of energy, and about sound and its transmission.

About the Photograph

In compound machines, one wheel is often used to turn another wheel. One way to have wheels turn one another is to fit them together as gears. The teeth on one gear fit into the notches between the teeth in another gear. Each gear is part of a wheel-and-axle machine. Gears are used to change the direction of a force and to change the distance and speed. When the small gear in the picture turns clockwise, the two larger gears will turn counterclockwise. The small gear will also go around several times for each time the larger gears go around. Gears are used in such things as watches, bicycles, and automobiles.

90

Teaching Options

Chapter 5 Preview

Properties of Matter

In this chapter students will explore the properties of matter and be able to describe three of its states—solid, liquid, and gas. They will also examine how atoms join to form different kinds of matter. Through observation, students will be able to identify and describe how matter can change physically and chemically.

Chapter 6 Preview

Work and Machines

Beginning with basic definitions of energy, work, and force, this chapter takes the student through the relationship of force and energy to friction, simple and compound machines, and gravity. Specific differences, such as fixed pulleys versus movable pulleys and simple machines versus compound machines, are also explored.

Unit 2

Physical Science

This picture shows some parts of a machine. Notice the jagged edges on the wheels. These edges help the wheels turn around to make the machine work.

Machines use energy to help make work easier for people. In this unit, you will learn about different kinds of energy, such as electricity. You also will discover what makes up all the objects around you.

SCIENCE IN THE NEWS During the next few weeks, look in newspapers or magazines for stories about electricity. Also look for pictures of machines that use electricity. Share the news stories and pictures with your class.

Teaching Tips

- Before starting the unit, lead a discussion about machines. Ask students to *name* examples of machines. List the examples of machines on the chalkboard.
- Direct students' attention to the photograph on page 90. Questions: **What is shown in the picture?** (gears) **What do you think the gears might be part of?** (Students might say a watch or clock. Accept any reasonable answers.) **How do the gears move?** (in circles)
- Explain to students how the gears move in relation to each other.

Chapter 7 Preview

Forms of Energy

This chapter focuses on light, electrical, mechanical, and sound energy. The concepts of electric charges, electric current, magnetism, and the visible spectrum are introduced and reinforced through concrete examples.

Chapter 8 Preview

Sound

This chapter gives students the opportunity to learn what causes sound, how sounds can vary, and how the body receives and interprets sound. Students will also learn about how sounds travel through different kinds of matter.

 Properties of Matter Planning Guide

TEACHING PLAN

Chapter Components	Skills	Materials
Chapter Opener/*DISCOVER*: Observing Different Materials pp. 92–93	*DISCOVER* p. 93 Science Process Skills *Observing, Inferring*	*DISCOVER* p. 93 (groups of 2) 30 plastic glasses, 1 box sugar, 1 box salt, 15 plastic spoons, water
Lesson 1 What Is Matter? pp. 94–98	Thinking Skills Challenge!: *Inferring* Find Out On Your Own: *Classifying, Organizing information*	Demonstration p. 94 objects of different size, color, smell, shape, and texture
Activity Observing that Air Has Volume p. 99	Science Process Skills *Observing, Collecting and interpreting data, Inferring*	(groups of 2) 15 clear-plastic cups, 1 roll paper towels, 15 large bowls
Lesson 2 What Makes Up Matter? pp. 100–102	Thinking Skills Challenge!: *Inferring* Find Out On Your Own: *Inferring*	Demonstration p. 100 clay, straws
Science and Technology Exploring the Inside of Atoms p. 103	Thinking Skills *Restating or explaining ideas, Inferring*	
Lesson 3 How Can Matter Change? pp. 104–108	Thinking Skills Challenge!: *Drawing conclusions* Find Out On Your Own: *Comprehending meaning*	Demonstration p. 104 spoon, bowl, ice cubes
Activity Observing Physical and Chemical Changes p. 109	Science Process Skills *Observing, Inferring*	(groups of 2) 30 plastic glasses, 1 box sugar, 1 box baking soda, 1 bottle vinegar, 30 plastic spoons
Skills for Solving Problems Using Equal Arm Balances and Pictographs pp. 110–111	Problem Solving Skills *Making decisions/Identifying and solving problems, Interpreting charts, maps, and graphs*	
Chapter Review pp. 112–113	Thinking Skills *Comparing, Contrasting, Restating or explaining ideas, Inferring, Sequencing*	

Teaching Options

Strategies	Extensions		Resource Masters
Cooperative Learning p. 92 (Also see p. T23.) Applying Whole Language p. 93 (Also see p. T30.)			Family Letter: *Resource Book* p. 51
Reading Strategies p. 94 (Also see pp. T26–T29.)	Special Education p. 95 Reinforcement pp. 95, 96, 97 Science and Art p. 96	Enrichment p. 97 Reteaching Suggestion p. 98 Game Suggestion p. 98	Vocabulary Preview: *Workbook* p. 25 Science and Social Studies: *Workbook* p. 26
			Activity Worksheet: *Resource Book* p. 55
Reading Strategies p. 100 (Also see pp. T26–T29.)	Enrichment p. 101 Special Education p. 101	Reinforcement p. 101 Reteaching Suggestion p. 102	Science Activity: *Workbook* p. 27
Reading Strategies p. 104 (Also see pp. T26–T29.)	Special Education p. 105 Reinforcement pp. 105, 107 Enrichment pp. 106, 107	Game Suggestion p. 106 Reteaching Suggestion p. 108	Science Skills: *Workbook* p. 28 Vocabulary Puzzle: *Workbook* p. 29
			Activity Worksheet: *Resource Book* p. 57
	Reteaching Suggestion p. 111		Science and Reading: *Resource Book* p. 59
Cooperative Learning p. 112 (Also see p. T23.)			Chapter Tests: Forms A and B *Test Book* pp. 45–58

Classroom Management

Advance Preparation

DISCOVER, page 93
Gather 2 small, plastic glasses, 1 plastic spoon, 1 hand lens, sugar, and salt for each group.

Activity, page 99
For each group provide a jar or a clean plastic glass, a large plastic bowl, and paper towels.

Demonstration, page 100
Bring colored clay and several drinking straws to class.

Demonstration, page 104
You will need a bowl of ice cubes for this demonstration.

Activity, page 109
Provide sugar, vinegar, baking soda, 2 small jars, and 2 plastic spoons for each group

Vocabulary Review

Use the following sentences with your students to review the meanings of the italicized words.

1. A *balance* is an instrument for measuring the mass of objects.
2. Ice is a *solid*, but water is a *liquid.*
3. Air bubbles in soda pop are a *gas.*
4. The wind blew small *particles* of dust in my eyes.
5. Ice is an example of matter in a solid *state.*
6. Liquid water becomes solid when its temperature drops below 0°C, the *freezing point* of water.

High-Potential Students

Ask students to find a copy of the story "The Golden Touch of King Midas" in the school or local library. Instruct students to make a list as they read of the objects King Midas touched that turned to gold. Have students use their lists to make a chart showing the properties of each object before it turned to gold. Each chart should include a column for the color, shape, size, and state of matter of each object. Allow students to share their charts with the class to initiate discussion about properties of objects.

Mainstreamed Students

Visually Impaired
To reinforce the concept that gas can spread into a large area, allow students to feel and hear the air escaping from a balloon.

Hearing Impaired
Allow hearing-impaired students to work with a partner when doing the activities in this chapter. The hearing student can aid the hearing-impaired student in making observations.

Science Fair Projects

The Experiment Skills on p. 356 and the Investigate feature in the chapter can be used for science fair projects. You also might encourage interested students to do one of the following projects:

1. To observe how various metals change chemically, collect different types of metal. Attach each piece of metal on a white, painted board. Use a plant sprayer to spray water on the metal pieces several times a day. Observe and record the chemical changes that occur. Such changes might include rust, color, brittleness, and pitting. Note how long it takes for each piece of metal to begin to change.
2. Find out how crystals form, using an encyclopedia or other reference sources for information. Compare the crystal structures of various solids. Make a drawing of the crystal patterns of each solid studied.

Bulletin Board

Ask students to give examples of how solids, liquids, and gases are important in their everyday lives.

FORMS OF MATTER

SOLID | LIQUID | GAS

MILK

Chapter 5 Poster

Science Discovery Center

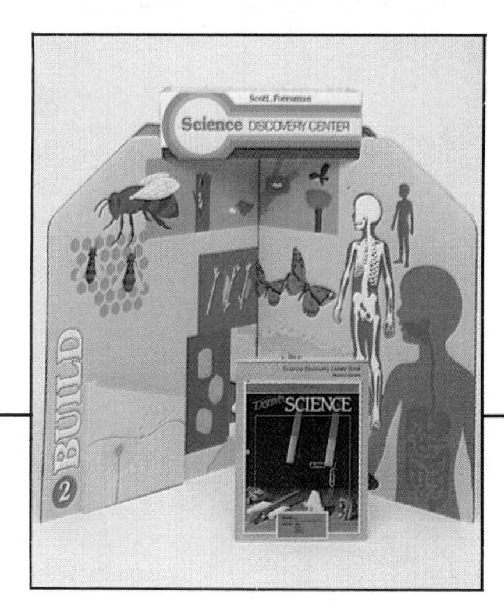

Use pages 55–60 from the *Science Discovery Center Book*. Place these worksheets in the appropriate pockets in the Science Discovery Center.

Overhead Transparencies

Use Transparencies 8 and 9 from the package of color overhead transparencies.

Teacher's Resource Book

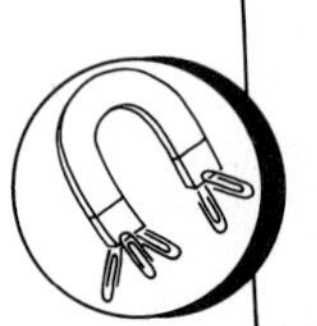

Dear Family,
Your student will be reading **Chapter 5: Properties of Matter** in *Discover Science*, published by Scott, Foresman. We will learn about the three states of matter—solid, liquid, and gas. Matter can change from one state to another, and has different properties in each state.
You and your student can do this activity together to learn more about the properties of matter.

Experimenting with Cornstarch

You will need a tablespoon measure; cornstarch; cup; water; dish; and metal, wood, and plastic objects (such as coins, toothpicks, buttons).

1. Put 1 tablespoon of cornstarch into a cup.
2. Add a few drops of water. Stir until a thick paste is formed. (If too much water is added, the paste will be too liquid, and you will have to start over.)
3. Pour the paste into a dish. Ask, "What happens?" (It should flow slowly. If it takes more than 5 minutes, add a few drops of water, stir, and try again.)
4. Put the metal, wood, or plastic objects on the paste. Ask, "What happens?" (They will float, as on a liquid.)
5. Remove a piece of the paste. Try to bend it. Try to break it. Ask, "What happens?" (Substance should both bend and break, like a solid. If it does not break, try again, using a quick bend or pull.
6. Ask, "Is the paste a solid or a liquid?" (It is neither, but has the properties of both solids and liquids at the same time.)

Workbook

Name ______

Use with Lesson 1: pages 94-98

Chapter 5

Vocabulary Preview

Properties of Matter

Chapter Vocabulary			
atom	evaporate	physical change	volume
chemical change	mass	property	water vapor
condense	matter	states of matter	

Vocabulary Cards

1. Write each word on a card.
2. Find each word in the glossary. Copy the pronunciation under the word on the card.
3. Practice saying the words with a partner.

Word Meaning

1. Which pair of words have the word *change* in them?

   a. **chemical change** b. **physical change**

2. Which two words have the word *vapor* in them?

   a. **water vapor** b. **evaporate**

3. Find the following words in sentences in Chapter 5 of your book. Copy one sentence for each word on a separate sheet of paper.
   atom condense mass matter volume
4. Now write your own definitions of these words on that same sheet of paper. Check your definitions in the glossary. Write the corrected definition of each word on the back of its vocabulary card.

At Home

1. Practice saying the words. Learn their meanings.
2. Look for the words and pictures of the words in newspapers and magazines. You can make a poster.

Teacher's Notes: Ask students if they can find any other words hidden in the term *states of matter*. Examples: mat, ate, test.

Workbook

Name ______

Use with Lesson 1: pages 94-98

Chapter 5

Science and Social Studies

Making Glass

Read the story. Then answer the questions.

Before people knew how to make glass, they found glass made by nature. When lightning strikes and melts sand, thin tubes of glass are formed. The heat from volcanoes turns sand into chunks of glass. People used this glass for jewelry, knives, and money.

About 5000 years ago, people began to make glass. The glass they made is called glaze. They used glaze to cover clay pots. People today still use glaze on clay.

People made the first all-glass pieces about 3500 years ago. Several solid materials are used to make the glass. Two main materials are sand and soda ash. These solids are ground up into very small pieces. Then they are heated over very hot fires. The sand and soda ash melt together. They form a liquid. The liquid is then poured into a shape. When the liquid cools, it becomes glass.

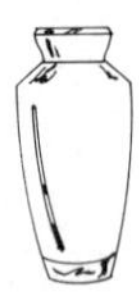

1. Before glass is made, in what state of matter is the sand and soda ash?
   Solid
2. How are the sand and soda ash changed into another state of matter?
   By heating them
3. During glass making, what state of matter do the sand and soda ash become?
   Liquid
4. Is the change that takes place when sand and soda ash are made into glass a chemical change or a physical change?
   Chemical change (Although the initial melting is a physical change)
5. List some items that you use that are made of glass.
   Answers may include windows, lightbulbs, mirrors

Teacher's Notes: Soda ash comes from a mineral trona or salt. Volcanic glass is called obsidian.

Teacher's Resource Book

Name ______

Use with Lesson 1: page 99

Observing That Air Has Volume

Activity Worksheet

Record Your Results

Towel after glass is pushed straight into water	
Towel after glass is turned over under water	

State Your Conclusion

1. What kept the water from reaching the paper towel in the glass when you pushed the glass straight into the water?
2. What happened to the paper towel when you turned the glass over under water. Explain.

Use What You Learned

How could people use the results of this experiment to help them work under water?

Chapter 5

Name ______________________

Use with Lesson 2: pages 100-102

Science Activity

How Do Gas Molecules Move?

Gather These Materials
- a bottle of perfume

Follow This Procedure
1. Work with the other students in your class. As everyone sits in their seats, have one person on one side of the room open the bottle of perfume.
2. Who do you think will smell the perfume first? Raise your hand when you can smell the perfume.

Record Your Results

Imagine that the box on the right is your classroom. Draw the perfume bottle on one side of the room. Use dots to stand for molecules of gas. Show how the gas molecules had moved when half the people in class could smell the perfume.	

State Your Conclusions
1. Where were those who first smelled the perfume sitting?
   They were in the seats closest to the perfume.
2. Why did some people smell the perfume before others?
   It took time for the molecules of perfume to move across the room.

Teacher's Notes: Put an open bottle of perfume in a box. In 15 minutes open the box. "Why is the smell in the box so strong?"

Chapter 5

Name ______________________

Use with Lesson 3: pages 104-108

Science Skills

Looking at Changes

Circle one word under each box to show whether a chemical or a physical change is taking place.

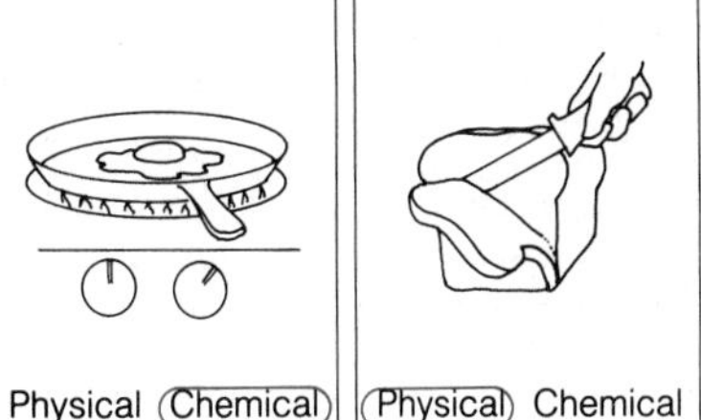

Egg cooking: Physical (Chemical) — Chemical circled

Slicing bread: (Physical) Chemical — Physical circled

Candle burning: Physical (Chemical) — Chemical circled

Car rusting: Physical (Chemical) — Chemical circled

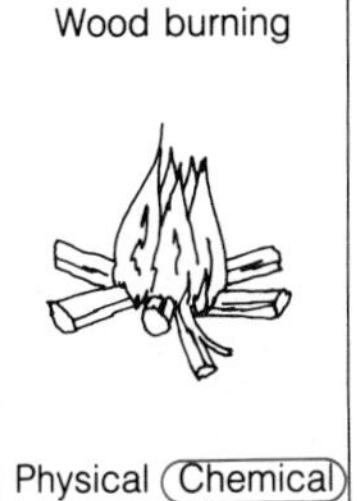

Wood burning: Physical (Chemical) — Chemical circled

Ice melting: (Physical) Chemical — Physical circled

1. How is an object different after a physical change?
   A physical change changes its shape or size.
2. How is an object different after a chemical change?
   A chemical change changes the matter in the object.

Teacher's Notes: Point out that cooking and burning are chemical changes.

Chapter 5

Name ______________________

Use with Lesson 3: pages 104-108

Vocabulary Puzzle

Cryptogram

Use the code to write the secret words.

1	2	3	4	5	6	7	8	9	10	11	12	13
A	B	C	D	E	F	G	H	I	J	K	L	M

14	15	16	17	18	19	20	21	22	23	24	25	26
N	O	P	Q	R	S	T	U	V	W	X	Y	Z

1. P H Y S I C A L C H A N G E
   16 8 25 19 9 3 1 12 3 8 1 14 7 5
2. S T A T E S O F M A T T E R
   19 20 1 20 5 19 15 6 13 1 20 20 5 18
3. C H E M I C A L C H A N G E
   3 8 5 13 9 3 1 12 3 8 1 14 7 5
4. E V A P O R A T E S
   5 22 1 16 15 18 1 20 5 19
5. W A T E R V A P O R
   23 1 20 5 18 22 1 16 15 18
6. P R O P E R T Y
   16 18 15 16 5 18 20 25
7. C O N D E N S E
   3 15 14 4 5 14 19 5
8. V O L U M E
   22 15 12 21 13 5
9. M A T T E R
   13 1 20 20 5 18
10. M A S S
    13 1 19 19
11. A T O M
    1 20 15 13

Teacher's Notes: Encourage students to use the vocabulary words during class discussions and in their writing.

Chapter 5

Name ______________________

Use with Lesson 3: page 109

Observing Physical and Chemical Changes

Activity Worksheet

Record Your Results

Sugar and vinegar	
Baking soda and vinegar	

State Your Conclusion
1. In each of these tests, did a physical change or chemical change take place?
2. What is one way you can tell that a chemical change takes place?

Use What You Learned
If you mix a spoonful of salt with 4 spoonfuls of water, the salt seems to disappear. Is this a chemical change or a physical change? Explain your answer.

Teacher's Resource Book

Name ______________________

Comprehension: cause and effect relationships

Chapter 5

Science and Reading

Planning Ahead

Read the story. Then answer the questions.

It had been raining hard all evening. Before going to bed, Mrs. Parker listened to the weather forecast on the radio and knew that the rain was expected to change to snow during the night. She planned to get up earlier than usual in the morning because if it snowed it would take longer to drive to work.

Sure enough, during the night the temperature dropped below freezing. By morning, several centimeters of snow had fallen. Several people who worked in Mrs. Parker's office got to work late, but she was able to get there on time.

1. How did Mrs. Parker know the rain was expected to change to snow?

   She listened to the weather forecast on the radio.

2. Why would Mrs. Parker need more time in the morning if it snowed?

   It would take longer to drive to work.

3. What happened during the night that caused the rain to change to snow?

   The temperature dropped below freezing.

4. Why did some people get to work late?

   Accept any answer that indicates they did not get up earlier, or it took longer to drive to work.

Test Book

Name ______________________

Chapter 5
Test A

Multiple Choice Choose the best answer.

1. The amount of matter in an object is that object's (1-1)
   - (a) mass.
   - b. space.
   - c. volume.
2. The color, size, and shape of an object are some of its (1-2)
   - (a) properties.
   - b. elements.
   - c. matter.
3. Solids, liquids, and gases are forms of (1-3)
   - a. mass.
   - (b) matter.
   - c. space.
4. The tiny particles that make up matter are called (2-1)
   - (a) atoms.
   - b. mass.
   - c. properties.
5. Which state of matter has particles that move around the most? (2-3)
   - a. a liquid
   - b. a solid
   - (c) a gas
6. What state of matter will fill any volume? (2-4)
   - a. a solid
   - b. a liquid
   - (c) a gas
7. Changes in such things as the size and shape of matter are called (3-1)
   - a. chemical changes.
   - (b) physical changes.
   - c. rusting.
8. What is water called when it is a gas? (3-2)
   - (a) water vapor
   - b. ice
   - c. water gas
9. The formation of drops of water on the outside of a cold pitcher is (3-2)
   - (a) condensation.
   - b. evaporation.
   - c. a chemical change.
10. What kind of change takes place when wood burns? (3-3)
    - (a) a chemical change
    - b. a physical change
    - c. a change of state

Numbers in parentheses after each question refer to the lesson number and the objective of that lesson.

Test Book

Name ______________________

Chapter 5
Test A

Short Answer Label the pictures above. Use these words:
evaporation liquid condensation solid

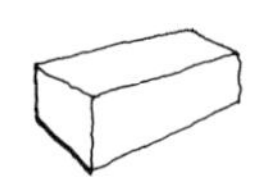

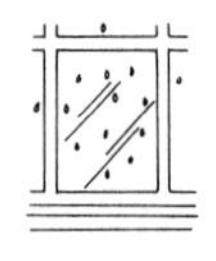

1. **liquid** 2. **evaporation** 3. **solid** 4. **condensation**

Short Essay Use complete sentences to answer each question.

1. What are evaporation and condensation? (3-3)

   Evaporation occurs when matter changes from a liquid to a gas.

   Condensation occurs when matter changes from a gas to a liquid.

2. Compare the shapes and volumes of solids, liquids and gases. (2-3, 2-4, 2-5)

   A solid has a certain shape and volume. A liquid has a certain volume but takes the shape of its container. A gas takes on the shape and volume of its container.

Numbers in parentheses after each question refer to the lesson number and the objective of that lesson.

Test Book

Name ______________________

Chapter 5
Test B

Multiple Choice Choose the best answer.

1. Everything that takes up space and has mass is called (1-1)
   - a. volume.
   - (b) matter.
   - c. weight.
2. To describe matter, it is necessary to name its (1-2)
   - (a) properties.
   - b. weight.
   - c. mass.
3. Three states of matter are (1-3)
   - a. solid, shape, mass.
   - (b) liquid, gas, solid.
   - c. gas, liquid, space.
4. The particles that make up matter are (2-1)
   - a. states.
   - (b) atoms.
   - c. properties.
5. Matter with a certain shape are
   - (a) solids
   - b. liquids
   - c. gases (2-2)
6. The measure of the amount of matter in an object is its (1-1)
   - (a) mass.
   - b. volume.
   - c. state.
7. Particles of matter that move about most freely form a (2-4)
   - (a) gas.
   - b. liquid.
   - c. solid.
8. A chemical change takes place when matter (3-1)
   - a. melts.
   - b. evaporates.
   - (c) rusts.
9. Liquid changed to a gas has (3-2)
   - a. condensed.
   - (b) evaporated.
   - c. melted.
10. Matter which changes from a gas to a liquid has (3-2)
    - (a) condensed.
    - b. evaporated.
    - c. boiled.
11. What does a balance measure? (1-1)
    - a. volume
    - b. properties
    - (c) mass

Numbers in parentheses after each question refer to the lesson number and the objective of that lesson.

Name ____________________

Chapter 5
Test B

Short Answer Label each picture. Use these words:
gas solid liquid condensation

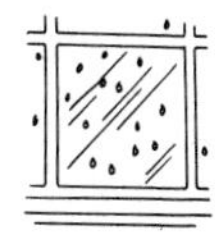
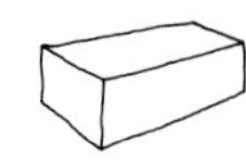

1. liquid 2. gas 3. condensation 4. solid

Short Essay Use complete sentences to answer each question.

1. Explain the difference between a physical change and a chemical change. **(3-2, 2-3)**

A physical change occurs when matter only changes its appearance. A chemical change occurs when matter changes into another kind of matter.

2. Describe how particles move in solids, liquids and gases. **(2-3, 2-4, 2-5)**

Particles in solids stay close together and do not change places. Particles in liquids are farther apart than those in solids and can change places. Gas particles do not pull together very strongly and move around more than solids or liquids.

Numbers in parentheses after each question refer to the lesson number and the objective of that lesson.

TEACHING PLAN

Major Concepts

Lesson 1 All things are made of matter.
Lesson 2 All matter is made up of particles.
Lesson 3 Some kinds of matter can change physically or chemically.

Chapter Vocabulary

atom, chemical change, condense, evaporate, mass, matter, physical change, property, states of matter, volume, water vapor

Getting Started

Point out that even though sand is actually many small particles, the particles can be put together to make something big, like the dragon. Ask students what other large objects they have seen that were actually made up of smaller pieces. (Examples include brick buildings, plastic foam cups, and tile floors.)

Chapter 5

Properties of Matter

Have you ever made an animal out of sand? If you look closely, you can see that many tiny parts of sand make up this dragon.

92

Teaching Options

Cooperative Learning ♦

Jigsaw Format (See page T23.)
Assign the following topics at random to your cooperative learning teams.
Topic A: How are solids and liquids alike? different?
Topic B: How are building blocks like atoms?
Topic C: What are the properties of an iron nail?
Topic D: How could you change the mass and volume of a balloon?
Have students search for information on their topic as they read the chapter. Then let all students with the same topic meet in an expert group to discuss the information. When students return to their teams, they may take turns presenting their topics to the team. Then give students a test covering all topics to complete individually (Chapter 5 Test A or B in the *Test Book*). Award Superteam certificates to teams whose average test scores exceed 90%, and Greatteam certificates to teams whose average test scores exceed 80%.

♦ ***Suitable as a language development activity***

Introducing the Chapter

Many objects around you look different from each other. In some ways, all the objects are alike. In this chapter you will learn about parts that make up all the objects in the world around you. In the activity below, you can observe and describe different objects.

Observing Different Materials

Notice that each glass in the picture has a small amount of white material in it. One glass has salt. The other glass has sugar.

You can get a glass with salt and a glass with sugar from your teacher. You will not know which material is in each glass. Look at the two materials. Add a few drops of water to each glass. Touch each material. Mix a small amount of each material in water. Compare the way the materials look when they are mixed in water. Notice the smell of each material. *CAUTION: Do not taste these materials. Never taste any materials, even if you think you know what they are.*

Talk About It

1. How were the materials alike?
2. How did you find out which material was salt and which material was sugar?

DISCOVER

Objective ♦

This optional *DISCOVER* activity will help students explore and build background information about the concept of identifying a substance from its properties. Later in the chapter students will be able to draw on this experience to help them assimilate the new content.

Science Process Skills

Observing, Inferring

Materials

For each group of 2: 1 plastic spoon, 2 small plastic glasses, spoonful of sugar, spoonful of salt, and 1 hand lens (if available)

Safety Tips (See page T24.)

- Remind the students to never taste an unkown substance, even if they think they know what it is.
- Wipe up spilled water immediately to prevent accidents.

Teaching Tip

- Provide labeled bottles for the students to *compare* with their unknown substances.

Answers

Talk About It

1. Both look white and grainy.
2. The sugar feels sticky.

Applying Whole Language ♦

Discuss the whole language framework with each Teaching Option you select. Here is an example applied to the Game Suggestion on p. 106.

1. Purpose: To play "Who Am I?" practicing awareness of properties and changes of matter

2. Context: Class game. Students take turns picking "mystery" objects. Others guess by asking questions.

3. Decisions: Specific rules and behaviors of players. Kinds of science questions to ask.

4. Evaluation: How did the game help us learn about matter? (See p. T30.)

Resource Book page 51

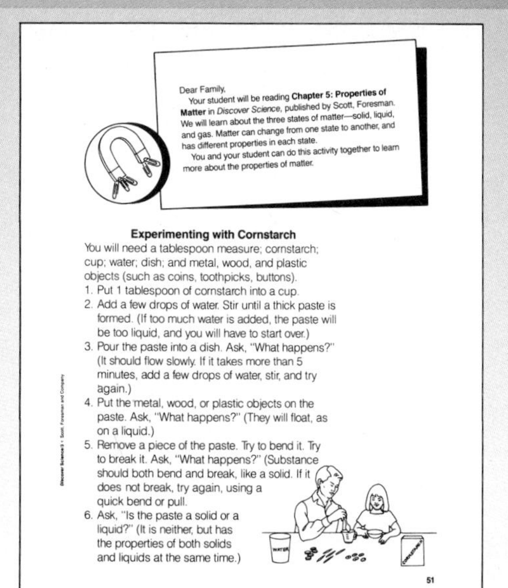

Dear Family,
Your student will be reading **Chapter 5: Properties of Matter** in *Discover Science*, published by Scott, Foresman. We will learn about the three states of matter—solid, liquid, and gas. Matter can change from one state to another, and has different properties in each state.
You and your student can do this activity together to learn more about the properties of matter.

Experimenting with Cornstarch

You will need a tablespoon measure; cornstarch; cup; water; dish; and metal, wood, and plastic objects (such as coins, toothpicks, buttons).

1. Put 1 tablespoon of cornstarch into a cup.
2. Add a few drops of water. Stir until a thick paste is formed. (If too much water is added, the paste will be too liquid, and you will have to start over.)
3. Pour the paste into a dish. Ask, "What happens?" (It should flow slowly. If it takes more than 5 minutes, add a few drops of water, stir, and try again.)
4. Put the metal, wood, or plastic objects on the paste. Ask, "What happens?" (They will float, as on a liquid.)
5. Remove a piece of the paste. Try to bend it. Try to break it. Ask, "What happens?" (Substance should both bend and break, like a solid. If it does not break, try again, using a quick bend or pull.
6. Ask, "Is the paste a solid or a liquid?" (It is neither, but has the properties of both solids and liquids at the same time.)

51

Science Background

Physical properties of matter can be observed and measured. Some examples of physical properties are state, color, shape, taste, odor, hardness or softness. Other physical properties include magnetic properties, malleability, and conductivity—electric and heat.

LESSON 1 pages 94–98

TEACHING PLAN

Lesson Objectives
- *Define* matter and mass.
- *List* some properties of matter.
- *Describe* three states of matter.

Lesson Vocabulary
mass, matter, property, states of matter, volume

1. Motivate

Demonstration Activity ♦
Display several objects with color, smell, shape, texture, and size differences. Allow students to examine the objects closely.

Discussion
Guide students to contrast the texture, smell, shape, color, and size of the objects. Questions: **How are these objects different?** (Answers will vary.) **How are these objects alike?** Point out that all the objects are alike because they are all made of matter. Explain that color, size, shape, smell, and texture of matter all describe what it is like.

LESSON GOALS

You will learn
- how objects have mass and are made of matter.
- how matter can be described by its properties.
- the three forms of matter.

volume (vol′yəm), the amount of space an object takes up.

[1]answer might include any large object in picture
[2]the desk
[3]answer might include desks or chairs
[4]answers might include pencils or erasers

All of these objects are made of matter.

1 What Is Matter?

Notice how the objects in the picture look different from each other. The desk looks larger than the pencil. The notebooks have different colors. Yet all the objects are alike in some ways.

Mass and Matter

These objects are alike because they all take up space. Notice how the paper clip takes up a small amount of space. Find an object that takes up more space than the paper clip.[1] Which object takes up the most space?[2]

The amount of space an object fills is its **volume.** Look at objects around you. Name an object in your classroom that has a large volume.[3] What object has a small volume?[4]

94

Teaching Options

Science Background

Everything in the universe is made of matter. Matter can exist in four states—solid, liquid, gas, and plasma. The fourth state, plasma, was recently classified by scientists. Plasma has the properties of a gas, but consists of charged particles rather than uncharged atoms or molecules. Plasma can be found in stars, nuclear explosions, and neon signs.

The mass of an object is not affected by gravity. Two pieces of matter cannot take up the same space at the same time. One piece of matter will displace the other.

Reading Strategies ♦

1. Guide students' pre-reading by asking: What parts of the lesson are familiar and which parts look hard and why?
2. Assign these strategies: Writing a Memory Sentence and Finding and Writing Sentences for Vocabulary Words (See pages T26–T29.)
3. Pair students to share what information is clear and unclear and initiate discussion using students' unanswered questions.

♦ ***Suitable as a language development activity***

All these objects are alike in another way. They all have **mass.** An object's mass is the measure of how much material makes up the object. You can use a balance to measure how much mass an object has. Find the picture of the apple on the balance. The apple has more mass than the eraser. The rock has more mass than the apple.

Everything that takes up space and has mass is called **matter.** All the objects around you are made of matter. You take up space and have mass. You are made of matter.

Imagine dreaming about your classroom. A dream does not take up space and have mass. A dream is not made of matter.

mass (mas), the measure of how much matter an object contains.

matter (mat′ər), anything that takes up space and has mass.

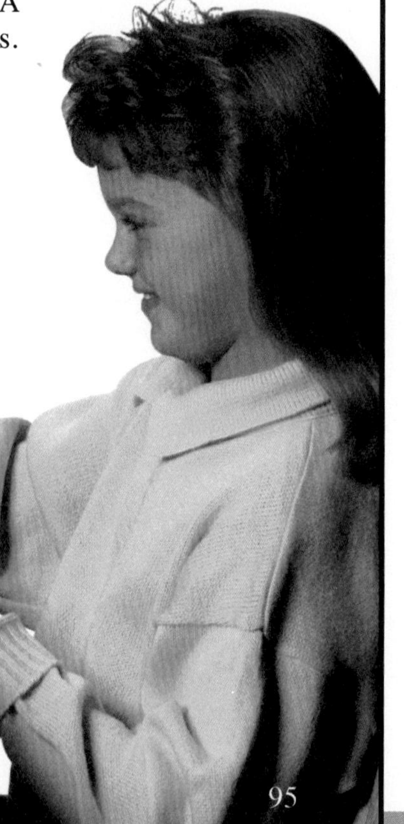

2. Teach

Teaching Tips

- **Possible Misconceptions:** Students might confuse the phrase "Is something the matter?" with "matter" in this chapter. Students might also confuse "property" with land, or property, that people own. Emphasize the scientific meanings of these words.
- Point out that some things do not have volume or mass. Guide students to *name* things that are not made of matter such as pain, sound, or gravity.)
- Explain that an object's mass can change. To illustrate this concept, have a volunteer bite an apple. Questions: **What lost mass?** (the apple) **What or who gained mass?** (the student) **Did the apple or the student change more noticeably in mass?** (the apple)

Workbook page 25 *

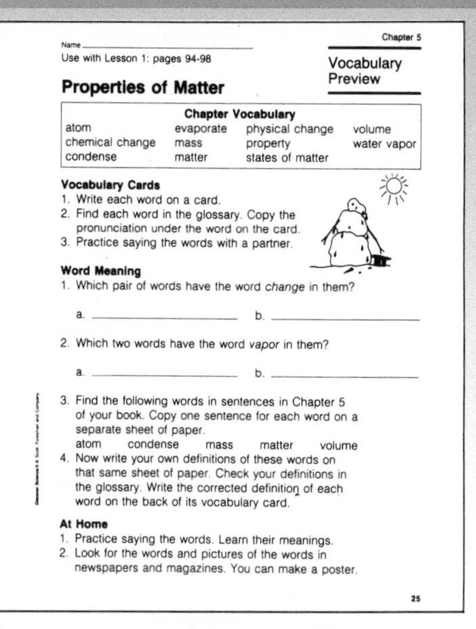

Name ______________

Use with Lesson 1: pages 94-98

Chapter 5

Vocabulary Preview

Properties of Matter

Chapter Vocabulary

atom	evaporate	physical change	volume
chemical change	mass	property	water vapor
condense	matter	states of matter	

Vocabulary Cards
1. Write each word on a card.
2. Find each word in the glossary. Copy the pronunciation under the word on the card.
3. Practice saying the words with a partner.

Word Meaning
1. Which pair of words have the word *change* in them?

   a. ______________ b. ______________

2. Which two words have the word *vapor* in them?

   a. ______________ b. ______________

3. Find the following words in sentences in Chapter 5 of your book. Copy one sentence for each word on a separate sheet of paper.
   atom condense mass matter volume
4. Now write your own definitions of these words on that same sheet of paper. Check your definitions in the glossary. Write the corrected definition of each word on the back of its vocabulary card.

At Home
1. Practice saying the words. Learn their meanings.
2. Look for the words and pictures of the words in newspapers and magazines. You can make a poster.

25

Answers to masters on pages 92E–92H

Special Education

Allow students with learning disabilities to measure the mass of various objects in the classroom with a balance. Have students *list* the mass of each object measured. Then have students use the lists to *compare* the objects.

Reinforcement

Obtain two balloons of the same size. Fill one with air and the other with water. Ask several volunteers to decide which balloon is heavier by holding each. Guide students to *infer* that the water balloon has more mass than the balloon filled with air. *CAUTION:* Students should wear cover goggles when working with balloons.

TEACHING PLAN

Teaching Tips

- Refer students to the *DISCOVER* on page 93 in which they observed and inferred physical properties. Point out that all matter has physical properties from which it can be identified.
- Have the students ***list*** the properties of each object in the picture. Then have students use the lists to ***compare*** the objects.
- Ask volunteers to choose an object in the classroom and have them ***describe*** the object by naming its properties. Encourage others to ***identify*** the object based upon the description of its properties.

You can describe these objects by naming their properties.

property (prop′ər tē), something about an object that can be observed, such as size, shape, color, or sound.

Properties of Matter

Think about how you might describe some of the objects in the picture. The dinner plate is large. The strawberries are small and juicy. The napkin is pink and soft. How can you describe the flowers?[1] When you describe an object, you tell about its **properties.** A property tells exactly what an object is like. You can describe matter by naming its properties.

Suppose you wanted to describe a slice of lemon. You probably would tell about its size, shape, and color. You also might describe its smell and sour taste. Size, shape, color, smell, and taste are properties of the lemon.

[1]yellow, pink, purple, soft

96

Teaching Options

Science Anecdote

Opening a can of carbonated beverage often causes a fizzing sound. Carbon dioxide gas is dissolved in the liquid. When the can is opened, the pressure on the liquid is lowered and this allows the gas to escape from the liquid.

Reinforcement

Have students write a description about the properties of a favorite object from home such as a bicycle, pet, food, or clothing. Allow the students to take turns reading their descriptions. Encourage others to guess the identity of the object described.

Science and Art

Have students each choose one property of matter such as color, smell, size, shape, or texture. Ask students to create a collage depicting the specific property of matter chosen. Students may use magazine and newspaper pictures, and actual objects on their collages. Display the collages in the classroom.

Matter can be solid, liquid, or gas.

States of Matter

Notice the diver in the picture. The diver is solid. The water in the picture is liquid. Find the air bubbles in the water. Air is a gas. Matter comes in three forms—solid, liquid, and gas. These forms are called the **states of matter.**

A solid keeps a certain shape and has a certain volume. Find some objects in the picture that are solid.[1]

Suppose you poured water into a glass. The water would take the shape of the glass. A liquid takes the shape of its container. A liquid keeps its volume when it changes shape. Imagine pouring water from a glass into a bowl. The volume of water would stay the same. How would the shape change?[2]

INVESTIGATE!

Find out if changing the shape of an object can change another property of that object. Write a hypothesis and test it with an experiment. Floating or sinking is a property of clay. You might observe how changing the shape of clay affects how it floats.

states of matter, the three forms of matter—solid, liquid, and gas.

[1]the coral, the diver, the diving mask
[2]take the shape of the bowl

Teaching Tips

- Explain to students that their bodies contain examples of all three states of matter. Question: **What are some examples of the three states of matter found in our bodies?** (Guide students to *infer* some of the following examples: solids—fingernails and bones; liquids—blood and saliva; gas—air in the lungs.)
- **Possible Misconception:** Some students may confuse states of matter with the states of the United States. Explain *states* as meaning forms or conditions.

Investigate!

Accept any testable hypothesis. Students' data will support various hypotheses. One possible hypothesis is: *Changing clay from a boat shape to a ball will cause the clay to sink.* Students should find out that a ball of clay will sink in water. Shaping the ball like a small boat can make the clay float in water.

Workbook page 26 *

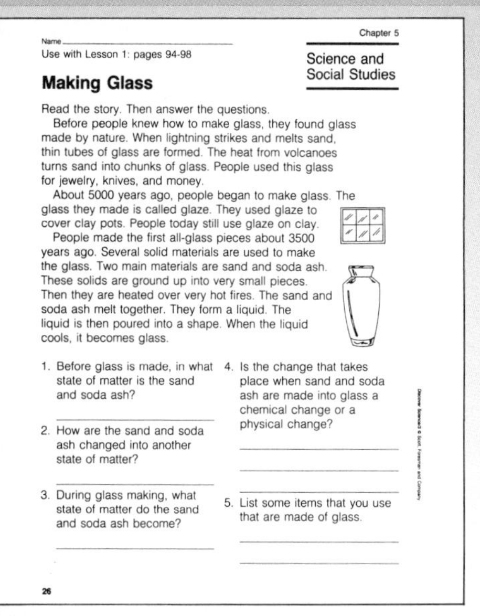

Name ____________

Use with Lesson 1: pages 94-98

Chapter 5

Science and Social Studies

Making Glass

Read the story. Then answer the questions.

Before people knew how to make glass, they found glass made by nature. When lightning strikes and melts sand, thin tubes of glass are formed. The heat from volcanoes turns sand into chunks of glass. People used this glass for jewelry, knives, and money.

About 5000 years ago, people began to make glass. The glass they made is called glaze. They used glaze to cover clay pots. People today still use glaze on clay.

People made the first all-glass pieces about 3500 years ago. Several solid materials are used to make the glass. Two main materials are sand and soda ash. These solids are ground up into very small pieces. Then they are heated over very hot fires. The sand and soda ash melt together. They form a liquid. The liquid is then poured into a shape. When the liquid cools, it becomes glass.

1. Before glass is made, in what state of matter is the sand and soda ash? ____________
2. How are the sand and soda ash changed into another state of matter? ____________
3. During glass making, what state of matter do the sand and soda ash become? ____________
4. Is the change that takes place when sand and soda ash are made into glass a chemical change or a physical change? ____________
5. List some items that you use that are made of glass. ____________

26

* Answers to masters on pages 92E–92H

Enrichment

Have the students *classify* the foods they ate for breakfast or lunch as solid, liquid, or gas. Students may be unaware that carbonated beverages contain a gas.

Reinforcement

Demonstrate forms of three states of matter. Obtain a plastic bag, one wooden block, and a glass of water. Put the wooden block into the plastic bag. Questions: **Does the block change shape?** (No) Blow air into the plastic bag and hold the end closed. **What shape does the gas take?** (the shape of the bag) Pour the water into the plastic bag. **How does the water change shape?** (changes from shape of glass to shape of the bag) Place the bag of water back into the glass. Let the students see that the volume has not changed.

LESSON 1 pages 94–98

TEACHING PLAN

3. Assess

Lesson Review

1. Anything that takes up space and has mass is matter.
2. Properties of matter tell exactly what an object is like. Size, shape, color, smell, and taste are some properties of matter.
3. solid, liquid, and gas
4. Challenge! The milk would spill out of the glass and take the shape of the space it spills on, because a liquid takes the shape of its container but keeps the same volume. **Thinking Skill:** *Inferring*

Find Out On Your Own

Answers will vary. The correlation between estimates and actual measurements should increase with practice. **Thinking Skill:** *Organizing information*

Solid, liquid, and gas

Notice the different shapes of the balloons in the picture. Air is a gas that fills the balloons. A gas takes the same shape as its container. Find the broken balloon. The air from the broken balloon spread out into the whole room. The same amount of gas can spread out to fill a larger volume.

Lesson Review

1. What is matter?
2. What are properties of matter?
3. What are the three states of matter?
4. **Challenge!** Suppose you added marbles to a full glass of milk. Would the shape and volume of the milk change? Explain your answer.

Study on your own, pages 324–325.

PHYSICAL SCIENCE

FIND OUT ON YOUR OWN

Find objects like those shown above to measure. Using metric measures, guess the length or height of each object. Also guess each object's mass. Then measure the objects.

Teaching Options

Science Anecdote

Hardness is a property of matter. A diamond is the hardest kind of matter that can be found in nature. A diamond can scratch any other substance.

Reteaching Suggestion ♦

Obtain two different objects such as a shoe and a piece of paper. Have a volunteer hold the objects in each hand. Point out that each of the objects takes up space. Questions: Which object has the greater mass? (the object with the greater amount of material in it) How do you know that each object is made of matter? (both objects take up space and have mass) Encourage students to list some of the properties of each object, including whether it is a solid, liquid, or gas.

Game Suggestion ♦

On separate index cards, list different properties of matter such as yellow, hard, small. Allow students to each pick a card. Challenge students to list as many objects as they can think of that have those properties. Instruct students to trade cards and repeat the activity. Compare the various objects chosen.

♦ ***Suitable as a language development activity***

Observing That Air Has Volume

Suggested grouping: pairs

Purpose

Observe that air takes up space.

Gather These Materials

• plastic glass • paper towels (or napkins) • large bowl • water

Follow This Procedure

1. Use a chart like the one shown to record your observations.
2. Loosely crumple a paper towel and put it in the bottom of the empty glass. Make sure the towel will not come out when the glass is turned upside down.
3. Run water into the large bowl. Turn the glass upside down. Push it straight down into the water as shown. Hold the glass there while you count to 30.
4. Lift the glass straight up out of the water. Pull the towel out of the glass. Record what you observe.
5. With the paper towel in the glass, turn the glass upside down and push it to the bottom of the water.
6. When the glass is at the bottom, slowly turn the glass over. Record what happens.

Record Your Results

Towel after glass is pushed straight into water	dry
Towel after glass is turned over under water	wet

State Your Conclusion

1. What kept the water from reaching the paper towel in the glass when you pushed the glass straight into the water?
2. What happened to the paper towel when you turned the glass over under water? Explain.

Use What You Learned

How could people use the results of this experiment to help them work under water?

99

Activity

Concept

Air has volume.

Objectives/Process Skills

- *Observe* that a paper towel remains dry in an inverted glass submerged in water.
- *Record* observations.
- *Infer* that air takes up space.

Time Allotment

Allow 20 minutes.

Safety Tips (See page T24.)

- Wipe up spilled water immediately to prevent accidents.
- If possible, use only clear plastic glasses and plastic bowls or storage boxes.

Teaching Tip

- Have students compare the flow of a liquid from a can with one hole and then with two holes. Explain that a liquid flows out more easily when a can has two holes because air enters one hole while liquid pours out the other.

Answers

State Your Conclusion

1. Air kept the water from reaching the paper towel.

2. The air moved out of the glass and water entered the glass making the paper wet.

Use What You Learned

If a large container is used, a person could bring enough air underwater to give him or her a place to breathe without returning to the surface. **Thinking Skill:** ***Applying information to new situations***

Resource Book page 55

Name
Use with Lesson 1: page 99
Observing That Air Has Volume

Chapter 5
Activity Worksheet

Record Your Results

Towel after glass is pushed straight into water	
Towel after glass is turned over under water	

State Your Conclusion

1. What kept the water from reaching the paper towel in the glass when you pushed the glass straight into the water?
2. What happened to the paper towel when you turned the glass over under water. Explain.

Use What You Learned

How could people use the results of this experiment to help them work under water?

55

Activity Results

TEACHING PLAN

Lesson Objectives

- *Define* atom, and *explain* how atoms join to form other particles of matter.
- *Compare* the arrangement and motion of the particles in solids, liquids, and gases.

Lesson Vocabulary

atom

1. Motivate

Demonstration Activity ♦

Roll colored clay into balls approximately 3 cm (about 1 inch). Explain that the balls represent the particles of matter (atoms) that make up solids, liquids, and gases. Cut soda straws in half. Connect several balls with the straws to represent how atoms are arranged in solids and gases.

Discussion

Have students compare and discuss the arrangement of the atoms in the models. Question: **What did you observe about the spacing of the atoms in the models?** (solid—atoms close together, liquid—atoms farther apart, gas—much space between atoms)

2 What Makes Up Matter?

LESSON GOALS

You will learn
- what makes up matter.
- how particles of matter move.

SCIENCE IN YOUR LIFE

Some matter has only one kind of atom. You breathe oxygen from the air. All the atoms of oxygen are the same. Other matter has more than one kind of atom. Two different kinds of atoms join together to make water.

atom (at′əm), a small particle that makes up matter.

Imagine looking at a pin like the one in the first picture. The tiny pin can be hard to see. Now imagine adding more pins like those in the second picture. You can see the group of pins more easily than one pin. Each pin is like a small particle of matter. The group of pins is like matter you can see.

Particles in Matter

The particles that make up matter are called **atoms.** You cannot see atoms because they are so small. Two or more atoms can join together to form larger particles of matter. Then many of these larger particles can join together to form the matter you see.

Scientists know about many kinds of atoms. These atoms can join together in different ways to make different kinds of matter.

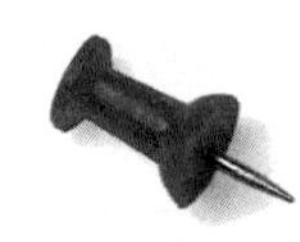

Small particles make up matter.

100

Teaching Options

Science Background

Elements are the basic substances that form matter. An element cannot be broken down into simpler substances by heat, light, or electricity. Scientists know of about 109 elements. The smallest particle of an element that still has all the properties of that element is an atom. Natural elements are those found in nature. Most scientists think all naturally occurring elements have been discovered. Scientists are able to make other elements in laboratories. Some elements made in laboratories last only for a short time. A compound is formed when atoms of two or more elements join. The smallest particle of a compound that still has all the properties of that compound is a molecule.

Reading Strategies ♦

1. Guide students' pre-reading by asking: Which parts of the lesson are familiar and which parts are new?
2. Assign these strategies: Writing Memory Sentences and Finding and Writing Sentences for Vocabulary Words (See pages T26–T29.)
3. Pair students to share what information is clear and unclear and initiate discussion using students' unanswered questions.

♦ *Suitable as a language development activity*

How Particles of Matter Move

Particles in matter are different in different states of matter. The pictures show how the particles in a solid look different from those in a liquid. Which state of matter has particles that are closer together?[1]

[1]solid

The particles in solids pull toward each other. A solid keeps a certain shape because its particles stay close together. The particles in matter are always moving. Particles in solids move back and forth, but they do not exchange places.

Notice how the particles in liquids are farther apart than those in solids. The pull between particles is weaker in liquids than in solids. Liquids change shape because the particles can move around each other.

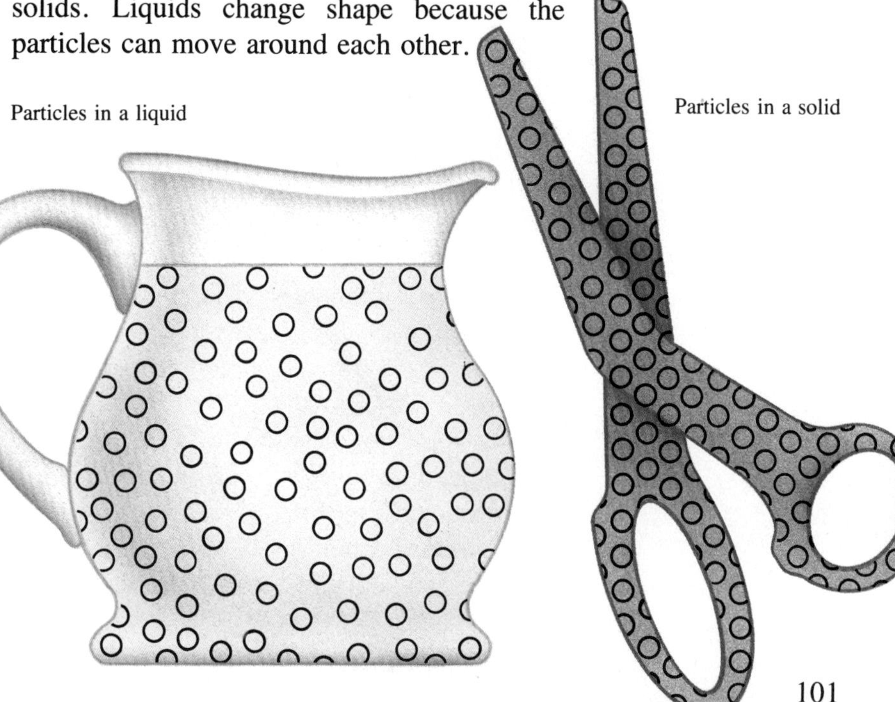

Particles in a liquid

Particles in a solid

2. Teach

Teaching Tips

- Have several students *name* the smallest thing they have ever seen. List student responses on the board. Explain that the types of matter listed on the board are huge compared to the tiny particles called atoms, which make up all matter.
- Instruct students to wave their hands in the air for a few seconds. Next, have them imagine waving their hands under water (liquid) and in sand (solid). Questions: **Through which substance air, water, or sand would it be easiest to move your hands? Why?** (Air. Guide students to *infer* that the particles of air—a gas—are spaced far apart, creating greater ease of movement.)
- **Possible Misconception:** Many students think that all of the atoms that make up an object must be of the same kind. Make sure students understand that objects can also be made up of different kinds of atoms.

Enrichment

Fill a clear, glass bowl with water and have students identify the states of matter in the bowl. (gas—air, liquid—water) Discuss how you might add another state in the bowl. (Add a solid such as sand or gravel.)

Special Education

To illustrate how atoms join together to form matter that we see, have students with learning disabilities use interlocking blocks to make models of atoms. Then work with students to join the atom models to form a house, car, box or other known object.

Reinforcement

Have students use toothpicks and clay to construct models showing the arrangement of atoms in solids, liquids, and gases. *CAUTION:* Tell students to use toothpicks carefully to avoid puncturing themselves.

LESSON 2 pages 100–102

TEACHING PLAN

Teaching Tip

- Point out that a person can sometimes smell food cooking in the kitchen when they are in another room in the house. Ask students to explain why this occurs. (Students should *infer* that the free moving particles of a gas spread the food odors through the air.)

3. Assess

Lesson Review

1. Atoms make up matter.
2. In solids, the particles move back and forth but stay close together. In liquids, the particles move around each other. The molecules in gases move freely.
3. Challenge! No. A gas will spread out to fill the container. **Thinking Skill:** *Inferring*

Find Out On Your Own

The color of the grape juice spreads throughout the water because the particles in a liquid are always moving. **Thinking Skill:** *Inferring*

Particles in a gas

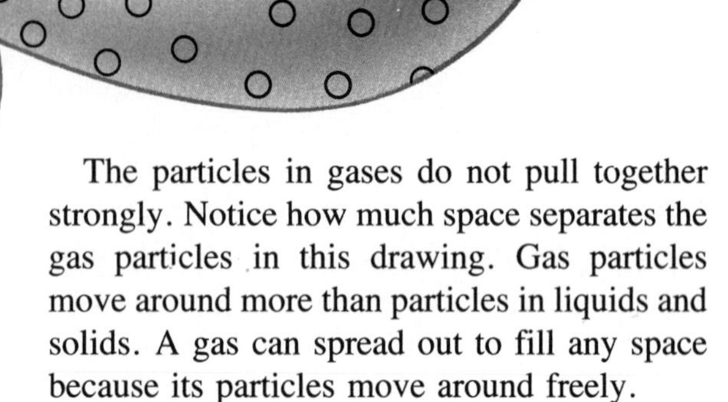

The particles in gases do not pull together strongly. Notice how much space separates the gas particles in this drawing. Gas particles move around more than particles in liquids and solids. A gas can spread out to fill any space because its particles move around freely.

Lesson Review

1. What makes up matter?
2. How do particles move in solids, liquids, and gases?
3. **Challenge!** Can a container of gas be only partly full? Explain your answer.

Study on your own, pages 324–325.

PHYSICAL SCIENCE
FIND OUT ON YOUR OWN

Draw a picture of a glass of water. Now imagine adding several drops of grape juice to the water. Color the picture to show how the liquid would look. Write a few sentences explaining what would cause the color to change.

102

Teaching Options

Science Anecdote

More than 2,000 years ago, a Greek scientist named Democritus felt that all things were made of tiny invisible particles. He called the particles "atomos," from the Greek word for "indivisible." Today, scientists know that atoms can be divided.

Reteaching Suggestion ♦

Ask a volunteer to come to the front of the room and stand in front of you. Ask the student to walk away. Explain that molecules in a gas can easily move away from each other. Ask the student to lightly hold your hand, and then pull away. Relate this to the molecules in a liquid, which are closer together but able to move. Next, ask the student to tightly grip both of your hands and try to pull away. Explain how this example is similar to how the molecules in a solid are close together and do not change places when they move.

Workbook page 27 *

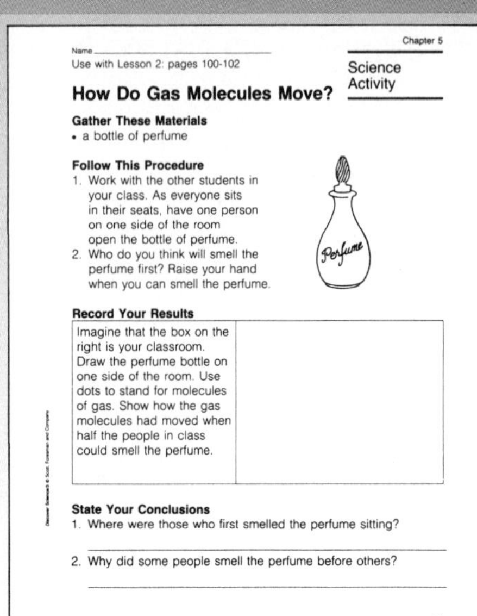

Chapter 5

Name ____________

Use with Lesson 2: pages 100-102

Science Activity

How Do Gas Molecules Move?

Gather These Materials
- a bottle of perfume

Follow This Procedure
1. Work with the other students in your class. As everyone sits in their seats, have one person on one side of the room open the bottle of perfume.
2. Who do you think will smell the perfume first? Raise your hand when you can smell the perfume.

Record Your Results

Imagine that the box on the right is your classroom. Draw the perfume bottle on one side of the room. Use dots to stand for molecules of gas. Show how the gas molecules had moved when half the people in class could smell the perfume.	

State Your Conclusions
1. Where were those who first smelled the perfume sitting?
____
2. Why did some people smell the perfume before others?
____

27

♦ *Suitable as a language development activity*

Exploring the Inside of Atoms

The Problem Our word *atom* comes from a Greek word, *atomos.* Over two thousand years ago, a Greek thinker wondered what the smallest particle of matter must be like. He called this pretend particle *atomos,* which means "not cuttable." Many years later, in 1803, an English scientist discovered the atom. He said it could not be split. However, in 1897, another English scientist found out that atoms are made of even smaller particles. Since that time, scientists have been trying to find out what the smallest particles of matter really are. You know an atom is tiny. How could scientists find out what is inside such a small particle?

The Breakthrough In 1930, Ernest O. Lawrence made the first "atom smasher." This machine helps scientists find new kinds of particles inside the atom. Scientists first make the smaller particles inside of atoms move very fast. Then they use these fast particles to smash apart even smaller particles inside the atom. Using atom smashers, scientists found many new types of particles inside the atom.

New Technology The picture shows part of a laboratory near Chicago, Illinois, called Fermilab. This laboratory has a tunnel that contains a giant atom smasher. Small particles from atoms move around the circle, going faster and faster as they move. Using this kind of atom smasher, scientists at Fermilab and at other laboratories around the world have learned much more about matter. Soon, the United States plans to build an even larger and more powerful atom smasher. It will help scientists find out even more about the smallest particles of matter.

Control room at Fermilab

What Do You Think?

1. How has the meaning of the word *atom* changed?
2. How is an atom smasher like a telescope and microscope?

Discussion

Discuss with students how the contents of a piñata (pot filled with candy, fruit, and so on, hung above the heads of children at Christmas time) are determined. (Children smash the piñata with a stick.) Explain that, in a similar way, atom smashers use small particles to hit and break open other particles. Scientists study what comes out of the collisions.

Teaching Tip

- Many students tend to think that scientists know all the answers. Point out that scientists are still very actively engaged in finding out more about the tiny particles that make up matter.

Answers

What Do You Think?

1. *Atom* comes from the Greek word that means "uncuttable." For many centuries, it was believed that atoms could not be broken down into smaller particles. The name *atom* described this belief. In 1987, however, it was found that atoms are made of even smaller particles, and we no longer think of atoms as being "uncuttable." **Thinking Skill:** *Restating or explaining ideas*

2. All three are used to study objects that otherwise could not be studied. **Thinking Skill:** *Inferring*

Science Background

An atom smasher has three main parts: a source of the particles that will be sped up; a vacuum chamber to speed them up in, and a source of an electric field to speed the particles up. Large, powerful magnets are used to bend the particles around the ring and keep the beam in focus.

LESSON 3 pages 104–108

TEACHING PLAN

Lesson Objectives

- *Define* physical change, and *explain* how some kinds of matter can change from one state to another.
- *Define* chemical change, and *explain* how some kinds of matter can change to become new matter.

Lesson Vocabulary

chemical change, condense, evaporate, physical change, water vapor

1. Motivate

Demonstration Activity ♦

Vigorously stir a bowl of ice cubes in front of the class. Have the students observe how the ice cubes change.

Discussion

Questions: **In which state of matter were the ice cubes?** (solid) **What state of matter did the ice cubes change into?** (liquid)

3 How Can Matter Change?

LESSON GOALS

You will learn
- what physical changes are and how matter can change from one state to another.
- how chemical changes can cause some kinds of matter to become new matter.

physical (fiz′ə kel) **change,** a change in the size, shape, state, or appearance of matter.

The pictures on these two pages show some ways you can change matter. Changes in matter happen around you every day. Some changes make matter look different. Other changes make one kind of matter become another kind of matter.

Physical Changes

Notice the different shapes of the clay. Changing the clay's shape does not change the clay into something else. The clay is still clay. **Physical changes** are changes in the way matter looks. Changes in size and shape, such as these changes in the clay, are physical changes. How is a person making a physical change in the piece of paper?[1]

[1]by cutting it

Physical changes

104

Teaching Options

Science Background

Physical changes can affect the size, shape, or state of a substance. These changes do not affect the chemical properties of a substance. Chemical changes result from chemical reactions. A chemical reaction is a process by which one or more substances are changed into one or more new substances. Such reactions can be molecules breaking apart into atoms, atoms joining to form molecules, or atoms changing places with other atoms to form new molecules.

Reading Strategies ♦

1. Guide students' pre-reading by asking: Which parts of the lesson are familiar and which parts are new?
2. Assign these strategies: Visualizing Information and Writing a Memory Sentence for each subhead. (See pages T26–T29.)
3. Pair students to share what information is clear and unclear and initiate discussion using students' unanswered questions.

♦ *Suitable as a language development activity*

Another kind of physical change happens when matter changes from one state to another state. Matter looks different when it changes state but it stays the same kind of matter.

Notice the liquid water in the glass. What happens if you put water in the freezer? Water changes to a solid when it is cooled to a temperature of 0° Celsius (0°C). This temperature is called the freezing point of water. How do particles in ice move differently from those in liquid water?[2]

Solids also can change to liquids. Heat speeds up the moving particles in ice. The particles move apart. Heat melts ice and changes it to liquid water.

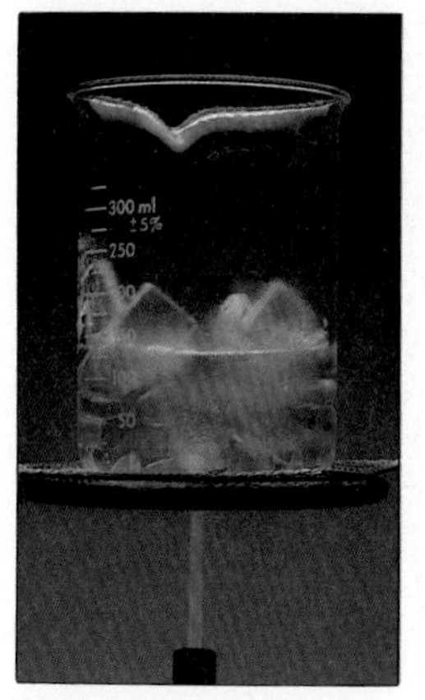
Liquid water and ice

[1]it freezes

[2]move back and forth, but do not move around each other like water

2. Teach

Teaching Tips

- **Metric-English Equivalent:**
0°C = 32°F
- Have the students color a sheet of paper. Question: **Is it still paper?** (Yes) Have the students shred the paper. Question: **Is it still paper?** (Yes) Explain that these are examples of physical changes.
- Have students *name* some physical changes they have observed. (sawing wood, crushing a sugar cube, crumpling paper, sharpening a pencil, stretching a rubber band, dissolving)

Special Education

Have students with learning disabilities tear, crumple, cut, and form sheets of paper into different shapes. Explain how each example is a physical change because the paper remains paper in every form.

Reinforcement

Display various objects such as a banana, clay, and paper. Ask students to describe ways to physically change the objects. (Students might ***describe*** mashing the banana, rolling the clay into a rope, and tearing the paper.) Guide students to observe that each object remained the same after the physical change.

TEACHING PLAN

Teaching Tips

- **Possible Misconception:** Some students might have difficulty understanding the difference between water vapor and air. For example, when water is heated, the first bubbles which appear are air being forced out of a solution. When the water boils, the bubbles are water vapor. Water usually boils at 100°C.
- **Metric-English Equivalent:** 100°C = 212°F
- Question: **Where have you seen water vapor condense, or change to a liquid?** (foggy mirrors or windows, dew)

Matter can change states.

SCIENCE IN YOUR LIFE

The strong smell of an onion comes from an oil. When you cut an onion, the oil changes into a gas. This gas evaporates and then spreads through the air.

evaporate (i vap′ə rāt′), to change from a liquid state to a gas.

water vapor (wô′ter vā′pər), water in the form of a gas.

condense (kən dens′), to change from a gas to a liquid state.

[1]the larger one

You probably have seen a puddle of water that disappears after a time. The water in the puddle changes into a gas. Matter **evaporates** when it changes from a liquid to a gas. Water in the form of a gas is called **water vapor.** Compare the two puddles in the picture. Water evaporates from the top part of each puddle. Which puddle will evaporate faster?[1]

Heat makes water particles move fast. Water quickly changes to a gas when water heats to a temperature of 100° Celsius. This temperature is the boiling point of water. Notice the bubbles in the pot of water. The gas in the bubbles is water vapor.

Cooling air causes water vapor to change to a liquid. Matter **condenses** when it changes from a gas to a liquid. This cold pitcher and the glasses cool the air around them. Then water vapor in the air condenses to small drops of water on the outsides of the containers.

106

Teaching Options

Science Anecdote

Dissolving sugar in tea is a physical change. The sugar cannot be seen but it still makes the tea taste sweet. Boiling the tea will cause the water to evaporate into water vapor. The sugar will remain.

Enrichment

Demonstrate how baking produces a chemical change. Make bread or muffins with the students. Have students *compare* the properties of the ingredients and the dough with the finished product. Point out that digesting food also causes chemical changes. The body changes the substances in food into other substances such as skin and muscles.

Game Suggestion ♦

Play "Who Am I." Begin with a volunteer. Have the student choose an object in the classroom. The other students then try to guess the identity of the object by asking questions relating to its properties or the way it can be changed either physically or chemically.

♦ ***Suitable as a language development activity***

Chemical Changes

You probably have seen different shapes carved out of wood. The wood does not change into a different material when it is carved. What kind of change happens to wood when carving changes its shape?[1]

The wood in the picture is burning. The fire changes the wood into ashes and smoke. The wood becomes a different kind of matter. A **chemical change** takes place when matter changes to a different kind of matter.

Notice the rust on these nails. The material that makes up the nails mixes with the air to form rust. Rusting is a chemical change.

chemical (kem′ə kəl) **change,** a change that causes matter to become a new kind of matter.

[1]physical change

Chemical changes

Teaching Tips

- Have students *list* the properties of the wood in the picture. Then have them list the properties of the ashes that result from the burning of the wood. Use the lists to contrast the different properties.
- Describe several physical and chemical changes and ask students to *identify* each change as physical or chemical. (Examples: physical—building a shelf, breaking a glass, tearing paper; chemical—baking bread, burning paper, and so on.)
- Have students *describe* where they have seen examples of rusting. (buildings, bicycles, roller or ice skates, old keys, door hinges, on old cars, and so on)

Workbook page 28 *

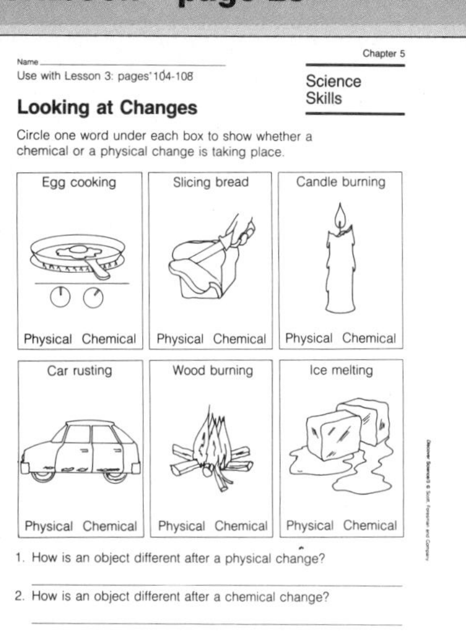

Name

Use with Lesson 3: pages 104-108

Chapter 5

Science Skills

Looking at Changes

Circle one word under each box to show whether a chemical or a physical change is taking place.

Egg cooking	Slicing bread	Candle burning
Physical Chemical	Physical Chemical	Physical Chemical

Car rusting	Wood burning	Ice melting
Physical Chemical	Physical Chemical	Physical Chemical

1. How is an object different after a physical change?

2. How is an object different after a chemical change?

28

* Answers to masters on pages 92E–92H

Enrichment

Obtain two identical iron nails. Paint one of the nails with house paint. Place both nails in a container of water. Ask students to *predict* what will happen. (Rust will form on the unpainted nail.) Allow the students to *observe* the nails daily for evidence of rust. When rust has formed on the unpainted nail, scrape some of it off and have students *compare* and *contrast* the properties of the nail before and after rusting occured. Ask students to *infer* why the painted nail did not rust. (The paint provided a protective coating which kept oxygen from joining with iron.)

Reinforcement

Provide students with a sample of iron wire. Allow them to change the sample physically. (bending, tearing, rolling, and so on) Then have students put their sample in a shallow dish of water to observe the formation of rust—a chemical change. *CAUTION:* To avoid punctures, remind students to handle the wire carefully around themselves and others.

TEACHING PLAN

Teaching Tip

- Have students *describe* where they have seen examples of tarnish on silver. (bowls, jewelry, spoons, forks, and so on)

3. Assess

Lesson Review

1. A physical change is a change in the size, shape, state, or appearance of matter.
2. Examples of a chemical change can include wood burning to ashes, rusting, and tarnishing of silver.
3. Challenge! The cool air causes the water vapor you breathe out to become a liquid. **Thinking Skill:** *Drawing conclusions*

Find Out On Your Own

Dry ice is a cold, white solid formed when carbon dioxide is greatly compressed and then cooled. Dry ice is used for cooling because it changes from a solid to a gas without becoming liquid. A physical change takes place when dry ice is formed. **Thinking Skill:** *Comprehending meaning*

Tarnish

[1]one looks silver colored, one looks black

Another sign of a chemical change can be a change in color. How do the two spoons in the picture look different? The spoons are made of silver. A gas in the air causes a black covering called tarnish to form on silver. A chemical change takes place because tarnish has different properties from the materials that form it.

Lesson Review

1. What is a physical change?
2. What are two examples of a chemical change?
3. **Challenge!** Why can you see your breath outside on a cold day?

Study on your own, pages 324–325.

LIFE SCIENCE
FIND OUT ON YOUR OWN
CONNECTION

Foods are often packed in dry ice to keep them safe to eat. Use library books to find out about dry ice. Write a paragraph telling how dry ice is formed. Tell whether a physical or chemical change happens when dry ice forms.

108

Teaching Options

Science Anecdote

Mixing water with wood ashes causes a chemical change to occur. The change produces lye—a chemical used in soap production.

Reteaching Suggestion ♦

Bring in two slices of bread and a toaster. Have a student cut or tear one slice. Question: **Is this a physical or chemical change?** (physical) Next, toast the other slice until it is very dark. Question: **Is this a physical or chemical change?** (chemical) Have students *compare* and *contrast* the color, texture, and strength of the bread and toast.

Workbook page 29 *

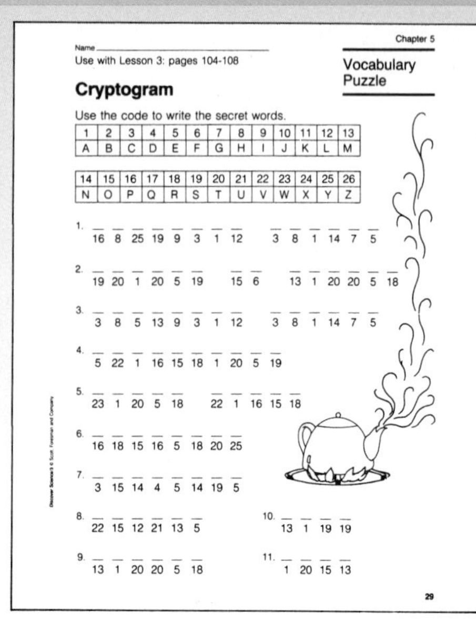

Name ______________

Use with Lesson 3: pages 104-108

Chapter 5

Vocabulary Puzzle

Cryptogram

Use the code to write the secret words.

1	2	3	4	5	6	7	8	9	10	11	12	13
A	B	C	D	E	F	G	H	I	J	K	L	M

14	15	16	17	18	19	20	21	22	23	24	25	26
N	O	P	Q	R	S	T	U	V	W	X	Y	Z

1. 16 8 25 19 9 3 1 12   3 8 1 14 7 5
2. 19 20 1 20 5 19   15 6   13 1 20 20 5 18
3. 3 8 5 13 9 3 1 12   3 8 1 14 7 5
4. 5 22 1 16 15 18 1 20 5 19
5. 23 1 20 5 18   22 1 16 15 18
6. 16 18 15 16 5 18 20 25
7. 3 15 14 4 5 14 19 5
8. 22 15 12 21 13 5
9. 13 1 20 20 5 18
10. 13 1 19 19
11. 1 20 15 13

29

♦ *Suitable as a language development activity*

Observing Physical and Chemical Changes

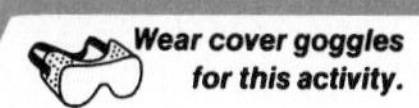

Suggested grouping: 2–4 students

Purpose
Observe a physical change and a chemical change.

Gather These Materials
- sugar • vinegar • baking soda
- 2 plastic cups • 2 plastic spoons
- 2 white labels

Follow This Procedure
1. Use a chart like the one shown to record your result.
2. Put one spoonful of sugar into a cup. Label the cup as shown.
3. Put 4 spoonfuls of vinegar into the cup with the sugar. Stir for one minute. Observe what happens. Record the result. *CAUTION: Do not taste.*

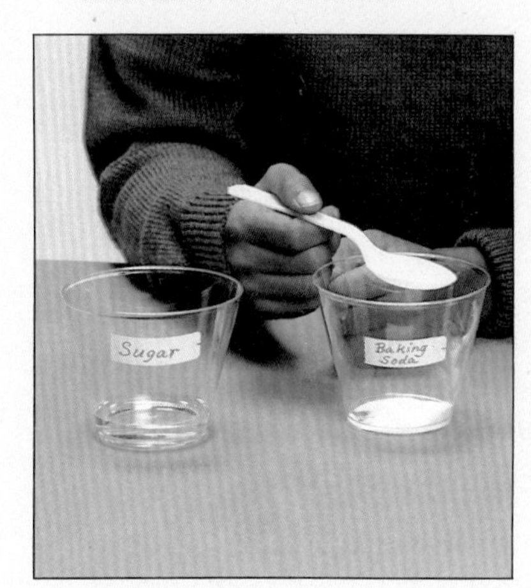

4. Use the other spoon to put one spoonful of baking soda into the other cup. Label the cup.
5. Put 4 spoonfuls of vinegar into the cup with the baking soda as shown in the picture. Observe what happens. Record the result. *CAUTION: Do not taste.*

Record Your Results

Sugar and vinegar	Sugar seems to disappear.
Baking soda and vinegar	Bubbling starts.

State Your Conclusion
1. In each of these tests, did a physical change or chemical change take place?
2. What is one way you can tell that a chemical change takes place?

Use What You Learned
If you mix a spoonful of salt with 4 spoonfuls of water, the salt seems to disappear. Is this a chemical change or a physical change? Explain your answer.

109

Resource Book page 57

Name ____
Use with Lesson 3: page 109
Observing Physical and Chemical Changes

Chapter 5
Activity Worksheet

Record Your Results

Sugar and vinegar	
Baking soda and vinegar	

State Your Conclusion
1. In each of these tests, did a physical change or chemical change take place?
2. What is one way you can tell that a chemical change takes place?

Use What You Learned
If you mix a spoonful of salt with 4 spoonfuls of water, the salt seems to disappear. Is this a chemical change or a physical change? Explain your answer.

57

* ***Answers to masters on pages 92E–92H***

Activity Results

Activity

Concept
Sugar and vinegar mixed together produce a physical change. Baking soda and vinegar produce a chemical change.

Objectives/Process Skills
- *Observe* what happens when sugar and baking soda are added to vinegar.
- *Record* observations.
- *Infer* which addition produces a chemical change and which produces a physical change.

Time Allotment
Allow 30 minutes.

Safety Tips (See page T24.)
- Instruct students not to place their faces over the mixture of baking soda and vinegar. The mixture could bubble up into their faces.
- Remind students never to taste chemicals used in science activities.
- Insist that the students wash their hands after finishing the activity.
- Wipe up spilled water immediately to avoid accidents.

Teaching Tips
- Allow water to evaporate from a salt water solution to demonstrate a physical change.
- Show students a rusty nail and explain that the rust is evidence of a chemical change.

Answers
State Your Conclusion
1. The sugar and vinegar produced a physical change. The baking soda and vinegar produced a chemical change.
2. The bubbling indicated a chemical change.

Use What You Learned
A physical change, the substances are still salt and water. **Thinking Skill:** ***Applying information to new situations***

TEACHING PLAN

Purpose
To develop the skills of collecting and organizing information using equal arm balances and pictographs to solve problems.

1. Motivate

Discussion
Question: **Where have you seen a balance or scale used?** (doctor's office, grocery store, hardware store, Post Office, and so on) Tell students that scientists also use balances. Discuss the importance to scientists of knowing the mass of objects precisely.

2. Teach

Teaching Tips
- Mass can be thought of in terms of density, the amount of matter in a certain volume.
- Point out that in the pictograph shown different-size symbols have different values.
- **Helpful Hint:** Demonstrate differences in density by letting the students lift a brick and a block of wood of the same size.

Using Equal Arm Balances and Pictographs

Problem: How can the mass of solids be measured and compared?

Part A. Using balances to collect information
1. Balances measure mass. A solid is placed on the left pan of the balance. Objects of known mass are placed on the right pan. When the pans balance, the mass of the things on the right pan is the same as the mass of the solid being measured. How can you tell the pans balance?
2. Look at the first picture. What is the total mass on the right pan? What is the mass of the wood cube?
3. What is the mass of the brick cube? the sponge cube? Are all the cubes the same size?

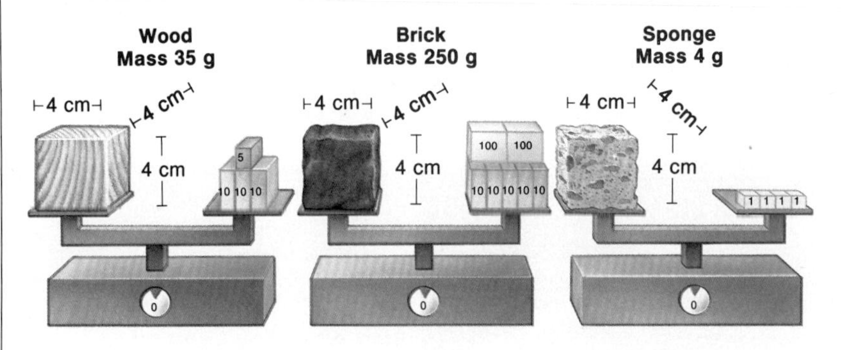

Part B. Using a Pictograph to Organize and Interpret Information.
4. This pictograph contains the information you collected about the mass of solids in Part A. What do the large objects stand for? The medium-sized objects? The small objects?

Teaching Options

Sample Pictograph for Part C

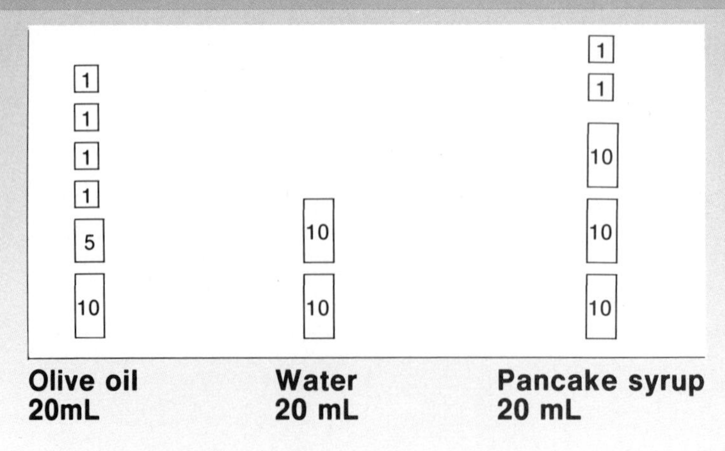

Olive oil 20mL | **Water 20 mL** | **Pancake syrup 20 mL**

♦ ***Suitable as a language development activity***

5. Look at the pictograph on the right. The solids in the pictograph are shown in order by mass. Which object has the least amount of mass? Which object has the most mass? Are all of the objects in the pictograph the same size?

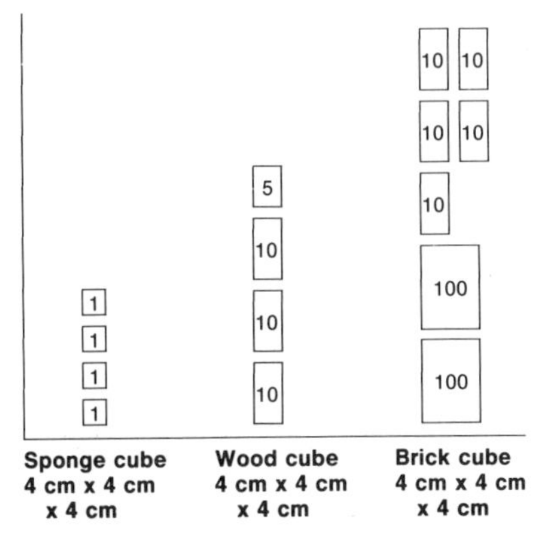

Part C. Using Balances and a Pictograph to Solve a Problem

Problem: How can liquids be arranged in order by mass?

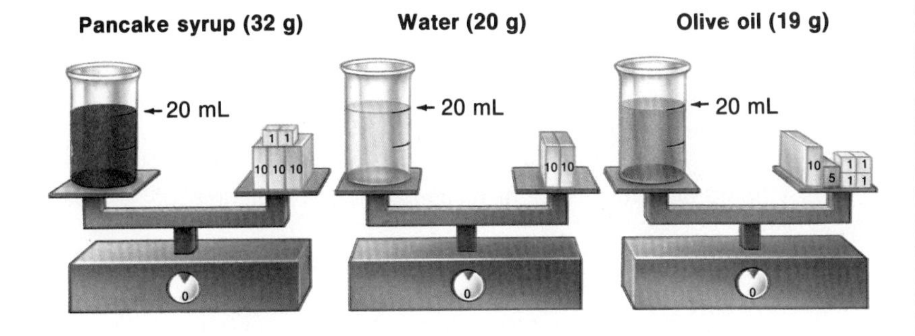

6. Use the balances to collect the information you need to solve the problem. Make a pictograph similar to the one shown in Part B to organize your information.
7. Look at your pictograph. How much liquid is in each jar? Which jar has the liquid with the greatest mass? the least mass?
8. If you mixed the oil with the water in the same jar, which liquid would float?

111

3. Assess

Part A

1. They are at the same level.
2. 35 g; 35 g
3. 250 g; 4 g; yes

Part B

4. Each large object stands for a mass of 100 g. Each medium-sized object stands for a mass of 10 g. Each small object stands for a mass of 1 g.
5. the sponge cube; the brick cube; yes

Part C

6. See Sample Pictograph for Part C in Teaching Options.
7. 20 mL; water; olive oil
8. oil

Reteaching Suggestion ♦

Find the mass of a book on a balance. Help students break the total mass into gram measurement blocks of 1g, 5g, and 10g each as represented in the illustration in Part A and the pictograph in Part B. Then have students decide in which place on the pictograph in Part B they would place the book, according to its mass. (Left to right, the masses are least to greatest.) On the chalkboard, have students draw the mass of the book in gram blocks, as it would appear in the pictograph.

Resource Book page 59 *

Name ______

Comprehension: cause and effect relationships

Chapter 5

Science and Reading

Planning Ahead

Read the story. Then answer the questions.

It had been raining hard all evening. Before going to bed, Mrs. Parker listened to the weather forecast on the radio and knew that the rain was expected to change to snow during the night. She planned to get up earlier than usual in the morning because if it snowed it would take longer to drive to work.

Sure enough, during the night the temperature dropped below freezing. By morning, several centimeters of snow had fallen. Several people who worked in Mrs. Parker's office got to work late, but she was able to get there on time.

1. How did Mrs. Parker know the rain was expected to change to snow?
2. Why would Mrs. Parker need more time in the morning if it snowed?
3. What happened during the night that caused the rain to change to snow?
4. Why did some people get to work late?

59

* Answers to masters on pages 92E–92H

REVIEW PLAN

Reviewing Science Words

1. property
2. volume
3. chemical change
4. mass
5. evaporates
6. states of matter
7. condense
8. physical change
9. water vapor
10. atom
11. matter

Reviewing What You Learned

1. b **3.** c **5.** d
2. c **4.** a **6.** b

Chapter 5 Review

Chapter Main Ideas

Lesson 1 • Everything that takes up space and has mass is matter. • Matter can be described by its properties. • The three states of matter are solid, liquid, and gas.

Lesson 2 • All matter is made of atoms. • Particles in solids are close together and move back and forth. The particles in liquids can move around each other. The particles in gases are far apart and move around freely.

Lesson 3 • Changes in the way matter looks are physical changes. A physical change takes place when matter changes from one state to another. • A chemical change takes place when matter changes to a different kind of matter.

Reviewing Science Words

atom	mass	states of matter
chemical change	matter	volume
condense	physical change	water vapor
evaporates	property	

Copy each sentence. Fill in the blanks with the correct word or words from the list.

1. Color is one ___ of matter.
2. A liquid takes the shape of its container and keeps the same ___.
3. A ___ takes place when fire changes wood into ashes and smoke.
4. The measure of how much matter makes up an object is called ___ .
5. Matter ___ when it changes from a liquid to a gas.
6. Solid, liquid, and gas are the three ___ .
7. Cooling the air causes water vapor to ___ .
8. A change in the shape of matter is a ___ .
9. Water in the form of a gas is called ___ .
10. A small particle that makes up matter is called an ___ .
11. All the objects around you are made of ___ .

112

Review Options

Cooperative Learning ♦

STAD Format (See page T23.)
Assign students to work in four- to five-member teams to study Chapter 5 Review. Students should work together to make sure that they and their teammates know the material in the chapter. After students have had enough time to study together, give them a test to complete individually (Chapter 5 Test A or B in the *Test Book*). Award Superteam certificates to teams whose average test scores exceed 90%, and Greatteam certificates to teams whose average test scores exceed 80%.

Test Book page 45 *

Name ____________

Chapter 5
Test A

Multiple Choice Choose the best answer.

1. The amount of matter in an object is that object's
a. mass.
b. space.
c. volume.

2. The color, size, and shape of an object are some of its
a. properties.
b. elements.
c. matter.

3. Solids, liquids, and gases are forms of
a. mass.
b. matter.
c. space.

4. The tiny particles that make up matter are called
a. atoms.
b. mass.
c. properties.

5. Which state of matter has particles that move around the most?
a. a liquid
b. a solid
c. a gas

6. What state of matter will fill any volume?
a. a solid
b. a liquid
c. a gas

7. Changes in such things as the size and shape of matter are called
a. chemical changes.
b. physical changes.
c. rusting.

8. What is water called when it is a gas?
a. water vapor
b. ice
c. water gas

9. The formation of drops of water on the outside of a cold pitcher is
a. condensation.
b. evaporation.
c. a chemical change.

10. What kind of change takes place when wood burns?
a. a chemical change
b. a physical change
c. a change of state

45

Reviewing What You Learned

Write the letter of the best answer.

1. A balance measures an object's
   (a) volume. (b) mass. (c) properties. (d) atoms.
2. A chemical change can take place when matter
   (a) evaporates. (b) condenses. (c) burns. (d) melts.
3. Which object has particles that move around the most?
   (a) a brick (b) water (c) air (d) a pencil
4. What makes up matter?
   (a) atoms (b) states (c) volume (d) properties
5. A physical change takes place when matter
   (a) rusts. (b) tarnishes. (c) burns. (d) condenses.
6. When you tell about an object's length, you describe a
   (a) volume (b) property (c) state (d) mass

Interpreting What You Learned

Write a short answer for each question or statement.

1. How are a flower and a book alike?
2. How is the shape of a gas different from the shape of a solid?
3. How can matter be described?
4. Explain how you would put these steps in the correct order:
   a. ashes
   b. log
   c. burning log

Extending Your Thinking

Write a paragraph to answer each question or statement.

1. Suppose a snowstorm covers the grass with snow. A week later, all of the snow is gone. Explain how this can happen.
2. When sugar is heated for a long time, it forms a solid black substance. What kind of change takes place? Explain your answer.

To explore scientific methods, see Experiment Skills on pages 356–357.

Interpreting What You Learned

1. Both are matter. **Thinking Skill:** *Comparing*

2. A gas takes the shape of its container; a solid keeps a certain shape. **Thinking Skill:** *Contrasting*

3. Matter can be described by its properties. **Thinking Skill:** *Restating or explaining ideas*

4. first, the log; second, the log burning; third, the ashes; representing a chemical change. **Thinking Skill:** *Sequencing*

Extending Your Thinking

1. Water molecules move faster as the temperature rises. The water molecules move farther apart. The snow melts and changes to liquid water in the form of puddles. Water molecules evaporate from the surface of the puddles to become water vapor. **Thinking Skill:** *Inferring*

2. A chemical change takes place. Two kinds of matter combine to form a new kind of matter. **Thinking Skill:** *Inferring*

Test Book page 46 *

Name ____________

Chapter 5
Test A

Short Answer Label the pictures above. Use these words:
evaporation liquid condensation solid

1. ______ 2. ______ 3. ______ 4. ______

Short Essay Use complete sentences to answer each question.

1. What are evaporation and condensation?

2. Compare the shapes and volumes of solids, liquids and gases.

46

Test Book page 47 *

Name ____________

Chapter 5
Test B

Multiple Choice Choose the best answer.

1. Everything that takes up space and has mass is called
   a. volume.
   b. matter.
   c. weight.
2. To describe matter, it is necessary to name its
   a. properties.
   b. weight.
   c. mass.
3. Three states of matter are
   a. solid, shape, mass.
   b. liquid, gas, solid.
   c. gas, liquid, space.
4. The particles that make up matter are
   a. states.
   b. atoms.
   c. properties.
5. Matter with a certain shape are
   a. solids
   b. liquids
   c. gases
6. The measure of the amount of matter in an object is its
   a. mass.
   b. volume.
   c. state.
7. Particles of matter that move about most freely form a
   a. gas.
   b. liquid.
   c. solid.
8. A chemical change takes place when matter
   a. melts.
   b. evaporates.
   c. rusts.
9. Liquid changed to a gas has
   a. condensed.
   b. evaporated.
   c. melted.
10. Matter which changes from a gas to a liquid has
    a. condensed.
    b. evaporated.
    c. boiled.
11. What does a balance measure?
    a. volume
    b. properties
    c. mass

47

Test Book page 48 *

Name ____________

Chapter 5
Test B

Short Answer Label each picture. Use these words:
gas solid liquid condensation

1. ______ 2. ______ 3. ______ 4. ______

Short Essay Use complete sentences to answer each question.

1. Explain the difference between a physical change and a chemical change.

2. Describe how particles move in solids, liquids and gases.

48

* *Answers to masters on pages 92E–92H*

Chapter 6 Work and Machines Planning Guide

TEACHING PLAN

Chapter Components	Skills	Materials
Chapter Opener/*DISCOVER:* Observing How People Move Objects pp. 114–115	*DISCOVER* p. 115 Science Process Skills *Observing, Experimenting*	*DISCOVER* p. 115 (groups of 3) 10 books, 10 boxes of crayons
Lesson 1 What Is Work? pp. 116–120	Thinking Skills Challenge!: *Applying information to new situations* Find Out On Your Own: *Explaining ideas*	Demonstration p. 116 1-meter-long cardboard sheet, toy car, string
Activity Observing How Friction Affects Motion p. 121	Science Process Skills *Observing, Measuring, Interpreting data*	(groups of 2) 1 ball of string, 15 large paper clips, 1 roll of waxed paper, 15 small terry towels, 15 hardcover books, 150 large metal washers
Lesson 2 What Are Simple Machines? pp. 122–127	Thinking Skills Challenge!: *Drawing conclusions* Find Out On Your Own: *Recognizing the main idea and supporting details*	Demonstration p. 122 ruler, cylindrical can or block, tape, clay, book
Activity Observing the Force Needed to Move an Object with a Lever p. 128	Science Process Skills *Observing, Making models, Measuring Interpreting data*	(individual) 30 tongue depressors, 30 pencils, 60 sugar cubes, 10 bottles of white glue, 1 roll of masking tape, 30 metric rulers
Science and Technology Teaching an Old Machine New Tricks p. 129	Thinking Skills *Recognizing cause and effect, Inferring*	
Lesson 3 What Are Compound Machines? pp. 130–131	Thinking Skills Challenge!: *Drawing conclusions* Find Out On Your Own: *Making generalizations, Making models*	Demonstration p. 130 2 pencils, pocket knife, pencil sharpener
Skills for Solving Problems Using Rulers and Bar Graphs pp. 132–133	Problem Solving Skills *Making decisions/Identifying and solving problems, Interpreting charts, maps, and graphs*	
Chapter Review pp. 134–135	Thinking Skills *Restating or explaining ideas, Recognizing cause and effect, Comprehending meaning, Drawing conclusions, Applying information to new situations, Organizing information*	

Teaching Options

Strategies	Extensions		Resource Masters
Cooperative Learning p. 114 (Also see p. T23.) Applying Whole Language p. 115 (Also see p. T30.)			Family Letter: *Resource Book* p. 63
Reading Strategies p. 116 (Also see pp. T26–T29.)	Reinforcement pp. 117, 118 Special Education p. 117 Enrichment pp. 118, 119	Science and Art p. 119 Game Suggestion p. 119 Reteaching Suggestion p. 120	Vocabulary Preview: *Workbook* p. 31 Science Activity: *Workbook* p. 32
			Activity Worksheet: *Resource Book* p. 67
Reading Strategies p. 122 (Also see pp. T26–T29.)	Reinforcement pp. 123, 124 Enrichment pp. 123, 125, 126 Special Education p. 124	Science and Music p. 125 Game Suggestion p. 126 Reteaching Suggestion p. 127	Science Skills: *Workbook* p. 33 Science and Social Studies: *Workbook* p. 34
			Activity Worksheet: *Resource Book* p. 69
Reading Strategies p. 130 (Also see pp. T26–T29.)	Special Education p. 131 Reteaching Suggestion p. 131		Vocabulary Puzzle: *Workbook* p. 35
	Reteaching Suggestion p. 133		Science and Reading: *Resource Book* p. 71
Cooperative Learning p. 134 (Also see p. T23.)			Chapter Tests: Forms A and B *Test Book* pp. 53–56

Classroom Management

Advance Preparation

Demonstration, p. 116
You will need a toy car and a piece of cardboard or posterboard about one meter (one yard) long.

Activity, page 121
Gather a small towel, a piece of waxed paper, and about ten metal washers for each group.

Demonstration, page 122
You will need a cylindrical can or block for this demonstration.

Activity, page 128
Each group will need a tongue depressor and 2 sugar cubes.

Vocabulary Review

Use the following sentences with your students to review the meanings of the italicized words.

1. A heavier object has more *mass* than a lighter object.
2. *Distance* is measured in centimeters or kilometers.
3. A *lubricant* helps things move by lessening friction.
4. A *machine* is a tool that makes it easier to do a certain job.
5. It is easier to push heavy objects up a *ramp* than it is to lift the same objects.

High-Potential Students

Encourage students to imagine they are living in prehistoric times, when everything a person needed came directly from nature. Challenge students to write a story about their family and how the necessities of life lead them to invent the six simple machines discussed in the chapter. Suggest that students discuss how the machines improved the ability to meet the following needs: shelter, food, water, and transportation. Ask students to write up their stories neatly and design a title page, illustrations, and a cover for the story. Encourage students to share their stories with the class.

Mainstreamed Students

Visually Impaired
Obtain examples of simple machines for students to examine and use.
Orthopedically Handicapped
Discuss the simple machines found in a wheelchair. These include the wheel and axle and the gears. Guide students to realize that a wheelchair ramp is an inclined plane.

Science Fair Projects

The Experiment Skills on p. 358 and the Investigate feature in the chapter can be used for science fair projects. You might also encourage interested students to do one of the following projects:

1. Collect working examples of as many levers as possible. Compare the amount of time and effort needed to complete the same job by hand with the amount of time and effort needed to complete the job using a lever. Make a chart showing the results. Experiment with changing the position of the fulcrum of each lever to determine the advantages and disadvantages of each position.
2. Use common objects to make a machine that will close a door, water plants, crack an egg, or open a window. Name the machine and design a box or container to package the machine. Make a diagram labeling the simple machines included in your new machine.

Bulletin Board

Encourage students to bring in pictures that illustrate the uses of the different kinds of machines.

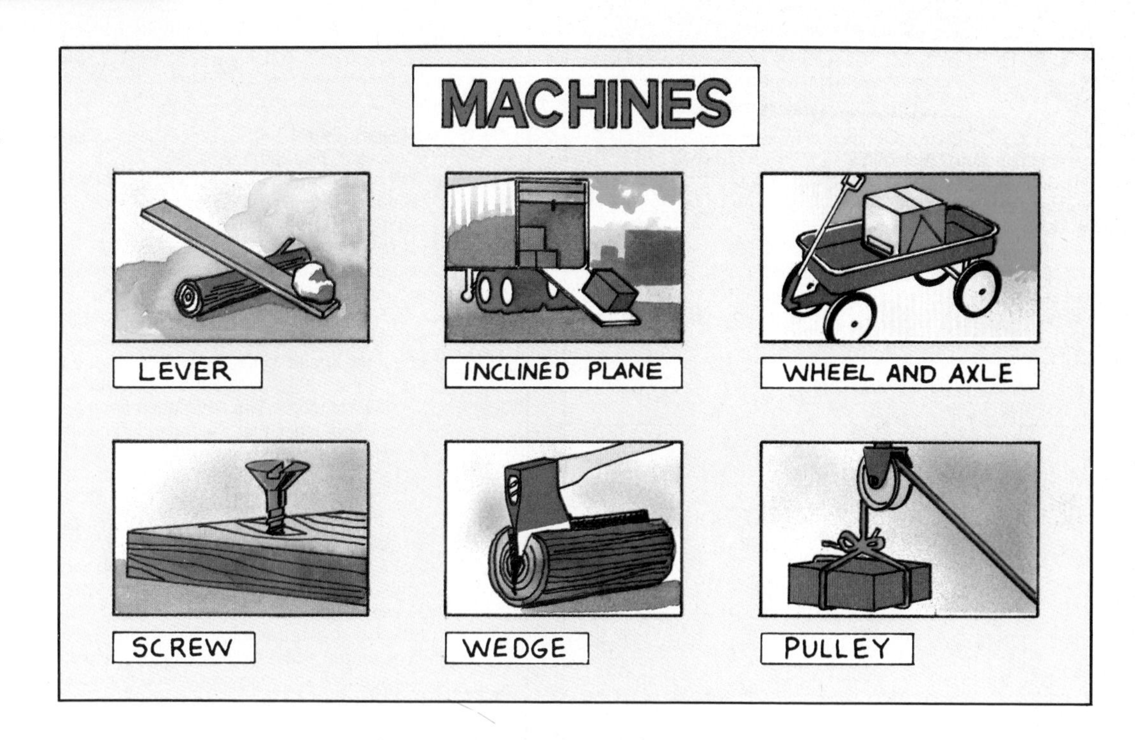

Chapter 6 Poster

Science Discovery Center

Use pages 65–70 from the *Science Discovery Center Book*. Place these worksheets in the appropriate pockets in the Science Discovery Center.

Overhead Transparencies

Use Transparency 10 from the package of color overhead transparencies.

CHAPTER 6 COPY MASTERS

Teacher's Resource Book

Dear Family,
Your student will be reading **Chapter 6: Work and Machines** in *Discover Science*, published by Scott, Foresman. We will learn that work is done when a force makes an object move through a distance. Machines help us do work. Simple machines include the lever, inclined plane, wedge, screw, pulley, and wheel and axle. A compound machine is made of two or more simple machines. You and your student can do this activity together to learn more about machines.

Visit an Auto Mechanic

Ahead of time, call a mechanic to find a time that you and your student may visit to see the machines and talk with him or her.

1. Ask the mechanic to show you and your student how a jack helps lift a car. Find the lever, fulcrum, and load. Explain that a person could not lift a car without a jack.
2. Look at some of the other machines. Look for simple and compound machines.
3. Ask the mechanic about safety precautions.

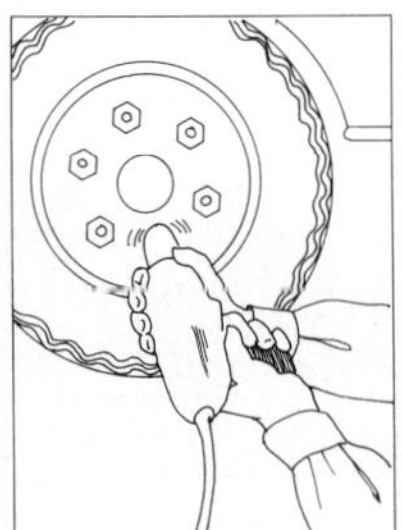

Workbook

Name ______________________

Use with Lesson 1: pages 116-120

Chapter 6

Vocabulary Preview

Work and Machines

Chapter Vocabulary			
(compound machine)	(fulcrum)	(lever)	(simple machine)
energy	(gear)	(load)	(wedge)
force	gravity	(pulley)	(wheel and axle)
friction	(inclined plane)	(screw)	work

Vocabulary Cards

1. Write each word on a card.
2. Find each word in the glossary. Copy the pronunciation under the word on the card.
3. Practice saying the words with a partner.

Word Meanings

1. Sort your vocabulary cards into the following two groups: things you can touch and things you cannot touch. You may use your glossary. Then compare your groups of cards with those of a classmate.
2. In the list above, circle the words that are things you can touch. Underline the words that are things you cannot touch. Two have been done for you.
3. Think about times that you have done work. Write a sentence using the word *work*.

Answers will vary.

To scientists, *work* means using a push or a pull to make something move. On your vocabulary card for *work*, write a sentence using the word as a scientist might.

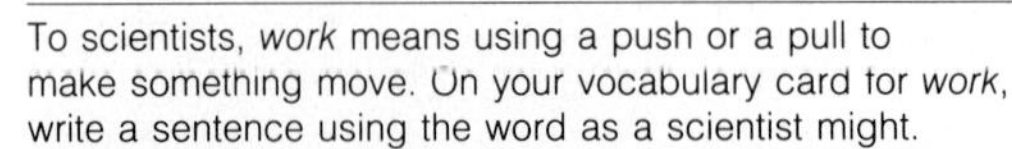

At Home

1. Practice saying the words. Learn their meanings.
2. Look for the words and pictures of the words in newspapers and magazines. You can make a poster.

Teacher's Notes: Ask students how the girl in the picture is doing work. (She is pushing the ball to make it move.)

Workbook

Name ______________________

Use with Lesson 1: pages 116-120

Chapter 6

Science Activity

Does Friction Help You Write?

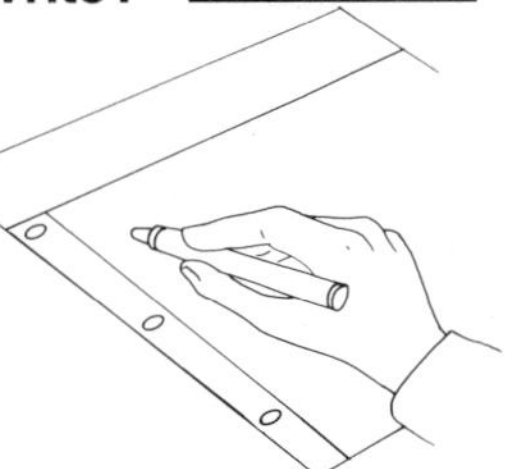

Gather These Materials

- crayon • piece of waxed paper
- piece of notebook paper • piece of fine sandpaper

Follow This Procedure

1. Without friction you could not write. Friction between your pencil and paper leaves a trail of pencil marks across a paper. Use the crayon to write your name on the piece of notebook paper. Notice how much friction there is between the paper and the crayon. Also notice how much the crayon is worn down.
2. Then write your name with the crayon on the waxed paper and the sandpaper. Notice how much friction there is between the papers and the crayon. Also notice how much the crayon is worn down each time.

Record Your Results

	Amount of Friction	How Much Crayon is Worn
Notebook paper	**some**	**some**
Waxed paper	**very little**	**very little**
Sandpaper	**much**	**much**

State Your Conclusions

1. List the papers in order from the least to most friction.

waxed paper, notebook paper, sandpaper

2. Explain how friction helps you write.

Friction between the paper and pencil causes bits of the pencil to stay on the paper.

Teacher's Notes: Tell students to use caution when writing on sandpaper, so they do not scrape their hands.

Teacher's Resource Book

Name ______________________

Use with Lesson 1: page 121

Observing How Friction Affects Motion

Chapter 6

Activity Worksheet

Record Your Results

Surface	Number of washers
Desk top	
Waxed paper	
Towel	

State Your Conclusion

1. On which surface did you need the most force to move the book?

____

2. Compare the amounts of friction caused by each of the materials under the book.

____

Use What Your Learned

Would it be easier to move a heavy carton across a wood floor or across a rug? Explain your answer.

____

Name ____________________

Use with Lesson 2: pages 122-127

Chapter 6

Science and Social Studies

Prehistoric Tools

Read the story.

Prehistoric people used stones and sticks to help them do work. Stick tools were used for digging and for hunting. Stone tools were made by hitting two stones together a few times. Flakes of stone were knocked away as the two stones hit. A sharp edge formed where the flakes of stone were knocked away. The sharp edge of the stone tool was used for chopping branches and bones. Because the stone was used for chopping, it is called a "chopper."

As time passed, new tools were made. One such tool was the hand ax. A hammer made of bone or wood was used to make the hand ax. The hammer was used to knock off flakes from the stone. The hand ax was shaped like a pear. It had sharp edges on both sides of the stone. The two sharp edges came together to form a point at the narrow end. The hand ax was used to cut branches and meat.

Answer the questions.

1. What was used to make a chopper?

   A stone

2. What was used to make a hand ax?

   A hammer

Look at the pictures of the stone tools below. Which one is a chopper? Which one is a hand ax? Write the name of each tool on the line under each picture.

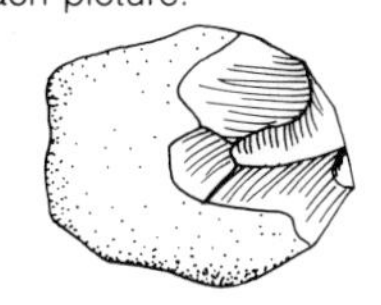

Chopper

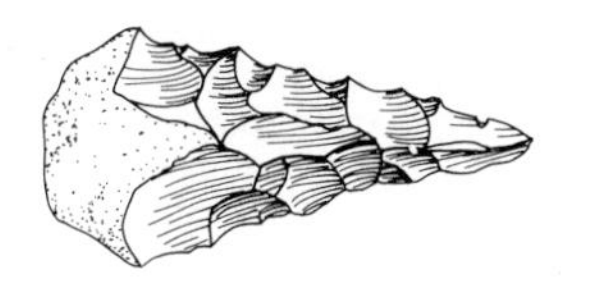

Hand ax

Teacher's Notes: Tell students that many prehistoric tools were made of flint. Later, flint was used to start fires.

Name ____________________

Use with Lesson 3: pages 130-131

Chapter 6

Science Skills

Identifying Machines

Six machines that help people do work are the lever, inclined plane, wedge, wheel and axle, screw, and pulley. Decide which machine is circled in each picture, and write its name on the line.

1. **wedge**

2. **screw**

3. **inclined plane**

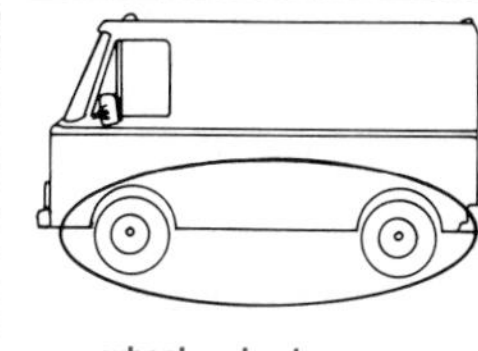

4. **wheel and axle**

5. **lever**

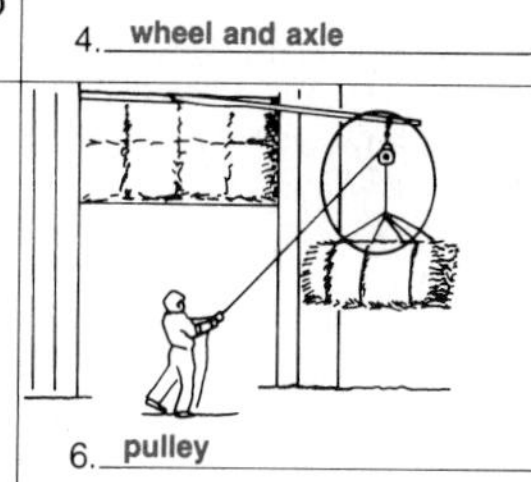

6. **pulley**

Teacher's Notes: Suggest that students make collages of simple machines from pictures cut from newspapers and magazines.

Name ____________________

Use with Lesson 2: page 128

Chapter 6

Activity Worksheet

Observing the Force Needed to Move an Object with a Lever

Record Your Results

Fulcrum	Line A	Line B	Line C
Force (easier, harder, hardest)			
Height: desk to tongue depressor end			

State Your Conclusion

1. Where was the fulcrum when you moved the sugar cubes the longest distance?

2. Compare the amount of force you used to move the sugar cubes when the fulcrum was moved from line A, to B, to C.

Use What You Learned

To move a heavy box with a lever, would you place the fulcrum close to the box or far away from the box? Explain.

Name ____________________

Use with Lesson 3: pages 130-131

Chapter 6

Vocabulary Puzzle

Wordsearch

Read the words in the word bank. Find and circle each word hidden in the puzzle. One is done for you.

Word Bank

compound machine	gravity	screw
energy	inclined plane	simple machine
force	lever	wedge
friction	load	wheel and axle
fulcrum	pulley	work
gear		

```
S C R E W Z A T G E A R O I N G I T
T Z A L C D Z W O A K C B E A R L T
U W I W O R K C O D V W Z L Y Z O O
G A F W L F G R A V I T Y E V F T U
R A R B L G P T F E R L L V X U R S
V S I M P L E M A C H I N E A L D Q
P E C P H O D A E M L P Q R B C F M
U I T K R A Q R K A W E D G E R P T
L E I L Z D C Z X V A A L M D U Z E
L C O M P L E X M A C H I N E M K N
E A N D N Q B C F M P N O A T L C E
Y W H E E L A N D A X L E Z R R L R
O U C C P Z T F O R C E N T L L O G
I N C L I N E D P L A N E V O A A Y
C O M P O U N D M A C H I N E Q X Z
```

Teacher's Notes: Play definition baseball during which students advance around the bases by defining the vocabulary words.

Teacher's Resource Book

Name ________________

Comprehension: drawing conclusions

Chapter 6

Science and Reading

The Sculptor

Read the story. Then circle the letter of the sentence that best answers each question.

Mike is a sculptor. He carves block of stones into works of art. He looks at the drawings and clay models he has made and draws crayon lines of the design onto the stone. He puts on his goggles, places the edge of a big steel tool against the stone, and hits it with a hammer. Chunks of stone fly everywhere. Mike turns the stone and works on all sides. He must work slowly and always think about what he is doing.

Mike uses grinding wheels to smooth the stone, and he uses files to smooth the hard-to-reach places. He rubs the stone with wet sandpaper and polishes it with a buffing wheel until it shines.

1. Why does Mike draw crayon marks on the stone?
   - a. It adds color to the sculpture.
   - (b.) It helps him konw what to chip away.
   - c. He likes to draw.

2. Why does Mike wear goggles when he works?
   - a. They help him see better.
   - b. They keep bright light out of his eyes.
   - (c.) They protect his eyes from flying chips.

3. Why must Mike work slowly and think about what he is doing?
   - (a.) He must not chip away too much stone.
   - b. He is not a very good sculptor.
   - c. He has never done this before.

Test Book

Name ________________

Chapter 6 Test A

Multiple Choice Choose the best answer.

1. What is needed to move an object? (1-1)
   - a. speed.
   - b. load.
   - (c.) force.

2. A force that makes objects pull toward each other is (1-2)
   - a. work.
   - b. friction.
   - (c.) gravity.

3. When an object moves through a distance, this is called (1-3)
   - a. energy.
   - b. gear.
   - (c.) work.

4. Whenever work is done, a person uses (1-4)
   - a. gravity.
   - (b.) energy.
   - c. ability.

5. Machines with a few or no moving parts are (2-1)
   - (a.) simple machines.
   - b. compound machines.
   - c. impossible to make.

6. A machine made of a bar that is supported underneath at some point is a (2-2)
   - a. inclined plane.
   - (b.) lever.
   - c. fulcrum.

7. A slanted road and a ramp are two examples of (2-2)
   - (a.) an inclined plane.
   - b. a wheel and axle.
   - c. a pulley.

8. A simple machine that is used to push objects apart is a (2-2)
   - (a.) wedge.
   - b. screw.
   - c. pulley.

9. What kind of machine is a bicycle? (3-1)
   - a. simple machine
   - b. lever
   - (c.) compound machine

10. A simple machine made of a rope and a wheel is a (2-2)
    - a. gear.
    - (b.) pulley.
    - c. wedge.

Numbers in parentheses after each question refer to the lesson number and the objective of that lesson.

Test Book

Name ________________

Chapter 6 Test A

Matching Put the letter of each object in the picture next to the term that describes the object.

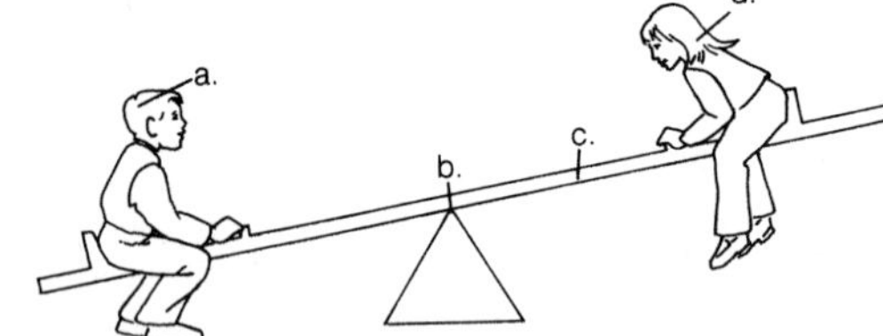

1. _a_ load
2. _c_ lever
3. _b_ fulcrum
4. _d_ force

Short Essay Use complete sentences to answer each question.

1. Name two ways to use machines safely. (3-2)

   Follow all safety rules. Ask an adult before using a machine. (Accept other reasonable answers.)

2. Describe a wheel and axle and explain how it works. (2-2)

   A wheel and an axle is a simple machine made of a rod attached to the center of a wheel. This machine is a lever that moves or turns objects when force is put on the wheel.

Numbers in parentheses after each question refer to the lesson number and the objective of that lesson.

Test Book

Name ________________

Chapter 6 Test B

Multiple Choice Choose the best answer.

1. A push or pull that changes the way an object moves is a (1-1)
   - a. lever.
   - (b.) force.
   - c. fulcrum.

2. The force caused by two objects rubbing together is
   - a. work.
   - (b.) friction.
   - c. gravity. (1-2)

3. The ability to work is (1-4)
   - a. gravity.
   - (b.) energy.
   - c. friction.

4. Gravity is a force that makes objects (1-2)
   - a. push apart.
   - (b.) pull together.
   - c. produce heat.

5. A pencil sharpener is a compound machine made up of several (3-1)
   - a. inclined planes.
   - (b.) simple machines.
   - c. fulcrums.

6. The point on which a lever is supported and turns is the (2-2)
   - a. load.
   - b. lever.
   - (c.) fulcrum.

7. A simple machine used to hold objects together is a (2-2)
   - a. wedge.
   - (b.) screw.
   - c. pulley.

8. Wheels with jagged edges like teeth are (2-2)
   - (a.) gears.
   - b. axle.
   - c. rod.

9. When two or more simple machines are put together, they form a (3-1)
   - a. lever.
   - b. simple machine.
   - (c.) compound machine.

10. Cars and roller skates both have a simple machine of a rod attached to a wheel that is a (2-2)
    - a. lever and fulcrum.
    - b. pulley.
    - (c.) wheel and axle.

Numbers in parentheses after each question refer to the lesson number and the objective of that lesson.

Name ____________________

Chapter 6
Test B

Matching Write the letter next to each simple machine in the blank next to the machine's name.

B.

A. C. D.

1. D lever
2. B inclined plane
3. A wedge
4. C wheel and axle

Short Essay Use complete sentences to answer each question.

1. What is an inclined plane and how does it help move objects? (2-2)

An inclined plane is a simple machine with a flat surface that is higher at one end. Inclined planes can help move an object to a higher or lower place.

2. Describe a pulley and name one way it can be used. (2-2)

A pulley is a simple machine made of a wheel and a rope. It can move things up and down like a flag on a pole.

Numbers in parentheses after each question refer to the lesson number and the objective of that lesson.

TEACHING PLAN

Major Concepts
Lesson 1 Work is done when a force uses energy to change the motion of an object.
Lesson 2 Six simple machines that help people do work are the lever, inclined plane, wedge, screw, wheel and axle, and pulley.
Lesson 3 Compound machines are combinations of simple machines.

Chapter Vocabulary
compound machine, energy, force, friction, fulcrum, gear, gravity, inclined plane, lever, load, pulley, screw, simple machine, wedge, wheel and axle, work

Getting Started
Have students look at the picture on page 114. Questions: **What tool is the man using to fix the bicycle?** (a wrench) **What do you think he might be doing with the wrench?** (turning a bolt or nut) **Why does he not just use his fingers?** (He can grip the bolt better with the wrench and turn the bolt tighter.)

Chapter 6

Work and Machines

People use tools to make work easier. This man is helping his grandson fix a bike. Find the tool the man is using.

114

Teaching Options

Cooperative Learning ♦

Jigsaw Format (See page T23.)
Assign the following topics at random to your cooperative learning teams.
Topic A: What are some simple machines, and how do they help people do work?
Topic B: What are some compound machines, and how do they make work easier?
Topic C: What is force, and how are gravity and friction related to force?
Topic D: How is energy related to work?
Have students search for information on their topic as they read the chapter. Then let all students with the same topic meet in an expert group to discuss the information. When students return to their teams, they may take turns presenting their topics to the team. Then give students a test covering all topics to complete individually (Chapter 6 Test A or B in the *Test Book*). Award Superteam certificates to teams whose average test scores exceed 90%, and Greatteam certificates to teams whose average test scores exceed 80%.

♦ ***Suitable as a language development activity***

Introducing the Chapter

The activity below will help you learn how people move objects. The lessons in this chapter will tell you what happens when people push and pull objects. You also will learn how people can move objects quickly and easily.

Observing How People Move Objects

A person is moving this car up a ramp that was made with a book. How is the person making the car move?[1] What are two ways a person could move the car down the ramp?[2]

Use a book to make a ramp like the one shown. Place a crayon box at the top of the ramp. Now try to move the box down the ramp without touching the box. You can use objects to help you.

Talk About It

1. How did you move the box down the ramp?
2. What would happen if you put a Ping-Pong ball at the top of the ramp?

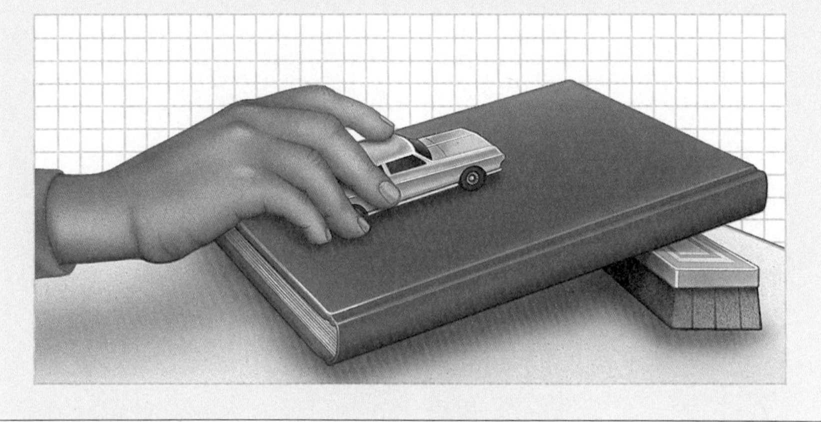

[1]by pushing

[2]pushing, pulling

DISCOVER

Objective ♦

This optional *DISCOVER* activity will help students explore and build background information about the concept concerning how forces cause objects to move. Later in the chapter students will be able to draw on this experience to help them assimilate the new content.

Science Process Skills

Observing, Experimenting

Materials

For each group of 3: book, object to support book, crayon box.

Teaching Tips

- After the students have moved the crayon box up and down the ramp, have them suggest how they could make the box easier to move. (greasing bottom of box, attaching wheels to the bottom)
- Have students describe ways they have observed people moving heavy objects up or down a ramp. (truckers' loading ramp, airplane ramp, furniture movers' ramp)

Answers

Talk About It

1. Answers will vary. Students might move box by pushing it with another object or blowing on it.

2. The ball would roll down the ramp.

Applying Whole Language ♦

Discuss the whole language framework with each Teaching Option you select. Here is an example applied to the Enrichment option on p. 126.

1. Purpose: To learn about machine hazards and write product safety rules to help children

2. Context: Team study, whole group sharing. Use printed and oral warnings as discussion starters.

3. Decisions: What are the parts of the machine? How can an accident happen?

4. Evaluation: Where is a good place to post our safety rules? (See p. T30.)

Resource Book page 63

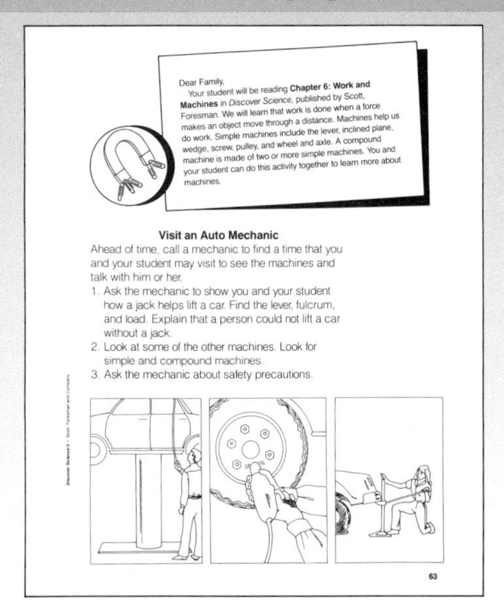

Dear Family,

Your student will be reading **Chapter 6: Work and Machines** in *Discover Science*, published by Scott, Foresman. We will learn that work is done when a force makes an object move through a distance. Machines help us do work. Simple machines include the lever, inclined plane, wedge, screw, pulley, and wheel and axle. A compound machine is made of two or more simple machines. You and your student can do this activity together to learn more about machines.

Visit an Auto Mechanic

Ahead of time, call a mechanic to find a time that you and your student may visit to see the machines and talk with him or her.

1. Ask the mechanic to show you and your student how a jack helps lift a car. Find the lever, fulcrum, and load. Explain that a person could not lift a car without a jack.
2. Look at some of the other machines. Look for simple and compound machines.
3. Ask the mechanic about safety precautions.

63

Science Background

Objects do not move unless a force, such as a push or pull, causes them to move. The force of gravity also can cause objects to move. Some objects, like the Ping-Pong ball, move more easily than others. Its smooth surface produces very little friction.

LESSON 1 pages 116–120

TEACHING PLAN

Lesson Objectives
- *Define* force and show how force relates to gravity and friction.
- *Define* work and *identify* situations involving work.
- *Explain* how energy and work are related.

Lesson Vocabulary
energy, force, friction, gravity, work

1. Motivate

Demonstration Activity ♦
Make a ramp on a table using a cardboard sheet 1 meter (about 3 feet) long. Raise one end of the ramp 30–40 cm (about 10–15 inches). Attach a string to a toy car. Allow a student to first pull the car up the ramp, then to lift the car to the top of the ramp.

Discussion
Discuss the fact that both methods achieve the same goal. Questions: **Which is easier, to pull the car up the ramp or to lift it?** (Pulling is easier.) **What caused the car to move?** (Explain that the student applied force to make the car move.)

1 What Is Work?

LESSON GOALS

You will learn
- how forces make objects move.
- how work is done.
- how work uses energy.

force (fôrs), a push or pull.

[1]the shopping cart
[2]pulling on rope
[3]book

You probably have seen a shopping cart like the one in the picture. The cart cannot move by itself. A person must push or pull the cart to make it move.

Force

A push or a pull is a **force.** Which picture shows a pushing force?[1] Which picture shows a pulling force?[2] A force changes the way an object moves. The shopping cart moves in the same direction the person pushes. Changing the direction of the force can make the shopping cart move in a different direction. You also can change the speed of an object. You can make an object move faster by using more force.

You need different amounts of force to move different objects. The heavier the object, the more force you need to move it. Would you need more force to move a book or a piece of paper?[3]

Pushing and pulling forces

Teaching Options

Science Background

Objects that start moving, stop moving, or change speed or direction change motion. Force is necessary for objects to change motion. Gravity is the force of one object pulling on another object. The earth's gravity pulls objects toward the center of the earth. The sun, the moon, and the other planets also have gravity. Friction is a force that opposes other forces on an object. Friction can be useful by helping people and objects slow down. Friction can also be undesirable when it interferes with the motion of an object or causes machine parts to wear out.

Reading Strategies ♦

1. Guide students' pre-reading by asking: What comes to mind when you think of the key words?
2. Assign these strategies: Writing a Memory Sentence and Finding and Writing Sentences for Vocabulary Words.(See pages T26–T29.)
3. Pair students to share what information is clear and unclear and initiate discussion using students' unanswered questions.

♦ ***Suitable as a language development activity***

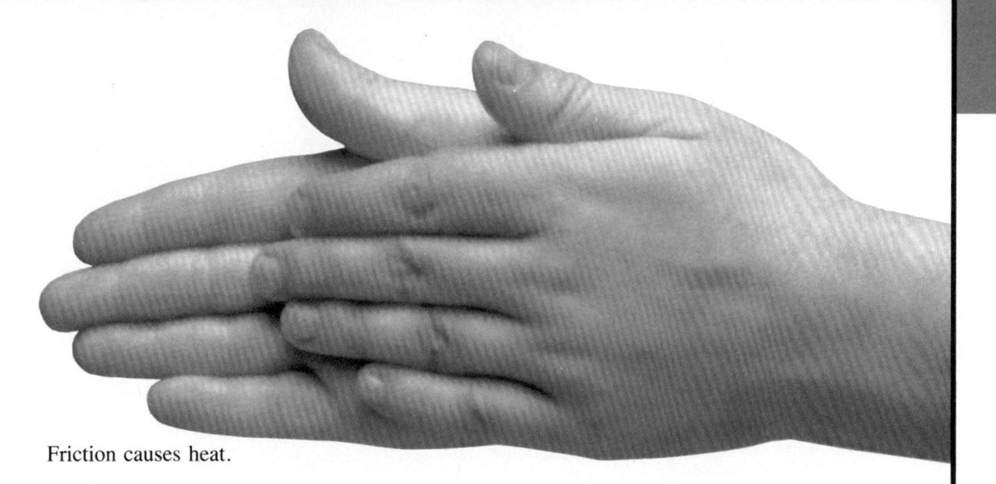

Friction causes heat.

What happens to a ball if you throw it into the air?[1] **Gravity** will pull the ball down. The earth's gravity is a force that pulls objects toward the center of the earth.

The more mass an object has, the more gravity pulls on it. How much an object weighs tells how much gravity pulls on the object. Suppose you weigh sixty pounds. Then the pull of the earth's gravity on you is sixty pounds.

Imagine rolling a ball across a floor. Think about how the ball would stop moving. **Friction** is a force that makes moving objects slow down or stop. Two objects rubbing against each other cause friction. The friction between the ball and floor would make the ball stop moving.

Look at the hands in the picture rubbing together. Rub your own hands together. Notice how your hands feel warm. Objects always become warm when they rub together. Friction causes heat.

INVESTIGATE!

Some objects have rough surfaces. Others have smooth. Find out how the roughness or smoothness of surfaces affects the amount of friction they produce. Write a hypothesis and test it with an experiment. You might compare how an object slides across a rough surface and a smooth surface.

gravity (grav′ə tē), the force that makes objects pull toward each other.

friction (frik′shən), the force caused by two objects rubbing together that slows down or stops moving objects.

[1]will fall down

2. Teach

Teaching Tips

- Have students place a book, a ruler, and a piece of paper on their desk tops. Ask them to move each item 12.5 cm (5 inches) using only their little finger. Have students *compare* the amount of force it took to move each item. Lead students to *conclude* that heavier objects require more force to move than lighter objects.
- Point out that gravity affects all things. Illustrate this point by asking students to draw what they think the classroom might look like if gravity did not hold things in place.
- Refer students to the *DISCOVER* on page 115. Question: **Why would the crayon box roll downward on the ramp, but we need to use force to move it upward?** (We need to overcome the force of gravity, which pulls the crayon box down.)

Investigate!

Accept any testable hypothesis. Most students' data will support various hypotheses. One possible hypothesis is: *Rough surfaces produce more friction than smooth surfaces.* Students should find out that less friction is produced when an object is rolled across a smooth surface than when it is rolled across a rough surface.

Workbook page 31 *

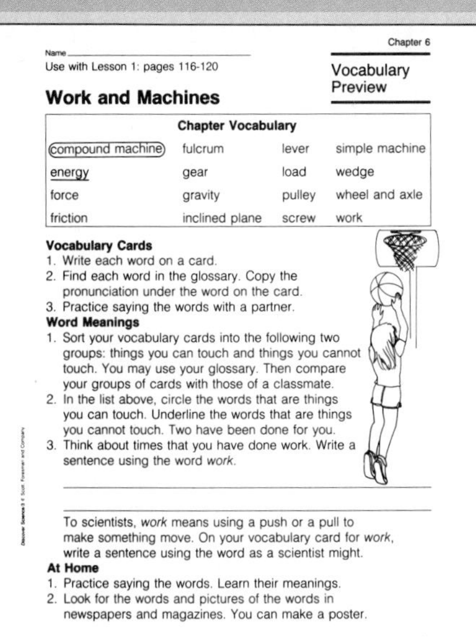

Name__________

Use with Lesson 1: pages 116-120

Chapter 6

Vocabulary Preview

Work and Machines

Chapter Vocabulary			
compound machine	fulcrum	lever	simple machine
energy	gear	load	wedge
force	gravity	pulley	wheel and axle
friction	inclined plane	screw	work

Vocabulary Cards
1. Write each word on a card.
2. Find each word in the glossary. Copy the pronunciation under the word on the card.
3. Practice saying the words with a partner.

Word Meanings
1. Sort your vocabulary cards into the following two groups: things you can touch and things you cannot touch. You may use your glossary. Then compare your groups of cards with those of a classmate.
2. In the list above, circle the words that are things you can touch. Underline the words that are things you cannot touch. Two have been done for you.
3. Think about times that you have done work. Write a sentence using the word *work*.

To scientists, *work* means using a push or a pull to make something move. On your vocabulary card for *work*, write a sentence using the word as a scientist might.

At Home
1. Practice saying the words. Learn their meanings.
2. Look for the words and pictures of the words in newspapers and magazines. You can make a poster.

31

**** Answers to masters on pages 114E–114H***

Reinforcement

Use the school gym or playground to play a game of kickball. As students play, direct their attention to the amount of force needed to move the ball, the amount of force needed to change the direction of the ball, and the amount of force needed to stop the ball. Return to class and ask students to describe when *force* and *gravity* were used during the game. Question: **Did the amount of force affect how the ball moved?** (yes) Ask students to explain how they used force to stop the ball and to change its direction.

Special Education

Allow students with learning disabilities to *observe* forces by rolling a ball in an area where it will continue rolling until it stops. Explain that students can use force to stop the ball from rolling, or the force of friction will slow down the ball gradually until it stops rolling.

LESSON 1 pages 116–120

TEACHING PLAN

Teaching Tips

- **Possible Misconception:** The term *work* has multiple meanings. Students might confuse phrases such as "going to work" or "doing homework" with the scientific use of the word *work*. Make sure different meanings of the term *work* are clear to students.
- To illustrate how you do more work when you use more force, have volunteers lift a science book in one hand and three books in the other hand at the same time. Have students compare the feeling of tiredness in their arms. (The arm holding one book feels less tired than the one holding three books because less force is being exerted.)

These students use force to lift the pumpkins.

work (wėrk), something done whenever a force makes an object move through a distance.

Work

Place a pencil on the top of your desk. Now move the pencil to the side of the desk. You need to use a force to make the pencil move. **Work** is done whenever a force makes an object move. You do work when you move the pencil across the desk.

The amount of work you do depends on how much force you use and how far the object moves. Suppose you lifted a pumpkin like one of those in the picture. You would be doing work when you used force to make the pumpkin move. Now suppose you lifted several pumpkins the same distance as you lifted the first pumpkin. You would use more force to lift several pumpkins. You do more work if you use more force to move something the same distance.

118

Teaching Options

Science Anecdote

In the late 1700s James Watt, a Scottish engineer, first used the term *horsepower* as a unit for measuring work. To convince people to use his engines, he needed to compare the amount of work his engines could do with the amount of work a horse could do.

Reinforcement

To demonstrate the concept of work, have an arm-wrestling match between two volunteers. Point out that without movement, there is no work. As the match progresses, ask the class to identify when work is and is not being done.

Enrichment

Provide students with two rubber band-powered toy airplanes. *CAUTION:* Instruct students to wear cover goggles. Load one airplane with a clay weight at the balance point (the point where the plane will balance on a pencil), and leave the other airplane unweighted. Wind each propeller the same number of winds. Let the planes fly and compare the distance traveled with the load each plane had to carry.

♦ ***Suitable as a language development activity***

The children in the picture on the left are using the same amount of force to make the swings move. Look for one person in the picture who is swinging higher than the other. This boy is doing more work because he is moving his swing a greater distance. The amount of work being done depends on how far the swing moves. More work is done when the same amount of force moves an object a greater distance.

Now look at the boy in the picture on the right. This boy is not doing work because he is not making the wall move through a distance.

Which students are doing work?[1]

[1]the students on the swings

119

Teaching Tips

- Call attention to the children in the picture. Questions: **Which child might have been swinging longer?** (Students should infer that the child swinging higher than the others might have had more opportunity to use more force.) **What might the children do to swing higher?** (Use more force by having someone push each of them.)
- Encourage students to describe instances when they have used force to do work. (Answers will vary.)

Science and Art

Have students make a collage showing energy-work relationships. Students may use magazines, newpapers, or student drawings. Examples: a girl eating an apple and then riding her bicycle.

Game Suggestion ♦

Divide students into three groups. Assign each group one-third of the letters of the alphabet in random order. Allow ten minutes for each group to make a list of objects that move, display work being done, or exert force causing something to move. The objects must each begin with a different letter of the alphabet as randomly assigned. For example, A-airplane, T-top, M-mousetrap, C-car, B-bicycle.

Enrichment

Have interested students *observe* their family for two days and *list* examples of force and work done each day. Examples include sweeping, bicycle riding, driving the automobile, putting dishes away, pushing the lawnmower, and so on. Review with students that work is done whenever a force makes an object move.

TEACHING PLAN

3. Assess

Lesson Review

1. A force is a push or a pull.
2. Work is done whenever a force makes an object move.
3. Energy is used whenever work is done.
4. Challenge! The smooth surface of ice produces less friction than the surface of grass. Therefore, slowing down or stopping can be difficult on ice. **Thinking Skill:** *Applying information to new situations*

Find Out On Your Own

Lubricants are substances that reduce friction. Some examples of lubricants are oil, grease, and graphite. A door would open easily if a lubricant were used on the hinges. **Thinking Skill:** *Explaining ideas*

energy (en′ər jē), the ability to do work.

Energy and Work

You can move around because you have **energy**—the ability to do work. You use energy whenever you use force to move an object through a distance. The more work you do, the more energy you need. Your body releases energy from food. You use this energy to do work. The boy in the pictures uses energy from his body to pull the dog. All objects that have energy can do work.

Lesson Review

1. What is force?
2. When do objects do work?
3. When do objects use energy?
4. **Challenge!** What makes walking on ice more difficult than walking on grass?

Study on your own, pages 326–327.

This boy is using energy to do work.

PHYSICAL SCIENCE
FIND OUT ON YOUR OWN

A lubricant is a material that can lessen friction between two objects. Look in library books to find out about lubricants. Write a few sentences telling about different kinds of lubricants. Explain how using a lubricant might help a door open easily.

120

Teaching Options

Science Anecdote

Energy cannot be made or destroyed. It can only change form. When you push an object, it moves because you give it energy.

Reteaching Suggestion ♦

Obtain two medium-sized boxes. Tape or glue sandpaper to the bottom of one box. Fill both boxes with books of equal weight. Place both boxes on the floor. Instruct a volunteer to push both boxes at the same time as far as possible. Have the class describe what is happening and the forces at work. (Gravity pulls the heavy boxes down. The student exerts force to overcome friction and moves the boxes which produces work.) Question: **Which box was more difficult to move?** (the box with sandpaper.) **Why?** (because the sandpaper produced more friction, and more force was needed to overcome the friction)

Workbook page 32 *

Name ____________
Use with Lesson 1: pages 116-120

Chapter 6
Science Activity

Does Friction Help You Write?

Gather These Materials
• crayon • piece of waxed paper • piece of notebook paper • piece of fine sandpaper

Follow This Procedure
1. Without friction you could not write. Friction between your pencil and paper leaves a trail of pencil marks across a paper. Use the crayon to write your name on the piece of notebook paper. Notice how much friction there is between the paper and the crayon. Also notice how much the crayon is worn down.
2. Then write your name with the crayon on the waxed paper and the sandpaper. Notice how much friction there is between the papers and the crayon. Also notice how much the crayon is worn down each time.

Record Your Results

	Amount of Friction	How Much Crayon is Worn
Notebook paper		
Waxed paper		
Sandpaper		

State Your Conclusions
1. List the papers in order from the least to most friction.
2. Explain how friction helps you write.

32

♦ *Suitable as a language development activity*

ACTIVITY

Observing How Friction Affects Motion

Suggested grouping: pairs

Purpose
Measure and *compare* the amount of force needed to drag a book across materials that cause different amounts of friction.

Gather These Materials
• string • large paper clip • book
• waxed paper • small towel
• metal washers

Follow This Procedure
1. Use a chart like the one shown to record your observations.
2. Tie the ends of the string together to make a loop. Put the string around the cover of the book.
3. Place the book on your desk top about 10 cm from the edge. Let the loop of the string hang over the edge of the desk.
4. Straighten a paper clip to form an S-shape. Hang the paper clip from the string. *CAUTION: Use the paper clip carefully.*
5. Put a metal washer on the paper clip as shown in the picture. Add metal washers one at a time, until the book begins to move. Record the number of washers needed.
6. Tape waxed paper to the desk top. Repeat steps 3–5.
7. Tape a towel to the desk top. Repeat steps 3–5.

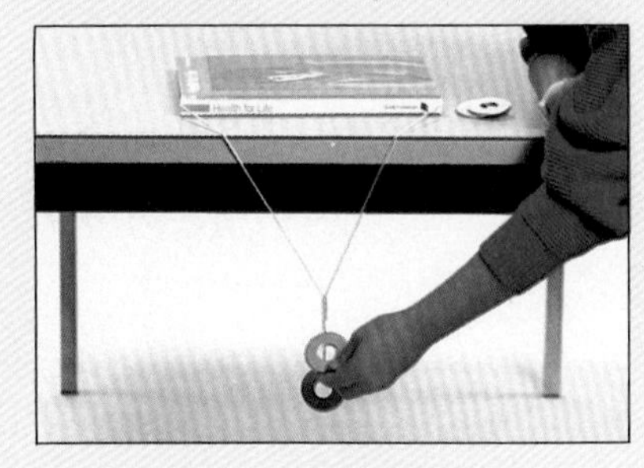

Record Your Results

Surface	Number of washers
Desk top	Results will vary. Most washers for towel, least for waxed paper
Waxed paper	
Towel	

State Your Conclusion
1. On which surface did you need the most force to move the book?
2. Compare the amounts of friction caused by each of the materials under the book.

Use What You Learned
Would it be easier to move a heavy carton across a wood floor or across a rug? Explain your answer.

121

Concept
More force is needed to move an object as the friction between two objects increases.

Objectives/Process Skills
• *Observe* the movement of a book across various surfaces.
• *Measure* the amount of force needed to pull a book across various surfaces.
• *Record* data.
• *Compare* the amounts of friction between the surfaces and the book.

Time Allotment
Allow 30 minutes.

Safety Tip (See page T24.)
• Remind students of the safe use of paper clips. Tell students to keep the clips away from the face, and to carefully straighten the clip by pulling on its side rather than on the pointed end.

Teaching Tip
• Students can perform the same tests using fine, medium, and coarse sandpaper under the books.

Answers
State Your Conclusions
1. The most force was needed to move the book across the towel.
2. The towel causes the most friction with the book. The waxed paper causes the least friction.

Use What You Learned
It would be easier to move the carton across the wood floor. There would be less friction between the smooth surfaces of the carton and the floor than between the carton and the rough rug.
Thinking Skill: *Applying information to new situations*

Resource Book page 67

Name ______
Use with Lesson 1: page 121
Observing How Friction Affects Motion

Chapter 6
Activity Worksheet

Record Your Results

Surface	Number of washers
Desk top	
Waxed paper	
Towel	

State Your Conclusion
1. On which surface did you need the most force to move the book?
2. Compare the amounts of friction caused by each of the materials under the book.

Use What Your Learned
Would it be easier to move a heavy carton across a wood floor or across a rug? Explain your answer.

67

Activity Results

See chart overprint above.

* *Answers to masters on pages 114E–114H*

TEACHING PLAN

Lesson Objectives

- *Describe* the lever and the inclined plane and *explain* how these machines can make work easier.
- *Explain* the functions of the wedge and the screw.
- *Describe* and give examples of the wheel and axle and the pulley.

Lesson Vocabulary

fulcrum, gear, inclined plane, lever, load, pulley, screw, simple machine, wedge, wheel and axle

1. Motivate

Demonstration Activity ♦

Use a ruler and a cylindrical can or block to demonstrate how a lever works. Tape the can to a flat surface and place the ruler on top of it. Using clay, secure a book at one end of the ruler, and then push down on the other end.

Discussion

Discuss the parts of the lever. Point out that the can is the fulcrum and the book is the load. Question: **How can this lever be used to lift a quarter?** (Place the quarter on one end of the stick and push down on the other end.)

2 What Are Simple Machines?

LESSON GOALS

You will learn
- how levers and inclined planes help make work easier.
- how wedges and screws help people do work.
- how people can use a wheel and axle and a pulley to do work.

simple machine (sim′ pəl mə shēn′), one of six kinds of tools with few or no moving parts that make work easier.

lever (lev′ər), a simple machine made of a bar that is supported underneath at some point.

fulcrum (ful′krəm), point on which a lever is supported and turns.

load (lōd), an object that is being moved.

Suppose you needed to move an object. You might use tools to help you. Machines are tools that use energy to do work. **Simple machines** are machines with a few or no moving parts. You probably have used several of the six kinds of simple machines.

Lever and Inclined Plane

Pretend you are lifting a heavy object like the one in the picture. Notice how these children use a board as a **lever.** A lever is a simple machine made of a board that is used to move objects. Find the point where the post holds up the board. The point on which a lever is held up is called the **fulcrum.** The lever moves back and forth on the fulcrum. The children push down on one end of the lever to move the object on the other end. The object being lifted is called the **load.**

Lifting a rock with a lever

122

Teaching Options

Science Background

Simple machines multiply the amount of force a person exerts. Other machines change the direction of movement. Simple machines transfer energy from a person to an object. The amount of work the person does equals the amount of work the machine produces, even though the machine increases the amount of force. A person using a lever exerts a small force to push the end of the lever a large distance. The other end of the lever exerts a large force to move an object a smaller distance.

Reading Strategies ♦

1. Guide students' pre-reading by asking: What ideas do you think you would find in the lesson to answer the question-title?
2. Assign these strategies: Visualizing Information and Finding and Writing Sentences for Vocabulary Words. (See pages T26–T29.)
3. Pair students to share what information is clear and unclear and initiate discussion using students' unanswered questions.

♦ *Suitable as a language development activity*

A lever moves back and forth on a fulcrum.

A hammer can be used as a lever.

[1]person on the left
[2]person on the right
[3]person on the right
[4]shortest part

This seesaw is really a lever. Which person on the seesaw is the force?[1] Which person is the load?[2] Notice how one person can lift the other person. Which person is closer to the fulcrum?[3] You can lift a load most easily by moving the fulcrum close to the load. Suppose the children moved the fulcrum away from the load. A person would need more force to lift the other person.

A hammer is another kind of lever. This hammer pulls a nail out of a piece of wood. Which part of the hammer is the fulcrum?[4]

2. Teach

Teaching Tips

- **Possible Misconception:** Students might think that a machine is a complicated piece of equipment. Some students might be surprised to learn that the six simple machines are really considered machines. Make sure students understand that a machine is any tool that can help a person do work or make work easier.
- Turn the light switch on and off several times. Point out that the light switch is a good example of a lever that makes a task easier and safer.
- Question: **What are some other things that are levers?** (Students might *name* the following: broom, shovel, crowbar, pliers, ice cube or sugar cube tongs, nutcracker, scissors, bottle cap opener, and tweezers.)

Reinforcement

Place a pencil under a metric ruler at the 25 cm (about 10 inches) mark. Place a book at the 3 cm (about 1 inch) mark of the ruler and ask a student to push down on the opposite end of the ruler to lift the book. Repeat with the pencil at the 6 cm (about 2.5 inches) mark and ask the student to *compare* the effort it took to lift the book. (Students should realize more effort was needed when the book was farther from the pencil—at the 3 cm mark.)

Enrichment

Have students look for examples of levers. Divide the students into groups and assign each group several tools. Have each group cooperatively locate the fulcrum, load, and force in each tool, and then report this information to the other groups. Tools include a broom, shovel, crowbar, fishing pole, pliers, hedge clippers, ice cube or sugar cube tongs, baseball bat, paper cutter, tennis racket, golf club, hockey stick, bottle cap opener, tweezer, canoe paddle, boat oars, wheel barrow, and nutcracker.

TEACHING PLAN

Teaching Tips

- Help students understand that a plane is any flat surface and that an inclined plane is a slanted or tilted flat surface.
- Question: **What could you do to reduce the amount of force needed to move an object up a path?** (Lead students to *infer* that the path could move diagonally back and forth up the hill. This route reduces the steepness of an incline, therefore reducing the force needed.)

inclined (in klīnd′) **plane**, a simple machine that is a flat surface with one end higher than the other.

The man in the picture is using a ramp to move a heavy object into a truck. A ramp is an **inclined plane** that helps move objects. An inclined plane is a simple machine with a flat surface that is higher at one end. You can use an inclined plane to help move an object to a higher or lower place.

You might have seen inclined planes in other places. A slanted road is an inclined plane. The person in the picture uses an inclined plane to go into the building. A path going up a hill also is an inclined plane. Notice that this path is steeper than the ramps in the picture. You would need more force to move an object up the path than up the ramps.

Inclined planes

Teaching Options

Science Anecdote

Some parts of the body are like simple machines. For example, arms are levers. The elbow is the fulcrum, the biceps provide the effort to pull the elbow, and the fist represents the resistance. Teeth are wedges that cut through food.

Reinforcement

To illustrate the advantage a screw has over a nail in terms of holding capacity, insert a small wood screw into a piece of wood and pound a nail into the same piece of wood. Allow students to compare the amount of force needed to pull the nail and the screw out of the wood using a claw hammer. (Students will find that more force was needed to pull out the screw, therefore the screw has greater holding ability.)

Special Education

To illustrate the concept of a screw, have students with learning disabilities make a screw using a pencil and a strip of paper. Have students wrap the paper in a spiral around the pencil. Refer students to the picture on page 125 for assistance. Point out that the edge of the paper is an inclined plane, and is also like the ridges of the screw.

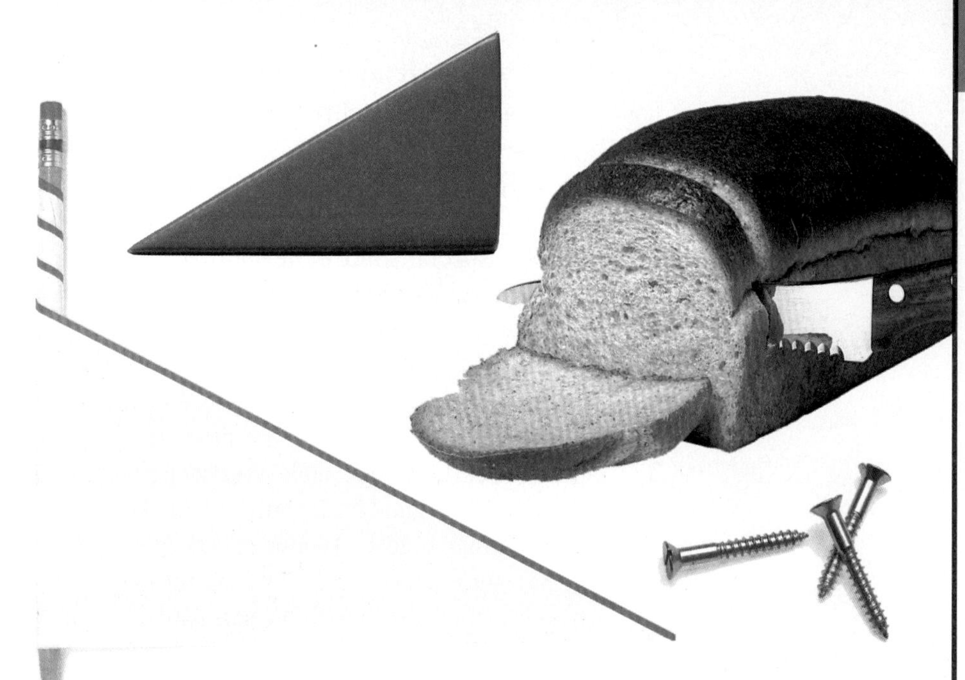

Find the wedges and the screws.

Wedge and Screw

Have you seen a doorstop like the one in the picture? This doorstop is a **wedge**—a simple machine used to push objects apart. Notice how two inclined planes come together to make a wedge. You might have seen other wedges. An ax is a wedge that splits wood. Find another wedge in the picture.[1]

A **screw** is a simple machine used to hold objects together. The picture shows that a screw really is an inclined plane wrapped around a rod. Find the colored edge of the inclined plane being wrapped around the pencil. This edge is like the ridges on a screw. What are some places you have seen screws?[2]

SCIENCE IN YOUR LIFE

A jar lid is a large screw. The ridges inside the lid of the jar are the ridges of the screw. These ridges hold the jar and the lid together.

wedge (wej), a simple machine used to cut or split an object.

screw (skrü), a simple machine used to hold objects together.

[1]the knife
[2]answers might include doors, tools, or furniture

Teaching Tips

- Display pictures of animals and have students identify body parts that are used as a wedge. (Students might ***name*** teeth, claws, and woodpecker's bill.)
- Point out that a screw can be used to lift objects. For example, screw-type jacks are used to help lift automobiles.

Workbook page 33 *

Name ____________
Use with Lesson 2: pages 122-127

Chapter 6

Science and Social Studies

Prehistoric Tools

Read the story.

Prehistoric people used stones and sticks to help them do work. Stick tools were used for digging and for hunting. Stone tools were made by hitting two stones together a few times. Flakes of stone were knocked away as the two stones hit. A sharp edge formed where the flakes of stone were knocked away. The sharp edge of the stone tool was used for chopping branches and bones. Because the stone was used for chopping, it is called a "chopper."

As time passed, new tools were made. One such tool was the hand ax. A hammer made of bone or wood was used to make the hand ax. The hammer was used to knock off flakes from the stone. The hand ax was shaped like a pear. It had sharp edges on both sides of the stone. The two sharp edges came together to form a point at the narrow end. The hand ax was used to cut branches and meat.

Answer the questions.

1. What was used to make a chopper?

____

2. What was used to make a hand ax?

____

Look at the pictures of the stone tools below. Which one is a chopper? Which one is a hand ax? Write the name of each tool on the line under each picture.

____

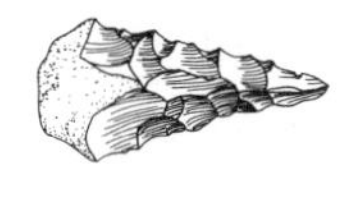

____

33

* **_Answers to masters on pages 114E–114H_**

Science and Music

Provide reference books with pictures of musical instruments. Have interested students locate the simple machines found on common musical instruments. (The school music teacher may be able to provide actual instruments.) Examples include: piano—levers, wedges, screws, and wheels and axles; French horn—the valves are a combination of levers, and wheels and axles)

Enrichment

Allow students to use reference books to locate the uses of as many simple machines as they can. Encourage students to make models or drawings of the various machines. Hold a contest for the most unusual use for a simple machine.

TEACHING PLAN

Teaching Tips

- Draw a skid (like a single ski) on the chalkboard. Explain that before wheels and axles were invented, the skid was one way that objects could be moved more easily. Ask students to ***name*** places in or around the classroom where wheels are, or could be, used instead of a skid. (chairs or desks)
- Point out that more work can be done with gears than with plain wheels.
- Question: **Why is it easier to pull down on a pulley than to pull up?** (Students should *explain* that it is easier because we can use our body weight to help or create force.)

[1]answers might include airplanes, bicycles, skateboards.

Wheel and Axle and Pulley

wheel and axle (hwēl ənd ak/səl), a simple machine that has a center rod attached to a wheel.

gear (gir), a wheel with jagged edges like teeth.

Wheel and axle

The picture shows different kinds of wheels. Find the rod attached to each wheel. A **wheel and axle** is a simple machine made of a rod attached to the center of a wheel. A wheel and axle is a special kind of lever that moves or turns objects. The axle, or rod, turns when you put force on the wheel.

You probably have seen a wheel and axle on cars, roller skates, and wagons. A doorknob also is a wheel and axle. What other machines with a wheel and axle have you seen?[1]

Notice the jagged edges of the wheels in the watch shown below. **Gears** are wheels with jagged edges like teeth. The teeth help the wheels turn each other. You can find gears in bicycles, cars, and many other machines.

126

Teaching Options

Science Anecdote

Most elevators use pulleys to move people up and down. An electric motor moves a pulley on the bottom of the elevator car. This pulley turns a pulley above the car. The pulley above the car turns a third pulley that moves the elevator up and down.

Enrichment

Encourage students to imagine what life would be like without rules. Have students draw or write a description of what it would be like to be on a road, in a game, in an airplane, at home, or in school if there were no rules for our safety. Then ask students to think of one safety rule for a machine discussed in the chapter.

Game Suggestion ♦

Twenty Questions: Divide the class into two teams. Think of a particular simple machine. Allow each team in turn to ask a question that can only be answered with *yes* or *no*. Continue the questions until a team correctly guesses the simple machine you are thinking of. Assign one point each time a team guesses correctly.

♦ ***Suitable as a language development activity***

A person pulls down on this rope to raise this paint can. The person is using a **pulley** to raise the can. A pulley is a simple machine with a wheel with a rope. The rope fits around the edge of the wheel. You can use a pulley to move a load up, down, or sideways.

A pulley can help move an object to a place that is hard to reach. For example, you might use a pulley to raise a flag to the top of a pole. A pulley also can help move a heavy load.

pulley (pu̇l′ē), a simple machine made of a wheel and a rope.

Lesson Review

1. How can levers and inclined planes help move objects?
2. How can a wedge and a screw help make work easier?
3. How do wheels and axles and pulleys help people do work?
4. **Challenge!** What kind of simple machine is a baseball bat?

Study on your own, pages 326–327.

PHYSICAL SCIENCE

FIND OUT ON YOUR OWN

Find pictures of simple machines and use the pictures to make a poster. Include at least two pictures of each of the six simple machines. Label the kind of simple machine used in each tool or object.

3. Assess

Lesson Review

1. A lever can help people lift a heavy load. An inclined plane can help move a load to a higher place.
2. A wedge cuts or splits an object. A screw holds objects together.
3. A wheel and axle helps people move and turn objects. A pulley can move a load up, down, or sideways.
4. Challenge! A baseball bat is an example of a lever. **Thinking Skill:** *Drawing conclusions*

Find Out On Your Own

The posters should illustrate all of the six simple machines. **Thinking Skill:** *Recognizing the main idea and supporting details*

Workbook page 34 *

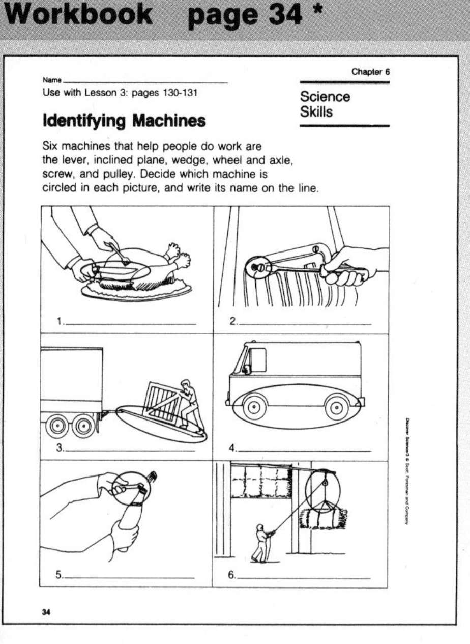

Name ____________

Use with Lesson 3: pages 130-131

Chapter 6

Science Skills

Identifying Machines

Six machines that help people do work are the lever, inclined plane, wedge, wheel and axle, screw, and pulley. Decide which machine is circled in each picture, and write its name on the line.

1. ____________
2. ____________
3. ____________
4. ____________
5. ____________
6. ____________

34

Reteaching Suggestion ♦

Point out that simple machines help people do work, and they also help people play. Question: **What are some machines used in sports?** (examples: levers—rowing oars, golf clubs, legs for kicking soccer balls; wedge—ice skate blades; pulleys—exercise machines) Encourage students to explain how these machines work.

Concept
The force needed to move an object that is closer to a fulcrum is less than the force needed to move the same object if it is farther from the fulcrum.

Objectives/Process Skills
- *Make a model* of a lever.
- *Observe* the amount of force needed to move an object.
- *Compare* the force needed when the fulcrum is moved from its previous position.
- *Record* observations.
- *Measure* the distance the force moves the lever when the fulcrum is in different positions.

Time Allotment
Allow 30 minutes.

Safety Tips (See page T24.)
- Use only nontoxic glue.
- Tell students not to put sugar cubes in their mouths; cubes might be contaminated.
- Warn students not to put washers or coins in their mouths; these objects might cause choking.
- Advise students to handle sharpened pencils carefully to avoid punctures.

Teaching Tips
- Thick wooden rulers also work well as levers.
- Washers or coins can be used as a forcé.
- A ball of clay can be used as a load.

Answers
State Your Conclusions

1. The sugar cube was moved a greater distance when the fulcrum was farther away from the load.

2. The most force was needed to move the sugar cube when the fulcrum was farthest away from the load. Less force was needed when the fulcrum was closer to the load, and the least force was needed when the fulcrum was closest to the load.

Use What You Learned

To lift a very heavy object, place the fulcrum closer to the load so that less force is needed. **Thinking Skill:** *Applying information to new situations*

Observe the Force Needed to Move an Object with a Lever

Suggested grouping: 1–2 students

Purpose

Observe that the closer the fulcrum is to the load, the less force is needed to lift an object.

Gather These Materials

• tongue depressor • pencil • 2 sugar cubes • glue • tape • metric ruler

Follow This Procedure

1. Use a chart like the one shown to record your observations.
2. Mark the tongue depressor with lines to divide it into 4 parts. Label the lines *A, B,* and *C.*
3. Glue the sugar cubes to one end of the tongue depressor.
4. Tape the pencil to your desk as a fulcrum. Place the tongue depressor on the pencil with the fulcrum at line *C* as shown.
5. Use the ruler to measure the height from the desk to the end of the tongue depressor.
6. Push on the end of the tongue depressor. Observe how much force you use to lift the sugar cubes.
7. Repeat steps 5–6 with the fulcrum at line *B* and line *A*. Compare the distances you moved the sugar cube. Compare the amount of force you needed to move the sugar cubes each time.

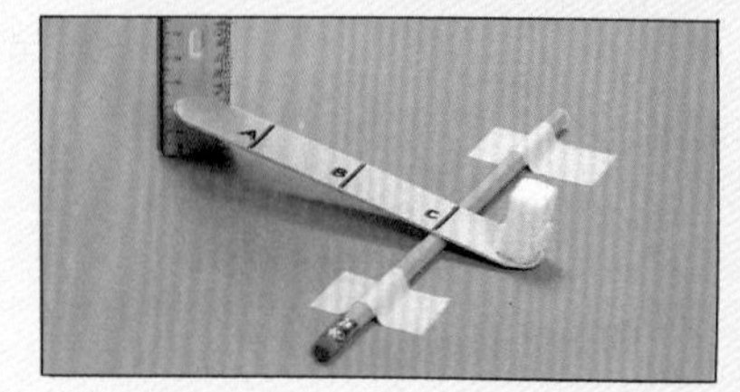

Record Your Results

Fulcrum	Line A	Line B	Line C
Force (easier, harder, hardest)	hardest	harder	easiest
Height: desk to tongue depressor end	Answers will vary.		

State Your Conclusion

1. Where was the fulcrum when you moved the sugar cubes the longest distance?
2. Compare the amount of force you used to move the sugar cubes when the fulcrum was moved from line *A*, to *B*, to *C*.

Use What You Learned

To move a heavy box with a lever, would you place the fulcrum close to the box or far away from the box? Explain.

128

Activity Results

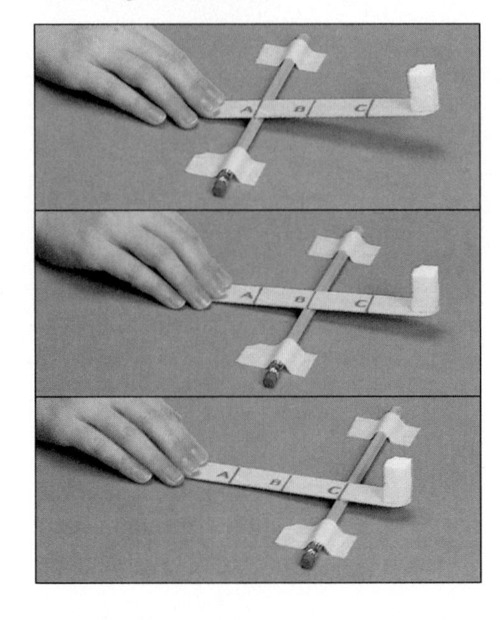

Resource Book page 69

Chapter 6

Name ______

Use with Lesson 2: page 128

Observing the Force Needed to Move an Object with a Lever

Activity Worksheet

Record Your Results

Fulcrum	Line A	Line B	Line C
Force (easier, harder, hardest)			
Height: desk to tongue depressor end			

State Your Conclusion

1. Where was the fulcrum when you moved the sugar cubes the longest distance?

______

2. Compare the amount of force you used to move the sugar cubes when the fulcrum was moved from line A, to B, to C.

______

Use What You Learned

To move a heavy box with a lever, would you place the fulcrum close to the box or far away from the box? Explain.

______

69

Teaching an Old Machine New Tricks

The Problem A windmill is a machine that uses energy to produce power. People have used windmills to do work for more than a thousand years. Early windmills often had four or five blades covered with cloth sails to catch the wind. These windmills pumped water, ground corn, and sawed timber. In 1890, for the first time, a windmill pumped water to make electricity. Soon, nearly six million windmills made electricity in the United States. Then large power companies began using oil to make less costly electricity. Many people took down their windmills. In the 1970s, supplies of oil were short. The cost of oil and other types of fuel rose. People had to find a way to make electricity and run machines without using fuels.

The Breakthrough Scientists and engineers began to work on the problem of energy needs. Windmills seemed to be part of the answer. Scientists designed new kinds of windmills, like the one in the picture. The new windmills have metal blades. The blades spin easily in wind from any direction. These windmills can work even when the wind is not blowing very hard.

New Technology Some of today's windmills have blades that measure 100 meters long from tip to tip. This is about the length of a football field! Some places have wind farms with several hundred large windmills. Wind farms can make as much electricity as a power plant. California has more than sixty-five wind farms that use five thousand windmills.

What Do You Think?

1. How did the low supply of oil cause people to start using windmills again?
2. Many years ago an eight-ton windmill blade tore loose from a windmill. Where should windmills be put to keep people safe?

129

Science and Technology

TEACHING PLAN

Discussion

Discuss the best locations for windmills. Question: **Is it better to put windmills in mountains or in flat places?** (in mountainous regions, because strong winds occur in mountain passes)

Teaching Tips

- Point out that a windmill is a complex machine made of several simple machines. Students will read about complex machines in the next lesson.
- Have students *list* the advantages and disadvantages of windmills. (Some advantages: making electricity with wind does not use scarce resources or pollute air or water; wind is free. Disadvantages: windmills can work only where wind is plentiful and constant; spinning blades might harm birds; falling blades can cause damage.)

Answers

What Do You Think?

1. The cost of oil and other fuels went up, so people wanted to produce energy without using fuels. **Thinking Skill:** *Recognizing cause and effect*

2. Windmills should not be too near areas with people in case a blade falls. **Thinking Skill:** *Inferring*

Teaching Options

Books to Read

Weitzman, David. *Windmills, Bridges, and Old Machines: Discovering Our Industrial Past.* Scribner's, 1982. (grades 5–6)

Hays, Dick, and Allen, Bill. *Windmills and Pumps of the Southwest.* Eakin, 1983. (for the teacher)

Hefner, Robert J. *The Windmills of Long Island.* Norton, 1984. (for the teacher)

Science Background

To generate electricity, windmills are connected to an electric generator through gears. Wind moves the vanes. Their motion moves a coil of wires in a magnetic field, which starts an electric current flowing. The installation of wind farms is as much a result of economics as of technological advances. As long as the cost of generating electricity through wind is lower than the cost of building other power plants, wind farms will likely exist. Government support has been important for the building of wind farms.

TEACHING PLAN

Lesson Objectives

- *Describe* and give examples of compound machines, and *explain* how compound machines make work easier.
- *Explain* the importance of using machines safely and give examples of the safe use of machines.

Lesson Vocabulary

compound machine

1. Motivate

Demonstration Activity ♦

Hold up two pencils. Sharpen one using a pocket knife. Point out that the knife is a simple machine—a wedge. Sharpen the second pencil in a pencil sharpener. Compare the efforts and the results when using the simple machine as opposed to using the compound machine.

Discussion

Have students discuss the advantages of using this compound machine. (Using a knife to sharpen a pencil can be dangerous; a pencil sharpener is safer and faster.) Point out that using the proper machine for a task is often the safer method.

3 What Are Compound Machines?

LESSON GOALS

You will learn

- how simple machines can be put together to make compound machines.
- that using machines carefully can help keep you safe.

compound machine (kom′pound mə shēn′), a machine made of two or more simple machines.

[1]wheel and axle
[2]screw

Look at the different machines that make up the paddle boat in the picture. The front of the boat is a wedge that pushes through the water. Screws hold parts of the boat together. What kind of simple machine is the paddle?[1]

Compound Machines

A **compound machine** is two or more simple machines put together. The paddle boat is a compound machine. You probably use many other compound machines. Look for the different parts of the pencil sharpener in the picture. The handle is a wheel that turns on an axle. Wedges sharpen the pencil. Find the gears that move the wedges. What kind of simple machine holds the pencil sharpener together?[2]

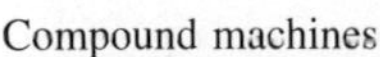

Compound machines

130

Teaching Options

Science Background

Compound machines are combinations of simple machines. Complex machines are combinations of simple and compound machines. Complex machines have many parts. All the movable parts produce friction. As a result of friction, large, complex machines often have low efficiency. Much of the work put into the machine is used to overcome friction.

Reading Strategies ♦

1. Guide students' pre-reading by asking: What ideas do you think you would find in the lesson to answer the question-title?
2. Assign these strategies: Writing a Memory Sentence and Finding and Writing a Sentence for the Vocabulary Words. (See pages T26–T29.)
3. Pair students to share what information is clear and unclear and initiate discussion using students' unanswered questions.

♦ ***Suitable as a language development activity***

Using Machines Safely

Using machines carefully is important. Asking an adult before using a machine and following safety rules can help keep you safe.

The picture shows compound machines you might use. A pair of scissors is made of two levers. The edge of each blade has the shape of a wedge. The screw that holds the two levers together is the fulcrum. You can use scissors safely by pointing them away from yourself.

What kind of simple machines make up the stapler in the picture?[1] How can you use a stapler safely?[2]

SCIENCE IN YOUR LIFE

A bicycle is a compound machine. Several different simple machines make up a bicycle. Handbrakes are levers. You can find a wheel and axle in the front and back wheels, as well as the pedals. The chain on the bicycle is a pulley. Screws hold parts of the bicycle together.

[1]lever, screw

People need to use machines safely.

Lesson Review

1. What is a compound machine?
2. How can you use machines safely?
3. **Challenge!** How is a shovel a compound machine?

Study on your own, pages 326–327.

HUMAN BODY

FIND OUT ON YOUR OWN

CONNECTION

Invent a compound machine that people can use to exercise properly. Draw a picture of it. Label each simple machine you use in your drawing. Write a few sentences explaining what your machine would do and how it would work.

[2]by being careful not to catch fingers between moving parts, and keeping fingers away from staples

131

2. Teach

Teaching Tip

• **Possible Misconception:** Many students think that garden tools, such as the spade, hoe, rake, and shovel, are simple machines. Make sure that students understand that the tools are compound machines; the ends are wedges and the handles are levers.

3. Assess

Lesson Review

1. A compound machine is a combination of one or more simple machines.
2. You can use machines safely by checking with an adult before using a machine, following safety rules, keeping body parts away from sharp edges and the moving parts of a machine.
3. Challenge! The blade is a wedge; the handle is a lever. **Thinking Skill:** *Drawing conclusions*

Find Out On Your Own

Answers and drawings will vary. Make sure students' compound machines consist of two or more simple machines. **Thinking Skills:** *Making generalizations, Making physical models*

Workbook page 35 *

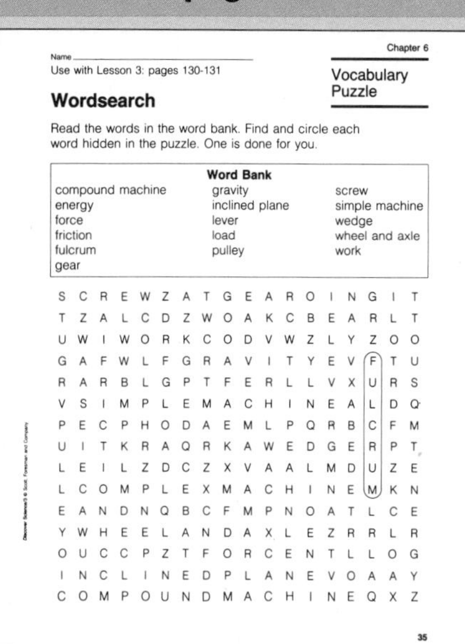

Name ____________________

Use with Lesson 3: pages 130-131

Chapter 6

Vocabulary Puzzle

Wordsearch

Read the words in the word bank. Find and circle each word hidden in the puzzle. One is done for you.

Word Bank

compound machine	gravity	screw
energy	inclined plane	simple machine
force	lever	wedge
friction	load	wheel and axle
fulcrum	pulley	work
gear		

S C R E W Z A T G E A R O I N G I T
T Z A L C D Z W O A K C B E A R L T
U W I W O R K C O D V W Z L Y Z O O
G A F W L F G R A V I T Y E V F T U
R A R B L G P T F E R L L V X U R S
V S I M P L E M A C H I N E A L D Q
P E C P H O D A E M L P Q R B C F M
U I T K R A Q R K A W E D G E R P T
L E I L Z D C Z X V A A L M D U Z E
L C O M P L E X M A C H I N E M K N
E A N D N Q B C F M P N O A T L C E
Y W H E E L A N D A X L E Z R R L R
O U C C P Z T F O R C E N T L L O G
I N C L I N E D P L A N E V O A A Y
C O M P O U N D M A C H I N E Q X Z

35

* *Answers to masters on pages 114E–114H*

Special Education

Take students with learning disabilities to the playground or bring a bicycle to class so that students may locate and move the various simple machines on the bicycle. Help students to understand that a bicycle is a compound machine because it consists of many simple machines put together.

Reteaching Suggestion ♦

Show students examples or pictures of compound machines. Ask students to identify the simple machines that make up each compound machine. Then have students explain how the compound machine works. (Example: scissors—two levers with a screw as the fulcrum and wedges on the blades.)

TEACHING PLAN

Purpose
To develop the skills of collecting and organizing information using metric rulers and bar graphs to solve problems.

1. Motivate

Discussion
Question: **What kinds of things have you measured with a ruler or meter stick?** (Answers will vary.) Display a metric ruler and review its division into centimeters and millimeters. Before beginning the activity, let students use metric rulers to measure objects in the classroom. Provide practice with measuring in millimeters.

2. Teach

Teaching Tip
- Point out that the pictures of inclined planes are not actual sizes.

Using Rulers and Bar Graphs

Problem: How does the length of the longest line of an inclined plane change when the bottom changes?

Part A. Using Rulers to Collect Information

1. The diagram shows three inclined planes that have the same slant. How do they differ?
2. The sides of the planes are measured with a metric ruler. The long lines on the ruler stand for centimeters. A centimeter is divided into 10 millimeters.
3. Look at the ruler below Plane A. Count the centimeters starting from the left end of the ruler. It measures 4 cm. How long is the bottom of Plane B? Plane C?
4. Measure the upright lines at the right end of each plane. How long is each one?
5. Measure the longest line of each plane. How long is each one?

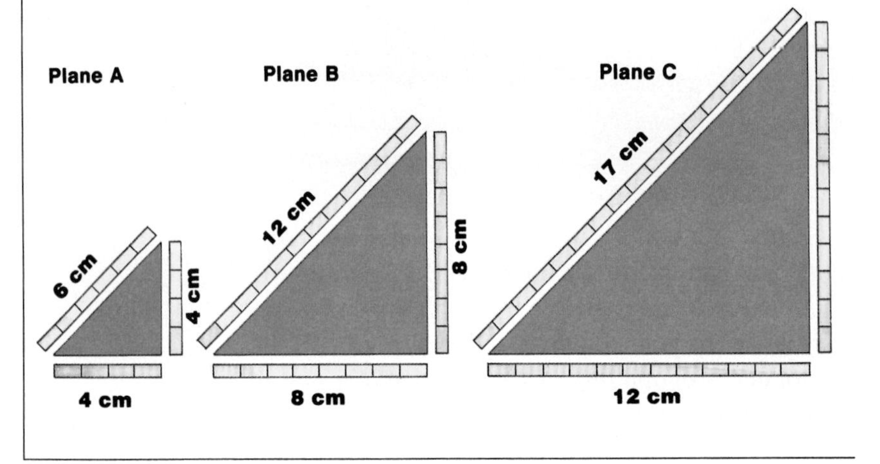

Teaching Options

Sample Bar Graph for Part C

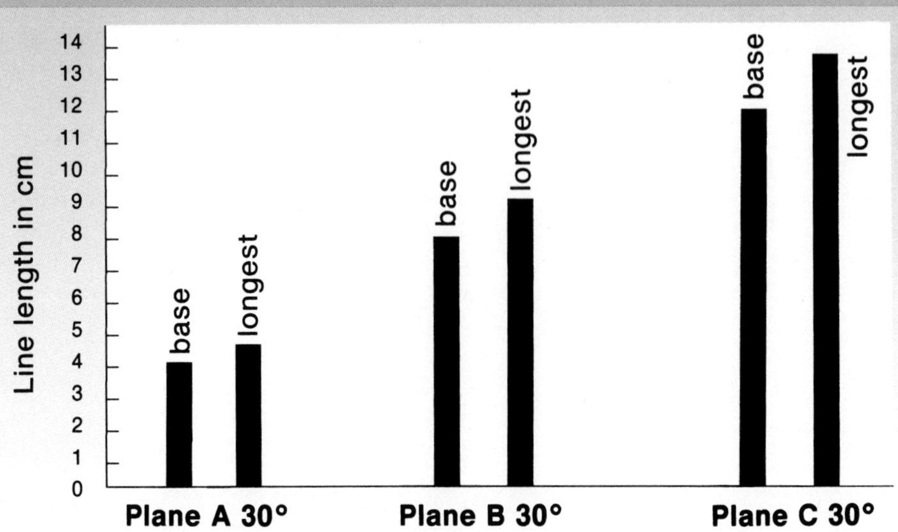

♦ ***Suitable as a language development activity***

Part B. Using Bar Graphs to Organize and Interpret Information

6. This bar graph contains the information you collected about inclined planes. The bottom line and the upright line are the same length in each plane. How long is the longest line of the plane with a 4 cm bottom? an 8 cm bottom? a 12 cm bottom?
7. If a plane had a bottom line 16 cm long, how would the length of its longest line compare with the longest lines of these planes?

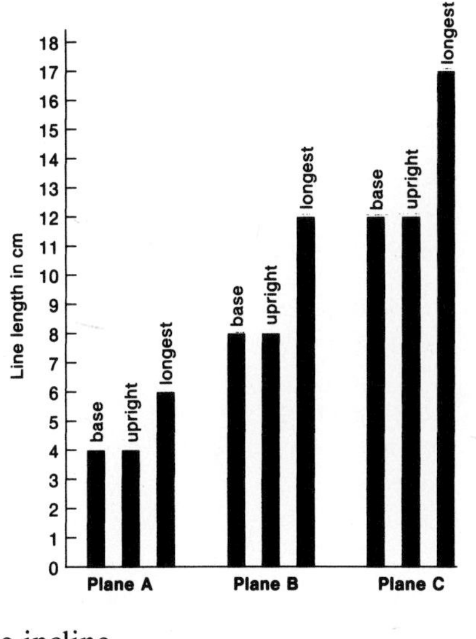

Part C. Using Rulers and Bar Graphs to Solve a Problem

Problem: How does the length of the incline change when the slant is smaller and the bottom stays the same length as in Part A?

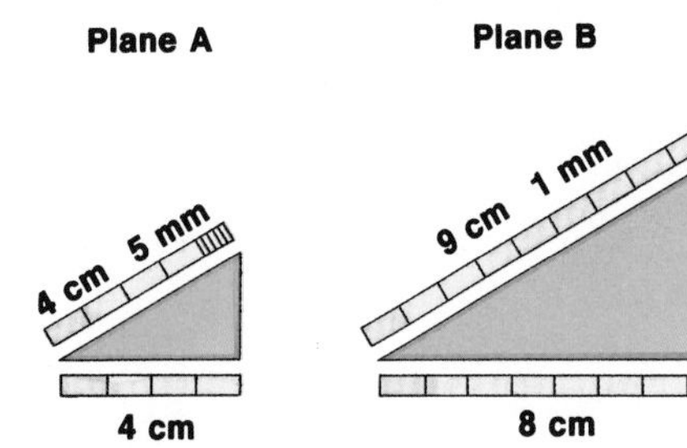

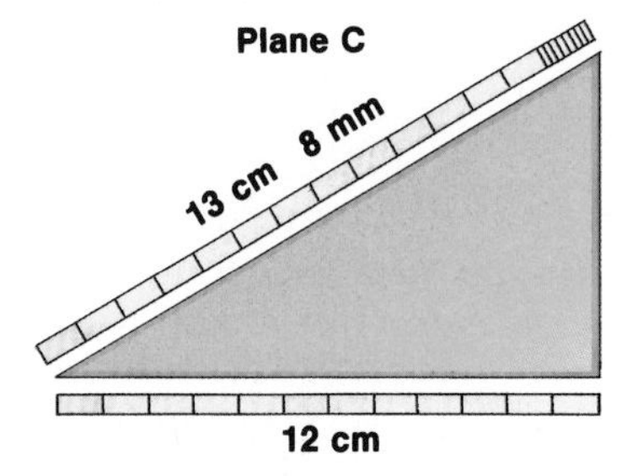

8. Use the rulers to collect the information you need to solve the problem. Make a bar graph like the one shown in Part B to organize your information.
9. Look at your bar graph. Compare it to the bar graph you made in Part B.. How does the length of the longest line change when the slant is smaller and the bottom stays the same length?

133

Teaching Tip

- Students might make cardboard models of the inclined planes. If so, note the measurements in the diagram have been rounded off. To make models, have students use the base and height measurements and 90° angles as shown. They can then modify the graph shown in Part B to match their models. Use this formula to find the actual length of the inclined plane, or hypotenuse: $base^2 + height^2 = hypotenuse^2$.

3. Assess

Part A

1. The length of their bases, their heights, and the length of their longest lines are different.
2. No answer is required.
3. 8 cm; 12 cm
4. Plane A—4 cm; Plane B—8 cm; Plane C—12 cm
5. Plane A—6 cm; Plane B—12 cm; Plane C—17 cm

Part B

6. 6 cm; 12 cm; 17 cm
7. It would be longer than the longest lines of the other planes.

Part C

8. See Sample Bar Graph for Part C.
9. The length of the longest line is smaller.

Reteaching Suggestion ♦

Have students use construction paper to make a square 6 cm on a side, a rectangle 6 cm × 8 cm, and another rectangle 8 cm × 12 cm. Have students draw diagonal slants within each quadrilateral, bisecting it. Then have them use metric rulers to measure the lengths of the slants. Students should construct a bar graph on which they record their findings.

Resource Book page 71 *

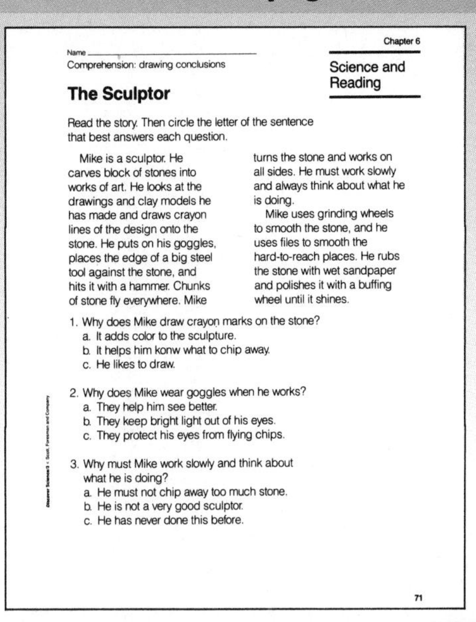

Name

Comprehension: drawing conclusions

Chapter 6

Science and Reading

The Sculptor

Read the story. Then circle the letter of the sentence that best answers each question.

Mike is a sculptor. He carves block of stones into works of art. He looks at the drawings and clay models he has made and draws crayon lines of the design onto the stone. He puts on his goggles, places the edge of a big steel tool against the stone, and hits it with a hammer. Chunks of stone fly everywhere. Mike turns the stone and works on all sides. He must work slowly and always think about what he is doing.

Mike uses grinding wheels to smooth the stone, and he uses files to smooth the hard-to-reach places. He rubs the stone with wet sandpaper and polishes it with a buffing wheel until it shines.

1. Why does Mike draw crayon marks on the stone?
 a. It adds color to the sculpture.
 b. It helps him konw what to chip away.
 c. He likes to draw.

2. Why does Mike wear goggles when he works?
 a. They help him see better.
 b. They keep bright light out of his eyes.
 c. They protect his eyes from flying chips.

3. Why must Mike work slowly and think about what he is doing?
 a. He must not chip away too much stone.
 b. He is not a very good sculptor.
 c. He has never done this before.

71

* Answers to masters on pages 114E–114H

REVIEW PLAN

Reviewing Science Words

1. gear
2. fulcrum
3. simple machines
4. force
5. wheel and axle
6. wedge
7. gravity
8. compound machine
9. friction
10. pulley
11. inclined plane
12. load
13. work
14. lever
15. energy
16. screw

Reviewing What You Learned

1. c **2.** d **3.** a **4.** d **5.** b **6.** b

Chapter 6 Review

Chapter Main Ideas

Lesson 1 • A force changes the way an object moves. • Work is done whenever force makes an object move. • Energy is used whenever work is done.

Lesson 2 • Levers and inclined planes help make work easier. • Wedges and screws help people do work. • A person can use a wheel and axle and a pulley to do work.

Lesson 3 • A compound machine is made of two or more simple machines. • Using machines safely is important.

Reviewing Science Words

compound machine	gravity	screw
energy	inclined plane	simple machines
force	lever	wedge
friction	load	wheel and axle
fulcrum	pulley	work
gear		

Copy each sentence. Fill in the blank with the correct word from the list.

1. A ___ has jagged edges like teeth.
2. A lever turns back and forth on a ___.
3. Six different ___ can help make work easier.
4. A push or a pull is a ___.
5. A machine that turns objects is a ___.
6. A machine that pushes objects apart is a ___.
7. ___ pulls objects toward the center of the earth.
8. A pencil sharpener is an example of a ___.
9. ___ causes heat.
10. A ___ is made of a wheel and a rope.
11. A ramp is an example of an ___.
12. An object being lifted or moved is a ___.
13. ___ is done whenever a force moves an object through a distance.

134

Review Options

Cooperative Learning ♦

STAD Format (See page T23.)
Assign students to work in four- to five-member teams to study Chapter 6 Review. Students should work together to make sure that they and their teammates know the material in the chapter. After students have had enough time to study together, give them a test to complete individually (Chapter 6 Test A or B in the *Test Book*). Award Superteam certificates to teams whose average test scores exceed 90%, and Greatteam certificates to teams whose average test scores exceed 80%.

Test Book page 53 *

Name ___

Chapter 6
Test A

Multiple Choice Choose the best answer.

1. What is needed to move an object?
   a. speed.
   b. load.
   c. force.
2. A force that makes objects pull toward each other is
   a. work.
   b. friction.
   c. gravity.
3. When an object moves through a distance, this is called
   a. energy.
   b. gear.
   c. work.
4. Whenever work is done, a person uses
   a. gravity.
   b. energy.
   c. ability.
5. Machines with a few or no moving parts are
   a. simple machines.
   b. compound machines.
   c. impossible to make.
6. A machine made of a bar that is supported underneath at some point is a
   a. inclined plane.
   b. lever.
   c. fulcrum.
7. A slanted road and a ramp are two examples of
   a. an inclined plane.
   b. a wheel and axle.
   c. a pulley.
8. A simple machine that is used to push objects apart is a
   a. wedge.
   b. screw.
   c. pulley.
9. What kind of machine is a bicycle?
   a. simple machine
   b. lever
   c. compound machine
10. A simple machine made of a rope and a wheel is a
    a. gear.
    b. pulley.
    c. wedge.

53

♦ ***Suitable as a language development activity***

14. A seesaw is a ____.
15. All objects that have ____ can do work.
16. A ____ is a machine used to hold objects together.

Reviewing What You Learned

Write the letter of the best answer.

1. The blade of a knife is an example of a
   (a) screw. (b) lever. (c) wedge. (d) force.
2. Gravity is one kind of
   (a) machine. (b) friction. (c) heat. (d) force.
3. A bicycle is a
   (a) compound machine. (b) pulley. (c) simple machine. (d) load.
4. A lever turns on a point called the
   (a) wedge. (b) force. (c) load. (d) fulcrum.
5. How much an object weighs measures the pull of
   (a) friction. (b) gravity. (c) energy. (d) work.
6. A simple machine sometimes used to raise a flag is a
   (a) wedge. (b) pulley. (c) gear. (d) axle.

Interpreting What You Learned

Write a short answer for each question or statement.

1. When is energy used?
2. Explain how using more force changes an object's speed.
3. Where can a fulcrum be moved to lift a load most easily?
4. Is more force needed to move an object the same distance across a level floor or up a hill?
5. What is one way a person can use scissors safely?

Extending Your Thinking

Write a paragraph to answer each question or statement.

1. Explain how the wheels of roller skates can become hot.
2. How do you use energy when you throw a ball?

To explore scientific methods, see Experiment Skills on pages 358–359.

Interpreting What You Learned

1. Energy is used whenever work is done. **Thinking Skill:** *Restating or explaining ideas*
2. Using more force can make an object move faster. **Thinking Skill:** *Recognizing cause and effect*
3. Move the fulcrum close to the load. **Thinking Skill:** *Comprehending meaning*
4. More force is needed to move an object up a hill. **Thinking Skill:** *Drawing conclusions*
5. Point the scissors away from oneself. **Thinking Skill:** *Organizing information*

Extending Your Thinking

1. When the wheels rub against the ground, they produce friction. Friction causes heat. **Thinking Skill:** *Applying information to new situations*
2. Throwing a ball is work because force is used to move the ball a distance. Energy is used whenever work is done. **Thinking Skill:** *Applying information to new situations*

Test Book page 54 *

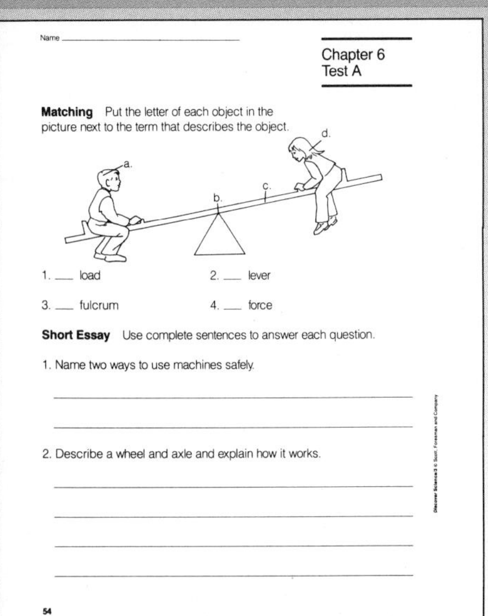

Name ____

Chapter 6
Test A

Matching Put the letter of each object in the picture next to the term that describes the object.

1. ___ load
2. ___ lever
3. ___ fulcrum
4. ___ force

Short Essay Use complete sentences to answer each question.

1. Name two ways to use machines safely.
2. Describe a wheel and axle and explain how it works.

54

Test Book page 55 *

Name ____

Chapter 6
Test B

Multiple Choice Choose the best answer.

1. A push or pull that changes the way an object moves is a
   a. lever.
   b. force.
   c. fulcrum.
2. The force caused by two objects rubbing together is
   a. work.
   b. friction.
   c. gravity.
3. The ability to work is
   a. gravity.
   b. energy.
   c. friction.
4. Gravity is a force that makes objects
   a. push apart.
   b. pull together.
   c. produce heat.
5. A pencil sharpener is a compound machine made up of several
   a. inclined planes.
   b. simple machines.
   c. fulcrums.
6. The point on which a lever is supported and turns is the
   a. load.
   b. lever.
   c. fulcrum.
7. A simple machine used to hold objects together is a
   a. wedge.
   b. screw.
   c. pulley.
8. Wheels with jagged edges like teeth are
   a. gears.
   b. axle.
   c. rod.
9. When two or more simple machines are put together, they form a
   a. lever.
   b. simple machine.
   c. compound machine.
10. Cars and roller skates both have a simple machine of a rod attached to a wheel that is a
    a. lever and fulcrum.
    b. pulley.
    c. wheel and axle.

55

Test Book page 56 *

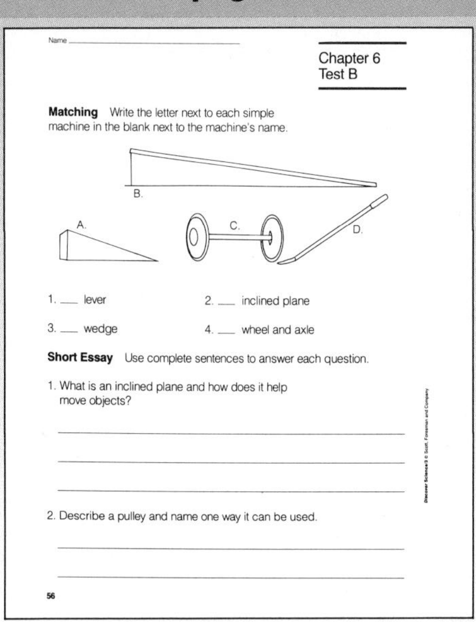

Name ____

Chapter 6
Test B

Matching Write the letter next to each simple machine in the blank next to the machine's name.

1. ___ lever
2. ___ inclined plane
3. ___ wedge
4. ___ wheel and axle

Short Essay Use complete sentences to answer each question.

1. What is an inclined plane and how does it help move objects?
2. Describe a pulley and name one way it can be used.

56

* *Answers to masters on pages 114E–114H*

Forms of Energy Planning Guide

TEACHING PLAN

Chapter Components	Skills	Materials
Chapter Opener/*DISCOVER:* Observing Magnets pp. 136–137	*DISCOVER* p. 137 Science Process Skills *Observing, Classifying*	*DISCOVER* p. 137 (groups of 2) each 15 pieces aluminum foil, sheets notebook paper, pieces cardboard, bar magnets, paper clips
Lesson 1 What Are Some Kinds of Energy? pp. 138–139	Thinking Skills Challenge!: *Applying information to new situations* Find Out On Your Own: *Drawing conclusions*	Demonstration p. 138 flashlight, toy car, radio, hair dryer
Lesson 2 What Is Electricity? pp. 140–145	Thinking Skills Challenge!: *Drawing Conclusions* Find Out On Your Own: *Recognizing the main idea and supporting details*	Demonstration p. 140 balloon, wool cloth
Activity Making an Electric Circuit p. 146	Science Process Skills *Observing, Collecting and interpreting data*	(groups of 2) 45 pieces thin insulated wire, 15 batteries (size D), 15 flashlight bulbs, masking tape, 15 sheets heavy paper
Science and People Finding Out About Energy p. 147	Thinking Skills *Recognizing cause and effect, Restating or explaining ideas, Comprehending meaning*	
Lesson 3 How Is Matter Heated? pp. 148–151	Thinking Skills Challenge!: *Applying information to new situations* Find Out On Your Own: *Collecting information, Communicating*	Demonstration p. 148 no materials needed
Lesson 4 How Does Light Travel? pp. 152–156	Thinking Skills Challenge!: *Drawing conclusions* Find Out On Your Own: *Recognizing the main idea and supporting details, Collecting information*	Demonstration p. 152 flashlight, aluminum foil
Activity Using a Prism p. 157	Science Process Skills *Observing, Collecting and interpreting data, Communicating*	(groups of 3) 10 prisms, 10 sheets heavy white paper, 10 sets crayons or colored markers
Skills for Solving Problems Using Thermometers and Bar Graphs pp. 158–159	Problem Solving Skills *Making decisions/Identifying and solving problems, Interpreting charts, maps, and graphs*	
Chapter Review pp. 160–161	Thinking Skills *Restating or explaining ideas, Drawing conclusions, Inferring*	

Teaching Options

Strategies	Extensions		Resource Masters
Cooperative Learning p. 136 (Also see p. T23.) Applying Whole Language p. 137 (Also see p. T30.)			Family Letter: *Resource Book* p. 75
Reading Strategies p. 138 (Also see pp. T26–T29.)	Special Education p. 139 Reteaching Suggestion p. 139		Vocabulary Preview: *Workbook* p. 37
Reading Strategies p. 140 (Also see pp. T26–T29.)	Science and Language Arts p. 141 Special Education p. 141 Reinforcement pp. 141, 142, 143, 144	Enrichment pp. 142, 143, 144 Game Suggestion p. 145 Reteaching Suggestion p. 145	Science Skills: *Workbook* p. 38
			Activity Worksheet: *Resource Book* p. 79
Reading Strategies p. 148 (Also see pp. T26–T29.)	Science and Health p. 149 Special Education p. 149 Reinforcement pp. 149, 150	Enrichment p. 150 Reteaching Suggestion p. 151	Science and Math: *Workbook* p. 39
Reading Strategies p. 152 (Also see pp. T26–T29.)	Science and Language Arts p. 153 Special Education pp. 153, 154 Reinforcement pp. 153, 154, 155	Science and Health p. 155 Reteaching Suggestion p. 156	Science Activity: *Workbook* p. 40 Vocabulary Puzzle: *Workbook* p. 41
			Activity Worksheet: *Resource Book* p. 81
	Reteaching Suggestion p. 159		Science and Reading: *Resource Book* p. 83
Cooperative Learning p. 160 (Also see p. T23.)			Chapter Tests: Forms A and B *Test Book* pp. 61–64

Classroom Management

Advance Preparation

DISCOVER, page 137
Provide a magnet for each group of students.

Demonstration, page 138
Bring in a flashlight, a toy car, a radio, and a hair dryer for this demonstration.

Demonstration, page 140
Bring a balloon and a piece of wool cloth to class.

Activity, page 146
For each group cut 3 pieces of thin, insulated wire to 20-cm lengths. Prepare each piece by scraping about 1 cm of insulation from each end, using a wire stripper. Also provide a size D battery, a flashlight bulb, and a sheet of heavy paper.

Demonstration, page 152
Bring 2 pieces of aluminum foil and a flashlight to class.

Activity, page 157
Provide a prism and heavy, white paper for each group.

Vocabulary Review

Use the following sentences with your students to review the meanings of the italicized words.

1. You use *energy* to move an object.
2. The *lenses* in glasses help people see objects more clearly.
3. A *compass* indicates direction with a magnetic needle that points north.
4. *Temperature* is a measure of how hot or cold something is.
5. A *thermometer* is used for measuring temperature.

High-Potential Students

Ask students to prepare a newspaper article on the electrical discovery of one of the following inventors: Benjamin Franklin, Thomas Edison, Samuel Morse, or Alexander Graham Bell. Encourage students to find information on the inventor's life and works. They should focus on the inventor's most famous inventions and describe how the inventor developed the devices. Ask the students to present their news reports to the class.

Mainstreamed Students

Visually Impaired
Give visually impaired students bar magnets to examine. You may wish to tape raised letters to the ends to designate which end is north and which end is south. Encourage students to feel the pull of the magnets when a north end and a south end are placed near each other. Direct them to note the push of the magnets when two north ends or two south ends are placed near each other.

Emotionally Handicapped
Ask the students to measure the temperature of the classroom in the morning and afternoon every day for a week. Direct them to note whether the sun was shining in the classroom when they measured the temperature. Have students record the temperature on a wall chart in the classroom. Encourage them to note correlations between temperature and weather conditions or other factors. At the end of the week ask them to share what they learned with the class.

Science Fair Projects

The Experiment Skills on p. 360 and the Investigate feature in the chapter can be used for science fair projects. You might also encourage interested students to do one of the following projects:

1. Take a flashlight apart and learn about its circuit. Make a diagram tracing the path of the current. Use the diagram to design a poster that explains how a flashlight works.
2. Use an encyclopedia to find out how a telephone works. If possible examine the receiver of an old telephone to identify the electromagnet and study the circuitry. Make a chart showing how the electromagnet, current wire, and metal disk work together to transmit sound.

lassroom Resources

Bulletin Board

Encourage students to draw or bring in pictures that show how they use different forms of energy.

HOW WE USE DIFFERENT FORMS OF ENERGY

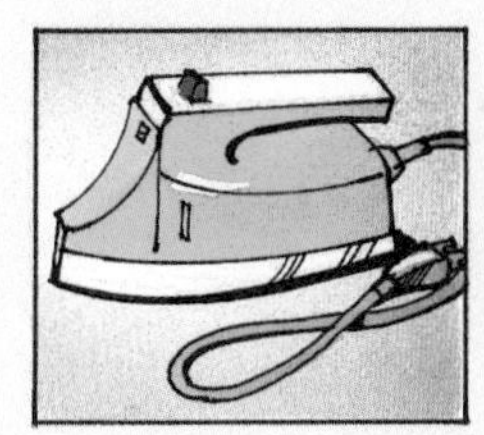

Chapter 7 Poster

Name the kinds of energy.

Science Discovery Center

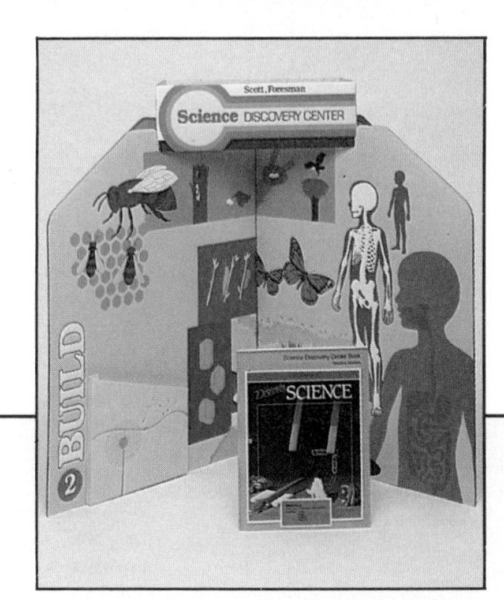

Use pages 75–80 from the *Science Discovery Center Book*. Place these worksheets in the appropriate pockets in the Science Discovery Center.

Overhead Transparencies

Use Transparencies 11, 12, and 13 from the package of color overhead transparencies.

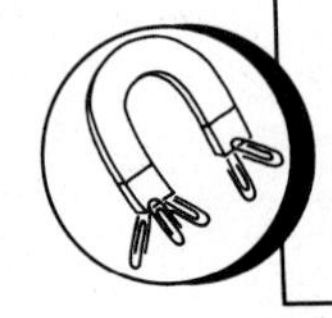

Dear Family,
Your student will be reading **Chapter 7: Forms of Energy,** in *Discover Science*, published by Scott, Foresman. We will learn about electricity, light, sound, and heat. Heat is the flow of energy from warmer places to cooler ones. Materials that carry heat easily are good conductors. Most metals are good heat conductors. Materials that do not carry heat easily are insulators.
You and your student can do this activity together to learn more about insulators.

Wool and Warmth

You will need two empty cans the same size, wool material large enough to wrap twice around one can (a scarf would work), rubber bands, hot tap water (***CAUTION:*** *Be sure the water is not hot enough to burn skin*).

1. Tightly wrap one can with two layers of wool. Use rubber bands to keep the wool in place.
2. Fill both cans with hot water.
3. Wait 30 minutes.
4. Have your student feel the water in each can. Ask, "How hot is the water?"
5. Discuss these questions:
   - Which can had the hotter water? (The one with the wool.)
   - Why do you think there was a difference?
   - Why do we wear wool when it is cold?

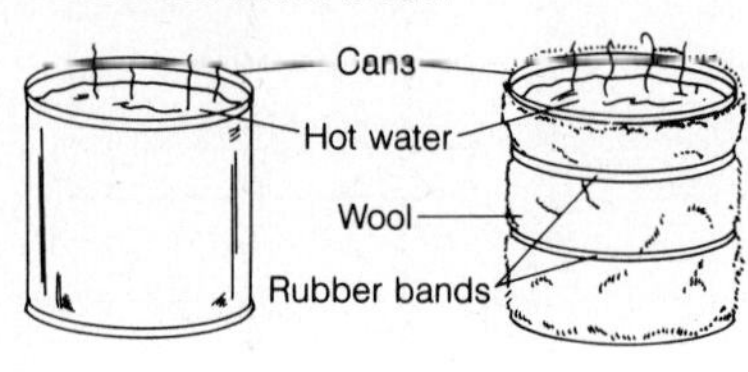

Name ____________________

Chapter 7

Use with Lesson 1: pages 138-139

Vocabulary Preview

Forms of Energy

Chapter Vocabulary

circuit	electromagnet	iris	reflects
conductor	energy of motion	magnetism	source
electric charges	image	prism	thermostat
electric current	insulator	pupil	visible spectrum

Vocabulary Cards

1. Write each word on a card.
2. Find each word in the glossary. Copy the pronunciation under the word on the card.
3. Practice saying the words with a partner.

The Pronunciation Game

1. With a partner, make two sets of six number cards. See the top picture. Put both sets of number cards in a paper bag.
2. Shuffle two sets of vocabulary cards together. Put these cards face up in a line. See the bottom picture.
3. Take turns during this part of the game. Pick a number card from the bag. The card tells you how many spaces to move a marker along the line of vocabulary cards. Try to pronounce the vocabulary word your marker lands on. If you can pronounce the word, leave your marker on that card. If you cannot pronounce the word, move your marker back to the start of the game.
4. The next player takes a turn. Continue until one player's marker reaches the finish.

1	2	3
4	5	6

Start Here

Finish

At Home

1. Practice saying the words. Learn their meanings.
2. Look for the words and pictures of the words in newspapers and magazines. You can make a poster.

Teacher's Notes: After finishing the chapter, students can play *The Definition Game*, using definitions rather than pronunciations.

Name ____________________

Chapter 7

Use with Lesson 2: pages 140-145

Science Skills

Looking at Circuits

Each picture shows a battery, a switch, and a bell. When these parts are connected correctly, they form a complete circuit. Electricity flows through the circuit and the bell rings. However, the bell cannot ring in any of the pictures below. Tell why each circuit is not complete.

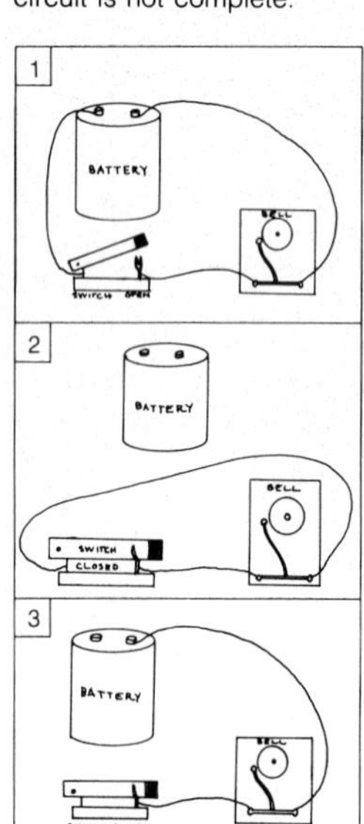

Picture 1: The switch is open, so the circuit is not complete.

Picture 2: The battery is left out of the circuit. (Need three wires: from bell to battery; from battery to switch; from switch to battery.)

Picture 3: Only one side of battery is connected to circuit. (Need a third wire—from switch to left side of battery.)

Teacher's Note: Have students compare these diagrams with the light-bulb circuit on page 141 in the text.

Name ____________________

Chapter 7

Use with Lesson 2: page 146

Activity Worksheet

Making an Electric Circuit

Record Your Results

	What happens to bulb
Wires not touched	
Battery first way	
Battery second way	

State Your Conclusion

1. Why does the bulb look different when the wires touch?

2. Explain what happens when you turn the battery backwards.

Use What You Learned

How do you think the switch on a flashlight works?

Name ________________________

Use with Lesson 3: pages 148-151

Chapter 7

Science and Mathematics

Thermostats

Most homes have at least one thermostat. Often, the thermostat is on the wall in the living room or hall. People use thermostats to start or stop their furnaces. The picture shows a type of thermostat found in many houses. Both sets of numbers on this thermostat tell the temperature in degrees Celsius.

The numbers along the bottom of the thermostat are part of a thermometer. They tell the air temperature in the room where the thermostat is located.

Above the top set of numbers is a dial that moves. If people want the room to be warmer, they push the dial to a higher number. This sets the thermostat at a higher temperature.

Moving the dial up causes the furnace to run. The furnace runs until the temperature shown on the thermometer is the same as the number next to the dial.

To cool the room off, the dial is pushed to a lower number. The lower setting causes the furnace to shut off. The furnace stays off until the thermometer reads the same as the number on the dial.

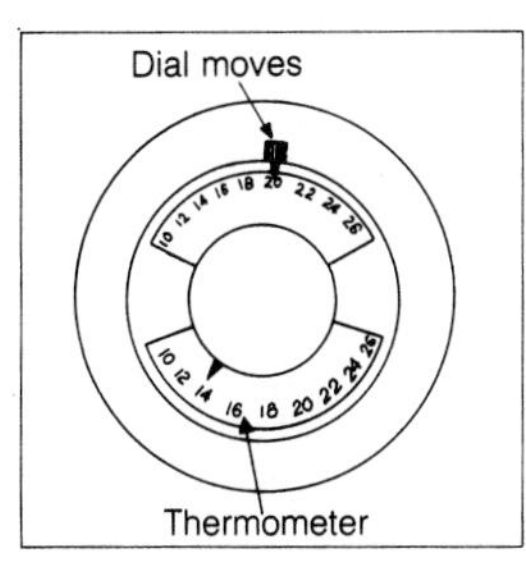

Use the thermostat drawing to answer these questions.

1. State the room temperature and the thermostat settings.

   room temperature: 14° C;

   thermostat setting: 20° C

2. Compare the room temperature with the setting on the thermostat.

   The room is 6° cooler than the thermostat setting.

3. What is the furnace doing?

   Furnace will run until the thermometer reads 20° C.

Teacher's Notes: Suggest that the students find thermostats in school and at home.

Name ________________________

Use with Lesson 4: pages 152-156

Chapter 7

Science Activity

How Does Your Eye See Things?

Gather These Materials

- piece of construction paper
- tape • piece of waxed paper
- shoe box • pin • scissors

Follow This Procedure

1. Roll the piece of construction paper to make a cone. Tape it together. Trim the big end.
2. On the waxed paper, trace around the big end of the cone. Cut the circle out. Tape it to the big end of the cone.
3. On one end of the shoe box, trace around the big end of the cone. Carefully cut out the circle. Tape the big end of the cone to the hole in the shoe box.
4. Make a pinhole in the other end of the shoe box.
5. Point the pinhole at something bright. *CAUTION: Do not point it at the sun.* Look through the cone.

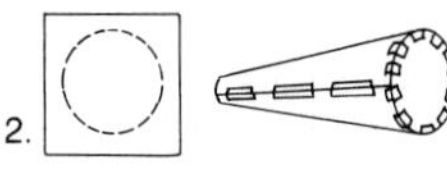

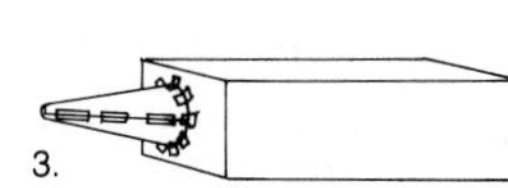

Record Your Results

Draw one thing that you saw through the cone.

Images will be drawn upside down.

State Your Conclusions

1. What do you see?

   an upside-down image

2. How is the image that you saw in the box like an image that forms on the back of your eye?

   Both the image in the box and on the back of your eye are upside down.

Teacher's Notes: Students can compare the box to a camera by placing photographic paper at the end opposite the pinhole.

Name ________________________

Use with Lesson 4: pages 152-156

Chapter 7

Vocabulary Puzzle

Wordsearch

Read the words in the word bank. Find and circle each word hidden in the puzzle. One is done for you.

Word Bank

circuit	electromagnet	iris	reflect
conductor	energy of motion	magnetism	source
electric charge	image	prism	thermostat
electric current	insulator	pupil	visible spectrum

```
E L E C T R I C C H A R G E A B C D V
L E G R Z M N V Z A R C P Q S H I A I
E N E R G Y O F M O T I O N O C C D S
C B S T W M F G B C D L N C U L O W I
T H E R M O S T A T T W C D R S N A B
R A O R E F L E C T E O O I C T D B L
O P W M K D B L J G I M A G E R U S E
M A G N E T I S M Q D B E S P J C K S
A Z U R S U L A A O I R I S U B T D P
G O C I R C U I T K J J M N P R O R E
N J A L I C E D O G B B R Q I P R S C
E C N E I N S U L A T O R P L N T Q T
T L E N S C A L L M Q Q P R I S M G R
E L E C T R I C C U R R E N T A K J U
Z A W V R S T L O M N B D E W R N M M
```

Teacher's Notes: Conduct a spelling bee, using the vocabulary words. In the second round, have students define the words.

Name ________________________

Use with Lesson 4: page 157

Chapter 7

Activity Worksheet

Using a Prism

Record Your Results

What you see	Names of colors

State Your Conclusion

1. What was the order of the colors in the spectrum?

2. How are the colors that you saw like a rainbow?

Use What You Learned

What do you think would happen if you passed the spectrum you made through a second prism?

Name ______

Comprehension: details and facts

Chapter 7

Science and Reading

Saving Electricity

Read the paragraph. Underline the main idea. Then answer the questions.

Some of the things we use can be recycled, or used again. Recycling paper, cans, and bottles saves electricity. It takes less electricity to make new paper out of old paper than it does to make it from wood pulp. It takes a lot of electric power to make aluminum for cans, so making aluminum cans from old cans saves a lot of electricity. It is easy to recycle returnable bottles. They can be washed and used again. Washing a bottle uses much less electricity than making a new one.

1. How does recycling paper save electricity?

   It takes less electricity to make paper from old paper than from wood pulp.

2. Why does recycling aluminum cans save a lot of electricity?

   It takes a lot of electricity to make aluminum.

3. How are returnable bottles recycled?

   They are washed and used again.

Name ______

Chapter 7 Test A

Multiple Choice Choose the best answer.

1. Plants use the energy from the sun to **(1-1)**
   (a.) make food.
   b. trap animals.
   c. stop growing.
2. Light, sound, and electricity are forms of **(1-2)**
   a. magnetism.
   b. sources.
   (c.) energy.
3. People get the energy they need to move their bodies from **(1-2)**
   (a.) food.
   b. sunlight.
   c. electricity.
4. The movement of electric charges from one place to another is called **(2-1)**
   a. sound.
   b. light.
   (c.) electric current.
5. All machines that have electric motors use **(2-2)**
   (a.) electromagnets.
   b. prisms.
   c. thermostats.
6. Some power plants use the energy produced by coal or oil to make **(2-3)**
   a. magnetism.
   (b.) electricity.
   c. conductors.
7. Objects that are made of metal are good heat **(3-1)**
   (a.) conductors.
   b. sources.
   c. insulators.
8. A tool that measures temperature is a **(3-2)**
   a. electromagnet.
   b. prism.
   (c.) thermometer.
9. A tool that controls the temperature in a home or building is a **(3-3)**
   a. thermometer.
   (b.) thermostat.
   c. heater.
10. A piece of glass that separates white light into the colors of the spectrum is a **(4-1)**
    a. pupil.
    (b.) prism.
    c. particle.

The numbers in parentheses after each question refer to the lesson number and the objective of that lesson.

Name ______

Chapter 7 Test A

Short Answer Fill in the blanks next to each part of the eye. Use these phrases: light enters here, this controls pupil size, this bends light to focus.

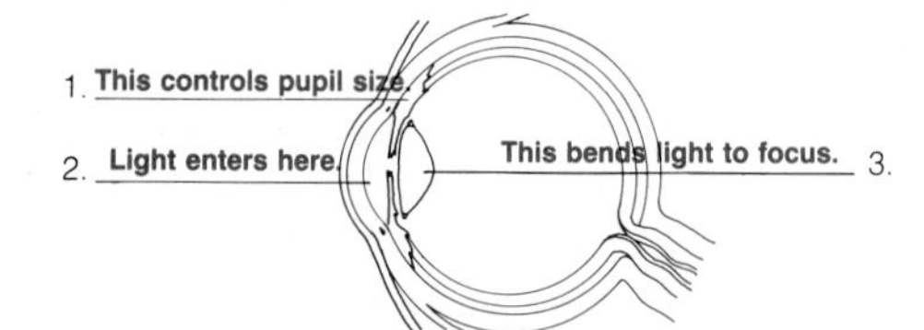

Short Essay Use complete sentences to answer each question.

1. Light waves bend when they travel through a lens. Explain why. **(4-2)**

   Light waves bend when they travel through a lens because a lens is a curved piece of glass. Lenses can be made of glass or other clear material.

2. How does the pupil of the eye react to light? **(4-2)**

   The pupil of the eye opens wide when the light is dim in order to let in more light. The pupil becomes smaller when light is bright to limit the amount of light entering the eye.

The numbers in parentheses after each question refer to the lesson number and the objective of that lesson.

Name ______

Chapter 7 Test B

Multiple Choice Choose the best answer.

1. Animals get energy to move their bodies from **(1-1)**
   (a.) food.
   b. sun.
   c. water.
2. When a light is turned on, energy from electricity changes into **(1-2)**
   a. sound energy.
   b. the energy of motion.
   (c.) light energy.
3. Plants use energy from the sun to make **(1-1)**
   (a.) food.
   b. electricity.
   c. light.
4. The tiny bits of electricity in all kinds of matter are known as **(2-1)**
   a. magnets.
   b. heat charges.
   (c.) electric charges.
5. People can make their supplies of coal and oil last longer by using **(2-3)**
   a. coal power plants.
   (b.) less electricity.
   c. more electricity.
6. A magnet attracts metal objects made out of **(2-2)**
   (a.) iron.
   b. aluminum.
   c. tin.
7. Most liquids and gases are not very good at conducting **(3-1)**
   (a.) heat.
   b. sources.
   c. sound.
8. A measure of how fast particles of matter are moving is called **(3-2)**
   a. pressure.
   b. volume.
   (c.) temperature.
9. A tool that controls the temperature in a home is a **(3-3)**
   a. thermometer.
   (b.) thermostat.
   c. heater.
10. What is formed when an object blocks light? **(4-1)**
    (a.) a shadow
    b. a rainbow
    c. an image

The numbers in parentheses after each question refer to the lesson number and the objective of that lesson.

Name ______________________

Chapter 7
Test B

Matching Match each part of the eye with its description. (4-3)

1. c pupil
2. d cells
3. a iris
4. b lens

a. the colored part of the eye
b. the part of the eye that bends light waves
c. the opening in the center of the eye that lets in light
d. the part of the eye that sends a message to the brain about what you see

Short Essay Use complete sentences to answer each question.

1. Name two ways to use energy wisely. (2-3)

Use less electricity. Use electricity safely. (Accept other good answers.)

2. What are three safety rules to follow when using electricity? (2-3)

Have dry hands when using things that use electricity. Keep things that use electricity away from water or shower. Do not touch parts which carry electric currents. (Accept other good answers.)

The numbers in parentheses after each question refer to the lesson number and the objective of that lesson.

TEACHING PLAN

Major Concepts

Lesson 1 Electricity, light, motion, and sound are kinds of energy.
Lesson 2 Electricity is a form of energy that results from movement of electrical charges.
Lesson 3 Heat is the flow of energy from warmer areas to cooler areas.
Lesson 4 Light waves can be separated into the colors of the spectrum and can be bent when they pass through lenses.

Chapter Vocabulary

circuit, conductor, electric charges, electric current, electromagnet, energy of motion, image, insulator, iris, lens, magnetism, prism, pupil, reflect, source, thermostat, visible spectrum

Getting Started

Have students pretend they are somewhere in this city scene. Ask them to describe the different places where the lights are coming from. Encourage students to think about how light and energy help people perform different tasks.

Chapter 7

Forms of Energy

Look at the many lights shining at night. Energy from electricity and light makes this city look bright.

136

Teaching Options

Cooperative Learning ♦

Jigsaw Format (See page T23.)
Assign the following topics at random to your cooperative learning teams.

Topic A: What is the main source of energy on the earth? How do people get and use this energy?
Topic B: How does electricity travel through matter?
Topic C: How is electricity like magnetism?
Topic D: What are some sources of heat? How does heat move through solid matter?

Have students search for information on their topic as they read the chapter. Then let all students with the same topic meet in an expert group to discuss the information. When students return to their teams, they may take turns presenting their topics to the team. Then give students a test covering all topics to complete individually (Chapter 7 Test A or B in the *Test Book*). Award Superteam certificates to teams whose average test scores exceed 90%, and Greatteam certificates to teams whose average test scores exceed 80%.

♦ ***Suitable as a language development activity***

Introducing the Chapter

In this chapter, you will read about electricity and magnets. You also will learn about heat and light. In the activity below, you will learn about the force of a magnet.

Observing Magnets

You can find out how the force of a magnet moves through different materials. Get a magnet, a paper clip, and pieces of paper, aluminum foil, and cloth.

Place the paper clip on your desk. Hold the magnet near the paper clip. Do not touch the clip with the magnet. What happens to the paper clip?[1] Does the magnet's force move through air?[2] Put the paper on top of the paper clip. Place the magnet against the top of the paper. Notice what happens to the paper clip. Place the aluminum foil and then the cloth between the magnet and the paper clip. Notice what happens.

Talk About It

1. How did you know the magnet had a force?
2. What happened when each of the materials was between the paper clip and magnet?

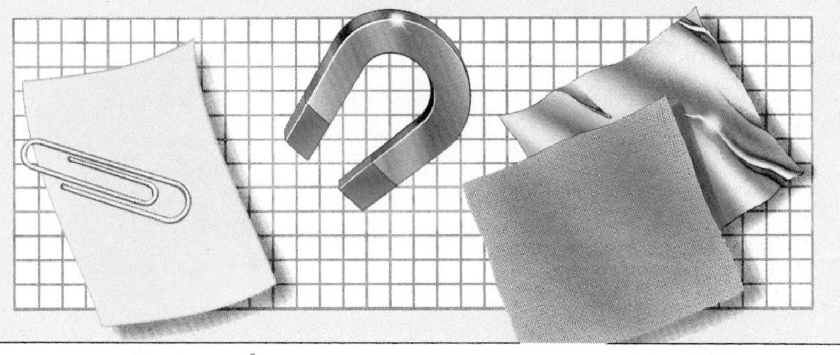

[1]moves toward magnet [2]yes

137

DISCOVER

Objective ♦

This optional *DISCOVER* activity will help students explore and build background information about the concept of magnetic attraction. Later in the chapter students will be able to draw on this experience to help them assimilate the new content.

Science Process Skills

Observing, Classifying

Materials

For each pair of students: magnet, piece of aluminum foil, paper clip, paper, piece of cloth

Teaching Tips

- Tell students to handle magnets carefully. Magnets might lose strength if they are dropped.
- Have students place other materials (cardboard, thin piece of plastic, thin piece of wool) between the magnet and paper clip. Predict through which objects the magnet will pull the clip.

Answers

Talk About It

1. Magnet pulled the paper clip.
2. The magnet attracts the paper clip through the paper, cloth, and foil.

Applying Whole Language ♦

Discuss the whole language framework with each Teaching Option you select. Here is an example applied to the Cooperative Learning option on p. 160.

1. Purpose: To make and play "Energy Password," a word game to review and expand vocabulary

2. Context: 4 × 4 grid, 16 words. Students make 2 or 3 clue cards for each word square.

3. Decisions: Kinds of clues: drawing, synonym, example, use, etc. Ways to play: alone, partner, trade cards with classmates.

4. Evaluation: How good was this game for review? (See p. T30.)

Resource Book page 75

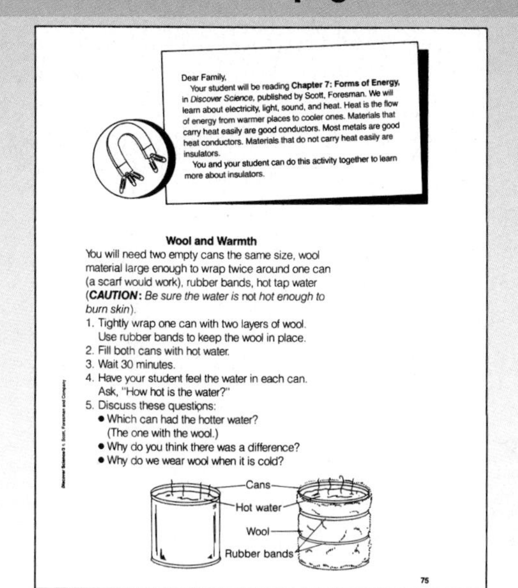

Dear Family,

Your student will be reading **Chapter 7: Forms of Energy,** in *Discover Science*, published by Scott, Foresman. We will learn about electricity, light, sound, and heat. Heat is the flow of energy from warmer places to cooler ones. Materials that carry heat easily are good conductors. Most metals are good heat conductors. Materials that do not carry heat easily are insulators.

You and your student can do this activity together to learn more about insulators.

Wool and Warmth

You will need two empty cans the same size, wool material large enough to wrap twice around one can (a scarf would work), rubber bands, hot tap water (***CAUTION:*** *Be sure the water is* not *hot enough to burn skin*).

1. Tightly wrap one can with two layers of wool. Use rubber bands to keep the wool in place.
2. Fill both cans with hot water.
3. Wait 30 minutes.
4. Have your student feel the water in each can. Ask, "How hot is the water?"
5. Discuss these questions:
   - Which can had the hotter water? (The one with the wool.)
   - Why do you think there was a difference?
   - Why do we wear wool when it is cold?

Cans
Hot water
Wool
Rubber bands

75

Science Background

Energy of motion is known as kinetic energy. The kinetic energy of an object increases rapidly as its speed increases. To determine an object's kinetic energy, square its speed. For example, if an object doubles its speed, its kinetic energy increases by four times. Kinetic energy also increases as mass increases.

TEACHING PLAN

Lesson Objectives

- *Explain* that electricity, light, motion, and sound are kinds of energy.
- *Explain* that light from the sun brings people most of their energy.

Lesson Vocabulary

energy of motion

1. Motivate

Demonstration Activity ♦

Display the following objects on a table in front of the students: flashlight, toy car, radio, and hairdryer. Push the toy car and say the word *energy*. Continue to turn on the remaining objects, repeating the word *energy* each time.

Discussion

Explain how each object produces energy. Questions: **What different kinds of energy do the objects produce?** (flashlight—light; toy car—motion; radio—sound; hairdryer— heat) **Do people use energy?** (Guide students to *infer* that people use energy to do work, such as running, lifting, carrying.)

LESSON GOALS

You will learn
- that energy comes in different forms.
- how people on earth get energy.

energy of motion (mō′shən), energy that moving objects have.

[1]Accept any answers that include electrical appliances.

1 What Are Some Kinds of Energy?

You use energy when you move objects from one place to another. You also use energy to do other things every day.

Different Kinds of Energy

How might you make this train move? You might push it with your hand. The moving train would have **energy of motion.** Moving objects have energy of motion.

You also could turn on electricity to make this train move. Electricity carries energy that can move the train. What other objects use electric energy?[1]

Light and sound are other kinds of energy. One kind of energy can change to another kind of energy. For example, energy from electricity changes to light energy when you turn on an electric lamp.

Electric train

Teaching Options

Science Background

Many different forms of energy exist. Mechanical energy, which is comprised of potential energy and kinetic energy, is the energy an object has from its motion and from forces acting on it. Potential energy is energy of position, while kinetic energy is energy of motion. Energy can be stored in matter as chemical energy, which is a form of potential energy. Particles of matter are in motion, and therefore have kinetic energy. Heat comes from the movement of the particles of matter. Energy that travels as waves in space is radiant energy, some types of which are light, radio waves, and X rays. Sound energy travels as waves through matter. Electricity is a convenient way to carry energy from one place to another.

Reading Strategies ♦

1. Guide students' pre-reading by asking: What ideas do you think you would find in the lesson to answer the question-title?
2. Assign these strategies: Visualizing Information and Mapping Examples. (See pages T26-T29.)
3. Pair students to share what information is clear and unclear and initiate discussion using students' unanswered questions.

♦ *Suitable as a language development activity*

How People Get Energy

A boy uses energy to kick the ball in the picture. Light from the sun brings people most of their energy on earth. How can the sun's energy help you work and play? You get energy to move your body from the food you eat. Your food comes from plants or from animals that eat plants. These plants get energy from the sun.

INVESTIGATE!

Find out how wind energy affects the motion of objects. Write a hypothesis and test your hypothesis with an experiment. You might try using a toy sailboat or a kite.

Lesson Review

1. What are some kinds of energy?
2. How do people on earth get energy?
3. **Challenge!** What happens to electric energy when you turn on a radio?

Study on your own, pages 328–329.

PHYSICAL SCIENCE

FIND OUT ON YOUR OWN

Energy from the sun is called solar energy. Find out how people can use solar energy to heat buildings. For what reason might people who live in Florida use more solar energy than people who live in Alaska? Write a few sentences explaining your answer.

2. Teach

Teaching Tip

- **Possible Misconception:** Some students might know that they need energy to work and play. However, electricity, light, and sound also are kinds of energy.

Investigate!

Accept any testable hypothesis. Most students' data will support various hypotheses. One possible hypothesis is: *Energy from the wind makes a kite able to move through the air.* Students should find out that wind energy can move various objects.

3. Assess

Lesson Review

1. Energy of motion, electricity, light, and sound are kinds of energy.
2. from sunlight
3. Challenge! Electric energy changes to sound energy. **Thinking Skill:** *Applying information to new situations*

Find Out On Your Own

Solar energy moves from the sun to the liquid in the solar panels, thus heating the liquid. People can use the heated liquid to heat their homes. Solar energy is most available in the parts of the country that have much sunny weather. **Thinking Skill:** *Drawing conclusions*

Workbook page 37 *

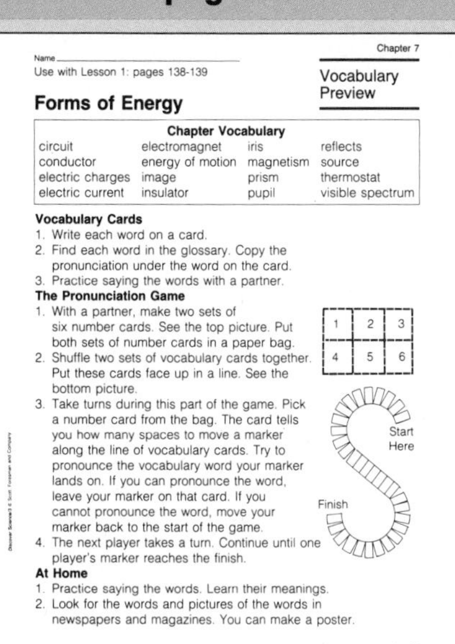

Name ____________

Use with Lesson 1: pages 138-139

Chapter 7

Vocabulary Preview

Forms of Energy

Chapter Vocabulary			
circuit	electromagnet	iris	reflects
conductor	energy of motion	magnetism	source
electric charges	image	prism	thermostat
electric current	insulator	pupil	visible spectrum

Vocabulary Cards

1. Write each word on a card.
2. Find each word in the glossary. Copy the pronunciation under the word on the card.
3. Practice saying the words with a partner.

The Pronunciation Game

1. With a partner, make two sets of six number cards. See the top picture. Put both sets of number cards in a paper bag.
2. Shuffle two sets of vocabulary cards together. Put these cards face up in a line. See the bottom picture.
3. Take turns during this part of the game. Pick a number card from the bag. The card tells you how many spaces to move a marker along the line of vocabulary cards. Try to pronounce the vocabulary word your marker lands on. If you can pronounce the word, leave your marker on that card. If you cannot pronounce the word, move your marker back to the start of the game.
4. The next player takes a turn. Continue until one player's marker reaches the finish.

At Home

1. Practice saying the words. Learn their meanings.
2. Look for the words and pictures of the words in newspapers and magazines. You can make a poster.

37

Answers to masters on pages 136E–136H

Special Education

Have students with learning disabilities find pictures in magazines of examples of each kind of energy discussed in the lesson.

Reteaching Suggestion ♦

Ask students to *list* examples of each kind of energy discussed in the lesson. Use students' lists to *compare* the various uses of energy.

TEACHING PLAN

Lesson Objectives
- *Define* electric charges and electric current.
- *Compare* magnetism and electricity.
- *Describe* wise uses of electricity.

Lesson Vocabulary
circuit, electric charges, electric current, electromagnet, magnetism

1. Motivate

Demonstration Activity ♦
Hold an inflated balloon near a volunteer's hair. Then rub the balloon against a wool cloth. Hold the balloon near the student's hair and slowly move the balloon away from the student's head. *CAUTION:* Cover goggles should be worn during this activity.

Discussion
Have students *describe* the results. (After the balloon was rubbed against the cloth, the hair was attracted to it.) Question: **What happened between the cloth and the balloon?** (Explain that electric charges moved between the balloon and the hair, causing the hair to be attracted to the balloon.)

2 What Is Electricity?

LESSON GOALS

You will learn
- about electric charges and electric current.
- how magnets work.
- how electricity can be used wisely.

Imagine walking into a dark room. What would you do first? You might turn on an electric light. Electric energy can help make your life easier. How have you used electricity today?[1]

[1] Accept any answers that indicate use of electricity.

SCIENCE IN YOUR LIFE

Try rubbing your feet on a wool rug and then reaching for a doorknob. You might see a spark. Sometimes electric charges jump between objects with unlike charges. These jumping charges make the sparks you see.

Electric Charge and Electric Current

Sometimes objects stick together. Look at the balloons in the picture. These balloons are sticking to the wall.

All matter has tiny bits of electricity called **electric charges.** Rubbing objects together can cause these electric charges to move from one object to another. These balloons were rubbed against wool cloth. The balloons picked up charges from the cloth. The balloons and the wall have unlike, or opposite, charges. Unlike charges pull on, or attract, each other. The pulling force between unlike charges makes the balloons stick to the wall.

electric charges, tiny bits of electricity in all matter.

140

Teaching Options

Science Background

Electrical charges are due to the movement of electrons. Electrons are negatively charged parts of atoms. When an object loses some negative electrons, it becomes positively charged. When an object gains electrons, it gains a negative charge. The term *static electricity* is sometimes used to describe the electrical phenomena that result from rubbing objects together or from the buildup of charges in other ways. *Static* means not moving. Since the electrons (or *charges*) are actually moving, the term is not really accurate. The real difference between phenomena that are called *static electricity* and electric current is that electrons flow steadily through matter to produce electric current.

Reading Strategies ♦

1. Guide students' pre-reading by asking: What two to three questions do you have for each subheading?
2. Assign these strategies: Visualizing Information and Writing a Memory Sentence for each vocabulary word. (See pages T26–T29.)
3. Pair students to share what information is clear and unclear and initiate discussion using students' unanswered questions.

♦ *Suitable as a language development activity*

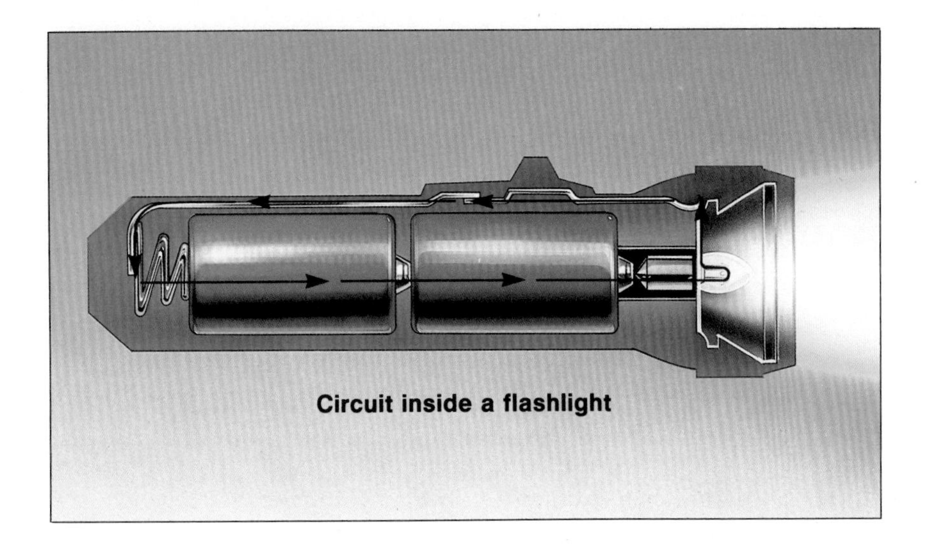

Circuit inside a flashlight

Suppose you rubbed two balloons against a piece of wool. If you held the two balloons near each other, they would push away from each other. The balloons would have charges that are alike. Like charges push away from, or repel, each other.

Think about times that you have used a flashlight. Electricity from a battery can make a flashlight work. A battery pushes electric charges from place to place. The moving of electric charges from one place to another is called **electric current.**

Follow the path of electric current shown in the picture. Notice how the current moves from one end of the battery through the wire to the bulb. Then the current moves from the bulb back to the battery. Electric current travels in a **circuit** when it travels in a path. This bulb lights because the circuit is complete.

electric current (kėr′ənt), the movement of electric charges from one place to another.

circuit (sėr′kit), the path along which electric current moves.

2. Teach

Teaching Tips

- **Possible Misconception:** Some students might not realize that electric current is the moving of electric charges. They might think that electric current originates from outlets in the wall.
- Rub two rubber or plastic combs with a wool cloth, as discussed with balloons in the text. Have a volunteer hold the combs lightly at their corners, and position them near each other. The student should feel a slight repelling between the combs. Draw two combs on the chalkboard. Draw double-sided arrows between the two combs on the chalkboard to illustrate how like charges repel.
- Explain that electric current is like a river. The flowing water carries rocks, fish, and leaves. An electric current carries electric charges from one place to another.

Science and Language Arts

Ask the students to imagine what a day without electricity would be like. To generate ideas ask: **How would you wake up, cook meals, read in the evening?** Encourage students to write a story about such a day. Allow students to share their stories in class.

Special Education

Allow students with learning disabilities to manipulate bar magnets to feel the like poles repel and the unlike poles attract. Also allow them to experiment with magnets and to *predict* which objects in the classroom would be attracted to magnets.

Reinforcement

Hold a sheet of newspaper against a wall. Release the paper and allow it to fall to the floor. Questions: **Did any charges move between the wall and the newspaper?** (no) Next, hold the newspaper against the wall and ask a volunteer to rub the paper with a wool cloth. Release the paper and *observe* the results. **Why does the newspaper stick to the wall?** (Students should *infer* that electric charges have moved from the cloth to the paper, and then moved between the wall and the newspaper.) **Do the wall and newspaper have like or unlike charges?** (unlike)

TEACHING PLAN

Teaching Tips

- Draw a complete circle and a broken circle on the board. Question: **Which drawing is like a complete circuit?** (the circle) **Which is like an incomplete circuit?** (the broken circle) Using colored chalk, illustrate the path of electrons around the complete circuit, and show how they cannot complete the path around the broken circle.
- Strip the insulation from a piece of old electric cord or insulated wire. Allow students to *identify* the material through which current can move easily, and the material through which current cannot move.
- Question: **Why is it important to have a rubber insulator on the outside of electrical wire?** (Guide students to *infer* that it prevents electric shock.)

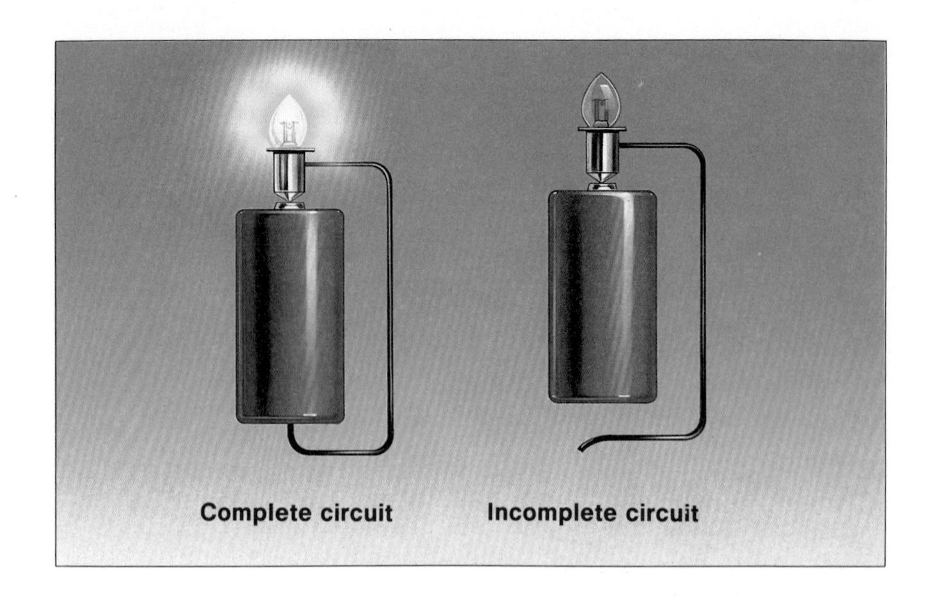

Compare these pictures of batteries. Follow the path of electric current shown in the picture on the left. Find the place where the path stops in the picture on the right. Notice that current cannot move from the battery through the wire back to the battery. The bulb does not light because the circuit is not complete.

Electric current can easily move through some materials, such as copper and other metals. Some other materials, such as rubber, plastic, and glass, do not carry electric current well. Look at the picture of the electric cord. Electric charges move through the copper wire inside the cord. The rubber covering keeps the charges from moving outside the cord.

142

Teaching Options

Science Anecdote

Too much current flowing through a wire can cause a fire. Buildings have fuses or circuit breakers that break the path of the electric current when too much current flows through the wire.

Reinforcement

To demonstrate the difference between a complete and an incomplete circuit, have students stand in a circle with legs apart and feet touching their neighbors'. Have the students pass a ball around the circle. Tell students that the electric charge (the ball) can move only when the circuit is complete (all feet touching). Allow students to pass the ball around the circle once. On the second time around, ask a student to bring his or her feet together to break the circuit. The ball must stop. Discuss how the circuit changed from complete to incomplete.

Enrichment

Invite a member of a local fire department to talk to the class about preventing home electrical fires. Discuss the proper uses of conductors and insulators in home appliances.

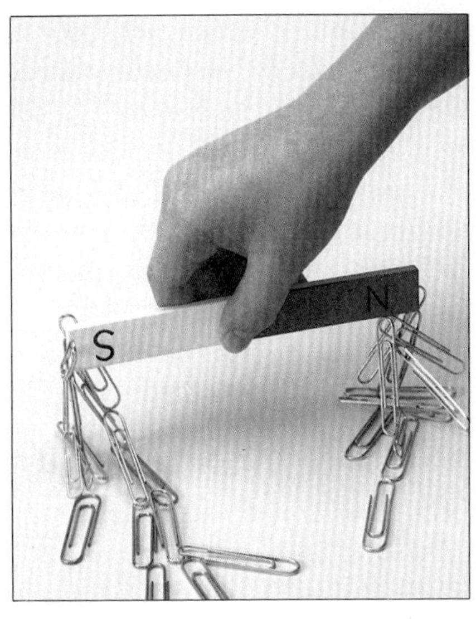

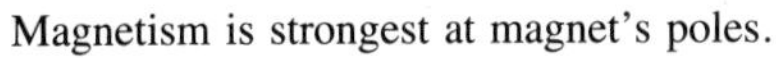
Magnetism is strongest at magnet's poles.

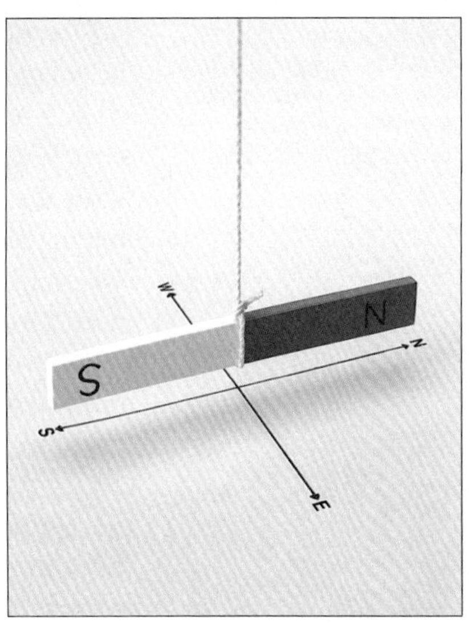

Poles of a magnet

Electricity and Magnets

Look at the picture of the magnet picking up paper clips. A magnet attracts objects with iron in them. Magnets have a force called **magnetism.** Does this magnet pick up more clips at its ends or at its center?[1] The poles of a bar magnet are usually at its ends. Magnetism is strongest at a magnet's poles.

The second picture shows a magnet at the end of a string. Find the marks *N* and *S* on the poles. The magnet turns until the pole marked with an *N* points north. The pole marked *S* points in the opposite direction, or south. A north pole and a south pole are unlike poles. If you hold unlike poles of two magnets near each other, they will attract each other. Like poles of magnets repel each other.

magnetism (mag′nə tiz′əm), the force around a magnet.

[1]at its ends

Teaching Tips

- Remind students of the *DISCOVER* activity on page 137 in which they experimented with magnets. Distribute magnets again and allow students to experiment further with them. Have students ***predict*** which objects the magnets will attract. Provide objects such as paper clips, pins, rubber bands, thumb tacks, nails, chalk, and so on.
- On the chalkboard, draw four pairs of bar magnets that are attracted together or are repelled apart. Have volunteers label the pictures with the letters *N* and *S* to show two like poles repelling and two unlike poles attracting. Ask the students to add arrows to show the direction of the magnetic force between each pair.

Enrichment

Have students make a list of all the uses of magnets in their homes. (to attach notes to the refrigerator, to hold cabinet doors closed, to hold paper clips or pins, to hold toys together, and so on) Compare the students' lists to discuss the many uses of magnets.

Reinforcement

Place a strong bar magnet on an overhead projector and turn on the lamp. Place a sheet of clear plastic over the magnet and slowly sprinkle iron filings over the paper. Allow students to ***observe*** how the iron filings collect in lines of force at the poles of the magnet. *CAUTION:* Tell students not to look directly into the lamp of the projector.

TEACHING PLAN

Teaching Tips

- Question: **How are electricity and magnetism alike?** (objects with like electric and magnetic charges repel one another; objects with unlike electric and magnetic charges attract one another)
- Review how the moving of electric charges from one place to another is called electric current. Have students use their fingers to trace the path of electric current in each illustration on page 144.
- Questions: **How does the machine in the picture use an electromagnet to pick up the heavy pieces of metal?** (Guide students to *infer* that when the current is on, the electromagnet picks up the metal, and when the current is off, it drops the metal.) **How is such a machine used?** (It can move large amounts of metals from place to place.)
- If possible, unscrew an old telephone receiver to allow students to *examine* the electromagnet in a telephone.
- Ask students to *list* other safety rules for using electricity. (Examples: turn off appliances before unplugging; never stick fingers or other objects into outlets.)

electromagnet (i lek′trō mag′nit), a wire coil that becomes a magnet when electric current moves through it.

Electric current can make a magnet. Look at the magnet this girl is using. Notice that loops of wire are wrapped around an iron nail. When electric current moves through the loops of wire, the nail becomes a magnet. This kind of magnet is called an **electromagnet.** You can make an electromagnet stronger by making more turns of wire in the loops.

Electromagnets are useful because their magnetism can be turned on and off. The machine in the picture uses a very strong electromagnet to lift heavy pieces of metal. Electromagnets are found in telephones, earphones, and machines that have electric motors.

Electricity can cause magnetism. Also, magnetism can cause electricity. Moving a magnet through loops of wire causes electric current to move through the wire. Power plants use magnets to make electricity. Wires carry the electricity to your home and school.

Machine with electromagnet

Using an electromagnet

Teaching Options

Science Anecdote

The ability of all atoms to respond to magnetism helps scientists detect certain diseases. In the magnetic resonance imaging (MRI) technique, a large magnet is used to give impulses to the atoms in a person's body. A computer analyzes the result and generates a picture of the body tissue.

Enrichment

Encourage interested students to research and report on the lives and discoveries of Michael Faraday and Hans Christian Oersted, two scientists who worked with electricity and magnetism.

♦ *Suitable as a language development activity*

How Electricity Can Be Used Wisely

Many power plants use energy from coal or oil to make electricity. Energy from the sun long ago was stored in coal and oil. People can make supplies of coal and oil last longer by using less electricity. How can you use less electricity?[1]

Using electricity wisely also means using it safely. Electricity can be harmful if it is not used properly. You can help keep yourself safe by following the safety rules in this chart.

Safety Rules

1. Always use dry hands to touch electric appliances and switches.
2. Keep electric appliances away from water.
3. Disconnect electric appliances when not in use.
4. Keep electric cords in places where the cords will not be damaged.
5. Check electric cords to make sure they are not worn.

Lesson Review

1. How are electric charges and electric current different?
2. How can electric current make a magnet?
3. How can electricity be used wisely?
4. **Challenge!** Many toys with batteries have switches to turn the toys on and off. What does a switch do to a circuit?

Study on your own, pages 328–329.

[1]Accept any answers that indicate that student would use electricity less.

EARTH SCIENCE
FIND OUT ON YOUR OWN
CONNECTION

The earth acts like a giant magnet. Magnetism is strongest near the north and south magnetic poles of the earth. A compass is a tool that helps people use the earth's magnetism to find directions. Find out more about compasses. On a piece of paper, draw a picture of a compass. Write a few sentences explaining how you could use a compass to find directions.

Teaching Tip

- Ask students to *describe* ways to use less electricity in school and at home.

3. Assess

Lesson Review

1. Electric charges are tiny bits of electricity in all matter. Electric current is movement of electric charges.
2. Moving electric current through loops of wire wrapped around an iron nail makes the nail become a magnet.
3. when it is not wasted and when it is used safely
4. Challenge! The switches connect and disconnect the circuit. When the circuit is not connected, electric charges cannot travel in a complete circuit, so the toy cannot work. **Thinking Skill:** *Drawing conclusions*

Find Out On Your Own

The needle of a compass is a magnet with north and south poles. One end of the compass needle points to the magnetic north pole of the earth. To find directions, turn a compass until the needle points to the *N*. Then you know which direction is north. **Thinking Skill:** *Recognizing the main idea and supporting details*

Workbook page 38 *

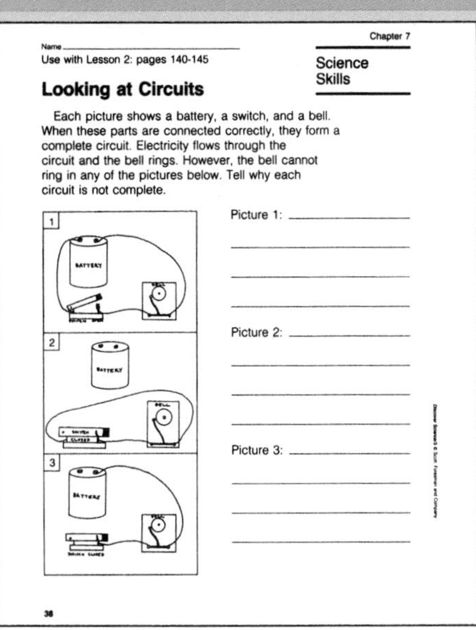
Name ____________
Use with Lesson 2: pages 140-145

Chapter 7

Science Skills

Looking at Circuits

Each picture shows a battery, a switch, and a bell. When these parts are connected correctly, they form a complete circuit. Electricity flows through the circuit and the bell rings. However, the bell cannot ring in any of the pictures below. Tell why each circuit is not complete.

Picture 1: ____________

Picture 2: ____________

Picture 3: ____________

38

* *Answers to masters on pages 136E–136H*

Game Suggestion ♦

Divide students into small teams. Provide each team with 20 paper clips, a bar magnet, and a large nail. Instruct students to make a magnet out of the nail by rubbing it with the magnet in one direction. Then challenge the teams to use the nail magnet to carry paper clips to a designated place in the classroom. Each team should try to carry as many paper clips as possible in 5 minutes.

Reteaching Suggestion ♦

Challenge students to create posters about electric charges, electric current, electricity and magnetism, or using electricity wisely. Point out that a poster should both attract attention and provide information. Encourage students to share their poster ideas with their classmates. Display the posters in the classroom.

Concept
A simple circuit consists of an energy source, a path for the current to flow along, and a means for opening and closing the circuit.

Objectives/Process Skills
- *Construct* an electric circuit.
- *Observe* what happens when the ends of the wires touch.
- *Record* observations.

Time Allotment
Allow 20 minutes.

Safety Tips (See page T24.)
- Instruct students to handle the sharp ends of the wires carefully.
- Warn students to not allow ends of one wire to remain in contact with both ends (terminals) of a battery.

Teaching Tips
- Try doing this activity using switches.
- Show students what happens if the bulb is loosened.

Answers
State Your Conclusion

1. Touching the wires closes the circuit, allowing the current to flow through the bulb so the light goes on.

2. Turning the battery changes the direction in which the current flows. The circuit is still closed and the light goes on.

Use What You Learned

The switch controls the current, which flows from the batteries through the bulb. Turning the switch to "on" closes the circuit and the bulb lights up; turning the switch to "off" breaks the circuit and the bulb does not light. **Thinking Skill:** *Inferring*

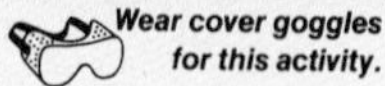

Making An Electric Circuit

Suggested grouping: pairs

Purpose

Make an electric circuit.

Gather These Materials

- three 20 cm pieces of thin insulated wire with about 1 cm of insulation scraped from the end
- flashlight battery (size D)
- flashlight bulb • masking tape
- heavy paper

Follow This Procedure

1. Use a chart like the one shown to record your observations.
2. Tape one end of one wire to the tip of the bulb's base. Tape one end of another wire to the metal on the side of the bulb's base.
3. Tape the other end of one of the wires to the point at one end of the battery. Tape one end of the third wire to the other end of the battery.
4. Tape the wires to the paper, like the picture shows, so that the free ends of the wires are close to each other but not touching. Leave enough wire free at the ends to make them touch in the next step. Observe the bulb.
5. Touch the free ends of the wires. Observe what happens.
6. Untape the battery, and tape it in backwards. Observe what happens when you touch the free ends of the wires.

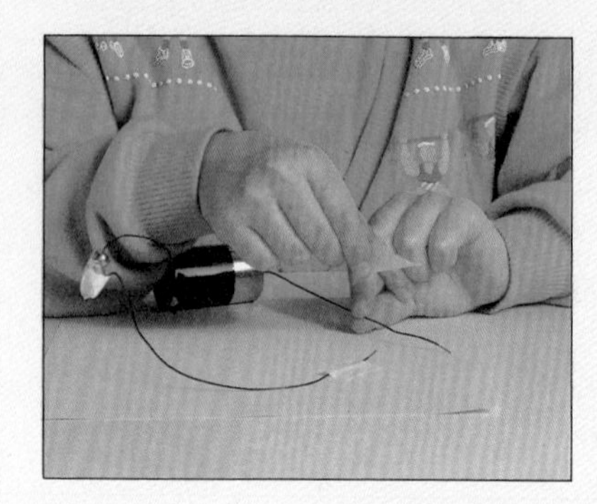

Record Your Results

	What happens to bulb
Wires not touched	bulb off
Battery first way	bulb lights
Battery second way	bulb lights

State Your Conclusion

1. Why does the bulb look different when the wires touch?
2. Explain what happens when you turn the battery backwards.

Use What You Learned

How do you think the switch on a flashlight works?

146

Activity Results

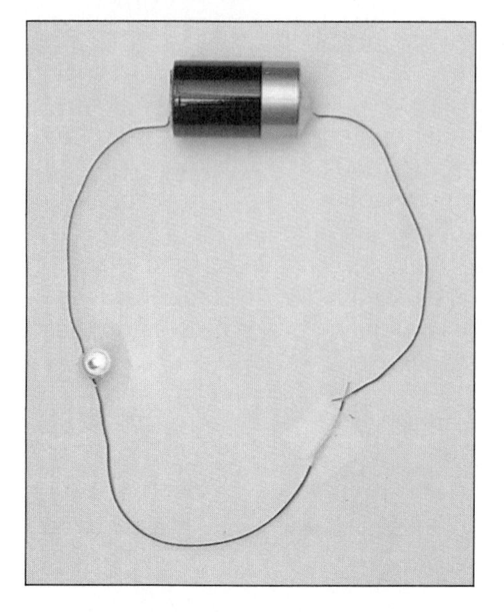

Resource Book page 79

Name ______

Use with Lesson 2: page 146

Chapter 7

Activity Worksheet

Making an Electric Circuit

Record Your Results

	What happens to bulb
Wires not touched	
Battery first way	
Battery second way	

State Your Conclusion

1. Why does the bulb look different when the wires touch?

2. Explain what happens when you turn the battery backwards.

Use What You Learned

How do you think the switch on a flashlight works?

79

Finding Out About Energy

Dr. Lise Meitner

Nuclear energy plant

In 1938, some scientists were experimenting with a special kind of matter called uranium. They tried shooting it with tiny particles. They found that it changed to form a different type of matter! How could this happen? A scientist named Lise Meitner, along with her nephew Otto Frisch, figured out what happened to the uranium.

Dr. Meitner found that the special treatment made each atom of uranium split into two new atoms. These new atoms were different kinds of matter. Splitting up the uranium had caused the change. Dr. Meitner called this splitting nuclear fission.

Dr. Meitner knew that energy is stored in each atom of uranium. She found that uranium releases a huge amount of energy when it splits apart. Dr. Meitner figured out just how much energy was released. This energy is called nuclear energy.

Like other forms of energy, nuclear energy can be used to do work. It can be used to make electricity in the same way coal and oil are used. However, just 1 gram of uranium can make the same amount of energy as 3 million grams of coal!

Electricity made by using nuclear energy provides light and heat to many homes and factories. Also, nuclear energy can be used to run submarines and spacecraft. The energy stored in a tiny bit of matter can make big things happen.

What Do You Think?

1. Describe what happens to a uranium atom when nuclear fission occurs.
2. Which can do more work, splitting one gram of uranium or burning one gram of coal?

Science and People

TEACHING PLAN

Discussion

Discuss with students what they already know about nuclear energy. Students might have negative misconceptions of all things "nuclear." Some of these ideas can be balanced by discussing some benefits of nuclear energy. For example, a much larger amount of coal or oil is required to do the same job as a small amount of uranium. Uranium can therefore be used where coal or oil would be impractical, such as in a submarine, which requires only a small amount of uranium as fuel for a long voyage.

Teaching Tip

- Tell students that nuclear energy provides about 17 percent of the electricity used in the United States.

Answers

What Do You Think?

1. It splits, forming two new atoms of a different type of matter. It also releases a large amount of energy. **Thinking Skills:** *Recognizing cause and effect, Restating or explaining ideas*

2. splitting one gram of uranium **Thinking Skill:** *Comprehending meaning*

Teaching Options

Science Background

Lise Meitner (1878–1968) was born in Austria and worked as a physicist in Germany until 1938, when she fled to Sweden. Otto Hahn and Fritz Strassmann first split the uranium atom, but their discovery was not recognized until Lise Meitner and her nephew, Otto Frisch, interpreted it in 1939. They determined that nuclear fission occurs when uranium atoms are bombarded with neutrons. A neutron strikes the uranium nucleus, combining with it to form two new, smaller nuclei (barium and krypton). Energy is released, along with one or two other neutrons, which can go on to cause further fission reactions, resulting in a chain reaction. In a nuclear power plant, the chain reaction is controlled to produce a steady supply of power.

LESSON 3 pages 148–151

TEACHING PLAN

Lesson Objectives
- *Name* some sources of heat and explain how heat moves through matter.
- *Explain* how temperature is measured.
- *Describe* ways to control home temperature.

Lesson Vocabulary
conductor, insulator, source, thermostat

1. Motivate

Demonstration Activity ♦
Rub your hands together rapidly, then place them on your face. Direct the students to do the same.

Discussion
Questions: **How did your hands feel as you rubbed them together?** (warm) **How did your face feel when you put your hands to your face?** (warm) **Where did the heat come from?** (the hands) Explain how heat moves from warm places and things to cooler ones. (warm hands to cool face)

3 How Is Matter Heated?

LESSON GOALS

You will learn
- how heat moves through matter.
- how you can measure temperature.
- how you can control home temperature.

source (sôrs), a place from which something comes.

Imagine walking outside on a hot summer day. You would feel warm. The sun is an energy **source** because it gives off energy that can warm you. A source is a place from which anything comes. What are some other energy sources?[1]

How Energy Moves Through Matter

Heat is the flow of energy from warmer places and objects to cooler ones. Pretend that you are holding an ice cube in your fist. Your hand is warmer than the ice cube. Energy from your hand moves to the ice cube. What happens to the ice cube?[2]

[1]fire, hot objects, friction
[2]it melts

Electric energy and most other forms of energy give off heat. Look at the hair dryer in the picture. When energy from electric current passes through the wires in this dryer, the wires become hot. This energy can warm air to dry the girl's hair.

Energy gives off heat.

148

Teaching Options

Science Background

The kinetic theory of matter explains energy transfer from one object to another. Heat (the amount of energy transferred) moves among objects in three ways. In conduction, particles of solids collide, transferring energy with each collision. In convection, heated particles of gas or liquid move faster than cooler particles, and move farther apart from each other. Heated gas or liquid, therefore, becomes less dense and rises above the cooler substance, generating a cycle of rising hot material and cooler sinking material. In radiation, waves of radiant energy from a source, such as the sun, increase the amount of energy in the particles of objects they touch. As the particles move faster, the objects heat up.

Reading Strategies ♦

1. Guide students' pre-reading by asking: Which parts of the lesson are familiar to you and which parts are new?
2. Assign these strategies: Writing a Memory Sentence and Mapping Examples for Vocabulary Words. (See pages T26-T29.)
3. Pair students to share what information is clear and unclear and initiate discussion using students' unanswered questions.

♦ ***Suitable as a language development activity***

Energy moves through matter.

You may have cooked food in a pan like this one. The burner on the stove is the energy source. Energy moves from the stove to the pan to the food. You know that the pan and the food are solids. Energy moves through one solid directly to another solid.

Materials that carry energy easily are called good **conductors.** Most metals are good conductors. Materials that do not carry energy easily are called **insulators.** Notice the wood handle on this pan. Wood is an insulator.

Energy also can move through liquids and gases. However, most liquids and gases are poor conductors.

The sun's energy can travel through empty space. When energy from the sun strikes an object, the sun's energy warms the object.

SCIENCE IN YOUR LIFE

You might have used an ice chest to keep food cold. An ice chest is made with materials that are good insulators. These materials keep energy from the air from moving through the ice chest to the cold food inside.

conductor (kən duk′tər), a material that easily carries heat.

insulator (in′sə lā′tər), a material through which energy cannot easily flow.

2. Teach

Teaching Tips

- Question: **How have you experienced heat today?** (sleeping in a warm bed, washing with warm water, eating warm toast or cereal, sitting in a heated classroom, and so on) Use students' examples to discuss how heat moves from warmer things to cooler things. (toaster to bread, furnace to classroom, warm water to face)
- Questions: **Why are cooking pots made of some type of metal?** (these metals are good conductors of heat and they transfer the heat from the stove to the food) **Why do some pots have wooden handles?** (wood is an insulator and does not carry the heat from the pot to your hand)
- **Possible Misconception:** Some students might think that the cold inside a refrigerator travels to foods. Actually, the heat leaves the foods until the temperature of the foods is the same as that of the refrigerator.

Science and Health

Ask students to bring in a favorite recipe which involves heat. Duplicate the recipes and make class cookbooks for all to enjoy. (Possible recipes include: pizza, wholegrain pancakes, blueberry muffins, and so on.)

Special Education

Allow students with learning disabilities to use alcohol thermometers to measure the temperatures of hot and cold liquids. Have them ***identify*** in which liquids the particles of matter are moving faster. (the warmer liquids) *CAUTION:* Wipe up any water spills immediately.

Reinforcement

Make a jar of "sun tea" to demonstrate how the sun's energy changes into heat when it strikes objects on the earth. Fill a gallon jar with cold water and 4 tea bags. *CAUTION:* Wipe up any water spills immediately. Replace the lid. Place the jar in a sunny window or outside in direct sunlight. After 3–4 hours, have students ***observe*** the color of the water. (light brown) Discuss how the sun's energy traveled through space to heat matter (jar, water, tea) on earth.

TEACHING PLAN

Teaching Tips

• Metric-English Equivalent
40°C = 104°F

• Obtain two pyrex measuring cups, some cold water, a pan, and a small burner. Pour some water into one cup and into the pan. Slowly heat the water in the pan over the burner. When the water is warm, pour it into the other cup. Question: **In which cup of water are the particles of matter moving faster? Explain your answer.** (the cup of warm water because the temperature of the water is higher) *CAUTION:* Have students stand at least 1 meter (about 3 feet) from the burner.

Measuring Temperature

You might listen to a weather report to find out how hot or cold a day is. The report tells about the temperature of the air. You learned in Chapter 5 that particles in matter are always moving. Temperature is a measure of how fast particles of matter are moving. The faster the particles of matter move, the higher the temperature of the matter.

A thermometer is a tool that measures temperature. The thermometer in the picture measures temperature in degrees Celsius. What temperature does this thermometer show?[1]

[1]24° Celsius

Thermometer

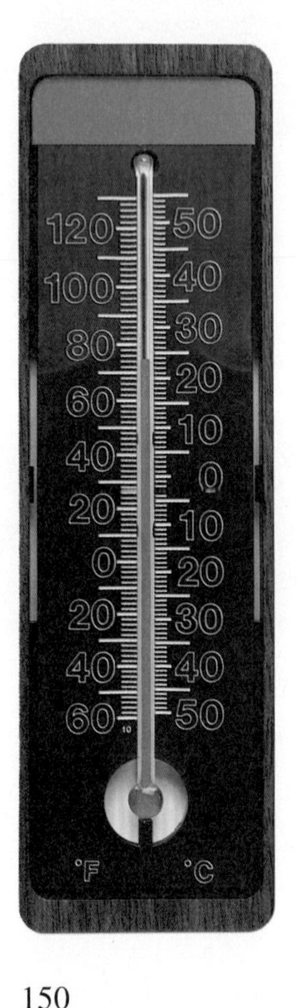

Thermostat

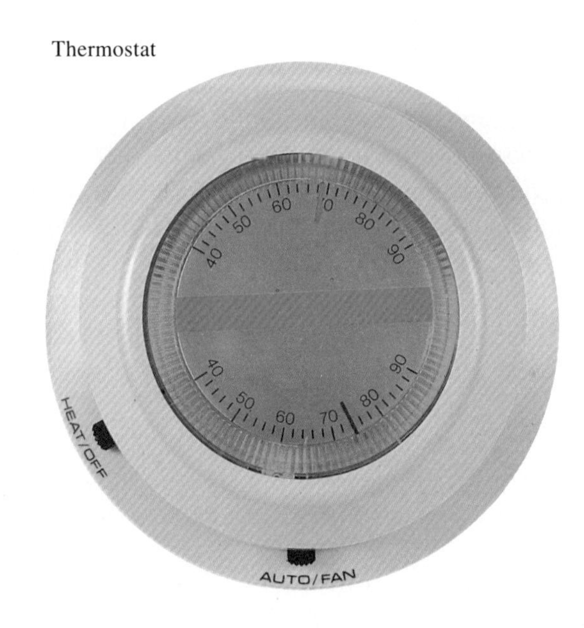

150

Teaching Options

Science Anecdote

The fur or hair on a mammal can trap air. Air is a good insulator and helps keep the mammal warm. Human beings often wear layers of clothing to trap air and keep them warm. Mammals without much hair, such as human beings, have layers of fat under their skin. Fat is also a good insulator.

Enrichment

Encourage interested students to research and report on how solar energy is used to heat homes and buildings.

Reinforcement

Demonstrate how temperature is a measure of how fast particles of matter are moving. Obtain two glass jars, warm water and cold water, and food coloring. Fill one jar with warm water and the other with cold water. Put two drops of food coloring into each jar. Do not stir the water. Have students *observe* the movement of the food coloring in each jar. (It spreads more rapidly in the warm water.) Ask students to explain why this occurs. (The particles of matter are moving faster in the warm water, which has a higher temperature.)

♦ ***Suitable as a language development activity***

Insulating an attic

Controlling Home Temperature

A **thermostat**, like the one shown on page 150, can control the temperature in a home. When the room temperature falls below the set temperature, the thermostat turns on the furnace. A thermostat also can control cooling systems.

thermostat
(thėr′mə stat), a tool that controls the temperature in a home or building.

People can help stop heat from moving out of a building. The man in the picture lines an attic with a good insulator to help keep heat in.

Lesson Review

1. How does energy move through solids?
2. What is used to measure temperature?
3. How can home temperature be controlled?
4. **Challenge!** How does touching a cold object make your fingers feel cold?

Study on your own, pages 328–329.

PHYSICAL SCIENCE

FIND OUT ON YOUR OWN

Look in library books to find out how feathers keep birds warm in the winter. Write a few sentences explaining what you learned.

Teaching Tips

- Explain that when energy from the furnace or the sun strikes your clothes and body the particles of matter in your body move faster and faster, and you begin to feel warm.
- Have students *describe* how the temperature in their homes is controlled. Have students ask their parents to show the students the thermostats and storm windows in their homes. Compare the various methods of heating homes.

3. Assess

Lesson Review

1. Energy moves through one solid directly to another solid.
2. A thermometer measures temperature.
3. A thermostat and the use of good insulators can control home temperature.
4. Challenge! Energy moves from warmer things to cooler ones. Energy moves from your warm fingers to the cold metal. **Thinking Skill:** *Applying information to new situations*

Find Out On Your Own

Feathers trap air. The air is a good insulator and helps keep birds warm. **Thinking Skill:** *Collecting information, Communicating*

Workbook page 39 *

Name ____________

Use with Lesson 3: pages 148-151

Chapter 7

Science and Mathematics

Thermostats

Most homes have at least one thermostat. Often, the thermostat is on the wall in the living room or hall. People use thermostats to start or stop their furnaces. The picture shows a type of thermostat found in many houses. Both sets of numbers on this thermostat tell the temperature in degrees Celsius.

The numbers along the bottom of the thermostat are part of a thermometer. They tell the air temperature in the room where the thermostat is located.

Above the top set of numbers is a dial that moves. If people want the room to be warmer, they push the dial to a higher number. This sets the thermostat at a higher temperature.

Moving the dial up causes the furnace to run. The furnace runs until the temperature shown on the thermometer is the same as the number next to the dial.

To cool the room off, the dial is pushed to a lower number. The lower setting causes the furnace to shut off. The furnace stays off until the thermometer reads the same as the number on the dial.

Dial moves

Thermometer

Use the thermostat drawing to answer these questions.

1. State the room temperature and the thermostat settings. ____________
2. Compare the room temperature with the setting on the thermostat. ____________
3. What is the furnace doing? ____________

39

Reteaching Suggestion ♦

Use index cards to make flash cards with the following terms from the lesson: temperature, thermometer, thermostat, conductor, insulator, matter, and heat. Hold up each flash card and ask the students to define the term in their own words or state an example for each one.

* *Answers to masters on pages 136E–136H*

TEACHING PLAN

Lesson Objectives

- *Explain* how white light can be separated into the visible spectrum's colors.
- *Explain* how light waves bend when they travel through a lens.
- *Explain* how the human eye functions.

Lesson Vocabulary

image, iris, lens, prism, pupil, reflect, visible spectrum

1. Motivate

Demonstration Activity ♦

Shine a flashlight down on a piece of aluminum foil. Hold your hand above the foil so students can *observe* the reflected light. Next, shine the light on some crumpled foil and have students *observe* the light reflected on your hand.

Discussion

Questions: **How did the light get from the flashlight to my hand?** (Let students *describe* how the foil reflected the light to your hand.) **Did the foil create light?** (no) Explain that the foil only reflects light. **Does smooth or crumpled foil seem to reflect more light?** (smooth) Note that smooth surfaces can focus the light better than uneven ones.

[1]the sun, fire, flashlight

LESSON GOALS

You will learn
- that white light is made of colors.
- how lenses work.
- how people see.

reflect (ri flekt′), to turn back.

SCIENCE IN YOUR LIFE

Perhaps you have seen fireflies on a clear summer evening. A firefly has a tiny taillight. Part of the firefly's tail has two different materials that can give off light when they mix together.

Notice how this light travels and reflects.

4 How Does Light Travel?

Suppose you tried to draw on paper when you were in a dark room. You would not be able to see. Paper does not give off light. This boy can see his paper because he turned on a lamp. The lamp is a light source. What are some other sources of light?[1]

The arrow shows how light travels in a straight line from the light source to the paper. Then light bounces off the paper to the boy's eyes. Light **reflects** when it bounces off an object. Light reflects best from smooth, shiny surfaces.

Light cannot go around objects. Suppose an object blocked the light from this lamp. The boy would see a shadow on his paper. A shadow forms whenever an object blocks light. Suppose your body blocks the sun's light when you are outside. What will you see?[2]

[2]your own shadow

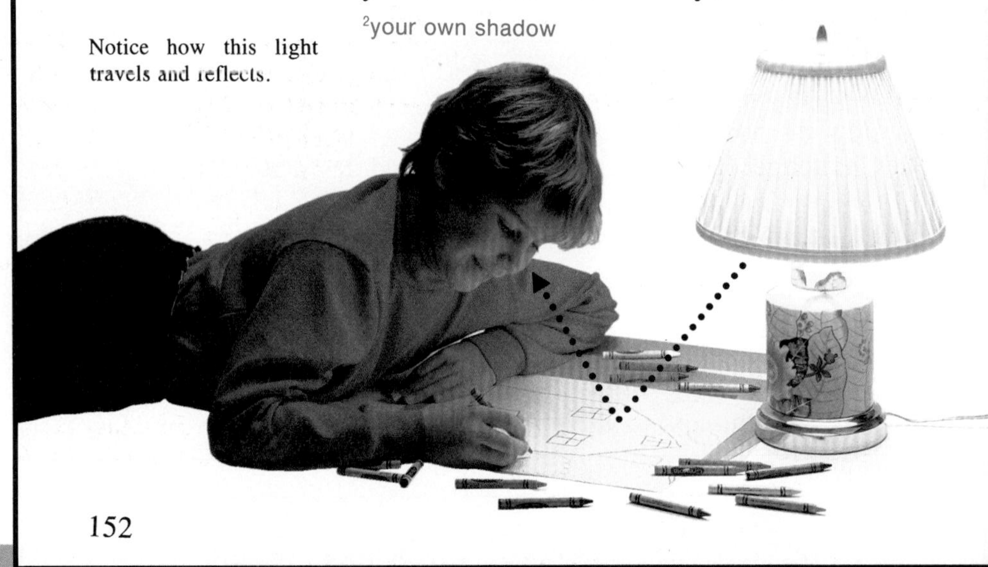

152

Teaching Options

Science Background

Lenses refract light to form images. A convex lens is thicker in the middle than at the edges. A concave lens is thinner in the middle than at the edges. In the human eye, the cornea and the lens work together as a convex lens to focus light from an object and form an image on the retina of the eye.

Reading Strategies ♦

1. Guide students' pre-reading by asking: What two or three questions do you have for each subheading?
2. Assign these strategies: Visualizing Information and Writing a Memory Sentence for the vocabulary words. (See pages T26-T29.)
3. Pair students to share what information is clear and unclear and initiate discussion using students' unanswered questions.

♦ ***Suitable as a language development activity***

A rainbow

Colors in White Light

Most of the light you see is white light. Have you ever seen a rainbow like this one? The rainbow shows how many different colors really make up the white light you see.

You might have seen a rainbow when the sun shines through rain. Light energy travels in waves. Raindrops in the air reflect and bend the light waves coming from the sun. Then the sunlight is separated into different colored parts. The band of colors you see in a rainbow is called the **visible spectrum.** The seven colors of the visible spectrum are red, orange, yellow, green, blue, indigo (in′də gō), and violet.

You also can use certain shapes of glass or plastic to bend white light. A **prism** is a piece of glass or plastic shaped like a triangle. A prism can separate white light into the colors of the visible spectrum.

visible spectrum (viz′ə bəl spek′trəm), the band of colors formed when a wave of white light is bent.

prism (priz′əm), a clear piece of glass or plastic that is shaped like a triangle and can be used for separating white light.

2. Teach

Teaching Tips

- Point out that the moon is not a source of light. Moonlight is reflected from the sun to the earth.
- Attach a large sheet of white paper to the wall. Shine light from a slide projector or flashlight onto the paper. Let students place various objects in front of the light to produce shadows.
- Ask students to *list* objects in the classroom that are light sources and objects that reflect light.
- Question: **When a rainbow occurs, what bends the light waves from the sun?** (Guide students to *infer* that the water drops bend the light.)

Science and Language Arts

Ask students to write a poem about how a rainbow happens. Encourage students to share their poems with the class.

Special Education

Provide a flashlight, white paper, and various objects to create shadows. Have students with learning disabilities trace the outline of the shadows and *compare* the tracings with the shape of the objects placed in front of the light. *CAUTION:* Warn students not to shine the light in other students' eyes.

Reinforcement

Use a prism and a slide projector to produce a spectrum. First shine the light from the projector onto a sheet of white paper. Have students observe the white light. Then shine the light through the prism. The colors will show better if you allow only a slit of light to be projected. Ask students to identify the colors of the spectrum.

TEACHING PLAN

Teaching Tips

• Point out that light rays passing through a lens always bend toward the thickest part of the lens.

• Introduce the terms concave and convex. Explain that a magnifying lens is a convex lens, which is thicker in the middle than at the edges. A concave lens is thinner in the center than at the edges.

• Have students *describe* any experiences with lenses. (eyeglasses, cameras, microscopes, and so on)

• Question: **Why is it sometimes helpful to use a lens that makes objects look bigger?** (Answers might include: to make details visible, to enable a person to read fine print, to allow someone to see how something is made, and so on.)

lens, a piece of glass or other clear material that bends light waves that pass through it.

How Lenses Work

You might have seen a **lens** like one of those in the picture. A lens is a curved piece of glass, plastic, or other clear material. Light waves bend when they pass through a lens.

Lenses can make objects look different. What you see depends on the kind of lens you use. Notice that lenses have different shapes. Some lenses are thick in the middle and thin at the edges. This kind of lens sometimes makes objects look bigger. Other lenses are thin in the middle and thick at the edges. This kind of lens always makes objects look smaller.

Both kinds of lenses are used in eyeglasses to help people see clearly. Lenses have many uses. The lens in this magnifying glass makes objects look bigger. Cameras, telescopes, and microscopes also have lenses.

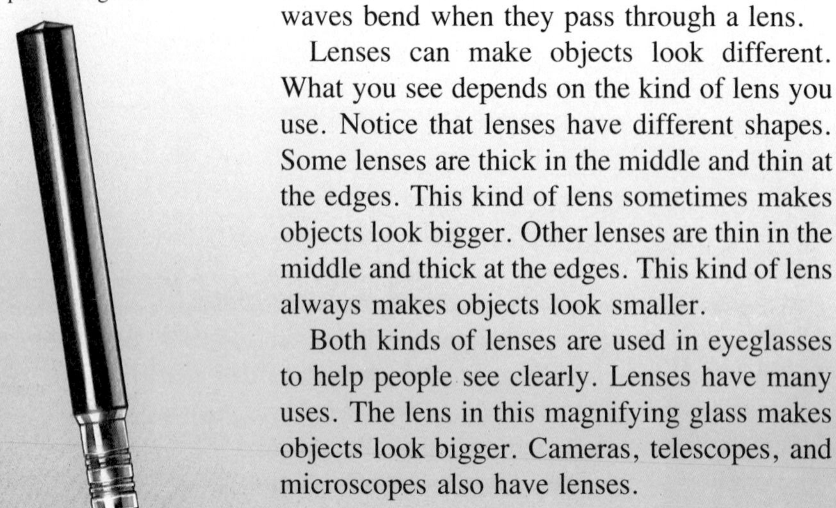

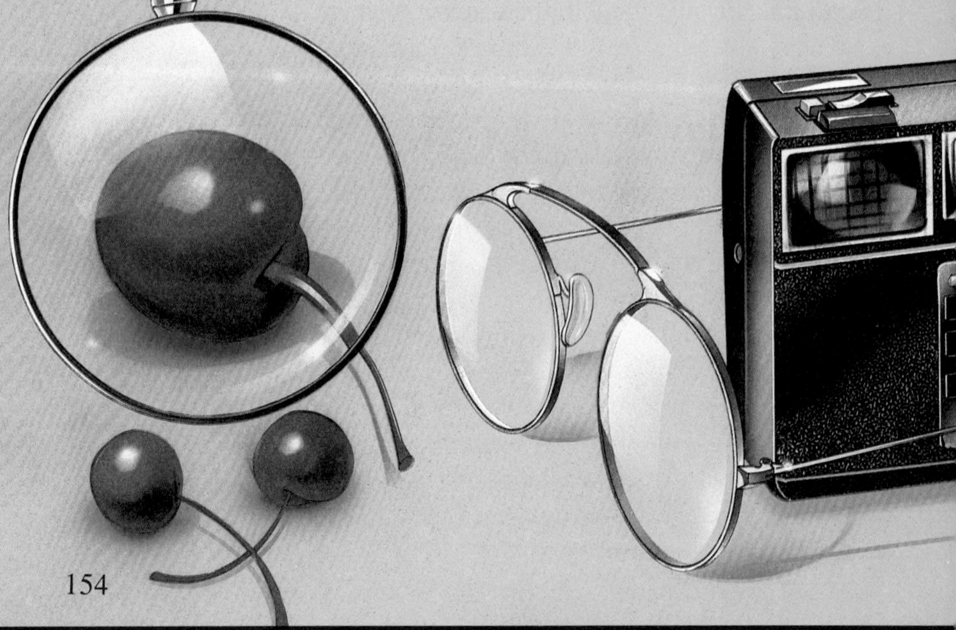

154

Teaching Options

Science Anecdote

Eyeglasses are probably the most familiar form of lens. People started wearing eyeglasses that were made especially to correct individual vision problems in the eighteenth century.

Enrichment

Encourage students to research an animal of their choice to find out how the eye structure of the animal differs from that of humans. For example, inside their eyeballs cats have a mirror-like surface that reflects light onto the retina, causing the cat's eyes to appear to glow.

Reinforcement

Provide students with hand lenses. Have students use the lenses to observe various objects in the classroom. Instruct students to draw each object as it appears through a hand lens. *CAUTION:* Remind students not to focus light rays on skin or paper.

How People See

Look at these pictures of a person's eye. How are the pictures different?[1] The black circle in the center of your eye is really an opening called the **pupil.** Light enters your eye through a clear layer that bends it. Then light passes through the pupil.

The colored part of your eye is the **iris.** Muscles in the iris control the size of the pupil. The pupil changes size to let in more or less light. The pupil opens wide when light is dim. The pupil becomes small when light is bright.

A lens is behind the pupil and iris of your eye. This lens bends light waves in much the same way as a glass lens does. The lens in your eye can change shape to help you see objects clearly. The lens becomes thicker when you look at a close object. The lens becomes thinner when you look at a faraway object.

[1]The pupils are different sizes.

pupil (pyü′pəl), the opening in the center of the eye that lets in light.

iris (ī′ris), the colored part of the eye.

Pupil in bright light

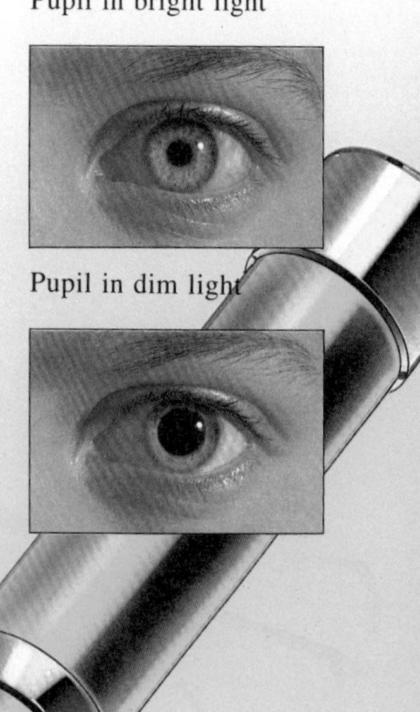

Pupil in dim light

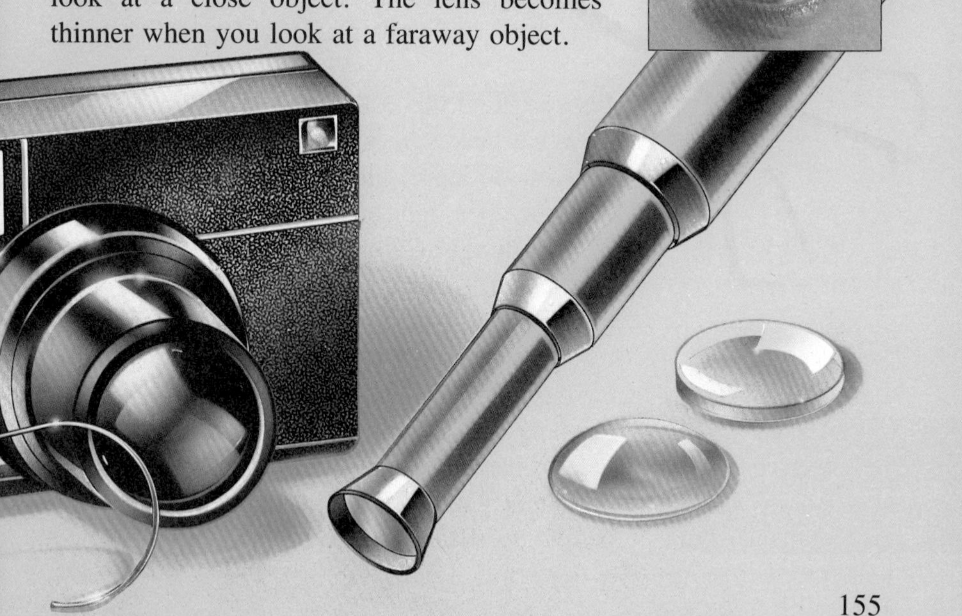

Teaching Tips

- Question: **How does the pupil adjust to changes in light?** (the iris opens the pupil wide when light is dim and the iris closes it down to a small size when light is bright)
- Point out that the eye muscles change the shape of the lens as you focus on an image.
- Question: **How would your eyes change if you went outside on a sunny day after being inside a dimly lit room?** (Guide students to *infer* that their pupils would get smaller.)

Workbook page 40 *

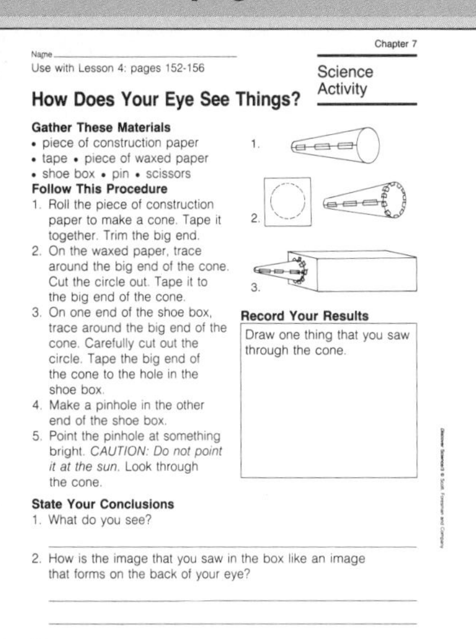

Name

Use with Lesson 4: pages 152-156

Chapter 7

Science Activity

How Does Your Eye See Things?

Gather These Materials
- piece of construction paper
- tape • piece of waxed paper
- shoe box • pin • scissors

Follow This Procedure
1. Roll the piece of construction paper to make a cone. Tape it together. Trim the big end.
2. On the waxed paper, trace around the big end of the cone. Cut the circle out. Tape it to the big end of the cone.
3. On one end of the shoe box, trace around the big end of the cone. Carefully cut out the circle. Tape the big end of the cone to the hole in the shoe box.
4. Make a pinhole in the other end of the shoe box.
5. Point the pinhole at something bright. *CAUTION: Do not point it at the sun.* Look through the cone.

Record Your Results

Draw one thing that you saw through the cone.

State Your Conclusions
1. What do you see?
2. How is the image that you saw in the box like an image that forms on the back of your eye?

40

*** *Answers to masters on pages 136E–136H***

Science and Health

Initiate a class discussion about proper eye care. Ask students to *list* ways to care for the eyes. Use students' lists to compile a class chart of proper eye care. Possible eye care tips include: do not rub your eyes; avoid playing with sharp objects; do not throw anything at a person's eyes; take a break from reading or close work to rest your eyes; work with a bright light that shines over your shoulder; never look directly into the sun.

Reinforcement

Point to various objects in the classroom that are near and far. Have students describe how the lenses in their eyes would change to see each object. (close objects: lens becomes thicker; far objects: lens becomes thinner)

TEACHING PLAN

Teaching Tips

- **Possible Misconception:** Some students might think that the image that forms on the back of the eye is right side up. Explain that the image is upside-down, and that nerve cells in the brain turn the image right side up.
- Have students *identify* the pupil, iris, lens, and retina shown in the diagram.

3. Assess

Lesson Review

1. The colors of the visible spectrum make up white light.
2. Lenses bend light waves.
3. Light enters the eye through the pupil.
4. Challenge! No, because the clouds can block the sun's light. You see sunlight that passes through raindrops when you see a rainbow. **Thinking Skill:** *Drawing conclusions*

Find Out On Your Own

Braille is a system of writing in which letters of the alphabet and numbers are represented by different arrangements of raised dots. People read by touching the dots. **Thinking Skills:** *Collecting information, Recognizing the main idea and supporting details*

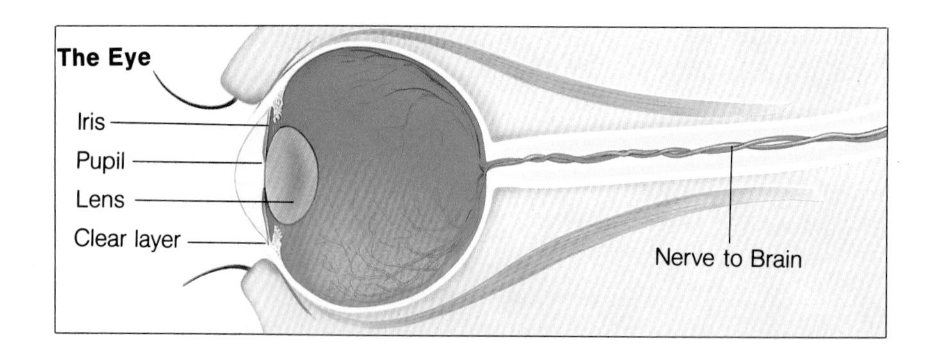

image (im′ij), a copy or likeness.

Use this picture to help you understand what happens when light waves bounce off an object and enter your eye. With your finger, trace the path of the light waves through the pupil and the lens. An **image,** or copy, of what you see forms on the back of your eye. Cells in the back of your eye send a message through a nerve to your brain. Your brain tells you what you see.

Lesson Review

1. What makes up white light?
2. How do lenses change light waves?
3. How does light enter the eye?
4. **Challenge!** Can you see a rainbow when the sun is blocked by clouds? Explain.

Study on your own, pages 328–329.

HUMAN BODY
FIND OUT ON YOUR OWN
CONNECTION

Many people who cannot see are still able to read. Look in library books to find out about Braille. Explain how blind people use Braille to read.

156

Teaching Options

Science Anecdote

Photographers can change the lenses on cameras. Some kinds of lenses can make distant objects look closer. The largest of these telephoto lenses magnifies 200X. Other kinds of camera lenses help people take pictures of large scenes. A fish-eye lens provides a 180° view.

Reteaching Suggestion ♦

Duplicate a diagram of the eye for each student. Ask students to label the pupil, iris, lens, and retina. Have the students trace the path of light through the pupil to the lens.

Workbook page 41 *

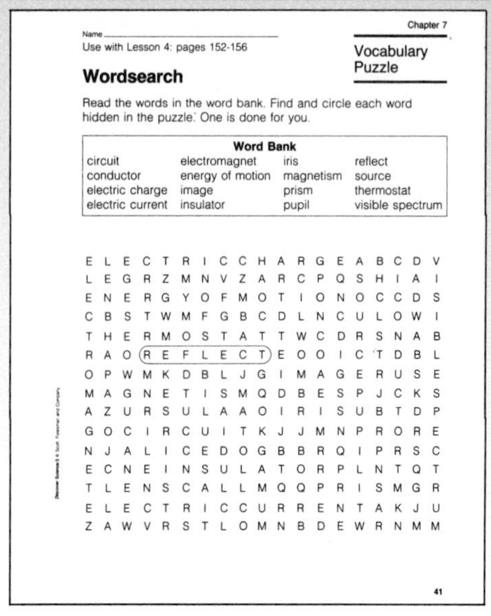
Name ____________________
Use with Lesson 4: pages 152-156

Chapter 7
Vocabulary Puzzle

Wordsearch

Read the words in the word bank. Find and circle each word hidden in the puzzle. One is done for you.

Word Bank

circuit	electromagnet	iris	reflect
conductor	energy of motion	magnetism	source
electric charge	image	prism	thermostat
electric current	insulator	pupil	visible spectrum

E L E C T R I C C H A R G E A B C D V
L E G R Z M N V Z A R C P Q S H I A I
E N E R G Y O F M O T I O N O C C D S
C B S T W M F G B C D L N C U L O W I
T H E R M O S T A T T W C D R S N A B
R A O R E F L E C T E O O I C T D B L
O P W M K D B L J G I M A G E R U S E
M A G N E T I S M Q D B E S P J C K S
A Z U R S U L A A O I R I S U B T D P
G O C I R C U I T K J J M N P R O R E
N J A L I C E D O G B B R Q I P R S C
E C N E I N S U L A T O R P L N T Q T
T L E N S C A L L M Q Q P R I S M G R
E L E C T R I C C U R R E N T A K J U
Z A W V R S T L O M N B D E W R N M M

41

♦ *Suitable as a language development activity*

Using a Prism

Suggested grouping: 2-3 students

Purpose
Observe how a prism separates white light.

Gather These Materials
• prism • white heavy paper
• colored markers or crayons

Follow This Procedure
1. Use a chart like the one shown to record your observations.
2. Hold a prism on end in direct sunlight as shown in the picture.

3. Turn the prism until a block of colored bands appears on the wall.
4. Have your partner hold the white paper so that the colored bands can be seen on it.
5. Observe the colors you see. Notice the order in which the colors appear.
6. In the chart, draw and color what you see. List the colors.

Record Your Results

What you see	Names of colors
red, orange, yellow, green, blue, indigo, violet	

State Your Conclusion
1. What was the order of the colors in the visible spectrum?
2. How are the colors that you saw like a rainbow?

Use What You Learned
What do you think would happen if you passed the visible spectrum you made through a second prism?

157

Activity

Concept
A prism breaks white light into its component colors, which are red, orange, yellow, green, blue, indigo, and violet.

Objectives/Process Skills
- *Observe* how a prism separates white light.
- *Record* observations.
- *Name* the colors of the spectrum in order.
- *Compare* the colors produced by the prism with a rainbow.

Time Allotment
Allow 20–30 minutes.

Safety Tips (See page T24.)
- Warn students not to look directly at the sun when attempting to focus the sun's rays through the prism.
- If students are using glass prisms, caution them to handle the prisms carefully to avoid breakage. Using plastic prisms would prevent the possibility of breakage and cuts.

Teaching Tips
- **Helpful Hint:** Turn off overhead lights for this activity. It works best in a darkened room.
- If your classroom does not have windows, the beam of light from a slide projector can be used as the source of white light.

Answers
State Your Conclusion
1. red, orange, yellow, green, blue, indigo, violet
2. A rainbow has the same colors, in the same order, as in the spectrum observed.

Use What You Learned
The colors would recombine to produce white light. **Thinking Skill:** *Inferring*

Resource Book page 81

Name ____________ Chapter 7
Use with Lesson 4: page 157 Activity Worksheet
Using a Prism

Record Your Results

What you see	Names of colors

State Your Conclusion
1. What was the order of the colors in the spectrum?
2. How are the colors that you saw like a rainbow?

Use What You Learned
What do you think would happen if you passed the spectrum you made through a second prism?

81

Activity Results

Students should see bands of color in the following order: red, orange, yellow, green, blue, indigo, and violet.

*** Answers to masters on pages 136E–136H**

TEACHING PLAN

Purpose
To develop the skills of collecting information and organizing information using thermometers and bar graphs to solve problems.

1. Motivate

Discussion
Explain that the colored liquid inside a thermometer changes size when the temperature changes. The liquid gets smaller and moves down the tube when the temperature is colder. It gets larger when the temperature is warmer. Question: **What happens to the level of the liquid in the tube when the temperature is warmer?** (It moves up the tube.)

2. Teach

Teaching Tip
- If students do the activity, they should be supervised to ensure that they handle light bulbs safely. Remind students that lit bulbs are hot. In Part C, students should allow light bulbs to cool before touching them. Care should also be taken to avoid breakage.

Skills for Solving Problems

Using Thermometers and Bar Graphs

Problem: How does temperature change when distance between a thermometer and a heat source is greater?

Part A. Using a Thermometer to Collect Information

1. Thermometers measure temperature in degrees Celsius. Spaces between the lines on a thermometer stand for 2°C. The colored liquid shows the temperature. Look at the top of the liquid in Thermometer A. What is the temperature? How far is the thermometer from the lamp?
2. What is the temperature on Thermometer B? How far is this thermometer from the lamp?
3. What is the temperature of the thermometer that is 6 cm from the lamp? 8 cm from the lamp?

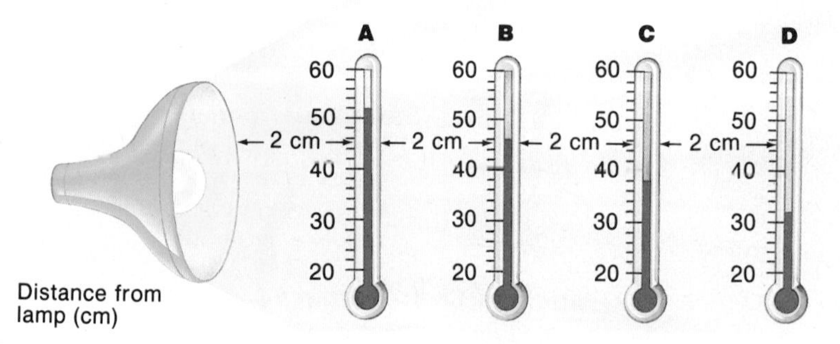

Part B. Using a Bar Graph to Organize and Interpret Information

4. The bar graph contains information you collected about how temperature changes with the distance from the heat source. The first bar shows the temperature of the thermometer that is 2 cm from the lamp. Move your finger

Teaching Options

Sample Bar Graph for Part C

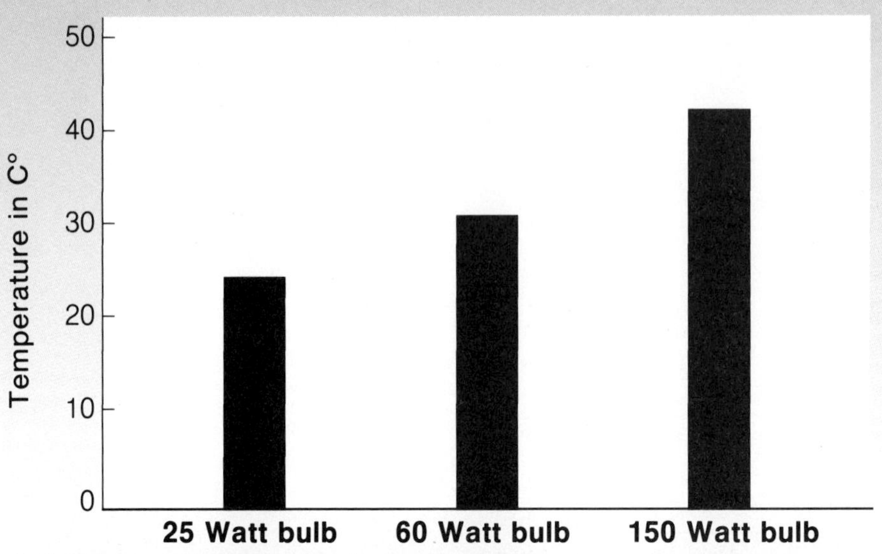

♦ ***Suitable as a language development activity***

from the top of the bar to the line at the left of the graph. This point tells the temperature on the thermometer. What is the temperature?

5. What temperature is on the thermometer that is 4 cm from the lamp? 6 cm? 8 cm? How far from the heat source is the temperature highest? lowest?
6. What would the temperature be on a thermometer that is 10 cm from the lamp?

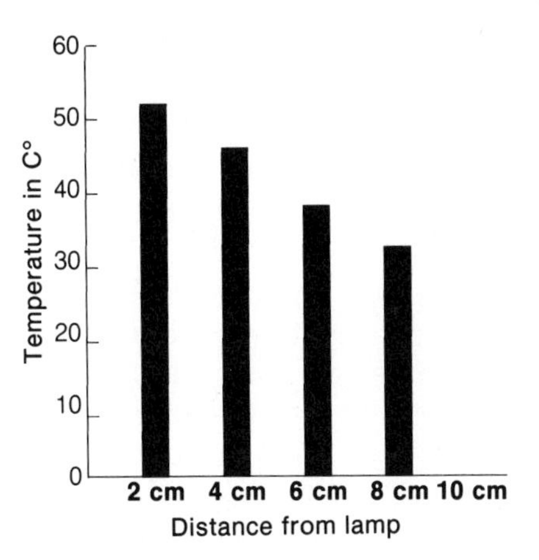

Part C. Using Thermometers and Bar Graphs to Solve New Problems

Problem: How does temperature change when the amount of electric power, or number of watts, of a heat source is greater?

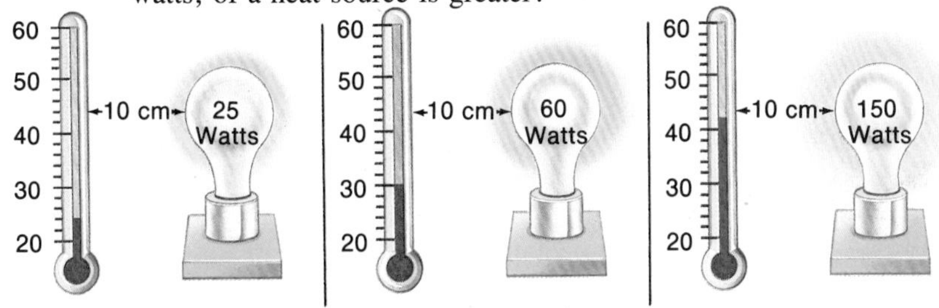

7. Use the picture to collect the information you need to solve the problem. Make a bar graph similar to the one in Part B to organize your information.
8. Look at the bar graph. How does temperature change when the amount of power, or number of watts, of a heat source increases?
9. You might want to do this experiment and use your own results to make a bar graph. *CAUTION: Light bulbs can get hot enough to cause burns. Do not touch hot bulbs.*

3. Assess

Part A

1. 52°C; 2 cm
2. 46°C; 4 cm
3. 38°C; 32°C

Part B

4. 52°C
5. 46°C; 38°C; 32°C; 2 cm; 8 cm
6. About 28°C

Part C

7. See Sample Bar Graph for Part C in Teaching Options
8. Temperature increases.
9. Use one bulb at a time. Turn a 45-watt bulb on for 30 seconds. Record the temperature and turn off the bulb. Let the bulb cool for at least one minute. Replace the bulb and repeat the procedure for the 60-watt and then for the 150-watt bulbs.

Reteaching Suggestion ♦

Turn on a 45-watt bulb for 30 seconds. Hold a thermometer 10 cm away and have students record the temperature. Turn it off for one minute. Then add another 45-watt bulb. Turn both of them on for thirty seconds. Then hold a thermometer so that it is 10 cm away from each bulb and have students record the temperature. Repeat the procedure with a third 45-watt bulb added. Question: **How does the temperature change as more light bulbs are added?** (It increases.) Have students make bar graphs of the information. *CAUTION:* No one should touch the hot bulbs.

Resource Book page 83 *

Name ____________
Comprehension: details and facts

Chapter 7
Science and Reading

Saving Electricity

Read the paragraph. Underline the main idea. Then answer the questions.

Some of the things we use can be recycled, or used again. Recycling paper, cans, and bottles saves electricity. It takes less electricity to make new paper out of old paper than it does to make it from wood pulp. It takes a lot of electric power to make aluminum for cans, so making aluminum cans from old cans saves a lot of electricity. It is easy to recycle returnable bottles. They can be washed and used again. Washing a bottle uses much less electricity than making a new one.

1. How does recycling paper save electricity?
2. Why does recycling aluminum cans save a lot of electricity?
3. How are returnable bottles recycled?

83

* *Answers to masters on pages 136E–136H*

REVIEW PLAN

Reviewing Science Words

1. electric charges
2. energy of motion
3. iris
4. circuit
5. reflects
6. thermostat
7. electromagnet
8. conductor
9. visible spectrum
10. magnetism
11. electric current
12. pupil
13. source
14. insulator
15. prism
16. image
17. lens

Reviewing What You Learned

1. b **3.** a **5.** c
2. d **4.** c

Chapter 7 Review

Chapter Main Ideas

Lesson 1 • Energy of motion, electricity, light, and sound are different kinds of energy. • People on earth get most of their energy from sunlight.

Lesson 2 • All matter has electric charges, and electric current travels in a circuit. • Electric current can make a magnet, and magnetism can cause electricity. • Electricity should be used wisely.

Lesson 3 • Heat is energy in motion. • Temperature is the measure of how fast particles of matter move. • People can control temperature in homes and buildings.

Lesson 4 • White light is made of colors. • A lens can change the way an object looks. • Light must travel through the eye for a person to see.

Reviewing Science Words

circuit	image	pupil
conductor	insulator	reflects
electric charges	iris	source
electric current	lens	thermostat
electromagnet	magnetism	visible spectrum
energy of motion	prism	

Copy each sentence. Fill in the blank with the correct word from the list.

1. All matter has ____.
2. Moving objects have ____.
3. The ____ is the colored part of the eye.
4. Electricity can travel in a path called a ____.
5. Light ____ when it bounces off an object.
6. A ____ can control home temperature.
7. An ____ can be made with wire and an iron nail.
8. A ____ is a material that carries heat easily.
9. The band of colors in a rainbow is the ____.
10. ____ is the force around a magnet.

160

Review Options

Cooperative Learning ♦

STAD Format (See page T23.)
Assign students to work in four- to five-member teams to study Chapter 7 Review. Students should work together to make sure that they and their teammates know the material in the chapter. After students have had enough time to study together, give them a test to complete individually (Chapter 7 Test A or B in the *Test Book*). Award Superteam certificates to teams whose average test scores exceed 90%, and Greatteam certificates to teams whose average test scores exceed 80%.

Test Book page 61 *

Name ____________

Chapter 7
Test A

Multiple Choice Choose the best answer.

1. Plants use the energy from the sun to
   a. make food.
   b. trap animals.
   c. stop growing.
2. Light, sound, and electricity are forms of
   a. magnetism.
   b. sources.
   c. energy.
3. People get the energy they need to move their bodies from
   a. food.
   b. sunlight.
   c. electricity.
4. The movement of electric charges from one place to another is called
   a. sound.
   b. light.
   c. electric current.
5. All machines that have electric motors use
   a. electromagnets.
   b. prisms.
   c. thermostats.
6. Some power plants use the energy produced by coal or oil to make
   a. magnetism.
   b. electricity.
   c. conductors.
7. Objects that are made of metal are good heat
   a. conductors.
   b. sources.
   c. insulators.
8. A tool that measures temperature is a
   a. electromagne*
   b. prism.
   c. thermometer.
9. A tool that controls the temperature in a home or building is a
   a. thermometer.
   b. thermostat.
   c. heater.
10. A piece of glass that separates white light into the colors of the spectrum is a
   a. pupil.
   b. prism.
   c. particle.

61

♦ ***Suitable as a language development activity***

11. ___ is produced when charges move along a path.
12. The ___ is an opening in the eye.
13. A lamp is a ___ of light.
14. An ___ does not carry heat easily.
15. A ___ can separate white light.
16. An ___ is a copy.
17. The glass in a magnifying glass is a ___.

Reviewing What You Learned

Write the letter of the best answer.

1. Which material is the best heat conductor?
   (a) wood (b) metal (c) water (d) air
2. What brings people most of their energy on earth?
   (a) gasoline (b) air (c) wind (d) sunlight
3. When moving particles in matter speed up, temperature
   (a) goes up. (b) goes down.
   (c) stays the same. (d) keeps going up and down.
4. Magnets can attract
   (a) plastic. (b) paper. (c) iron. (d) glass.
5. What forms when an object blocks light?
   (a) a prism (b) a rainbow (c) a shadow (d) an image

Interpreting What You Learned

Write a short answer for each question.

1. What are three kinds of energy?
2. What are two ways to control home temperature?
3. What is a complete circuit?

Extending Your Thinking

Write a paragraph to answer each question or statement.

1. What happens if a hot piece of metal is attached to a cold piece of metal?
2. Suppose the lens in your eye could not change shape. How would you have to move in order to see objects that were close and far away?

To explore scientific methods, see Experiment Skills on pages 360–361.

Interpreting What You Learned

1. Kinds of energy include electricity, light, sound, sun's energy, energy of motion. **Thinking Skill:** *Restating or explaining ideas*
2. Home temperature can be controlled by using a thermostat, using insulators in roof, using storm windows. **Thinking Skill:** *Restating or explaining ideas*
3. A complete circuit is a complete path followed by electric current. **Thinking Skill:** *Restating or explaining ideas*

Extending Your Thinking

1. energy moves from the hot piece to the cold piece; cold piece then becomes hot **Thinking Skill:** *Drawing conclusions*
2. You would need to move forward or backward to focus on the objects. **Thinking Skill:** *Inferring*

Test Book page 62 *

Name

Chapter 7
Test A

Short Answer Fill in the blanks next to each part of the eye. Use these phrases: light enters here, this controls pupil size, this bends light to focus.

1.
2.
3.

Short Essay Use complete sentences to answer each question.

1. Light waves bend when they travel through a lens. Explain why.

2. How does the pupil of the eye react to light?

62

Test Book page 63 *

Name

Chapter 7
Test B

Multiple Choice Choose the best answer.

1. Animals get energy to move their bodies from
   a. food.
   b. sun.
   c. water.
2. When a light is turned on, energy from electricity changes into
   a. sound energy.
   b. the energy of motion.
   c. light energy.
3. Plants use energy from the sun to make
   a. food.
   b. electricity.
   c. light.
4. The tiny bits of electricity in all kinds of matter are known as
   a. magnets.
   b. heat charges.
   c. electric charges.
5. People can make their supplies of coal and oil last longer by using
   a. coal power plants.
   b. less electricity.
   c. more electricity.
6. A magnet attracts metal objects made out of
   a. iron.
   b. aluminum.
   c. tin.
7. Most liquids and gases are not very good at conducting
   a. heat.
   b. sources.
   c. sound.
8. A measure of how fast particles of matter are moving is called
   a. pressure.
   b. volume.
   c. temperature.
9. A tool that controls the temperature in a home is a
   a. thermometer.
   b. thermostat.
   c. heater.
10. What is formed when an object blocks light?
   a. a shadow
   b. a rainbow
   c. an image

63

Test Book page 64 *

Name

Chapter 7
Test B

Matching Match each part of the eye with its description.

1. ___ pupil
2. ___ cells
3. ___ iris
4. ___ lens

a. the colored part of the eye
b. the part of the eye that bends light waves
c. the opening in the center of the eye that lets in light
d. the part of the eye that sends a message to the brain about what you see

Short Essay Use complete sentences to answer each question.

1. Name two ways to use energy wisely.

2. What are three safety rules to follow when using electricity?

64

* *Answers to masters on pages 136E–136H*

 Sound Planning Guide

TEACHING PLAN

Chapter Components	Skills	Materials
Chapter Opener/*DISCOVER:* Inferring What Makes Sounds pp. 162–163	*DISCOVER* p. 163 Science Process Skills *Observing, Inferring*	*DISCOVER* p. 163 (groups of 2) 15 plastic rulers
Lesson 1 What Is Sound? pp. 164–166	Thinking Skills Challenge!: *Inferring* Find Out On Your Own: *Sequencing, Making generalizations*	Demonstration p. 164 tape recorder with side speakers, tissue paper, tape, marker
Activity Predicting Sounds p. 167	Science Process Skills *Predicting, Identifying and controlling variables, Observing*	(groups of 2) 15 large rubber bands, 15 books (24 cm long), 15 metric rulers, 30 pencils
Lesson 2 How Does Sound Travel? pp. 168–170	Thinking Skills Challenge!: *Inferring* Find Out On Your Own: *Inferring*	Demonstration p. 168 large hard rubber or plastic comb, audio-visual cart, lunch cart, or metal tray
Activity Listening to Sound Through Different Materials p. 171	Science Process Skills *Observing, Collecting and interpreting data, Inferring*	(groups of 2) 15 zip-type plastic bags, 15 pencils with erasers, 15 plastic cups, 15 small wood blocks, water
Lesson 3 How Do People Make Sound? pp. 172–174	Thinking Skills Challenge!: *Inferring* Find Out On Your Own: *Recognizing cause and effect*	Demonstration p. 172 tuning fork
Science and Technology Sending Voices Over the Telephone p. 175	Thinking Skills *Summarizing, Restating or explaining ideas*	
Skills for Solving Problems Using Rulers and Pictographs pp. 176–177	Problem Solving Skills *Making decisions/Identifying and solving problems, Interpreting charts, maps, and graphs*	
Chapter Review pp. 178–179	Thinking Skills *Inferring, Comparing, Restating ideas, Contrasting, Recognizing cause and effect*	

Teaching Options

Strategies	Extensions	Resource Masters
Cooperative Learning p. 162 (Also see p. T23.) Applying Whole Language p. 163 (Also see p. T30.)		Family Letter: *Resource Book* p. 87
Reading Strategies p. 164 (Also see pp. T26–T29.)	Reinforcement p. 165 Special Education p. 165 Reteaching Suggestion p. 166	Vocabulary Preview: *Workbook* p. 43 Science Skills: *Workbook* p. 44
		Activity Worksheet: *Resource Book* p. 91
Reading Strategies p. 168 (Also see pp. T26–T29.)	Reinforcement p. 169 Special Education p. 169 Reteaching Suggestion p. 170	Science Activity: *Workbook* p. 45
		Activity Worksheet: *Resource Book* p. 93
Reading Strategies p. 172 (Also see pp. T26–T29.)	Special Education p. 173 Enrichment p. 173 Reteaching Suggestion p. 174	Science and Language Arts: *Workbook* p. 46 Vocabulary Puzzle: *Workbook* p. 47
	Reteaching Suggestion p. 177	Science and Reading: *Resource Book* p. 95
Cooperative Learning p. 178 (Also see p. T23.)		Chapter Tests: Forms A and B *Test Book* pp. 69–72

Classroom Management

Advance Preparation

DISCOVER, page 163
Each group of students will need a plastic ruler for this activity.

Demonstration, page 164
You will need a tape recorder with side speakers, a piece of tissue paper, and a tape with recorded sounds.

Demonstration, page 168
You will need a large, hard rubber or plastic comb and a metal cart or tray.

Activity, page 171
Provide each group of students with a sealable plastic bag and a wood block.

Demonstration, page 172
Obtain a tuning fork for the class.

Vocabulary Review

Use the following sentences with your students to review the meanings of the italicized words.

1. *Matter* comes in three forms—solid, liquid, and gas.
2. *Energy* is the ability to move objects.
3. A *wave* has an up-and-down motion.
4. When you breathe, air passes through a tube in your neck called your *windpipe.*

High-Potential Students

Ask high-potential students to set aside fifteen minutes every day for two weeks to listen to sounds around them and record what they hear. They should choose a different listening location each day, close their eyes, and sit quietly. Ask the students to list the sounds they hear in a daily log. After two weeks, have the students summarize their experiences in oral reports to the class. Ask students to tell how their listening skills changed over the two week period.

Mainstreamed Students

Hearing Impaired
Have students who are hearing impaired share with the class the sounds they can hear. They might also discuss types of equipment that help them stay in touch with current events (such as closed caption television) and how they know that a phone is ringing, or a siren is wailing. If students feel uncomfortable sharing this information, they might draw a picture or write a story about these subjects. Respect the privacy of students who do not want to participate in such activities.

Visually Impaired
Obtain a model of an ear and let students who are visually impaired feel the parts of the ear. If a model is not available, help the class make one out of modeling clay.

Science Fair Projects

The Experiment Skills on p. 362 and the Investigate feature in the chapter can be used for science fair projects. You might also encourage interested students to do one of the following projects:

1. Find out how porpoises and whales are able to navigate in the ocean. Compare this form of nagivation to that used by bats.
2. Use plastic cups for receivers and different types of string and wire as telephone lines to set up a telephone system. Attach one end of each line to a cup and the other ends of the lines to a single wire ring. Compare the ability of each type of line to carry sound.
3. Obtain or make several different types of whistles. Organize the whistles from the one with the highest pitch to the one with the lowest pitch. If a tuned instrument is available, try to match each whistle's sound to a known pitch.

Classroom Resources

Bulletin Board

Encourage students to list or describe how sounds can help them stay safe.

HOW ARE SOUNDS DIFFERENT?

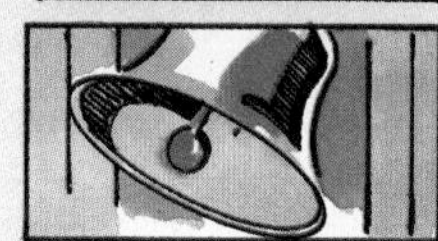

Chapter 8 Poster

How does the harp make music?

Science Discovery Center

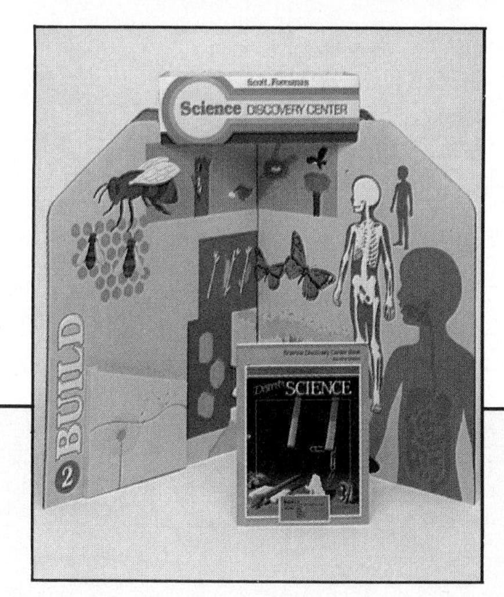

Use pages 85–90 from the *Science Discovery Center Book*. Place these worksheets in the appropriate pockets in the Science Discovery Center.

Overhead Transparencies

Use Transparency 14 from the package of color overhead transparencies.

CHAPTER 8 COPY MASTERS

Dear Family,
Your student will be reading **Chapter 8: Sound** in *Discover Science*, published by Scott, Foresman. We will learn that sound is made by vibrations. Sound has volume (loud to soft) and pitch (high to low). Sound travels in waves through air, water, and solids. We use our vocal cords to make sounds, speak, and sing. Our ears and eardrums help us hear.
You and your student can do this activity together to learn more about sound.

Listen Closely!

You will need paper, pencils, and four different places that you and your student can visit.

1. Go to the first location. Sit down and close your eyes. Concentrate on what you hear.
2. Listen for 10 minutes. You and your student each write down the name of each sound you hear.
3. When the 10 minutes are up, compare your lists.
4. Repeat steps 1 to 3 at the other locations.
5. Discuss the many different sounds you heard, how the sounds were made, which sounds were heard at all four places, which sounds were loud or soft, high or low, pleasant or unpleasant.

Name ________________

Use with Lesson 1: pages 164-166

Chapter 8

Vocabulary Preview

Sound

Chapter Vocabulary			
(eardrum)	(nerve)	vibrate	volume
echo	pitch	(vocal cords)	

Vocabulary Cards

1. Write each word on a card.
2. Find each word in the glossary. Copy the pronunciation under the word on the card.
3. Practice saying the words with a partner.

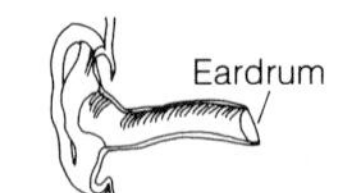

Word Meanings

1. Find the following words in sentences in Chapter 8 of your book. Copy one sentence for each word.
   a. echo ________________
   b. vibrate ________________
   c. volume ________________
2. Now write your own definition of these words on a separate sheet of paper. Check your definitions in the glossary. Write the corrected definition of each word on the back of its vocabulary card.
3. Which vocabulary words are names of parts of the body? Circle those words in the list.

At Home

1. Practice saying the words. Learn their meanings.
2. Look for the words and pictures of the words in newspapers and magazines. You can make a poster.

Teacher's Notes: Ask students to write the words *eardrum, nerve,* and *vocal cords* in sentences.

Name ________________

Use with Lesson 1: pages 164-166

Chapter 8

Science Skills

Musical Instruments

Read about musical instruments.

Most musical instruments can be grouped as stringed, wind, or percussion (per kush′ en) instruments. Each group makes sounds in different ways.

In stringed instruments, strings vibrate to make sounds. The length of the string determines the pitch. Short strings make high sounds. Long strings make low sounds. All the strings can make soft and loud sounds.

In wind instruments, wind makes sounds. The wind is blown into a tube. In some wind instruments, a piece of wood vibrates to make sounds. In others, the player's lips vibrate to make sounds.

In percussion instruments, parts of the instruments are hit to make sound. Drums are percussion instruments. So are gongs and bells. The part of the instrument that is hit vibrates to make sounds.

Study each picture. Then circle the type of instrument.

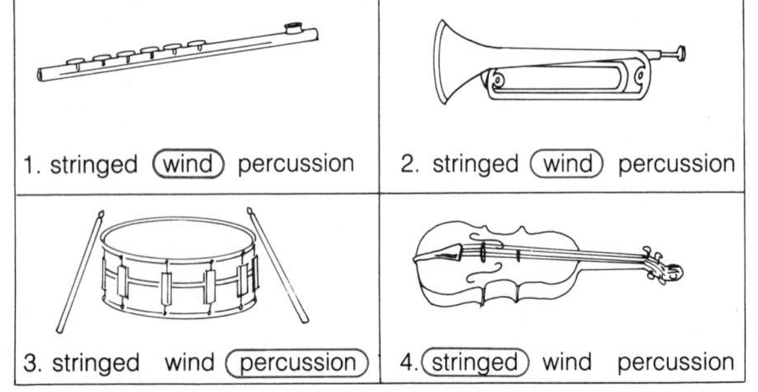

Teacher's Notes: Point out to advanced students that wind instruments can be grouped as woodwind (flute) or brass (bugle).

Name ________________

Use with Lesson 1: page 167

Chapter 8

Activity Worksheet

Predicting Sounds

Record Your Results

	Prediction	Result
Long band		
Short band		

State Your Conclusion

1. How does the length of a vibrating object change the pitch of the sound produced?
   ____
2. At which length did the vibrating part of the rubber band vibrate the fastest.
   ____

Use What You Learned

The strings of a cello are longer than the strings of a violin. Which instrument would you use to play music with low sounds?

____

Workbook

Chapter 8

Name ____________________

Use with Lesson 2: pages 168-170

Science Activity

How Can You See Sound?

Wear cover goggles for this activity.

Gather These Materials

• balloon • scissors • an orange juice can with the ends removed • rubber bands • tape • a small mirror to fit inside the can • glue • paper • a flashlight

ORANGE JUICE

Follow This Procedure

1. Work with a partner. Cut off the end of the balloon. Fit the balloon tightly over one end of the juice can.
2. Put a rubber band over the balloon to keep it in place. Tape down the balloon edges.
3. Glue the back side of the mirror to the balloon.
4. Tape the can to your desk. Have your partner hold the paper about two meters from the balloon end of the can.
5. Shine a flashlight onto the mirror. Aim the light so that its reflection hits the paper.
6. Take turns with your partner talking and shouting into the open end of the can. Watch what happens to the light reflecting on the paper.

Record Your Results

Record what moved when you spoke into the can.

The light moved, or vibrated.

State Your Conclusion

1. What happened to the light when you spoke into the can?
   The light moved, or vibrated.
2. What made this happen?
   Vibrations from my voice vibrated the balloon, which in turn vibrated the light reflecting off the mirror.

Teacher's Note: Have students compare the balloon with their own ears. "What part of your ear is like the balloon?" (eardrum)

Teacher's Resource Book

Chapter 8

Name ____________________

Use with Lesson 2: page 171

Activity Worksheet

Listening to Sound Through Different Materials

Record Your Results

	Observation
Sounds through air	
Sounds through water	
Sounds through wood	

State Your Conclusion

1. Through which material did you hear the loudest sound? the quietest sound?
2. Does sound travel most easily through solids, liquids, or gases? Explain your answer.

Use What You Learned

Suppose you were trying to hear footsteps. Would you put your ear next to the ground to hear the sounds better or would you hold your head up in the air? Explain.

Workbook

Chapter 8

Name ____________________

Use with Lesson 3: pages 172-174

Science and Language Arts

Using Sound Words

Work with a partner. Read the following words aloud to each other. Think about how the words sound.

buzz	hoot	pop
fizz	meow	quack
hiss	moo	tick

Working by yourself, read the following sentences. Fill in each blank with a sound word from the list.

1. The **tick** of the clock kept Yoshi awake.
2. Samuel heard the **buzz** of the bee.
3. The kittens will **meow** when they are hungry.
4. Michael heard the low **moo** of the cow in the field.
5. The snake warned the dog with a **hiss**.
6. The **hoot** of the owl surprised them.
7. She could hear the soda **fizz** in the bottle.
8. If you listen carefully, you can hear the popcorn **pop**.
9. Whenever Miguel walks near the ducks they **quack**.

Teacher's Notes: Ask students to make a list of other words that imitate sounds.

Workbook

Chapter 8

Name ____________________

Use with Lesson 3: pages 172-174

Vocabulary Puzzle

Word Scramble

Unscramble the letters in each pie to write a word. Write on the line below the pie. One is done for you.

L C C D A S R O O V — 1. VOCAL CORDS

H C O E — 2. **ECHO**

H I C T P — 3. **PITCH**

R N V E E — 4. **NERVE**

M V E O U L — 5. **VOLUME**

V E I T B A R — 6. **VIBRATE**

E M U R R A D — 7. **EARDRUM**

Teacher's Notes: Encourage the students to increase their vocabularies by learning the names of other parts of the ear.

Name ____________

Comprehension: predicting outcomes

Chapter 8

Science and Reading

What Will Happen Next?

Read each paragraph. Circle the letter of the sentence that tells what will happen next. Then answer the question.

Jason's Wish

Jason wished he could have a cat. One day his mother said to him, "There's a surprise for you in your room." As he walked toward his room, he heard a meow.

a. Jason will not like his surprise.
(b.) Jason will find a cat in his room.
c. Jason will not go to his room.

What clues in the story make you think as you do?

Jason wished he had a cat; his mother had a surprise for him; he heard a meow.

Jane's Wish

On the way home, Jane's dad told her that he had a surprise for her. As they turned into their driveway, she heard piano music but thought it was coming from a radio. Jane wished that she had a piano so she could learn to play it as well as her mother played.

(a.) Jane will see a piano in her house.
b. Jane's parents will give her a radio.
c. Jane's mother will learn to play the piano.

What clues in the story make you think as you do?

Jane's dad had a surprise for her; she heard piano music; she wished she could learn to play the piano.

Name ____________

Chapter 8
Test A

Multiple Choice Choose the best answer.

1. Sound happens when matter moves back and forth very quickly and (1-1)
a. echoes.
(b.) vibrates.
c. bounces.

2. The loudness or softness of a sound is called (1-2)
(a.) volume.
b. pitch.
c. tone.

3. The highness or lowness of a sound is called the (1-2)
a. loudness.
b. volume.
(c.) pitch.

4. Slow vibrations cause sounds that have a (1-2)
a. high pitch.
(b.) low pitch.
c. low volume.

5. How does sound move through a liquid that has particles that are far apart? (1-2)
(a.) slowly.
b. quickly.
c. not at all.

6. Sound bouncing back from an object causes (2-2)
a. a low volume.
(b.) an echo.
c. a high pitch.

7. The thin flaps at the top of the windpipe are called (3-1)
a. eardrums.
b. sound waves.
(c.) vocal cords.

8. The thin skin that covers the middle part of the ear is called the (3-2)
(a.) eardrum.
b. nerve.
c. pitch.

9. A part of the body that carries messages to the brain is called (3-2)
a. an eardrum.
(b.) a nerve.
c. a pitch.

10. The liquid in the ear carries soundwaves to (3-2)
a. the eardrum.
b. bones.
(c.) nerves.

Numbers in parentheses after each question refer to the lesson number and the objective of that lesson.

Name ____________

Chapter 8
Test A

Matching Write the letter of each ear part in the picture next to its name.

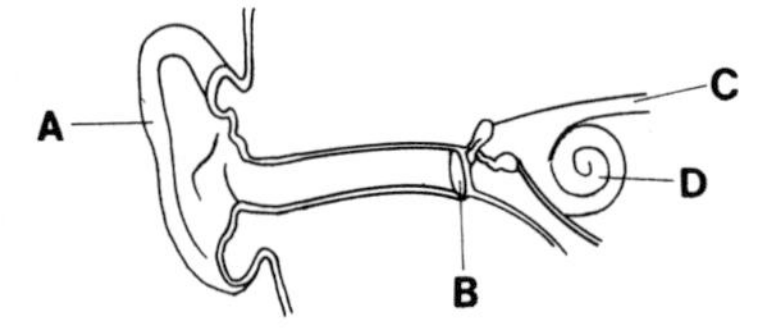

1. B eardrum
2. A outer ear
3. C middle ear
4. D nerve

Short Essay Use complete sentences to answer each question.

1. Explain how people make sound when they talk. (3-1)

When people talk, their vocal cords vibrate. Vocal cords are thin flaps at the top of the windpipe. Air comes from the lungs and passes through the cords making them vibrate. The vibrations make sounds.

2. What causes an echo? (2-2)

When sound waves bounce back from a surface, the sound can be heard again. Echoes can be heard the best when sound bounces back from a hard, smooth surface.

Numbers in parentheses after each question refer to the lesson number and the objective of that lesson.

Name ____________

Chapter 8
Test B

Multiple Choice Choose the best answer.

1. Matter moving back and forth very quickly causes (1-1)
(a.) sound.
b. light.
c. echoes.

2. The pitch of a sound is lowest when vibrations are (1-2)
a. very fast.
(b.) slow.
c. fast.

3. What happens when air from the lungs passes between the vocal cords? (3-1)
(a.) Vocal cords vibrate.
b. The windpipe closes.
c. A nerve sends messages.

4. When an object vibrates harder, its sound will be (1-2)
(a.) louder.
b. softer.
c. lower.

5. Because particles in gases are very far apart, sound moves through gases (2-1)
a. quickly.
(b.) slowly.
c. very fast.

6. Echoes can be heard best when sound bounces back from (2-2)
a. hard, rough surfaces.
b. soft, thick surfaces.
(c.) hard, smooth surfaces.

7. The vocal cords are located at the top of the (3-1)
(a.) windpipe.
b. ear.
c. lungs.

8. A special nerve inside your head carries messages from your ear to your (3-2)
a. eardrum.
(b.) brain.
c. outer ear.

9. The eardrum is made of (3-2)
(a.) thin skin.
b. small bones.
c. liquid.

10. The sound of a ticking clock can be heard the best when the sound passes through (2-2)
(a.) a solid.
b. a gas.
c. a liquid.

Numbers in parentheses after each question refer to the lesson number and the objective of that lesson.

Name ____________________

Chapter 8 Test B

Matching Write the letter of each picture next to its description.

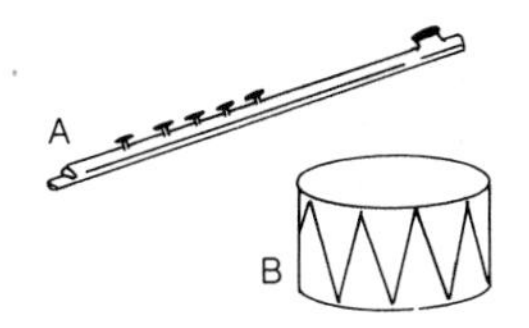

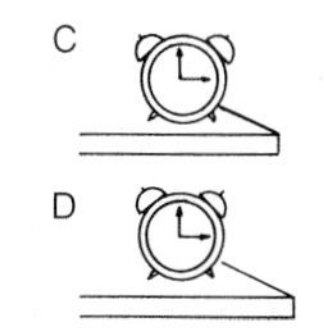

1. A has a high pitch
2. C The sound waves are softer.
3. B has a low pitch
4. D The sound waves are louder.

Short Essay Use complete sentences to answer each question.

1. What can cause hearing loss? (3-2)

Listening to loud noises over and over for a long time can cause hearing loss.

2. Why does sound travel faster through solids than through liquids? (2-1)

Sound moves from one particle to another. Particles in solids are closer together than they are in liquids. Sound moves more quickly when particles are closer together.

Numbers in parentheses after each question refer to the lesson number and the objective of that lesson.

Name ____________________

Unit Test 2

Multiple Choice Choose the best answer.

1. The tiny particles that make up matter are called (5-2-1)
   a. mass.
   (b) atoms.
   c. properties.

2. The color and shape of an object are some of its (5-1-2)
   (a) properties.
   b. elements.
   c. matter.

3. Three states of matter are
   a. solid, shape, mass.
   (b) liquid, gas, solid.
   c. gas, liquid, space. (5-1-3)

4. The change from one type of matter into another is a (5-3-2)
   (a) chemical change.
   b. physical change.
   c. atomic change.

5. A chemical change takes place when matter (5-3-2)
   a. melts.
   b. evaporates.
   (c) rusts.

6. A simple machine made of a rope and a wheel is a (6-2-3)
   a. gear.
   (b) pulley.
   c. wedge.

7. A force that makes objects pull toward each other is (6-1-1)
   a. work.
   b. friction.
   (c) gravity.

8. The force caused by two objects rubbing together is
   a. energy.
   (b) friction.
   c. gravity. (6-1-1)

9. The ability to work is (6-1-3)
   a. gravity.
   (b) energy.
   c. friction.

10. What kind of machine is a bicycle? (6-3-1)
    a. simple machine
    b. lever
    (c) compound machine

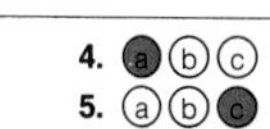
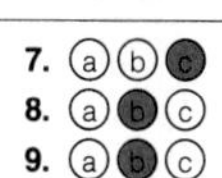

1. ⓐ ● ⓒ
2. ● ⓑ ⓒ
3. ⓐ ● ⓒ
4. ● ⓑ ⓒ
5. ⓐ ⓑ ●
6. ⓐ ● ⓒ
7. ⓐ ⓑ ●
8. ⓐ ● ⓒ
9. ⓐ ● ⓒ
10. ⓐ ⓑ ●

Numbers in parentheses after each question refer to the chapter and lesson numbers and the objective of that lesson.

Name ____________________

Unit Test 2

Multiple Choice Choose the best answer.

11. All machines that have electric motors use (7-2-2)
    (a) electromagnets.
    b. prisms.
    c. thermostats.

12. Sound bouncing back from an object causes (8-2-2)
    a. a low volume.
    (b) an echo.
    c. a high pitch.

13. Plants use energy from the sun to make (7-1-2)
    (a) food.
    b. electricity.
    c. light.

Use the picture to answer question 14.

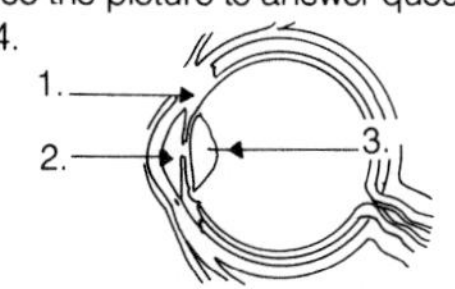

14. Which part of the eye is the lens?
    a. 1 b. 2 (c) 3 (7-4-3)

15. Slow vibrations cause sounds that have a (8-1-2)
    a. high pitch.
    (b) low pitch.
    c. low volume.

16. A tool that measures temperature is a (7-3-2)
    (a) thermometer.
    b. prism.
    c. electromagnet.

17. A magnet attracts metal objects made out of (7-2-2)
    (a) iron.
    b. aluminum.
    c. tin.

18. The eardrum is made of
    (a) thin skin.
    b. small bones.
    c. liquid. (8-3-2)

19. What is formed when an object blocks light? (7-4-1)
    (a) a shadow
    b. a rainbow
    c. an image

11. ● ⓑ ⓒ
12. ⓐ ● ⓒ
13. ● ⓑ ⓒ
14. ⓐ ⓑ ●
15. ⓐ ● ⓒ
16. ● ⓑ ⓒ
17. ● ⓑ ⓒ
18. ● ⓑ ⓒ
19. ● ⓑ ⓒ

Numbers in parentheses after each question refer to the chapter and lesson numbers and the objective of that lesson.

TEACHING PLAN

Major Concepts

Lesson 1 Sound is a form of energy produced when matter vibrates.
Lesson 2 Sound travels in all directions from a source.
Lesson 3 The vocal cords vibrate to make sound, and the ear receives sound.

Chapter Vocabulary

eardrum, echo, nerve, pitch, vibrate, vocal cords, volume

Getting Started

Explain to students that dolphins are mammals. Even though dolphins live in water, they have lungs and breathe air just like people. Dolphins emit the clicking sounds, and then wait for the sounds to echo off objects that might be in their way. This process is called echolocation or sonar.

Chapter 8

Sound

Look at the dolphins in the picture. Dolphins make clicking sounds as they swim. They can use sounds to send messages to each other. They also listen to sounds to help them find their way.

162

Teaching Options

Cooperative Learning ♦

Jigsaw Format (See page T23.)
Assign the following topics at random to your cooperative learning teams.
Topic A: What is sound?
Topic B: If someone is calling you from outside, how can you hear them inside your house?
Topic C: Explain two ways in which sounds can differ.
Topic D: How can a doctor hear your heart beat through a stethoscope?
Have students search for information on their topic as they read the chapter. Then let all students with the same topic meet in an expert group to discuss the information. When students return to their teams, they may take turns presenting their topics to the team. Then give students a test covering all topics to complete individually (Chapter 8 Test A or B in the *Test Book*). Award Superteam certificates to teams whose average test scores exceed 90%, and Greatteam certificates to teams whose average test scores exceed 80%.

♦ ***Suitable as a language development activity***

Introducing the Chapter

In this chapter, you will learn how sounds are alike and how sounds are different. You will learn how sound travels through matter. You also will read about how people make and hear sound. The activity below will help you learn what makes sound.

[1]back and forth quickly
[2]sound
[3]the sound stops

Inferring What Makes Sounds

You hear many different sounds every day. You can make sound with objects in your classroom. Place a plastic ruler on your desk so that part of it hangs over the edge. Hold the ruler tightly on the desk with one hand as shown in the picture. Gently press down on the free end of the ruler with your other hand. Let go of the free end of the ruler. How does the ruler move?[1] What do you hear?[2] What happens when the ruler stops moving?[3] Slide the ruler on the desk to make the part that moves longer and then shorter. Listen to the different sounds you can make.

Talk About It

1. What was happening to the ruler when sounds were made?
2. How did the sounds change when you moved the ruler?

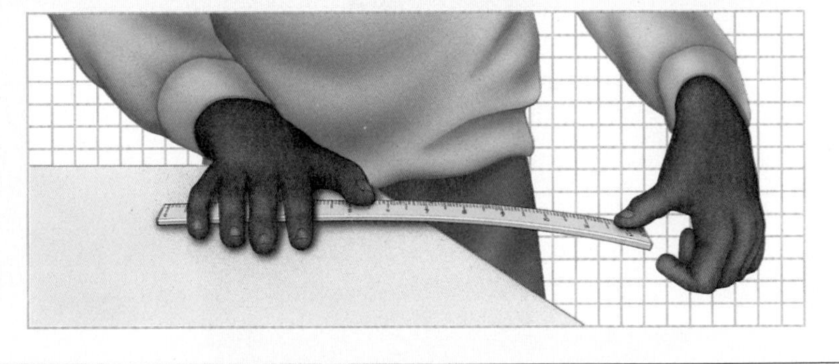

DISCOVER

Objective ♦

This introductory *DISCOVER* activity will help students explore and build background information about the concept that vibrations produce sound. Later in the chapter, students will be able to draw on this experience to help them assimilate the new content.

Science Process Skills

Observing, Inferring

Materials

For each pair of students: a plastic ruler

Safety Tip (See page T24.)

- Tell students not to bend ruler so much that it might break.

Teaching Tips

- Ask students to ***name*** different types of musical instruments. Question: **Which part of the instruments vibrate to make sound?** (Help students ***identify*** the vibrating parts of the instruments.)
- Encourage students to create a band by using rubber bands, combs, and other objects as musical instruments.

Answers

Talk About It

1. the vibrating ruler
2. higher or lower

Applying Whole Language ♦

Discuss the whole language framework with each Teaching Option you select. Here is an example applied to the Special Education option on p. 173.

1. Purpose: To make posters and compare hearing abilities of animals.
2. Context: Individual project, whole group sharing. References: books, pictures, film, visits.
3. Decisions: What ear parts should we study/compare? How can we group results?
4. Evaluation: What did we learn that we did not expect? (See p. T30.)

Resource Book page 87

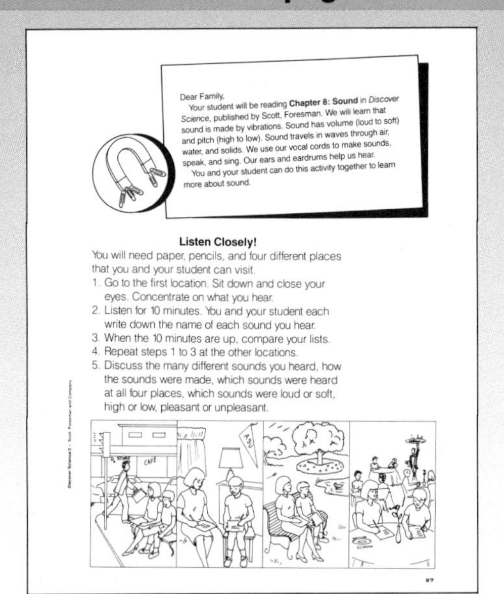

Dear Family,
Your student will be reading **Chapter 8: Sound** in *Discover Science*, published by Scott, Foresman. We will learn that sound is made by vibrations. Sound has volume (loud to soft) and pitch (high to low). Sound travels in waves through air, water, and solids. We use our vocal cords to make sounds, speak, and sing. Our ears and eardrums help us hear.
You and your student can do this activity together to learn more about sound.

Listen Closely!

You will need paper, pencils, and four different places that you and your student can visit.

1. Go to the first location. Sit down and close your eyes. Concentrate on what you hear.
2. Listen for 10 minutes. You and your student each write down the name of each sound you hear.
3. When the 10 minutes are up, compare your lists.
4. Repeat steps 1 to 3 at the other locations.
5. Discuss the many different sounds you heard, how the sounds were made, which sounds were heard at all four places, which sounds were loud or soft, high or low, pleasant or unpleasant.

Science Background

Sound is produced when matter vibrates. Sound waves travel through matter as compressional waves. Some of the characteristics of sound are quality, pitch, frequency, intensity, and duration. The intensity of a sound decreases as it moves farther away from its source.

LESSON 1 pages 164–166

TEACHING PLAN

Lesson Objectives
- *Explain* how sound is produced.
- *Describe* how volume and pitch are determined.

Lesson Vocabulary
pitch, vibrate, volume

1. Motivate

Demonstration Activity ♦
Obtain a tape recorder with side speakers. Turn the bass level all the way up. Tape a sample sheet of tissue paper (on the top and both sides, allowing some slack) over each speaker. Use a marker to draw lines on the tissue paper. Turn the recorder on. Then have the students *observe* the tissue paper.

Discussion
Question: **What happened to the tissue paper when the tape recorder was turned on?** (The paper vibrated.) Explain that the sound pushed matter and made it vibrate. (Electrical energy was turned into sound energy that made the air vibrate, causing the tissue paper to vibrate.)

1 What Is Sound?

LESSON GOALS

You will learn
- how sounds are alike.
- how sounds are different.

vibrate (vī′brāt), move quickly back and forth.

You learned in Chapter 7 that you use many different kinds of energy each day. Electricity and light are some kinds of energy. Sound is another kind of energy.

How Sounds Are Alike

Suppose you took a walk outside. You might hear people, animals, or cars making sounds. All these sounds are different. Yet all sounds are alike in some ways.

Sound happens when matter **vibrates,** or moves back and forth very quickly. Look at the drum in the picture. If you hit a drum, parts of the drum will vibrate. The drum makes sound. All the sounds around you happen only when objects vibrate.

164

Teaching Options

Science Background

Sound waves are a series of compressions and rarefactions caused by vibrating objects. The distance a vibrating object moves from its resting position, is its amplitude. The larger the amplitude of a sound wave, the more energy it has, and the louder the sound produced. The unit that is used to measure the strength or loudness of a sound is the decibel. The human ear can hear decibels of 0–85 without sustaining any damage. Decibel levels between 85 and 115 are dangerous, but do not generally cause pain to the listener. Decibel levels above this range can cause permanent hearing loss if experienced over a long period of time. A rocket being launched creates about 180 decibels.

Reading Strategies ♦

1. Guide students' pre-reading by asking: Which parts of the lesson look easy, which parts look hard, and why?
2. Assign these strategies: Visualizing Information and Writing a Memory Sentence (See pages T26–T29.)
3. Pair students to share what information is clear and unclear and initiate discussion using students' unanswered questions.

♦ *Suitable as a language development activity*

How Sounds Are Different

Suppose you lightly tapped a drum. You would hear a soft sound. Suppose you hit the drum harder. How would the sound change?[1] The loudness or softness of a sound is called **volume.** The harder you hit the drum, the more it will vibrate. The more an object vibrates, the louder the sound will be. A loud sound has more volume than a soft sound.

Pretend you are listening to music. You probably would hear high sounds and low sounds. **Pitch** describes how high or low a sound is. Objects that vibrate slowly, such as the drum, make sounds with a low pitch. This recorder makes air particles vibrate quickly. The recorder has a high pitch. Different volumes and pitches can be combined to make pleasant music.

INVESTIGATE!

Find out if animals hear the same sounds as people. Write a hypothesis and test your hypothesis with an experiment. You might blow on a whistle with a very high pitch. See if both people and dogs can hear the sounds.

volume (vol′yəm), the loudness or softness of a sound.

pitch (pich), how high or low a sound is.

[1]get louder

2. Teach

Teaching Tips

- **Possible Misconception:** Students might think that sound, like light, can travel through vacuums and space. Tell them sound can travel only through matter.
- Tell students that the thickness of an object also affects how fast or slow vibrations occur, which affects the pitch. Let students try testing this idea by using a thick rubber band and a thin rubber band of about the same length. Have students ***compare*** the sounds that the two rubber bands make. *CAUTION:* Students should wear safety goggles during this activity.
- If possible, have students ***examine*** a guitar and ***compare*** the lengths and thicknesses of the strings.

Investigate!

Accept any testable hypothesis. Most students' data will support this hypothesis: *Dogs hear sounds that people cannot hear.* Students should find out that dogs can hear high-pitched sounds that people are unable to hear.

Workbook page 43 *

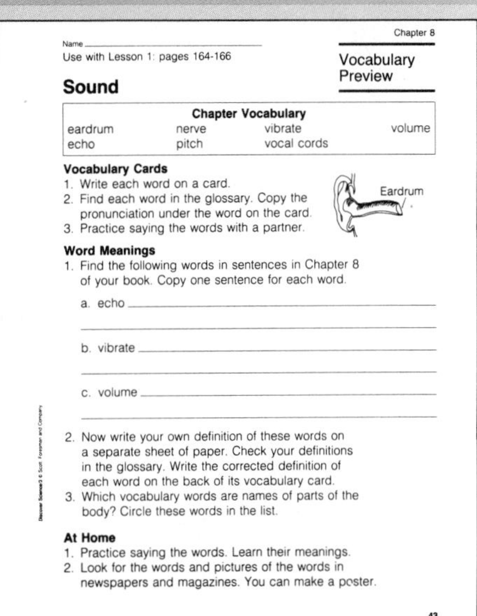

Chapter 8

Name ____________

Use with Lesson 1: pages 164-166

Vocabulary Preview

Sound

Chapter Vocabulary

eardrum, nerve, vibrate, volume, echo, pitch, vocal cords

Vocabulary Cards

1. Write each word on a card.
2. Find each word in the glossary. Copy the pronunciation under the word on the card.
3. Practice saying the words with a partner.

Word Meanings

1. Find the following words in sentences in Chapter 8 of your book. Copy one sentence for each word.
   a. echo ____________
   b. vibrate ____________
   c. volume ____________
2. Now write your own definition of these words on a separate sheet of paper. Check your definitions in the glossary. Write the corrected definition of each word on the back of its vocabulary card.
3. Which vocabulary words are names of parts of the body? Circle these words in the list.

At Home

1. Practice saying the words. Learn their meanings.
2. Look for the words and pictures of the words in newspapers and magazines. You can make a poster.

43

* ***Answers to masters on pages 162E–162H***

Special Education

Give each student a tuning fork and a large beaker half-filled with water. Instruct the students to tap the fork lightly against the heel of their hand and then to place the tips of the vibrating fork into the beaker. Students should note the size of the splash. Have the students continue this procedure, but they should increase the force of the tap each time. Three or four trials should establish the concepts. (*CAUTION:* To prevent slipping accidents, wipe up any water spills immediately.)

Reinforcement

Obtain a tuning fork from the school music teacher or a music store. Tap the tuning fork. Immediately dip the fork tips barely into water. Allow the students to observe what happens. Let students take turns tapping the fork and placing it on the small bone behind their ears. Explain that the sound they hear is caused when the tuning fork vibrates.

TEACHING PLAN

3. Assess

Lesson Review

1. Sound is made when particles of matter vibrate.
2. Sounds can differ in volume and pitch.
3. Challenge! You can change the pitch of a vibrating rubber band by changing the length of the rubber band. **Thinking Skill:** *Inferring*

Find Out On Your Own

Students should obtain and organize pictures of objects that make sounds. **Thinking Skills:** *Sequencing*, *Making generalizations*

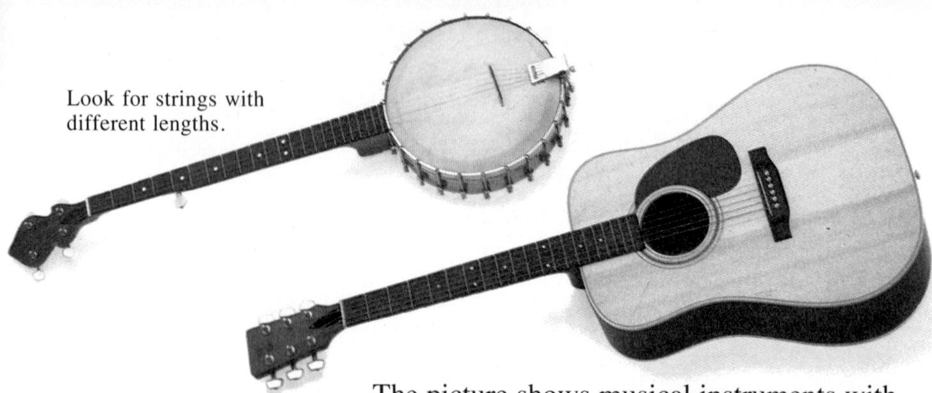

Look for strings with different lengths.

The picture shows musical instruments with strings. You can make different sounds by plucking the strings. Notice how some strings are shorter than others. As a string is shortened, it vibrates more quickly. How does making a string shorter change the pitch of the sound?[1]

[1]gets higher

Lesson Review

1. How is sound made?
2. How are some sounds different?
3. **Challenge!** How can you change the pitch of a sound made by a vibrating rubber band?

Study on your own, pages 330–331.

PHYSICAL SCIENCE
FIND OUT ON YOUR OWN

Look through magazines for pictures of objects that can make sound. Use these pictures to make a poster about sound. Place the pictures in order from those that make the softest sound to those that make the loudest sound. Write the name of each object under its picture.

166

Teaching Options

Science Anecdote

On the Hawaiian Island of Kauai, (kou ī′), when the wind blows over the sand in a particular place, the vibrations make a sound like a barking dog. The place is called *Barking Sands*.

Reteaching Suggestion ♦

Bring a thin glass and some lemon juice or vinegar to class. Wash your hands with soap to rid them of any oils. Place the lemon juice or vinegar around the rim of the glass and rub the glass gently in a circular motion. (The glass will begin to "sing.") Rub faster and slower so students can detect the differences in the volume of the sound that is made. Adding water to the glass will change the pitch produced. Question: **What caused the sound?** (the glass vibrating as it was rubbed)

Workbook page 44 *

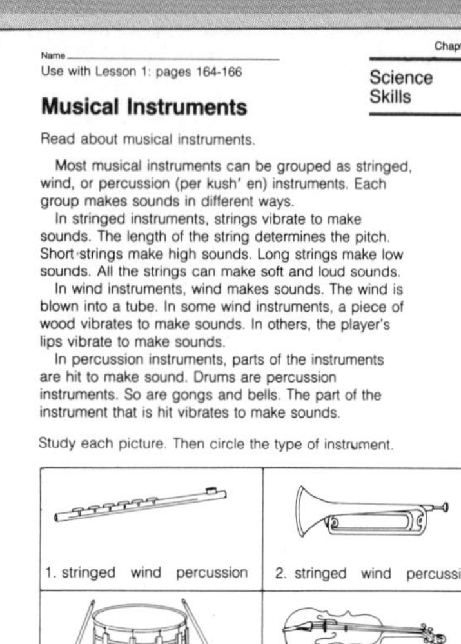

Name ____________

Use with Lesson 1: pages 164-166

Chapter 8

Science Skills

Musical Instruments

Read about musical instruments.

Most musical instruments can be grouped as stringed, wind, or percussion (per kush′ en) instruments. Each group makes sounds in different ways.

In stringed instruments, strings vibrate to make sounds. The length of the string determines the pitch. Short strings make high sounds. Long strings make low sounds. All the strings can make soft and loud sounds.

In wind instruments, wind makes sounds. The wind is blown into a tube. In some wind instruments, a piece of wood vibrates to make sounds. In others, the player's lips vibrate to make sounds.

In percussion instruments, parts of the instruments are hit to make sound. Drums are percussion instruments. So are gongs and bells. The part of the instrument that is hit vibrates to make sounds.

Study each picture. Then circle the type of instrument.

1. stringed wind percussion	2. stringed wind percussion
3. stringed wind percussion	4. stringed wind percussion

44

♦ *Suitable as a language development activity*

Predicting Sounds

Wear cover goggles for this activity.

Suggested grouping: pairs

Purpose
Predict how the length of a vibrating rubber band affects the sound the rubber band makes.

Gather These Materials
• large rubber band • book about 24 cm long • centimeter ruler • 2 pencils

Follow This Procedure
1. Use a chart like the one shown to record your observations.
2. Put the rubber band the long way around the book. Place the pencils under the band about 10 cm apart. *CAUTION: Handle the rubber band carefully.*
3. Ask a partner to hold a finger over the band and each pencil as shown. Gently pluck the band between the pencils. Listen to the pitch of the sound.

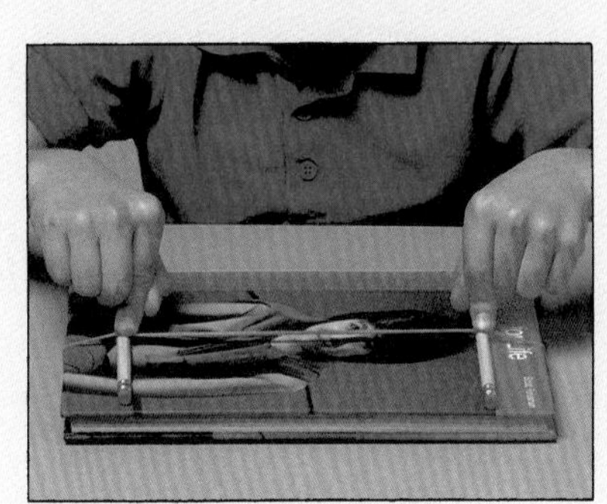

4. Predict how the pitch will change if you make the vibrating part of the band longer. Record your prediction.
5. Hold the rubber band and pencils 19 cm apart. Pluck the rubber band. Record the result.
6. Predict how the pitch will change if you shorten the vibrating part of the band. Record your prediction.
7. Hold the rubber band and pencils 5 cm apart. Pluck the rubber band. Record the result.

Record Your Results

	Prediction	Result
Long band	Predictions might	low pitch
Short band	vary.	high pitch

State Your Conclusion
1. How does the length of a vibrating object change the pitch of the sound produced?
2. At which length did the vibrating part of the rubber band vibrate the fastest?

Use What You Learned
The strings of a cello are longer than the strings of a violin. Which instrument would you use to play music with low sounds?

167

Concept
The length of a vibrating object affects the pitch of the sound produced.

Objectives/Process Skills
- *Observe* the pitch of the sound produced by a section of vibrating rubber band.
- *Predict* how the pitch will change when the vibrating section is made longer or shorter.
- *Record* if the predictions were correct.
- *Conclude* that the longer vibrating rubber band produces a lower pitch.

Time Allotment
Allow 20 minutes.

Safety Tips (See page T24.)
- Remind students to be careful when using sharpened pencils.
- Tell students to carefully stretch the rubber bands.
- Collect the rubber bands when students are finished using them.
- Cover goggles should be worn at all times during this activity.

Teaching Tips
- Use rubber bands of different thicknesses to demonstrate how the thickness of a vibrating object affects the pitch of the sound produced.
- Correlate the frequency of vibration to the length of the rubber band and the pitch of the sound.

Resource Book page 91

Chapter 8

Name ______

Use with Lesson 1: page 167

Activity Worksheet

Predicting Sounds

Record Your Results

	Prediction	Result
Long band		
Short band		

State Your Conclusion
1. How does the length of a vibrating object change the pitch of the sound produced?

2. At which length did the vibrating part of the rubber band vibrate the fastest.

Use What You Learned
The strings of a cello are longer than the strings of a violin. Which instrument would you use to play music with low sounds?

91

* *Answers to masters on pages 162E–162H*

Activity Results

The shorter the vibrating object, the higher the pitch of the sound produced.

Answers
State Your Conclusion
1. The longer the vibrating object, the lower the pitch.
2. The vibrating part of the rubber band vibrated the fastest when it was 5 cm (about 2 in.) long. At this length, the rubber band produced the highest pitch.

Use What You Learned
Because the strings of a cello are longer than that of a violin, a cello can produce lower sounds than a violin. **Thinking Skill:** ***Applying information to new situations***

TEACHING PLAN

Lesson Objectives

- *Compare* how sound waves travel through different kinds of matter.
- *Explain* how an echo is formed.

Lesson Vocabulary

echo

1. Motivate

Demonstration Activity ♦

Obtain a large hard rubber or plastic comb. Stand where the students can see you, and pluck the teeth of the comb. Next, take the same comb and place the end of it on an empty cart, or metal tray. Pluck the comb again.

Discussion

Question: **How did the sound change after the comb was placed on metal?** (Students should *compare* the two sounds they hear. The sound obtained after the comb was placed against an object with a large surface should have been louder.) Point out that after the comb was placed against the cart or tray, the cart or tray also vibrated, which pushed more air molecules around and created a louder sound.

2 How Does Sound Travel?

LESSON GOALS

You will learn
- how sound travels through matter.
- what makes an echo.

Pretend that your friend is talking to you. Sound from your friend's voice moves through the air. You learned in Chapter 5 that particles make up matter. Sound moves from place to place by making the particles in matter vibrate.

How Sound Moves Through Matter

Sound travels in waves. The picture shows how sound waves might look if you could see them. First, the school bell vibrates when it rings. Then the air particles next to the bell begin to vibrate. The sound waves keep moving from one air particle to another. Notice how sound waves spread out in all directions.

You would hear a loud sound if you stood close to the school bell. The sound waves are strongest near the bell. These sound waves get weaker as they move away from the bell. You might hear only a soft sound from far away.

Sound waves

168

Teaching Options

Science Background

Sound waves travel through matter as compressional waves. Compression is the place in a sound wave where particles of matter are squeezed together. Rarefaction is the place in a sound wave where particles of matter are spread farthest apart. The compressions and rarefactions in sound waves are evenly spaced. The distance between two compressions or rarefactions is the wavelength. The speed of a sound wave can be determined by the equation: wavelength × frequency = speed.

Reading Strategies ♦

1. Guide students' pre-reading by asking: What two or three questions do you have for each subheading?
2. Assign these strategies: Writing a Memory Sentence and Mapping Examples (See pages T26–T29.)
3. Pair students to share what information is clear and unclear and initiate discussion using students' unanswered questions.

♦ *Suitable as a language development activity*

Suppose you placed a ticking clock in one part of a room. Now suppose you walked away from the clock. You could still hear the clock ticking. Sound waves from the clock would move through the air.

The girl in the picture is listening to a ticking clock. Notice that she places her ear on the table. Sound waves also can move through wood.

You can only hear sound when it travels through matter. Sound moves through solids, liquids, and gases. Particles of matter in solids are the closest together. For this reason, sound moves fastest and most easily through solids. Particles in liquids are farther apart. Sound moves more slowly through liquids. The particles in gases are the farthest apart. Which state of matter carries sound most slowly?[1]

SCIENCE IN YOUR LIFE

Imagine a flash of lightning in the sky. Seconds later, the sound of thunder follows. You see lightning before you hear thunder because light travels faster than sound. Light travels about 300,000,000 meters a second. Sound travels about 300 meters a second.

[1]gas

Sound waves can travel through solid matter.

2. Teach

Teaching Tips

- Place a clear dish such as a glass pie plate on an overhead projector. Fill the pan 1/2 full of water, and add a few wood shavings from the pencil sharpener. As students ***observe***, drop a bean into the water. Point out the waves traveling in all directions. Question: **What moved when the bean dropped into the pan?** (the water and the wood shavings) Relate the motion of the water waves to the sound waves caused by any vibrating object. Explain that all objects that are struck by molecules in motion will move or vibrate (as much as their makeup allows).
- When discussing how vibrating objects create waves, refer students to the *DISCOVER* activity on page 163. Question: **What vibrated?** (the ruler and the rubber band) **What did the vibrations cause?** (sound waves) **What type of matter did the sound waves travel through?** (gas)
- **Possible Misconception:** Students might not understand why echoes do not occur when they face the walls in their homes and talk. Explain that sound can be absorbed by furniture, rugs, and other items in a room, and that if their homes were empty, they would hear echoes when they talked.

Special Education

Help students with learning disabilities use reference books to learn what ultrasound is and how it is used. Some students might want to find out how ultrasound produces images.

Reinforcement

Have students write a story or make a flow chart about how sound is produced. Elements in the chart or story should include: objects that can make sounds, particles of matter, and the human ear. Students might want to be creative and write a story such as *I am a sound.* (Students could write the story in first person, and tell about their travels.)

TEACHING PLAN

3. Assess

Lesson Review

1. Sound travels through all kinds of matter. It travels fastest through solids and slowest through gases.
2. An echo is sound that bounces back from an object.
3. Challenge! No. Sound waves cannot be heard unless they travel through matter. **Thinking Skill:** *Inferring*

Find Out On Your Own

Bats make high-pitched clicking sounds. The sounds bounce off of objects and make certain kinds of echoes that help the bats avoid flying into objects. Bats also use this method to locate the insects they use for food. **Thinking Skill:** *Inferring*

echo (ek′ō), a sound that bounces back from an object.

Echoes

Pretend you see a ball like this one bouncing against a smooth wall. Use your finger to trace the path of the ball in the picture. Now imagine calling out your name in a large empty room. You might hear sound even after you stop speaking. Like the ball, sound waves can bounce off different surfaces. An **echo** is a sound bouncing back from an object. Echoes can best be heard when sound bounces from hard, smooth surfaces.

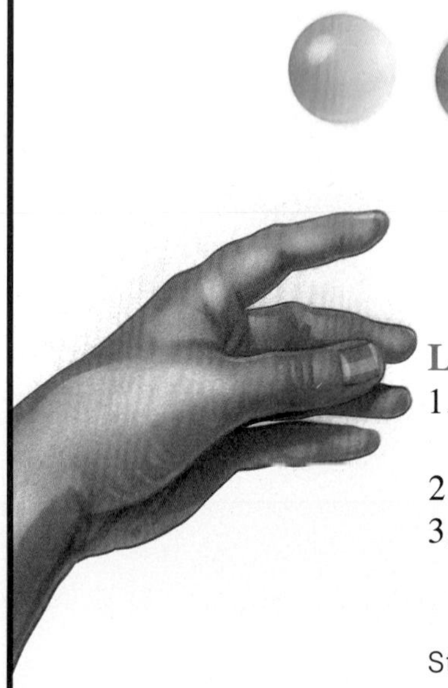

Lesson Review

1. How does sound move through different kinds of matter?
2. What is an echo?
3. **Challenge!** Could sound waves from a radio travel through a room with no air? Explain your answer.

Study on your own, pages 330–331.

LIFE SCIENCE — FIND OUT ON YOUR OWN — CONNECTION

Bats are animals that fly in the dark. Look in library books to find out what kind of sounds bats hear. Write a few sentences telling how these sounds help bats.

170

Teaching Options

Science Anecdote

Sound travels at about 1,191 kilometers (738 miles) per hour through air, but travels at about 18,000 kilometers (11,160 miles) per hour through steel. Years ago people used to listen with one ear against the railroad tracks to tell if a train was coming.

Reteaching Suggestion ♦

Direct the students to draw a train with its horn blowing. Instruct them to draw the sound waves as they would travel from the horn, to the houses near the train tracks. Have them draw at least one bounce in the sound waves. You might wish to repeat the activity in the *Teaching Tips* on page 169 to show students how sound waves travel.

Workbook page 45 *

Name ______________________ Chapter 8

Use with Lesson 2: pages 168-170

Science Activity

How Can You See Sound?

Wear cover goggles for this activity.

Gather These Materials
• balloon • scissors • an orange juice can with the ends removed • rubber bands • tape • a small mirror to fit inside the can • glue • paper • a flashlight

ORANGE JUICE

Follow This Procedure
1. Work with a partner. Cut off the end of the balloon. Fit the balloon tightly over one end of the juice can.
2. Put a rubber band over the balloon to keep it in place. Tape down the balloon edges.
3. Glue the back side of the mirror to the balloon.
4. Tape the can to your desk. Have your partner hold the paper about two meters from the balloon end of the can.
5. Shine a flashlight onto the mirror. Aim the light so that its reflection hits the paper.
6. Take turns with your partner talking and shouting into the open end of the can. Watch what happens to the light reflecting on the paper.

Record Your Results
Record what moved when you spoke into the can.

State Your Conclusion
1. What happened to the light when you spoke into the can?
2. What made this happen?

♦ *Suitable as a language development activity*

Listening to Sound Through Different Materials

Suggested grouping: pairs

Purpose

Test different materials to learn which substances sound travels through most easily.

Gather These Materials

• sealable bag • pencil with eraser • water • cup • wood block

Follow This Procedure

1. Use a chart like the one shown to record your observations.
2. Fill the bag with air by blowing into the bag. Seal the bag closed.
3. Hold the bag next to your ear. Cover your other ear with your hand. Listen while your partner taps the bag lightly with the pencil eraser. Record whether the sounds are loud or quiet.
4. Use the cup to fill the bag with water. Seal the bag.
5. Repeat step 3. Record whether the sounds are louder or quieter than before.
6. Hold the block next to your ear, and cover the other ear. Ask your partner to tap lightly on the block. Record whether the sounds are louder or quieter than before.

Record Your Results

	Observation
Sounds Through Air	quiet
Sounds Through Water	louder
Sounds Through Wood	loudest

State Your Conclusion

1. Through which material did you hear the loudest sound? the quietest sound?
2. Does sound travel most easily through solids, liquids, or gases? Explain your answer.

Use What You Learned

Suppose you were trying to hear footsteps. Would you put your ear next to the ground to hear the sounds better or would you hold your head up in the air? Explain.

171

Activity

Concept

Sound travels most easily through solids and least easily through gases.

Objectives/Process Skills

- *Observe* how sound travels through air, water, and wood.
- *Record* observations.
- *Infer* that sound travels best through solids.

Time Allotment

Allow 30 minutes.

Safety Tips (See page T24.)

- Wipe any spilled water to prevent accidents.
- Remind students to be careful with sharpened pencils.

Teaching Tips

- **Helpful Hint:** Make sure students hold the bag and wooden block next to their ears.
- Direct students to tap the eraser with the same amount of force each time.
- Allow students to test other materials, such as sand, soil, oil, and so on.

Answers

State Your Conclusion

1. wooden block; air

2. Sound travels most easily through solids. The tapping sound was the loudest through the wooden block.

Use What You Learned

The sound of the footsteps could be heard more loudly by placing the ear next to the ground because sound travels more easily through solids than through gases. **Thinking Skill:** *Inferring*

Resource Book page 93

Name ______

Use with Lesson 2: page 171

Listening to Sound Through Different Materials

Chapter 8

Activity Worksheet

Record Your Results

	Observation
Sounds through air	
Sounds through water	
Sounds through wood	

State Your Conclusion

1. Through which material did you hear the loudest sound? the quietest sound?
2. Does sound travel most easily through solids, liquids, or gases? Explain your answer.

Use What You Learned

Suppose you were trying to hear footsteps. Would you put your ear next to the ground to hear the sounds better or would you hold your head up in the air? Explain.

93

Activity Results

The tapping sound will be the loudest through the wood and the quietest when heard through the air.

*** Answers to masters on pages 162E–162H***

TEACHING PLAN

Lesson Objectives

- *Explain* that vibrations of the vocal cords make sounds.
- *Explain* how the ear receives sound waves and sends signals to the brain about the sound.

Lesson Vocabulary

eardrum, nerve, vocal cords

1. Motivate

Demonstration Activity ♦

Have students tap tuning forks and place them on the small bones behind their ears. Then direct them to place their fingers on their throats and hum softly.

Discussion

Questions: **What did you hear?** (low, soft sounds) **What caused the sound from the tuning fork?** (vibrations of the metal) Lead students to *infer* that structures in their throats must be vibrating.

3 How Do People Make Sound?

LESSON GOALS

You will learn
- how people make sound.
- how people hear sound.

vocal cords (vō′kəl kôrdz), two pairs of thin, ropelike flaps at the top of the windpipe.

[1]vibrating movement in throat

Think about sounds you can make with your voice. You can whisper or talk. You make sound when you laugh. You might even enjoy singing.

Making Sound

Place your fingers lightly on the front of your throat like the girl in the picture is doing. Now say your name out loud. What do you feel with your fingertips?[1]

You probably can feel your **vocal cords** moving back and forth. Vocal cords are thin flaps at the top of your windpipe. When you talk, air comes from your lungs and passes between your vocal cords. The air makes your vocal cords vibrate. This vibrating movement makes sound.

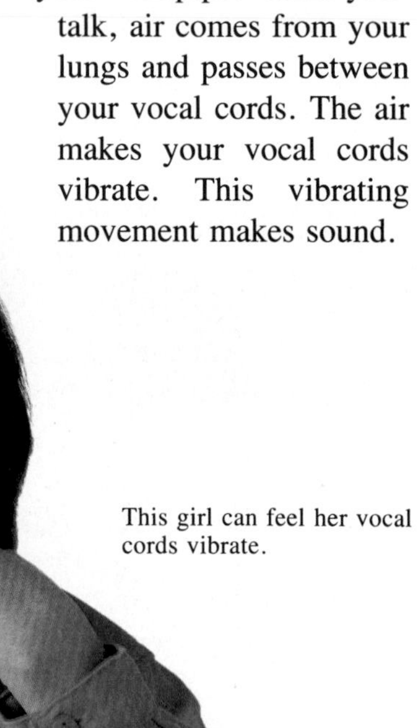

This girl can feel her vocal cords vibrate.

Teaching Options

Science Background

Many parts of the body help people communicate verbally. The diaphragm provides the force for expelling air from the lungs. The air from the lungs causes the vocal cords to vibrate. Different vibrations of the vocal cords cause different sounds. When a person makes a high-pitched sound, the vocal cords vibrate more quickly, and when he or she makes low-pitched sounds, the vocal cords vibrate slowly. The lips and tongue help form the sounds of speech. Different languages combine different sounds and vocal pitches, and use the tongue and lips differently.

Reading Strategies ♦

1. Guide students' pre-reading by asking: What comes to mind when you think of the key words?
2. Assign these strategies: Visualizing Information and Mapping Examples (See pages T26–T29.)
3. Pair students to share what information is clear and unclear and initiate discussion using students' unanswered questions.

♦ *Suitable as a language development activity*

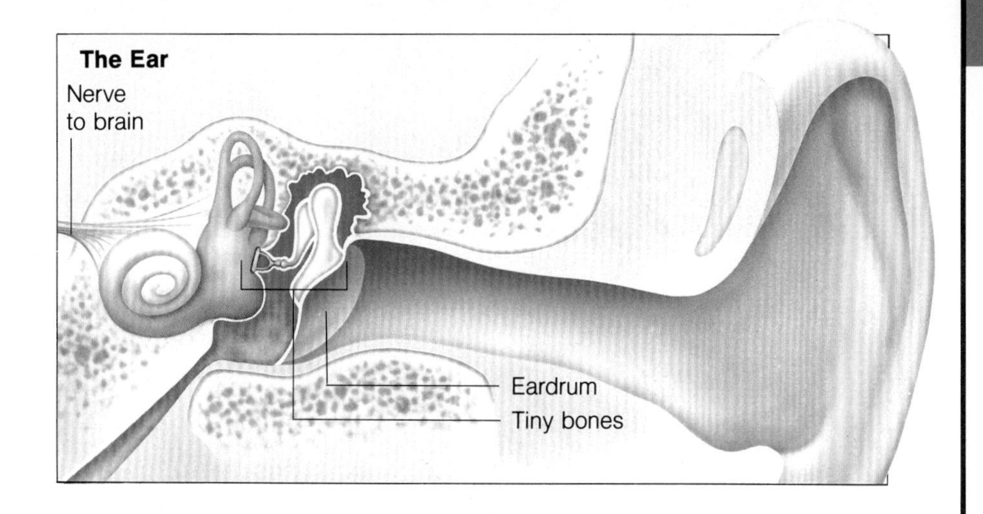

Hearing Sound

Imagine trying to catch a ball. You might cup your hands together. Shaping your hands this way helps you catch the ball. Now think about the shape of the outer part of your ear. This shape helps catch sound waves in the air.

Look at the picture to follow the path of the sound waves through the ear. First, the part of the ear you can see moves the sound waves to the part of the ear inside the head. Find the **eardrum**—the thin skin that covers the middle part of the ear. Sound waves make your eardrum vibrate. Then the tiny bones in the middle of your ear begin to vibrate. Find the part of the ear that is shaped like a shell. A liquid fills this part of your ear. This liquid carries sound waves to a special **nerve**—a part of the body that carries messages to the brain. Your brain tells you about the sound you hear.

SCIENCE IN YOUR LIFE

A rabbit has very long ears that can move around. These large ears can catch many sound waves. If another animal tries to sneak up on a rabbit, the rabbit can hear the sounds from far away. Then the rabbit has time to run away.

eardrum, thin skin that covers the middle part of the ear and vibrates when sound waves reach it.

nerve (nėrv), a part of the body that carries messages to the brain.

173

2. Teach

Teaching Tips

• **Possible Misconception:** Students might have unusual ideas about why they sometimes "lose" their voices, since they know that air from the lungs causes the vocal cords to move and make sound. Explain that the vocal cords are located in the larynx, an area of the windpipe that sometimes becomes infected by germs that cause laryngitis—a temporary loss of one's voice.

• Copy the sketch of the inner and outer ear on page 173. Place the copy on an overhead projector, and use a colored marker to show how sound enters the ear canal, vibrates the eardrum, and moves up the bones to the nerve. Question: **When do you know what type of sound you are hearing?** (when the brains interprets the message from the nerve in the ear)

• Obtain a pair of ear protectors for students to examine. Ask students to *name* some places where it would be helpful to wear ear protectors. (Students should *name* places where loud noises are usually present, such as at airports, in factories, and near drilling and construction sites.)

Workbook page 46 *

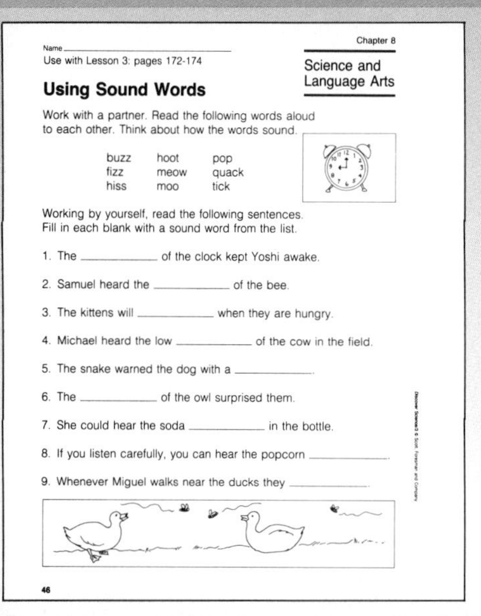
Name ____________
Use with Lesson 3: pages 172-174

Chapter 8

Science and Language Arts

Using Sound Words

Work with a partner. Read the following words aloud to each other. Think about how the words sound.

buzz	hoot	pop
fizz	meow	quack
hiss	moo	tick

Working by yourself, read the following sentences. Fill in each blank with a sound word from the list.

1. The ________ of the clock kept Yoshi awake.
2. Samuel heard the ________ of the bee.
3. The kittens will ________ when they are hungry.
4. Michael heard the low ________ of the cow in the field.
5. The snake warned the dog with a ________.
6. The ________ of the owl surprised them.
7. She could hear the soda ________ in the bottle.
8. If you listen carefully, you can hear the popcorn ________.
9. Whenever Miguel walks near the ducks they ________.

46

* Answers to masters on pages 162E–162H

Special Education

Have students with learning disabilities cut out magazine pictures of animals' ears to place on a poster. Discuss which animals have ears that are designed especially to capture sound.

Enrichment

Invite a trained vocalist to visit the classroom to sing for the students. Have him or her tell what types of training were helpful in learning how to sing, and what part of the body is primarily used in breathing exercises.

TEACHING PLAN

3. Assess

Lesson Review

1. People make sound when air from the lungs passes between the vocal cords and causes them to vibrate.
2. Sound waves travel from the outer part of the ear, to the eardrum, to the bones in the ear, to the liquid, to the nerve, to the brain.
3. Challenge! Sound waves move faster through the bones in the ear because the bones are solid and sound waves move fastest through solids. **Thinking Skill:** *Inferring*

Find Out On Your Own

A cricket makes sound by rubbing its wings together. A fly makes sound by rapidly vibrating its wings. **Thinking Skill:** *Recognizing cause and effect*

Protecting hearing

Many sounds can help you. For example, a siren can warn you of danger. Sounds also can be harmful. Listening to loud noises over and over or for a long time can cause you to lose hearing. This hearing loss might last for a short time or it might last forever. How is the person in the picture protecting his hearing?[1]

[1]adjusting volume of radio

Lesson Review

1. How do people make sounds with their voices?
2. What path do sound waves follow through the ear?
3. **Challenge!** Do sound waves move faster through the outer part of the ear or through the tiny bones in the ear? Explain your answer.

Study on your own, pages 330–331.

LIFE SCIENCE FIND OUT ON YOUR OWN CONNECTION

Many animals make sounds using different body parts. Look in library books to find out how crickets and flies make sound. Write a few sentences telling what you learned.

174

Teaching Options

Science Anecdote

Physicians often treat people who have tinnitus (ringing in the ears) by using a hearing aid that makes sounds similar to the sound of the wind or the ocean.

Reteaching Suggestion ♦

Have students create a skit in which they are the parts of the ear. Ask the students to act out how the ear receives and transfers sound waves. One student might want to be the brain and tell what type of sound was heard and transferred.

Workbook page 47 *

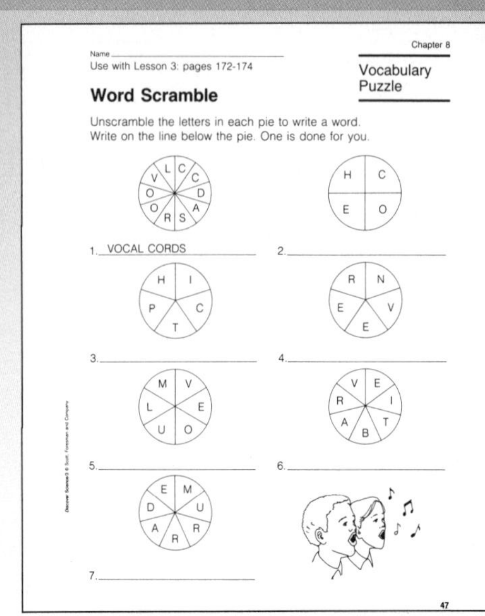

Name

Use with Lesson 3: pages 172-174

Chapter 8

Vocabulary Puzzle

Word Scramble

Unscramble the letters in each pie to write a word. Write on the line below the pie. One is done for you.

1. VOCAL CORDS
2.
3.
4.
5.
6.
7.

47

♦ ***Suitable as a language development activity***

Sending Voices over the Telephone

The Problem By the late 1800s, the telegraph made it possible to send messages over long distances. The telegraph changed a code into electric signals, dots, and dashes. However, only one telegraph message could be sent over the wire at one time. Also, telegraph wires could not carry human voices.

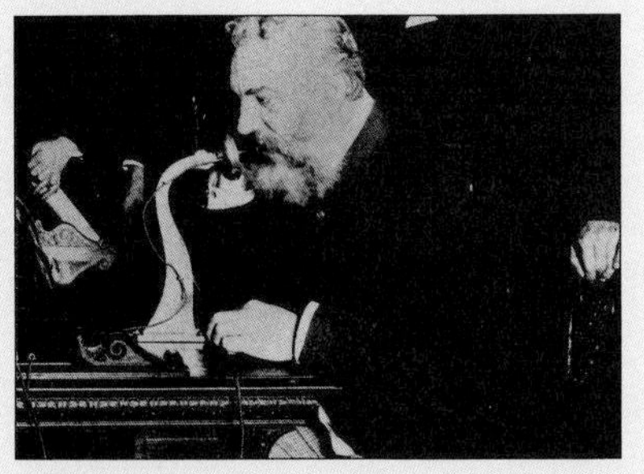
Alexander Graham Bell

The Breakthrough Alexander Graham Bell was born into a family very interested in sound. His grandfather was an actor. His mother was a musician. His father taught deaf people to speak. Bell also became a teacher of the deaf. He became interested in electricity while working on ways to teach the deaf to imitate speech. Bell found a good partner for his experiments. Thomas A. Watson was a repair mechanic.

Bell and Watson worked together to develop the telephone. They were working with an instrument made of several metal reeds and electromagnets. A line ran from one room into the next room. In March 1876, Watson was at the end of the line in the other room. He heard Bell say, "Mr. Watson, come here. I want you!" This famous sentence was the first ever carried by a telephone. The sound waves from Bell's voice traveled as electrical signals along the line. More than one message could travel along the line at once.

New Technology Today new technology lets us send more messages than ever before over long distances. This is possible because of fiber optics. Scientists discovered that beams of light can carry messages. Thin threads of clear glass or plastic can carry light beams for more than 30 kilometers before they fade away. These thin threads are called optical fibers.

What Do You Think?
1. In what way was teamwork an important part of the development of the telephone?
2. How does fiber optics improve the telephone?

Science and Technology

Discussion
Question: **Have you ever become interested in a second topic while learning about something else?** Tell students that Alexander Graham Bell first thought of sending speech through wires while researching teaching speech to the deaf.

Teaching Tips
- **Metric-English Equivalent:**
30 kilometers = about 19 miles
- Point out that Alexander Graham Bell maintained his interest in the problems of people who were deaf throughout his lifetime. He confided to his family that he would prefer to be remembered as a teacher of the deaf than as the inventor of the telephone.

Answers
What Do You Think?
1. Teamwork led to the development of the telephone because neither Watson nor Bell had all of the skills necessary to develop the invention on his own. Bell contributed his ideas and Watson contributed his mechanical know-how. Their cooperative efforts resulted in the telephone. **Thinking Skill:** *Summarizing*
2. Fiber optics allow more messages to be sent at one time. **Thinking Skills:** *Restating or explaining ideas*

Science Background

The telephone consists of two main parts, the transmitter and the receiver. The transmitter changes the voice into an electrical signal while the receiver changes the electrical signal back to sound waves. When a person talks into a telephone, the sound waves of the voice vibrates the air. The air then vibrates a round metal disk in the transmitter. The disk presses against a cup containing carbon grains. Depending on the pressure on the carbon grains, an electric current flowing through them is changed. At the receiver, the electric current goes through an electromagnet. The electromagnet's attraction for a metal disk varies. The vibrating disk reproduces sound waves.

* ***Answers to masters on pages 162E–162H***

TEACHING PLAN

Purpose
To develop the skills of collecting and organizing information using rulers and pictographs to solve problems.

1. Motivate

Discussion
Display or show pictures of several articles of clothing. Questions: **How could you find out how long these articles of clothing are?** (Measure them using a ruler or tape measure.) **Why are their lengths important to you?** (Guide students to *conclude* that for a proper fit, a person needs to know length or size.) Explain to students that knowing the length of a sound wave is also important, because wavelength is related to the pitch of the sound.

2. Teach

Teaching Tip
- Be sure students understand that a slow vibration produces a low-pitched sound. As the rate of vibration increases, pitch becomes higher.

Skills for Solving Problems

Using Rulers and Pictographs

Problem: How does the rate an object vibrates affect the pitch of a sound?

Part A. Using Rulers to Collect Information

1. Rulers measure distance in a straight line. The marks on a ruler stand for centimeters. Move your finger up a ruler in the picture. Read the number of centimeters when your finger is even with the top of the straw. How long is the longest straw?
2. Objects make sounds when they vibrate. A straw vibrates when you blow through it. The shorter the straw, the more rapidly it vibrates. Is the pitch of the longest straw high or low?
3. How long is the shortest straw? Is its pitch high or low?

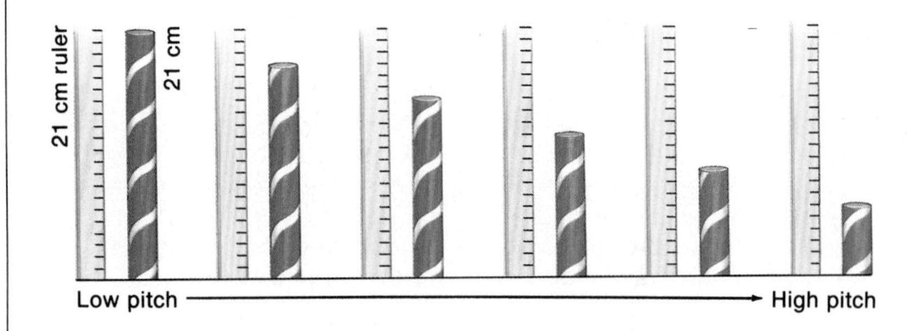

Part B. Using a Pictograph to Organize and Interpret Information

4. The pictograph contains the information you collected about how the rate an object vibrates affects the pitch of a sound. How are the lengths of the straws different?

176

Teaching Options

Sample Pictograph for Part C

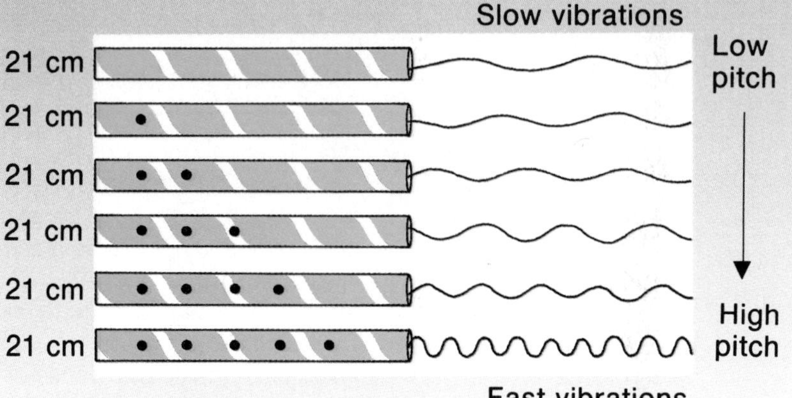

♦ ***Suitable as a language development activity***

5. How does the rate the straws vibrate change with the differences in length? How does the pitch of the sound change?

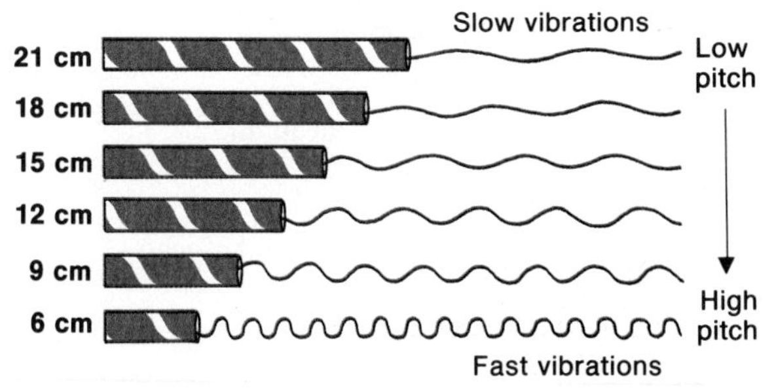

Part C. Using Rulers and Pictographs to Solve a Problem

Problem: Can the pitch change even if the length of an object that vibrates stays the same?

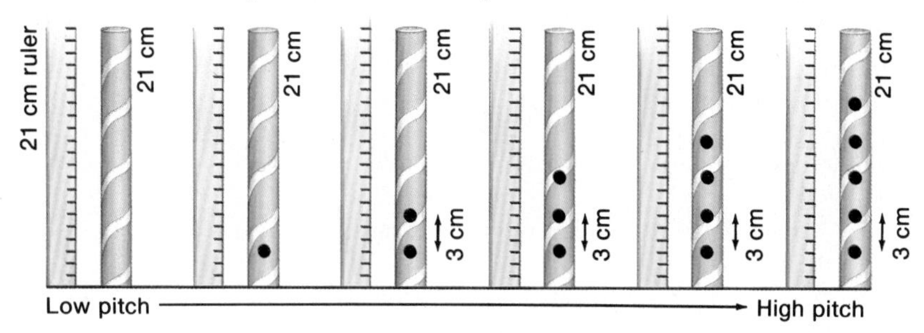

6. Use the picture to collect the information you need to solve the problem. Make a pictograph similar to the one in Part B to organize your information.
7. Look at your pictograph. Can pitch change if the length of a vibrating object stays the same?
8. Are the straws all the same length? How are the straws different from the straws in Part A? How are the straws different from each other?
9. You might want to do this experiment and use your own results to make a pictograph.

177

3. Assess

Part A

1. 21 cm
2. low
3. 6 cm; high

Part B

4. Each straw is 3 cm longer than the one below it.
5. The vibrations get faster as the straws get shorter. The pitch gets higher as the vibrations are faster in shorter straws.

Part C

6. See Sample Pictograph for Part C in Teaching Options.
7. Yes, holes affect the pitch and vibrations of a straw in the same way shortening the straw does.
8. The straws are all the same length. Most of the straws in Part C have holes in them. The straws in Part C have different numbers of holes.
9. If students are doing the activity, flatten one end of the straw and trim the outer angles of the flat part to resemble a reed. Students might want to experiment with the placement of holes in the straw and try to produce musical notes.

Reteaching Suggestion ♦

Obtain a bottle with a long neck and narrow mouth. Blow gently across the mouth to emit a low pitched tone. Then pour some water into the bottle. Have students measure the water level with a centimeter ruler and record it. Then blow gently across the mouth again. Question: **How has the sound changed?** (The pitch has gotten higher.) Repeat the process several times, making sure students measure and record the water level. Help them develop a pictograph that shows the relationship between the water level and the pitch.

Resource Book page 95 *

Name

Comprehension: predicting outcomes

Chapter 8

Science and Reading

What Will Happen Next?

Read each paragraph. Circle the letter of the sentence that tells what will happen next. Then answer the question.

Jason's Wish

Jason wished he could have a cat. One day his mother said to him, "There's a surprise for you in your room." As he walked toward his room, he heard a meow.

a. Jason will not like his surprise.
b. Jason will find a cat in his room.
c. Jason will not go to his room.

What clues in the story make you think as you do?

Jane's Wish

On the way home, Jane's dad told her that he had a surprise for her. As they turned into their driveway, she heard piano music but thought it was coming from a radio. Jane wished that she had a piano so she could learn to play it as well as her mother played.

a. Jane will see a piano in her house.
b. Jane's parents will give her a radio.
c. Jane's mother will learn to play the piano.

What clues in the story make you think as you do?

95

* *Answers to masters on pages 162E–162H*

REVIEW PLAN

Reviewing Science Words

1. nerve
2. echo
3. pitch
4. vibrate
5. volume
6. vocal cords
7. eardrum

Reviewing What You Learned

1. a **3.** b **5.** d **7.** a
2. c **4.** c **6.** b **8.** d

Interpreting What You Learned

1. Any sound will occur when matter vibrates. **Thinking Skill:** *Comparing*
2. Sounds that are too loud can cause hearing loss. **Thinking Skill:** *Restating ideas*
3. The particles of matter in solids are closer together than the particles of matter in air. **Thinking Skill:** *Contrasting*

Chapter 8 Review

Chapter Main Ideas

Lesson 1 • Sound happens when particles of matter vibrate. • How much an object vibrates controls the volume of a sound. How fast an object vibrates controls the pitch of a sound.

Lesson 2 • Sound travels in waves that spread out in all directions. Sound travels through all states of matter—solids, liquids, and gases. The speed of sound through matter is fastest through solids and slowest through gases. • An echo is sound bouncing back from an object.

Lesson 3 • People make sound when the vocal cords vibrate. • Sound waves travel through all the parts of the ear. People hear when the nerve in the ear carries messages to the brain about sound.

Reviewing Science Words

eardrum pitch vocal cords
echo vibrate volume
nerve

Copy each sentence. Fill in the blank with the correct word from the list.

1. A special ___ in the ear carries messages about sound to the brain.
2. An ___ is a sound bouncing back from an object.
3. Objects that move back and forth slowly make sounds with a low ___.
4. The air particles next to a bell begin to ___ when the bell rings.
5. The more an object vibrates, the louder the ___ will be.
6. When a person talks, air comes from the lungs and passes between the ___.
7. The ___ is a skin that covers the middle part of the ear.

178

Review Options

Cooperative Learning ♦

STAD Format (See page T23.)
Assign students to work in four- to five-member teams to study Chapter 8 Review. Students should work together to make sure that they and their teammates know the material in the chapter. After students have had enough time to study together, give them a test to complete individually (Chapter 8 Test A or B in the *Test Book*). Award Superteam certificates to teams whose average test scores exceed 90%, and Greatteam certificates to teams whose average test scores exceed 80%.

Test Book page 69 *

Name ______

Chapter 8
Test A

Multiple Choice Choose the best answer.

1. Sound happens when matter moves back and forth very quickly and
   a. echoes.
   b. vibrates.
   c. bounces.
2. The loudness or softness of a sound is called
   a. volume.
   b. pitch.
   c. tone.
3. The highness or lowness of a sound is called the
   a. loudness.
   b. volume.
   c. pitch.
4. Slow vibrations cause sounds that have a
   a. high pitch.
   b. low pitch.
   c. low volume.
5. How does sound move through a liquid that has particles that are far apart?
   a. slowly.
   b. quickly.
   c. not at all.
6. Sound bouncing back from an object causes
   a. a low volume.
   b. an echo.
   c. a high pitch.
7. The thin flaps at the top of the windpipe are called
   a. eardrums.
   b. sound waves.
   c. vocal cords.
8. The thin skin that covers the middle part of the ear is called the
   a. eardrum.
   b. nerve.
   c. pitch.
9. A part of the body that carries messages to the brain is called
   a. an eardrum.
   b. a nerve.
   c. a pitch.
10. The liquid in the ear carries soundwaves to
   a. the eardrum.
   b. bones.
   c. nerves.

69

♦ ***Suitable as a language development activity***

Reviewing What You Learned

Write the letter of the best answer.

1. Sound waves travel most quickly through
   (a) solids. (b) gases. (c) air. (d) liquids.
2. Which part of the ear collects sound waves?
   (a) eardrum (b) nerve (c) outer part (d) bones
3. How fast an object vibrates controls
   (a) volume. (b) pitch. (c) echoes. (d) hearing.
4. What makes the vocal cords vibrate?
   (a) sounds (b) the lungs (c) air (d) the windpipe
5. Hitting an object harder will make the volume
   (a) softer. (b) the same. (c) lower. (d) louder.
6. The liquid in the ear carries sound waves to the
   (a) bones. (b) nerve. (c) eardrum. (d) brain.
7. Echoes can be heard best when sound bounces from surfaces that are
   (a) smooth. (b) rough. (c) soft. (d) bumpy.
8. Objects that vibrate slowly have a pitch that is
   (a) high. (b) loud. (c) soft. (d) low.

Interpreting What You Learned

Write a short answer for each question or statement.

1. How are all sounds alike?
2. How can sounds be harmful?
3. How does sound move through solids more easily than it moves through air?
4. What happens when sound waves reach the eardrum? What makes the vocal cords vibrate?

Extending Your Thinking

Write a paragraph to answer each question or statement.

1. How can guitar strings with different lengths make music that has different sounds?
2. Explain how covering the ears can keep a person from hearing a sound.

To explore scientific methods, see Experiment Skills on pages 362–363.

179

4. The eardrum vibrates when it receives sound waves.
5. Air from the lungs makes the vocal cords vibrate by passing between them. **Thinking Skill:** *Recognizing cause and effect*
6. Sounds can be different in volume and pitch. **Thinking Skill:** *Contrasting*

Extending Your Thinking

1. Pressing on the strings of a guitar can change the length of the part of the strings that vibrate. The different lengths of the strings produce different pitches (or notes) in the music. **Thinking Skill:** *Inferring*
2. Covering the ears can prevent the outer part of the ear from catching sound waves. People cannot hear unless sound waves enter and travel through the ear. **Thinking Skill:** *Inferring*

Test Book page 70 *

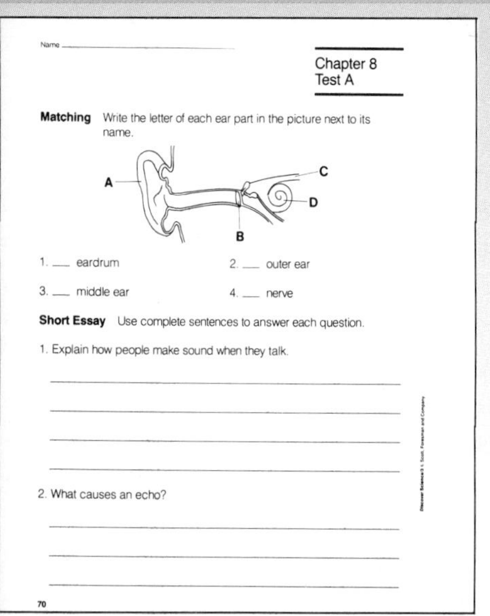

Name

Chapter 8
Test A

Matching Write the letter of each ear part in the picture next to its name.

1. ___ eardrum 2. ___ outer ear
3. ___ middle ear 4. ___ nerve

Short Essay Use complete sentences to answer each question.

1. Explain how people make sound when they talk.

2. What causes an echo?

70

Test Book page 71 *

Name

Chapter 8
Test B

Multiple Choice Choose the best answer.

1. Matter moving back and forth very quickly causes
   a. sound.
   b. light.
   c. echoes.
2. The pitch of a sound is lowest when vibrations are
   a. very fast.
   b. slow.
   c. fast.
3. What happens when air from the lungs passes between the vocal cords?
   a. Vocal cords vibrate.
   b. The windpipe closes.
   c. A nerve sends messages.
4. When an object vibrates harder, its sound will be
   a. louder.
   b. softer.
   c. lower.
5. Because particles in gases are very far apart, sound moves through gases
   a. quickly.
   b. slowly.
   c. very fast.
6. Echoes can be heard best when sound bounces back from
   a. hard, rough surfaces.
   b. soft, thick surfaces.
   c. hard, smooth surfaces.
7. The vocal cords are located at the top of the
   a. windpipe.
   b. ear.
   c. lungs.
8. A special nerve inside your head carries messages from your ear to your
   a. eardrum.
   b. brain.
   c. outer ear.
9. The eardrum is made of
   a. thin skin.
   b. small bones.
   c. liquid.
10. The sound of a ticking clock can be heard the best when the sound passes through
    a. a solid.
    b. a gas.
    c. a liquid.

71

Test Book page 72 *

Name

Chapter 8
Test B

Matching Write the letter of each picture next to its description.

A B C D

1. ___ has a high pitch 2. ___ The sound waves are softer.
3. ___ has a low pitch 4. ___ The sound waves are louder.

Short Essay Use complete sentences to answer each question.

1. What can cause hearing loss?

2. Why does sound travel faster through solids than through liquids?

72

* **Answers to masters on pages 162E–162H**

TEACHING PLAN

Purpose
The careers discussed in this feature give students background information on some familiar careers in physical science, and introduce careers students might not have heard about. Several careers that involve working with matter and energy are investigated.

Teaching Tips
- Ask students to *name* some ways that the work of chemists has benefitted their lives.
- Discuss electrical safety with the class. Go over some safety rules: stay away from power lines; never use an electrical appliance when near or in water; never stick objects into outlets; and do not use appliances with frayed electrical cords.
- Ask if any students have watched an electrician or carpenter work. Have students tell what they would like and dislike about these careers.
- Ask how many students play musical instruments or sing in a choir. Discuss some of the different styles of music, such as classical, folk, rock, country, and jazz.

Careers

Your books, your clothes, and even your body are made up of atoms. **Chemists** are the people who study atoms. They learn about the matter that forms when different atoms join together. Some chemists study the atoms and particles inside living things. Other chemists help make useful products, such as clothing, tape,

Chemist

and medicines. To become a chemist, you need to go to college for at least four years.

Electricians work with electric energy. Electricians put electric wires, switches, and outlets into buildings. They also repair old electric wiring. Electricians work carefully so they do not get an electric shock from the wires. To become an electrician, you need to take special classes. Then you train with an electrician for about four years.

Like electricians, **carpenters** work on buildings such as houses. To build a house, carpenters first make a wooden frame. Then they put in doors and windows. Inside the house they put up walls, cabinets, and wood trim. Carpenters use many different tools, such as saws, hammers, and drills. They must measure everything closely so that all parts of the house fit together. Carpenters can learn their skills on the job or by taking classes at a special school.

Musicians create sounds that please the ear. When a musician performs music, he or she makes sound energy. Some musicians play musical instruments. Others make music by singing, writing music, or leading groups of musicians. Many musicians start their training at a very early age. Some study music in college. Others become good musicians by practicing and performing with other musicians.

Carpenter

180

Teaching Options

Enrichment
Ask a local chemist or chemistry teacher to visit the class and do a few simple demonstrations. Invite a local electrician or carpenter to talk with the class. If possible, take the class to a musical performance.

Where To Write
For more information about these careers contact:

Chemist: American Chemical Society, Career Services, 1155 Sixteenth Street NW, Washington, DC 20036.

Electrician: National Electrical Contractors Association, 7315 Wisconsin Avenue, Bethesda, MD 20814.

Carpenter: Associated Builders and Contractors, Inc., 729 Fifteenth Street NW, Washington, DC 20005.

Musician: National Association of Schools of Music, 11250 Roger Bacon Drive, Reston, VA 22090. Ask for their brochure, *Careers in Music.*

The Violin

A violin can make beautiful music. It can also teach a lot about sounds. A violin has 84 parts. All these parts work together to make the sound that you can hear. Together, these parts vibrate and send sound waves into the air.

The picture shows the main parts of a violin. Every part adds something to the sound the violin makes.

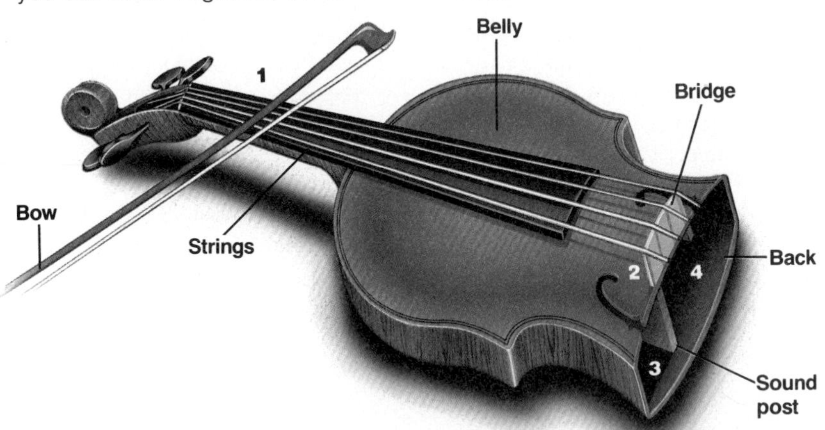

1 The sound begins with the strings. A person draws a bow across a string. The bow makes the string vibrate. A string vibrates in many ways at once, making many different sounds. You hear all these sounds mixed together.

2 A vibrating string does not make very loud sounds. The other parts of the violin help make the sound louder. When a string vibrates, it causes the bridge to vibrate too. The bridge passes the vibrations to the belly of the violin.

3 Vibrations from the bridge also pass down the sound post to the back of the violin. The sound post and the back vibrate along with the string and the belly. Special kinds of wood are used to make the violin, because different kinds of wood make different sounds.

4 All of the vibrating parts cause the air inside the violin to vibrate too. Each vibrating part sends out sound waves and makes its own special sound. Musicians say that the violin has a rich sound. They mean that the violin has many different kinds of sounds that go together well.

181

How it Works

Teaching Tips

- Students can make a one-stringed instrument by stretching a piece of string over two pencils taped on a rigid piece of cardboard or wood. The instrument can be used to illustrate many of the characteristics of the violin described in the feature. Included are the effect of tension on the string, of finger placement, and of plucking versus bowing the string.
- Ask students to select one other musical instrument to study. Have them find out how the musical sounds are produced in that instrument.
- Perhaps a student would be willing to bring his or her own violin to class and demonstrate how it produces musical sounds.

Science Background

The construction of a violin is as much a science as it is an art. Much research has been done to determine why the sounds of some violins, such as those made by Antonio Stradivari (1644–1737), are so much finer than those of other violins. Although a great deal has been learned from the research, the question still has not been answered. Almost every aspect of a violin appears to contribute to the quality of the sound it produces. Included in these aspects are the composition and thickness of the strings; the composition, thickness, size, and shape of the body; the type of varnish used as a coating; and the size and shape of the space inside the violin body.

Unit 2 Review

Complete the Sentence

1. thermostat
2. volume
3. nerve
4. force
5. work
6. evaporates
7. conductors
8. pitch
9. electromagnet
10. condenses
11. echo
12. friction

Short Answer

1. The water particles move faster, evaporate, and become water vapor; a physical change. *(Restating ideas)*

2. The volume will remain the same. Gas takes the shape of its container. *(Recognizing cause and effect)*

3. Gravity causes the ball to fall to the ground. Friction of the ball against the air and ground cause it to stop rolling. *(Cause and effect)*

4. Gears are wheels with jagged edges like teeth, found in bicycles, cars, and machines. *(Explaining ideas)*

5. magnetism; at a magnet's poles *(Restating or explaining ideas)*

6. A current must flow from a source to an appliance and back again to complete a circuit. A cord has one wire to carry electricity to the appliance and one wire to carry it back. *(Making inferences)*

7. The wood handle is a poor conductor of heat and is insulated from the hot pan. *(Making inferences)*

8. The air particles do not vibrate as fast when the pitch is lower. *(Making inferences)*

9. Vocal cords are thin muscles that vibrate like the strings on a violin. Tightening raises the pitch in the same way that tightening violin strings does. *(Making analogies)*

Essay

1. A lever is made of a bar that is supported at the fulcrum. A heavy object can be lifted by placing the fulcrum close to the load. *(Restating ideas)*

2. A lens bends the light that passes through it. If the lens is thick in the middle and thin at the edges, it can make objects look bigger. If the lens is thin in the middle and thick at the edges, the object will look smaller. *(Restating ideas)*

Unit 2 Review

Complete the Sentence

Fill in the blank with the correct word or words from the list.

condenses, friction, conductors, nerve, echo, pitch, electromagnet, thermostat, evaporates, volume, force, work

1. The ___ controls the temperature in a home.
2. An object's ___ is the amount of space it takes up.
3. Sound is carried by a ___ from the ear to the brain.
4. An object can be made to move faster by using more ___.
5. ___ is done when an object is moved through a distance.
6. Water ___ when it changes from a liquid to a gas.
7. Energy is carried easily by materials that are good ___.
8. Objects create a sound with a high ___ when they vibrate very quickly.
9. A wire coil becomes an ___ when an electric current moves through it.
10. Water vapor ___ to a liquid on a cold glass.
11. A sound produces an ___ when it bounces back from an object.
12. Two objects rubbing together cause the force of ___.

Short Answer

Write a short answer for each question or statement.

1. Describe the change that takes place when water is boiled. Is this change a physical change or a chemical change?
2. A person removes some of the gas from a container. What happens to the volume of the gas in the container? Why?
3. Describe the forces that make a thrown ball stop moving.
4. Describe gears and tell where you might find them.
5. What kind of force do magnets have? Where is this force the strongest?
6. Why does an electric cord need to have two wires inside it?
7. How can a wood handle on a hot pot protect you from burns?
8. How does the movement of air particles change when the pitch of a flute becomes lower?
9. How are vocal cords like the strings of a violin?

Essay

Write a paragraph for each question or statement.

1. Describe a lever and explain how a lever can help lift a heavy object.
2. How can lenses make objects seem larger or smaller?

182

Test Book page 77 *

Name ______________

Unit Test 2

Multiple Choice Choose the best answer.

1. The tiny particles that make up matter are called
   a. mass.
   b. atoms.
   c. properties.
2. The color and shape of an object are some of its
   a. properties.
   b. elements.
   c. matter.
3. Three states of matter are
   a. solid, shape, mass.
   b. liquid, gas, solid.
   c. gas, liquid, space.
4. The change from one type of matter into another is a
   a. chemical change.
   b. physical change.
   c. atomic change.
5. A chemical change takes place when matter
   a. melts.
   b. evaporates.
   c. rusts.
6. A simple machine made of a rope and a wheel is a
   a. gear.
   b. pulley.
   c. wedge.
7. A force that makes objects pull toward each other is
   a. work.
   b. friction.
   c. gravity.
8. The force caused by two objects rubbing together is
   a. energy.
   b. friction.
   c. gravity.
9. The ability to work is
   a. gravity.
   b. energy.
   c. friction.
10. What kind of machine is a bicycle?
    a. simple machine
    b. lever
    c. compound machine

1. (a)(b)(c) 4. (a)(b)(c) 7. (a)(b)(c) 10. (a)(b)(c)
2. (a)(b)(c) 5. (a)(b)(c) 8. (a)(b)(c)
3. (a)(b)(c) 6. (a)(b)(c) 9. (a)(b)(c)

77

Test Book page 78 *

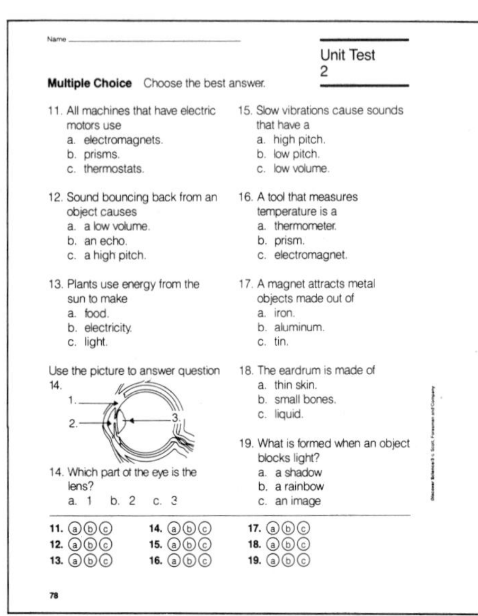

Name ______________

Unit Test 2

Multiple Choice Choose the best answer.

11. All machines that have electric motors use
    a. electromagnets.
    b. prisms.
    c. thermostats.
12. Sound bouncing back from an object causes
    a. a low volume.
    b. an echo.
    c. a high pitch.
13. Plants use energy from the sun to make
    a. food.
    b. electricity.
    c. light.

Use the picture to answer question 14.

14. Which part of the eye is the lens?
    a. 1 b. 2 c. 3
15. Slow vibrations cause sounds that have a
    a. high pitch.
    b. low pitch.
    c. low volume.
16. A tool that measures temperature is a
    a. thermometer.
    b. prism.
    c. electromagnet.
17. A magnet attracts metal objects made out of
    a. iron.
    b. aluminum.
    c. tin.
18. The eardrum is made of
    a. thin skin.
    b. small bones.
    c. liquid.
19. What is formed when an object blocks light?
    a. a shadow
    b. a rainbow
    c. an image

11. (a)(b)(c) 14. (a)(b)(c) 17. (a)(b)(c)
12. (a)(b)(c) 15. (a)(b)(c) 18. (a)(b)(c)
13. (a)(b)(c) 16. (a)(b)(c) 19. (a)(b)(c)

78

Unit Projects and Books

Science Projects

1. Compare the sizes of different objects. Use paper clips to find the perimeter of an envelope. Perimeter is the distance around the envelope. Use small squares to find the envelope's area. The number of squares needed to cover the envelope is its area. Then use the paper clips and squares to find the perimeter and area of a sheet of paper.

   Next find the perimeter and area of the envelope in metric measures. Use a ruler to find its perimeter in centimeters. Find the envelope's area by using a sheet of 1 cm graph paper. Place the envelope on top of the graph paper. Trace the outline of the envelope. Then count how many square centimeters it takes to cover the envelope. Do the same with the paper.
2. Make your own telephone. Use two paper cups. Punch a small hole through the bottom of each cup. Thread the end of a long piece of string through each hole. Fasten the string by tying large knots that will not slip through the holes. With the help of a friend, stretch the string in a straight line. Speak into one of the cups. Your friend should hold his ear close to the second cup. Can your friend hear you? Now have your friend talk to you.

Books About Science

Wheels at Work: Building and Experimenting with Models of Machines by Bernie Zubrowsky. Morrow, 1986. Learn to make a model of a pulley and a gear using readily available materials. Gr. 4-7

The Science of Music by Melvin Berger. Crowell, 1989. Learn about sound and the science behind music.

Science and Society

Noise Pollution More people want to fly in and out of Barnes County Airport. The airport needs to build another runway to handle more airplanes. This would bring money and jobs to the county. Neighbors of the airport object. The added noise might break their windows and damage their hearing. The airport manager agrees that there will be more noise. But she says it will only last a few hours each day. She has to think about the whole county, not just those who live near the airport. The county will buy the homes of people who want to move away. What are the arguments for a new runway? What are the arguments against the runway?

183

Unit Projects and Books

TEACHING PLAN

Science Projects

1. Students should observe that the objects have both a perimeter and an area and that different units of measurement can be used to determine perimeter and area.

2. When a student speaks, the air in front of his or her mouth vibrates, making the bottom of the cup vibrate. These are transmitted to the string and travel through it to the bottom of the second cup, then to the air in the cup. These vibrations are picked up by the second student's ear.

Science and Society

Encourage students to explore both sides of the issue in class or on their own. In class, you may wish to set up a debate on the issue. On their own, students might do further research and choose one side of the issue to support. Throughout this exercise, emphasize the importance of respecting the opinions of others. **Thinking Skill:** *Judging and evaluating*

Books in Spanish

Ardley, Neil, *Experimentos con agua.* Watts, 1985. Explores properties of water. (grades 4–5)

Ardley, Neil. *Cosas en movimiento.* Watts, 1985. Explores the effects of gravity and friction. (grades 4–6)

Ardley, Neil. *Musica y sonido.* Watts, 1985. Explores sound and music. (grades 4–6)

Teaching Options

Science Background

Science and Society Many people are not aware of the effects of loud sounds. For many young people loud music, television, and conversations are an everyday experience. Studies suggest that loud noises are distracting. In most cases, students learn less in a noisy environment than in a quiet one. Research also shows that loud sounds can cause physiological damage. People who listen to very loud music or work around loud machinery tend to show more hearing loss than those in quieter environments. The critical factor in noise pollution is sound intensity, measured in decibels.

*Answers to masters on pages 162E–162H

3 UNIT 3/Earth Science

Planning Guide

Unit Components			Pages
Unit Overview/*SCIENCE IN THE NEWS*			184–185
Chapter 9 Rocks and Soil	**Lesson 1** How Are Rocks Formed? **Lesson 2** What Is Soil Made Of?	**Lesson 3** How Do People Use Rocks and Soil?	186–205
Chapter 10 Changes in the Earth	**Lesson 1** What Is the Inside of the Earth Like? **Lesson 2** How Do Water and Wind Change the Earth's Crust?	**Lesson 3** How Do Earthquakes and Volcanoes Change the Earth's Crust? **Lesson 4** How Do Living Things Change the Earths Crust?	206–227
Chapter 11 Clouds and Storms	**Lesson 1** How Do Clouds Form? **Lesson 2** What Happens to Water in the Clouds?	**Lesson 3** What Causes Storms?	228–245
Chapter 12 The Sun, Moon, and Planets	**Lesson 1** How Do the Earth and Moon Move? **Lesson 2** What Are the Sun and the Planets Like?	**Lesson 3** What Do Scientists Learn from Space Travel?	246–267
Careers	surveyor, aerial photographer, volcanologist, hurricane hunter, aerospace assembler		268
How It Works	Space suits		269
Unit Review			270
Unit Projects and Books	Science Projects Science and Society Books About Science		271

Science Process Skills Book

The Science Process Skills Book contains worksheets that can be used to teach and then assess student mastery of the basic science process skills. In addition, other worksheets in this book teach students the manipulative skills they will need to use basic science equipment. You might use some or all of these worksheets

- at the beginning of the year to teach and assess skills.
- directly before students encounter a particular skill in an activity.
- as a remedial tool when students begin to have trouble with specific skills.

cience Resources for the Unit

Resource Books

douze, Jean, and Israel, Guy, eds. *The Cambridge Atlas of Astronomy*. Cambridge Univ Pr, 1985. Summarizes current knowledge about the sun, moon, planets, stars, and comets.

ll, Pat. *Rocks and Minerals*. Collier, 1985. Identifies more than 400 rocks and minerals and explains how rocks form.

llant, Roy A. *Our Restless Earth*. Watts, 1986. Discusses and explains the formation of the solar system and the beginnings of the planets.

uber, Patricia. *Volcano: The Eruption and Healing of Mount St. Helens*. Bradbury, 1986.

nan, Colin A., ed. *The Sky-Watcher's Handbook: Night and Day What to Look for in the Heavens Above*. Crown, 1985. Gathers information about weather, the sun, and stars.

Community Resources

sit a nearby farm to study the use d conservation of soil.
Visit a recycling center.
Invite a representative from a cal oil, gas, or mining company to scuss uses of minerals.
Arrange a visit to a local evision station to learn how ather reports are produced.

Audio-Visual Resources

Basic Facts About the Earth, Sun, Moon, and Stars. Sterling. Series of Films, 9 minutes each. Discusses stars, constellations, and galaxies.

The Earth's Changing Crust. General. Filmstrip, 57 frames. Explains many changes on the earth's surface.

An Introduction to Forecasting the Weather. National Geographic. Wonders of Learning Kit (30 booklets, teacher's guide, activity sheets, cassette). Explores methods of forecasting the weather.

The Night Sky. Britannica. Film, 11 minutes. Explores the broad scope of inquiry into the sun, moon, stars, planets, constellations, and galaxies.

The Solar System and Beyond. National Geographic. Filmstrips with cassettes, 15 minutes each. *Sun, Earth, and Moon; Exploring the Solar System; Comets, Meteoroids, and Asteroids; and Beyond the Solar System.*

The Wrinkled Ball. Macmillan. Film, 7 minutes. Shows the irregularities in the earth's surface, the distribution of land and water, the great variations in landforms, and the composition of the earth's interior.

Computer Software

Astronomy for Everyone. C.C. Publications. Apple IIe+, IIe, IIc, 48K. 15–20 minutes, 4–6 activities. Basic astronomy concepts including Halley's Comet.

The Earth and Its Composition. Right-On. Apple II Series, 48K disk or cassette, or Commodore 64, 64K. Discusses the basic components of earth, water, air, and volcanoes.

The Earth Moves: Folds & Faults. Aquarius. Apple IIe+, IIe, IIc, 48K. 5+ minutes, 9 activities. Students can graphically simulate the earth's faults.

Explorations in Science: Weather Wise. Jostens. Apple II+, IIe, IIc, 48K. Take a simulated ride in a hot-air balloon.

Mickey's Space Adventure. Walt Disney. Apple II+, IIe, IIc, 64K. 20–40 minutes, 2–3 activities. Using planning and map-making skills, learn facts about the planets.

Rocks: Properties and Uses of Minerals. Spectrum Software. Apple II Series, 48K disk or cassette or TRS80 I, III 32K. Introduces properties of rocks.

The Water Cycle. Educational Computing Systems. Apple II Series, 48K. Graphics teach meteorological concepts.

TEACHING PLAN

Unit Overview

This unit focuses on the earth and its place in the solar system. It explores the structure of the earth; changes in the earth's crust, soil and rocks; weather conditions on earth; and the sun, moon, and other planets.

About the Photograph

The photograph shows a scene in Torres del Paine National Park in the region of southern Chile known as Patagonia. Lakes, glaciers, forests, rivers, mountains, and glacial fiords form a unique park ecosystem practically unspoiled by humans. Few places in the world have as many glaciers. The lakes in the park are fed by water from melting glaciers and are interconnected by the Rio Paine. The rock formations which give their name to the park are three granite towers with sheer faces. Frequent showers and storms occur in the park, but high in the mountains the forest thins out and masses of fallen rocks abound. The tops of the towers are often shrouded in mist.

184

Teaching Options

Chapter 9 Preview

Rocks and Soil

This chapter describes the three types of rocks and discusses how they are formed. The chapter explores the relationship between rocks and soil, and examines the capacity of soil to transfer nutrients to plants. Students will also study the various uses of minerals and the importance of conserving and recycling these minerals.

Chapter 10 Preview

Changes in the Earth

In this chapter students will discover some of the factors that change the earth, such as volcanoes, earthquakes, people, and animals. The chapter discusses the three layers of the earth and explores how scientists study the earth. It also examines the effects of wind and water in weathering and erosion, and identifies examples of weathering and erosion. The formation and effects of volcanoes and earthquakes are explored.

Unit 3

Earth Science

Over the years, the land in the picture has slowly changed. Notice the shapes of the mountains. Look at the snow on the mountain tops.

The earth is always changing. In this unit, you will discover what causes changes in land and changes in weather. You also will learn about places far away from earth.

SCIENCE IN THE NEWS During the next few weeks, look in newspapers or magazines for news about storms. Also, look for stories that tell how storms changed the land or changed the lives of people. Share the news with your class.

Teaching Tips

- Direct students' attention to the photograph on page 184. Questions: **What do you see in the picture?** (mountains, snow, water, trees and other plants) **What do you notice about the shapes of the mountains?** (Students should *observe* that they have sharp, jagged peaks and steep sides.) **Do you think it is warm or cold on the mountains? Why?** (Students should *infer* that it would be cold, because the mountain tops are snow covered.)
- Explain to students that rivers of ice, called glaciers, are also found in these mountains. Point out the lake in the foreground of the picture. Tell students that the water in the lake comes from the melting of the glaciers. Question: **Do you think you would enjoy swimming in this lake? Why?** (Students should *infer* that the water would be too cold for swimming.)

Chapter 11 Preview

Clouds and Storms
This chapter introduces basic weather conditons, cloud formations, and the concepts of evaporation and condensation. Students will learn about some types of severe weather, such as thunderstorms, hurricanes, and tornadoes. The chapter also discusses safety practices to follow during severe weather.

Chapter 12 Preview

Sun, Moon, and Planets
This chapter describes the rotation and revolution of the earth and how rotation causes day and night. It also describes the moon's features, movements, and phases. Students will learn about the sun and planets that make up our solar system. How scientists have learned about the moon and other planets is also discussed.

Rocks and Soil Planning Guide

TEACHING PLAN

Chapter Components	Skills	Materials
Chapter Opener/*DISCOVER:* Observing Rocks pp. 186–187	*DISCOVER* p. 187 Science Process Skills *Observing, Communicating*	*DISCOVER* p. 187 (groups of 2) 45 assorted rocks
Lesson 1 How Are Rocks Formed? pp. 188–191	Thinking Skills Challenge!: *Applying information to new situations* Find Out On Your Own: *Collecting information, Sequencing*	Demonstration p. 188 White bread and wheat or rye bread, book
Activity Testing Objects for Hardness p. 192	Science Process Skills *Observing, Collecting and interpreting data*	(groups of 2) 15 school pencils, 15 pieces of chalk, 15 pennies
Science and Technology Entering the Diamond Age p. 193	Thinking Skills *Restating or explaining ideas, Recognizing cause and effect, Inferring*	
Lesson 2 What Is Soil Made Of? pp. 194–196	Thinking Skills Challenge!: *Applying information to new situations, Making inferences* Find Out On Your Own: *Collecting information, Communicating*	Demonstration p. 194 glass jar with lid, soil, water
Activity Observing Soil Samples p. 197	Science Process Skills *Observing, Classifying, Collecting and interpreting data*	(groups of 2) 1 bag potting soil, 1 bag sand, 1 bucket garden soil, 1 bucket clay soil, 15 sheets white construction paper, 60 paper cups, 1 box toothpicks, 15 hand lenses, 15 plastic spoons
Lesson 3 How Do People Use Rocks and Soil? pp. 198–201	Thinking Skills Challenge!: *Applying information to new situations, Making inferences* Find Out On Your Own: *Suggesting alternatives, Communicating*	Demonstration p. 198 stalk of celery wrapped in aluminum foil, salt shaker, salt
Skills for Solving Problems Using Graduated Cylinders and Bar Graphs pp. 202–203	Problem Solving Skills *Making decisions/Identifying and solving problems, Interpreting charts, maps, and graphs*	
Chapter Review pp. 204–205	Thinking Skills *Recognizing the main idea and supporting details, Restating or explaining ideas, Sequencing, Contrasting, Applying information to new situations, Inferring*	

Teaching Options

Strategies	Extensions		Resource Masters
Cooperative Learning p. 186 (Also see p. T23.) Applying Whole Language p. 187 (Also see p. T30.)			Family Letter: *Resource Book* p. 99
Reading Strategies p. 188 (Also see pp. T26–T29.)	Science and Social Studies p. 189 Special Education p. 189 Reinforcement p. 190	Enrichment p. 190 Game Suggestion p. 191 Reteaching Suggestion p. 191	Vocabulary Preview: *Workbook* p. 49 Science Activity: *Workbook* p. 50
			Activity Worksheet: *Resource Book* p. 103
Reading Strategies p. 194 (Also see pp. T26–T29.)	Science and Social Studies p. 195 Special Education p. 195	Reinforcement p. 195 Reteaching Suggestion p. 196	Science Skills: *Workbook* p. 51
			Activity Worksheet: *Resource Book* p. 105
Reading Strategies p. 198 (Also see pp. T26–T29.)	Science and Art p. 199 Reinforcement p. 199 Special Education p. 199	Enrichment p. 200 Game Suggestion p. 201 Reteaching Suggestion p. 201	Science and Social Studies: *Workbook* p. 52 Vocabulary Puzzle: *Workbook* p. 53
	Reteaching Suggestion p. 177		Science and Reading: *Resource Book* p. 107
Cooperative Learning p. 204 (Also see p. T23.)			Chapter Tests: Forms A and B *Test Book* pp. 81–84

Classroom Management

Advance Preparation

DISCOVER, page 187
Ask students to bring rocks from their neighborhood or from rock collections to class

Demonstration, page 188
Bring several slices of white bread and several slices of wheat or rye bread to class.

Activity, page 197
Place 2 teaspoons of soil samples in a paper cup for each group of students or ask students to bring soil samples to class. Provide a plastic spoon for each group.

Demonstration, page 198
Bring a stalk of celery, aluminum foil, and salt to class.

Vocabulary Review

Use the following sentences with your students to review the meanings of the italicized words.

1. The color and hardness of a rock are two of its *properties*.
2. If a suitcase is very full, you might have to put *pressure* on it to close it.
3. Some statues are carved out of a hard stone called *marble.*
4. The soil you see at a park, on a lawn, or on a farm is called *topsoil.*
5. Rocks are made of one or more *minerals.*
6. Sometimes farmers add fertilizers to the soil to give the plants extra *nutrients* to grow.

High-Potential Students

Ask students to compare how well plants grow in three types of soil—sand, clay, and loam. Tell students to obtain three potted plants of the same variety. They should pot one plant in sand, one in clay, and one in loam. Encourage them to measure and record the height of the plants when they are placed in the pots. Each plant should receive the same amount of water and sunlight throughout the investigation. Direct students to keep accurate records of the measurements and dates for several weeks. Have them display their plants, and write reports following this sample outline:

Title: ________(Plant growth in different soils)
Hypothesis: __(I think that . . .)
Materials: ___(List materials used)
Results: ______(Record in a graph or chart)
Conclusion: __(Check hypothesis)

Mainstreamed Students

Visually Impaired
Allow visually impaired students to touch the samples of igenous, sedimentary, and metamorphic rock. Discuss the different properties of the rocks with them.

Emotionally Handicapped
Encourage students to put together a rock collection. Ask students to bring in their collections and show samples from their rock collection to the class.

Science Fair Projects

The Experiment Skills on p. 364 and the Investigate feature in the chapter can be used for science fair projects. You might also encourage interested students to do one of the following projects:

1. To test the hardness of rocks in your area, collect samples of different rocks. Find a copy of Moh's scale of hardness in a reference book. Using Moh's scale, rank the order of hardness of the rock samples. Prepare a chart of the findings.
2. Make a soil profile of your area by obtaining soil samples from different areas in your neighborhood using the following method: remove the lid from a tin can and clean out the can; push the can into some soil, when it is full of soil pull the can up. Analyze the contents of the soil and prepare a display showing the location and type of soil found in each area.

lassroom Resources

Bulletin Board

Ask students to bring in pictures that show different ways we use rocks, such as in buildings, roads, statues, gravestones, and landscaping

ROCKS FORM IN DIFFERENT WAYS

IGNEOUS ROCKS

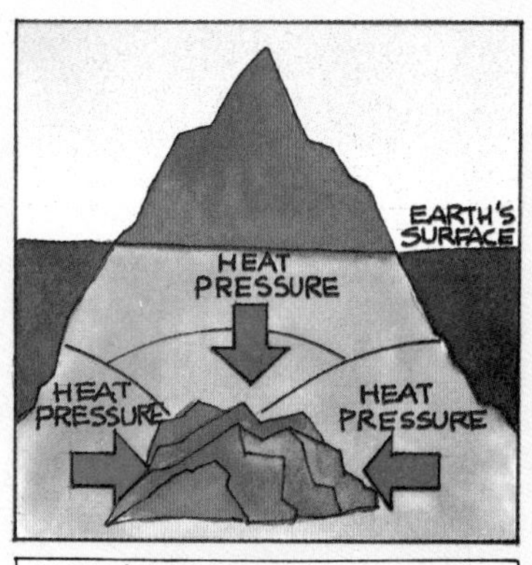

METAMORPHIC ROCKS

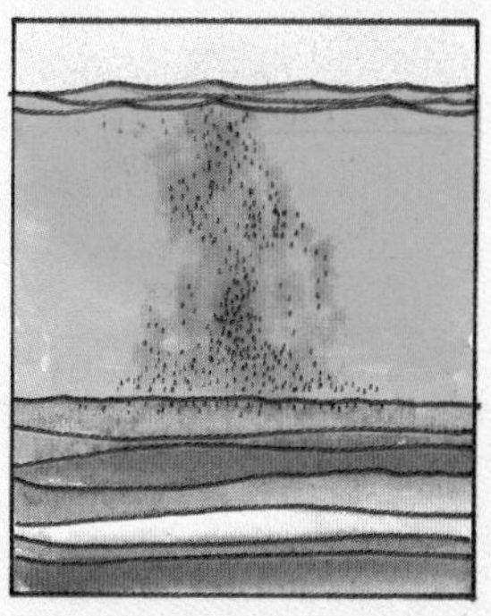

SEDIMENTARY ROCKS

Chapter 9 Poster

Why does this rock need to be repaired?

Science Discovery Center

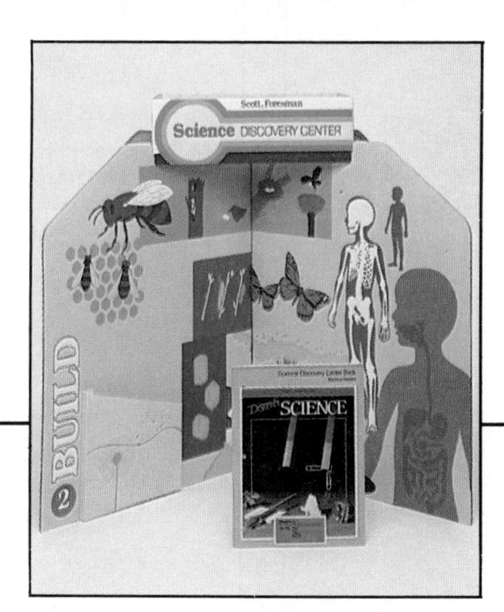

Use pages 95–100 from the *Science Discovery Center Book.* Place these worksheets in the appropriate pockets in the Science Discovery Center.

Overhead Transparencies

Use Transparency 15 from the package of color overhead transparencies.

Teacher's Resource Book

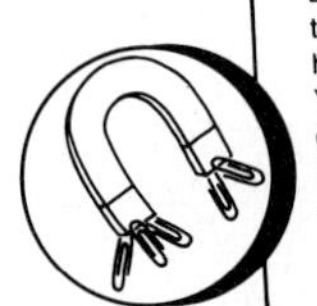

Dear Family,
Your student will be reading **Chapter 9: Rocks and Soil** in *Discover Science*, published by Scott, Foresman. We will learn that rocks are made of one or more minerals. Each mineral has its own special properties—color, hardness, shininess. You can see different minerals in some rocks. Rocks can be grouped by size, shape, color, or how they were formed.
You and your student can do this activity together to learn more about rocks.

Rock Collection

You will need egg cartons and a marker.

1. Ask your student to collect 12 rocks.
2. Together, study each rock. Ask, "What shape is it? What colors can you see? Can you see crystals? Are they large or small? Is the rock shiny or dull, rough or smooth?
3. Use the egg carton to group the rocks. For example, you can group all the shiny rocks at one end, and all the dull rocks at the other. Put the label on the inside of the carton lid.
4. Go to the library. Find books on rocks and minerals. Use the books to help you identify the rocks. Try to group your rocks as igneous, sedimentary, or metamorphic.

Workbook

Name ______________________

Use with Lesson 1: pages 188-191

Chapter 9

Vocabulary Preview

Rocks and Soil

Chapter Vocabulary

clay soil	igneous rock	natural resource	recycle
decay	loam	nutrient	sandy soil
humus	metamorphic rock	ore	sedimentary rock

Vocabulary Cards

1. Write each word on a card.
2. Find each word in the glossary. Copy the pronunciation under the word on the card.
3. Practice saying the words with a partner.

Word Meanings

1. Sort your vocabulary cards into the following three groups: rock words, soil words, and other words. You may use your glossary. Then compare your groups of cards with those of a classmate.
2. Circle the rock words in the list above. Underline the soil words.
3. Find the following words in Chapter 9 of your book. Copy one sentence for each word.

decay: ______________________

nutrient: ______________________

recycle: ______________________

natural resource: ______________________

At Home

1. Practice saying the words. Learn their meanings.
2. Look for the words and pictures of the words in newspapers and magazines. You can make a poster.

Teacher's Notes: Have students write sentences with vocabulary words omitted. Have them exchange papers and fill in the words.

Workbook

Name ______________________

Use with Lesson 1: pages 188-191

Chapter 9

Science Activity

How Do Sedimentary Rocks Form?

Gather These Materials

• a slice of white bread • a slice of wheat bread • waxed paper • several heavy books

Follow This Procedure

1. Put the slice of white bread on top of the slice of wheat bread. Wrap the two together in waxed paper.
2. Place the wrapped bread on your desk. Set heavy books on the bread. Keep them there for two hours.
3. Unwrap the bread. Draw what you see.

Record Your Results

Draw what the bread looks like after having pressure on it for several hours.

Students will draw flattened layers of bread.

State Your Conclusions

1. What happened to the slices of bread?
   Pressure forced the slices together.
2. What do you think happens to rocks that have pressure on them?
   Pressure forces the rocks together.
3. What do you think happens as sediment settles on the bottom of a lake?
   Pressure from the water and the new sediment changes previous sediments, forming sedimentary rock.

Teacher's Notes: Point out that rivers carry sediment, such as sand, mud, gravel, and clay, into lakes, seas, and the ocean.

Teacher's Resource Book

Name ______________________

Use with Lesson 1: page 192

Testing Objects for Hardness

Chapter 9

Activity Worksheet

Record Your Results

	Object to be tested	Chalk	Pencil lead	Penny
Scratched by	Fingernail			
	Chalk			
	Pencil lead			
	Penny			

State Your Conclusion

1. Which is harder, pencil lead or chalk?

   ____

2. List the objects you tested in order of their hardness, with the softest first.

   ____

Use What You Learned

A diamond is the hardest mineral. If you had a real diamond and a fake diamond made of glass, how could you tell which one was real?

____

Workbook

Name

Use with Lesson 2: pages 194-196

Chapter 9

Science Skills

Looking at a Log

The log once was part of a living tree. Now it provides other living things with food and shelter. In time, the log will become part of the soil.

What causes the log to break down into soil? The plants that grow on the log help break it up. Their roots make holes in the wood. Also, small animals chew holes in the wood. Rainwater fills these holes.

During cold weather, the water in the holes changes to ice. As water freezes, it swells and takes up more space. As a result, ice breaks up the log even more.

As the log breaks down, it adds nutrients to the soil. Other plants can get what they need from that soil. Perhaps one day another tree will grow where the log now lies.

Use the picture and the paragraphs above to answer these questions.

1. What living things do you see on the log? **insects, mouse, worms, moss, other plants**
2. What helps turn the log into soil? **animals chewing; plant roots; rainwater, ice**

Teacher's Notes: Discuss living things found on logs: beetles, worms, and other small animals; moss, other plants. Larger animals, such as bears, squirrels, and raccoons, make their homes in some logs.

Teacher's Resource Book

Name

Use with Lesson 2: page 197

Chapter 9

Activity Worksheet

Observing Soil Samples

Record Your Results

Sample Part	a	b	c	d
Color				
Hard or soft				
Size (large, medium, or small)				
Shape				

State Your Conclusion

1. How many different kinds of bits did you find in the first soil sample?
2. Did one of your piles have humus? How do you know?

Use What You Learned

Which kinds of soil were in your samples? Would plants be able to grow well in your soil samples?

Workbook

Name

Use with Lesson 3: pages 198-201

Chapter 9

Science and Social Studies

Using Soil to Make Houses

Read the story. Then answer the questions.

In some places, people live in houses made from soil. The soil is made into bricks. The house is made from these bricks.

Making bricks from soil is not a new idea. Thousands of years ago, ancient Egyptians used bricks made from soil. Indians in the American Southwest used bricks from soil to build homes, called pueblos (pweb′ lōs).

Today, many houses in the Southwest and in Mexico are made with adobe (e dō′ bē) bricks. Adobe is the Spanish name for bricks made of soil. Adobe houses are cooler than wood or stone houses. Adobe houses help people keep cool in the hot Southwest and Mexico.

Making adobe bricks takes time. First, sandy clay is mixed with water and straw or grass. Then this mixture is poured into wooden frames. When the bricks are dry, the frames are taken away. The bricks are then placed in the hot sun to bake for two weeks.

1. Who used bricks made of soil to build homes long ago? **Ancient Egyptians and Indians in the Southwest**
2. Today, where might you see a house made from adobe bricks? **The Southwest and Mexico**
3. What materials are used to make adobe bricks? **Sandy clay, water, and straw or grass**
4. How are adobe bricks made hard? **They are baked in the sun.**

Teacher's Notes: Encourage students to find out how sod is used to make homes.

Workbook

Name

Use with Lesson 3: pages 198-201

Chapter 9

Vocabulary Puzzle

Fill-In

Use the clues to write the missing letters.

1. Useful materials that come from the earth
   N A T U R A L R E S O U R C E S
2. Rock that was changed by heat or pressure
   M E T A M O R P H I C R O C K
3. Rock that forms from layers pressed together
   S E D I M E N T A R Y R O C K
4. Rock that forms from melted minerals
   I G N E O U S R O C K
5. Needed by plants and animals to live and grow
   N U T R I E N T S
6. Tightly packed soil
   C L A Y S O I L
7. Loose soil with large grains
   S A N D Y S O I L
8. Mixture of clay, sand, and humus
   L O A M
9. To break down slowly
   D E C A Y
10. To change something so it can be used again
    R E C Y C L E
11. Rock rich in minerals
    O R E
12. Decayed matter in soil
    H U M U S

Teacher's Notes: Encourage students to identify kinds of soil.

Teacher's Resource Book

Name ____________

Comprehension: details

Chapter 9

Science and Reading

Only the Facts

Read the article. Then write a sentence to answer each question.

Landfills are places where garbage is dumped. Today many of our landfills are full or nearly full. Some towns are helping by having garbage recycling programs. Materials such as glass, cans, and newspaper can be recycled and used again.

There are two kinds of recycling programs. Some towns have recycling centers. People drop off their materials to be recycled at these centers. Other towns have curbside recycling programs. People are given bins in which to collect materials for recycling. Then the bins are put out by the curb, and the materials are collected.

1. How are towns helping with the landfill problem?

They are helping by having recycling programs.

2. What are two kinds of recycling that towns have?

Towns have recycling centers or curbside recycling programs.

3. How does a curbside recycling program work?

People put materials to be recycled in bins, and then the materials are collected.

Test Book

Name ____________

Chapter 9
Test A

Multiple Choice. Choose the best answer.

1. Rocks are made of one or more (1-1)
   a. carbons.
   (b) minerals.
   c. nutrients.
2. Rocks formed from melted minerals that have cooled are (1-2)
   a. sedimentary rocks.
   (b) igneous rocks.
   c. metamorphic rocks.
3. Igneous or sedimentary rocks that are changed by heat or pressure are (1-2)
   a. recycled rocks.
   (b) metamorphic rocks.
   c. fossils.
4. Freezing and melting water causes rocks to break slowly into small pieces and forms (2-1)
   a. carbons.
   (b) soil.
   c. fossils.
5. Materials from the soil that plants need to grow are (2-2)
   a. clays.
   b. sediments.
   (c) nutrients.
6. Soil that is a mixture of clay, sand, and humus is called (2-3)
   a. clay soil.
   b. sandy soil.
   (c) loam.
7. Rocks that have large amounts of useful minerals within them are (3-1)
   a. mines.
   (b) ores.
   c. loams.
8. A mineral used to make film, jewelry, and mirrors is (3-3)
   a. copper.
   (b) silver.
   c. granite.
9. Minerals and soil are two things people use that are (3-4)
   (a) natural resources.
   b. igneous rocks.
   c. nutrients.
10. What kind of soil has few nutrients? (2-2)
    a. loam.
    b. humus.
    (c) sandy soil.

Numbers in parentheses after each question refer to the lesson number and the objective of that lesson.

Test Book

Name ____________

Chapter 9
Test A

Matching Match the correct definition and term.

Short Answer Use the pictures to answer the questions.

marble sedimentary granite graphite

1. Which rock is a igneous rock used to make buildings? **granite**
2. Which rock is used to make pencil tips? **graphite**
3. Which rock formed when layers of material were pressed together? **sedimentary**
4. Which rock is a metamorphic rock? **marble**

Short Essay Use complete sentences to answer each question.

1. Explain why soil is important for people. (3-1)

Plants grow and reproduce in soil. People eat the plants that grow in soil. People get meat from animals that eat plants. Peoplealso use plants for clothing and other things.

2. Explain how miners reclaim, or restore, the land after it has been mined. (3-3)

Miners must fill in the mining pit so that land can be used again. Plants must be replaced too.

Numbers in parentheses after each question refer to the lesson number and the objective of that lesson.

Test Book

Name ____________

Chapter 9
Test B

Multiple Choice. Choose the best answer.

1. The soft material in a pencil tip comes from (1-1)
   a. a plant.
   (b) a rock.
   c. soil.
2. The colors of rocks come from different (1-1)
   a. kinds of soil.
   (b) minerals.
   c. plants.
3. When melted minerals move to the earth's surface and harden, they are (1-2)
   (a) igneous rocks.
   b. sedimentary rocks.
   c. metamorphic rocks.
4. Rocks formed when layers of material are pressed together are called (1-2)
   a. igneous rocks.
   (b) sedimentary rocks.
   c. metamorphic rocks.
5. Large rocks can be slowly broken into small pieces by water or (2-1)
   a. sandy soil.
   (b) plant roots.
   c. decayed matter.
6. Decayed matter that becomes part of the soil is called (2-2)
   a. sand.
   b. sediments.
   (c) humus.
7. Tightly packed soil with tiny grains is (2-3)
   (a) clay soil.
   b. sandy soil.
   c. loam.
8. A mineral used to make some kinds of pipe and used in batteries is (3-2)
   a. granite.
   b. limestone.
   (c) lead.
9. Materials that are used over and over again are (3-4)
   a. wasted.
   (b) recycled.
   c. decayed.
10. Changing the kinds of plants grown in the soil every few years replaces lost (3-4)
    (a) nutrients.
    b. ores.
    c. sand.

Numbers in parentheses after each question refer to the lesson number and the objective of that lesson.

Name ______________________

Chapter 9
Test B

Short Answer Use the pictures to answer the questions.

1. Which rock is a metamorphic rock? marble
2. Which rock is made of soft carbon? graphite
3. Which rock was formed in layers under water? sedimentary
4. Which rock is a hard igneous rock? granite

marble sedimentary granite graphite

Short Essay Use complete sentences to answer each question.

1. Name two ways people use objects that come from the soil. (3-1)

People eat plants that grow in the soil. Other kinds of plants are used for clothing, paper, and many other things.

2. Define a natural resource. Give two examples. (3-4)

A natural resource is something people use that comes from the earth. Minerals and soils are two examples of natural resources. (Accept other examples.)

Numbers in parentheses after each question refer to the lesson number and the objective of that lesson.

TEACHING PLAN

Major Concepts

Lesson 1 Rocks are formed in three different ways: from melted minerals inside the earth; as a result of layers of materials pressing together; from changes of existing rocks due to heat or pressure.
Lesson 2 Soil is a mixture of rock particles, organic matter, air, and water.
Lesson 3 Rocks and soil are useful to people.

Chapter Vocabulary

clay soil, decay, humus, igneous rock, loam, metamorphic rock, natural resource, nutrient, ore, recycle, sandy soil, sedimentary rock

Getting Started

Ask students what colors they see in the rock in the picture. Then ask them how they think the rock came to be this shape. Lead students to *infer* that the hole was hollowed out by wind or water. Encourage students to describe other unusual rocks they have seen.

Chapter 

Rocks and Soil

Have you ever seen a rock shaped like this? Notice that you can see different colors in the rock.

186

Teaching Options

Cooperative Learning ♦

Jigsaw Format (See page T23.)
Assign the following topics at random to your cooperative learning teams.
Topic A: What senses are needed to study rocks and soil?
Topic B: Describe the ways in which rocks and soil are formed.
Topic C: What are the properties of rocks and soil?
Topic D: In what ways can rocks and soil be helpful to humans?
Have students search for information on their topic as they read the chapter. Then let all students with the same topic meet in an expert group to discuss the information. When students return to their teams, they may take turns presenting their topics to the team. Then give students a test covering all topics to complete individually (Chapter 9 Test A or B in the *Test Book).* Award Superteam certificates to teams whose average test scores exceed 90%, and Greatteam certificates to teams whose average test scores exceed 80%.

♦ ***Suitable as a language development activity***

Introducing the Chapter

Different kinds of rocks are made of different materials. In this chapter, you will learn about how rocks and soil are formed. You will read about ways people use rocks and soil. The activity below will help you learn about different properties of rocks.

Observing Rocks

Put the rocks your teacher gives you on your desk. Observe the rocks carefully. On a piece of paper, write some words that describe one of the rocks. Give the paper to a classmate to read. Ask your classmate to point out the rock that you described.

Talk About It

1. How are the rocks you collected alike?
2. How was the rock you described different from the others?

DISCOVER

Objective ♦

This optional *DISCOVER* activity will help students explore and build background information concerning the classification of rocks. Later in the chapter students will be able to draw on this experience to help them assimilate the new content.

Science Process Skills

Observing, *Communicating*

Materials

For each pair of students: assorted rocks

Teaching Tips

- Bring rocks to school, or ask students to bring in rocks from their neighborhoods or collections.
- Group similar rocks together and have students *list* the characteristics of the rocks in each group. Use the lists to *classify* rocks.

Answers

Talk About It

1. Answers will vary but might include that the rocks were different or similar in size, shape, or color.

2. Students might describe rocks by their appearance, texture, or odor. Some students might have struck the rocks against a hard surface and described the differences in sound.

Applying Whole Language ♦

Discuss the whole language framework with each Teaching Option you select. Here is an example applied to the Game Suggestion on p. 201.

1. Purpose: To invent a new use for a recycled material

2. Context: The role of inventors, and using technology for the good of people

3. Decisions: How the material can be used to help people. What are the steps of recycling the material from its previous use? How can we show the steps?

4. Evaluation: Did we come up with new ideas? Were they new to us? (See p. T30.)

Resource Book page 99

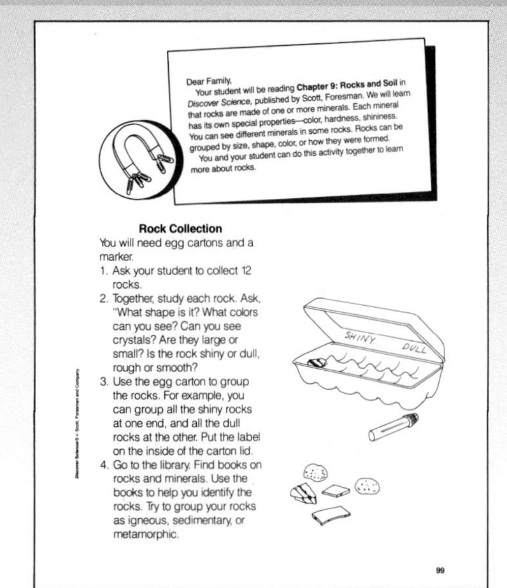

Dear Family,

Your student will be reading **Chapter 9: Rocks and Soil** in *Discover Science*, published by Scott, Foresman. We will learn that rocks are made of one or more minerals. Each mineral has its own special properties—color, hardness, shininess. You can see different minerals in some rocks. Rocks can be grouped by size, shape, color, or how they were formed.

You and your student can do this activity together to learn more about rocks.

Rock Collection

You will need egg cartons and a marker.

1. Ask your student to collect 12 rocks.
2. Together, study each rock. Ask, "What shape is it? What colors can you see? Can you see crystals? Are they large or small? Is the rock shiny or dull, rough or smooth?
3. Use the egg carton to group the rocks. For example, you can group all the shiny rocks at one end, and all the dull rocks at the other. Put the label on the inside of the carton lid.
4. Go to the library. Find books on rocks and minerals. Use the books to help you identify the rocks. Try to group your rocks as igneous, sedimentary, or metamorphic.

SHINY DULL

99

Science Background

Several characteristics can be used to classify rocks, including size, shape, and color. However, these characteristics are not sufficient to clearly distinguish among rocks. Geologists further classify rocks according to their composition. Geologists also use characteristics such as hardness, shape of crystals, and reactions with acids.

TEACHING PLAN

Lesson Objectives
- *State* that rocks are made of minerals.
- *Explain* how igneous rock, sedimentary rock, and metamorphic rock form.

Lesson Vocabulary
igneous rock, metamorphic rock, sedimentary rock

1. Motivate

Demonstration Activity ♦
Alternately layer slices of white and wheat or rye bread. Place a book on top of the slices.

Discussion
Ask students what will happen to the bread slices if left under the book. (they will stick together) Point out how the individual layers of bread are still visible. Question: **What causes the slices of bread to stick together?** (pressure from the book) Point out that sedimentary rocks form in a similar way.

1 How Are Rocks Formed?

LESSON GOALS

You will learn
- that rocks are made of minerals.
- how the three kinds of rocks are formed.

Suppose you scratched the tip of a pencil with your fingernail. Part of the tip would come off. The soft, black material in a pencil tip came from a rock.

Minerals and Rocks

The tip of a pencil is made of graphite (graf′īt), a form of the mineral carbon. You learned in Chapter 1 that a mineral forms in the earth from matter that was never alive. The pictures show a few of the 2,500 known minerals. Each mineral has its own special properties. A mineral can be any color, shiny or dull, and soft or hard.

Rocks are made of one or more minerals. If you look closely at a rock, you might see different colors. Substances in the minerals cause the different colors.

Minerals

Selenite

Calcite with fluorite

Silver

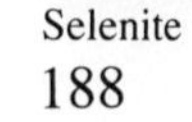

Teaching Options

Science Background

The three kinds of rock—igneous, sedimentary, and metamorphic—are parts of a cycle in which rocks change slowly, but continuously, from one kind into another. Heat and pressure can change igneous and sedimentary rocks into metamorphic rock. Weathering can break down igneous, sedimentary, and metamorphic rocks into sediments. Layers of sediments fuse together to form sedimentary rocks. Deep in the earth, heat, pressure, and liquids can change sedimentary or igneous rocks into metamorphic rocks. If the heat and pressure are high enough, metamorphic rocks can melt and form magma, which cools into igneous rock. As one type of rock is destroyed, another type forms.

Reading Strategies ♦

1. Guide students' pre-reading by asking: What ideas do you think you would find in the lesson to answer the question-title?
2. Assign these strategies: Writing a Memory Sentence and Finding and Writing Sentences for Vocabulary Words: igneous rock, metamorphic rock, and sedimentary rock. (See pages T26-T29.)
3. Pair students to share what information is clear and unclear and initiate discussion using students' unanswered questions.

♦ ***Suitable as a language development activity***

Kinds of Rocks

How would you group rocks in a rock collection? You might group them by their size, shape, color, or what they are made of. You also can group rocks by how they were formed.

igneous (ig′nē əs) **rock,** a rock that forms from melted minerals.

The earth is so hot deep inside that minerals melt. Some rocks form from melted minerals inside the earth. Melted minerals sometimes move to the earth's surface. Then the minerals cool and harden into rock. **Igneous rocks** are rocks that form from melted minerals. The igneous rock in the first picture formed from melted minerals that poured out of an opening in a mountain.

Granite (gran′it) is an igneous rock that is very hard. People use granite to make buildings. What do you notice about the color of this piece of granite?[1] Each color you see is a different mineral.

[1] has different colors

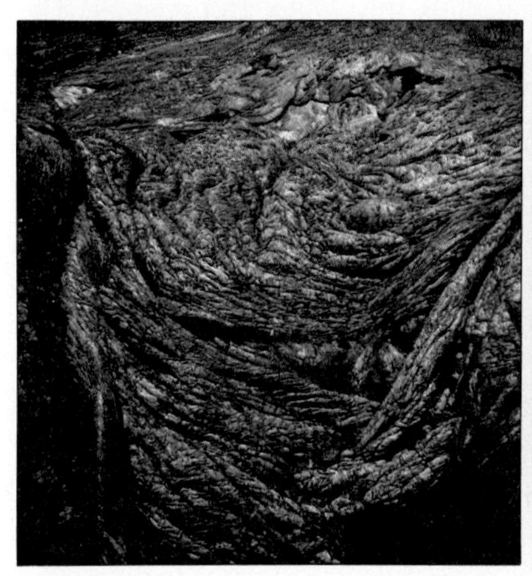
Igneous rock

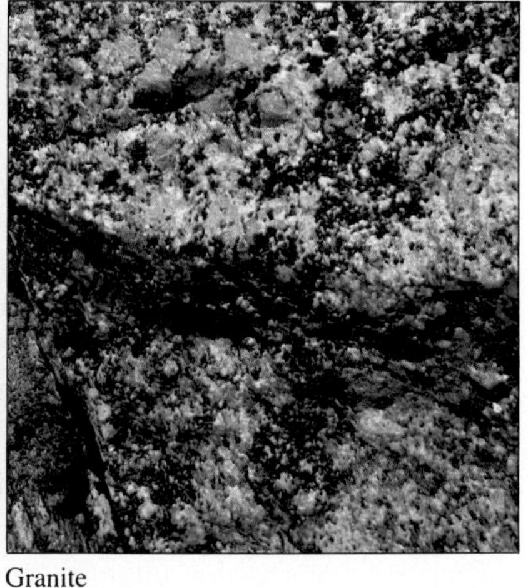
Granite

2. Teach

Teaching Tips

- Ask students to *compare* and *contrast* the three minerals shown in the photograph on this page.
- Allow students to *observe* the different colors in granite (pink, tan, gray, black). Question: **Do you think granite is made of more than one mineral?** (Guide students to *infer* that it consists of more than one mineral.)
- Provide hand lenses and samples of salt, sugar, granite, and so on. Allow students to use the lenses to *observe* the crystal structure of each sample. *CAUTION:* Direct students not to use the lenses to concentrate the sun's rays onto skin or paper.

Workbook page 49 *

Chapter 9

Name ______

Use with Lesson 1: pages 188-191

Vocabulary Preview

Rocks and Soil

Chapter Vocabulary			
clay soil	igneous rock	natural resource	recycle
decay	loam	nutrient	sandy soil
humus	metamorphic rock	ore	sedimentary rock

Vocabulary Cards

1. Write each word on a card.
2. Find each word in the glossary. Copy the pronunciation under the word on the card.
3. Practice saying the words with a partner.

Word Meanings

1. Sort your vocabulary cards into the following three groups: rock words, soil words, and other words. You may use your glossary. Then compare your groups of cards with those of a classmate.
2. Circle the rock words in the list above. Underline the soil words.
3. Find the following words in Chapter 9 of your book. Copy one sentence for each word.

decay: ______

nutrient: ______

recycle: ______

natural resource: ______

At Home

1. Practice saying the words. Learn their meanings.
2. Look for the words and pictures of the words in newspapers and magazines. You can make a poster.

49

Science and Social Studies

Have the students research a state or country of their choice to determine the minerals found there. Ask students to make a map to indicate the location and uses of the mineral deposits.

Special Education

Have students with learning disabilities each bring in a rock. Let them *list* the properties of the rock.

***** ***Answers to masters on pages 186E–186H***

TEACHING PLAN

Teaching Tips

- **Possible Misconception:** Most students are not aware of the rock cycle. They often think that since rocks are not living things, they never change. Emphasize the process of the rock cycle throughout the study of the chapter.
- Have students *infer* what natural force, other than water, can move sediments. (wind) Discuss how sediments settle down to the earth when the wind slows down.
- Question: **What is needed in order for metamorphic rocks to form?** (heat and pressure)
- Refer students to the *DISCOVER* activity on page 187 in which they ***described*** rocks. Have students ***observe*** the rocks with a magnifying lens, and then ***classify*** them as sedimentary or igneous. *CAUTION:* Instruct students not to use lenses to concentrate the sun's rays onto skin or paper.

Layers in sedimentary rock

sedimentary (sed′ə men′tər ē) **rock,** rock that forms when layers of material are pressed together.

metamorphic (met′ə môr′fik) **rock,** an igneous or sedimentary rock that was changed by heat or pressure.

Another kind of rock can form under water. Rivers and streams carry bits of rock, shells, and other materials into lakes and oceans. The materials sink and form layers at the bottom of the water. The top layers press down on the bottom layers. Over millions of years, the bottom layers harden into **sedimentary rock.** You can see layers of sedimentary rock in the picture.

Sometimes, igneous or sedimentary rock can be changed into a third kind of rock—**metamorphic rock.** Great heat inside the earth can change a rock's minerals into other minerals. Pressure also can change the minerals in rocks. When the minerals in a rock change, the rock becomes metamorphic rock.

190

Teaching Options

Science Anecdote

Diamond is the hardest mineral known. It can be scratched only with another diamond. The largest diamond ever found weighed about 3,000 carats, or had a mass of about 600 grams (21 ounces). It was found in South Africa in 1905.

Reinforcement

Demonstrate how sediments form into layers by putting sand, gravel, soil, and water in a tightly-covered jar. *CAUTION:* Immediately wipe up any water that spills on the floor. Shake the jar and let it stand for three days. Have students ***observe*** the layers of sediment. Question: **What would have to happen to turn this to rock?** (Other materials would have to be deposited on top.)

Enrichment

Take the students on a walk through the neighborhood to ***identify*** buildings made of sandstone or limestone (brownstone is a type of limestone). Have students ***observe*** the layers of sediment in these rocks.

◆ ***Suitable as a language development activity***

Limestone is a sedimentary rock. When limestone is heated and squeezed inside the earth, it changes into marble—a metamorphic rock. Marble is much harder than limestone. People can use very sharp tools to carve objects out of marble.

Lesson Review

1. What are rocks made of?
2. How does each of the three kinds of rock form?
3. **Challenge!** How can an igneous rock become part of a sedimentary rock?

Study on your own, pages 332–333.

SCIENCE IN YOUR LIFE

Sometimes scientists find fossils in sedimentary rocks. Fossils are signs of past life. Some fossils are bones, teeth, footprints, or whole plants or animals that lived many years ago.

PHYSICAL SCIENCE — **FIND OUT ON YOUR OWN** — *CONNECTION*

A mineral can scratch a softer mineral. For example, the mineral graphite can scratch the mineral talc because graphite is harder than talc. However, talc cannot scratch graphite.

Read about the properties of the minerals shown here. Decide which mineral is the softest and which is the hardest. On a piece of paper, list the names of the minerals from the softest to the hardest.

Diamond can scratch quartz and gypsum.

Gypsum cannot scratch diamond or quartz.

Quartz can scratch gypsum but not diamond.

3. Assess

Lesson Review

1. Rocks are made of minerals.
2. Igneous rock forms when minerals in the earth melt, then cool and harden. Sedimentary rock forms when layers of material are pressed together over a long time. Heat and pressure change or rearrange the minerals in igneous or sedimentary rock to form metamorphic rock.
3. Challenge! Pieces of igneous rock might break off and be washed away into a body of water. The pieces sink into sediment layers under the water. Over time, the layers, including the pieces of igneous rock, are pressed into sedimentary rock. **Thinking Skill:** *Applying information to new situations*

Find Out On Your Own

In order of softest to hardest: gypsum—quartz—diamond. **Thinking Skills:** *Collecting information, Sequencing*

Workbook page 50 *

Name ______________________
Use with Lesson 1: pages 188-191

Chapter 9

Science Activity

How Do Sedimentary Rocks Form?

Gather These Materials
• a slice of white bread • a slice of wheat bread • waxed paper • several heavy books

Follow This Procedure
1. Put the slice of white bread on top of the slice of wheat bread. Wrap the two together in waxed paper.
2. Place the wrapped bread on your desk. Set heavy books on the bread. Keep them there for two hours.
3. Unwrap the bread. Draw what you see.

Record Your Results
Draw what the bread looks like after having pressure on it for several hours.

State Your Conclusions
1. What happened to the slices of bread?
____
2. What do you think happens to rocks that have pressure on them?
____
3. What do you think happens as sediment settles on the bottom of a lake?
____

50

**Answers to masters on pages 186E–186H*

Game Suggestion ♦

Write the name, classification, and description of various igneous, sedimentary, and metamorphic rocks on separate index cards. Let a student choose a card and read the description to the class. The other students try to guess the type of rock by asking questions that can be answered with a *yes* or a *no*. The student who correctly guesses the rock goes next. Continue until all students have had a turn.

Reteaching Suggestion ♦

Display samples of sandstone, pumice, limestone, and marble. Challenge students to select the rocks from the samples which best match the following clues: a) formed when limestone is placed under heat and pressure (marble); b) formed when many layers of sand are pressed together (sandstone); c) formed from bubbly surface lava that cools (pumice); d) formed from seashells and sea animals pressed together for many years (limestone).

Concept
One object is harder than another if it scratches the surface of that object.

Objectives/Process Skills
- *Test* and *observe* objects for hardness by performing scratch tests.
- *Record* observations.
- *Classify* objects by their hardness.

Time Allotment
Allow 30 minutes.

Safety Tips (See page T24.)
- Tell students to scratch the chalk gently with the pencil. Students should lay the chalk down, not hold it, while they scratch it.
- Remind students to handle sharpened pencils safely. They should not poke themselves or other students.
- Warn students not to put any of the objects into their mouths.

Teaching Tip
- If mineral samples are available, let students test for hardness of various minerals using their fingernails, a penny, and a steel nail.

Answers
State Your Conclusion
1. Pencil lead
2. Chalk, pencil lead, fingernail, penny

Use What You Learned
Use the scratch test. The real diamond would scratch the glass. **Thinking Skill:** ***Applying information to new situations***

Testing Objects for Hardness

Suggested grouping: pairs

Purpose
Observe and test objects for hardness by doing a scratch test.

Gather These Materials
• pencil • chalk • penny

Follow This Procedure
1. Use a chart like the one shown to record your observations.
2. One substance is harder than another if it scratches the surface of that object. A mark that rubs off is not a scratch. Test the lead of a pencil to see if it can be scratched with your fingernail as shown in the picture. Write *yes* or *no* in the chart.
3. Test a piece of chalk and a penny to see if they can be scratched with your fingernail. Record your results.
4. Test the pencil lead and penny to see if they can be scratched by the chalk. Record your results.
5. Test the chalk and penny to see if they can be scratched by the pencil lead. Record your results.
6. Test the chalk and pencil lead to see if they can be scratched by the penny. Record your results.

Record Your Results

	Object to be tested	Chalk	Pencil lead	Penny
Scratched by	Fingernail	yes	yes	no
	Chalk	x	no	no
	Pencil lead	yes	x	no
	Penny	yes	yes	x

State Your Conclusions
1. Which is harder, pencil lead or chalk?
2. List the objects you tested in order of their hardness, with the softest first.

Use What You Learned
A diamond is the hardest mineral. If you had a real diamond and a fake diamond made of glass, how could you tell which one was real?

Activity Results
Chalk is scratched by the fingernail, pencil lead, and penny. Pencil lead is scratched by the fingernail and penny, but not by the chalk. The penny is not scratched by the fingernail, the pencil lead, or the chalk.

Resource Book page 103

Name ______
Use with Lesson 1: page 192
Testing Objects for Hardness

Chapter 9
Activity Worksheet

Record Your Results

	Object to be tested	Chalk	Pencil lead	Penny
Scratched by	Fingernail			
	Chalk			
	Pencil lead			
	Penny			

State Your Conclusion
1. Which is harder, pencil lead or chalk?

2. List the objects you tested in order of their hardness, with the softest first.

Use What You Learned
A diamond is the hardest mineral. If you had a real diamond and a fake diamond made of glass, how could you tell which one was real?

103

Entering the Diamond Age

Diamonds

The Problem Diamonds are valuable minerals made of carbon. People dig diamonds out of the ground. Diamonds make beautiful jewelry. You might be surprised to learn that diamonds are more than beautiful. They are also the hardest substance found in nature. Their hardness makes them very useful in industry. Diamonds cut, grind, and drill holes through hard metal. Diamonds also conduct heat and electricity well. The problem is that industry needs more diamonds than are found in nature.

The Breakthrough In 1955, a team of American scientists made the first artificial diamonds. They used a special press to squeeze carbon under great heat and pressure. Most of the artificial diamonds were very small. They worked well in industrial uses. In 1970, American scientists also made larger artificial diamonds.

New Technology Recently, scientists learned how to coat things with thin films of diamonds. The clear films are just as hard as real diamonds. They also have other properties of diamonds. Cutting tools coated with diamond films will work better than present tools made with diamonds. New tools with diamond films will last longer and are harder all over than present diamond tools. In the future, many products may have diamond films. Windows in airplanes might be covered with a diamond film. Raindrops hitting the windows at high speed would not scratch these windows. Diamond-coated parts inside the hottest parts of cars would not melt. They could keep track of how much gas the car used. Computer parts coated with diamond film would make computers work faster. Coating computer disks with diamond film would make disks last longer.

What Do You Think?

1. Why are diamonds useful for cutting in industry?
2. How could diamond films improve eyeglasses?

Science and Technology

TEACHING PLAN

Discussion

Hold a pencil next to a picture of a diamond. Question: **How are these two items alike?** (The "lead" in the pencil really is graphite, which also is a form of carbon.) Explain that the way the particles of carbon link together determines whether graphite or diamond forms.

Teaching Tips

- Explain that when researchers make diamond films, they carefully control conditions so that the carbon particles will form diamonds, not graphite.
- Point out that diamonds are excellent conductors of heat, so they are well suited for computer chips. Currently, chips are made of silicon. Diamond-coated chips could be placed closer together than silicon ones, thus speeding up the rate of computing.

Answers

What Do You Think?

1. Diamonds are hard enough to cut through metal. **Thinking Skills:** ***Restating or explaining ideas, Recognizing cause and effect***

2. Glasses could not be scratched. **Thinking Skill:** *Inferring*

Teaching Options

Science Background

Most diamond-coating processes use methane, a hydrocarbon that is the main component of natural gas. A methane molecule consists of a carbon atom surrounded by four hydrogen atoms. The different processes for diamond-coating involve stripping away the hydrogen atoms. Then the carbon atoms can link together on the surface of the object being coated in the precise crystalline structure of a true diamond film. In 1987, tiny diamonds were discovered in rocks from outer space (meteorites). Studying these diamonds may show us how to make diamonds more easily on the earth.

LESSON 2 pages 194–196

TEACHING PLAN

Lesson Objectives
- *Explain* how soil is formed.
- *Describe* clay soil, sandy soil, and loam.

Lesson Vocabulary
clay soil, decay, humus, loam, nutrient, sandy soil

1. Motivate

Demonstration Activity ♦
Shake some soil with water in a glass jar, and allow the mixture to settle. *CAUTION:* Immediately wipe up any water that spills on the floor to prevent slipping accidents. Have students ***observe*** the layers.

Discussion
Question: **What layers formed as the soil settled?** (Students should recognize heavier particles such as pebbles and sand on the bottom, finer particles of clay or silt above the sand, and muddy water with plant or animal matter on the top layer.)

2 What Is Soil Made Of?

LESSON GOALS

You will learn
- how soil is formed.
- how the three kinds of soil differ.

What do you notice about the size of this rock?[1] Over thousands of years, the rock will slowly crumble and become part of the soil.

How Soil Is Formed

How can a large solid rock break into small pieces? If you look closely at a rock, you will see many tiny holes and cracks. When it rains or snows, water fills the holes and cracks. This water freezes and pushes against the rock. As the water continues to freeze and melt, the rock slowly breaks apart. Over many years, the water causes the rock to break into small pieces and form soil.

[1]large size

194

Teaching Options

Science Background

Several factors affect the formation of soil in a given area. For example, the type of bedrock from which the soil forms determines the mineral composition of the soil. The amount of time during which weathering has taken place affects how much soil is formed. The climate and slope of the land also affect how fast and how thick of a layer the soil forms. Soil forms faster in areas with warm, humid climates, and forms thicker layers on flat land than in mountainous regions, generally speaking.

Reading Strategies ♦

1. Guide students' pre-reading by asking: What ideas do you think you would find in the lesson to answer the question-title?
2. Assign these strategies: Writing a Memory Sentence and Mapping Examples of three kinds of soil. (See pages T26-T29.)
3. Pair students to share what information is clear and unclear and initiate discussion using students' unanswered questions.

♦ ***Suitable as a language development activity***

What can you notice about the rock in the picture on the right?[1] Soil can collect in the cracks of a rock. Plants can grow in this soil. Then plant roots can break rocks apart. Over many years, the roots can help crumble a large rock.

Soil is made of more than rock. Soil also has air, water, and matter that was once alive. When plants and animals die, they **decay,** or rot. The decayed matter—**humus**—becomes part of the soil. Humus gives soil a dark color and adds **nutrients** to the soil. Nutrients are materials plants need to live and grow.

Kinds of Soil

The soil where you live might be very different from soil in other places. You might have noticed that soil can have different colors. Soil also can have pieces of different sizes. Soil can feel very soft or very hard.

SCIENCE IN YOUR LIFE

When a rock breaks apart, its minerals become part of the soil. Some minerals can give the soil a certain color. The mineral iron can give soil a red color.

decay (di kā′), to slowly break down, or rot.

humus (hyü′məs), the decayed matter in soil.

nutrient (nü′trē ənt), a material that plants and animals need to live and grow.

[1] plants growing in it

Plant growing in cracks in rock

Rocks help form soil.

2. Teach

Teaching Tips

- **Possible Misconception:** Some students may think soil extends deep into the earth. Point out that soil is a relatively thin layer that lies above a very thick layer of rock.
- Place several samples of "soft" rock, such as shale or sandstone, into a can with a tight lid. Have students pass the can around and ask each student to shake the can vigorously. Then, open the can and allow students to ***observe*** the rocks. Question: **Why is shaking the rocks like the action of water and wind on rocks?** (both actions wear down rocks)
- If possible, have a sample of humus available for students to ***observe***.

Science and Social Studies

Ask students to each choose a different country to research the types of soil—good farm land, rocky soil, sand—found around the world. Help students use reference materials to find out the types of soil found in each country. Encourage students to make maps using shading or different colors to indicate the different soils in various areas.

Special Education

Obtain samples of clay soil, sandy soil, and loam. Allow students with learning disabilities to feel and dig into each sample to ***compare*** the differences between the various types. Have students pour water on the samples to ***observe*** how each type of soil holds water. *CAUTION:* Immediately wipe up any water that spills on the floor. Direct students to wash their hands thoroughly after this activity.

Reinforcement

Obtain a jar large enough for the students to see from around the room. Place rocks in the bottom, then sand, some potting soil, grass or other plants. Shake the jar gently several times so that the different materials blend slightly. Hold the jar up for the students to see. Point out that the soil in the jar begins as rock, becomes smaller, is mixed with plant matter called humus, and finally becomes soil.

TEACHING PLAN

3. Assess

Lesson Review

1. Soil forms as rocks break apart. As dead plants and animals decay, they also become part of the soil.
2. Clay soil has tiny grains, holds water, and has nutrients. Sandy soil has larger grains, does not hold water, and has few nutrients. Loam is a mixture of clay, sand, and humus. It holds water and has nutrients. Plants grow well in loam but not in clay or sandy soil.
3. Challenge! The roots might push against the sidewalk and break the sidewalk apart. **Thinking Skills:** *Applying information to new situations, Making inferences*

Find Out On Your Own

The uppermost layer, topsoil, is rich in humus. Topsoil holds water long enough for plants to absorb moisture, but it also allows excess moisture to drain away. Plants grow well in topsoil. **Thinking Skills:** *Collecting information, Communicating*

Clay soil

Sandy soil

Loam

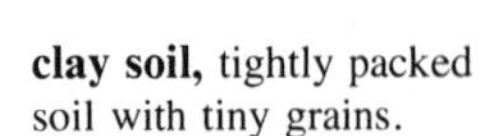

clay soil, tightly packed soil with tiny grains.

sandy soil, loose soil with large grains.

loam (lōm), good planting soil that is a mixture of clay, sand, and humus.

Compare the three kinds of soil in the pictures. Tiny grains in **clay soil** make it feel smooth. Clay soil has many nutrients and holds water well. Many plants cannot grow in clay soil because the grains are so close together.

Sandy soil is loose and easy to dig. Most plants do not grow well in sandy soil. Sand does not hold water, and it has few nutrients.

Loam is a mixture of clay, sand, and humus. Many plants grow well in this soil. Loam holds water and has many nutrients for plants.

Lesson Review

1. How is soil formed?
2. How do the three kinds of soil differ?
3. **Challenge!** What might happen to a sidewalk that has a tree growing near it?

Study on your own, pages 332–333.

PHYSICAL SCIENCE
FIND OUT ON YOUR OWN
CONNECTION

The part of the soil you usually see is called topsoil. Look in library books to find out about the properties of topsoil. Write a few sentences explaining why topsoil is important.

196

Teaching Options

Science Anecdote

Terra cotta is a brownish-red clay that is used to make pottery, statues, and building decorations. The use of terra cotta has been dated to prehistoric times. Terra cotta figurines about 5,000 years old were found in Greece.

Reteaching Suggestion ♦

Divide students into small groups. Provide each group with samples of clay soil, sandy soil, and loam in clay pots. Allow each group to plant one plant in each pot. Instruct the groups to label their pots and place them near sunshine. Ask students to *infer* in which soil the plants will grow best. (loam) Have students *measure* and *compare* the plant growth in each type of soil. Discuss the results.

Workbook page 51 *

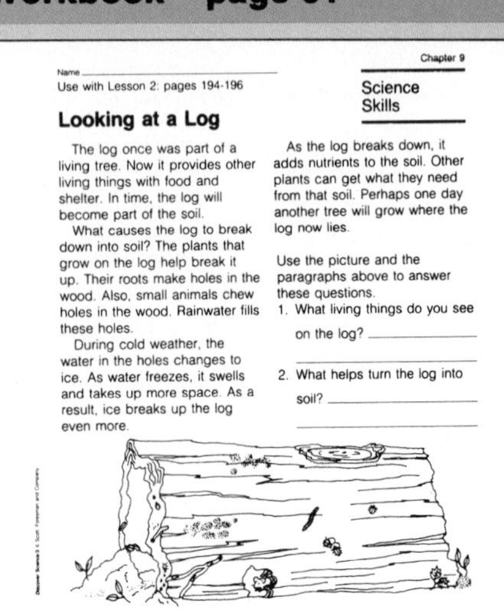
Chapter 9

Name ____________

Use with Lesson 2: pages 194-196

Science Skills

Looking at a Log

The log once was part of a living tree. Now it provides other living things with food and shelter. In time, the log will become part of the soil.

What causes the log to break down into soil? The plants that grow on the log help break it up. Their roots make holes in the wood. Also, small animals chew holes in the wood. Rainwater fills these holes.

During cold weather, the water in the holes changes to ice. As water freezes, it swells and takes up more space. As a result, ice breaks up the log even more.

As the log breaks down, it adds nutrients to the soil. Other plants can get what they need from that soil. Perhaps one day another tree will grow where the log now lies.

Use the picture and the paragraphs above to answer these questions.

1. What living things do you see on the log? ____________
2. What helps turn the log into soil? ____________

51

♦ *Suitable as a language development activity*

Observing Soil Samples

Suggested grouping: 1–2 students

Purpose
Observe soil samples to see what they are made of.

Gather These Materials
- 4 soil samples in paper cups
- white construction paper
- toothpick • hand lens • plastic spoon

Follow This Procedure
1. Use a chart like the one shown to record your observations.
2. Put two spoonfuls of soil from each sample on the white paper.
3. Use the toothpick to separate one soil sample into different materials. Look for light pieces, dark pieces, and soft material. Pile like pieces together.
4. Label each pile with a letter. Carefully look at each pile with a hand lens, as shown in the picture. Look at the color, hardness or softness, size, and shape of the bits in each pile. Record your observations.
5. Repeat steps 3 and 4 with the other three soil samples.

Record Your Results

Sample part	a	b	c	d
Color				
Hard or soft				
Size (large, medium, or small)				
Shape				

Answers will vary, depending on types of soil.

State Your Conclusions
1. How many different kinds of bits did you find in the first soil sample?
2. Did one of your piles have humus? How do you know?

Use What You Learned
Which kinds of soil were in your samples? Would plants be able to grow well in your soil samples?

197

Concept
Different kinds of soil are composed of different kinds of particles.

Objectives/Process Skills
- *Observe* soil samples.
- *Sort* particles in each sample.
- *Compare* soil samples.

Time Allotment
Allow 20–30 minutes.

Safety Tips (See page T24.)
- Instruct students to wash their hands thoroughly after completing the activity.
- Advise students to handle the pointed toothpicks safely, keeping them away from the face and from other students.

Teaching Tips
- Students could be asked to bring in soil samples from home, or teacher can provide samples.
- Put soil samples in small paper cups for ease in handling.

Answers
State Your Conclusion
1. Answers will vary according to the soil sample.
2. Answers will vary, but most likely the sample will contain humus. Dark soil would indicate the presence of humus.

Use What You Learned
Answers will vary. Loam soil would allow plants to grow well. Sandy soil would not allow most plants to grow well. Many plants would not grow well in clay soil.
Thinking Skill: *Predicting*

Resource Book page 105

Name ______ Chapter 9
Use with Lesson 2: page 197
Activity Worksheet
Observing Soil Samples

Record Your Results

Sample Part	a	b	c	d
Color				
Hard or soft				
Size (large, medium, or small)				
Shape				

State Your Conclusion
1. How many different kinds of bits did you find in the first soil sample?
2. Did one of your piles have humus? How do you know?

Use What You Learned
Which kinds of soil were in your samples? Would plants be able to grow well in your soil samples?

105

Activity Results

Results will vary according to soil samples students have. The piles might include large or small particles, hard or soft pieces, clay soil, sandy soil, or loam.

* ***Answers to masters on pages 186E–186H***

TEACHING PLAN

Lesson Objectives

- *Describe* how people use rocks and soil.
- *Explain* the importance of using natural resources wisely.

Lesson Vocabulary

natural resource, ore, recycle

1. Motivate

Demonstration Activity ♦

Obtain a stalk of celery wrapped in aluminum foil and a salt shaker. Unwrap the celery and lightly salt it.

Discussion

Point out that everything used in the demonstration either came from minerals (aluminum foil and the salt shaker—if it is glass), is a mineral (salt), or grows in soil that contains minerals (celery). Question: **What are some other ways you use soil, rocks, and minerals?** (Accept any reasonable answers.)

3 How Do People Use Rocks and Soil?

LESSON GOALS

You will learn
- how people use rocks and soil.
- the importance of using natural resources wisely.

ore (ôr), rock that is rich in useful minerals.

Objects that came from rocks and soil

This glass is made of minerals that came from rocks. The juice and the salad came from plants that grew in soil. Every day, people use objects that came from rocks and soil.

How Rocks and Soil Are Important

People eat plants that grow in soil. People get meat from animals that eat plants. Some materials in clothing, such as cotton, also come from plants. The paper in this book was made from trees that grew in soil. What other ways do you use objects that came from soil?[1]

Rocks are important to people too. Some rocks—called **ores**—have large amounts of useful minerals. People can separate the minerals from the rocks. The ore in the picture has the mineral copper. People use copper to make coins, electric wires, and pots and pans.

Copper

[1]Answers might include cotton clothing, wooden furniture, maple syrup or other foods.

198

Teaching Options

Science Background

Deposits of metal ores and other important minerals are formed in many ways. Sometimes, water and air break down rocks, changing their chemical composition. Water can dissolve minerals from rocks, and dissolved minerals from scattered areas can collect in underground streams. If the groundwater concentrates the minerals in a small area and then dries up, ore deposits remain. Salty seas also can evaporate and leave behind minerals. Deposits of rock salt, gypsum, and other minerals have resulted from the evaporation of salty seas.

Reading Strategies ♦

1. Guide students' pre-reading by asking: What ideas do you think you would find in the lesson to answer the question-title?
2. Assign these strategies: Writing a Memory Sentence and Finding and Writing Sentences for the Vocabulary Words. (See pages T26-T29.)
3. Pair students to share what information is clear and unclear and initiate discussion using students' unanswered questions.

♦ *Suitable as a language development activity*

Open-pit mine

Some ores have more than one kind of useful mineral. For example, an ore might have the minerals lead and silver. Lead is used to make batteries and some kinds of pipes. People use silver to make film, jewelry, and mirrors.

Sometimes ores are found deep under the ground. Miners must dig deep tunnels to get the ores out. Other times ores are found near the earth's surface. Miners remove the soil and top layers of rock to reach the ores. As you can see in the picture, this kind of mining leaves a large pit in the ground. Miners must fill in the pit after the ores are mined. Then, people can use the land again.

2. Teach

Teaching Tips

- Question: **What items in the classroom are made from things grown in soil?** (wood desks, chairs, cotton clothing, paper, and so on) Have students *identify* items in the classroom that are made from minerals. (jewelry, cans, other metal objects)
- If available, show students other pictures of mining methods and machinery, such as open pit sandstone or limestone quarries, or coal mines.
- If any minerals are mined in your area, trace the ore from the mine to the final product.

Science and Art

Challenge students to create recycling posters. Remind students that posters should both attract attention and provide information. Display students' posters in the classroom and school building.

Reinforcement

Ask students to *list* items found in their homes that either come from minerals or from things grown in soil. Use students' lists to discuss the importance of minerals and soil in daily life.

Special Education

Have students with learning disabilities make a collage of pictures of items that illustrate how rocks and soil are used daily.

TEACHING PLAN

Teaching Tips

- Have students name items that they threw away today. (Students may mention paper, food, plastic, broken metal toy parts, and so on.) List students' responses on the board. Use the list to *compare* and *contrast* which items can be recycled.
- Question: **How many students have taken part in recycling projects, such as paper, aluminum can, or glass drives?** Review the processes involved in recycling each of these materials.

Investigate!

Accept any testable hypothesis. Most students' data will support this hypothesis: *Adding fertilizers to soil helps plants grow.* Students should find out that plants grow well in soil that has large amounts of nutrients.

natural resource (nach′ər əl ri sôrs′), something people use that comes from the earth.

recycle (rē si′kəl), to change something so it can be used again.

Using Natural Resources Wisely

A **natural resource** is something people use that comes from the earth. Minerals and soil are two kinds of natural resources. The earth has a limited amount of natural resources. People must be careful not to waste them.

Some materials can be used over and over again—or **recycled.** People use the machines in these pictures to recycle aluminum (ə lü′mə nəm) cans. Aluminum is a metal found in some minerals. The machine chops the used cans into small pieces. Then, the pieces are melted and rolled into sheets. People use the aluminum sheets to make new cans.

People can recycle some materials

200

Teaching Options

Science Anecdote

The United States is the world's largest manufacturer of fertilizer. Every year, it manufactures over 5 billion dollars worth of fertilizer.

Enrichment

Encourage interested students to locate recycling projects in your area. Have students find out the procedures involved in collecting and delivering materials to these centers.

Workbook page 52 *

Name ________________

Use with Lesson 3: pages 198-201

Chapter 9

Science and Social Studies

Using Soil to Make Houses

Read the story. Then answer the questions.

In some places, people live in houses made from soil. The soil is made into bricks. The house is made from these bricks.

Making bricks from soil is not a new idea. Thousands of years ago, ancient Egyptians used bricks made from soil. Indians in the American Southwest used bricks from soil to build homes, called pueblos. Pueblos are like small apartment buildings.

Today, many houses in the Southwest and in Mexico are made with adobe bricks. Adobe is the Spanish name for bricks made of soil. Adobe houses are cooler than wood or stone houses. Adobe houses help people keep cool in the hot Southwest and Mexico.

Making adobe bricks takes time. First, sandy clay is mixed with water and straw or grass. Then this mixture is poured into wooden frames. When the bricks are dry, the frames are taken away. The bricks are then placed in the hot sun to bake for two weeks.

1. Who used bricks made of soil to build homes long ago?

____

2. Today, where might you see a house made from adobe bricks?

____

3. What materials are used to make adobe bricks?

____

4. How are adobe bricks made hard?

____

52

♦ *Suitable as a language development activity*

Field of peas

People can grow plants in ways that use soil wisely. Some plants, such as these peas, add nutrients to the soil. Different plants use up different nutrients in the soil. People can change the kinds of plants they grow each year. Then the lost nutrients have time to build up again in the soil.

INVESTIGATE!

Fertilizers have nutrients that plants need. Find out how fertilizers affect the growth of plants. Write a hypothesis and test your hypothesis with an experiment. You might try growing one plant with fertilizer and another one without fertilizer.

Lesson Review

1. How are rocks and soil important?
2. Why should people not waste natural resources?
3. **Challenge!** Paper is made from trees. How do people use natural resources wisely when they take newspapers to be recycled?

Study on your own, pages 332–333.

EARTH SCIENCE

FIND OUT ON YOUR OWN

Think of one way you can avoid wasting minerals or soil. Write your idea on a piece of paper. Then, make a poster that shows your idea for using minerals or soil wisely.

3. Assess

Lesson Review

1. Answers may include: people use the soil when they grow plants or animals for food, or use paper. People use minerals from rocks and soils to make useful items such as coins, jewelry, scissors, and so on.

2. The earth has a limited amount of natural resources.

3. Challenge! When people return newspapers, the newspapers are reused. Newspapers are made of plants that use minerals from the soil. Reusing newspapers is one way of not wasting minerals. **Thinking Skills:** *Applying information to new situations, Making inferences*

Find Out On Your Own

Answers might include reusing aluminum foil wrap, saving newspapers for recycling, and reusing planting soil and adding fertilizer to it. **Thinking Skills:** *Suggesting alternatives, Communicating*

Workbook page 53 *

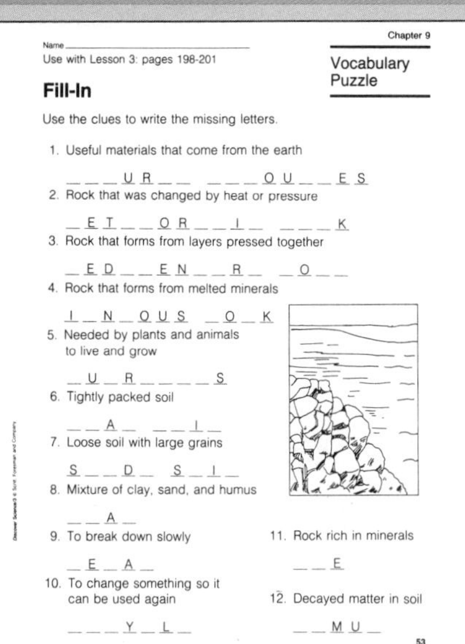
Name ____________

Use with Lesson 3: pages 198-201

Chapter 9

Vocabulary Puzzle

Fill-In

Use the clues to write the missing letters.

1. Useful materials that come from the earth
   _ _ _ U R _ _   _ _ _ O U _ _ E S
2. Rock that was changed by heat or pressure
   _ E T _ _ O R _ _ I _   _ _ _ K
3. Rock that forms from layers pressed together
   _ E D _ _ E N _ _ R _   _ O _ _
4. Rock that forms from melted minerals
   I _ N _ O U S   _ O _ K
5. Needed by plants and animals to live and grow
   _ U _ R _ _ _ _ S
6. Tightly packed soil
   _ _ A _   _ _ I _
7. Loose soil with large grains
   S _ _ D _   S _ I _
8. Mixture of clay, sand, and humus
   _ _ A _
9. To break down slowly
   _ E _ A _
10. To change something so it can be used again
    _ _ _ Y _ L _
11. Rock rich in minerals
    _ _ E
12. Decayed matter in soil
    _ _ M U _

53

*** Answers to masters on pages 186E–186H***

Game Suggestion ♦

Divide students into small groups. Assign a material, such as glass, aluminum, paper, wood, iron, steel, and so on, to each group. Challenge each group to invent a new use for their recycled material.

Reteaching Suggestion ♦

Work with students to *describe* the changes an aluminum can goes through from mining to recycling. (The ore is mined in an underground mine, processed into metal, manufactured into a can at a factory, and recycled into aluminum to be reused to make more cans.)

TEACHING PLAN

Purpose
To develop the skills of collecting and organizing information using graduated cylinders and bar graphs to solve problems.

1. Motivate

Discussion
Display graduated cylinders and other containers used to measure liquids, such as measuring cups. Discuss how these items are used. Question: **What do you use to measure a liquid?** (measuring cups or containers of known volume)

2. Teach

Teaching Tips

- Demonstrate how to measure precisely using a graduated cylinder. Hold the cylinder at eye level and read the line even with the bottom of the curved surface (meniscus) of the liquid.
- To keep the soil in the funnel, plug the neck of the funnel with cotton.

Using Graduated Cylinders and Bar Graphs

Problem: How much water will pass through different kinds of soil?

Part A. Using Graduated Cylinders to Collect Information

1. Graduated cylinders measure volume. Each line on these cylinders stands for 10 mL.
2. Each funnel has 100 cc of soil in it. When 100 mL of water is poured into a funnel, the water soaks the soil. Extra water goes through the soil into the cylinder. Look at the line on the cylinder that measures the level of the water. How many mL of water passed through the sandy soil? the clay soil? the loam soil?

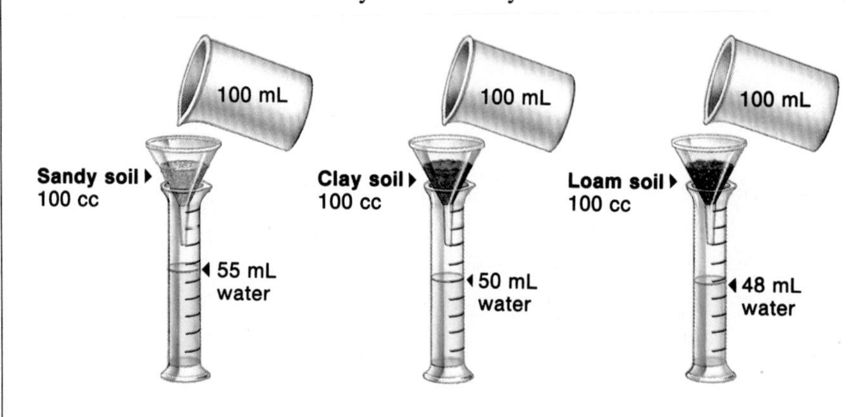

Part B. Using a Bar Graph to Organize and Interpret Information

3. This bar graph contains the information you collected about how much water will pass through soil. Each bar shows how much water passed through one kind of soil. What does each line on the scale on the left stand for?

Teaching Options

Sample Bar Graph for Part C

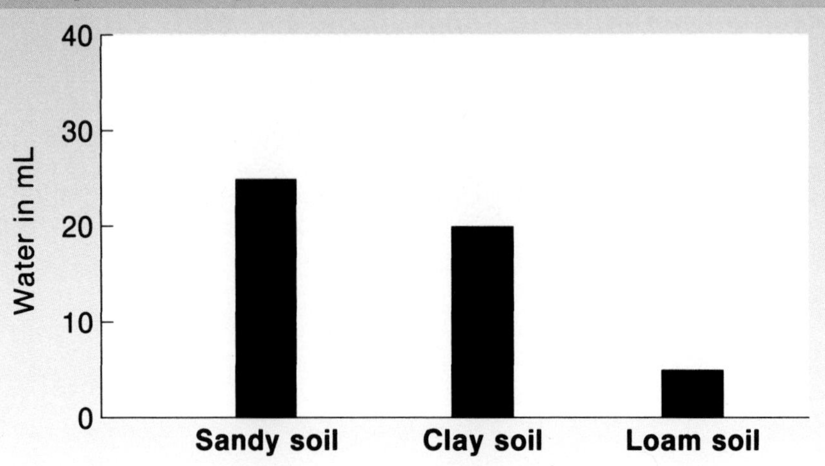

♦ *Suitable as a language development activity*

4. The first bar shows how much water passed through the sandy soil. Trace with your finger from the top of the bar across to the scale on the left. The bar shows that 55 mL of water ran into the cylinder. How much water ran into the second cylinder? the third?

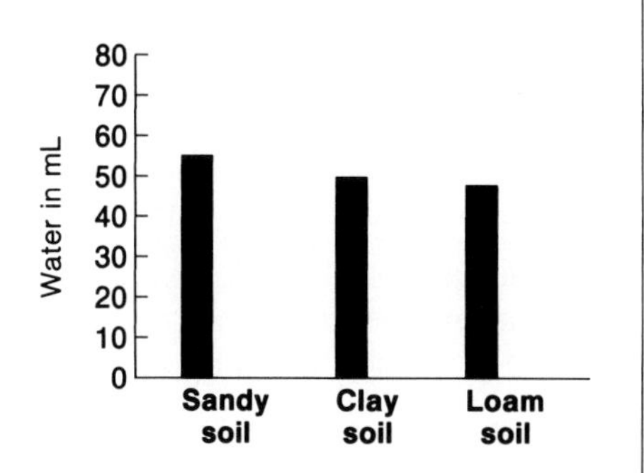

Part C. Using Graduated Cylinders and a Bar Graph to Solve a Problem

Problem: Does changing the amount of soil in the funnel affect how much water passes through it?

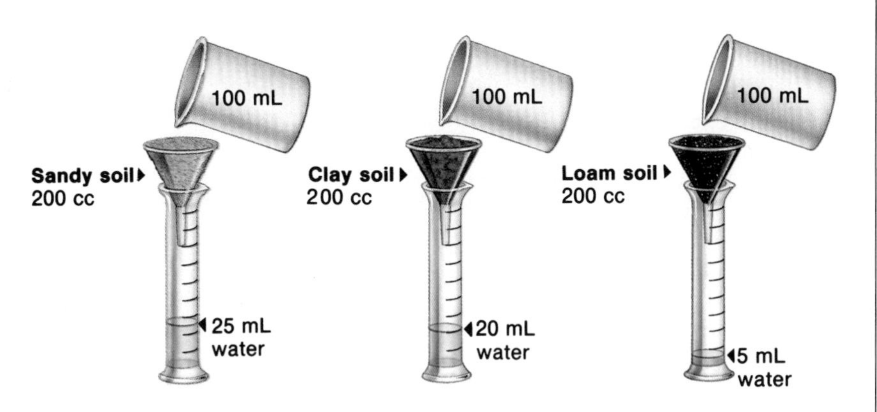

5. Use the graduated cylinders to collect the information you need to solve the problem. Make a bar graph similar to the one shown in Part B to organize your information.
6. Look at your bar graph. Does changing the amount of soil in the funnel affect how much water passes through it?
7. You might want to do this experiment and use your own results to make a bar graph.

203

Teaching Tip

- Explain to students that volume measures the number of cubic units an object will hold. In these experiments, the volume of soil is measured in cubic centimeters, which is abbreviated cc.

3. Assess

Part A

1. No answer is required.
2. 55 mL; 50 mL; 48 mL

Part B

3. 10 mL of water
4. 50 mL; 48 mL

Part C

5. See Sample Bar Graph for Part C in Teaching Options.
6. Increasing the amount of soil in the funnel results in less water passing through it.
7. You might also allow students to try the activity using different amounts of the same kinds of soil in the funnel.

Reteaching Suggestion ♦

Use graduated cylinders and funnels to test how much water passes through soil collected from outside or from potting soil you bring to class. Pour 100 ml of water through 100 cc, 200 cc, and 300 cc of soil. Draw a bar graph on the board similar to the one that appears on page 203, but mark each specimen according to its cubic volume. Have students draw the bars according to the measurements. Then help them *conclude* that less water passes through greater volumes of soil.

Resource Book page 107 *

Name ____________

Comprehension: details

Chapter 9

Science and Reading

Only the Facts

Read the article. Then write a sentence to answer each question.

Landfills are places where garbage is dumped. Today many of our landfills are full or nearly full. Some towns are helping by having garbage recycling programs. Materials such as glass, cans, and newspaper can be recycled and used again.

There are two kinds of recycling programs. Some towns have recycling centers. People drop off their materials to be recycled at these centers. Other towns have curbside recycling programs. People are given bins in which to collect materials for recycling. Then the bins are put out by the curb, and the materials are collected.

1. How are towns helping with the landfill problem?

____

2. What are two kinds of recycling that towns have?

____

3. How does a curbside recycling program work?

____

107

* **Answers to masters on pages 186E–186H**

REVIEW PLAN

Reviewing Science Words

1. metamorphic rock
2. sandy soil
3. decays
4. loam
5. nutrient
6. recycle
7. igneous rock
8. clay soil
9. sedimentary rock
10. natural resources
11. humus
12. ore

Reviewing What You Learned

1. d	**3.** a	**5.** b	**7.** d
2. b	**4.** c	**6.** a	**8.** d

Chapter 9 Review

Chapter Main Ideas

Lesson 1 • Rocks are made of one or more minerals. • Igneous rocks, sedimentary rocks, and metamorphic rocks are three kinds of rocks that formed in different ways.

Lesson 2 • Rock can break apart into small pieces and form part of the soil. Soil also has air, water, and decayed matter. • Three kinds of soil are clay soil, sandy soil, and loam.

Lesson 3 • Many objects people use come from rocks and soil. • People should use natural resources wisely.

Reviewing Science Words

clay soil	loam	ore
decays	metamorphic rock	recycle
humus	natural resources	sandy soil
igneous rock	nutrient	sedimentary rock

Copy each sentence. Fill in the blank with the correct word from the list.

1. A rock becomes a ___ when heat or pressure changes the minerals in an igneous or sedimentary rock.
2. ___ is loose soil with large grains.
3. A plant ___ when it slowly breaks down, or rots.
4. ___ is a mixture of clay, sand, and humus.
5. A ___ is a material in the soil that a plant needs to grow.
6. People can ___ some kinds of minerals so that they can be used again.
7. A rock that is formed from melted minerals is an ___.
8. Smooth soil with tiny grains is ___.
9. ___ is formed when layers of material are pressed together.
10. Minerals and soil are two kinds of ___.
11. ___ is decayed matter in soil.
12. A rock that has a large amount of useful minerals is an ___.

204

Review Options

Cooperative Learning ♦

STAD Format (See page T23.)
Assign students to work in four- to five-member teams to study Chapter 9 Review. Students should work together to make sure that they and their teammates know the material in the chapter. After students have had enough time to study together, give them a test to complete individually (Chapter 9 Test A or B in the *Test Book*). Award Superteam certificates to teams whose average test scores exceed 90%, and Greatteam certificates to teams whose average test scores exceed 80%.

Test Book page 81 *

Name ______________

Chapter 9
Test A

Multiple Choice. Choose the best answer.

1. Rocks are made of one or more
   a. carbons.
   b. minerals.
   c. nutrients.
2. Rocks formed from melted minerals that have cooled are
   a. sedimentary rocks.
   b. igneous rocks.
   c. metamorphic rocks.
3. Igneous or sedimentary rocks that are changed by heat or pressure are
   a. recycled rocks.
   b. metamorphic rocks.
   c. fossils.
4. Freezing and melting water causes rocks to break slowly into small pieces and forms
   a. carbons.
   b. soil.
   c. fossils.
5. Materials from the soil that plants need to grow are
   a. clays.
   b. sediments.
   c. nutrients.
6. Soil that is a mixture of clay, sand, and humus is called
   a. clay soil.
   b. sandy soil.
   c. loam.
7. Rocks that have large amounts of useful minerals within them are
   a. mines.
   b. ores.
   c. loams.
8. A mineral used to make film, jewelry, and mirrors is
   a. copper.
   b. silver.
   c. granite.
9. Minerals and soil are two things people use that are
   a. natural resources.
   b. igneous rocks.
   c. nutrients.
10. What kind of soil has few nutrients?
    a. loam.
    b. humus.
    c. sandy soil.

81

♦ ***Suitable as a language development activity***

Reviewing What You Learned

Write the letter of the best answer.

1. Most plants grow well in
   (a) sandy soil. (b) clay soil. (c) ores. (d) loam.
2. The soft black material in a pencil tip comes from
   (a) a plant. (b) a rock. (c) sandy soil. (d) humus.
3. The grains in clay soil are
   (a) close together. (b) far apart. (c) very large.
   (d) loose and easy to dig.
4. Miners remove soil and top layers of rock to reach
   (a) igneous rocks. (b) loam. (c) ores. (d) clay.
5. What kind of rocks can form in layers under water?
   (a) igneous (b) sedimentary (c) humus (d) loam
6. Which plant parts often break rocks apart?
   (a) roots (b) stems (c) flowers (d) leaves
7. Which kind of rock is marble?
   (a) sedimentary (b) ore (c) igneous
   (d) metamorphic
8. Which kind of soil has few nutrients?
   (a) loam (b) clay soil (c) humus (d) sandy soil

Interpreting What You Learned

Write a short answer for each question or statement.

1. How does humus help plants grow?
2. Explain how water can break a large rock apart.
3. Describe the steps in recycling an aluminum can.
4. How can a farmer grow plants to use soil wisely?
5. How does clay soil differ from sandy soil?

Extending Your Thinking

Write a paragraph to answer each question or statement.

1. Which would dry faster after a rainstorm, a sandy beach or a field with many different plants? Explain your answer.
2. Most minerals are mined only in places where they are found in large amounts. Give one possible reason for this.

To explore scientific methods, see Experiment Skills on pages 364–365.

Interpreting What You Learned

1. Humus has nutrients that plants need to live and grow. **Thinking Skill:** *Recognizing main idea and supporting details*
2. Water from rain or melted snow can get into the cracks and holes of a rock. Water expands as it freezes, and can break apart the rock. **Thinking Skill:** *Restating or explaining ideas*
3. A machine chops the can into pieces, the pieces are melted and rolled into sheets, the sheets are made into new cans. **Thinking Skill:** *Sequencing*
4. If the farmer changes the kinds of plants every few years, the soil will not always lose the same nutrients. Some plants add nutrients to soil. **Thinking Skill:** *Restating or explaining ideas*
5. Clay soil is smooth, has tiny grains that are close together, holds water well, and has many nutrients. Sandy soil has large grains, is loose, does not hold water well, and has few nutrients. **Thinking Skill:** *Contrasting*

Extending Your Thinking

1. The sandy beach would dry faster because sand does not hold water well. A field with many plants has good planting soil that holds water. **Thinking Skills:** *Applying information to new situations, Inferring*
2. It would cost too much money and involve too much work to dig a mine for only a small amount of most minerals. **Thinking Skill:** *Inferring*

Test Book page 82 *

Name ______________

Chapter 9
Test A

Matching Match the correct definition and term.

Short Answer Use the pictures to answer the questions.

marble sedimentary granite graphite

1. Which rock is a igneous rock used to make buildings? ________
2. Which rock is used to make pencil tips? ________
3. Which rock formed when layers of material were pressed together? ________
4. Which rock is a metamorphic rock? ________

Short Essay Use complete sentences to answer each question.

1. Explain why soil is important for people.
2. Explain how miners reclaim, or restore, the land after it has been mined.

82

Test Book page 83 *

Name ______________

Chapter 9
Test B

Multiple Choice. Choose the best answer.

1. The soft material in a pencil tip comes from
   a. a plant.
   b. a rock.
   c. soil.
2. The colors of rocks come from different
   a. kinds of soil.
   b. minerals.
   c. plants.
3. When melted minerals move to the earth's surface and harden, they are
   a. igneous rocks.
   b. sedimentary rocks.
   c. metamorphic rocks.
4. Rocks formed when layers of material are pressed together are called
   a. igneous rocks.
   b. sedimentary rocks.
   c. metamorphic rocks.
5. Large rocks can be slowly broken into small pieces by water or
   a. sandy soil.
   b. plant roots.
   c. decayed matter.
6. Decayed matter that becomes part of the soil is called
   a. sand.
   b. sediments.
   c. humus.
7. Tightly packed soil with tiny grains is
   a. clay soil.
   b. sandy soil.
   c. loam.
8. A mineral used to make some kinds of pipe and used in batteries is
   a. granite.
   b. limestone.
   c. lead.
9. Materials that are used over and over again are
   a. wasted.
   b. recycled.
   c. decayed.
10. Changing the kinds of plants grown in the soil every few years replaces lost
    a. nutrients.
    b. ores.
    c. sand.

83

Test Book page 84 *

Name ______________

Chapter 9
Test B

Short Answer Use the pictures to answer the questions.

1. Which rock is a metamorphic rock? ________
2. Which rock is made of soft carbon? ________
3. Which rock was formed in layers under water? ________
4. Which rock is a hard igneous rock? ________

marble sedimentary granite graphite

Short Essay Use complete sentences to answer each question.

1. Name two ways people use objects that come from the soil.
2. Define a natural resource. Give two examples.

84

* *Answers to masters on pages 186E–186H*

Chapter 10 Changes in the Earth Planning Guide

TEACHING PLAN

Chapter Components	Skills	Materials
Chapter Opener/*DISCOVER*: Observing Drops of Water pp. 206–207	*DISCOVER* p. 207 Science Process Skills *Observing, Making models*	*DISCOVER* p. 207 (groups of 2) 1 roll waxed paper, 1 bag sand, 15 medicine droppers
Lesson 1 What Is the Inside of the Earth Like? pp. 208–210	Thinking Skills Challenge!: *Inferring* Find Out On Your Own: *Recognizing the main idea and supporting details*	Demonstration p. 208 peach, knife
Activity Inferring Shapes from Sounds p. 211	Science Process Skills *Observing, Collecting and interpreting data, Inferring*	(groups of 2) 15 small boxes with lids, assorted small objects to place in boxes, 1 roll masking tape
Lesson 2 How Do Water and Wind Change the Earth's Crust? pp. 212–215	Thinking Skills Challenge!: *Making generalizations* Find Out On Your Own: *Collecting information, Communicating, Restating or explaining ideas*	Demonstration p. 212 sugar cubes, small container with fitted lid
Activity Making a Model of Erosion p. 216	Science Process Skills *Observing, Collecting and interpreting data, Communicating*	(groups of 5) 180 small rough stones, 12 coffee cans with lids, 12 large clear-plastic cups, 1 roll paper towels, 1 roll masking tape, 6 marking pens, water
Science and Technology Predicting Earthquakes p. 217	Thinking Skills *Comprehending meaning, Drawing conclusions*	
Lesson 3 How Do Earthquakes and Volcanoes Change the Earth's Crust? pp. 218–221	Thinking Skills Challenge!: *Recognizing cause and effect* Find Out On Your Own: *Interpreting charts, maps, and graphs; Collecting information*	Demonstration p. 218 pail, water, light-weight floating object
Lesson 4 How Do Living Things Change the Earth's Crust? pp. 222–223	Thinking Skills Challenge!: *Recognizing cause and effect* Find Out On Your Own: *Visualizing, Collecting information, Drawing conclusions*	Demonstration p. 222 clear-plastic tray, rocks, soil, small object to represent a house
Skills for Solving Problems Using Balances and Line Graphs pp. 224–225	Problem Solving Skills *Making decisions/Identifying and solving problems, Interpreting charts, maps, and graphs*	
Chapter Review pp. 226–227	Thinking Skills *Recognizing cause and effect, Inferring, Drawing conclusions, Predicting*	

Teaching Options

Strategies	Extensions		Resource Masters
Cooperative Learning p. 206 (Also see p. T23.) Applying Whole Language p. 207 (Also see p. T30.)			Family Letter: *Resource Book* p. 111
Reading Strategies p. 208 (Also see pp. T26–T29.)	Game Suggestion p. 209 Special Education p. 209 Reteaching Suggestion p. 210		Vocabulary Preview: *Workbook* p. 55 Science Activity: *Workbook* p. 56
			Activity Worksheet: *Resource Book* p. 115
Reading Strategies p. 212 (Also see pp. T26–T29.)	Enrichment pp. 213, 215 Reinforcement p. 213 Science and Geography p. 210	Special Education p. 214 Reteaching Suggestion p. 215	Science Skills: *Workbook* p. 57
			Activity Worksheet: *Resource Book* p. 117
Reading Strategies p. 218 (Also see pp. T26–T29.)	Enrichment pp. 219, 220, 221 Special Education p. 219	Reinforcement pp. 219, 220 Reteaching Suggestion p. 221	Science and Reading: *Workbook* p. 58
Reading Strategies p. 222 (Also see pp. T26–T29.)	Special Education p. 223 Reteaching Suggestion p. 223		Vocabulary Puzzle: *Workbook* p. 59
			Science and Reading: *Resource Book* p. 119
Cooperative Learning p. 226 (Also see p. T23.)			Chapter Tests: Forms A and B *Test Book* pp. 89–92

Classroom Management

Advanced Preparation

Activity, page 211
Collect enough shoeboxes with lids to have one for each group of students.

Activity, page 216
Collect 30 small, rough stones for each group of students. Limestone or sandstone will work well.

Vocabulary Review

Use the following sentences with your students to review the meanings of the italicized words.

1. Some cakes have two or three *layers.*
2. Knives, forks, and spoons are usually made of *metal.*
3. A thermometer is an *instrument* used to measure temperature.
4. A person breathes in oxygen and breathes out *carbon dioxide.*
5. Mammoth Cave is a large *cavern* carved out of the ground by water.
6. *Scientists* try to find out how and why things happen.
7. An *ore* contains material that is mined.

High-Potential Students

Encourage students to make a mural titled, "The Changing Earth." Illustrate the three layers of the earth and how wind, water, weathering, people, earthquakes, and volcanoes change the earth's surface. Use a roll of shelf paper for the mural. Instruct students to include labels near each illustration. Display the mural in the classroom or school hallway.

Instruct students to bring pictures from vacations or areas near their homes that show the outdoors. Challenge the students to point out and explain ways in which the pictures show how the earth is constantly changing. They might point out pictures of rivers, soil, mountains, and changes made by people.

Mainstreamed Students

Visually Impaired
Allow students to feel stones which have been weathered to understand how this process changes rocks.

Emotionally Handicapped
Have students collect recent newspaper and magazine articles about the environment, volcanoes, or earthquakes. Allow the students to share the information from the articles with the class.

Allow students to help collect the shoeboxes and stones for use in the activities.

Science Fair Projects

The Experiment Skills on p. 366 and the Investigate feature in the chapter can be used for science fair projects. You might also encourage interested students to do one of the following projects:

1. Using clay or papier-mâché, construct a model of the earth showing the three layers; the crust, mantle, and core. Include facts about each layer with your presentation.
2. Experiment with the effects of flowing water on a slope of exposed soil, a slope of soil covered with grass, and a third slope of soil covered with plant debris, such as dead leaves and stems. Paint trays can be used for the models. A sprinkling can might be used to help simulate rain. Observe the effects of the water on each of the surfaces. Also, measure the amount of soil washed off the slope for each surface. Discuss the applications of this experiment for soil erosion and farming practices.
3. Prepare a world map showing the locations of all known active and inactive volcanoes. Refer to encyclopedias, geology books, and other reference sources for the information.
4. Prepare a model of a landscape or cityscape on top of a sheet of strong cardboard, such as that used for large cardboard boxes. Various materials could be used for the model including building blocks, paper cutouts, plaster of Paris, clay, and sugar cubes. Simulate earthquakes of various strengths by tapping the cardboard. After each trial, the damage should be assessed and recorded in a chart. Make a scale of perhaps ten different tapping strengths and the associated damage.

lassroom Resources

Bulletin Board

Ask students to draw pictures of the places they have observed where the earth's crust is changing or has been changed.

Chapter 10 Poster

Science Discovery Center

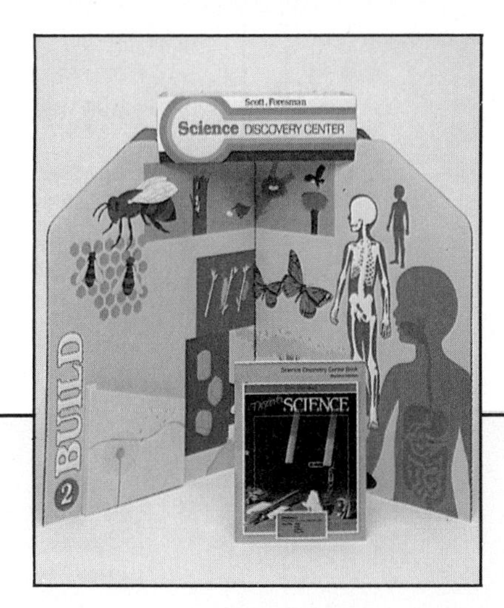

Use pages 105–110 from the *Science Discovery Center Book.* Place these worksheets in the appropriate pockets in the Science Discovery Center.

Overhead Transparencies

Use Transparencies 16 and 17 from the package of color overhead transparencies.

Teacher's Resource Book

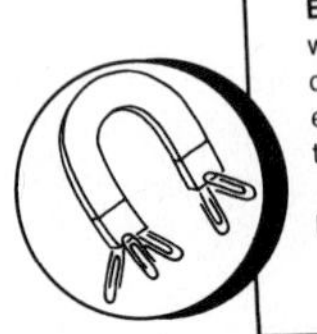

Dear Family,
Your student will be reading **Chapter 10: Changes in the Earth** in *Discover Science*, published by Scott, Foresman. We will learn that the earth has a crust, mantle, and core. The crust changes because of weathering. When water freezes, it expands. If water freezes inside a rock, the water will crack the rock.
By doing this activity together, you and your student can learn more about weathering.

The Power of Water

You will need a balloon, water, plaster, and a large, clear plastic bag.

1. With your student, fill a small balloon with water. Do not expand the balloon, but fill it full.
2. Tie a knot at the end of the balloon to close it.
3. Mix the plaster according to the directions.
4. Cover the balloon with the plaster. Allow it to dry. You now have a plaster rock.
5. Place your plaster rock in a large, clear plastic bag. Place the bag in the freezer for 24 hours.
6. Remove the bag and observe the rock. Ask, "What happened to the plaster? What happened to the water inside the balloon? How is this activity similar to what happens in nature? How is this like the process of weathering?"

Workbook

Name ____________

Use with Lesson 1: pages 208-210

Chapter 10

Vocabulary Preview

Changes in the Earth

Chapter Vocabulary				
core	earthquake	erupt	magma	volcano
crust	erosion	lava	mantle	weather

Vocabulary Cards

1. Write each word on a card.
2. Find each word in the glossary. Copy the pronunciation under the word on the card.
3. Practice saying the words with a partner.

Word Meanings

1. If the *crust* of a pie is on the outside, what do you think the *crust* of the earth is?
   The crust is the outside part of the earth.
2. If the *core* of an apple is on the inside, where do you think the *core* of the earth is?
   The core is in the center of the earth.
3. If the ground can be covered by a *mantle* of snow, what do you think the *mantle* does to the earth's core?
   The mantle covers the earth's core.
4. What is the meaning of each of the two words that make up the word *earthquake*?
   Earth is the planet on which we live.
   Quake means to shake or tremble.

At Home

1. Practice saying the words. Learn their meanings.
2. Look for the words and pictures of the words in newspapers and magazines. You can make a poster.

Teacher's Notes: Ask students to identify the vocabulary associated with the volcano drawing. Answers include magma, lava, erupt, volcano.

Workbook

Name ____________

Use with Lesson 1: pages 208-210

Chapter 10

Science Activity

What Are the Layers of Earth?

Gather These Materials

- three different colors of clay
- 30 cm dental floss
- two pencils

Follow This Procedure

1. Make a ball the size of a marble. Use one color of clay.
2. Cover the ball with a layer of another color of clay. The layer should be 1 cm thick.
3. Use the third color of clay to cover the ball. This layer should be thin.
4. Use dental floss to cut the ball in half. See the picture.

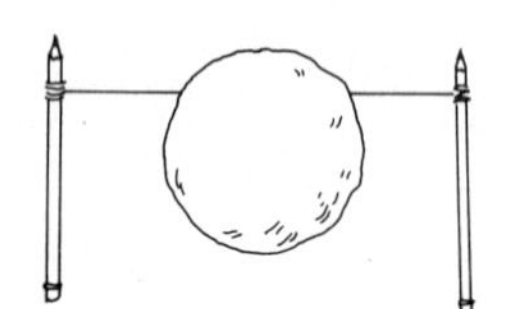

Record Your Results

Draw a picture of what the layers look like. Label the core, the mantle, and the crust.

crust
mantle
core

State Your Conclusions

1. Which layer is thickest?
   The core of the earth is the thickest layer.
2. Which layer is thinnest?
   The crust is the thinnest layer.
3. Why do scientists study the earth's layers?
   Answers will vary. Example: Scientists study the layers to learn about how the earth was formed.

Teacher's Notes: Invite a geologist to talk with students about the layers that make up the crust in or near your community.

Teacher's Resource Book

Name ____________

Use with Lesson 1: page 211

Chapter 10

Activity Worksheet

Inferring Shapes From Sounds

Record Your Results

Sounds made	Objects you think are in box	Objects in box

State Your Conclusion

1. How did the sounds help you to tell what was in the box?
2. How did you tell how many objects were in the box?

Use What You Learned

What are some things you could not learn about the hidden objects by shaking the box? Explain your answer.

Workbook

Name ______________________

Use with Lesson 2: pages 212-215

Chapter 10

Science Skills

After a Forest Fire

Once, the hills in the picture were covered by a forest. Many trees, smaller plants, and animals lived in the forest. These plants provided homes and food for the animals. The plants also soaked up rainwater. In this way, they kept soil from moving down hillsides.

The picture shows how the hills looked shortly after a fire. Changes have taken place. Soon the hills will show many signs of erosion.

Answer the questions.

1. What did the fire do to the forest animals? **Some died in the fire, the rest ran away**
2. How will erosion change the hills? **Paths or holes form in the hillsides; soil washed to bottom of hill; hills lowered**
3. How could people help save the soil and the hills? **Plant seeds to hold the soil in place.**

Teacher's Note: Ask students to name places where they see rapid erosion taking place near the school or in the neighborhood.

Teacher's Resource Book

Name ______________________

Use with Lesson 2: page 216

Chapter 10

Activity Worksheet

Making a Model of Erosion

Record Your Results

	Water	Stones
Can A		
Can B		
Pile C	XXX	

State Your Conclusion

1. Compare the way the water in can A looks with the way the water in can B looks.

____

____

2. How are the stones in piles A, B, and C different from each other after the shaking?

____

____

Use What You Learned

How are the stones moving in the cans like the rocks moving in a stream of water?

____

____

____

Teacher's Resource Book

Name ______________________

Use with Lesson 3: pages 218-221

Chapter 10

Science and Reading

A Volcano Erupts

Read the story. Then put the pictures in order.

Mount Saint Helens was a quiet volcano. It had not erupted for 123 years. Evergreen forests grew on its slopes. Many plants, animals, and birds lived there. Then, on May 18, 1980, the volcano erupted. The hot rock and ash made leaves and plants burn. The explosions knocked down millions of trees. Plants and animals died.

Several years have passed. The gray ash has formed rich soil. First, small plants started to grow. Then tiny animals came back to live. Next, the berry bushes grew. When the berries were ripe, the animals that eat the berries came back to live. Then, the trees started to get big. The birds came back to live in the trees.

Now squirrels and deer live on Mount Saint Helens again. Now the beautiful song of the mountain bluebird fills the air.

Look at the pictures. What happened first, second, third, and fourth? Write the numbers 1, 2, 3, or 4 in the proper boxes to show the proper order.

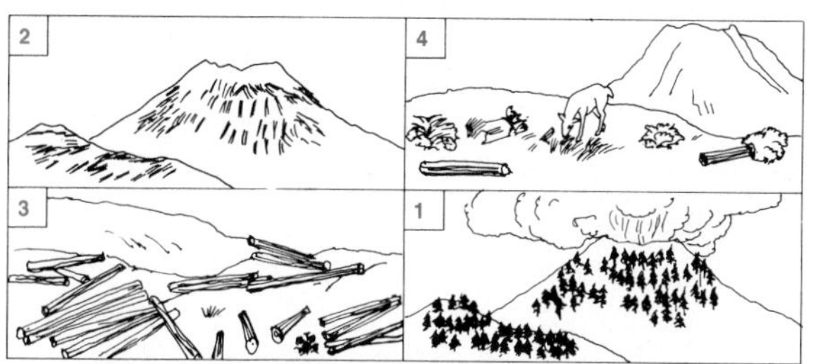

Teacher's Notes: Ask students if they think that Mount Saint Helens will be covered with trees as it once was. Ask them why.

Workbook

Name ______________________

Use with Lesson 4: pages 222-223

Chapter 10

Vocabulary Puzzle

Crossword

Write the words in the puzzle. Use the word bank and the clues below.

Word Bank	
core	lava
crust	magma
earthquake	mantle
erosion	volcano
erupt	weather

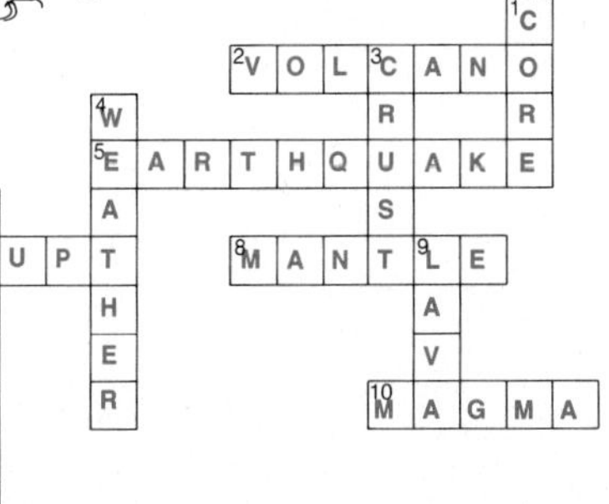

Across

2. a mountain with an opening through which lava and other materials come out
5. a shaking or sliding of the earth's crust
7. to burst out
8. the middle layer of the earth
10. melted rock inside the earth

Down

1. the center part of the earth
3. the outer layer of the earth
4. to wear down rocks
6. the movement of soil or rocks by wind or water
9. hot, melted rock that flows from a volcano

Teacher's Notes: Encourage interested students to increase their vocabularies by learning the names of different kinds of rocks.

Name ______________________

Comprehension: time sequence

Chapter 10

Science and Reading

The Order of Events

Read the story. Then number the events in the order that they happened in the story.

Julie needed to know more about earthquakes for her science project. After school, she went to the library to find out about earthquakes. She found a book and sat down to look at it.

Suddenly, the building began to shake. Julie thought it was just a train passing by. Then someone shouted, "Earthquake!" In a few seconds, the small earthquake was over.

Julie went back to her work and closed her book. She had more information about earthquakes now than she could ever find in a book. She wrote about how it felt to be in an earthquake.

3 Someone shouted, "Earthquake!"

4 Julie wrote how it felt to be in an earthquake.

2 The building began to shake.

1 Julie went to the library.

Name ______________________

Chapter 10 Test A

Multiple Choice Choose the best answer.

1. The layer of the earth made of rocks and soil is called the
   a. mantle.
   (b.) crust.
   c. core. **(1-1)**

2. The layer of the earth made mostly of iron is called the **(1-1)**
   a. mantle.
   b. crust.
   (c.) core.

3. How do scientists study the mantle and core of the earth? **(1-2)**
   a. by digging
   b. by going inside the earth
   (c.) with special instruments

4. The wearing down or breaking apart of rocks is called **(2-1)**
   (a.) weathering.
   b. mining.
   c. an eruption.

5. When the earth's crust is changed because soil is moved by wind, this is **(2-2)**
   a. weathering.
   (b.) wind erosion.
   c. an eruption.

6. Plants can help change the earth's **(4-1)**
   a. core.
   b. magma.
   (c.) crust.

7. A shaking or sliding of the earth's crust is **(3-1)**
   a. erosion.
   (b.) an earthquake.
   c. weathering.

8. A mountain with an opening where lava, ash, rocks, and other materials come out is a **(3-2)**
   a. magma.
   b. mantle.
   (c.) volcano.

9. How can strong earthquakes cause damage or hurt people? **(3-2)**
   a. hot magma
   (b.) falling objects
   c. strong winds

10. A bulge in a volcano might mean that the volcano **(3-2)**
    (a.) is about the erupt.
    b. will never erupt.
    c. has stopped erupting.

Numbers in parentheses after each question refer to the lesson number and the objective of that lesson.

Name ______________________

Chapter 10 Test A

Short Answer Write the name of each part of the volcano in the correct blank. Use these words: magma, eruption, lava, new crust.

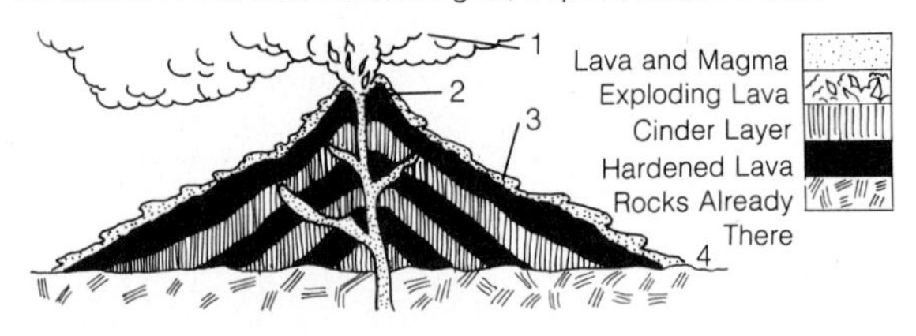

1. eruption 2. magma 3. lava 4. new crust

Short Essay Use complete sentences to answer each question.

1. Name two ways to protect yourself during an earthquake. **(3-1)**

   If indoors, stay away from windows and furniture that could fall over. If outdoors, stay away from high buildings and other objects that could fall.

2. Explain how wind and water can cause erosion. **(2-2)**

   Wind and water can move tons of soil and rocks. Erosion usually takes a long time. Sometimes, heavy rains and strong winds move soil quickly.

Numbers in parentheses after the questions refer to the lesson number and the objective of that lesson.

Name ______________________

Chapter 10 Test B

Multiple Choice Choose the best answer.

1. The earth's crust is mostly the land we walk on and the **(1-1)**
   a. rocks of the mantle.
   (b.) land under oceans.
   c. iron in the core.

2. Scientists cannot see the inside of the earth, but they can study it from **(3-1)**
   a. pictures of the core.
   (b.) movements of the crust.
   c. trips inside deep holes.

3. The middle layer of the earth is called the **(1-1)**
   (a.) mantle.
   b. core.
   c. crust.

4. When the earth's crust is changed because soil is moved by water, this is called **(2-2)**
   (a.) water erosion.
   b. weathering.
   c. an earthquake.

5. Freezing and thawing, plant roots, and rainwater all can make rocks **(2-1)**
   a. grow larger.
   (b.) weather.
   c. erupt.

6. When lava cools, it forms **(3-2)**
   a. sand.
   (b.) rocks.
   c. soil.

7. The earth's crust is usually changed very slowly by **(3-1)**
   a. earthquakes.
   (b.) erosion and weathering.
   c. volcanoes.

8. Plants can help change the earth's **(4-2)**
   a. mantle.
   (b.) crust.
   c. center.

9. People who build buildings, cut down trees, and mine for ore help **(4-1)**
   (a.) change the earth's crust.
   b. change the earth's mantle.
   c. prevent erosion.

10. Large blocks of rock in the earth's crust move and form cracks during **(3-1)**
    a. a volcano.
    b. a flood.
    (c.) an earthquake.

Numbers in parentheses after each question refer to the lesson number and the objective of that lesson.

Name ________________________

Chapter 10
Test B

Short Answer Write the name of each part of the volcano in the correct blank. Use these words: magma, eruption, lava.

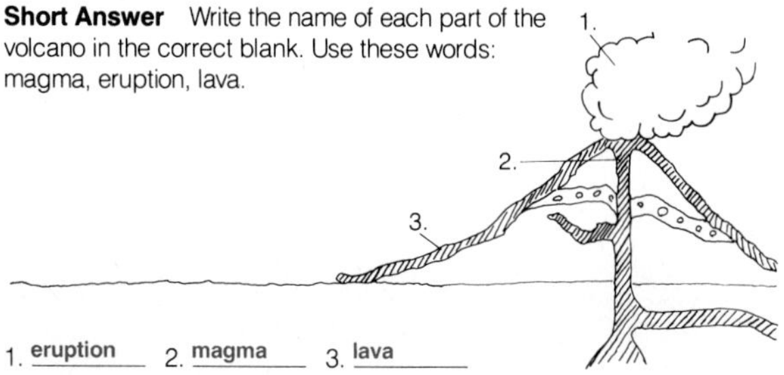

1.
2.
3.

1. eruption 2. magma 3. lava

Short Essay Use complete sentences to answer each question.

1. Name three ways rocks are weathered. **(2-1)**

Three types of weathering are water, wind, and freezing and thawing.

2. Name two ways to protect yourself during an earthquake. **(3-1)**

Stay away from windows and tall furniture, if you are indoors. When outdoors, stay away from high buildings.

Numbers in parentheses after each question refer to the lesson number and the objective of that lesson.

TEACHING PLAN

Major Concepts
Lesson 1 The earth is a sphere with three layers: the crust, the mantle, and the core.
Lesson 2 Over time, water and wind can weather rocks.
Lesson 3 Earthquakes and volcanoes can change the earth's crust quickly.
Lesson 4 People, plants, and animals change the earth's crust.

Chapter Vocabulary
crust, core, earthquake, erosion, erupt, lava, magma, mantle, volcano, weather

Getting Started
Questions: **How was the gorge in the photograph carved out?** (The running water loosened and carried away rocks and soil.) **What do you think the gorge will look like next year?** (It will probably look the same, because erosion of rock occurs very slowly.) **What do you think it will look like in a few hundred years?** (It will probably be wider and longer.) Tell students that running water is only one of the forces that changes the earth's surface. In this chapter they will be learning about other forces as well.

Chapter 10

Changes in the Earth

Notice how the water flows over these rocks. Over the years, the water has worn away some of the rock.

206

Teaching Options

Cooperative Learning ♦

Jigsaw Format (See page T23.)
Assign the following topics at random to your cooperative learning teams.
Topic A: List and describe the three layers of the earth.
Topic B: What changes occur on the earth as rock weathers and as erosion occurs?
Topic C: What are earthquakes and volcanoes and how do they change the earth's crust?
Topic D: Give four examples showing how people, plants, or animals change the earth's crust.
Have students search for information on their topic as they read the chapter. Then let all students with the same topic meet in an expert group to discuss the information. When students return to their teams, they may take turns presenting their topics to the team. Then give students a test covering all topics to complete individually (Chapter 10 Test A or B in the *Test Book*). Award Superteam certificates to teams whose average test scores exceed 90% and Greatteam certificates to teams whose average test scores exceed 80%.

♦ ***Suitable as a language development activity***

Introducing the Chapter

In this chapter, you will read about the inside and outside of the earth. You will learn how the earth can change. In the activity below, you will learn how a drop of water can change the land.

Observing Drops of Water

Suppose you dug a small hole in a field. This would not change the land much. Suppose you and your classmates spent a week digging holes in the field. Then the field would look very different.

Place a small mound of sand on a piece of waxed paper. Use a medicine dropper to let a drop of water fall on the sand, as shown in the picture. Notice what happens to the sand. Now hold the dropper higher above the sand. Squirt all the water in one place on the sand. Refill the dropper several times. Squirt all the water in the same place.

Talk About It

1. What did one drop of water do to the mound of sand?
2. What happened when you let a lot of water fall quickly on the sand in the same place?

DISCOVER

Objective ♦
This optional *DISCOVER* activity will help students explore and build background information about the concept of erosion. Later in the chapter students will be able to draw on this experience to help them assimilate the new content.

Science Process Skills
Observing, Making models

Materials
For each pair of students: mound of sand, sheet of waxed paper, a medicine dropper

Safety Tip (See page T24.)
- Immediately wipe up any sand or water spills to prevent slipping accidents.

Teaching Tip
- Have students *apply* their observations by asking how a heavy rainfall might change the earth's surface. (wearing away soil on hillsides)

Answers
Talk About It
1. Did not change the mound greatly
2. Water quickly flowed down the mound and left a dug-out path.

Applying Whole Language ♦

Discuss the whole language framework with each Teaching Option you select. Here is an example applied to the Cooperative Learning option on p. 226.
1. Purpose: To organize study by answering lesson title questions through picture analysis.
2. Context: Teams. Share ideas contained within graphic aids. Form summaries using text to supplement. Write summaries.
3. Decisions: What purposes does each graphic aid serve?
4. Evaluation: How did pictures help us understand the text? (See p. T30.)

Resource Book page 111

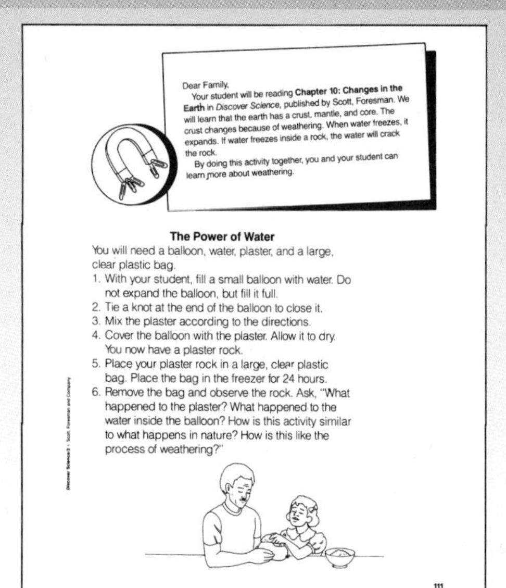

Dear Family,
Your student will be reading **Chapter 10: Changes in the Earth** in *Discover Science*, published by Scott, Foresman. We will learn that the earth has a crust, mantle, and core. The crust changes because of weathering. When water freezes, it expands. If water freezes inside a rock, the water will crack the rock.
By doing this activity together, you and your student can learn more about weathering.

The Power of Water
You will need a balloon, water, plaster, and a large, clear plastic bag.
1. With your student, fill a small balloon with water. Do not expand the balloon, but fill it full.
2. Tie a knot at the end of the balloon to close it.
3. Mix the plaster according to the directions.
4. Cover the balloon with the plaster. Allow it to dry. You now have a plaster rock.
5. Place your plaster rock in a large, clear plastic bag. Place the bag in the freezer for 24 hours.
6. Remove the bag and observe the rock. Ask, "What happened to the plaster? What happened to the water inside the balloon? How is this activity similar to what happens in nature? How is this like the process of weathering?"

111

Science Background

For erosion to occur, material must be picked up and carried away. One of the most important and noticeable ways erosion is accomplished is by running water. The amount of running water, its speed, the type of surface over which it runs, and the length of time it flows are important factors in the amount of erosion that occurs.

TEACHING PLAN

Lesson Objectives

- *Describe* the three layers of the earth.
- *Explain* how scientists study the inside of the earth.

Lesson Vocabulary

crust, core, mantle

1. Motivate

Demonstration Activity ♦

Cut a peach in half and show it to the class. Ask students to *observe* the peach closely and determine its main layers.

Discussion

Lead students to *observe* that the peach has three main layers: the skin, the meaty part of the fruit, and the pit. Tell them that the earth also has three main layers.

LESSON GOALS

You will learn
- that the earth has three layers.
- how scientists study the inside of the earth.

1 What Is the Inside of the Earth Like?

People used to think the earth was flat. They believed that if you sailed a boat far out into the ocean, the boat would fall off the earth. Today people know this is not true. The earth is not flat. It is shaped like a ball. The earth only seems flat because it is so large. When you look around you, you are seeing only a small part of the earth.

Layers of the Earth

Compare the picture of the peach with the picture of the earth. You can see that they both have three layers.

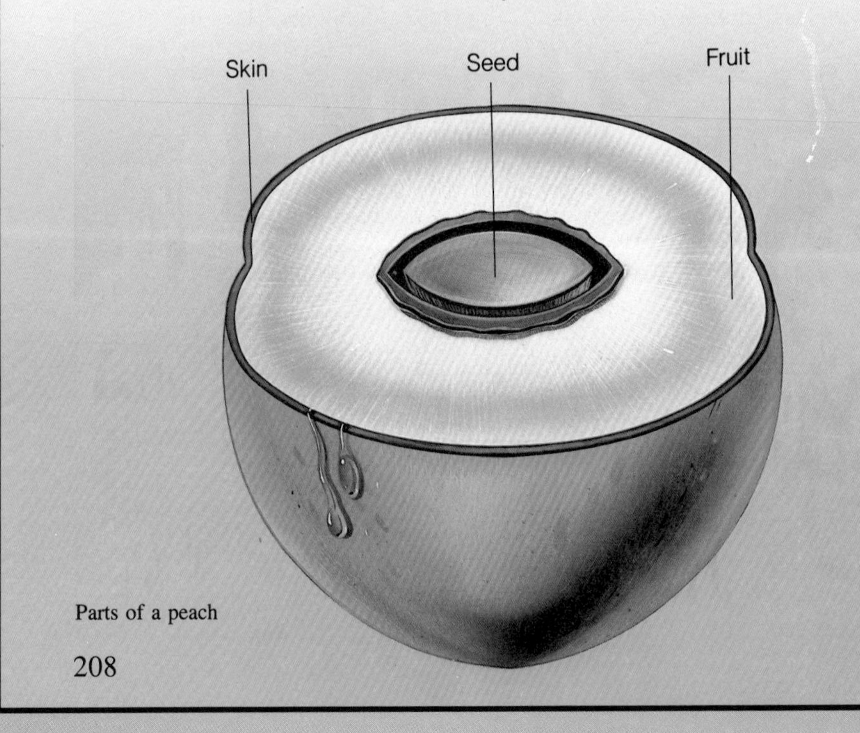

Parts of a peach

Teaching Options

Science Background

Rocks of the earth's crust consist mostly of oxygen and silicon. The crust varies in thickness from 8 km (5 mi) under the ocean to about 40 km (25 mi) under the continents. The mantle contains many heavier elements, such as iron and magnesium. The core is mostly iron and nickel. To learn about the earth's interior, scientists study seismic waves produced by earthquakes and underground explosions. These waves travel at different speeds through different materials. Seismic waves travel relatively quickly through dense, solid material. The waves slow down when traveling through partly melted material, which leads geologists to conclude that a top layer of the mantle is partly (1 to 10 percent) melted.

Reading Strategies ♦

1. Guide students' pre-reading by asking: What ideas do you think you would find in the lesson to answer the question-title?
2. Assign these strategies: Visualizing Information and Writing a Memory Sentence. (See pages T26-T29.)
3. Pair students to share what information is clear and unclear and initiate discussion using students' unanswered questions.

♦ *Suitable as a language development activity*

Find the outer layer of the earth. This layer is like the skin of the peach. The outer layer, or **crust,** of the earth is made up of rocks and soil. The land you walk on and the land under the oceans are part of the crust.

What part of the earth is like the middle part of the peach?[1] The middle layer of the earth is the **mantle.** It is mostly made of rock. Some of the rock in the mantle is partly melted.

Notice that the center part of the earth is like the peach seed. The center of the earth—the **core**—is mostly iron. The outside part of the core has liquid iron. The inside part has solid iron. The core is the hottest part of the earth. The temperature of the core is almost as hot as the surface of the sun!

SCIENCE IN YOUR LIFE

Some parts of the top of the earth's crust are stronger than others. When workers build a very tall building, they first test the ground. The workers need to find out if the crust is strong enough to hold up the building.

crust (krust), the outer layer of the earth.

mantle (man′tl), the middle layer of the earth.

core (kôr), the center part of the earth.

[1] the middle layer

Mantle
Core
Crust

Layers of the earth

2. Teach

Teaching Tips

- On a large ball, such as a beach ball, mark off a small circle, perhaps 2 centimeters (about an inch) in diameter. Show students that this small circle by itself looks relatively flat even though it is actually curved and part of a sphere. Question: **How would the curved shape of the circle change if the ball became bigger?** (The shape would become flatter.)
- Use a toy boat or other small object to ***demonstrate*** the appearance of a ship coming over the horizon on a spherical surface as opposed to a flat or cubical surface.
- Direct students' attention to the illustration on page 209. Ask which of the three layers of the earth is the thinnest? (crust) the thickest? (mantle)
- **Possible Misconception:** Students might visualize the partly melted portion of the mantle as being like a syrup or slush. Actually, this part of the mantle, called the asthenosphere, is evidenced to be only 1–10 percent melted and flows only a few centimeters per year.

Workbook page 55 *

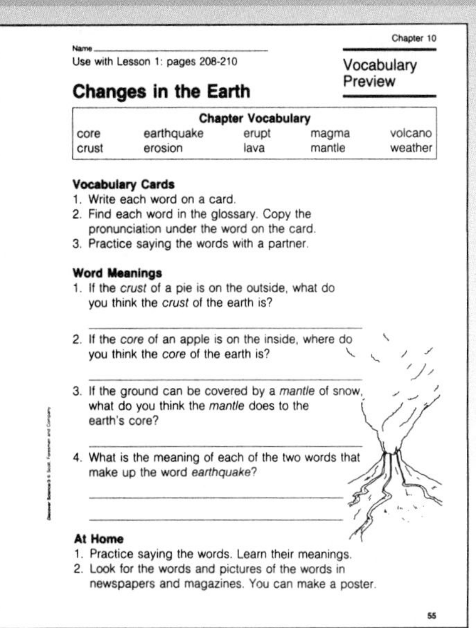
Chapter 10

Name ______

Use with Lesson 1: pages 208-210

Vocabulary Preview

Changes in the Earth

Chapter Vocabulary

core	earthquake	erupt	magma	volcano
crust	erosion	lava	mantle	weather

Vocabulary Cards
1. Write each word on a card.
2. Find each word in the glossary. Copy the pronunciation under the word on the card.
3. Practice saying the words with a partner.

Word Meanings
1. If the *crust* of a pie is on the outside, what do you think the *crust* of the earth is?
2. If the *core* of an apple is on the inside, where do you think the *core* of the earth is?
3. If the ground can be covered by a *mantle* of snow, what do you think the *mantle* does to the earth's core?
4. What is the meaning of each of the two words that make up the word *earthquake*?

At Home
1. Practice saying the words. Learn their meanings.
2. Look for the words and pictures of the words in newspapers and magazines. You can make a poster.

55

**** Answers to masters on pages 206E–206H***

Game Suggestion ♦

Create a board game entitled *Voyage to the Center of the Earth.* Draw a path on an open file folder or posterboard. The background and path could be drawn so that the path appears to lead from the earth's surface to the core. Divide the path into squares. On each square, write a question from the lesson. Using a number cube, students move colored chips or other kinds of markers around the path answering the questions on the squares. A wrong answer results in a missed turn or backward jump. The goal is to reach the center of the earth.

Special Education

Have students with learning disabilities illustrate the three layers of the earth using crayons or paint. Ask them to label each layer.

TEACHING PLAN

3. Assess

Lesson Review

1. The crust is made up of rocks and soil. The mantle is made of solid and partly melted rock. The core is made of liquid and solid iron.
2. Scientists use special instruments and study movements of the crust to find out about the inside of the earth.
3. Challenge! Scientists know more about the crust because they can dig into the crust to observe it. The mantle and core are too deep to dig into. **Thinking Skill:** *Inferring*

Find Out On Your Own

A seismograph amplifies and records small movements of the ground. Scientists can determine the location and size of earthquakes from these records. **Thinking Skill:** *Recognizing the main idea and supporting detail*

Studying the inside of the earth

How Scientists Study the Inside of the Earth

The earth's mantle and core are too far down to reach. Scientists study the inside of the earth from the outside. Movements of the earth's crust give scientists information about the inside of the earth. Scientists also use instruments to find out about the inside of the earth.

Lesson Review

1. What makes up each layer of the earth?
2. How do scientists know what the inside of the earth is like?
3. **Challenge!** Why do scientists know more about the crust than the other layers of the earth?

Study on your own, pages 334–335.

PHYSICAL SCIENCE *FIND OUT ON YOUR OWN* CONNECTION

A seismograph is one machine that scientists can use to learn about the earth's crust. Look in library books to find out how a seismograph works. Write a few sentences telling what you have learned.

210

Teaching Options

Science Anecdote

Seismic waves are used in offshore oil exploration. A ship tows a sound source and telephones. Sound waves made by the source bounce off different rock layers beneath the ocean floor and are intercepted by the telephones. This method is used to map geologic features such as faults, folds, and domes, which are often traps for oil.

Reteaching Suggestion ♦

Provide students with three different colors of clay. Instruct them to use the clay to make a model of the layers of the earth. Call on students to *describe* the different parts of the models.

Workbook page 56 *

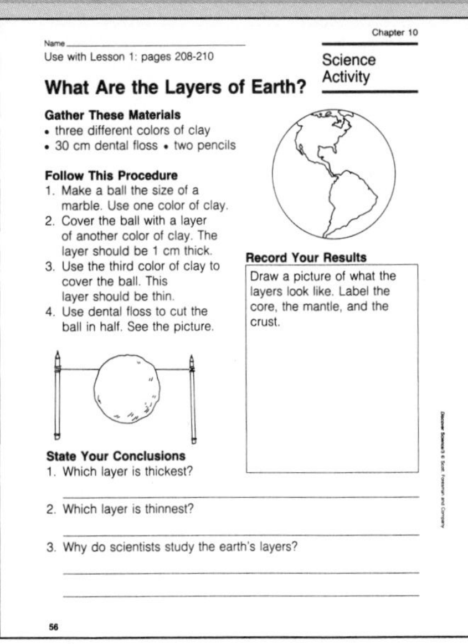

Chapter 10

Name ____

Use with Lesson 1: pages 208-210

Science Activity

What Are the Layers of Earth?

Gather These Materials
- three different colors of clay
- 30 cm dental floss • two pencils

Follow This Procedure
1. Make a ball the size of a marble. Use one color of clay.
2. Cover the ball with a layer of another color of clay. The layer should be 1 cm thick.
3. Use the third color of clay to cover the ball. This layer should be thin.
4. Use dental floss to cut the ball in half. See the picture.

Record Your Results

Draw a picture of what the layers look like. Label the core, the mantle, and the crust.

State Your Conclusions
1. Which layer is thickest?
2. Which layer is thinnest?
3. Why do scientists study the earth's layers?

56

♦ *Suitable as a language development activity*

Inferring Shapes From Sounds

Suggested grouping: pairs

Purpose

Describe the sounds hidden objects make when they move in a box and *infer* the shapes and types of the objects.

Gather These Materials

• covered box containing several objects • tape

Follow This Procedure

1. Use a chart like the one shown to record your observations.
2. Tilt, turn, and shake the box. Listen to the sounds the objects make. Do they sound like they are sliding or rolling? What do they sound like when they hit the sides of the box? How many objects do you think are in the box?
3. Record your observations. Write the names of the objects you think are in the box.
4. Open the box. Write the names of the objects in the chart.
5. Make up another mystery box like the one in the picture. Use objects that you get from your teacher. Seal the box with tape. Trade boxes with a partner.
6. Repeat steps 2, 3, and 4.

Record Your Results

Sounds Made	Objects you think are in box	Objects in box
Answers will vary depending on objects in box.		

State Your Conclusion

1. How did the sounds help you to tell what was in the box?
2. How did you tell how many objects were in the box?

Use What You Learned

What are some things you could not learn about the hidden objects by shaking the box? Explain your answer.

211

Activity

Concept

The size and shape of an object can be determined by the sounds it makes when shaken around the inside of a closed box.

Objectives/Process Skills

- *Observe* sounds made by objects in a closed box when the box is tilted or shaken.
- *Record* and *interpret* observations.
- *Infer* the contents of a closed box.

Time Allotment

Allow 30 minutes.

Safety Tip (See page T24.)

- Be sure no sharp objects are placed in the boxes.

Teaching Tips

- Make a variety of boxes, some having only one object and some having several objects.
- Have a collection of small objects available for students to make mystery boxes for their partners.
- As a variation on this activity, have one member of each group tape a few objects to the bottom of the box. The partner must feel the objects without looking and draw a map of how the objects are arranged.
- Poke small holes into the sides and top of the box. Allow students (wearing cover goggles) to probe the inside of the box using a very thin dowel rod or pencil.

Answers

State Your Conclusion

1. Answers will vary. Students might mention sliding, rolling, or bouncing sounds.

2. Answers might include the use of sound, feel of movement, and weight.

Use What You Learned

Answers might include any physical properties that could not be identified by sound or mass, such as color, newness of object, or what it is made of. **Thinking Skill:** *Inferring, Predicting*

Resource Book page 115

Name ______
Use with Lesson 1: page 211

Chapter 10

Activity Worksheet

Inferring Shapes From Sounds

Record Your Results

Sounds made	Objects you think are in box	Objects in box

State Your Conclusion

1. How did the sounds help you to tell what was in the box?

2. How did you tell how many objects were in the box?

Use What You Learned

What are some things you could not learn about the hidden objects by shaking the box? Explain your answer.

115

*** Answers to masters on pages 206E–206H**

Activity Results

Answers will vary depending on the objects in the box. Students might hear sliding, rolling, or bouncing sounds. Students might be able to determine size and approximate mass of objects.

TEACHING PLAN

Lesson Objectives
- *Explain* how water and wind can weather rocks.
- *Define* erosion.

Lesson Vocabulary
erosion, weather

1. Motivate

Demonstration Activity ♦
Place several sugar cubes in a small container and cover it tightly, either with a fitted lid or with your hand. Shake the container vigorously for a few minutes. Remove the lid and display the sugar cubes on a table or desk next to a few sugar cubes that were not shaken. Allow the students to *observe* and *compare* the sugar cubes.

Discussion
Discuss how the sugar cubes that were shaken look different from those that were not. (Those that were shaken might be broken, cracked, or rounded.) State that the changes they observed in the cubes are similar to changes that occur to rocks in rivers and on the ground.

2 How Do Water and Wind Change the Earth's Crust?

LESSON GOALS

You will learn
- how rocks weather.
- how water and wind cause erosion.

Notice how smooth the rocks in the picture look. These rocks used to have very sharp edges. Over many years, these sharp edges have worn away.

Weathering

weather (weᴛʜ′ər), to wear down or break apart rocks.

You learned in Chapter 9 that rocks crumble and wear away. When rocks break down, they **weather.** Most rocks weather slowly, over many years. Rocks can weather in several different ways.

Freezing water weathers rocks. You know that water can fill the holes and cracks in a rock. What happens to the rock as the water continues to freeze and melt?[1]

[1] it cracks

Moving water can wear away rocks.

212

Teaching Options

Science Background

One product of weathering is soil, which contains bits of weathered rock, organic remains, water, and air. Rock, sand, and soil that is eroded from one place and dropped elsewhere is called sediment. Besides wind and running water (rivers) discussed in the text, other agents of erosion include glaciers, water waves and currents, groundwater, and mass movements due to gravity, such as landslides. Running water is considered the most effective agent of erosion. Satellite or aerial photographs show how valleys eroded by creeks, streams, and rivers dominate the landscape. Running water erodes the land in two ways: by breaking up rock along the riverbed and banks and by removing weathered materials.

Reading Strategies ♦

1. Guide students' pre-reading by asking: What questions do you have about the picture of the rocks in the water?
2. Assign these strategies: Visualizing Information and Writing a Memory Sentence. (See pages T26-T29.)
3. Pair students to share what information is clear and unclear and initiate discussion using students' unanswered questions.

♦ *Suitable as a language development activity*

Plants also help weather rocks. You learned that plant roots can grow in soil that collects in the cracks of rocks. The roots can break rocks apart as they grow. The roots of a large plant, such as a tree, can break a boulder apart.

Sometimes, rainwater and gases in the air can weather rocks. You learned in Chapter 1 that carbon dioxide is a gas in the air. Carbon dioxide can mix with water, such as rain, to form a new material. This material can turn hard minerals in rocks to soft clay. Then the rocks slowly break apart. Sometimes, materials in the water eat away the minerals in rocks and leave hollow spaces. Some of these hollow spaces are beautiful underground caverns, like the one in the picture.

INVESTIGATE!

Find out how moving water affects the beaches along lakes and oceans. Write a hypothesis and test it with an experiment. You might make a model to show wave action using a pan, water, and sand.

Weathered rocks can form underground caverns.

2. Teach

Teaching Tips

- Make a clear distinction between the term *weather* as it is used in this chapter and its more familiar meaning related to meteorology.
- Emphasize that many episodes of freezing and thawing usually occur before a rock breaks or before potholes form in streets.
- Questions: **What can happen to metal objects left outside for many weeks?** (They can rust.) **Is the rusty object weaker or stronger than the original object?** (weaker) Explain that rust is the result of another kind of weathering that occurs when oxygen combines with the iron in some metal objects. This weathering takes place faster in the presence of water. Bring in a rusted nail or other object for students to ***observe.*** Break the object or flake off some of the rust to show how weathering weakens rocks. *CAUTION:* Handle rusty objects with gloves, and wear cover goggles if you break the objects.

Investigate!

Accept any testable hypothesis. One possible hypothesis is: *Moving water changes the beaches along lakes and oceans*. Students should find out that moving water can expand or erode beaches.

Enrichment

Inform the class that calcium is a mineral found in bones and eggshells and that calcium makes these substances hard. Demonstrate the weathering power of weak acids on calcium by placing a chicken bone in a jar of vinegar—a weak acid. After several days, the bone will become rubbery as the vinegar dissolves the calcium. Let students investigate the effect of vinegar on an eggshell. *CAUTION:* Students should wear cover goggles while handling vinegar. Remind students to wash their hands thoroughly after handling chicken bone and egg shells.

Reinforcement

Demonstrate the power of expanding water in breaking apart rocks. Fill a plastic bottle to the top with water, cap the bottle, and freeze it. As the water freezes, the bottle will either crack or the cap will pop off and ice will protrude from the top as the water freezes. *CAUTION:* Do not use a glass bottle. Immediately wipe up any water that spills on the floor to prevent slipping accidents.

TEACHING PLAN

Teaching Tips

- Show the class a picture of a river. Question: **What is flowing in the river besides water?** (soil and rocks)
- Tell students to ***recall*** the *DISCOVER* activity on page 207 in which they ***observed*** water running down a hill of sand, making paths. Ask what the wearing away of the sand was called. (erosion)
- Show students satellite or aerial photographs of an area in which valleys and ridges dominate the scene. Such pictures are available in geology or earth science textbooks. Point out the valleys, which most likely were all caused by erosion by rivers at the bottom of the valleys.
- Point out that erosion by a river is caused not only by the water but also by the rocks, sand, and bits of soil hitting against the banks and river bottom.

Erosion

erosion (i rô′zhən), the movement of soil or rocks by wind or water.

Every year, wind and water move tons of soil and rocks. **Erosion** is the moving of soil and rocks by wind or water. Erosion changes the earth's crust. Most of the time these changes happen very slowly, over many years.

Look at the paths on the hill. The paths were made by erosion. As rainwater rolled down the hill, it picked up some of the soil. The water carved out the paths as it moved the soil down the hill. The soil has collected at the bottom of the hill.

Soil erosion

214

Teaching Options

Science Anecdote

New Orleans is located on the Mississippi River Delta, which is made of eroded materials deposited by the Mississippi River. The delta is expanding out into the ocean at a rate of nearly 110 m (360 ft) per year. Fossil evidence and ancient sediments indicate that the position of the delta has changed several times over the last few thousand years.

Science and Geography

Help students locate the Mississippi River on a map of the United States. Point out that the river flows toward the Gulf of Mexico. Call on volunteers to trace some of the major tributaries of the Mississippi River to recognize that waters of a tributary flow into a larger river. Have students trace the tributaries backwards toward the beginning of the rivers to find the extent of the Mississippi River system. Lead students to understand that some water and eroded material from such states as Montana, Minnesota, and New York eventually reaches the Mississippi delta.

Special Education

Obtain several rounded pebbles from a stream, beach, or gardening store to show students with learning disabilities. Point out that the pebbles came from a stream or a beach. Question: **Why are they rounded?** (Tell students that the roundness is caused by the collision of rocks with other rocks and sand particles within a stream or along a beach.) You may wish to compare the shapes of the pebbles to the shapes of the sugar cubes in the Demonstration Activity.

♦ ***Suitable as a language development activity***

Wind erosion

Wind erosion formed the large piles of sand in the picture. Strong desert winds blew the sand over long distances. When the winds calmed down, the sand stopped moving. Over time, the large piles of sand were formed.

Lesson Review

1. How do rocks weather?
2. What is erosion?
3. **Challenge!** How can erosion be both harmful and helpful?

Study on your own, pages 334–335.

LIFE SCIENCE — FIND OUT ON YOUR OWN — CONNECTION

Plants help keep soil from being eroded by protecting the soil from wind and water. The plants' roots help to hold down the soil. Use an encyclopedia to find out how farmers use plants to stop erosion on their land. On a sheet of paper, describe two ways that farmers fight erosion.

Teaching Tip

- **Possible Misconception:** Some students might think the terms *weathering* and *erosion* can be used interchangeably. Although both processes may loosen rock, erosion involves its removal.

3. Assess

Lesson Review

1. Rocks weather by the freezing and melting of water, by plant roots breaking rocks apart, and by carbon dioxide mixing with water to form a material that softens minerals in rocks or eats them away.
2. Erosion is the moving of soil or rocks by wind or water.
3. Challenge! Erosion can be harmful when it removes soil from the land or destroys property along a riverbank. It can be helpful when the soil is dropped somewhere else, building up new land.
Thinking Skill: *Making generalizations*

Find Out On Your Own

Methods include strip cropping, in which water-absorbing grasses are planted between strips of crops; contour plowing, in which furrows are made along a hill rather than up and down it; and the use of tree lines to act as barriers against wind erosion. **Thinking Skills:** *Collecting information; Communicating; Restating or explaining ideas*

Workbook page 57 *

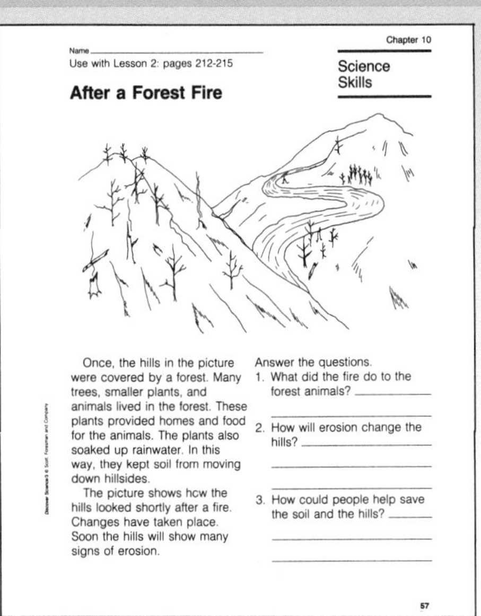

Name ____

Use with Lesson 2: pages 212-215

Chapter 10

Science Skills

After a Forest Fire

Once, the hills in the picture were covered by a forest. Many trees, smaller plants, and animals lived in the forest. These plants provided homes and food for the animals. The plants also soaked up rainwater. In this way, they kept soil from moving down hillsides.

The picture shows hcw the hills looked shortly after a fire. Changes have taken place. Soon the hills will show many signs of erosion.

Answer the questions.

1. What did the fire do to the forest animals? ____
2. How will erosion change the hills? ____
3. How could people help save the soil and the hills? ____

57

* **_Answers to masters on pages 206E–206H_**

Enrichment

Conduct a tour around the schoolyard, looking for signs of weathering and erosion. In an urban setting, look for such things as worn building stone, cracked and pitted sidewalks, rounded pebbles used for decorative stone, and rivulets carved in dirt by water flowing along street gutters.

Reteaching Suggestion ♦

Show students pictures of the Grand Canyon. Discuss the effects of weathering and erosion in terms of the Grand Canyon. Point out the depth of the canyon—1.6 km (1 mi)—and reinforce the idea that such features made by slow, steady processes require a long time to form—about 10 million years in this case.

Concept
Moving water erodes rocks.

Objectives/Process Skills
- *Observe* rocks that have been shaken in water 2000 times and 1000 times.
- *Record* observations.
- *Compare* the two piles of shaken rocks and the control group of rocks.
- *Describe* how the three piles of rocks differ and how the water in the two cans differ.

Time Allotment
Allow 30 minutes.

Safety Tips (See page T24.)
- Wipe up spilled water immediately.
- Remind students to be certain that the lids are tight.
- Tape the lids of the coffee cans securely.
- Students should use both hands—one on the lid and the other on the bottom of the can.
- Students should allow themselves plenty of room before shaking the cans.

Teaching Tips
- Have the students mark their cans with their names as well as their control letter.
- Have students feel as well as observe the three piles of stones in Step 7.
- Use limestone or sandstone for this activity.

Answers
State Your Conclusion

1. The water in can B is more cloudy. Small bits of rock can be seen in the water in can B.

2. The stones in pile B have the smoothest edges. The edges of the stones in piles A and B are smoother than those in pile C.

Use What You Learned

Moving rocks in a stream bump into each other, causing small pieces to break off the rocks. **Thinking Skill:** *Applying information to new situations*

Making A Model of Erosion

Suggested grouping: 5 students

Purpose

Observe how rocks erode in moving water.

Gather These Materials

• 30 small, rough stones • 2 coffee cans, half filled with water, with lids • 2 large, clear plastic cups • 3 sheets of paper towel • masking tape • marking pen

Follow This Procedure

1. Use a chart like the one shown to record your observations.
2. Divide the stones into three equal piles. Use the tape and the pen to label the piles A, B, and C.
3. Put pile A into one coffee can and pile B into the other can. Label the cans A and B.
4. Hold can A with both hands as shown in the picture. Shake the can 100 times. Ask another person in the group to shake the can 100 times.
5. Pour the water from can A into a cup. Look at the water.
6. Put the stones from can A on the paper towel. Look at the stones. Record what you see.
7. Repeat steps 4, 5, and 6 with can B. Have ten people shake the can 100 times. Compare the three piles of stones.

Record Your Results

	Water	Stones
Can A	cloudy	smoother
Can B	more cloudy	smoothest
Pile C		roughest

State Your Conclusion

1. Compare the way the water in can A looks with the way the water in can B looks.
2. How are the stones in piles A, B, and C different from each other after the shaking?

Use What You Learned

How are the stones moving in the cans like the rocks moving in a stream of water?

216

Activity Results

Shaking the can erodes the rocks, and bits of rock give the water a cloudy appearance.

Resource Book page 117

Name ______________

Chapter 10

Use with Lesson 2: page 216

Making a Model of Erosion

Activity Worksheet

Record Your Results

	Water	Stones
Can A		
Can B		
Pile C	XXX	

State Your Conclusion

1. Compare the way the water in can A looks with the way the water in can B looks.

2. How are the stones in piles A, B, and C different from each other after the shaking?

Use What You Learned

How are the stones moving in the cans like the rocks moving in a stream of water?

117

Predicting Earthquakes

The Problem One of the clearest signs that the earth is always changing is an earthquake. Over a hundred years ago, an earthquake shook the city of Los Angeles. In 1906, an earthquake caused a lot of damage in San Francisco. In 1987, a big earthquake hit Los Angeles. Scientists have been studying the causes of earthquakes. They know that large pieces of crust called plates cover the earth. Plates move only a few millimeters each year. Earthquakes occur when two plates hit each other or move apart. If scientists could tell when the earthquakes might happen, they might help save lives.

The Breakthrough Scientists looked for ways to keep track of small movements of the earth's surface. Several discoveries in the 1960s gave them the tools they needed. Scientists learned how to send satellites into space. They also learned how to make a special kind of light, called a laser. Laser light travels long distances in a very straight line. Lasers can make measurements over long distances.

New Technology Today, scientists shoot laser light at satellites from ground stations. These stations are on each edge of a plate in the earth's surface. Scientists time how long the laser light takes to reflect back from the satellite. Computers record the time and the position of the satellite. If the light travels back faster or slower, the station must have moved. If the station moved, then the crust must have moved. This helps people measure shifts in the earth's crust. Then they can predict that an earthquake is likely to occur. However, no one can tell exactly where or when an earthquake will hit.

What Do You Think?

1. What steps would you take if you lived where earthquakes were likely to occur?
2. Why is being prepared for earthquakes just as important as knowing when they might happen?

217

Science and Technology

TEACHING PLAN

Discussion

Question: **What is an earthquake?** (a shaking or sliding of the earth's crust) Encourage students to discuss what they already know about earthquakes. Question: **What kinds of problems might earthquakes cause?** (Accept reasonable responses.)

Teaching Tips

- Point out that scientists knew in advance that Mexico City was likely to experience the earthquake that occurred in September, 1985. They did not know, however, precisely when the earthquake would occur.

Answers

What Do You Think?

1. Students might respond that they would live in houses built to withstand an earthquake, and learn emergency procedures. **Thinking Skill:** *Comprehending meaning*

2. Predicting earthquakes is very inexact, and it is not clear what the best way to react to a specific prediction would be. People can prepare for earthquakes and thus be safer when an earthquake actually occurs. **Thinking Skill:** *Drawing conclusions*

Teaching Options

Books to Read

Paananen, Eloise. *Tremor: Earthquake Technology in the Space Age*. Messner, 1982. (for the teacher)

Science Background

The plates involved in California along the San Andreas Fault are the Pacific and North American plates. Some satellites revolve in orbits that change little. They appear to hover overhead. A laser is beamed to such a satellite from two ground stations, one on each side of a plate boundary. If the time it takes for the laser beam to reflect back to one of the stations changes, the location of the ground station must have changed. In other words, the plate must have shifted in location. Scientists can now barely detect ordinary continental drift direction.

TEACHING PLAN

Lesson Objectives

- *Describe* how earthquakes change the earth's crust.
- *Explain* how volcanoes erupt.

Lesson Vocabulary

earthquake, erupt, lava, magma, volcano

1. Motivate

Demonstration Activity ♦

Obtain a pail of water. Place a lightweight, floating object on the water. Put your hand in the water at the bottom of the pail. Allow the water to become still. Then shake your hand. *CAUTION:* Wipe up water spills immediately.

Discussion

Question: **What happens when I shake my hand at the bottom of the pail?** (Waves form at the top of the water, making the object bob.) Explain that waves traveled from your hand throughout the pail. Compare this action to movements of rock underground and waves of energy that travel outward from the movement, causing the shaking of an earthquake.

3 How Do Earthquakes and Volcanoes Change the Earth's Crust?

LESSON GOALS

You will learn
- how earthquakes change the earth's crust.
- how volcanoes change the earth's crust.

Weathering and erosion usually change the earth's crust very slowly. Sometimes, however, the crust changes very quickly. The picture shows where a large crack occurs in the earth's crust. Land on either side of this crack can suddenly move a couple meters in a minute.

earthquake (ėrth′kwāk′), a shaking or sliding of the earth's crust.

Earthquakes

Earthquakes change the earth's crust.

An **earthquake** is a shaking or sliding of the earth's crust. Earthquakes happen when rocks inside the earth move. As the rocks move, they may form new cracks in the crust. Sometimes, the movement of the rocks makes the land move up, down, or sideways. The picture shows land after an earthquake. How did the earthquake change the land?[1]

[1] formed a large crack

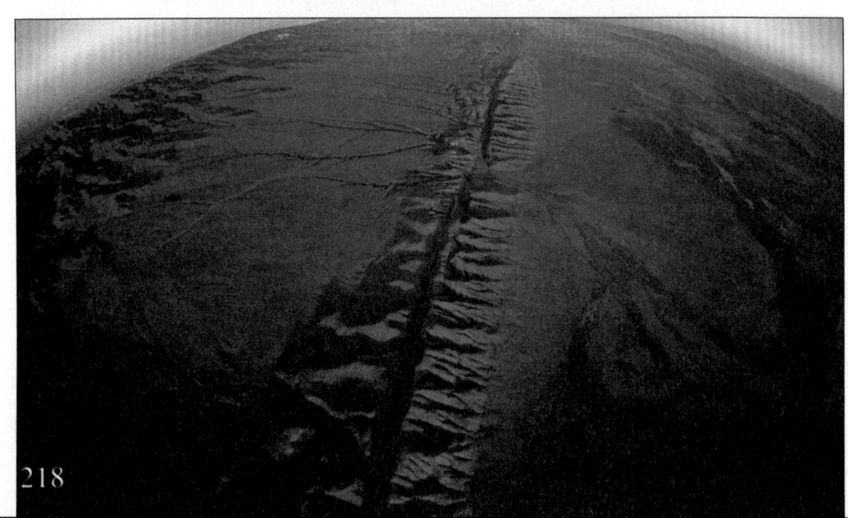

218

Teaching Options

Science Background

Earthquakes and volcanoes are most common near the boundaries of tectonic plates—large sections of the earth's crust that slowly move over partially melted rock in the upper mantle. Earthquakes and volcanoes occur when two plates move into each other, away from each other, or past each other. The San Andreas Fault in California, where many earthquakes occur, is part of the boundary between two plates that slide past each other. Faults are cracks in rock along which movement occurs. Faults vary in length from a few meters to over 1,000 km (620 mi).

Reading Strategies ♦

1. Guide students' pre-reading by asking: What words or ideas would you expect to find under each subheading?
2. Assign these strategies: Visualizing Information and Writing a Memory Sentence. (See pages T26–T29.)
3. Pair students to share what information is clear and unclear and initiate discussion using students' unanswered questions.

♦ *Suitable as a language development activity*

Most earthquakes are too weak for people to notice. However, some earthquakes are strong enough to shake buildings. The walls of buildings might fall down during a strong earthquake. The picture shows the kind of damage an earthquake can do. The movement of the ground twisted these railroad tracks as if they were a rubber band. Earthquakes can injure or kill people. Many people are hurt by falling objects.

Scientists try to predict earthquakes. Knowing when an earthquake will happen can save lives. People would have time to go to a safe place.

If you are indoors during an earthquake, stay away from windows and furniture that could slide or fall over. If outdoors, stay away from high buildings and other objects that could fall.

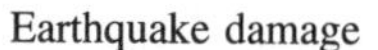

Earthquake damage

2. Teach

Teaching Tips

- **Possible Misconception:** Some students might think that, during an earthquake, huge cracks open up in the ground, swallowing people, cars, and houses. Point out that most cracks caused by the shaking during an earthquake are only a few meters (yards) deep.
- Point out that scientists use an instrument called a seismograph to locate and measure the strength of an earthquake.
- Question: **Why would a person probably be safer in an open field during an earthquake than in a building?** (The building might collapse during the earthquake. Also point out that much damage from earthquakes is from the fires that result when gas lines are broken and electrical equipment is damaged.)

Enrichment

Compare the large cracks (faults) in the crust to cracks formed in a sidewalk where a growing tree root or a slumping of the soil has offset the pavement. Look for such a place on the school grounds and lead a discussion on the site. Perhaps take the class to the general area and challenge them to find something that reminds them of the movement of the earth's crust, which would cause an earthquake.

Special Education

Have students with learning disabilities experiment with wooden blocks to visualize how the earth's crust moves when an earthquake occurs.

Reinforcement

Instruct the students to hold their palms together, with fingers pointing away, while they balance an unsharpened pencil across the index fingers. Tell them to slide their palms slowly and steadily past each other and to *observe* the pencil. It will move but should stay balanced on the fingers. Now tell them to come back to the original position, press their palms together, and jerk their palms a little past each other. The pencil will likely fall off their fingers. Compare this action to the buildup of energy in the earth's crust and the sudden movements of the crust that cause earthquakes.

TEACHING PLAN

Teaching Tips

- Have students use their fingers to trace the upward movement of magma through the top of the volcano illustrated on page 220.
- Ask students to *compare* and *contrast* magma and lava. (Both are melted rock but magma is inside the earth whereas lava is on the earth's surface.)
- Provide students with an *analogy* of an erupting volcano by squeezing toothpaste from a tube. Explain that the weight of surrounding rock provides the force that squeezes the magma through the crust, just as the hand provides the force that squeezes toothpaste through the tube.
- Provide samples of volcanic rock for the students to touch and *observe.* Such rocks as pumice and obsidian are relatively inexpensive at rock shops.
- On a world map, display the locations of famous volcanic eruptions. Point out Hawaii on the map. Tell students that these islands were formed from cooled lava that built up from the ocean floor due to volcanic eruptions over millions of years. On the chalkboard, draw a simple sideview of a volcano building up from the ocean floor, eventually forming an island.

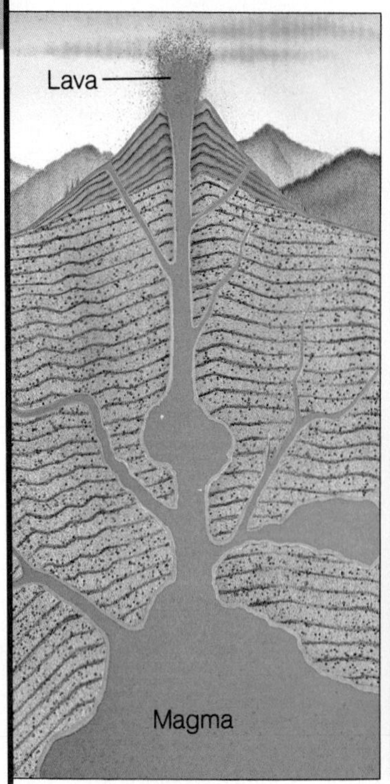

A volcano

Kilauea Volcano in Hawaii

volcano (vol kā′nō), a mountain with an opening through which lava, ashes, rocks, and other materials come out.

erupt (i rupt′), to burst out.

magma (mag′mə), hot, melted rock deep inside the earth.

Volcanoes

Hot, melted rock is shooting out from the volcano in the picture. A **volcano** is a mountain with an opening at the top. When a volcano erupts, melted rock, ashes, and other materials burst out from its opening.

What causes a volcano to **erupt**? The picture shows that hot, melted rock—or **magma**—lies deep inside the earth. The magma is squeezed up through the crust. Sometimes, the magma is forced up through the volcano, and the volcano erupts.

220

Teaching Options

Science Anecdote

The Hawaiian Islands formed as the plate on which they rest moved northwestward over a stationary "hot spot," or small melting region, in the earth's mantle. As the plate has continued to move, a submarine volcano has formed just southeast of Hawaii and may become another Hawaiian Island in as little as 100 years.

Reinforcement

Make a model of a volcano using modeling clay or plaster of Paris. Allow students to add twigs, grass, and rocks for realism. Place a large test tube or a small, porcelain crucible in the crater of the volcano. *CAUTION:* Have students wear cover goggles. Immediately wipe up any liquid that spills on the floor to prevent accidents. Place 1–2 tablespoons of baking soda in the container. Pour about 3 mL (1/2 oz.) of vinegar into the container. Have students *observe* and *describe* the flow of the "lava" and *compare* it to that of a real volcano.

Enrichment

Read newspaper or magazine accounts of volcanic eruptions to the class. These may be recent eruptions or historical ones. Discuss the geological conditions of the eruption as well as the effects of the eruption.

♦ ***Suitable as a language development activity***

Magma that comes out of a volcano is called **lava.** When the lava cools, it hardens into solid rock. The rock forms new crust. Like earthquakes, volcanoes can change the earth's crust very quickly.

When a volcano erupts, it can hurt people and damage property. Hot lava that flows down a volcano can burn homes.

Scientists look for warning signals to predict when a volcano will erupt. Small earthquakes near a volcano might mean that magma is rising inside. A bulge on a volcano also might mean that the volcano is about to erupt.

SCIENCE IN YOUR LIFE

A volcano can form new land. The Hawaiian Islands were formed by underwater volcanoes. The islands are the tops of volcanoes in the Pacific Ocean.

lava (lä′və), hot, melted rock that flows from a volcano.

Lesson Review

1. How can earthquakes change the earth's crust?
2. What causes a volcano to erupt?
3. **Challenge!** Why do earthquakes occur when magma rises in a volcano?

Study on your own, pages 334–335.

EARTH SCIENCE

FIND OUT ON YOUR OWN

Look in an encyclopedia to find out where earthquakes occur most often on the earth. Then, write the answers to the following questions.

1. What part of North America has the most earthquakes?
2. What parts of the earth have the fewest earthquakes?
3. Do more earthquakes occur under water or on land?

Teaching Tips

- Discuss how a series of small earthquakes or a bulge forming on a volcano might be warning signs of an eruption.
- Point out that a bulge on Mount St. Helens was detected and monitored for weeks before the volcano erupted. Another important indicator of a coming eruption is the type of gas and ash that often escapes from cracks in the crater.

3. Assess

Lesson Review

1. Earthquakes can cause cracks in the earth's crust and cause the land to move up, down, or sideways.
2. Magma being forced up through a volcano causes the volcano to erupt.
3. Challenge! The rising magma pushes on the surrounding rock and causes it to move. **Thinking Skill:** *Recognizing cause and effect*

Find Out On Your Own

1. The West Coast has the most earthquakes. **2.** Australia and Antarctica have the fewest earthquakes. **3.** More earthquakes occur underwater. **Thinking Skills:** *Interpreting charts, maps, and graphs; Collecting information*

Workbook page 58 *

Name ______
Use with Lesson 3: pages 218-221

Chapter 10
Science and Reading

A Volcano Erupts

Read the story. Then put the pictures in order.

Mount Saint Helens was a quiet volcano. It had not erupted for 123 years. Evergreen forests grew on its slopes. Many plants, animals, and birds lived there. Then, on May 18, 1980, the volcano erupted. The hot rock and ash made leaves and plants burn. The explosions knocked down millions of trees. Plants and animals died.

Several years have passed. The gray ash has formed rich soil. First, small plants started to grow. Then tiny animals came back to live. Next, the berry bushes grew. When the berries were ripe, the animals that eat the berries came back to live. Then, the trees started to get big. The birds came back to live in the trees.

Now squirrels and deer live on Mount Saint Helens again. Now the beautiful song of the mountain bluebird fills the air.

Look at the pictures. What happened first, second, third, and fourth? Write the numbers 1, 2, 3, or 4 in the proper boxes to show the proper order.

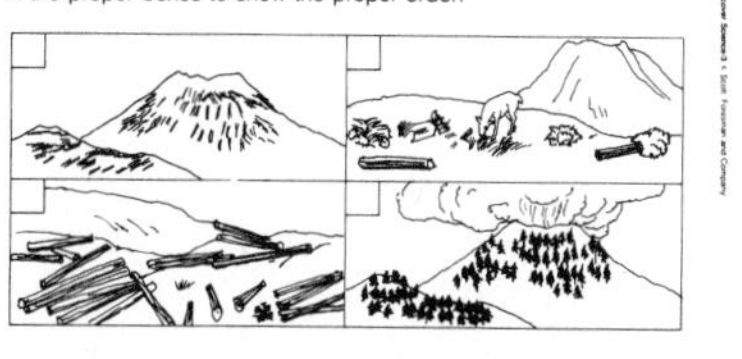

58

* *Answers to masters on pages 206E–206H*

Enrichment

Ask if any students have been to Yellowstone National Park. Let these students relate their experiences and *describe* what they saw. Discuss the various features that are due to a shallow region of magma beneath the surface. These features include geysers and a variety of features at hot springs. Consult a book about national parks or volcanoes for an explanation of these features.

Reteaching Suggestion ♦

Use colored construction paper to make two models: one of a cutaway view of a volcano and another of a side-view of a block of the earth's crust. Attach the models to the board. Ask volunteers to use the models to explain what happens during a volcanic eruption and an earthquake to change the earth's crust.

TEACHING PLAN

Lesson Objectives

- *Explain* how animals and plants change the earth's crust.
- *Explain* how people change the earth's crust.

1. Motivate

Demonstration Activity ♦

Fill a clear plastic tray with rocks and soil. Place the rocks on the bottom and the soil on top. Place a small object on the soil surface to represent a house or other building.

Discussion

Ask how the building of the house changes the crust. (digging in the ground to build the basement and/or lay the foundation, digging in the yard to install a septic system or sewer pipes, installing gas pipes into the house) As each of these items is mentioned, include them in your model, preferably near the outer edge of the tray so that students can see them. You may want to use straws for pipes and a thread spool for a septic tank.

4 How Do Living Things Change the Earth's Crust?

LESSON GOALS

You will learn
- how people change the earth's crust.
- how plants and animals change the earth's crust.

Whenever you dig a hole in the ground or split a rock in half, you are changing the earth's crust. People, plants, and animals change the earth's crust every day.

How Animals and Plants Change the Earth's Crust

Some animals, such as earthworms, live in the earth's crust. These animals change the earth's crust as they dig through the soil. Other animals look for food in the soil. How is the animal in the picture changing the land?[1]

Plants also help change the earth's crust. Plants can protect the land from erosion by holding the soil down. How can plants help change rocks into soil?[2]

[1]digging holes
[2]growing roots crack rocks

A prairie dog

Teaching Options

Science Background

The Surface Mining Control and Reclamation Act of 1977 was passed to control the environmental impact of coal strip-mining activities. Mining companies must prove they can reclaim—restore—the land before receiving a mining permit. After mining, the land must be restored so that it can be used for the purpose it served before mining. Restoration includes filling holes, contouring the land, removing wastes, and replanting vegetation. Mining companies must use the best available technology to prevent the pollution of streams and groundwater from toxic and acidic materials that wash off the mining wastes. Most experts agree that this law can help protect the environment if it is properly enforced.

Reading Strategies ♦

1. Guide students' pre-reading by asking: Which parts of the lesson are familiar and which parts are new?
2. Assign these strategies: Visualizing Information and Writing a Memory Sentence. (See pages T26-T29.)
3. Pair students to share what information is clear and unclear and initiate discussion using students' unanswered questions.

♦ *Suitable as a language development activity*

How People Change the Earth's Crust

When people build things, they change the land. People dig rocks and soil out of the ground to build roads and buildings. People cut down plants. Then wind and water can easily erode the soil. You learned in Chapter 9 that people dig mines to reach ores. Digging mines changes the earth's crust. Look at the picture. How are these workers changing the land?[1]

Lesson Review

1. How can plants and animals change the earth's crust?
2. How do people change the land when they build things?
3. **Challenge!** Why do plants grow better in soil in which animals live or dig?

Study on your own, pages 334–335.

SCIENCE IN YOUR LIFE

You might have seen cracked sidewalks near trees. As the trees' roots grew, they pushed up on the sidewalks. The pushing caused the sidewalks to break apart.

Building a road

[1]moving soil to build a road

EARTH SCIENCE

FIND OUT ON YOUR OWN

Long ago, there were no living things on the earth. Try to imagine what the earth might have looked like then. How was the earth's surface different from the surface today? Write a short paragraph that describes what you think the earth's crust used to be like. Then, use library books to find out what scientists think the earth was like at that time in history.

223

2. Teach

Teaching Tips

- Question: **How do you change the earth's crust?** (Lead students to understand that simple activities such as digging soil and kicking a stone change the crust.) Emphasize that these changes are not necessarily bad.
- **Possible Misconception:** Students may think that the crust refers only to the top layer of the earth underground. The crust also, however, includes the surface and near-surface.

3. Assess

Lesson Review

1. Animals dig in the soil and plant roots help break rocks apart.
2. People change the land by digging rocks and soil and cutting down trees.
3. Challenge! Plant roots grow more easily in loose soil. **Thinking Skill:** *Recognizing cause and effect*

Find Out On Your Own

Students might say the surface looked rocky because there were no plants or animals. **Thinking Skills:** *Visualizing; Collecting information; Drawing conclusions*

Workbook page 59 *

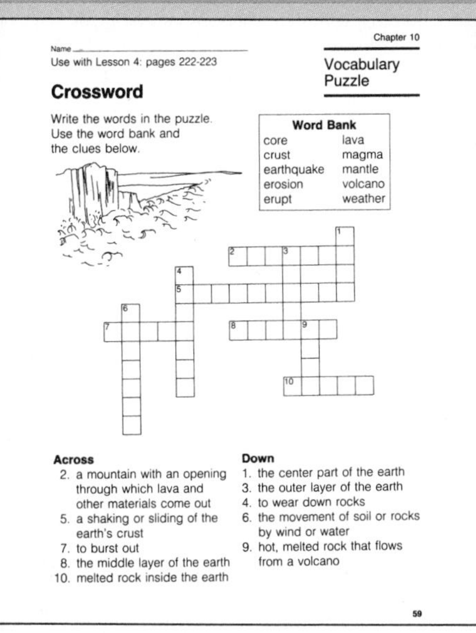

Chapter 10

Name

Use with Lesson 4: pages 222-223

Vocabulary Puzzle

Crossword

Write the words in the puzzle. Use the word bank and the clues below.

Word Bank

core	lava
crust	magma
earthquake	mantle
erosion	volcano
erupt	weather

Across
2. a mountain with an opening through which lava and other materials come out
5. a shaking or sliding of the earth's crust
7. to burst out
8. the middle layer of the earth
10. melted rock inside the earth

Down
1. the center part of the earth
3. the outer layer of the earth
4. to wear down rocks
6. the movement of soil or rocks by wind or water
9. hot, melted rock that flows from a volcano

59

* *Answers to masters on pages 206E–206H*

Special Education

Ask students with learning disabilities to *describe* ways that the earth's crust has been changed near their homes or near the school.

Reteaching Suggestions ♦

Have students look through magazines to find pictures of ways that people, animals, and plants have changed the earth's crust. Discuss what was changed in each picture.

TEACHING PLAN

Purpose
To develop the skills of collecting and organizing information using balances and line graphs to solve problems.

1. Motivate

Discussion
Question: **How do people change as they get older?** (Humans grow, mature, and become old. They look different at different stages.)

2. Teach

Teaching Tip
• Point out that, like humans and other living things, the earth changes as it gets older. Erosion is one process that changes the surface of the earth.

Skills for Solving Problems

Using Balances and Line Graphs

Problem: What effect do plants have on soil erosion?

Part A. Using a Balance to Collect Information

1. The pictures below show three trays of soil. Tray A is fully covered with grass plants. Tray B has only half the number of plants as Tray A. How many plants does Tray C have?
2. Each tray had the same amount of water sprinkled over it. The soil that washed away from each tray was collected and put on a balance. Each balance shows the mass of soil that was eroded from each tray. Tray A had 4 grams of soil eroded from it. How much soil was eroded from Tray B? Tray C?

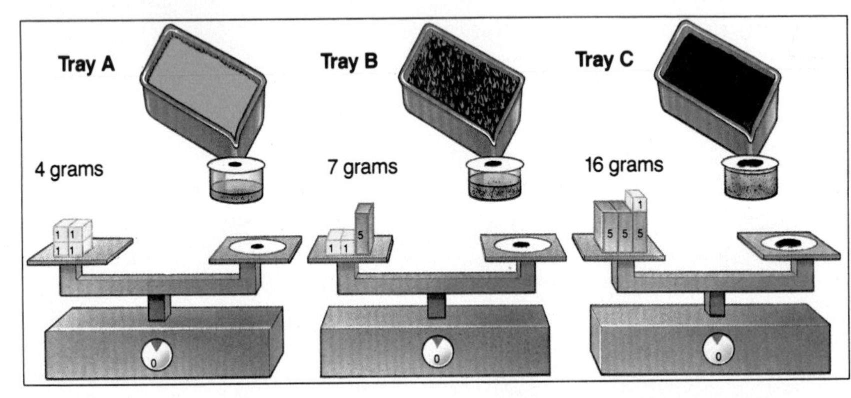

Part B. Using a Line Graph to Organize and Interpret Information

3. The line graph contains the information you collected in Part A. The dots on the graph show how many grams of soil were eroded from each tray. The line connecting the dots shows that erosion in each tray increased. If the line between two dots is not steep, the

Teaching Options

Sample Line Graph for Part C

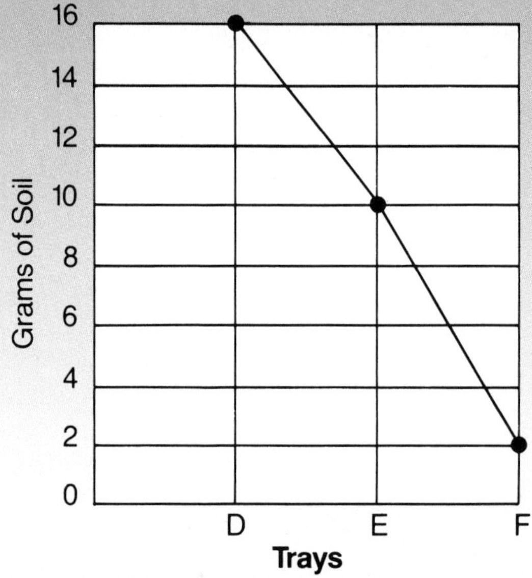

♦ ***Suitable as a language development activity***

increase in the amount of erosion was small. If the line is steep, the increase in the amount of erosion was great. Where is the line the steepest on the graph? Between which two trays did erosion increase the most? Between which two trays did erosion increase the least?

4. What effect do plants have on soil erosion?

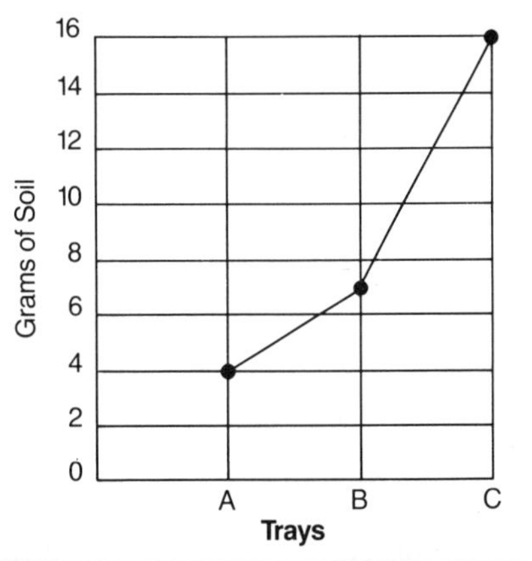

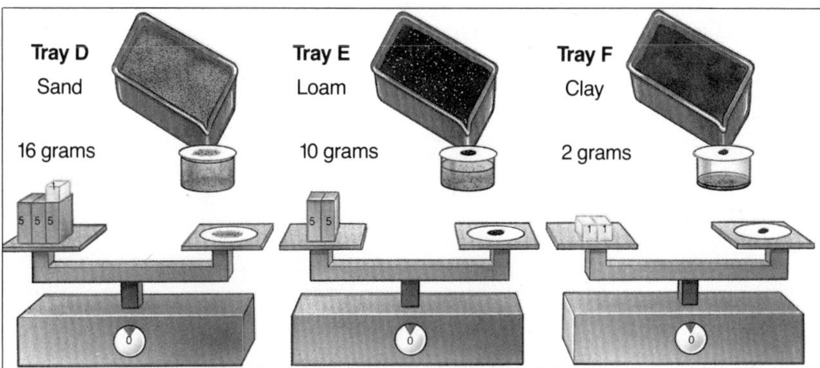

Part C. Using a Balance and Line Graph to Solve a Problem

Problem: How are different types of soil affected by erosion?

5. Use the pictures above to collect the information you need to solve the problem. Make a line graph similar to the one shown in Part B to organize your information.
6. Look at your line graph. Which kind of soil eroded the most? The least?

3. Assess

Part A

1. none

2. 7 g; 16 g

Part B

3. between lines B and C; A and B; A and B

4. Plants decrease the amount of erosion.

Part C

5. See Sample Line Graph for Part C in Teaching Options.

6. Sand eroded the most; clay eroded the least.

Reteaching Suggestion ♦

Present three soil trays: *Tray 1, covered by grass plants; Tray 2, covered by large rocks;* and *Tray 3, covered by small rocks.* Indicate amounts of soil erosion from each tray as follows: *Tray 1, 4g; Tray 2, 13g; Tray 3, 9g.* Have students place 5- and 1-gram masses on a balance to show the amount of eroded soil from each tray. Then have them draw a line graph on the board with the information. Questions: **From which tray did the most soil erode?** (Tray 2) **the least?** (Tray 1)

Resource Book page 119 *

Chapter 10

Name ______

Comprehension: time sequence

Science and Reading

The Order of Events

Read the story. Then number the events in the order that they happened in the story.

Julie needed to know more about earthquakes for her science project. After school, she went to the library to find out about earthquakes. She found a book and sat down to look at it.

Suddenly, the building began to shake. Julie thought it was just a train passing by. Then someone shouted, "Earthquake!" In a few seconds, the small earthquake was over.

Julie went back to her work and closed her book. She had more information about earthquakes now than she could ever find in a book. She wrote about how it felt to be in an earthquake.

__ Someone shouted, "Earthquake!"

__ Julie wrote how it felt to be in an earthquake.

__ The building began to shake.

__ Julie went to the library.

119

* Answers to masters on pages 206E–206H

REVIEW PLAN

Reviewing Science Words

1. magma
2. erosion
3. mantle
4. volcano
5. weather
6. crust
7. erupts
8. earthquake
9. core
10. lava

Reviewing What You Learned

1. b	**3.** a	**5.** c	**7.** a
2. c	**4.** d	**6.** b	**8.** d

Interpreting What You Learned

1. Three ways include by digging in the soil, by mining, and by cutting trees. **Thinking Skill:** *Restating or explaining ideas*

2. They study the inside of the earth by using special instruments and by studying the movements of the earth's crust. **Thinking Skill:** *Restating or explaining ideas*

3. The changes caused by an earthquake take place rapidly. The changes caused by erosion occur slowly. **Thinking Skill:** *Contrasting*

Chapter 10 Review

Chapter Main Ideas

Lesson 1 • The earth has three layers—the crust, the mantle, and the core. • Scientists learn about the inside of the earth by studying movements of the earth's crust and by using instruments.

Lesson 2 • Water, plants, and air help weather rocks. • Water and wind help cause erosion of the soil.

Lesson 3 • Earthquakes cause fast changes in the earth's crust. • Volcanoes cause the earth's crust to change very quickly.

Lesson 4 • Plants change the earth's crust by breaking rocks apart; animals change the earth's crust by digging in the soil. • People change the earth's crust by building roads and buildings, cutting down trees, and mining.

Reviewing Science Words

crust	erupts	mantle
core	lava	volcano
earthquake	magma	weather
erosion		

Copy each sentence. Fill in the blank with the correct word from the list.

1. ___ is hot, melted rock deep inside the earth.
2. The movement of soil or rocks by wind or water is ___.
3. The ___ is the middle layer of the earth.
4. A ___ is a mountain with an opening through which lava, ashes, rocks, and other materials come out.
5. Rocks ___ when they wear down or break apart over a long time.
6. The outer layer of the earth is the ___.
7. A volcano ___ when melted rock, ashes, and other materials burst out from its opening.
8. A shaking or sliding of the earth's crust is an ___.
9. The ___ is the inner layer of the earth.
10. ___ is hot, melted rock that flows from a volcano.

226

Review Options

Cooperative Learning ♦

STAD Format (See page T23.)
Assign students to work in four- to five-member teams to study Chapter 10 Review. Students should work together to make sure that they and their teammates know the material in the chapter. After students have had enough time to study together, give them a test to complete individually (Chapter 10 Test A or B in the *Test Book*). Award Superteam certificates to teams whose average test scores exceed 90% and Greatteam certificates to teams whose average test scores exceed 80%.

Test Book page 89 *

Name ______________________

Chapter 10
Test A

Multiple Choice Choose the best answer.

1. The layer of the earth made of rocks and soil is called the
   a. mantle.
   b. crust.
   c. core.
2. The layer of the earth made mostly of iron is called the
   a. mantle.
   b. crust.
   c. core.
3. How do scientists study the mantle and core of the earth?
   a. by digging
   b. by going inside the earth
   c. with special instruments
4. The wearing down or breaking apart of rocks is called
   a. weathering.
   b. mining.
   c. an eruption.
5. When the earth's crust is changed because soil is moved by wind, this is
   a. weathering.
   b. wind erosion.
   c. an eruption.
6. Plants can help change the earth's
   a. core.
   b. magma.
   c. crust.
7. A shaking or sliding of the earth's crust is
   a. erosion.
   b. an earthquake.
   c. weathering.
8. A mountain with an opening where lava, ash, rocks, and other materials come out is a
   a. magma.
   b. mantle.
   c. volcano.
9. How can strong earthquakes cause damage or hurt people?
   a. hot magma
   b. falling objects
   c. strong winds
10. A bulge in a volcano might mean that the volcano
    a. is about the erupt.
    b. will never erupt.
    c. has stopped erupting.

89

♦ ***Suitable as a language development activity***

Reviewing What You Learned

Write the letter of the best answer.

1. The mantle is mostly made of
   (a) water. (b) rock. (c) metal. (d) soil.
2. Which of the following makes slow changes in the earth's crust?
   (a) volcano (b) earthquake (c) erosion (d) lava
3. Plants help change rocks into
   (a) soil. (b) magma. (c) volcanoes. (d) lava.
4. Plants can help change the earth's
   (a) core. (b) mantle. (c) magma. (d) crust.
5. The earth's core is mostly
   (a) cold. (b) water (c) iron. (d) soil.
6. What mixes with carbon dioxide to turn hard minerals into clay?
   (a) air (b) water (c) lava (d) rock
7. When large blocks of rock in the earth's crust move, they form
   (a) an earthquake. (b) erosion. (c) lava. (d) weathering.
8. When lava cools, it forms
   (a) humus. (b) sandy soil. (c) clay. (d) rock.

Interpreting What You Learned

Write a short answer for each question or statement.

1. What are three ways people can change the earth's crust?
2. How do scientists study the inside of the earth?
3. How does the way an earthquake changes the land differ from the way erosion changes the land?
4. Explain how water can cause rocks to weather.
5. How can people protect themselves during an earthquake?
6. What is one warning signal that a volcano is about to erupt?

Extending Your Thinking

Write a paragraph to answer each question or statement.

1. How could you get an idea of how deep a hole is without really looking into it?
2. In which place would you find more erosion, a thick forest or a bare hill?

To explore scientific methods, see Experiment Skills on pages 366–367.

4. Water that freezes in the cracks of rocks can cause the rock to break apart. Water mixed with carbon dioxide makes a material that can turn hard minerals into soft clay. **Thinking Skill:** *Recognizing cause and effect*

5. Indoors, people should stay away from windows and furniture that could fall over or slide. Outdoors, people should stay away from buildings and objects that could fall. **Thinking Skill:** *Restating or explaining ideas*

6. Answers may include many small earthquakes near the volcano or a bulge on the volcano. **Thinking Skill:** *Restating or explaining ideas*

Extending Your Thinking

1. Students may suggest tossing a rock into the hole and measuring how long the rock takes to hit bottom. Another suggestion may be to tie a heavy object to a rope, lower the rope into the hole until it hits bottom, and measure how much of the rope is in the hole. **Thinking Skill:** *Inferring, identifying and suggesting alternatives*

2. A bare hill would have more erosion due to the slope and the lack of protective vegetation to cover and hold the soil in place. **Thinking Skills:** *Drawing conclusions, Predicting, Inferring*

Test Book page 90 *

Name ____________

Chapter 10
Test A

Short Answer Write the name of each part of the volcano in the correct blank. Use these words: magma, eruption, lava, new crust.

Lava and Magma
Exploding Lava
Cinder Layer
Hardened Lava
Rocks Already There

1. ______ 2. ______ 3. ______ 4. ______

Short Essay Use complete sentences to answer each question.

1. Name two ways to protect yourself during an earthquake.

2. Explain how wind and water can cause erosion.

90

Test Book page 91 *

Name ____________

Chapter 10
Test B

Multiple Choice Choose the best answer.

1. The earth's crust is mostly the land we walk on and the
   a. rocks of the mantle.
   b. land under oceans.
   c. iron in the core.
2. Scientists cannot see the inside of the earth, but they can study it from
   a. pictures of the core.
   b. movements of the crust.
   c. trips inside deep holes.
3. The middle layer of the earth is called the
   a. mantle.
   b. core.
   c. crust.
4. When the earth's crust is changed because soil is moved by water, this is called
   a. water erosion.
   b. weathering.
   c. an earthquake.
5. Freezing and thawing, plant roots, and rainwater all can make rocks
   a. grow larger.
   b. weather.
   c. erupt.
6. When lava cools, it forms
   a. sand.
   b. rocks.
   c. soil.
7. The earth's crust is usually changed very slowly by
   a. earthquakes.
   b. erosion and weathering.
   c. volcanoes.
8. Plants can help change the earth's
   a. mantle.
   b. crust.
   c. center.
9. People who build buildings, cut down trees, and mine for ore help
   a. change the earth's crust.
   b. change the earth's mantle.
   c. prevent erosion.
10. Large blocks of rock in the earth's crust move and form cracks during
    a. a volcano.
    b. a flood.
    c. an earthquake.

91

Test Book page 92 *

Name ____________

Chapter 10
Test B

Short Answer Write the name of each part of the volcano in the correct blank. Use these words: magma, eruption, lava.

1. ______ 2. ______ 3. ______

Short Essay Use complete sentences to answer each question.

1. Name three ways rocks are weathered.

2. Name two ways to protect yourself during an earthquake.

92

* *Answers to masters on pages 206E–206H*

 Clouds and Storms Planning Guide

TEACHING PLAN

Chapter Components	Skills	Materials
Chapter Opener/*DISCOVER:* Observing How Water Changes pp. 228–229	*DISCOVER* p. 229 Science Process Skills *Observing, Inferring*	*DISCOVER* p. 229 (individual) 1 roll paper towels, 30 straws, 30 plastic bags
Lesson 1 How Do Clouds Form? pp. 230–232	Thinking Skills Challenge!: *Drawing conclusions* Find Out On Your Own: *Making models*	Demonstration p. 230 30 paper towels, water
Activity Measuring Temperature at Which Dew Forms p. 233	Science Process Skills *Observing, Collecting and interpreting data, Measuring, Inferring*	(groups of 2) 15 soup cans, 1 bag ice cubes or crushed ice, water, 15 stirring sticks, 15 Celsius thermometers (alcohol only)
Lesson 2 What Happens to Water in the Clouds? pp. 234–236	Thinking Skills Challenge!: *Inferring* Find Out On Your Own: *Making models*	Demonstration p. 234 spray bottle, plate of glass or sauce pan, water
Activity Observing Part of the Water Cycle p. 237	Science Process Skills *Observing, Making models, Collecting and interpreting data*	(groups of 2) 15 large jars, hot tap water, 1 bag ice cubes, 1 roll plastic wrap, 15 rubber bands
Lesson 3 What Causes Storms? pp. 238–240	Thinking Skills Challenge!: *Drawing conclusions* Find Out On Your Own: *Restating or explaining ideas*	Demonstration p. 238 tall clear plastic or glass container, water, sand, spoon or stick
Science and People Mapping Tornadoes p. 241	Thinking Skills *Inferring, Applying information to new situations*	
Skills for Solving Problems Using Rain Gauges and Charts pp. 242–243	Problem Solving Skills *Making decisions/Identifying and solving problems, Interpreting charts, maps, and graphs*	
Chapter Review pp. 244–245	Thinking Skills *Restating or explaining ideas Recognizing the main idea and supporting details, Sequencing, Recognizing patterns and relationships, Inferring, Drawing conclusions*	

Teaching Options

Strategies	Extensions		Resource Masters
Cooperative Learning p. 228 (Also see p. T23.) Applying Whole Language p. 229 (Also see p. T30.)			Family Letter: *Resource Book* p. 123
Reading Strategies p. 230 (Also see pp. T26–T29.)	Science and Language Arts p. 231 Special Education p. 231	Reteaching Suggestion p. 232	Vocabulary Preview: *Workbook* p. 61
			Activity Worksheet: *Resource Book* p. 127
Reading Strategies p. 234 (Also see pp. T26–T29.)	Special Education p. 235 Reinforcement p. 235	Reteaching Suggestion p. 236	Science and Social Studies: *Workbook* p. 62 Science Activity: *Workbook* p. 63
			Activity Worksheet: *Resource Book* p. 129
Reading Strategies p. 238 (Also see pp. T26–T29.)	Special Education p. 239 Game Suggestion p. 239	Reteaching Suggestion p. 240	Science Skills: *Workbook* p. 64 Vocabulary Puzzle: *Workbook* p. 65
	Reteaching Suggestion p. 243		Science and Reading: *Resource Book* p. 131
Cooperative Learning p. 244 (Also see p. T23.)			Chapter Tests: Forms A and B *Test Book* pp. 97–100

Classroom Management

Advance Preparation

Activity, page 233
Collect enough aluminum cans for each group of students. Soup cans or fruit cans with the labels removed work well. Have large containers of water prepared ahead of time. Water should be at room temperature to start the activity.

Vocabulary Review

Use the following sentences with your students to review the meanings of the italicized words.

1. Invisible *water vapor* in the air *condenses* into tiny drops of water that form clouds.
2. A puddle on the sidewalk disappears as the water *evaporates* into the air.
3. Heat is one kind of *energy*.
4. The shock that you sometimes feel when touching metal is an *electrical charge*.

High-Potential Students

Have students make poetry booklets about clouds. Introduce students to various poetry styles such as haiku, cinquain, free verse, limerick, and triplet. Provide resources to assist students in locating information about clouds and cloud terms. Encourage students to illustrate their poems and bind their booklets using yarn, fasteners, or staples.

Mainstreamed Students

Emotionally Handicapped
Ask students to relate their experiences with water vapor condensing on a cold container, such as beads of water on a can of soda or on a bathroom mirror.

Hearing Impaired
Be sure hearing impaired students are aware of the television station bulletins that flash across the bottom of the television screen before or during a storm.

Visually Impaired
Provide visually impaired students with cotton balls to make the three types of clouds discussed in this chapter.

Ask students to describe a thunderstorm, rainfall, or snowfall using senses other than sight. Students might refer to the sounds of distant thunder, the muffled or crunching sounds of cars and people during a snowfall, and the musty smell that sometimes occurs during a light rain.

Science Fair Projects

The Experiment Skills on p. 368 and the Investigate feature in the chapter can be used for science fair projects. You might also encourage interested students to do one of the following projects:

1. Measure the weekly rainfall in your area with a simple rain gauge. Place a coffee can outside during a rainfall. After the rain has stopped, measure the depth of the water with a ruler that measures centimeters and inches. Record the measurements over a one monthy period and make a chart showing the results.
2. Research how animals change when the weather changes. Refer to encyclopedias and other reference sources. Illustrate your findings with pictures and drawings.
3. Take photographs of different cloud types and arrange them in a chart. On the chart include the characteristics of each cloud type.
4. Make posters related to storm safety for thunderstorms, hurricanes, and tornadoes.

lassroom Resources

Bulletin Board

Ask students to imagine they are water drops going through the water cycle. Have them write a short story or describe their experiences as they go through the cycle.

Chapter 11 Poster

Science Discovery Center

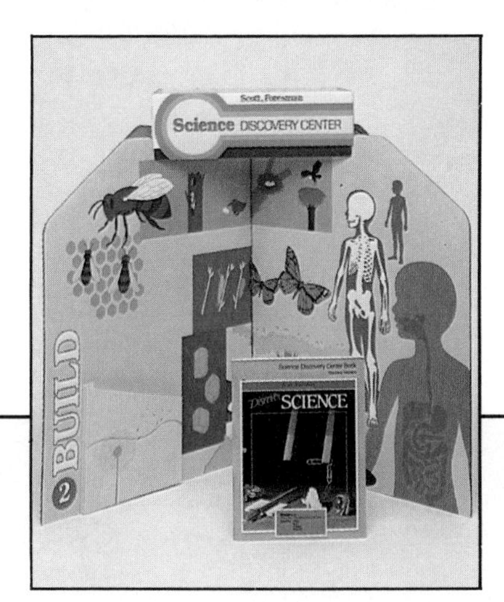

Use pages 115–120 from the *Science Discovery Center Book.* Place these worksheets in the appropriate pockets in the Science Discovery Center.

Overhead Transparencies

Use Transparencies 18 and 19 from the package of color overhead transparencies.

CHAPTER 11 COPY MASTERS

Teacher's Resource Book

Dear Family,
Your student will be reading **Chapter 11: Clouds and Storms** in *Discover Science*, published by Scott, Foresman. We will learn that when air cools, the water vapor in it condenses and makes tiny droplets of water. These droplets join together to make clouds. Three kinds of clouds are cirrus, cumulus, and stratus clouds. Fog is a cloud near the ground.
By doing this activity together, you and your student can learn more about fog.

Making Fog

You will need a clear glass jar or bottle, hot water, ice cubes, and cheesecloth.

1. Fill the jar with hot water. Then pour out about three-fourths of the water.
2. Wrap in cheesecloth enough ice cubes to cover the mouth of the jar. Hold the ice cubes in the mouth.
3. Hold the bottle up to the light. Thin streams of fog moving down into the bottle should be visible.
4. Explain to your student that the ice cubes cool the hot air, which is full of water vapor. The coolness causes the water vapor to condense into fog.

Workbook

Name ______________________

Use with Lesson 1: pages 230-232

Chapter 11

Vocabulary Preview

Clouds and Storms

	Chapter Vocabulary	
cirrus clouds	fog	stratus clouds
cumulus clouds	hurricane	tornado
dew	precipitation	water cycle

Vocabulary Cards

1. Write each word on a card.
2. Find each word in the glossary. Copy the pronunciation under the word on the card.
3. Practice saying the words with a partner.

cirrus clouds

Word Meanings

1. Cirrus clouds are sometimes called *mares tails*. These clouds are often thin and feathery like the tails of horses. On your vocabulary card marked *cirrus clouds*, draw cirrus clouds as shown.
2. Cumulus clouds are thick, white, and puffy. On your card marked *cumulus clouds*, draw a soft cumulus cloud as shown.
3. Stratus clouds are like a blanket. They spread out over the sky. On your card marked *stratus clouds*, draw even-layered clouds as shown.
4. Write your own definitions of the following words. Use the backs of the cards. Check your definitions in the glossary.
   dew fog hurricane tornado

cumulus cloud

stratus clouds

At Home

1. Practice saying the words. Learn their meanings.
2. Look for the words and pictures of the words in newspapers and magazines. You can make a poster.

Teacher's Notes: Ask students to draw pictures for the words they have defined in the *Word Meanings*.

Teacher's Resource Book

Name ______________________

Use with Lesson 1: page 230

Chapter 11

Activity Worksheet

Measuring Temperature at Which Dew Forms

Record Your Results

Air Temperature	
Water Temperature	
Temperature at which water forms on can	

State Your Conclusion

1. How did the temperature at which water formed on the can compare with room temperature?
____
____
2. What caused water to form on the outside of the can?
____
____

Use What You Learned

What can happen to water vapor in the air when it rises and cools?
____
____
____
____

Workbook

Name ______________________

Use with Lesson 2: pages 234-236

Chapter 11

Science and Social Studies

Mapping Rainfall

The map shows how much it usually rains in one year.

1. Find the areas marked ▩. These areas have more than 150 cm of rain each year. Color these areas green.
2. Find the areas marked ▨. These areas have 125 cm to 150 cm of rain each year. Color these areas blue.
3. Find the areas marked ░. These areas have 40 cm to 125 cm of rain each year. Color these areas red.
4. Find the areas marked □. These areas have 0 to 40 cm of rain each year. Color these areas yellow.
5. Circle the amount of rain that most areas get.
   Over 150 cm 125 to 150 cm (40 to 125 cm) 0 to 40 cm

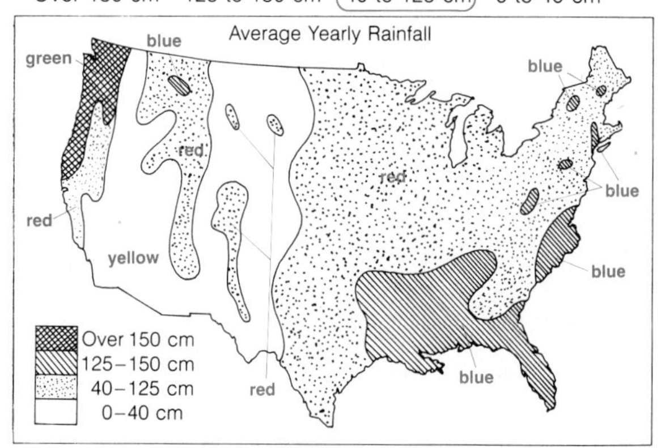

Teacher's Notes: You might want to point out that both rain and snow are included when figuring yearly rainfall. Ask why.

Name

Use with Lesson 2: pages 234-236

Chapter 11

Science Activity

How Is a Rainbow Made?

Gather These Materials

• a small bowl • water • a small mirror • a piece of white paper

Follow This Procedure

1. Fill a small bowl with water. Place a mirror in the bowl.
2. Turn the bowl so that the sun passes through the water and hits the mirror. Hold a sheet of white paper so that the sunlight shines from the mirror onto the paper.

Record Your Results

Draw in color what you saw.
Students will draw rainbows.

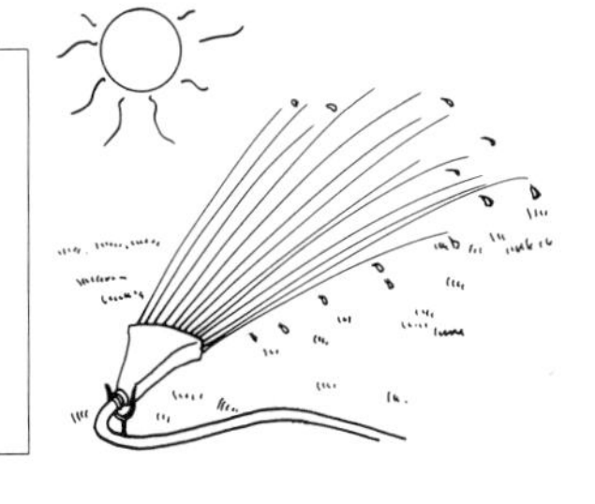

State Your Conclusions

1. What happened when sunlight was bent by the water?
   A rainbow formed.
2. What other ways does light bend to make a rainbow?
   Answers will vary. Example: Sunlight passes through crystal or cut glass, through a water fountain or a water sprinkler.

Teacher's Notes: Point out that red, orange, yellow, green, blue, and violet are the main colors in a rainbow.

Name

Use with Lesson 2: page 237

Observing Part of the Water Sample

Chapter 11

Activity Worksheet

Record Your Results

After 5 minutes	
After 10 minutes	

State Your Conclusion

1. What happened to some of the hot water in the jar?
2. What did putting ice on the plastic wrap do to the warm air in the jar?

Use What You Learned

What part or parts of the water cycle were shown in this activity?

Name

Use with Lesson 3: pages 238-240

Chapter 11

Science Skills

Looking at Storms

A. The pictures show a hurricane, a tornado, and a thunderstorm. Decide which storm is in each picture, and write its name in the picture.

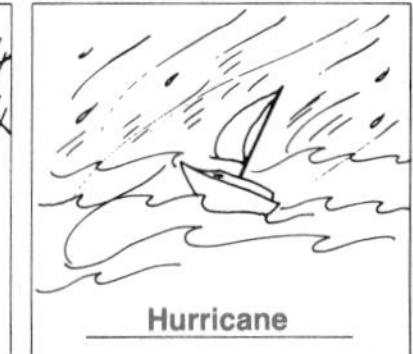

B. Decide which kind of storm is described in each sentence below. Write the number for the sentence in the correct box below. You will write some numbers in more than one box.

1. This storm covers the largest region.
2. The land or water gets heavy rainfall.
3. You see lightning and hear thunder.
4. A funnel-shaped cloud appears.
5. This storm covers the smallest area of the three.
6. It moves through air, touching ground sometimes.
7. It grows when warm moist air quickly moves upward.
8. It has the strongest winds on earth.
9. When it moves to land, it begins to disappear.
10. It is strong enough to tear down buildings.

Hurricane	Tornado	Thunderstorm
1 2 7 9 10	4 5 6 7 8 10	2 3 7

Teacher's Notes: Have students compare these storms.

Name

Use with Lesson 3: pages 238-240

Chapter 11

Vocabulary Puzzle

Cryptogram

Each number in the code stands for a letter. Use the code to write the words below. One is done for you.

1	2	3	4	5	6	7	8	9	10	11	12	13
A	B	C	D	E	F	G	H	I	J	K	L	M

14	15	16	17	18	19	20	21	22	23	24	25	26
N	O	P	Q	R	S	T	U	V	W	X	Y	Z

1. T O R N A D O
   20 15 18 14 1 4 15
2. H U R R I C A N E
   8 21 18 18 9 3 1 14 5
3. W A T E R C Y C L E
   23 1 20 5 18 3 25 3 12 5
4. S T R A T U S C L O U D S
   19 20 18 1 20 21 19 3 12 15 21 4 19
5. C U M U L U S C L O U D S
   3 21 13 21 12 21 19 3 12 15 21 4 19
6. C I R R U S C L O U D S
   3 9 18 18 21 19 3 12 15 21 4 19
7. P R E C I P I T A T I O N
   16 18 5 3 9 16 9 20 1 20 9 15 14
8. F O G
   6 15 7
9. D E W
   4 5 23

Teacher's Note: Discuss the local weather. Encourage students to use the chapter vocabulary during the discussion.

Teacher's Resource Book

Name ____________

Comprehension: cause and effect relationships

Chapter 11

Science and Reading

Why Did They Do It?

Read the story and the questions. Circle the letter of the sentence that best answers each question.

Joey and his sister, Lisa, went to play golf because it was such a beautiful day. Joey was teaching Lisa how to play golf. Since Joey was such a good golfer, Lisa was happy to have him for a teacher.

Joey and Lisa had only played four holes when clouds began to fill the sky. Then they heard thunder in the distance. Joey and Lisa were near the clubhouse, so they ran inside for shelter. Joey told Lisa that it is not safe for golfers to stay on a golf course during a thunderstorm. They may get struck by lightning.

1. Why was Lisa happy to have Joey for a teacher?
   (a.) Joey was a good golfer.
   b. It was a beautiful day.
   c. Joey didn't like to play golf.

2. Why did Joey and Lisa run for shelter inside?
   a. Joey wanted to show Lisa the clubhouse.
   b. They wanted to get something to drink.
   (c.) A golf course is not a safe place during a storm.

3. Why is it not safe for golfers to stay on a golf course during a thunderstorm?
   a. They may get their clothes all wet.
   (b.) They may get struck by lightning.
   c. They may forget what their score is.

Test Book

Name ____________

Chapter 11 Test A

Multiple Choice Choose the best answer.

1. Water vapor that condenses on cool surfaces is (1-1)
   (a.) dew.
   b. rain.
   c. snow.

2. Clouds are formed by (1-2)
   a. cool air rising.
   (b.) water vapor condensing.
   c. lightning.

3. Stratus clouds form in (1-3)
   a. fog.
   b. balls.
   (c.) layers.

4. Clouds made of tiny ice crystals are called (1-3)
   (a.) cirrus.
   b. cumulus.
   c. stratus.

5. A puffy cloud that looks like cotton is called (1-3)
   a. cirrus.
   (b.) cumulus.
   c. stratus.

6. Water droplets that are too heavy to float make (2-1)
   a. fog.
   b. clouds.
   (c.) rain.

7. The changing and moving of the earth's water is called (2-2)
   (a.) the water cycle.
   b. a hurricane.
   c. precipitation.

8. Most water evaporates from
   a. lakes.
   b. ponds.
   (c.) oceans. (2-2)

9. How does wind change the speed that water evaporates? (2-2)
   (a.) makes it faster
   b. no change
   c. makes it slower

10. When electric charges jump between clouds, we see (3-1)
    a. rain.
    (b.) lightning.
    c. thunder.

11. A hurricane forms over a warm
    (a.) ocean.
    b. city.
    c. desert. (3-2)

12. Tornadoes can form during a
    a. snowstorm.
    b. heavy fog.
    (c.) thunderstorm. (3-2)

Numbers in parentheses after each question refer to the lesson number and the objective of that lesson.

Test Book

Name ____________

Chapter 11 Test A

Short Answer Label the pictures of the water cycle. Use these words: (2-2)

evaporation precipitation condensation

1. precipitation 2. evaporation 3. condensation

Match each meaning in Column I with a word in Column II. Write the correct letter in the blank. (3-2)

Column I	Column II
b 4. a funnel-shaped cloud that forms over land during a thunderstorm	a. hurricane
c 5. formed from electric charges	b. tornado
a 6. a storm that can form over a warm ocean	c. lightning

Short Essay Use complete sentences to answer the question.

1. What is one way hurricanes and tornadoes are alike? How are they different? (3-2)

   Both have strong winds. Hurricanes are huge storms. Tornadoes move along a narrow path.

Numbers in parentheses after each question refer to the lesson number and the objective of that lesson.

Test Book

Name ____________

Chapter 11 Test B

Multiple Choice Choose the best answer.

1. What causes water to evaporate faster? (1-1)
   a. rain
   b. clouds
   (c.) sun

2. Clouds that form just above the ground are called (1-2)
   a. dew.
   (b.) fog.
   c. droplets.

3. The layered clouds often observed during a light rain are called (1-3)
   a. cirrus.
   b. cumulus.
   (c.) stratus.

4. Feathery clouds floating high in the air on a mostly clear day are (1-3)
   (a.) cirrus.
   b. cumulus.
   c. stratus.

5. Clouds that appear puffy and cottonlike are (1-3)
   a. cirrus.
   (b.) cumulus.
   c. stratus.

6. Most of the earth's water is (2-2)
   a. in the clouds.
   (b.) in the ocean.
   c. under the ground.

7. Heavy droplets of water that can no longer float in the clouds fall as (2-1)
   a. fog.
   b. snow.
   (c.) rain.

8. The fastest winds on earth are in a (3-2)
   (a.) hurricane.
   b. tornado.
   c. stratus cloud.

9. Electricity that jumps from one cloud to another during a storm is called (3-1)
   (a.) lightning.
   b. thunder.
   c. heat.

10. A storm that forms over the ocean and has heavy rains is a (3-2)
    (a.) hurricane.
    b. tornado.
    c. wind storm.

Numbers in parentheses after each question refer to the lesson number and the objective of that lesson.

Name ______________________

Chapter 11
Test B

Short Answer Label the storms.
Use these words: tornado thunderstorm hurricane (3-2)

1. hurricane 2. tornado 3. thunderstorm

Fill in the blanks with the correct answers.

4. A storm with strong, destructive winds and heavy rains that can cause floods is a hurricane. (3-2)

5. During a thunderstorm, large cumulus clouds form, winds begin to blow, and rain falls. (3-2)

6. Tornadoes make the most powerful winds on earth. (3-2)

Short Essay Use complete sentences to answer the question.
1. Explain how water moves through the water cycle. (2-2)

Rain falls to the ground and collects in streams. Energy from the sun makes some water evaporate. Then water vapor forms in the air. Some water vapor condenses into clouds and rain falls again.

Numbers in parentheses after each question refer to the lesson number and the objective of that lesson.

TEACHING PLAN

Major Concepts

Lesson 1 Rising warm, moist air causes the development of clouds.
Lesson 2 Water in clouds can form large raindrops and fall back to earth as precipitation.
Lesson 3 Thunderstorms, hurricanes, and tornadoes are caused by rapidly rising air.

Chapter Vocabulary

cirrus clouds, cumulus clouds, dew, fog, hurricane, precipitation, stratus clouds, tornado, water cycle

Getting Started

Ask students to ***describe*** what they see, hear, and feel during a thunderstorm. Ask them why they think a thunderstorm occurs. Have them ***explain*** what they would do if they found themselves in a thunderstorm like the one they see in the picture.

Chapter 11

Clouds and Storms

Think about the last time you saw a storm. You might have seen large clouds like these in the sky. You also might have seen lightning.

Teaching Options

Cooperative Learning ♦

Jigsaw Format (See page T23.)
Assign the following topics at random to your cooperative learning teams.

Topic A: What is evaporation and condensation?
Topic B: How do clouds form?
Topic C: Describe the travels of a drop of water as it goes through the water cycle.
Topic D: How do hurricanes and tornadoes form?

Have students search for information on their topic as they read the chapter. Then let all students with the same topic meet in an expert group to discuss the information. When students return to their teams, they may take turns presenting their topics to the team. Then give students a test covering all topics to complete individually (Chapter 11 Test A or B in the *Test Book).* Award Superteam certificates to teams whose average test scores exceed 90%, and Greatteam certificates to teams whose average test scores exceed 80%.

♦ ***Suitable as a language development activity***

Introducing the Chapter

In this chapter, you will find out how clouds form. You also will learn what causes rainy and stormy weather. The air always has some water in it—even on a clear day. This activity will help you learn about water in the air.

Observing How Water Changes

Wear cover goggles for this activity.

Wet a paper towel and squeeze out the extra water. Use the paper towel to make a wet mark on a chalkboard. Look at the chalkboard as it dries.

Place one end of a straw inside a small plastic bag. Hold the edges of the bag tightly around the sides of the straw. Now blow through the straw. How does the plastic bag look?[1] Open the plastic bag. Run your finger along the inside of the bag. How does the inside of the bag feel?[2]

Talk About It

1. Where did the water go as the chalkboard dried?
2. How did the inside of the plastic bag become wet?

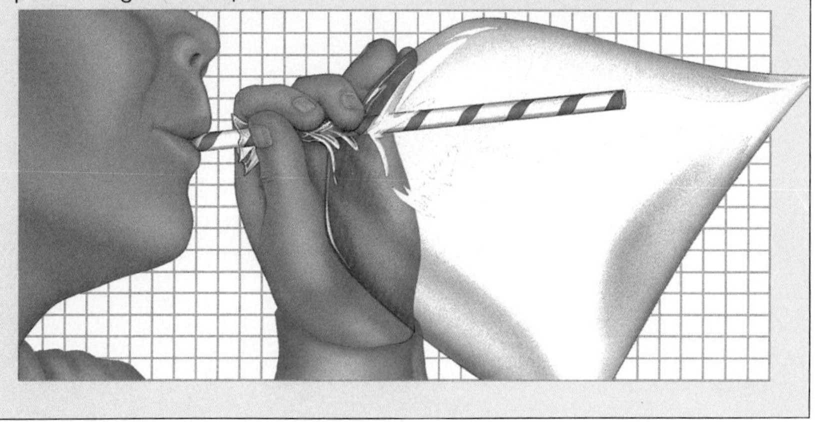

[1]cloudy [2]wet

DISCOVER

Objective ♦

This optional *DISCOVER* activity will help students explore and build background information about the concepts of evaporation and condensation. Later in the chapter students will be able to draw on this experience to help them assimilate the new content.

Science Process Skills

Observing, Inferring

Materials

For each student: paper towel, straw, small plastic bag

Safety Tips (See page T24.)

- Remind students to wipe up any spills on the floor immediately.
- Tell students to keep plastic bag away from their faces.

Teaching Tip

- **Helpful Hint:** The first part of this activity may best be done as a demonstration using one student to wet a large portion of the board.

Answers

Talk About It

1. The water went into the air.
2. The bag became wet from water in a person's breath.

Applying Whole Language ♦

Discuss the whole language framework with each Teaching Option you select. Here is an example applied to the Special Education option on p. 239.

1. Purpose: To make and display posters describing storms in 3 stages: development, action, after-effects

2. Context: Small group. Use photos, drawings, captions, warnings, descriptions, weather data.

3. Decisions: Type of storm. Library resources. Sequence.

4. Evaluation: How did our posters show sequence? How are our posters useful? (See p. T30.)

Resource Book page 123

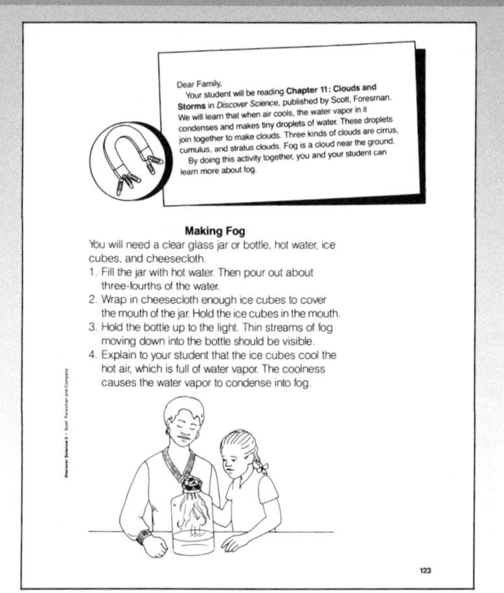

Dear Family,

Your student will be reading **Chapter 11: Clouds and Storms** in *Discover Science*, published by Scott, Foresman. We will learn that when air cools, the water vapor in it condenses and makes tiny droplets of water. These droplets join together to make clouds. Three kinds of clouds are cirrus, cumulus, and stratus clouds. Fog is a cloud near the ground.

By doing this activity together, you and your student can learn more about fog.

Making Fog

You will need a clear glass jar or bottle, hot water, ice cubes, and cheesecloth.

1. Fill the jar with hot water. Then pour out about three-fourths of the water.
2. Wrap in cheesecloth enough ice cubes to cover the mouth of the jar. Hold the ice cubes in the mouth.
3. Hold the bottle up to the light. Thin streams of fog moving down into the bottle should be visible.
4. Explain to your student that the ice cubes cool the hot air, which is full of water vapor. The coolness causes the water vapor to condense into fog.

123

Science Background

During evaporation, molecules leave a water surface and become gas molecules in water vapor. Solid, liquid, and gas are the three phases of matter. Any change from one phase to another requires either an input of energy or a disposal of energy.

TEACHING PLAN

Lesson Objectives

- *State* examples of evaporation and condensation.
- *Explain* how clouds form.
- *Compare* different types of clouds.

Lesson Vocabulary

cirrus clouds, cumulus clouds, dew, fog, stratus clouds

1. Motivate

Demonstration Activity ♦

Provide each student with a damp paper towel. Instruct the students to moisten the back of each hand. Tell them to blow on the back of one hand for about thirty seconds and *observe* the results.

Discussion

Questions: **Which hand dried the quickest?** (the one students blew on) **Why?** (Students should *infer* that blowing on their hand caused it to dry quicker.) **How did your hands feel?** (Both hands should feel cool as they dry, but the one that was blown on should feel cooler since it is drying faster.) Point out that heat and wind (the blowing) speed up evaporation.

1 How Do Clouds Form?

LESSON GOALS

You will learn

- how water evaporates and condenses.
- how clouds form.
- how three kinds of clouds differ.

dew (dü), water vapor that condenses on cool surfaces, usually during the night.

[1]dry

[2]answers might include plants, car windows, or other cool surfaces

Imagine climbing out of a swimming pool like the girl in the picture will do. Your skin would feel wet. Suppose you sat in the sun for a few minutes. How would your skin feel?[1]

How Water Changes Forms

You learned in Chapter 5 that water can change from one state to another. Water on your skin evaporates and becomes water vapor. Heat from the sun helps water evaporate. Wind also helps water evaporate.

Sometimes you might notice water on the grass in the morning. When air cools during the night, water vapor in the air condenses into small drops of water. These drops of water —called **dew**—settle on the grass. What other places might you see dew?[2]

Water evaporates from your skin.

230

Teaching Options

Science Background

Clouds are made of tiny water droplets, ice crystals, or combinations of both. Each droplet condenses around a microscopic particle of dust, smoke, or salt in the air. Cloud droplets or ice crystals are so small—0.02 to 0.06 mm (0.0008 to 0.0024 in) in diameter—that the ever-present movements of air are enough to suspend them in the atmosphere. Clouds are classified on the basis of altitude and shape. This classification is largely reflected in cloud names, which are often compound words. Thus altostratus clouds are those of the stratus that form at intermediate altitudes. Other compound cloud names refer to two shapes of clouds. Thus stratocumulus clouds have shapes characteristic of both stratus and cumulus.

Reading Strategies ♦

1. Guide students' pre-reading by asking: What comes to mind when you think of the vocabulary words?
2. Assign these strategies: Visualizing Information and Writing a Memory Sentence. (See pages T26–T29.)
3. Pair students to share what information is clear and unclear and initiate discussion using students' unanswered questions.

♦ *Suitable as a language development activity*

[1]water vapor from breath condenses on window

How Clouds Form

You probably have seen your breath in the air on a cold day. Water vapor is in your warm breath. This water vapor condenses when it meets the cold air. What happens if you breathe out onto a cold window?[1]

Water vapor in the air condenses in much the same way as water in your breath does. As warm air rises, it cools. As air cools, the water vapor in the air condenses into tiny droplets of water. The picture shows that many of these droplets come together to form clouds.

Sometimes the air just above the ground cools quickly. Then the water vapor in the air condenses to form **fog**—a cloud near the ground. You might notice that the air around you feels wet when fog forms.

INVESTIGATE!

Find out how a low temperature affects water vapor. Write a hypothesis and test it with an experiment. You might place a metal pan in a freezer for several hours. Take it out and observe what happens.

fog, a cloud that forms just above the surface of the earth.

Forming Clouds

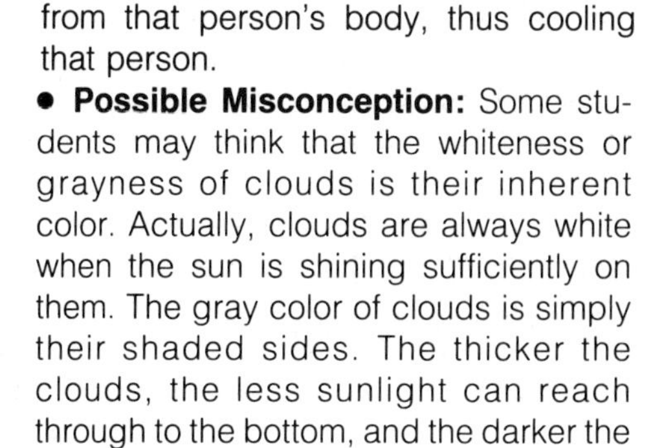

2. Teach

Teaching Tips

- Refer students to the *DISCOVER* activity on page 229 as you discuss evaporation and condensation.
- Question: **How does your skin feel when you come out of a pool or shower?** (cool or cold) Explain that when a person steps out of a shower, the evaporating water uses some of the heat from that person's body, thus cooling that person.
- **Possible Misconception:** Some students may think that the whiteness or grayness of clouds is their inherent color. Actually, clouds are always white when the sun is shining sufficiently on them. The gray color of clouds is simply their shaded sides. The thicker the clouds, the less sunlight can reach through to the bottom, and the darker the bottom appears.

Investigate!

Accept any testable hypothesis. Most students' data will support various hypotheses. One possible hypothesis is: *A temperature below freezing turns water vapor in the air to ice.* Students should find out that frost forms on a cold metal pan when it is exposed to warm air. The pan cools the air near it causing the water vapor in the air to turn to ice.

Workbook page 61 *

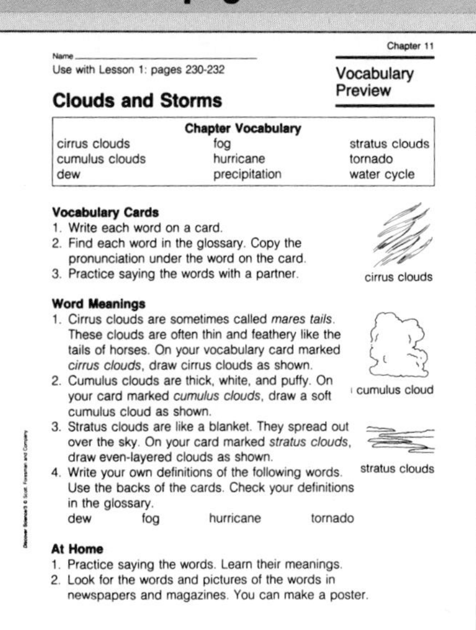

Name ______ Chapter 11

Use with Lesson 1: pages 230-232

Vocabulary Preview

Clouds and Storms

Chapter Vocabulary

cirrus clouds	fog	stratus clouds
cumulus clouds	hurricane	tornado
dew	precipitation	water cycle

Vocabulary Cards

1. Write each word on a card.
2. Find each word in the glossary. Copy the pronunciation under the word on the card.
3. Practice saying the words with a partner.

cirrus clouds

Word Meanings

1. Cirrus clouds are sometimes called *mares tails*. These clouds are often thin and feathery like the tails of horses. On your vocabulary card marked *cirrus clouds*, draw cirrus clouds as shown.
2. Cumulus clouds are thick, white, and puffy. On your card marked *cumulus clouds*, draw a soft cumulus cloud as shown.
3. Stratus clouds are like a blanket. They spread out over the sky. On your card marked *stratus clouds*, draw even-layered clouds as shown.
4. Write your own definitions of the following words. Use the backs of the cards. Check your definitions in the glossary.
   dew fog hurricane tornado

cumulus cloud

stratus clouds

At Home

1. Practice saying the words. Learn their meanings.
2. Look for the words and pictures of the words in newspapers and magazines. You can make a poster.

61

Science and Language Arts

Take the class outside to *observe* clouds. Encourage students to write descriptive sentences, essays, or poems about the shape, color, and movement of the clouds. Allow students to illustrate their sentences and display the descriptions in class.

Special Education

Show students with learning disabilities pictures of heavy fog, such as the fogs that form in San Francisco Bay and off the coast of New England. Discuss their similar appearance to that of clouds higher in the air. Ask students to relate their experiences with fog. Be sure students understand that fog is not a cloud that descends to the ground but one that forms at the ground.

*** Answers to masters on pages 228E–228H**

TEACHING PLAN

Teaching Tips

- Have students pronounce the names of the major cloud types. Ask them to state one characteristic of each type.
- Ask students to *classify* the clouds that appear outside.

3. Assess

Lesson Review

1. Water can evaporate to form a gas or condense to form a liquid.
2. Water vapor in the air condenses to tiny droplets of water that come together to form clouds.
3. Cirrus clouds are high, feathery clouds made of tiny pieces of ice. Cumulus clouds look like cotton. Large ones can cause storms. Stratus clouds form in layers and often bring light rains.
4. Challenge! You can speed up evaporation by putting it in the sun. **Thinking Skill:** *Drawing conclusions*

Find Out On Your Own

Generally, the diary should show that cirrus clouds and small cumulus clouds are associated with fair weather, large cumulus clouds are associated with storms, and stratus clouds are associated with light rain or snow. **Thinking Skill:** *Making physical models*

Cirrus clouds

Cumulus clouds

Stratus clouds

cirrus (sir′əs) **clouds,** high, feathery white clouds made of tiny pieces of ice.

cumulus (kyü′myə ləs) **clouds,** fluffy clouds that look like puffs of cotton.

stratus (strā′təs) **clouds,** clouds that form in sheets or layers and spread out over the sky.

Kinds of Clouds

The pictures show three kinds of clouds. **Cirrus clouds** are made of tiny pieces of ice. These feathery clouds float high in the air. You often see cirrus clouds in good weather.

Cumulus clouds look like cotton. You usually see these clouds in good weather. Weather can become stormy when cumulus clouds become large.

Stratus clouds form in layers. These clouds usually spread across the sky. You often see stratus clouds during a rain.

Lesson Review

1. How can water change forms?
2. How do clouds form?
3. How do three kinds of clouds differ?
4. **Challenge!** How can you make water in a glass evaporate faster?

Study on your own, pages 336–337.

EARTH SCIENCE
FIND OUT ON YOUR OWN

Keep a cloud diary for a week. On a piece of paper, write down the kinds of clouds you see every day. Next to the name of each kind of cloud, write a sentence describing the weather.

Teaching Options

Science Anecdote

Water in such tiny particles as those comprising clouds remain liquid far below the normal freezing point. Such water is called supercooled. Clouds are made entirely of water droplets down to about −12°C (10°F). Below −40°C (−40°F) the clouds are all ice crystals. Such is the case with cirrus clouds.

Reteaching Suggestion ♦

Prepare a simple drawing showing warm air rising, cooling, and water vapor condensing to form clouds. Provide each student with a copy of the drawing. Then write the phrases *warm air, heated air rises and cools,* and *water vapor condenses into water droplets that form clouds.* Have the students write the phrases in the correct places on the drawing. Review the process and allow the students to color their drawings when completed.

♦ ***Suitable as a language development activity***

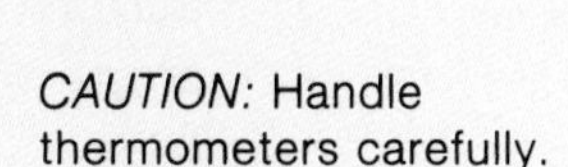

Measuring Temperature at Which Dew Forms

Suggested grouping: pairs

Purpose

Measure the temperature at which dew forms.

Gather These Materials

• shiny can • water • ice cubes or crushed ice • stirring stick • thermometer

Follow This Procedure

1. Use a chart like the one shown to record your observations.
2. Use the thermometer to measure the temperature in your room.

   CAUTION: Handle thermometers carefully.
3. Half fill the can with water. Place the thermometer into the can of water as shown in the picture. Record the temperature of the water.
4. Add ice, a little at a time, to the water. Slowly stir the ice and water with the stirring stick.
5. Carefully watch the can. Read the thermometer when water forms on the outside of the can. Record the temperature.

Record Your Results

Air temperature	Answers will vary, depending on humidity.
Water temperature	
Temperature at which water forms on can	

State Your Conclusion

1. How did the temperature at which water formed on the can compare with room temperature?
2. What caused water to form on the outside of the can?

Use What You Learned

What can happen to water vapor in the air when it rises and cools?

233

Resource Book page 127

Name
Use with Lesson 1: page 230
Measuring Temperature at Which Dew Forms

Chapter 11
Activity Worksheet

Record Your Results

Air Temperature	
Water Temperature	
Temperature at which water forms on can	

State Your Conclusion

1. How did the temperature at which water formed on the can compare with room temperature?

2. What caused water to form on the outside of the can?

Use What You Learned

What can happen to water vapor in the air when it rises and cools?

127

Activity Results

See chart overprint above. Results will vary depending on relative humidity.

Concept

The dew point is the temperature at which water vapor in the air begins to condense.

Objectives/Process Skills

- *Observe* the formation of dew on the can.
- *Measure* the temperature at which dew begins to form.
- *Record* and interpret data.
- *Infer* where the water on the can came from.

Time Allotment

Allow 20-25 minutes.

Safety Tips (See page T24.)

- Tape the top edge of the cans to prevent cuts.
- Tell students to let you know if any glass objects break.
- Remind students to wipe up any spills on the floor immediately.

Teaching Tip

- Have students relate their experiences involving dew point, such as the appearance of water on the outside of a cold can of soda or glass of water and the appearance of dew on grass in the evening or morning. Discuss the causes of this dew. (Air is chilled below the dew point.)

Answers

State Your Conclusion

1. The temperature at which water formed on the can was lower than room temperature.

2. The water formed when the water vapor in the air around the can condensed.

Use What You Learned

When water vapor in the air rises and cools, it can condense and form clouds.

Thinking Skill: *Applying information to new situations*

LESSON 2 pages 234–236

TEACHING PLAN

Lesson Objectives
- *Explain* how rain and snow form.
- *Describe* the relationships between evaporation, condensation, and precipitation in the water cycle.

Lesson Vocabulary
precipitation, water cycle

1. Motivate

Demonstration Activity ♦
Spray a mist of water on a smooth surface held at an incline. Ask a volunteer to run a finger through the water droplets that cling to the surface. *CAUTION:* Immediately wipe up any water that spills on the floor.

Discussion
Questions: **What happened as the finger ran through the water on the surface?** (Streams of water ran down from the finger.) **What caused these streams of water?** (Let students *hypothesize* about what they saw. Some students may correctly say the finger pushed some of the tiny droplets together, making larger drops that were heavy enough to run down the surface.)

2 What Happens to Water in the Clouds?

LESSON GOALS

You will learn
- how rain and snow form.
- how all the water on earth keeps moving through the water cycle.

precipitation
(pri sip′ə tā′shən), moisture that falls to the ground.

The Water Cycle

Think about water vapor condensed on a mirror. What happens if you push the tiny drops of water together? The drops become large. They get heavy and run down the mirror. Rain drops form in much the same way.

Forming Rain and Snow

The tiny drops of water in clouds are light enough to float in the air. The drops can bump into each other and form larger drops. The large drops are too heavy to float in the air. These drops fall as rain. The rain that falls from clouds is called **precipitation.** Snow is another kind of precipitation. Snowflakes are feathery bits of ice formed in clouds where air is freezing.

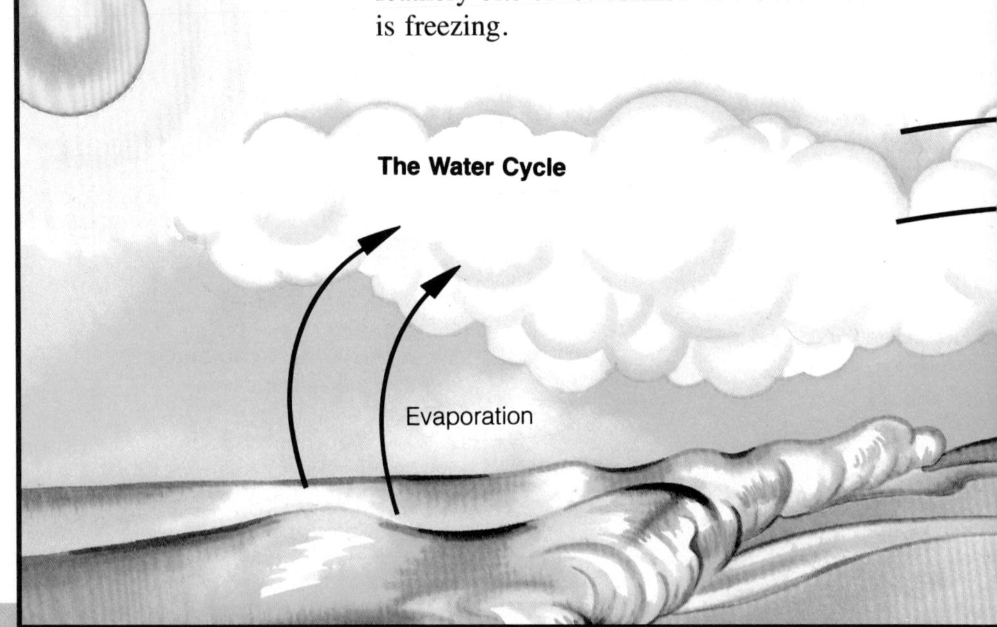

The Water Cycle

Teaching Options

Science Background

Average raindrops have diameters of 1 to 2 mm (1/25 to 1/10 in.). In the tropics, rain formed by water drops coming together is common. But in the middle and high latitudes, rain usually results from the melting of snow as it falls from high, cold layers of air through lower, warm layers. Snow forms where the air is well below freezing, and a mixture of ice crystals and supercooled water droplets make up the clouds. The water adheres to the ice crystals and freezes, increasing the size of the snowflakes and making them fall more rapidly from the cloud. The precipitation stays in the form of snow only if the underlying layers of air are below freezing. Otherwise, it melts and reaches the ground as rain.

Reading Strategies ♦

1. Guide students' pre-reading by asking: What ideas do you think you would find in the lesson to answer the question-title?
2. Assign these strategies: Visualizing Information and Mapping Examples of precipitation. (See pages T26–T29.)
3. Pair students to share what information is clear and unclear and initiate discussion using students' unanswered questions.

♦ *Suitable as a language development activity*

The Water Cycle

All the water on earth keeps moving and changing forms. The movement of water on the earth by evaporation, condensation, and precipitation is called the **water cycle.**

Use your finger to trace the raindrop's path through the water cycle in the picture. Suppose a raindrop falls on a hillside. Imagine the drop of water running down the hillside into a stream. Where does the water in the stream flow? Energy from the sun makes some of the water evaporate. Then the drop of water changes to water vapor. Some of the water vapor might condense into a tiny drop and form part of a cloud. Rain or snow might fall. Then the whole water cycle repeats itself.

[1]to the ocean

SCIENCE IN YOUR LIFE

Hail is a kind of precipitation made up of lumps of ice. These icy lumps are formed inside storm clouds. Most hailstones are about the size of a pea. Some hailstones can get as big as a baseball.

water cycle (sī′kəl), the movement of water from the ground to the air and back to the ground by evaporation, condensation, and precipitation.

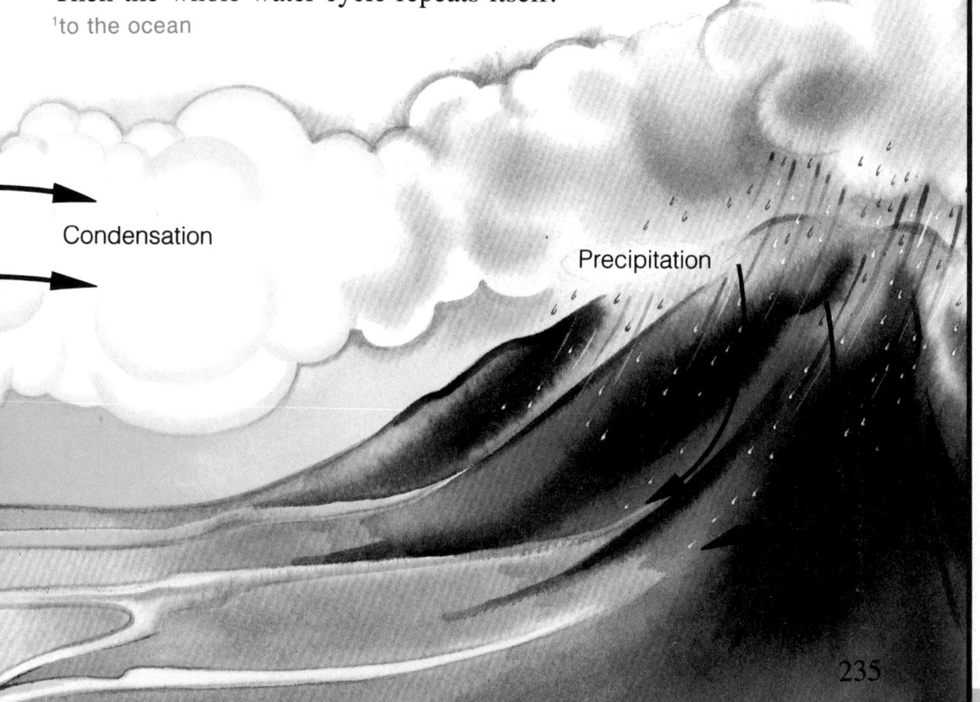

235

2. Teach

Teaching Tips

- Refer students to the suggested demonstration at the beginning of this lesson as you discuss page 234.
- **Possible Misconception:** Some students might think that all clouds bring precipitation. Be sure students understand that the droplets of water must attain a certain size before they can become heavy enough to fall as precipitation. Most clouds, such as cumulus of fair weather, do not have enough moisture for droplets to attain a large size.
- Question: **What would happen to snow if it fell through a layer of air that was above freezing?** (Guide students to *infer* that the snow would melt and continue falling as rain.)
- Have a student look up the word *cycle* in the dictionary. (a period of time or complete process that repeats itself in the same order) Ask students to name other cycles. (seasons, day and night, washing machine cycles, and so on)
- Point out that the movement of water in the ground and evaporation from plants is also part of the water cycle.

Workbook page 62 *

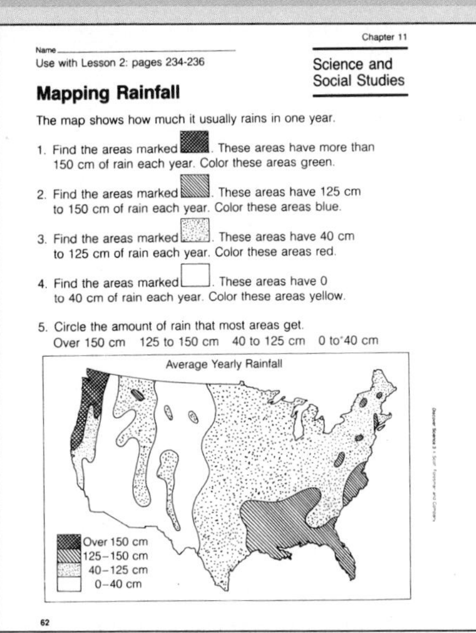

Chapter 11

Name

Use with Lesson 2: pages 234-236

Science and Social Studies

Mapping Rainfall

The map shows how much it usually rains in one year.

1. Find the areas marked ▇. These areas have more than 150 cm of rain each year. Color these areas green.
2. Find the areas marked ▧. These areas have 125 cm to 150 cm of rain each year. Color these areas blue.
3. Find the areas marked ░. These areas have 40 cm to 125 cm of rain each year. Color these areas red.
4. Find the areas marked □. These areas have 0 to 40 cm of rain each year. Color these areas yellow.
5. Circle the amount of rain that most areas get.
   Over 150 cm   125 to 150 cm   40 to 125 cm   0 to 40 cm

62

* *Answers to masters on pages 228E–228H*

Special Education

Have students with learning disabilities make a chart which is divided into three columns labeled *Evaporation, Condensation,* and *Precipitation.* Students should then draw pictures to illustrate each step of the water cycle. Encourage students to refer to the charts as they learn about the water cycle.

Reinforcement

Have students write a paragraph describing three different paths that a particular particle of water might take through the water cycle. Point out that cycles in nature provide several ways for a material to move from one point to another.

TEACHING PLAN

Teaching Tips

• On a map or globe, show the class that the earth's surface consists mostly of water.

• On a local, state, or national map, call on students to find a river and follow it downstream. Let students discover where the river leads (most likely to a larger river, a lake, or an ocean)

3. Assess

Lesson Review

1. Tiny drops of water in clouds come together to form heavier drops that fall as rain. If the air is cold enough, bits of ice form and fall as snow.

2. Evaporation, condensation, and precipitation are three stages in the water cycle.

3. Challenge! Water from the ocean evaporates, condenses, and precipitates. This water could fall on a body of water being used as a source of household water. **Thinking Skill:** *Inferring*

Find Out On Your Own

Students might mention that each snowflake has a different shape and pattern. However, all snowflakes have six sides and six shapes. **Thinking Skill:** *Making physical models*

All the water on earth is part of the water cycle.

Some water evaporates from land and small bodies of water like the one in the picture. Much of the water from rivers and streams reaches lakes and oceans. Most of the earth's water is in oceans, so most water evaporates from oceans. Water evaporates, condenses, and falls as precipitation over and over again.

Lesson Review

1. How do rain and snow form?
2. What are the parts of the water cycle?
3. **Challenge!** Explain how water from your faucet might once have been in the ocean.

Study on your own, pages 336–337.

PHYSICAL SCIENCE
FIND OUT ON YOUR OWN
CONNECTION

Look in library books to find out about shapes and patterns of snowflakes. Use white paper to cut out different kinds of snowflakes. Write a few sentences explaining how snowflakes are different and how they are alike.

Teaching Options

Science Anecdote

People have tried to modify weather patterns by cloud seeding—spraying clouds with particles that act as nuclei around which water condenses. This action has been partially successful for increasing rainfall, lessening the severity of storms, and clearing fog.

Reteaching Suggestion ♦

Prepare sentence strips describing the steps of the water cycle and pictures which correspond to each step. Randomly arrange the sentences and pictures on the bulletin board. Ask the students to arrange the sentences and pictures in the correct order. One group or person can do the sentences while another does the pictures.

Workbook page 63 *

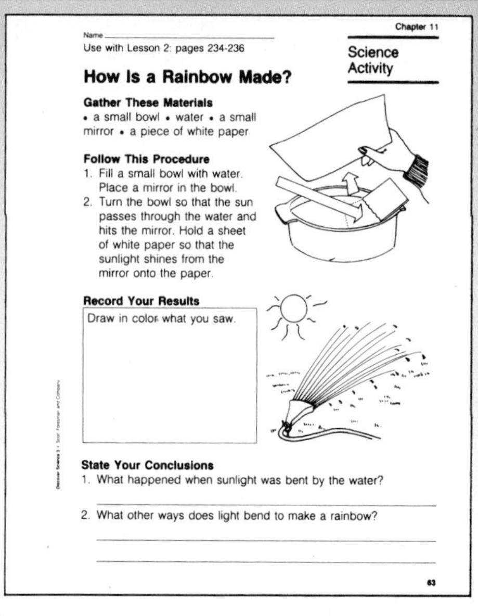

Chapter 11

Name ______

Use with Lesson 2: pages 234-236

Science Activity

How Is a Rainbow Made?

Gather These Materials

• a small bowl • water • a small mirror • a piece of white paper

Follow This Procedure

1. Fill a small bowl with water. Place a mirror in the bowl.
2. Turn the bowl so that the sun passes through the water and hits the mirror. Hold a sheet of white paper so that the sunlight shines from the mirror onto the paper.

Record Your Results

Draw in color what you saw.

State Your Conclusions

1. What happened when sunlight was bent by the water?
2. What other ways does light bend to make a rainbow?

63

♦ *Suitable as a language development activity*

Observing Part of the Water Cycle

Suggested grouping: pairs

Purpose

Observe how water moves in the water cycle.

Gather These Materials

• large jar • hot tap water • ice cubes • plastic wrap • rubber band

Follow This Procedure

1. Use a chart like the one shown to record your observations.
2. Half fill the jar with hot water.
3. Place the plastic wrap over the opening of the jar. Let the plastic wrap droop to make a bowl shape.
4. Put a rubber band around the top of the jar to hold the plastic wrap in place.
5. Place 2 or 3 ice cubes on the plastic wrap, as shown in the picture. Wait 5 minutes.
6. Remove the ice cubes. Observe the plastic wrap. Record what you observe.
7. Replace the ice cubes on the plastic wrap. Wait 5 more minutes. Remove the plastic wrap from the jar. Observe the bottom of the plastic wrap. Record what you observe.

Record Your Results

After 5 minutes	some water
After 10 minutes	more water

State Your Conclusions

1. What happened to some of the hot water in the jar?
2. What did putting ice on the plastic wrap do to the warm air in the jar?

Use What You Learned

What part or parts of the water cycle were shown in this activity?

237

Resource Book page 129

Name

Use with Lesson 2: page 237

Observing Part of the Water Sample

Chapter 11

Activity Worksheet

Record Your Results

After 5 minutes	
After 10 minutes	

State Your Conclusion

1. What happened to some of the hot water in the jar?

2. What did putting ice on the plastic wrap do to the warm air in the jar?

Use What You Learned

What part or parts of the water cycle were shown in this activity?

129

Activity Results

Students should see condensation on bottom of plastic wrap after 5 minutes, more condensation after 10 minutes.

Activity

Concept

Water evaporates and condenses as it goes through the water cycle.

Objectives/Process Skills

- *Observe* the results of evaporation and condensation.
- *Record* observations.
- *Recognize the cause and effect* of cooling warm, moist air.
- *Identify* evaporation and condensation as parts of the water cycle.

Time Allotment

Allow 25–30 minutes.

Safety Tips (See page T24.)

- Remind students to wipe up any spills on the floor immediately.
- Plastic containers should be used.

Teaching Tips

- Aluminum foil can be substituted for plastic wrap.
- Have students look in magazines for pictures that illustrate the processes of evaporation, condensation, and precipitation.
- Have students work in small groups to draw and color murals illustrating the water cycle. Encourage them to show several different paths.

Answers

State Your Conclusion

1. Some of the hot water evaporated into the air above the cup.

2. The ice cooled the air beneath and above the plastic wrap.

Use What You Learned

Evaporation and condensation occurred. **Thinking Skill:** *Drawing conclusions*

***** *Answers to masters on pages 228E–228H*

TEACHING PLAN

Lesson Objectives
- *Explain* how a thunderstorm can form when warm, moist air rises.
- *Explain* how rapidly rising air causes hurricanes and tornadoes.
- *State* ways people can protect themselves during thunderstorms, hurricanes, and tornadoes.

Lesson Vocabulary
hurricane, tornado

1. Motivate

Demonstration Activity ♦
Half-fill with water a tall, clear plastic container. Carefully place a thin layer of sand on the bottom of the container. Use a spoon or stick to stir the water vigorously, creating a whirlpool to simulate movement within a tornado. Do not touch the bottom of the container with the stirrer, but rather, let the sand be drawn eventually into the whirlpool.

Discussion
Questions: **What kind of violent weather occurrence does this water remind you of?** (tornado) **What does the sand represent?** (soil and objects on the ground)

[1]answers might include a rumbling noise or a loud crack

3 What Causes Storms?

LESSON GOALS

You will learn
- how a thunderstorm forms.
- how hurricanes and tornadoes form.
- how people can keep safe during storms.

Imagine walking outside on a sunny day. Suddenly, the sky gets dark. The wind blows and rain begins to fall. Lightning flashes in the sky. A thunderstorm is beginning.

Thunderstorms

The picture shows what happens when a thunderstorm forms. Warm, moist air quickly moves up in the sky. As cool air replaces the rising air, winds begin to blow. Large cumulus clouds form. Water droplets in the clouds become large and heavy. The rain begins to fall.

You learned in Chapter 7 that electric charges can jump between objects. You see lightning when electric charges jump between clouds or from a cloud to the ground. You often can hear thunder after lightning flashes. How can you describe the sound of thunder?

How Storms Form

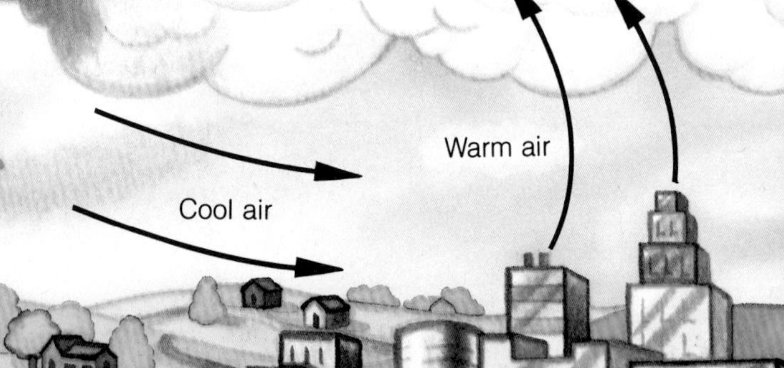

238

Teaching Options

Science Background

Thunderstorms are common in the United States, partly because cool polar air clashes with warm tropical air in the middle latitudes. Thunderstorms are characterized by strong up and down air movements. The rapid heating of air along the path of a lightning bolt makes the air expand quickly, producing intense sound waves, which we hear as thunder. Light travels more quickly than sound, so we see the lightning before we hear the thunder. Tornadoes and hurricanes are both areas of extremely low air pressure. The spinning motion in both of these phenomena result from air spinning from an area of high pressure into an area of low pressure. Winds within a tornado may reach 400 km (250 mi) per hour.

Reading Strategies ♦

1. Guide students' pre-reading by asking: What information do you get from the pictures?
2. Assign these strategies: Visualizing Information and Writing a Memory Sentence. (See pages T26–T29.)
3. Pair students to share what information is clear and unclear and initiate discussion using students' unanswered questions.

♦ *Suitable as a language development activity*

A tornado

Hurricanes and Tornadoes

A **hurricane** is a huge storm that forms when warm, moist air begins to rise very quickly from a warm ocean. The water vapor condenses to form clouds. Cool air moves in to replace the warm air that rises. The air and the clouds begin to spin around and around. As more water vapor rises from the ocean, more air begins to spin and the storm grows stronger. A hurricane has powerful winds that can destroy buildings and heavy rains that can cause floods. When a hurricane moves over land, it dies out.

Look at the **tornado** in the picture. A tornado can form during a thunderstorm when air rises very quickly. The air twists into a funnel-shaped cloud that moves along a narrow path. A tornado has the fastest winds on earth. Usually, tornadoes stay in the air. A tornado that touches the ground can destroy objects in its path.

hurricane (hėr′ə kān), a huge storm that forms over a warm ocean and has strong winds and heavy rains.

tornado (tôr nā′dō), a funnel cloud that has very strong winds and moves along a narrow path.

2. Teach

Teaching Tips

- Review how condensation and evaporation are involved in the formation of clouds and precipitation.
- Explain that during a thunderstorm some parts of clouds become positively charged and other parts become negatively charged. Also, the base of the cloud is negatively charged while the ground becomes positively charged. The electric current that passes between these opposite charges is lightning.
- Ask students if they have ever received an electric shock, such as when touching metal after walking across a rug on a dry day. Lead students to ***make an analogy*** between lightning and the electric shock.
- **Possible Misconception:** Some students may think that lightning and thunder are two unrelated phenomena, except that they both occur during thunderstorms. Point out the relationship between lightning and thunder. As an example showing the difference in travel times of light waves and sound waves, ask a student to stand at the end of a long hall and drop a book. Students will *see* the book hit the floor before they *hear* it. Also point out how an outfielder hears the "crack of the bat" after the bat hits the ball.
- Ask students to *compare* and *contrast* hurricanes and tornadoes.

Workbook page 64 *

Chapter 11

Name ____________

Use with Lesson 3: pages 238-240

Science Skills

Looking at Storms

A. The pictures show a hurricane, a tornado, and a thunderstorm. Decide which storm is in each picture, and write its name in the picture.

B. Decide which kind of storm is described in each sentence below. Write the number for the sentence in the correct box below. You will write some numbers in more than one box.

1. This storm covers the largest region.
2. The land or water gets heavy rainfall.
3. You see lightning and hear thunder.
4. A funnel-shaped cloud appears.
5. This storm covers the smallest area of the three.
6. It moves through air, touching ground sometimes.
7. It grows when warm moist air quickly moves upward.
8. It has the strongest winds on earth.
9. When it moves to land, it begins to disappear.
10. It is strong enough to tear down buildings.

Hurricane	Tornado	Thunderstorm

64

Answers to masters on pages 228E–228H

Special Education

Have students with learning disabilities draw and label pictures illustrating how a thunderstorm, tornado, or hurricane forms. Use the illustrations to review the sequence of events involved in the formation of these storms.

Game Suggestion ♦

Using vocabulary words, play a variation of password, called "Weather Word." Divide the students into two teams. Tell one student in each team the "weather word." Those students then give clues to their teammates who must try to guess the word. The student who correctly guesses the word becomes the next student on that team to give clues for the next word.

TEACHING PLAN

Teaching Tip

• **Possible Misconception:** Many students will think that standing under a tree during a thunderstorm is safe. Tall trees are subject to being struck by lightning. Rapid heating of the sap makes parts of the tree explode.

3. Assess

Lesson Review

1. Warm, moist air rises quickly, forming large cumulus clouds. Water droplets in the clouds become large and heavy and fall as rain.
2. Hurricanes form when warm, moist air rises quickly over an ocean. Tornadoes form when air rises quickly and twists forming a funnel cloud of strong winds.
3. People should go inside a building or a car. If outside, lie in a low place and stay away from trees.
4. Challenge! It no longer has a source of water vapor—the ocean. **Thinking Skill:** *Drawing conclusions*

Find Out On Your Own

Changes in air pressure cause the mercury in a barometer to rise or fall. Storms are areas of low pressure. An approaching storm is usually indicated by a declining air pressure. **Thinking Skill:** *Restating or explaining ideas*

A tornado drill

SCIENCE IN YOUR LIFE

A lightning rod is a metal rod placed on top of a building. The rod is connected to the ground. If lightning strikes a lightning rod, the electric charges move through the rod and into the ground. Then, the electric charges do not harm the building.

Keeping Safe During Storms

You can help keep yourself safe during a thunderstorm. Going inside a building or a car with a hard top can help protect you. If you cannot go indoors, lie down in a low place. Stay away from trees and water.

Learn about the emergency plans in your community for hurricanes and tornadoes. These children are having a tornado drill at school. They stay indoors away from windows.

Lesson Review

1. How does a thunderstorm form?
2. How do hurricanes and tornadoes form?
3. How can people keep safe during thunderstorms?
4. **Challenge!** Why does a hurricane die out when it moves over land? Explain.

Study on your own, pages 336–337.

PHYSICAL SCIENCE
FIND OUT ON YOUR OWN
CONNECTION

A barometer measures air pressure. Look in library books to learn more about barometers. Write a few sentences explaining how a barometer works. Tell how a barometer can be useful for learning about weather.

Teaching Options

Science Anecdote

The power of tornadoes has been documented by many strange events. For example, tornadoes have plucked chickens of their feathers, stripped the leaves and bark from trees, sucked blankets out of houses through chimneys, driven straws into telephone poles, and popped corks out of bottles.

Reteaching Suggestion ♦

Have students make a chart comparing thunderstorms, hurricanes, and tornadoes. They should include characteristics of the storms and how the storms form. This is a good time to review the tornado, hurricane, or other weather-related school emergency plans.

Workbook page 65 *

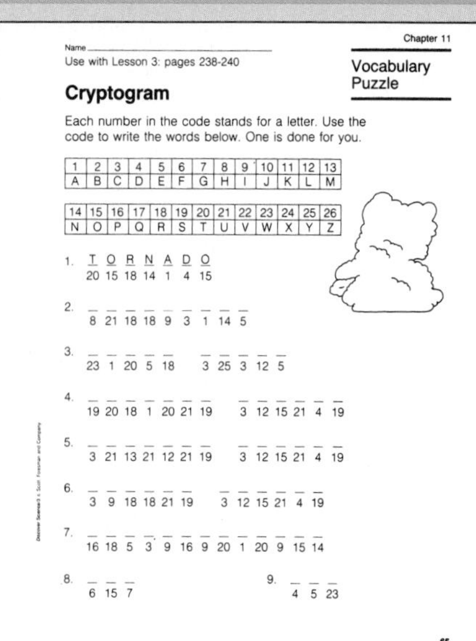

Name ____________
Use with Lesson 3: pages 238-240

Chapter 11
Vocabulary Puzzle

Cryptogram

Each number in the code stands for a letter. Use the code to write the words below. One is done for you.

1	2	3	4	5	6	7	8	9	10	11	12	13
A	B	C	D	E	F	G	H	I	J	K	L	M

14	15	16	17	18	19	20	21	22	23	24	25	26
N	O	P	Q	R	S	T	U	V	W	X	Y	Z

1. T O R N A D O
20 15 18 14 1 4 15
2. 8 21 18 18 9 3 1 14 5
3. 23 1 20 5 18 3 25 3 12 5
4. 19 20 18 1 20 21 19 3 12 15 21 4 19
5. 3 21 13 21 12 21 19 3 12 15 21 4 19
6. 3 9 18 18 21 19 3 12 15 21 4 19
7. 16 18 5 3 9 16 9 20 1 20 9 15 14
8. 6 15 7
9. 4 5 23

65

♦ *Suitable as a language development activity*

Mapping Tornadoes

"The United States has the biggest and the best," says Dr. Ted Fujita. "That's why I came here from Japan." Dr. Fujita is not talking about mountains or baseball games. He is talking about tornadoes. He has been interested in these storms his whole life. Now as a scientist, he knows more about tornadoes than almost anyone in the world.

Dr. Fujita knows that in the middle of the United States tornadoes form easily. Cold winds from Canada blow south and east. Warm, moist winds from the Gulf of Mexico blow north. When these winds meet, they can twist together to make a tornado.

No one can tell just when a tornado will strike. However, Dr. Fujita found a way to help people make a good guess. He put together a map to show where tornadoes occurred from 1930 to 1978. In that time, about 23,000 tornadoes occurred. By using this map, Dr. Fujita and other scientists can tell where tornadoes are most likely to happen.

Most tornadoes last a short time—less than one hour. They usually travel just a short distance. However, tornadoes are so powerful that they can toss railroad cars and flatten houses.

Dr. Ted Fujita

Dr. Fujita thinks that the more people learn about tornadoes, the more lives can be saved.

What Do You Think?

1. Why do you think a tornado is sometimes called a twister?
2. Why do you think tornadoes are not common in Oregon?

Science and People

Discussion

Ask students what they would do if they heard a tornado warning while at home or in a car. Point out that in a house, the safest place is on the lowest floor, in the smallest room, and in the most interior space. A car is not a safe place to be in a tornado, even though a car would be a safe place to be in a thunderstorm. In a tornado, people should leave their cars and lie down in low areas.

Teaching Tip

- Point out that tornadoes usually occur in the late afternoon. Most tornadoes occur in April, May, or June. They appear to be grey or black because of the dust and debris that the winds pick up. Some of the biggest tornadoes make a roaring sound that can be heard for miles.

Answers

What Do You Think?

1. Because warm and cold winds twist together to make a tornado. **Thinking Skill:** *Inferring*

2. Because the cold Canadian winds and the warm, moist Gulf winds do not meet there. **Thinking Skill:** *Applying information to new situations*

Science Background

When cold Canadian air collides with warm, moist Gulf air, the cold air sinks and the warm air rises. As a result, cooler air is pulled in from the sides to take the place of the warm air. The rotation of the earth makes the rising warm air spin. The more it spins inward, the faster it goes. A low pressure area is created, causing a tornado. Tornadoes rotate counterclockwise north of the equator and clockwise south of the equator. Dr. Fujita says that at the equator, the winds "get confused and don't know which way to go. That's why there are so few tornadoes at the equator."

TEACHING PLAN

Purpose
To develop the skills of collecting and organizing information using rain gauges and charts to solve problems.

1. Motivate

Discussion
Display a variety of containers used to measure liquids, and discuss the uses of each. Questions: **When is it important to measure liquids?** (Accept reasonable responses.) Point out that meteorologists measure rainfall using rain gauges. **Why might meteorologists be interested in the amount of rainfall in an area?** (Knowing the amount of rainfall can help them identify trends and predict weather.)

2. Teach

Teaching Tip
- If possible, obtain a rain gauge (available at many hardware stores.) Place the rain gauge in an area that is sheltered from strong wind and is unlikely to be disturbed. Help students measure rainfall over a period of time.

Skills for Solving Problems

Using Rain Gauges and Charts

Problem: How can rainfall be measured?

Part A. Using Rain Gauges to Collect Information

1. Rain gauges are containers that are used to collect and measure rainfall. Look at the rain gauges below. Each line on these rain gauges is 1 centimeter. What is the abbreviation for centimeter?
2. The rain gauge in example A collected 1 centimeter of rain. How much rain was collected in examples B, C, and D?

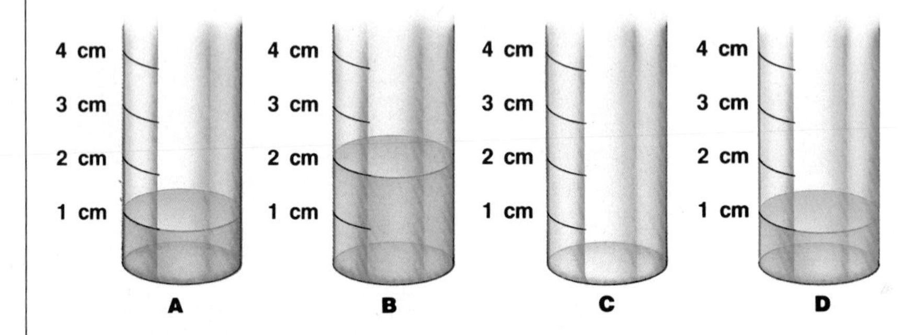

Part B. Using a Chart to Organize and Interpret Information

3. This chart contains the information you collected from the four rain gauges. Which rain gauge collected the most rainfall? Which rain gauge collected no rainfall?

Rain Gauge	Rainfall collected
A	1 cm
B	2 cm
C	0 cm
D	1 cm
Total rainfall	4 cm

242

Teaching Options

Sample Chart for Part C

Rain gauge	Rainfall collected
Sunday	0 cm
Monday	2 cm
Tuesday	2 cm
Wednesday	1 cm
Thursday	0 cm
Friday	0 cm
Saturday	3 cm
Total rainfall	8 cm

♦ *Suitable as a language development activity*

4. Look again at the rain gauges. Which rain gauges collected the same amount of rainfall?

5. What is the total amount of rainfall that was collected by all of the rain gauges?

Part C. Using Rain Gauges and Charts to Solve a Problem

Problem: What was the total rainfall for one week?

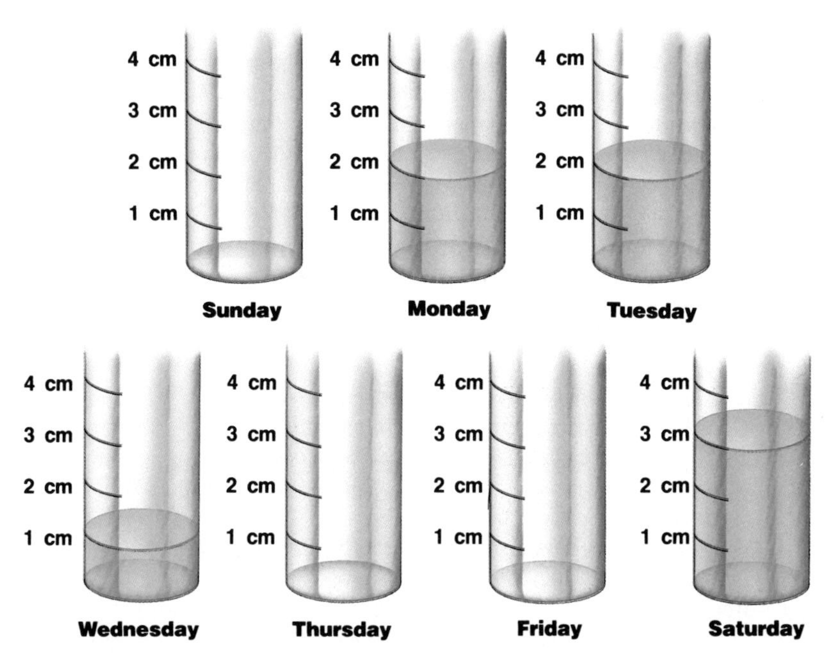

6. Use the picture to collect the information you need to solve the problem. Make a chart similar to the one in Part B to organize your information.
7. Look at your chart. How many days of the week did it rain?
8. What was the total rainfall for the week?

243

3. Assess

Part A

1. cm
2. B—2 cm; C—0 cm; D—1 cm

Part B

3. B
4. A and D
5. 4 cm

Part C

6. See Sample Chart for Part C in Teaching Options.
7. 4
8. 8 cm

Reteaching Suggestion ♦

Indicate hypothetical amounts of rain for each day of the week as follows: *Monday, 3 cm; Tuesday, 0 cm; Wednesday, 0 cm; Thursday, 1 cm; Friday, 0 cm; Saturday, 2 cm; Sunday, 1 cm.* Have students draw rain gauges filled with water to represent the rainfall for each day, or fill actual rain gauges. Then have them make charts that compare and contrast the rainfall for each day of the week. Question: **What is the week's total rainfall?** (7 cm)

Resource Book page 131 *

Name ______________________

Chapter 11

Comprehension: cause and effect relationships

Science and Reading

Why Did They Do It?

Read the story and the questions. Circle the letter of the sentence that best answers each question.

Joey and his sister, Lisa, went to play golf because it was such a beautiful day. Joey was teaching Lisa how to play golf. Since Joey was such a good golfer, Lisa was happy to have him for a teacher.

Joey and Lisa had only played four holes when clouds began to fill the sky. Then they heard thunder in the distance. Joey and Lisa were near the clubhouse, so they ran inside for shelter. Joey told Lisa that it is not safe for golfers to stay on a golf course during a thunderstorm. They may get struck by lightning.

1. Why was Lisa happy to have Joey for a teacher?
   a. Joey was a good golfer.
   b. It was a beautiful day.
   c. Joey didn't like to play golf.

2. Why did Joey and Lisa run for shelter inside?
   a. Joey wanted to show Lisa the clubhouse.
   b. They wanted to get something to drink.
   c. A golf course is not a safe place during a storm.

3. Why is it not safe for golfers to stay on a golf course during a thunderstorm?
   a. They may get their clothes all wet.
   b. They may get struck by lightning.
   c. They may forget what their score is.

131

* Answers to masters on pages 228E—228H

REVIEW PLAN

Reviewing Science Words

1. precipitation
2. cirrus clouds
3. hurricane
4. water cycle
5. cumulus clouds
6. fog
7. dew
8. stratus clouds
9. tornado

Reviewing What You Learned

1. b
2. a
3. b
4. d
5. c
6. a
7. c
8. d

Chapter 11 Review

Chapter Main Ideas

Lesson 1 • Water changes forms when it evaporates and condenses. • Water vapor in the air condenses to form clouds. • Cirrus clouds, cumulus clouds, and stratus clouds are three kinds of clouds.

Lesson 2 • Rain and snow are forms of precipitation. • Water on earth evaporates, condenses, and falls as precipitation over and over again as it goes through the water cycle.

Lesson 3 • A thunderstorm forms when warm, moist air rises. • A hurricane forms when warm ocean water evaporates and rises very quickly; a tornado can form during a thunderstorm when air rises quickly. • People can help keep themselves safe during storms.

Reviewing Science Words

cirrus clouds	fog	stratus clouds
cumulus clouds	hurricane	tornado
dew	precipitation	water cycle

Write a short answer for each question or statement.

1. Rain that falls from the clouds is one kind of ___.
2. ___ are high, feathery clouds made of pieces of ice.
3. A storm that forms over a warm ocean and has strong winds and heavy rains is a ___.
4. The ___ is the movement of water between the ground and the air by evaporation, condensation, and precipitation.
5. Clouds that look like puffs of cotton are ___.
6. A cloud that forms just above the surface of the earth is called ___.
7. ___ is water vapor that condenses on cool surfaces, usually during the night.
8. ___ are clouds that form in layers and spread out over the sky.
9. A funnel cloud that has strong winds and moves along a path is called a ___.

244

Review Options

Cooperative Learning ♦

STAD Format (See page T23.)
Assign students to work in four- to five-member teams to study Chapter 11 Review. Students should work together to make sure that they and their teammates know the material in the chapter. After students have had enough time to study together, give them a test to complete individually (Chapter 11 Test A or B in the *Test Book*). Award Superteam certificates to teams whose average test scores exceed 90% and Greatteam certificates to teams whose average test scores exceed 80%.

Test Book page 97 *

Name ______

Chapter 11
Test A

Multiple Choice Choose the best answer.

1. Water vapor that condenses on cool surfaces is
   a. dew.
   b. rain.
   c. snow.
2. Clouds are formed by
   a. cool air rising.
   b. water vapor condensing.
   c. lightning.
3. Stratus clouds form in
   a. fog.
   b. balls.
   c. layers.
4. Clouds made of tiny ice crystals are called
   a. cirrus.
   b. cumulus.
   c. stratus.
5. A puffy cloud that looks like cotton is called
   a. cirrus.
   b. cumulus.
   c. stratus.
6. Water droplets that are too heavy to float make
   a. fog.
   b. clouds.
   c. rain.
7. The changing and moving of the earth's water is called
   a. the water cycle.
   b. a hurricane.
   c. precipitation.
8. Most water evaporates from
   a. lakes.
   b. ponds.
   c. oceans.
9. How does wind change the speed that water evaporates?
   a. makes it faster
   b. no change
   c. makes it slower
10. When electric charges jump between clouds, we see
   a. rain.
   b. lightning.
   c. thunder.
11. A hurricane forms over a warm
   a. ocean.
   b. city.
   c. desert.
12. Tornadoes can form during a
   a. snowstorm.
   b. heavy fog.
   c. thunderstorm.

97

♦ ***Suitable as a language development activity***

Reviewing What You Learned

Write the letter of the best answer.

1. The fastest winds on earth are in a
   (a) hurricane. (b) tornado. (c) rainstorm. (d) stratus cloud.
2. How does wind change the speed at which water evaporates?
   (a) makes it faster (b) makes no change
   (c) makes it slower (d) stops water from evaporating
3. One kind of precipitation is
   (a) fog. (b) snow. (c) dew. (d) water vapor.
4. Clouds form when water vapor
   (a) warms up. (b) freezes. (c) evaporates. (d) condenses.
5. A hurricane forms over a warm
   (a) desert. (b) river. (c) ocean. (d) lake.
6. Which kind of clouds can usually be seen during a thunderstorm?
   (a) cumulus (b) icy (c) stratus (d) cirrus
7. Most of the earth's water is in
   (a) rivers. (b) lakes. (c) oceans. (d) soil.
8. What forms when air just above the ground cools quickly?
   (a) a storm (b) dew (c) rain (d) fog

Interpreting What You Learned

Write a short answer for each question.

1. What should you do if you are outdoors during a thunderstorm?
2. What causes lightning and thunder?
3. Suppose a raindrop falls. What steps in the water cycle would the raindrop then follow?
4. How is a cloud like condensed water vapor on a mirror?

Extending Your Thinking

Write a paragraph to answer each question.

1. How could you explain to someone that you take in water every time you breathe?
2. Which place would be safer during a tornado, the top floor of a building with many windows or a basement with no windows? Explain your answer.

To explore scientific methods, see Experiment Skills on pages 368–369.

Interpreting What You Learned

1. You should go indoors. If you cannot go indoors, lie in a low place away from trees. Do not stay in water. **Thinking Skill:** *Restating and explaining ideas*

2. Lightning is caused by electric charges that jump between clouds or between clouds and the ground. The sudden heating of the air by lightning causes the sound of thunder. **Thinking Skill:** *Recognizing the main idea and supporting details*

3. It might flow into a river leading to a lake or ocean. It might later evaporate and become water vapor. The water vapor might condense into cloud droplets and later fall again as rain. **Thinking Skill:** *Sequencing*

4. In both cases, water vapor cools and condenses into tiny drops of water. **Thinking Skill:** *Recognizing patterns and relationships*

Extending Your Thinking

1. Water is in the air you breathe in the form of a gas called water vapor. **Thinking Skill:** *Inferring*

2. A basement with no windows would be safer because there would be little danger of being hit by broken window glass or other objects. **Thinking Skill:** *Drawing conclusions*

Test Book page 98 *

Name

Chapter 11
Test A

Short Answer Label the pictures of the water cycle. Use these words:
evaporation precipitation condensation

1. ______ 2. ______ 3. ______

Match each meaning in Column I with a word in Column II. Write the correct letter in the blank.

Column I	Column II
___ 4. a funnel-shaped cloud that forms over land during a thunderstorm	a. hurricane
___ 5. formed from electric charges	b. tornado
___ 6. a storm that can form over a warm ocean	c. lightning

Short Essay Use complete sentences to answer the question.

1. What is one way hurricanes and tornadoes are alike? How are they different?

98

Test Book page 99 *

Name

Chapter 11
Test B

Multiple Choice Choose the best answer.

1. What causes water to evaporate faster?
   a. rain b. clouds c. sun
2. Clouds that form just above the ground are called
   a. dew. b. fog. c. droplets.
3. The layered clouds often observed during a light rain are called
   a. cirrus. b. cumulus. c. stratus.
4. Feathery clouds floating high in the air on a mostly clear day are
   a. cirrus. b. cumulus. c. stratus.
5. Clouds that appear puffy and cottonlike are
   a. cirrus. b. cumulus. c. stratus.
6. Most of the earth's water is
   a. in the clouds. b. in the ocean. c. under the ground.
7. Heavy droplets of water that can no longer float in the clouds fall as
   a. fog. b. snow. c. rain.
8. The fastest winds on earth are in a
   a. hurricane. b. tornado. c. stratus cloud.
9. Electricity that jumps from one cloud to another during a storm is called
   a. lightning. b. thunder. c. heat.
10. A storm that forms over the ocean and has heavy rains is a
    a. hurricane. b. tornado. c. wind storm.

99

Test Book page 100 *

Name

Chapter 11
Test B

Short Answer Label the storms.
Use these words: tornado thunderstorm hurricane

1. ______ 2. ______ 3. ______

Fill in the blanks with the correct answers.

4. A storm with strong, destructive winds and heavy rains that can cause floods is a ______.
5. During a ______, large cumulus clouds form, winds begin to blow, and rain falls.
6. ______ make the most powerful winds on earth.

Short Essay Use complete sentences to answer the question.

1. Explain how water moves through the water cycle.

100

* *Answers to masters on pages 228E—228H*

Chapter 12 The Sun, Moon, and Planets Planning Guide

TEACHING PLAN

Chapter Components	Skills	Materials
Chapter Opener/*DISCOVER:* Observing Light pp. 246–247	*DISCOVER* p. 247 Science Process Skill *Observing*	*DISCOVER* p. 247 (groups of 2) 15 flashlights with batteries, 15 dark-colored balls
Lesson 1 How Do the Earth and Moon Move? pp. 248–252	Thinking Skills Challenge!: *Inferring* Find Out On Your Own: *Recognizing the main idea and supporting details, Making models*	Demonstration p. 248 polystyrene ball, pencil, chart paper
Activity Comparing the Sun and Moon p. 253	Science Process Skills *Making models, Measuring, Collecting and interpreting data, Inferring*	(groups of 3) 10 large balls, 10 small balls, 10 meter sticks, 10 pieces of chalk
Lesson 2 What Are the Sun and the Planets Like? pp. 254–257	Thinking Skills Challenge!: *Inferring* Find Out On Your Own: *Restating or explaining ideas, Communicating*	Demonstration p. 254 2 flashlights
Activity Making Models of Planets p. 258	Science Process Skills *Making models, Collecting and interpreting data*	(groups of 2) 15 large sheets lightweight cardboard, 15 compasses (for drawing circles), 15 metric rulers, 3 balls of string, 15 scissors, 15 sets of crayons or colored markers, 15 coat hangers
Science and People Doing Experiments in Space p. 259	Thinking Skills *Inferring, Predicting*	
Lesson 3 What Do Scientists Learn from Space Travel? pp. 260–263	Thinking Skills Challenge!: *Inferring* Find Out On Your Own: *Restating or explaining ideas*	Demonstration p. 260 desk, chair
Skills for Solving Problems Using Calendars and Charts pp. 264–265	Problem Solving Skills *Making decisions/Identifying and solving problems, Interpreting charts, maps, and graphs*	
Chapter Review pp. 266–267	Thinking Skills *Contrasting, Comparing, Restating or explaining ideas, Inferring*	

Teaching Options

Strategies	Extensions		Resource Masters
Cooperative Learning p. 246 (Also see p. T23.) Applying Whole Language p. 247 (Also see p. T30.)			Family Letter: *Resource Book* p. 135
Reading Strategies p. 248 (Also see pp. T26–T29.)	Science and Social Studies p. 249 Special Education p. 249 Reinforcement pp. 250, 251	Enrichment pp. 250, 251 Science and Language Arts p. 251 Reteaching Suggestion p. 252	Vocabulary Preview: *Workbook* p. 67 Science Activity: *Workbook* p. 68
			Activity Worksheet: *Resource Book* p. 139
Reading Strategies p. 254 (Also see pp. T26–T29.)	Science and Health p. 255 Special Education p. 255 Reinforcement pp. 255, 256	Enrichment p. 256 Game Suggestion p. 257 Reteaching Suggestion p. 257	Science Skills: *Workbook* p. 69
			Activity Worksheet: *Resource Book* p. 141
Reading Strategies p. 260 (Also see pp. T26–T29.)	Science and Language Arts p. 261 Special Education p. 261 Reinforcement p. 262	Enrichment p. 262 Game Suggestion p. 263 Reteaching Suggestion p. 263	Science and Social Studies: *Workbook* p. 70 Vocabulary Puzzle: *Workbook* p. 71
	Reteaching Suggestion p. 265		Science and Reading: *Resource Book* p. 143
Cooperative Learning p. 266 (Also see p. T23.)			Chapter Tests: Forms A and B *Test Book* pp. 105–108

Classroom Management

Advance Preparation

DISCOVER, page 247
Provide a flashlight and a dark-colored ball for each group of students.

Demonstration, page 248
Obtain a polystyrene ball for this demonstration.

Activity, page 253
Each group of students will need a large ball, such as a basketball, a volleyball, or an inflatable rubber ball, and a small ball, such as a tennis ball. You might be able to use the balls from the *DISCOVER* activity on page 247 and from the demonstration on page 248.

Demonstration, page 254
You will need two flashlights for this demonstration. You might use the flashlights from the *DISCOVER* activity on page 247.

Activity, page 258
Each group of students will need a compass and a coat hanger.

Vocabulary Review

Use the following sentences with your students to review the meanings of the italicized words.

1. A *moon* is a large, rocky object that orbits—or circles—a planet.
2. We see most objects because light *reflects*—or bounces off—their surfaces.
3. During a *solar eclipse*, the moon passes between the sun and the earth.
4. *Energy* is the ability to move objects.
5. *Temperature* is a measure of how hot or cold something is.
6. With *telescopes* we can see more stars than we can see with our eyes alone.
7. A *volcano* is a mountain that can spew melted rock and hot gases.

High-Potential Students

Ask students to begin a scrapbook about the solar system to which they can add information as the year progresses. Have the students use reference materials to gain information, and then ask them to draw, color, and label the planets and their satellites. Throughout the year, have the students add information to their scrapbooks on current research on the sun, the planets, and space travel.

Mainstreamed Students

Emotionally Handicapped
Have students who are emotionally handicapped choose a planet to report on and have them look in the library to find pictures and information about the planet they choose. Let them give oral reports about what they learn or make charts and displays that provide information about the planets.

Visually Impaired
Help students who are visually impaired make clay models of the planets and the sun. Have the sighted students help visually impaired students put the planets and the sun in the proper order to represent the solar system.

Science Fair Projects

The Experiment Skills on p. 370 and the Investigate feature in the chapter can be used for science fair projects. You might also encourage interested students to do one of the following projects:

1. Obtain some cardboard or paper plates and paint the phases of the moon on them. Label each phase and make a chart that shows the position of the moon in relation to the sun and the earth, during each phase.
2. Find out about food and items that were invented or used for space travel that are now being used on Earth. Obtain and display some of these items along with a written explanation of how they were useful in space.
3. Obtain information about the space missions Mercury, Gemini, and Apollo. Use cards to list the dates of the missions, the names of the astronauts on board, and some of the highlights during each mission. Display this information under pictures of each spacecraft or mission crew.

lassroom Resources

Bulletin Board

Divide the class into nine groups. Assign each group a planet. Ask each group to find out similar kinds of facts about their planet and add them to the bulletin board.

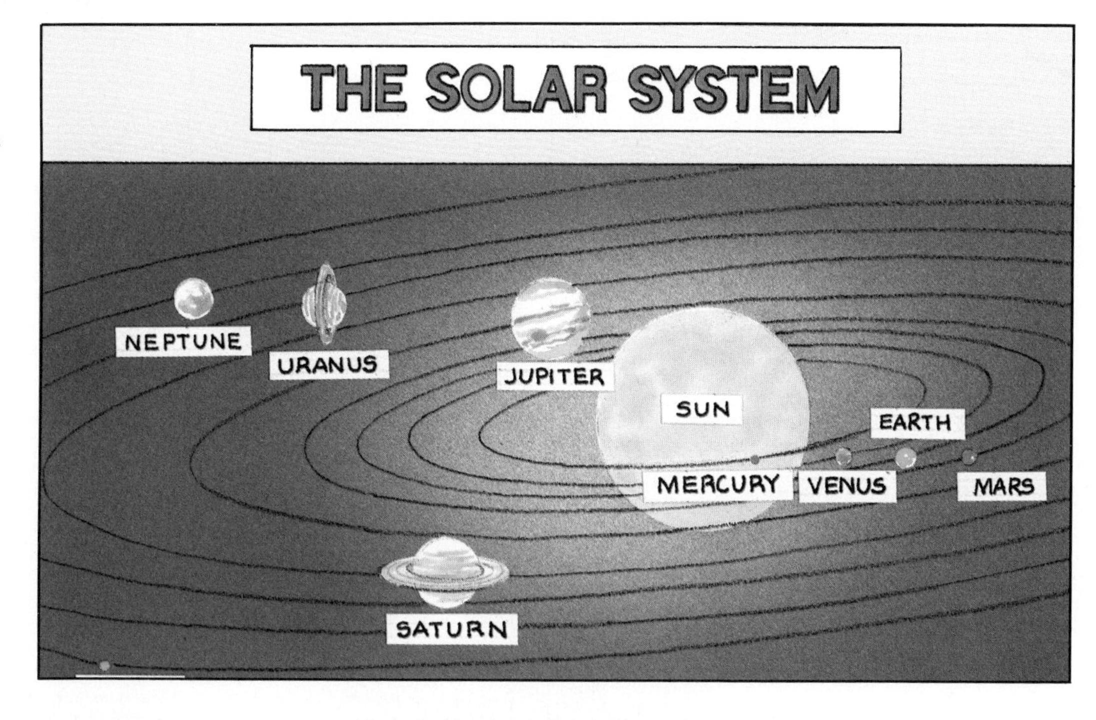

Chapter 12 Poster

Science Discovery Center

Use pages 125–130 from the *Science Discovery Center Book.* Place these worksheets in the appropriate pockets in the Science Discovery Center.

Overhead Transparencies

Use Transparencies 20 and 21 from the package of color overhead transparencies.

Teacher's Resource Book

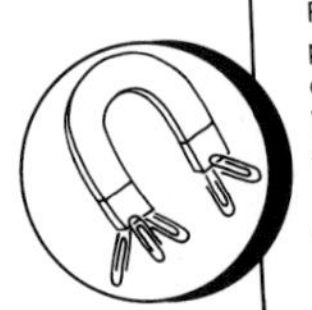

Dear Family,
Your student will be reading **Chapter 12: The Sun, Moon, and Planets** in *Discover Science*, published by Scott, Foresman. We will learn about the solar system and the planets and their satellites. The moon revolves around the earth. We can see the moon because it reflects the sun's light. We see different phases (shapes) of the moon because of the way the moon moves.
You and your student can do this activity together to learn more about the phases of the moon.

Phases of the Moon

You will need a calendar for one month with spaces large enough to draw the moon, a pencil or black crayon, and a flashlight.

1. With your student, look for the moon in the morning, afternoon, or evening sky. A newspaper or astronomy magazine can tell you when the moon will rise. The best time to start this project is an evening when the moon is full.
2. Take the calendar, pencil, and a flashlight. (On the day of the new moon, you will see nothing.) On the calendar, your student can draw the shape of the moon. Ask, "Can you see dark shapes in the moon?" (the craters) Help your student make the picture accurate.
3. Try to go out to look for the moon every day for a month. (Tell your student that the word "month" comes from the Old English word mona, meaning moon.) Record the shape of the moon every day.
4. At the end of the month, ask, "How did the shape of the moon change?"
5. Find a book about the moon and learn more about the phases. Label the phases of the moon on your calendar.

Workbook

Name ______________________

Use with Lesson 1: pages 248-252

Chapter 12

Vocabulary Preview

The Sun, Moon, and Planets

Chapter Vocabulary		
astronaut	phase	rotation
axis	planet	satellite
crater	revolution	solar system
orbit		

Vocabulary Cards

1. Write each word on a card.
2. Find each word in the glossary. Copy the pronunciation under the word on the card.
3. Practice saying the words with a partner.

1	2	3
4	5	6

The Pronunciation Game

1. Work with a partner. Together, make two sets of six number cards. See the top picture. Put both sets of number cards in a paper bag.
2. Shuffle two sets of vocabulary cards together. Put them face up in a line. See bottom picture.
3. Now take turns. Pick a number card from the bag. The card tells you how many steps to move a marker along the line of vocabulary cards. Say the vocabulary word your marker lands on. If you pronounce the word correctly, leave your marker on the card. If you pronounce the word incorrectly, move your marker back to start.
4. The next player takes a turn. Continue until one player's marker reaches the finish.

Finish

Start

At Home

1. Practice saying the words. Learn their meanings.
2. Look for the words and pictures of the words in newspapers and magazines. You can make a poster.

Teacher's Notes: After finishing the chapter, students can play *The Definition Game*, using definitions rather than pronunciations.

Workbook

Name ______________________

Use with Lesson 1: pages 248-252

Chapter 12

Science Activity

How Does the Moon Move?

Gather These Materials

• paper • marker or crayon

Follow This Procedure

1. Work with a friend. Write *Front of the Moon* on the paper. Hold the paper in front of you. You will pretend to be the moon.
2. Have your friend stand about a meter in front of you. Your friend will pretend to be the earth.
3. Walk around your friend so that the front of the moon always faces earth. Notice how you move.
4. Now walk around your friend so that the front of the moon always faces the same wall.

Record Your Results

1. What did you have to do so that the front of the moon always faced the earth?
I had to turn around (rotate) while I walked around the earth.
2. How was this different from the way you walked when the front always faced a wall?
I did not rotate, but different sides of the moon faced the earth.

State Your Conclusions

1. What did you do to show that the moon revolves around the earth?
I walked around my friend in a circle.
2. What did you do to make the front of the moon always face the earth?
I turned around and rotated on my axis.
3. How does the moon move?
The moon rotates on its axis so that the same side always faces the earth. If the moon did not rotate or rotated faster, different sides would face earth.

Teacher's Notes: Have the students do the activity again, but switch roles, so that each has a chance to be the moon.

Teacher's Resource Book

Name ______________________

Use with Lesson 1: page 253

Comparing the Sun and Moon

Chapter 12

Activity Worksheet

Record Your Results

Distance to Moon	
Distance to Sun	

State Your Conclusion

1. Was the moon closer to you or closer to the sun when it looked the same size as the sun?
____
2. In what way does distance affect how large an object looks?
____

Use What You Learned

Why do the sun and moon in the sky appear to be the same size?
____

Name ____________________

Use with Lesson 2: pages 254-257

Chapter 12

Science Skills

The Solar System

A *fact* is something that is true. A fact can be tested. An *opinion* explains how someone feels. Often, an opinion cannot be tested.

Read the statements below. Decide whether each is a fact or an opinion. Write *F* in front of each fact. Write *O* in front of each opinion.

The Moon

F 1. The moon is a satellite of the earth.

O 2. The full moon is a very beautiful sight.

F 3. The sun is a star.

O 4. We think there is no life in the solar system except on earth.

F 5. Many other planets have moons, just as the earth does.

F 6. Jupiter is the largest planet in the solar system.

F 7. The planet Venus is so bright that you can see it during the day.

O 8. People will never live on the planet Mars.

F 9. Jupiter has sixteen moons, and Mars has two moons.

F 10. Moonlight is really reflected sunlight.

O 11. I think being an astronaut must be exciting.

Teacher's Notes: Encourage the students to share facts they have learned about and their own opinions of the solar system.

Name ____________________

Use with Lesson 2: page 258

Chapter 12

Activity Worksheet

Making Models of Planets

Record Your Results

Picture of Planet				
Name of Planet				

State Your Conclusion

1. How do Mercury, Venus, Earth, and Mars compare in size?

2. What is the order of the first four planets from the sun?

Use What You Learned

Which of these planets takes the longest to orbit the sun? Explain.

Name ____________________

Use with Lesson 3: pages 260-263

Chapter 12

Science and Social Studies

Living in Space

Someday people will be able to live in space. Instead of living in houses, the people will live in space stations. The picture shows how a space station might look.

Everything people need for life must be present in the station. The spacecraft must hold plants, animals, and water. People can use the plants and animals for food. Living things also need oxygen. Plants can provide fresh supplies of oxygen. In the space station, water and oxygen are used, then cleaned, and reused.

People need a place to live too. Suppose the wheel part of the station contains apartments. From their windows, people could see the earth, as well as the moon, sun, and stars. Imagine the view!

The station can get its energy directly from sunlight. This energy, called solar energy, can run machines and provide heat and light for the spacecraft.

Today, scientists are making plans to build a space station. However, it takes years of work and planning to do this job.

Answer these questions.

1. What things must people in a space station have?
   food, water, oxygen, space, energy
2. How is energy useful?
   run machines, heat, light
3. Why is solar energy the best energy for a spacecraft?
   lots of available sunlight in space

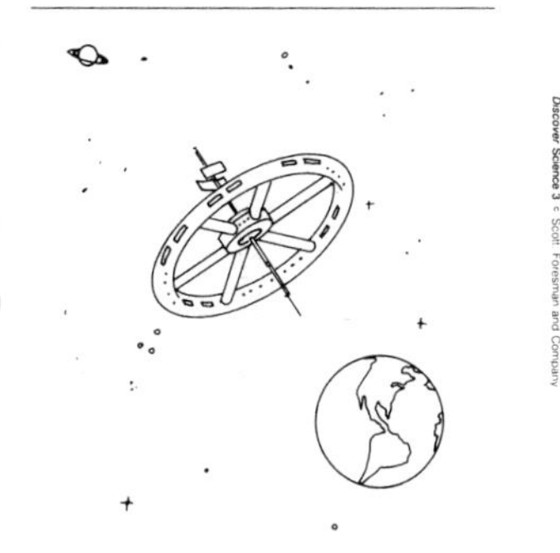

Teacher's Notes: Discuss the items needed for life in space.

Name ____________________

Use with Lesson 3: pages 260-263

Chapter 12

Vocabulary Puzzle

Word Scramble

Unscramble the letters in each pie. Write the word on the line, as shown.

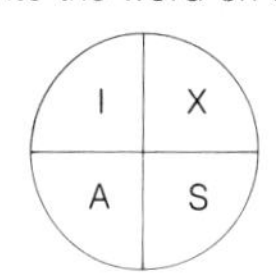

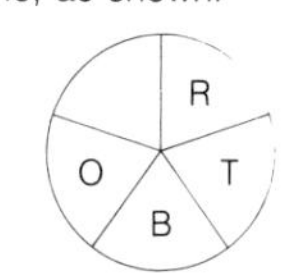

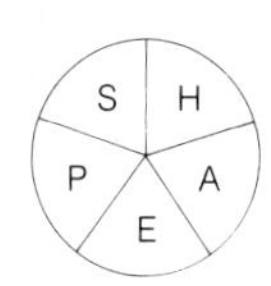

1. AXIS 2. ORBIT 3. PHASE

A P T N L E

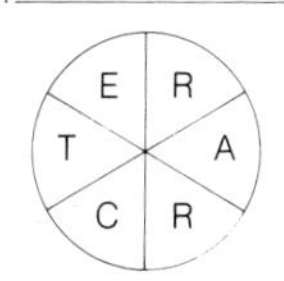

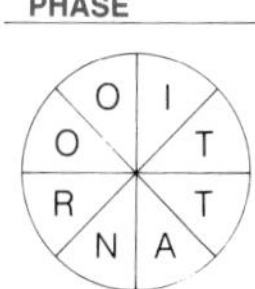

4. PLANET 5. CRATER 6. ROTATION

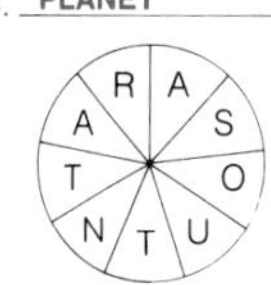

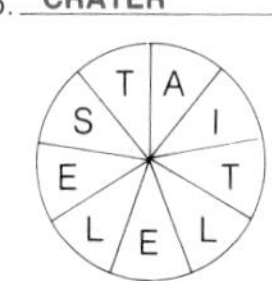

7. ASTRONAUT 8. SATELLITE 9. REVOLUTION

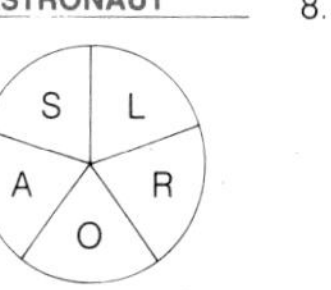

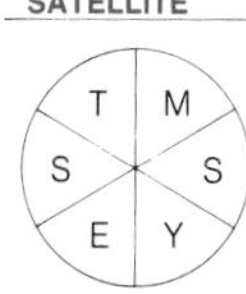

10. SOLAR SYSTEM

Teacher's Notes: Ask volunteers to define the vocabulary words.

Name ______________________

Comprehension: predicting outcomes

Chapter 12

Science and Reading

What Do You Predict?

Read each story. Then write a sentence that tells what will happen next.

1. Rosa reads everything she can about space. She hopes to be an astronaut someday. One day, Rosa reads in the newspaper that an astronaut will speak at the library. Rosa runs to find her father.

   Possible answer: She asks him if he will take her to see the astronaut speak.

2. John's class has been studying the moon. One day, John's teacher tells the class that there will be an eclipse of the moon that night. John runs home to tell his parents about the eclipse. Then he remembers that his neighbor has a telescope.

   Possible answer: John runs to ask his neighbor if he may use his telescope to see the eclipse.

3. Sue has been studying for her science test. She still does not understand what causes the phases of the moon. She decides to go to the kitchen for a snack and finds her father there with a friend. He introduces Sue and tells her that his friend is a science teacher at the high school.

   Possible answer: Sue asks her father's friend to help her understand the phases of the moon.

Name ______________________

Chapter 12
Test A

Multiple Choice. Choose the best answer.

1. The earth rotates around an imaginary line called the (1-1)
   a. Equator.
   (b.) axis.
   c. phase.

2. The movement of an object in a complete orbit around another object is a (1-1)
   (a.) revolution.
   b. rotation.
   c. crater.

3. The moon revolves around the earth in about (1-2)
   a. one year.
   b. one day.
   (c.) one month.

4. The moon reflects light from the (1-2)
   a. earth.
   (b.) sun.
   c. planets.

5. The shape of the lighted part of the moon as it is seen from the earth is the moon's (1-3)
   (a.) phase.
   b. path.
   c. stage.

6. The sun looks bigger than other stars because it (2-1)
   a. is hotter.
   b. really is bigger.
   (c.) is closer to the earth.

7. Venus, Mercury, and Mars are three of the sun's (2-2)
   a. moons.
   (b.) planets.
   c. systems.

8. The large holes that are found on the moon are called (3-1)
   (a.) craters.
   b. cones.
   c. scoops.

9. All the planets, their moons, and the sun make up the (2-2)
   a. sun system.
   (b.) solar system.
   c. planet system.

10. The moon is the earth's (1-2)
    a. spacecraft.
    (b.) satellite.
    c. astronaut.

11. What is the largest planet?
    (a.) Jupiter
    b. Earth
    c. Mars (2-2)

Numbers in parentheses after each question refer to the lesson number and the objective of that lesson.

Name ______________________

Chapter 12
Test A

Matching Write the letter of each planet in the blank next to its name. (2-2)

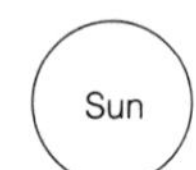

a. b. c. d.   e.   f.   g.   h.   i.

1. Earth c   2. Venus b

3. Saturn f   4. Pluto i

Short Essay Use complete sentences to answer each question.

1. Name two ways scientists study the planets. (3-2)

   Scientists use spacecraft with cameras to take pictures of the planets. Special machines are also used to collect soil and rock samples from the planets. These samples are then studied by scientists here on earth.

2. Explain where the moon gets its light. (1-2)

   The moon has no light of its own. It reflects light from the sun.

Numbers in parentheses after each question refer to the lesson number and the objective of that lesson.

Name ______________________

Chapter 12
Test B

Multiple Choice. Choose the best answer.

1. Every 24 hours the earth makes one complete (1-1)
   a. phase.
   b. revolution.
   (c.) rotation.

2. The path an object follows as it moves around another object is called (1-1)
   a. a revolution.
   (b.) an orbit.
   c. an axis.

3. The sun keeps the earth in orbit by the force of (1-2)
   a. magnetism.
   (b.) gravity.
   c. rotation.

4. The length of time it takes the earth to complete one revolution is (1-2)
   (a.) one year.
   b. one day.
   c. one month.

5. When all the lighted part of the moon can be seen, the phase is called the (1-3)
   a. new moon.
   b. half moon.
   (c.) full moon.

6. The sun is made of (2-1)
   a. liquids.
   (b.) gases.
   c. volcanoes.

7. The solar system is made up of objects that revolve around the (2-2)
   a. earth.
   (b.) sun.
   c. moon.

8. A planet that is only a little cooler than the earth is (2-2)
   a. Pluto.
   (b.) Mars.
   c. Venus.

9. What instrument have scientists used for four hundred years to study the sky? (3-1)
   a. satellite.
   (b.) telescope.
   c. camera.

10. Scientists have landed spacecraft with cameras and special machines on the (3-2)
    a. sun.
    b. stars.
    (c.) planets.

Numbers in parentheses after each question refer to the lesson number and the objective of that lesson.

Name ______________________

Chapter 12 Test B

Matching Match each planet with its description.
Look at the drawing of the planets to help with your answers. (2-2)

Sun a. b. c. d. e. f. g. h. i.

C 1. the planet between Venus and Mars — A. Mercury

I 2. the smallest planet — E. Jupiter

E 3. the planet with sixteen moons — C. Earth

A 4. the planet closest to the sun — I. Pluto

Short Essay Use complete sentences to answer each question.

1. Why does the moon appear to have different shapes? (1-2)

As the moon revolves around the earth, different amounts of its lighted part can be seen. These shapes are called the moon's phases.

2. Name two things scientists know about Saturn. (2-2)

Scientists know that Saturn is made mostly of gases and that it is cold. Saturn also has rings made of ice.

Numbers in parentheses after each question refer to the lesson number and the objective of that lesson.

Name ______________________

Unit Test 3

Multiple Choice Choose the best answer.

1. Rocks are made of one or more (9-1-1)
   a. carbons.
   (b.) minerals.
   c. nutrients.

2. Minerals and soil that people use are (9-3-2)
   (a.) natural resources.
   b. igneous rocks.
   c. nutrients.

3. Materials from the soil that plants need are (9-2-1)
   a. clays.
   b. sediments.
   (c.) nutrients.

4. The wearing down or breaking apart of rocks is called (10-2-1)
   a. an eruption.
   b. mining.
   (c.) weathering.

5. Which is a mineral? (9-3-1)
   (a.) copper
   b. wood
   c. cotton

Use the picture to answer questions 6 and 7.

3. 1. 2.

6. In this picture of a volcano, magma is (10-3-2)
   a. 1. b. 2. (c.) 3.

7. Which number shows lava from the volcano? (10-3-2)
   (a.) 1 b. 2 c. 3

8. Plants can help change the earth's (10-4-1)
   a. mantle.
   b. center.
   (c.) crust.

9. The layer of the earth with soil is called the (10-1-1)
   (a.) crust.
   b. mantle.
   c. core.

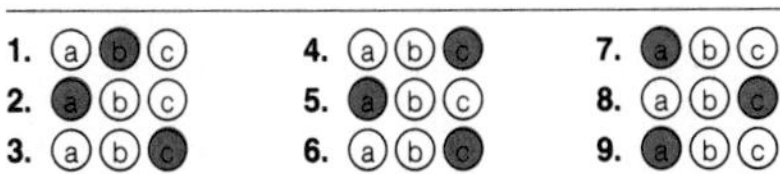

Numbers in parentheses after each question refer to chapter and lesson numbers and the objective of that lesson.

Name ______________________

Unit Test 3

Multiple Choice Choose the best answer.

10. The changing and moving of the earth's water is called
    (a.) the water cycle.
    b. a hurricane.
    c. precipitation. (11-2-2)

11. When electric charges jump between clouds, we see
    a. rain.
    (b.) lightning.
    c. thunder. (11-3-1)

12. Clouds are formed by (11-1-2)
    a. cool air rising.
    (b.) water vapor condensing.
    c. lightning.

13. Layered clouds seen during a light rain are called (11-1-3)
    a. cirrus.
    b. cumulus.
    (c.) stratus.

14. A hurricane forms over a warm
    (a.) ocean.
    b. city.
    c. desert. (11-3-2)

15. The sun is made of (12-2-1)
    a. liquids.
    (b.) gases.
    c. volcanoes.

16. The time the earth takes for one revolution is (12-1-1)
    (a.) one year.
    b. one day.
    c. one month.

17. The moon is the earth's
    a. source of light.
    b. spacecraft.
    (c.) satellite. (12-1-2)

18. Scientists have landed spacecraft with cameras and special machines on the
    (a.) planets.
    b. stars.
    c. sun. (12-3-2)

19. All the planets, their moons, and the sun make up the
    a. sun system.
    (b.) solar system.
    c. planet system. (12-2-2)

10. **a** b c
11. a **b** c
12. a **b** c
13. a b **c**
14. **a** b c
15. a **b** c
16. **a** b c
17. a b **c**
18. **a** b c
19. a **b** c

Numbers in parentheses after each question refer to the chapter and lesson numbers and the objective of that lesson.

TEACHING PLAN

Major Concepts

Lesson 1 The earth rotates on its axis and revolves around the sun; the moon rotates and revolves around the earth.
Lesson 2 The sun and the planets and their moons make up most of the solar system.
Lesson 3 Space travel can provide information about the moon and the planets.

Chapter Vocabulary

astronaut, axis, crater, orbit, phase, planet, revolution, rotation, satellite, solar system

Getting Started

Show students a softball representing the earth and a golf ball representing the moon. Ask students to place the golfball at a distance from the softball to approximate a scale model of the earth and moon. The golf ball should be about 2.3 meters (2.5 yards) away from the softball. Tell students this distance represents 384,403 kilometers (238,857 miles). Explain that a rocket would take about 6 days to make a round-trip journey to the moon.

Chapter 12

The Sun, Moon, and Planets

For hundreds of years people wondered what caused parts of the moon to look light or dark. Scientists now know that mountains and plains make the shades you see on the moon.

246

Teaching Options

Cooperative Learning ♦

Jigsaw Format (See page T23.)
Assign the following topics at random to your cooperative learning teams.

Topic A: What are some ways that the earth moves and how much time does each way take?
Topic B: How does the moon appear to us as it moves through the sky?
Topic C: What have scientists learned from spacecraft, and what have astronauts brought back to the earth from space?
Topic D: What star is closest to the earth and what objects revolve around it?

Have students search for information on their topic as they read the chapter. Then let all students with the same topic meet in an expert group to discuss the information. When students return to their teams, they may take turns presenting their topics to the team. Then give students a test covering all topics to complete individually (Chapter 12 Test A or B in the *Test Book*). Award Superteam certificates to teams whose average test scores exceed 90%, and Greatteam certificates to teams whose average test scores exceed 80%.

♦ ***Suitable as a language development activity***

Introducing the Chapter

Think about the moon shining on a clear night. Light from the sun causes the moonlight you see. The activity below will help you learn what causes the earth's moon to shine. In this chapter, you will learn about the sun and the planets and their moons. You also will read about space travel.

Observing Light

Hold a flashlight in one hand and a dark-colored ball in the other hand. This picture shows you how to shine the flashlight directly on the ball. Pretend that the light from the flashlight is light from the sun and the ball is the earth's moon.

Move the ball slowly to the left. Notice the change in the light on the ball. Now continue to move the ball until it is completely out of the light.

Talk About It

1. Describe how the ball looked when the flashlight was shining on it.
2. How did the light on the ball change when you moved the ball away from the light?

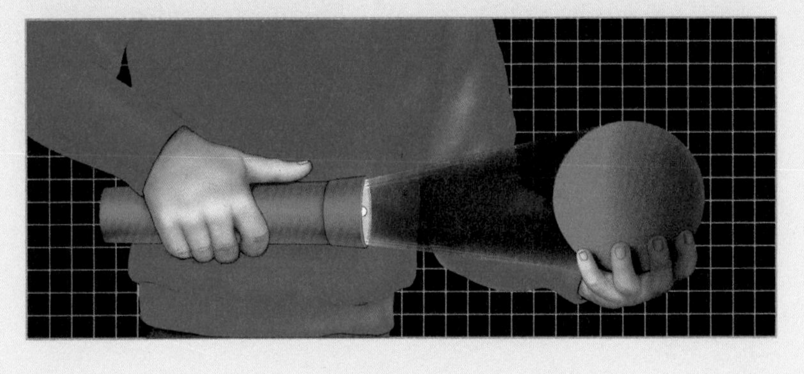

DISCOVER

Objective ♦

This optional *DISCOVER* activity will help students explore and build background information about the concept of reflecting light. Later in the chapter students will be able to draw on this experience to help them assimilate the new content.

Science Process Skill

Observing

Materials

For each pair of students: flashlight, dark-colored ball

Teaching Tips

- Question: **What does the flashlight represent?** (the sun) Have students *compare* the way light reflects off the ball to the way sunlight reflects off the moon. Question: **Which side of the moon or the ball is lit?** (the side that faces the light source).
- Warn students not to shine the flashlight in other people's eyes.

Answers

Talk About It

1. The side of the ball towards the flashlight looked light.

2. The part of the ball in the direct light of the flashlight looked light; the rest of the ball looked dark.

Applying Whole Language ♦

Discuss the whole language framework with each Teaching Option you select. Here is an example applied to the Science and Language Arts option on p. 251.

1. Purpose: To write a poem or poems about the moon

2. Context: The role of poets

3. Decisions: Individual, group, or whole class? What key words should we include? What key ideas? What feelings to express?

4. Evaluation: How did writing our poem make us feel? How does it make others feel? Do our poems teach? (See p. T30.)

Resource Book page 135

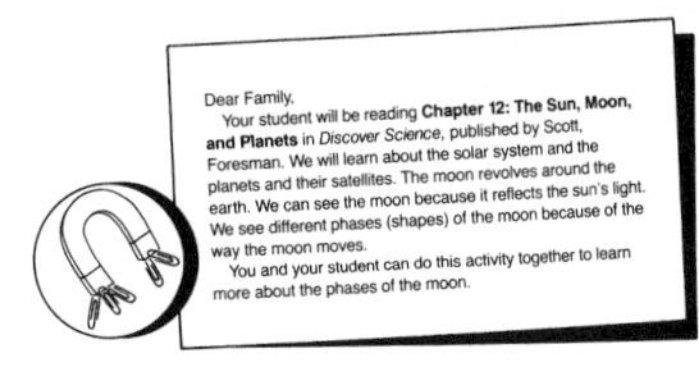

Dear Family,

Your student will be reading **Chapter 12: The Sun, Moon, and Planets** in *Discover Science*, published by Scott, Foresman. We will learn about the solar system and the planets and their satellites. The moon revolves around the earth. We can see the moon because it reflects the sun's light. We see different phases (shapes) of the moon because of the way the moon moves.

You and your student can do this activity together to learn more about the phases of the moon.

Phases of the Moon

You will need a calendar for one month with spaces large enough to draw the moon, a pencil or black crayon, and a flashlight.

1. With your student, look for the moon in the morning, afternoon, or evening sky. A newspaper or astronomy magazine can tell you when the moon will rise. The best time to start this project is an evening when the moon is full.
2. Take the calendar, pencil, and a flashlight. (On the day of the new moon, you will see nothing.) On the calendar, your student can draw the shape of the moon. Ask, "Can you see dark shapes in the moon?" (the craters) Help your student make the picture accurate.
3. Try to go out to look for the moon every day for a month. (Tell your student that the word "month" comes from the Old English word mona, meaning moon.) Record the shape of the moon every day.
4. At the end of the month, ask, "How did the shape of the moon change?"
5. Find a book about the moon and learn more about the phases. Label the phases of the moon on your calendar.

135

Science Background

The moon is a natural satellite of the earth. The surface of the moon has a pattern of light and dark parts. The light parts are highland areas that contain many craters. The dark parts are the places where, billions of years ago, molten material flowed out of the moon's interior.

TEACHING PLAN

Lesson Objectives
- *Describe* rotation and revolution of the earth and the time required for each.
- *Describe* the moon and explain how it rotates on its axis and revolves around the earth.
- *Describe* the phases of the moon.

Lesson Vocabulary
axis, orbit, phase, revolution, rotation, satellite

1. Motivate

Demonstration Activity ♦
Poke a pencil through a polystyrene ball. Trace a large circle on chart paper with the pencil, explaining that it is moving like the earth revolving around the sun. Then spin the ball on the pencil, to represent earth rotating. Retrace the circle while spinning the ball. Explain that the earth revolves and rotates at the same time.

Discussion
Question: **What would happen to the light on earth if the earth revolved, but did not rotate?** (One side of the earth would always be light; one side would be dark.)

1 How Do the Earth and Moon Move?

LESSON GOALS

You will learn
- how the earth moves during 24 hours and 1 year.
- what the moon looks like and how the moon moves.
- about phases of the moon.

axis (ak′sis), an imaginary straight line through the center of the earth around which the earth rotates.

rotation (rō tā′shən), the act of spinning on an axis.

Have you ever seen a top spinning around and around? The earth spins around in much the same way as a top does.

How the Earth Moves

The earth spins around an imaginary line that runs through its center. This line is called the **axis.** Every twenty-four hours, the earth makes one complete spin, or a **rotation.**

Follow what happens to point A as the earth rotates. How does the time change? The rotation of the earth causes day and night. The part of the earth toward the sun has daylight. The part away from the sun has night.

248

Teaching Options

Science Background

The Foucault pendulum demonstrates the rotation of the earth. This pendulum consists of a swinging metal weight suspended from a very long wire. In 1851, J.B.L. Foucault showed that instead of constantly swinging back and forth in one direction, this pendulum's swing changes direction. This occurs because the earth turns under the pendulum. The rate at which the direction changes varies according to latitude. At the poles, the pendulum makes a complete rotation in 24 hours. At the equator, where the earth does not turn under the pendulum, the direction does not change. In the northern hemisphere, the plane of swing moves clockwise; in the southern hemisphere, it moves counterclockwise.

Reading Strategies ♦

1. Guide students' pre-reading by asking: Which parts of the lesson are familiar and which parts are new?
2. Assign these strategies: Visualizing Information and Writing a Memory Sentence for each subheading. (See pages T26–T29.)
3. Pair students to share what information is clear and unclear and initiate discussion using students' unanswered questions.

♦ *Suitable as a language development activity*

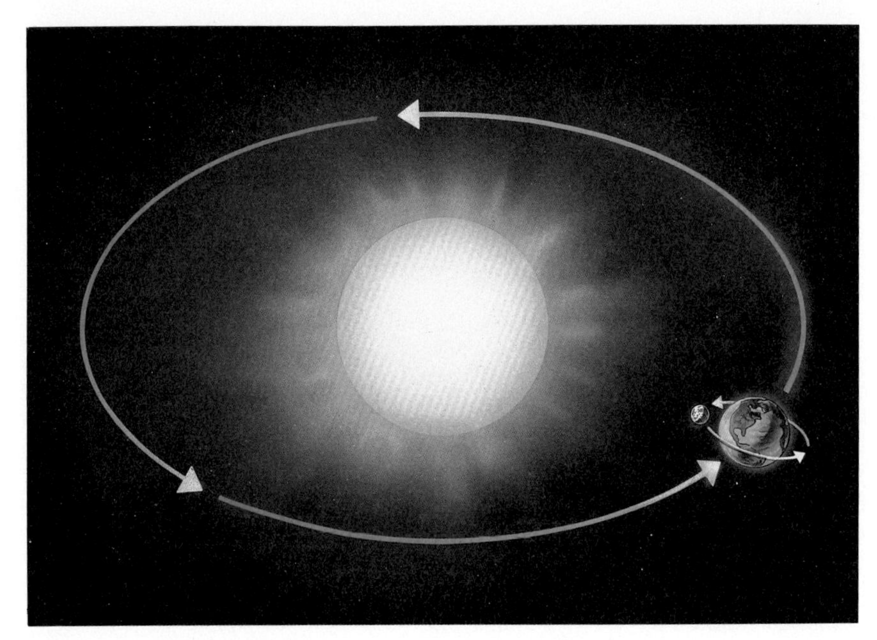

The earth's orbit around the sun

This morning you might have seen the sun in one part of the sky. Will you see the sun in the same part of the sky in the late afternoon?[1] The sun really is not moving across the sky. The way the earth rotates makes the sun appear to move.

The earth also moves in a path around the sun. This path is called an **orbit.** Notice in the picture that the path is shaped almost like a circle. Use your finger to trace the orbit of the earth around the sun. The force of gravity from the sun pulls on the earth and keeps the earth in its orbit. Each complete orbit around the sun is called one **revolution.** The earth takes about 365 days, or one year, to make one revolution.

orbit (ôr′bit), a closed, curved path that an object follows as the object moves around another object.

revolution (rev′ə lü′shən), movement of an object in an orbit around another object.

[1]No

2. Teach

Teaching Tips

- Refer students to the *Demonstration Activity*. Point out that the pencil represents the earth's axis. Question: **How long does the earth take to rotate on its axis?** (24 hours, or one day)
- Explain that, years ago, people used to determine what time it was by using a sun dial showing the position of the shadow of an object in relation to the sun. Help students make a crude sundial. Tape a ruler to the side of a box, leaving most of the ruler sticking up at an angle, above the box (towards the sky). Place white paper on top of the box and place the box outside. Place the raised end of the ruler toward the north. In half hour intervals for two to four hours, assign different students to go outside and mark where the ruler's shadow falls on the paper. Display the box in the classroom with a sign that reads, "When the earth rotates, it changes its position in relation to the sun."
- **Possible Misconception:** Make sure students understand that the sun does not move. Emphasize that the sun rises in the east and sets in the west because the earth is rotating, which makes the sun *appear* to move.

Workbook page 67 *

Name ____________

Use with Lesson 1: pages 248-252

Chapter 12

Vocabulary Preview

The Sun, Moon, and Planets

Chapter Vocabulary

astronaut	phase	rotation
axis	planet	satellite
crater	revolution	solar system
orbit		

Vocabulary Cards

1. Write each word on a card.
2. Find each word in the glossary. Copy the pronunciation under the word on the card.
3. Practice saying the words with a partner.

1	2	3
4	5	6

The Pronunciation Game

1. Work with a partner. Together, make two sets of six number cards. See the top picture. Put both sets of number cards in a paper bag.
2. Shuffle two sets of vocabulary cards together. Put them face up in a line. See bottom picture.
3. Now take turns. Pick a number card from the bag. The card tells you how many steps to move a marker along the line of vocabulary cards. Say the vocabulary word your marker lands on. If you pronounce the word correctly, leave your marker on the card. If you pronounce the word incorrectly, move your marker back to start.
4. The next player takes a turn. Continue until one player's marker reaches the finish.

Finish

Start

At Home

1. Practice saying the words. Learn their meanings.
2. Look for the words and pictures of the words in newspapers and magazines. You can make a poster.

67

* **Answers to masters on pages 246E–246H**

Science and Social Studies

Tell the students that satellites in space help cartographers (map makers) draw more accurate maps of the earth. Help the students use reference materials in the library to find out how cartographers measured parts of the earth before satellites were in space.

Special Education

Have students with learning disabilities act out the movements of the sun, moon, and earth. Have one student pretend to be the earth, rotating and revolving around the sun. Have another student pretend to be the moon, revolving around the earth. Another student might be the sun. As the earth and moon change their positions, point out the different phases of the moon.

TEACHING PLAN

Teaching Tips

- Use a model or diagram to show students that we always see the same side of the moon. The side of the moon that faces us is completely illuminated when it is in the opposite part of the sky from the sun, and that same side is completely dark when the moon is between the earth and the sun. (Make sure students understand that the side of the moon that does not face the earth is light when the side facing the earth is dark. The side of the moon facing away from the earth is dark when the side we see is light.)
- Questions: **How is the moon like the earth?** (Students should *compare* the two and state that both the moon and the earth have rocks, soil, hills, and valleys. The moon and the earth both get light from the sun.) **How are the moon and the earth different?** (Students should state that the moon lacks air and water.)

Investigate!

Accept any testable hypothesis. Most students' data will support this hypothesis: *The earth rotates in a west-to-east direction.* Students should find out that the earth's rotation causes the sun to appear to rise in the east and set toward the west. Only a west to east rotation produces this effect.

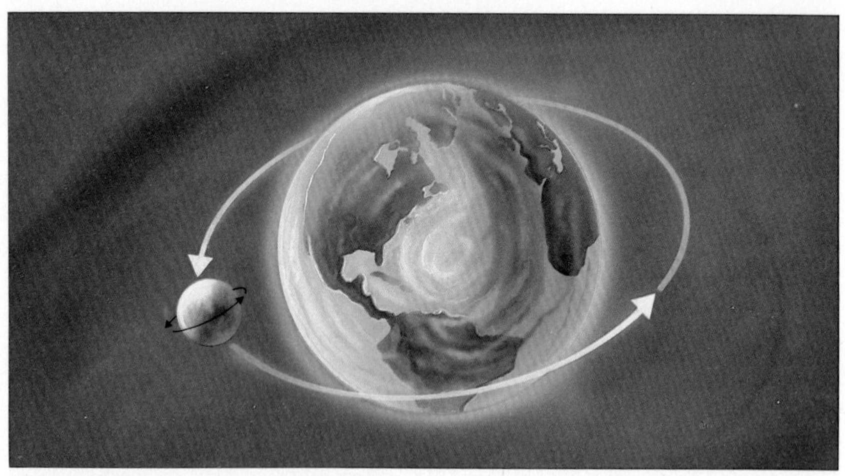

The moon's orbit around the earth

INVESTIGATE!

Find out if the earth rotates in an east-to-west or a west-to-east direction. Write a hypothesis and test it with an experiment. You might observe the directions in which the sun rises and sets. Then compare this to a model of the earth and the sun using two balls.

satellite (sat′l īt), object that revolves around another object.

How the Moon Moves

Find the earth's moon in the picture. What is its shape?[1] Like the earth, the moon has rocks and soil. The moon also rotates on its axis. Yet the earth's moon is different from the earth. This moon has no water or air.

The moon is called a **satellite** of the earth because it revolves around the earth. Use your finger to trace the moon's orbit. The moon takes about one month to make one revolution around the earth.

The moon does not give off its own light. The light you see comes from the sun's light reflecting off the moon. As it rotates, the part of the moon facing the sun reflects sunlight. The other part of the moon is dark. From the earth, you only see the lighted part of the moon that faces you. As the moon revolves, you see different amounts of its lighted part.

[1]round

Teaching Options

Science Anecdote

A person who weighs about 132 pounds on earth will weigh only about 22 pounds on the moon.

Reinforcement

Ask students to pretend that they are astronauts and have them write a story about landing on the moon. Ask them to *describe* the moon's appearance and physical features, and tell how the moon is moving in relation to the earth and the sun.

Enrichment

Write to or call NASA to get samples of moon rocks for students to examine. (NASA certifies teachers to receive these rocks.) Tell students that scientists believe the rocks are about 4 billion years old.

Half moon

Full moon

Phases of the Moon

You probably have seen different shapes of the moon. These shapes are the **phases** of the moon. You can see all the moon's phases as the moon makes one revolution around the earth.

phase (fāz), the shape of the lighted part of the moon as it is seen from the earth.

During the new moon phase, the lighted part of the moon faces away from the earth and you cannot see it. As the moon revolves, you can see more and more of its lighted part. About one week after the new moon, the moon looks like a half circle. It is sometimes called a half moon. Find the half moon in the picture.

As the moon continues to revolve, you can see more and more of its lighted part. About a week after the half moon, you can see a complete circle. This phase is called a full moon, shown in the picture. After the full moon, you can see less and less of the moon each night. About two weeks after the full moon, the new moon phase comes again.

Teaching Tips

- Make sure students understand that the dark part of the moon is the part on which the sun is not shining; the shadow is *not* caused by earth's shadow.
- To help them understand how the different phases of the moon occur, remind students of the *DISCOVER* activity on page 247, in which a ball moved close to and away from a light. You might repeat the activity, and explain that the ball represents the moon. This time, stand in front of a sunny window, or have a student point a flashlight at the ball. Move the ball around your head, and have the students ***describe*** which phase of the moon is being represented at different points. Questions: **Does the moon ever change shape?** (no) **Does the sun ever move?** (no) **Does earth cast a shadow on the moon?** (no) If students seem to have difficulty answering the last question, remind them that you represent the earth. **Where is my shadow?** (Students should state that your shadow falls behind you when light is in front.)
- Explain that if the moon passes in the earth's shadow, the moon will be dark. Point out that the moon usually passes above or below the earth's shadow, but that occasionally the moon does pass through the earth's shadow. Explain that this occurrence results in an eclipse of the moon, or a lunar eclipse.

Science and Language Arts

Have students look through library books, music books, and magazines to find lyrics, literature, and articles written about the moon. Some of the information students might look for could include romantic or mystical powers that were thought to be associated with the moon, music or literature that mentions a particular phase of the moon, and literature that gives the moon a personality, such as that of a man.

Enrichment

Have students use reference materials to find out about other parts of the solar system, such as the planets and their satellites, stars, comets, asteroids, and meteors.

Reinforcement

Have students keep a log of the phases of the moon, until they have recorded the whole cycle. Students might want to obtain binoculars or a telescope to help them ***observe*** the moon more closely, even though they can study the phases without optical aids. Students might wish to ***examine*** the terminator line (the line separating the dark part of the moon from the light part).

TEACHING PLAN

3. Assess

Lesson Review

1. The earth rotates on its axis once each twenty-four hours. It revolves around the sun once each year.
2. The moon rotates on its axis and revolves around the earth.
3. The phases of the moon are the different shapes of the moon that can be seen as the moon revolves. The phases include the new moon, quarter moon, half moon, and full moon.
4. Challenge! A day would be ten hours. **Thinking Skill:** *Inferring*

Find Out On Your Own

Students should indicate that, during a solar eclipse, the moon is between the sun and the earth so that the moon blocks out the sun's light. **Thinking Skills:** *Recognizing the main idea and supporting details*; *Making physical models*

Sometimes you might see the moon during the daytime. You can see a full moon only at night. All the other phases of the moon can be seen during the day as well as at night.

Lesson Review

1. How does the earth move every twenty-four hours? every year?
2. How does the moon move?
3. What are the phases of the moon?
4. **Challenge!** Suppose the earth made a rotation around its axis every ten hours. How many hours would pass from the start of one day to the start of another day?

Study on your own, pages 338–339.

EARTH SCIENCE

FIND OUT ON YOUR OWN

This picture was taken during a solar eclipse. Look in a book about space to find out about a solar eclipse. Draw a picture of the sun, earth, and moon during a solar eclipse. Write a paragraph that explains your picture.

252

Teaching Options

Science Anecdote

Astronauts who went on the Apollo missions that landed on the moon brought back 400 kg (about 882 pounds) of moon rocks to study.

Reteaching Suggestion ♦

Draw several circles on the chalkboard. Assign different students to come to the board and shade in the appropriate part of the circle to show a phase of the moon that you have determined. You also might ask students to fill in the blanks (orally) to sentences that you read. For example, you might say, "The moon revolves around ___." (Students should say, "the earth.")

Workbook page 68 *

Name ______________________
Use with Lesson 1: pages 248-252

Chapter 12

Science Activity

How Does the Moon Move?

Gather These Materials
• paper • marker or crayon

Follow This Procedure
1. Work with a friend. Write *Front of the Moon* on the paper. Hold the paper in front of you. You will pretend to be the moon.
2. Have your friend stand about a meter in front of you. Your friend will pretend to be the earth.
3. Walk around your friend so that the front of the moon always faces earth. Notice how you move.
4. Now walk around your friend so that the front of the moon always faces the same wall.

Record Your Results
1. What did you have to do so that the front of the moon always faced the earth? ______________________
2. How was this different from the way you walked when the front always faced a wall? ______________________

State Your Conclusions
1. What did you do to show that the moon revolves around the earth? ______________________
2. What did you do to make the front of the moon always face the earth? ______________________
3. How does the moon move? ______________________

68

♦ ***Suitable as a language development activity***

Comparing the Sun And Moon

Suggested grouping 3–4 students

Purpose

Infer how the sun and moon can look the same size.

Gather These Materials

• large ball • small ball • meter stick • chalk

Follow This Procedure

1. Use a chart like the one shown to record your observations.
2. Ask a classmate to hold up a large ball. Pretend that the large ball is the sun.
3. Ask another classmate to hold up a small ball in front of the sun. Pretend that the small ball is the moon.
4. Stand in front of the classmate holding the moon, as shown.

The distance between you and the sun should be about 5 meters.

5. Ask the classmate holding the moon to move back and forth between you and the sun. When the moon and the sun appear to be the same size, tell the classmate to stop moving.
6. Use chalk to mark an *X* where you are standing. Measure the distance between the *X* and the moon. Record the distance.
7. Measure the distance between the *X* and the sun. Record the distance.

Record Your Results

Distance to Moon	Distances
Distance to Sun	will vary.

State Your Conclusion

1. Was the moon closer to you or closer to the sun when it looked the same size as the sun?
2. In what way does distance affect how large an object looks?

Use What You Learned

Why do the sun and moon in the sky appear to be the same size?

253

Activity

Concept

The sun and the moon appear to be the same size because of their different distances from the earth.

Objectives/Process Skills

- *Make a model* of the sun and moon.
- *Measure* the distances to each ball when they appear to be the same size.
- *Record* data.
- *Infer* how the sun and moon can look the same size.

Time Allotment

Allow 10 minutes.

Safety Tips (See page T24.)

- Warn students to handle meter sticks carefully. Explain that people can get hurt if sticks are waving about.
- Tell students not to look directly into the bright light of the projector.

Teaching Tip

- Use the balls and light from an overhead projector to show how the moon blocks out the sun during a solar eclipse.

Answers

State Your Conclusion

1. The moon was closer to the student.
2. The shorter the distance, the larger the object looks.

Use What You Learned

The sun seems to be the same size as the moon because the sun is much farther away from the earth. **Thinking Skill:** ***Recognizing and using space or time relationships***

Resource Book page 139

Name
Use with Lesson 1: page 253
Comparing the Sun and Moon

Chapter 12
Activity Worksheet

Record Your Results

Distance to Moon	
Distance to Sun	

State Your Conclusion

1. Was the moon closer to you or closer to the sun when it looked the same size as the sun?
2. In what way does distance affect how large an object looks?

Use What You Learned

Why do the sun and moon in the sky appear to be the same size?

139

Activity Results

Things that are close appear larger; things that are farther away appear smaller.

*** Answers to masters on pages 246E–246H**

TEACHING PLAN

Lesson Objectives

- *Describe* the sun and *explain* how the sun affects the earth.
- *Describe* the planets and *identify* their positions in the solar system.

Lesson Vocabulary

planet, solar system

1. Motivate

Demonstration Activity ♦

Darken the room and turn on two flashlights of the same size. Move one flashlight close to the students so that it appears to be brighter and larger than the other flashlight. Point out that a similar situation occurs when people view stars. Some small stars that are close to the earth appear brighter and bigger than larger stars that are farther away.

Discussion

Mention that the sun is a star. Question: **How and why does the sun look different from other stars?** (It looks bigger and brighter because it is closer to the earth.)

2 What Are the Sun and the Planets Like?

LESSON GOALS

You will learn
- what the sun is like and how the sun produces energy.
- that the sun and the planets and their moons make up most of the solar system.

Scientists have learned many things about the sun. They know the sun is much larger than the earth. They also know that the sun is the most important source of energy for the earth.

The Sun

Have you ever watched twinkling stars at night? Each star is made of hot, glowing gases. The sun is a star. The sun looks much larger than other stars because it is closer to the earth. Objects that are closer to you often seem to be larger. You cannot see other stars during the day because the sun is so bright.

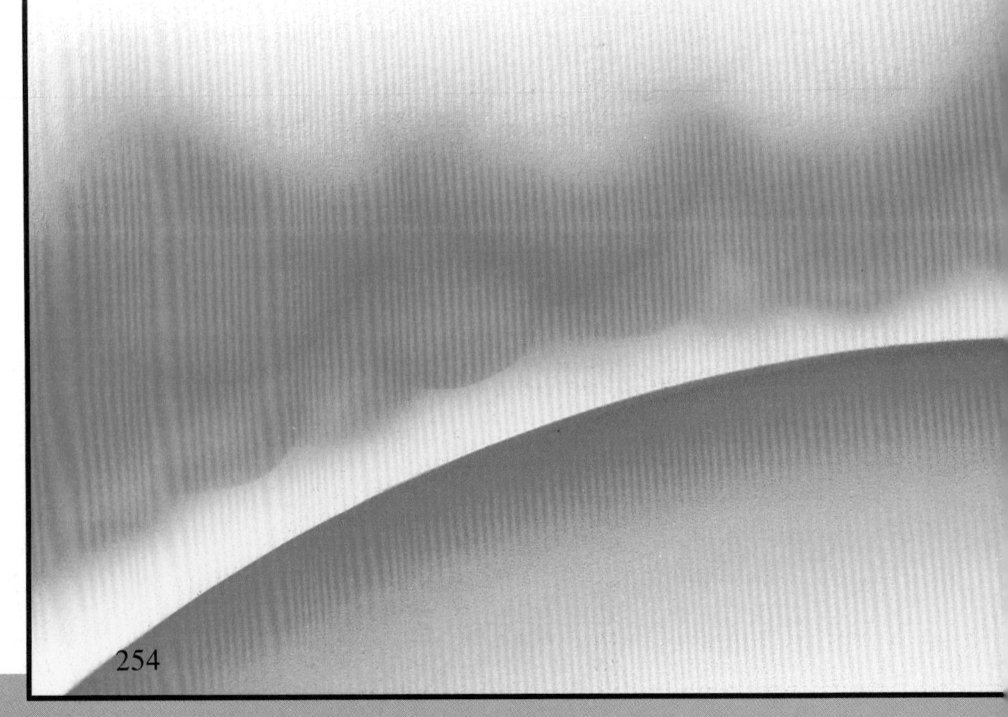

254

Teaching Options

Science Background

The dark areas on the sun's surface are sunspots. Sunspots have strong magnetic fields that affect the sun's gases. The number of sunspots increases and decreases regularly in a pattern called the sunspot cycle. When the number of sunspots is high, tremendous explosions, known as solar flares, erupt from the sun's surface. Solar flares shoot particles and radiation into space, and some of them reach the earth. Scientists know that nuclear fusion within the stars changes small amounts of gas into large amounts of energy. This energy makes the stars hot and makes them shine. Scientists think that the sun's energy will last for another 5 billion years.

Reading Strategies ♦

1. Guide students' pre-reading by asking: Which parts of the lesson are familiar and which parts are new?
2. Assign these strategies: Visualizing Information and Writing a Memory Sentence for each subheading. (See pages T26–T29.)
3. Pair students to share what information is clear and unclear and initiate discussion using students' unanswered questions.

♦ *Suitable as a language development activity*

This picture shows that the sun is shaped like a ball. Notice how the sun glows. The temperature at the sun's surface is about twelve times hotter than the hottest temperature you need to cook food. The temperature in the center of the sun is even hotter. Changes in matter take place in the center of the sun. Energy comes from these changes. This energy produces the sun's heat and light.

Even though the sun is the closest star to earth, it is still very far away. The light from the sun you see now really left the sun eight and one-half minutes ago.

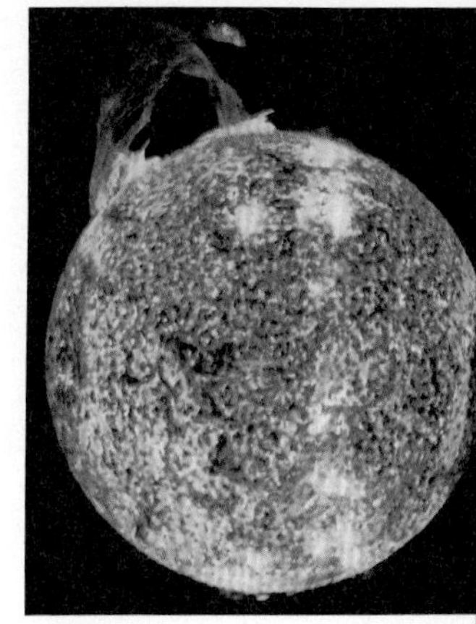

The sun

255

2. Teach

Teaching Tips

- Questions: **If you had not read anything about the sun, how would you know that the sun is very hot?** (Students might ***describe*** how the sun feels on their skin, how heat from the sun causes some things to melt, or how too much sun might damage the skin.)
- **How can you tell that the sun gives off energy?** (Students might ***describe*** how the sun makes water, metal, and other objects hot when they are exposed to the sun's direct rays.)
- **Possible Misconception:** Some students might think that the sun gets hotter in summer and in the middle of the day, and is cooler in the winter and in the early morning and evening. Explain that the sun's temperature does not vary according to times of the day. Temperatures on the earth vary according to season and time of day.

Science and Health

Ask students to find out what ultraviolet rays are and how they can damage the skin and the eyes. Ask the students also to find out if ultraviolet rays have good effects. Students might also want to learn what the ozone layer is and how it helps protect us from harmful ultraviolet radiation.

Special Education

Engage students with learning disabilities in a discussion of what the earth would be like if it did not have the sun to heat it. Help the students realize that earth probably would be too cold for people and most animals and plants to live on. Point out that some scientists think that dinosaurs might have become extinct when the sunlight was partially blocked for a time, causing the earth to become colder. Then, plants that the dinosaurs ate died, and the dinosaurs died because they did not have enough food.

Reinforcement

Ask students to ***list*** words that describe the sun. Encourage the students to think of as many words as possible. Then have the students draw some of the ways the sun helps people.

TEACHING PLAN

Teaching Tips

- Tell students that the nine planets will change order in 1999 when Pluto swings out after being inside Neptune's orbital path.
- Explain that the planets are not really arranged in a row, as the diagram indicates. Tell students, too, that distances on the diagram are only approximate.
- Encourage students to gather interesting information about the planets. Students might also organize the planets by their sizes or other common characteristics. Ask students to *list* information about each planet on a grid or chart. Students might list information such as: *cold, 16 moons,* or *rings* under the names of the appropriate planets.
- Mention that Jupiter, Uranus, and Neptune also have small rings. Some students might want to do independent study to learn more about the particular planets they are interested in.

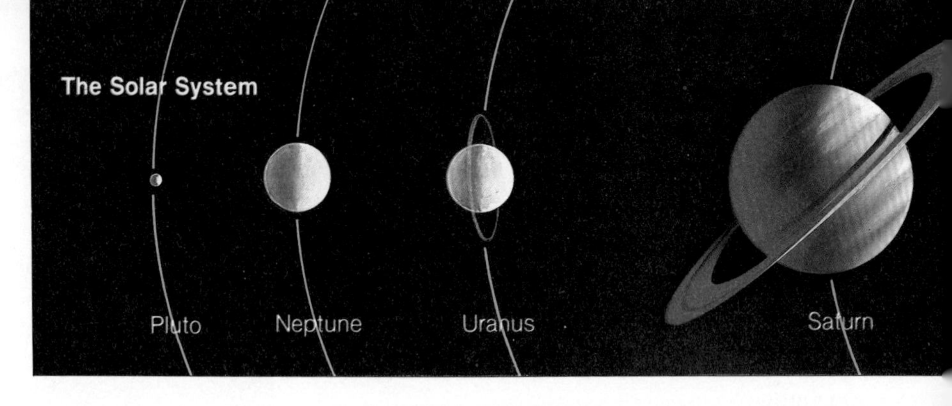

planet (plan′it), a large body of matter revolving around the sun.

solar system (sō′lər sis′təm), the sun, the planets and their moons, and other objects that revolve around the sun.

The Planets and Their Moons

You know that the earth revolves around the sun. Earth is a **planet**—a large body of matter that revolves around the sun. The other planets also revolve around the sun. The nine planets are Mercury, Venus, Earth, Mars, Jupiter, Saturn, Uranus, Neptune, and Pluto. The sun and the planets and their moons make up most of the **solar system.**

Look at the picture of the solar system. Notice that Mercury, Venus, Earth, and Mars are the closest planets to the sun. All of these planets are rocky. Mercury and Venus are very hot. The temperatures on Mars are a little cooler than those on Earth.

Different planets have different numbers of moons. Scientists know about two moons that revolve around Mars. They know about sixteen moons that revolve around Jupiter. How many moons revolve around Earth?[1]

Compare the sizes of the planets. Jupiter is the largest planet. More than one thousand planets the size of Earth could fit into Jupiter. What is the smallest planet?[2]

[1]one
[2]Pluto

256

Teaching Options

Science Anecdote

Saturn's rings are extremely thin. Imagine a phonograph record 4 km (2.5 mi) across but the same thickness as a record. These dimensions are similar to those of Saturn's rings.

Reinforcement

Assign nine students to be the nine planets, and one student to be the sun. Have each student stand in his or her planet's correct position from the sun. Starting with the sun, have each of the students *describe* the celestial body they represent. Give all students a chance to participate.

Enrichment

Obtain a copy of the song *The Family of the Sun* from the Smithsonian Institution's National Air and Space Museum. Teach the song to the students. The song is sung to the tune of *The Farmer in the Dell* and it reviews all nine planets in order, as well as some of their major features.

♦ *Suitable as a language development activity*

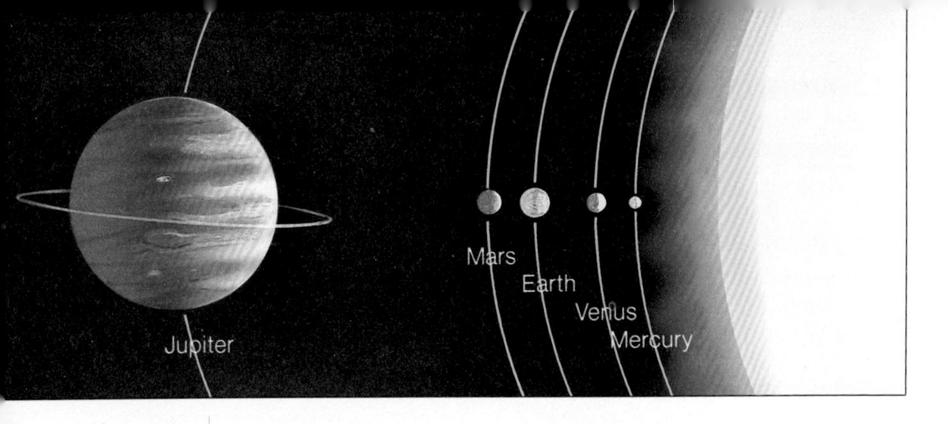

Jupiter, Saturn, Uranus, and Neptune are made mostly of gases. These planets are cold because they are far away from the sun. Notice the rings around Saturn. Scientists have learned that these rings are made of millions of pieces of ice. Scientists are still learning about the planets and their moons.

Lesson Review

1. What is the sun made of?
2. What makes up the solar system?
3. **Challenge!** What is one reason why Mercury is hotter than Earth?

Study on your own, pages 338–339.

SCIENCE IN YOUR LIFE

The time a planet takes to revolve around the sun is called a year. Mercury makes one revolution around the sun every 88 days. Therefore, a year on Mercury is only 88 days long. Jupiter takes twelve of Earth's years to revolve around the sun. A year on Jupiter is twelve Earth years long.

PHYSICAL SCIENCE — *FIND OUT ON YOUR OWN* — *CONNECTION*

Look in recent books about space to find out more about Mars. Draw a picture of Mars. Next to your picture, tell what minerals give Mars its color.

3. Assess

Lesson Review

1. The sun is a star made of hot, glowing gases.
2. The sun, the planets, and their moons make up most of the solar system. (Students might also mention other parts of the solar system that they have learned about in their research.)
3. Challenge! Because it is closer to the sun. **Thinking Skill:** *Inferring*

Find Out On Your Own

Students' pictures should indicate Mars' reddish appearance. Mars was named the *red planet* because of this appearance. Rustlike dust on the surface of Mars causes this color. Winds on Mars blow the dust about. Information from space probes indicated that the color is due to iron compounds. **Thinking Skills:** *Restating or explaining ideas, Communicating*

Workbook page 69 *

Name ____________
Use with Lesson 2: pages 254-257

Chapter 12
Science Skills

The Solar System

A *fact* is something that is true. A fact can be tested. An *opinion* explains how someone feels. Often, an opinion cannot be tested.

Read the statements below. Decide whether each is a fact or an opinion. Write *F* in front of each fact. Write *O* in front of each opinion.

The Moon

_F_ 1. The moon is a satellite of the earth.
___ 2. The full moon is a very beautiful sight.
___ 3. The sun is a star.
___ 4. We think there is no life in the solar system except on earth.
___ 5. Many other planets have moons, just as the earth does.
___ 6. Jupiter is the largest planet in the solar system.
___ 7. The planet Venus is so bright that you can see it during the day.
___ 8. People will never live on the planet Mars.
___ 9. Jupiter has sixteen moons, and Mars has two moons.
___ 10. Moonlight is really reflected sunlight.
___ 11. I think being an astronaut must be exciting.

69

*** *Answers to masters on pages 246E–246H***

Game Suggestion ♦

Divide the students into nine teams. Secretly assign each team a planet. Allow each team ten minutes to write down all of the information they can find about their planet. Then have the teams take turns going to the front of the room. The team members must answer at least three questions about their planet and give one hint as to which planet they represent. The object of the game is for the class to ask as many questions as possible before the planet's identity is discovered.

Reteaching Suggestion ♦

Have students make a model of the solar system. Give each student a supply of salt dough and nine thick threads. Have students roll a dough ball for each planet, and roll the end of a piece of thread into each ball. Allow the dough to dry, and let students paint it. Then have the students use needles to poke holes through the top of a shoe box, and hang each planet in position by running its thread through a hole and tying a knot. Cut a small hole in one end of the box to view the solar system, and make a hole at the other end to let light in. You also might have the class make a scale model in the hallway.

Activity

Concept
Mercury, Venus, Earth, and Mars are similar in size.

Objectives/Process Skills
- *Make models* of Mercury, Venus, Earth, and Mars.
- *Compare* the appearances of the four planets.
- *Record* data.

Time Allotment
Allow 30 minutes.

Safety Tips (See page T24.)
- Tell students to handle scissors and compasses carefully. Use round-tipped scissors.
- Use only nontoxic markers, crayons, and glue.
- Warn students to handle the hangers carefully. They should not wave the hangers about, and should be careful when attaching their drawings to the hangers. Explain that eye injuries or puncture wounds could result if students are not careful. You may want to tape the sharp ends of the coat hanger.

Teaching Tips
- Students might make more detailed models of planets by gluing cotton batting to their circles.
- Students can use papier-mâché to build their models.
- Have students compose a class story about an imaginary trip to Mercury, Venus, and Mars.

Answers
State Your Conclusion

1. The planets, in order of size from the smallest to largest, are Mercury, Mars, Venus, and Earth.

2. The closest planet to the sun is Mercury, followed by Venus, Earth, and Mars.

Use What You Learned

Mars takes the longest time to orbit the sun because, of these four planets, it is the farthest away from the sun. **Thinking Skill:** *Inferring*

Making Models of Planets

Suggested grouping: 2 to 4 students

Purpose

Compare the appearances of the four planets closest to the sun.

Gather These Materials

• cardboard • compass • ruler • string • scissors • crayons or markers • coat hanger

Follow This Procedure

1. Use a chart like the one shown to record your observations.
2. Cut out 2 circles for each planet you make. The table shows how wide to make the circles for each planet.
3. Use the ruler to draw a line from the center of each circle to the edge. Cut each line.
4. Use crayons or markers to color your circles.

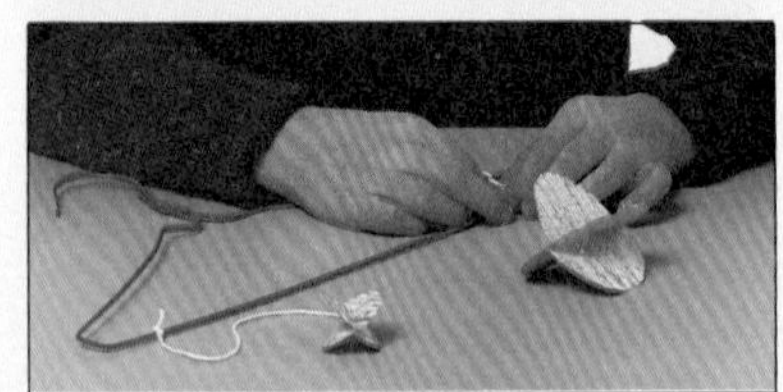

5. Using the two circles for each planet, push the circle together at the slits as shown.
6. Tape string to each planet model. Attach the models to the coat hanger, as shown. Attach the models in the correct order as shown in the table.
7. In the chart, draw a picture of each planet model you made. Write the name of the planet under the picture.

Planet	Width of Circle
Mercury	3 cm
Venus	7 cm
Earth	8 cm
Mars	4 cm

Record Your Results

Picture of Planet	Mercury,	Venus,	Earth,	Mars.
Name of Planet				

State Your Conclusion

1. How do Mercury, Venus, Earth, and Mars compare in size?
2. What is the order of the first four planets from the sun?

Use What You Learned

Which of these planets takes the longest to orbit the sun? Explain.

258

Activity Results

See chart overprint.

Resource Book page 141

Name

Use with Lesson 2: page 258

Chapter 12

Activity Worksheet

Making Models of Planets

Record Your Results

Picture of Planet				
Name of Planet				

State Your Conclusion

1. How do Mercury, Venus, Earth, and Mars compare in size?

2. What is the order of the first four planets from the sun?

Use What You Learned

Which of these planets takes the longest to orbit the sun? Explain.

141

Doing Experiments in Space

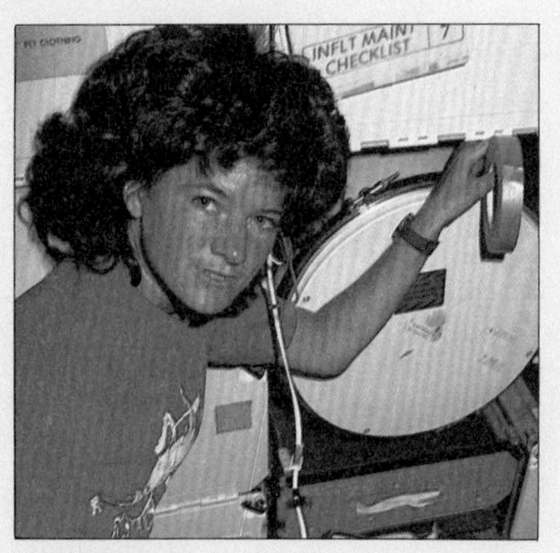

Dr. Sally Ride

Sally Ride saw an advertisement in the newspaper. It said that the space program needed new astronauts, and they wanted scientists. Dr. Ride was a scientist who was interested in stars and planets. She decided to try to get the job. More than 8,000 other people also wanted to become astronauts, but only thirty-five were chosen. When Dr. Ride found out she was one of them, she was so excited she wanted to go up in space the next day! However, she and the other new astronauts needed to spend several years training before they were ready to fly. The picture shows Dr. Ride training for her flight.

Finally, Dr. Ride got her chance. She became the first American woman to fly in space in 1983. On her flight, the spacecraft made one orbit around the earth every 90 minutes. In her six days in space, she went around the earth ninety-eight times!

One of Dr. Ride's main jobs on the flight was to use a robot arm to reach out from the spacecraft. Dr. Ride and another astronaut used the arm to practice releasing and picking up satellites in space. One of the satellites was a space lab. This type of satellite could be left in space to do experiments and take pictures. Later, another spacecraft could pick it up and bring it back to Earth.

Dr. Ride also did experiments to see how plants grow when they do not feel the pull of gravity. What she learned could help scientists plan for future farms in space.

The work that Sally Ride did helps everyone understand more about space. Dr. Ride and the other astronauts know that a great deal is still left to discover.

What Do You Think?

1. Why do you think some experiments are done on a space lab satellite and not by astronauts in flight?
2. How would farms in space be useful?

Science and Technology

TEACHING PLAN

Discussion

Question: **What do you think astronauts need to learn before they are ready to fly in space?** (how to move, how to eat, how to work in a weightless environment)

Teaching Tips

- Question: **Have you ever had to work hard for a long time before you were able to do something that you wanted to do?** (Students will probably answer yes.) Point out that Sally Ride had to train for years before going on her first flight.
- Point out that satellites can be used to receive voice and picture signals from the earth and send them to another place on the earth. Some telephone calls and television programs are transmitted by means of satellites.

Answers

What Do You Think?

1. Some experiments might take a long time to do, and astronauts might not want to stay in space for a long time. **Thinking Skill:** *Inferring*

2. If there is not enough land for farming on Earth, it would be useful to have farms in space, where there is plenty of sunlight. **Thinking Skill:** *Predicting*

Teaching Options

Books To Read

Blacknall, Carolyn. *Sally Ride: America's First Woman in Space*. Dillon Press, 1984. (for grade 3 and up)

Briggs, Carole S. *Women in Space: Reaching the Last Frontier*. Lerner, 1987. (for grade 5 and up)

Science Background

Sally Ride's first space flight took place in June, 1983. Part of her training involved adjusting to a weightless environment. To do this, she and other astronauts practiced in a specially prepared jet plane. This jet was empty in the passenger area, and it had padded walls, floor, and ceiling. It would repeatedly fly to a high altitude, and then dive straight down before climbing up again. During each dive, the astronauts experienced weightlessness for less than a minute. They floated around the padded area in a manner similar to that experienced in space. While they floated, the astronauts practiced activities that they would need to carry out in space, such as eating, drinking, putting on space suits, and so on.

LESSON 3

TEACHING PLAN

Lesson Objectives
- *Explain* that astronauts explored the surface of the moon.
- *Describe* how spacecraft have provided information about Mars, Venus, and other planets.

Lesson Vocabulary
astronaut, crater

1. Motivate

Demonstration Activity ♦
While you hold the back of a chair, have a volunteer sit on the chair and put his or her feet on a desk in front of the chair. Tell the students that the volunteer is in a position similar to that taken by the Apollo astronauts when their spacecraft was launched.

Discussion
Question: **What problems might you have if you stayed in a small space for a week?** (Students should *describe* difficulties they think they might encounter.) Compare the size of the Apollo capsule to the size of a space shuttle's cargo bay (which might be large enough to hold your classroom).

3 What Do Scientists Learn From Space Travel?

LESSON GOALS

You will learn
- how space travel helped scientists learn about the moon.
- how spacecraft helped scientists learn about Mars, Venus, and other planets.

People have always wondered about objects in the sky. For many years, people thought everything in the sky revolved around the earth. Almost four hundred years ago, a man named Galileo first used a telescope to study distant objects in the sky. Galileo found mountains and valleys on the moon. He found that Venus has phases like the moon. Galileo also saw four moons revolving around Jupiter. The telescope helped scientists prove that the earth was not the center of the solar system.

Today, people use large telescopes like this one to study objects in space. Scientists also use spacecraft that carry special cameras and other tools that help us find out about the solar system.

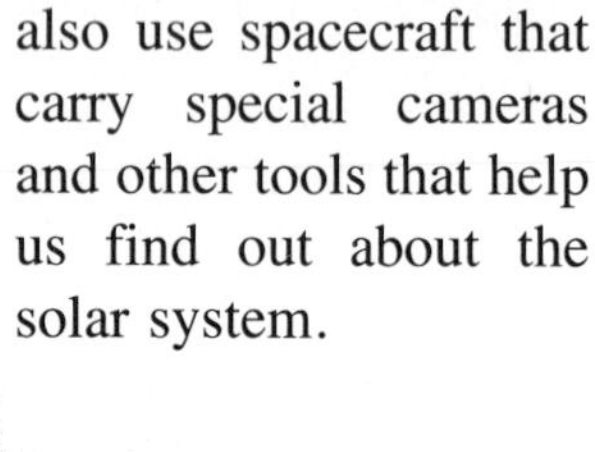

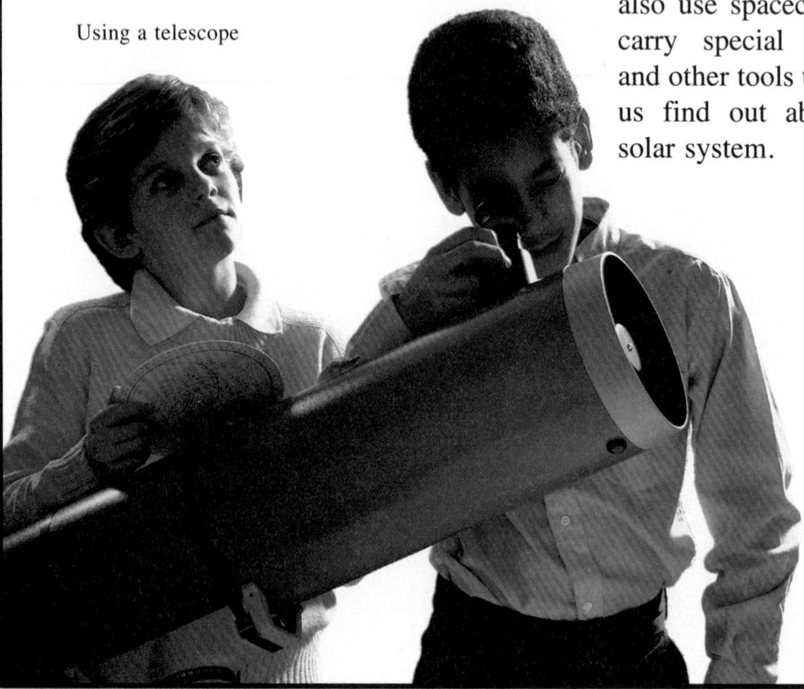

Using a telescope

Teaching Options

Science Background

Prior to the space program, scientists used observational data to develop theories about the planets and their moons. For example, before unmanned space vehicles had landed on the moon, some scientists believed that the moon was covered with thick layers of dust. There actually is a layer of dust on the moon, but it is not nearly as thick as scientists first thought. Flights of space probes continue to provide information about the planets and their moons, which could not be obtained through ground-based observations. In 1989, Project Galileo began its journey toward Jupiter, where it will send a small probe through the atmosphere and send information back to Earth.

Reading Strategies ♦

1. Guide students' pre-reading by asking: What ideas do you think you would find in the lesson to answer the question-title?
2. Assign these strategies: Visualizing Information and Writing a Memory Sentence for each subheading. (See pages T26–T29.)
3. Pair students to share what information is clear and unclear and initiate discussion using students' unanswered questions.

♦ *Suitable as a language development activity*

Astronaut James Irwin

Rocket on a launch pad

Learning About the Moon

Rockets, like the one in the picture, have carried people to the moon. **Astronauts** are people who travel in space. Astronauts Neil Armstrong and Buzz Aldrin were the first people to land on the moon and return to the earth. They traveled in the spacecraft named Apollo 11.

The picture shows another astronaut on the moon. American astronauts landed on the moon six times. They visited different parts of the moon. They saw large **craters** on the moon. A crater is a hole in the ground that is shaped like a bowl. Most scientists think that rocks from space crashed into the moon and caused these craters. Scientists on earth study rocks that astronauts collected to learn about the moon.

SCIENCE IN YOUR LIFE

People on earth use some foods and some kinds of medical equipment that were first made for astronauts. Other objects used on earth, such as solar batteries, also were first made for space travel.

astronaut (as′trə nôt), a person who travels in space.

crater (krā′tər), a large hole in the ground that is shaped like a bowl.

261

2. Teach

Teaching Tips

- Bring a telescope to class and allow students to examine it. Instruct students to look at objects that are fairly close and objects that are far away. Ask students to *compare* how the objects appear when viewed with and without the telescope.
- **Possible Misconception:** Students might not realize that craters also exist on the earth.
- Explain that the moon has no air or water. Questions: **What will happen to the footprints of the men who walked on the moon?** (Allow students to *make inferences.*) **Is there any water on the moon to wash them away, or any wind to blow them away?** (Lead students to *draw the conclusion* that the footprints could be there for millions of years.)
- Question: **If you dropped a hammer and a feather on earth, which would hit the ground first, and why?** (the hammer; air pushing on the feather would make it float instead of fall straight down) Tell students that an astronaut dropped the same two items on the moon in 1971, and both items hit the ground at the same time.

Workbook page 70 *

Name ____________

Use with Lesson 3: pages 260-263

Chapter 12

Science and Social Studies

Living in Space

Someday people will be able to live in space. Instead of living in houses, the people will live in space stations. The picture shows how a space station might look.

Everything people need for life must be present in the station. The spacecraft must hold plants, animals, and water. People can use the plants and animals for food. Living things also need oxygen. Plants can provide fresh supplies of oxygen. In the space station, water and oxygen are used, then cleaned, and reused.

People need a place to live too. Suppose the wheel part of the station contains apartments. From their windows, people could see the earth, as well as the moon, sun, and stars. Imagine the view!

The station can get its energy directly from sunlight. This energy, called solar energy, can run machines and provide heat and light for the spacecraft.

Today, scientists are making plans to build a space station. However, it takes years of work and planning to do this job.

Answer these questions.

1. What things must people in a space station have?

____

____

2. How is energy useful?

____

3. Why is solar energy the best energy for a spacecraft?

____

____

70

* **Answers to masters on pages 246E–246H**

Science and Language Arts

Have the students write letters about what they would like to see the space program do or make better. Mention that the letters do not have to be shared with the class, but allow interested students to send the letters to a NASA educational office or a NASA representative.

Special Education

Bring in a selection of library books that contain photographs of moon missions. Have students with learning disabilities find pictures of the spacecraft that were used on missions to the moon. Ask the students to describe what space travel conditions were like for the astronauts.

TEACHING PLAN

Teaching Tips

- Draw the orbital path of Mars on the chalkboard, with the orbital path of Earth inside it. Show a path from Earth to Mars that represents the path taken by the Viking spacecraft. Mention that the Viking spacecraft was only a little over 16 seconds late getting to Mars.
- Point out that Mars is the proposed target for the first manned exploration of another planet. Question: **Why might an astronaut explore Mars rather than Venus?** (Venus is very hot, but the temperature on Mars is close to the temperature on Earth.) Mention that Venus is hot enough to melt lead.
- Question: **Do you think it is important for scientists to continue to gather information about other planets? Why or why not?** (Guide students to *judge and evaluate* the benefits and drawbacks of the space program. Students will not necessarily agree with one another or with you, but they should begin to consider the reasons for starting and continuing the space program.)

Scientists learn about planets.

Learning About the Planets

Scientists use spacecraft to study planets. The Viking spacecraft landed on the planet Mars. No astronauts traveled on this spacecraft. Cameras on the Viking took many pictures of Mars. Scientists learned that Mars has many hills and valleys. The pictures also showed large volcanoes.

Scientists sent spacecraft into orbit around Venus. Spacecraft also have landed on Venus. Clouds always cover Venus. Spacecraft also have taken pictures of rocks under the clouds.

The scientists in the picture are studying photographs of Venus to learn more about this planet.

Teaching Options

Science Anecdote

The word *astronaut* means *sailor among the stars*. The first time astronauts were sent to the moon, the trip from Earth to the moon took about four days. The return trip took less than three days.

Reinforcement

Write the following words on the board and ask students to *list* their meanings or *describe* (in writing) their importance: *Galileo, telescope, astronaut, crater, Viking.* When all of the students have completed the assignment, ask for volunteers to share their answers with the class.

Enrichment

Students might enjoy hearing excerpts from the radio program *The War of the Worlds*. Ask students to use what they now know about Mars to write a speech about why people on Earth need not fear being invaded by creatures from Mars. Explain that when the radio show was aired in 1938, many people did not know much about the planet Mars, and believed that the fictitious invasion was real.

♦ ***Suitable as a language development activity***

Spacecraft also have taken pictures of Jupiter, Saturn, Uranus, and Neptune. Scientists will keep learning about the planets as they send more spacecraft into space.

Lesson Review

1. How has space travel helped scientists learn about the moon?
2. How have spacecraft helped scientists learn about the planets?
3. **Challenge!** Suppose you stood on the surface of Venus. Would you be able to see the earth and the stars? Explain your answer.

Study on your own, pages 338–339.

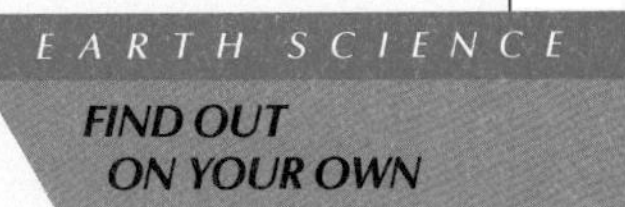

The pictures show satellites made by people. Scientists have sent such satellites into orbit around the earth. Look in library books to find out how some of these satellites help scientists learn about weather on earth. Write a few sentences explaining what you learned.

3. Assess

Lesson Review

1. Astronauts visited different parts of the moon, did experiments, and brought back pieces of moon rock for study.
2. Spacecraft have taken pictures of the planets, collected soil from Mars, and performed tests.
3. Challenge! No, you could not see the sky because Venus is covered by clouds. **Thinking Skill:** *Inferring*

Find Out On Your Own

Weather satellites can measure temperatures, measure the amount of moisture in the air, and send pictures of clouds and storms back to the earth. **Thinking Skill:** *Restating or explaining ideas*

Workbook page 71 *

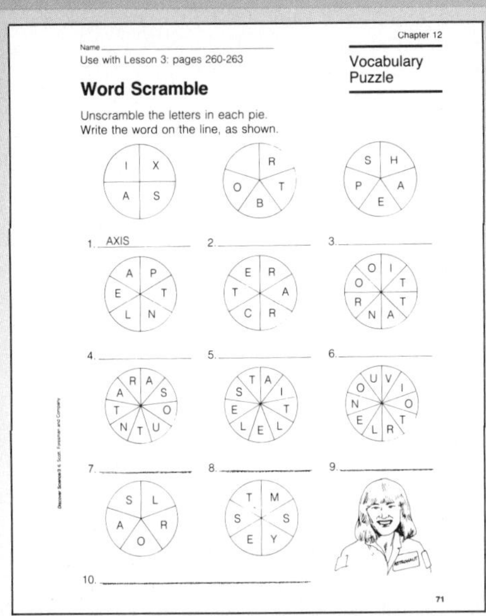

Name ______ Chapter 12
Use with Lesson 3: pages 260-263 Vocabulary Puzzle

Word Scramble

Unscramble the letters in each pie.
Write the word on the line, as shown.

1. AXIS
2. ______
3. ______
4. ______
5. ______
6. ______
7. ______
8. ______
9. ______
10. ______

71

*** Answers to masters on pages 246E–246H**

Game Suggestion ♦

Make a game board with Earth in the center and the other planets surrounding it. Mark paths from Earth to the planets, and from the planets to one another. Have the students make hazard cards, such as *Low fuel—skip a turn, Spacecraft out of order,* and so on. Students should also make cards such as *Perfect launch—advance twice!* Students should roll a pair of dice, and attempt to travel to all of the planets and back to Earth without encountering any hazards.

Reteaching Suggestion ♦

Provide reference materials about the unmanned spacecraft and satellites that have been sent into orbit to gather information. Have students make a map of the solar system and mark the paths of some of the different spacecraft and satellites that have sent back to Earth information about other planets. (Make sure students include the *Viking* landers on their map.)

TEACHING PLAN

Purpose
To develop the skills of collecting and organizing information using calendars and charts to solve problems.

1. Motivate

Discussion
Display various types of calendars. If possible, include a calendar that indicates the dates of the phases of the moon. Question: **How do you and your families use a calendar?** (Answers will vary, but might include for recording appointments, keeping track of time, and so on.)

2. Teach

Teaching Tips
- Review with students the appearance of each of the phases of the moon.
- Point out some of the idiosyncrasies of the calendar, such as the variation of the number of days in each month.

Skills for Solving Problems

Using Calendars and Charts

Problem: How many days pass between full moons?

Part A. Using Calendars to Collect Information

1. A calendar divides time into years, months, and days. Calendars tell about time. How many months does this calendar show?
2. A full moon is when you see all of the lighted side of the moon. Between each full moon, the moon revolves once around the earth. The dates of the full moons are circled below. When was the full moon in January? in February? How many days passed between the two dates?
3. How many days passed between full moons in February and in March? in March and in April? in April and in May? in May and in June?

January

S	M	T	W	T	F	S
					1	2
3	4	5	6	(7)	8	9
10	11	12	13	14	15	16
17	18	19	20	21	22	23
24	25	26	27	28	29	30
31						

February

S	M	T	W	T	F	S
	1	2	3	4	(5)	6
7	8	9	10	11	12	13
14	15	16	17	18	19	20
21	22	23	24	25	26	27
28	29					

March

S	M	T	W	T	F	S
		1	2	3	4	5
(6)	7	8	9	10	11	12
13	14	15	16	17	18	19
20	21	22	23	24	25	26
27	28	29	30	31		

April

S	M	T	W	T	F	S
					1	2
3	4	(5)	6	7	8	9
10	11	12	13	14	15	16
17	18	19	20	21	22	23
24	25	26	27	28	29	30

May

S	M	T	W	T	F	S
1	2	3	(4)	5	6	7
8	9	10	11	12	13	14
15	16	17	18	19	20	21
22	23	24	25	26	27	28
29	30	31				

June

S	M	T	W	T	F	S
			1	2	(3)	4
5	6	7	8	9	10	11
12	13	14	15	16	17	18
19	20	21	22	23	24	25
26	27	28	29	30		

Part B. Using Charts to Organize and Interpret Information

4. The chart contains the information you collected about the number of days between the full moons. What do the numbers in the column on

Teaching Options

Sample Chart for Part C

Times	Number of days
January to February	29 days
February to March	30 days
March to April	29 days
April to May	29 days
May to June	30 days

♦ ***Suitable as a language development activity***

the right stand for? What periods of time are shown in the left column?

5. What might be the reason that the number of days between full moons is not always exactly the same?

Times	Number of days
January to February	29 days
February to March	30 days
March to April	30 days
April to May	29 days
May to June	30 days

Part C. Using Calendars and Charts to Solve a Problem

Problem: How many days pass between new moons?

January

S	M	T	W	T	F	S
					1	2
3	4	5	6	7	8	9
10	11	12	13	14	15	16
17	18	19	20	21	22	23
24	25	26	27	28	29	30
31						

February

S	M	T	W	T	F	S
	1	2	3	4	5	6
7	8	9	10	11	12	13
14	15	16	17	18	19	20
21	22	23	24	25	26	27
28	29					

March

S	M	T	W	T	F	S
		1	2	3	4	5
6	7	8	9	10	11	12
13	14	15	16	17	18	19
20	21	22	23	24	25	26
27	28	29	30	31		

April

S	M	T	W	T	F	S
					1	2
3	4	5	6	7	8	9
10	11	12	13	14	15	16
17	18	19	20	21	22	23
24	25	26	27	28	29	30

May

S	M	T	W	T	F	S
1	2	3	4	5	6	7
8	9	10	11	12	13	14
15	16	17	18	19	20	21
22	23	24	25	26	27	28
29	30	31				

June

S	M	T	W	T	F	S
			1	2	3	4
5	6	7	8	9	10	11
12	13	14	15	16	17	18
19	20	21	22	23	24	25
26	27	28	29	30		

6. Use the calendar to collect the information you need to solve the problem. Make a chart similar to the one in Part B to organize your information.
7. Look at your chart. How many days pass between new moons?
8. Compare your chart with the chart in Part B. How are the numbers of days between full moons and the numbers between new moons alike or different?

3. Assess

Part A

1. six

2. January 7; February 5; twenty-nine days

3. thirty days; thirty days; twenty-nine days; thirty days

Part B

4. the number of days from the full moon of one month to the full moon of the next; months

5. The moon revolves around the earth in a 29 1/2 day period. Since the calendar does not show half days, the number of days between full moons varies to adjust for the half days.

Part C

6. See Sample Chart for Part C in Teaching Options.

7. three periods of twenty-nine days and three of thirty days

8. All are twenty-nine or thirty days.

Reteaching Suggestion ♦

Show students a calendar in which no national holidays have been marked. Ask them to help you mark the following holidays: *New Year's Day; Valentine's Day; Memorial Day; July 4th; Labor Day; Halloween;* and *Thanksgiving.* Have them count the number of days from one holiday to the next, starting with New Year's Day and ending with Thanksgiving. Then have them create a chart showing the time period and the number of days. (The number of days will vary, depending on the year.)

Resource Book page 143 *

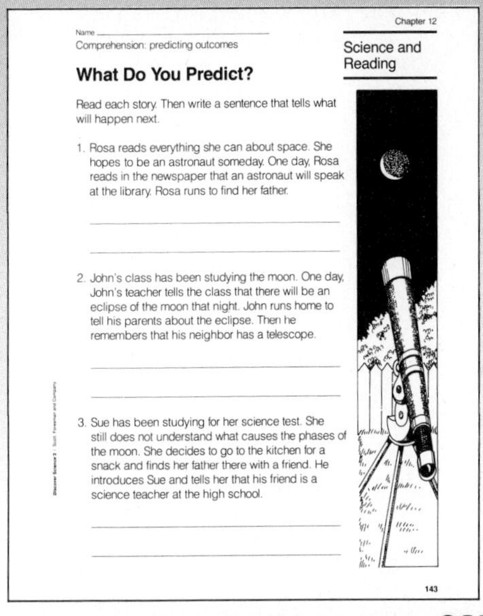
Chapter 12

Name ____________

Comprehension: predicting outcomes

Science and Reading

What Do You Predict?

Read each story. Then write a sentence that tells what will happen next.

1. Rosa reads everything she can about space. She hopes to be an astronaut someday. One day, Rosa reads in the newspaper that an astronaut will speak at the library. Rosa runs to find her father.

2. John's class has been studying the moon. One day, John's teacher tells the class that there will be an eclipse of the moon that night. John runs home to tell his parents about the eclipse. Then he remembers that his neighbor has a telescope.

3. Sue has been studying for her science test. She still does not understand what causes the phases of the moon. She decides to go to the kitchen for a snack and finds her father there with a friend. He introduces Sue and tells her that his friend is a science teacher at the high school.

143

* Answers to masters on pages 246E–246H

REVIEW PLAN

Reviewing Science Words

1. axis
2. crater
3. phase
4. rotation
5. satellite
6. revolution
7. astronaut
8. planet
9. solar system
10. orbit

Reviewing What You Learned

1. d **2.** b **3.** c **4.** a **5.** c **6.** d **7.** b **8.** a

Chapter 12 Review

Chapter Main Ideas

Lesson 1 • The earth rotates on its axis and revolves in an orbit around the sun. • The moon rotates on its axis and revolves around the earth. • All the phases of the moon can be seen as the moon makes one revolution.

Lesson 2 • The sun is a ball of hot, glowing gases and is the most important source of energy for the earth. • The sun and the planets and their moons make up most of the solar system.

Lesson 3 • Space travel has helped scientists learn about the moon. • Spacecraft have helped scientists learn about Mars, Venus, and other planets.

Reviewing Science Words

astronaut	phase	rotation
axis	planet	satellite
crater	revolution	solar system
orbit		

Copy each sentence. Fill in the blank with the correct word from the list.

1. The earth rotates around an imaginary straight line called its ___.
2. A ___ is a hole in the ground that is shaped like a bowl.
3. The full moon is one ___ of the moon.
4. The earth makes one ___ when it makes one complete spin around its axis.
5. The moon is called a ___ of the earth because it revolves around the earth.
6. Each complete orbit the earth makes around the sun is called a ___.
7. An ___ is a person who travels in space.
8. A large body of matter that revolves around the sun is called a ___.
9. The sun and the planets and their moons make up the ___.
10. An ___ is a path an object follows as it moves around another object.

266

Review Options

Cooperative Learning ♦

STAD Format (See page T23.) Assign students to work in four- to five-member teams to study Chapter 12 Review. Students should work together to make sure that they and their teammates know the material in the chapter. After students have had enough time to study together, give them a test to complete individually (Chapter 12 Test A or B in the *Test Book*). Award Superteam certificates to teams whose average test scores exceed 90%, and Greatteam certificates to teams whose average test scores exceed 80%.

Test Book page 105 *

Name ______________

Chapter 12 Test A

Multiple Choice. Choose the best answer.

1. The earth rotates around an imaginary line called the
   a. Equator.
   b. axis.
   c. phase.
2. The movement of an object in a complete orbit around another object is a
   a. revolution.
   b. rotation.
   c. crater.
3. The moon revolves around the earth in about
   a. one year.
   b. one day.
   c. one month.
4. The moon reflects light from the
   a. earth.
   b. sun.
   c. planets.
5. The shape of the lighted part of the moon as it is seen from the earth is the moon's
   a. phase.
   b. path.
   c. stage.
6. The sun looks bigger than other stars because it
   a. is hotter.
   b. really is bigger.
   c. is closer to the earth.
7. Venus, Mercury, and Mars are three of the sun's
   a. moons.
   b. planets.
   c. systems.
8. The large holes that are found on the moon are called
   a. craters.
   b. cones.
   c. scoops.
9. All the planets, their moons, and the sun make up the
   a. sun system.
   b. solar system.
   c. planet system.
10. The moon is the earth's
    a. spacecraft.
    b. satellite.
    c. astronaut.
11. What is the largest planet?
    a. Jupiter
    b. Earth
    c. Mars

105

♦ ***Suitable as a language development activity***

Reviewing What You Learned

Write the letter of the best answer.

1. The earth rotates once every
   (a) 15 hours. (b) 18 hours. (c) 20 hours. (d) 24 hours.
2. The earth revolves around the sun once every
   (a) month. (b) 365 days. (c) day. (d) 29 days.
3. Which phase of the moon looks like a bright circle?
   (a) new moon (b) complete moon
   (c) full moon (d) half moon
4. The spacecraft Apollo 11 landed on
   (a) the moon. (b) Mars. (c) Pluto. (d) Venus.
5. What is the most important source of energy on earth?
   (a) moon (b) solar system (c) sun (d) planets
6. The planet closest to the sun is
   (a) Pluto. (b) Earth. (c) Neptune. (d) Mercury.
7. Which is the largest planet?
   (a) Neptune (b) Jupiter (c) Earth (d) Mars
8. The Viking landed on
   (a) Mars. (b) Uranus. (c) Saturn. (d) Venus.

Interpreting What You Learned

Write a short answer for each question or statement.

1. Name two planets that are colder than Earth.
2. Name three ways Mars and Earth are alike.
3. What makes the moon shine?

Extending Your Thinking

Write a paragraph to answer each question or statement.

1. Neptune takes longer to revolve around the sun than Mercury does. What can explain this difference?
2. Suppose the moon revolved completely around the earth in one week. How long would it take to see all the phases of the moon?

To explore scientific methods, see Experiment Skills on pages 370–371.

267

Interpreting What You Learned

1. Answers could include any two of the following: Jupiter, Saturn, Uranus, Mars, Neptune, Pluto. **Thinking Skill:** *Contrasting*

2. Answers could include any three of the following: both are shaped like a ball; both revolve around the sun; both have the same kind of temperatures; both have hills, valleys, and volcanoes. **Thinking Skill:** *Comparing*

3. The moon reflects light from the sun. **Thinking Skill:** *Restating or explaining ideas*

Extending Your Thinking

1. Neptune has a longer distance to travel in its orbit around the sun because it is farther from the sun. **Thinking Skill:** *Inferring*

2. We would see all the phases in one week because we see all the phases as the moon makes one revolution around the earth. **Thinking Skill:** *Inferring*

Test Book page 106 *

Name

Chapter 12
Test A

Matching Write the letter of each planet in the blank next to its name.

Sun a. b. c. d. e. f. g. h. i.

1. Earth ___ 2. Venus ___
3. Saturn ___ 4. Pluto ___

Short Essay Use complete sentences to answer each question.

1. Name two ways scientists study the planets.

2. Explain where the moon gets its light.

106

Test Book page 107 *

Name

Chapter 12
Test B

Multiple Choice. Choose the best answer.

1. Every 24 hours the earth makes one complete
   a. phase.
   b. revolution.
   c. rotation.
2. The path an object follows as it moves around another object is called
   a. a revolution.
   b. an orbit.
   c. an axis.
3. The sun keeps the earth in orbit by the force of
   a. magnetism.
   b. gravity.
   c. rotation.
4. The length of time it takes the earth to complete one revolution is
   a. one year.
   b. one day.
   c. one month.
5. When all the lighted part of the moon can be seen, the phase is called the
   a. new moon.
   b. half moon.
   c. full moon.
6. The sun is made of
   a. liquids.
   b. gases.
   c. volcanoes.
7. The solar system is made up of objects that revolve around the
   a. earth.
   b. sun.
   c. moon.
8. A planet that is only a little cooler than the earth is
   a. Pluto.
   b. Mars.
   c. Venus.
9. What instrument have scientists used for four hundred years to study the sky?
   a. satellite.
   b. telescope.
   c. camera.
10. Scientists have landed spacecraft with cameras and special machines on the
    a. sun.
    b. stars.
    c. planets.

107

Test Book page 108 *

Name

Chapter 12
Test B

Matching Match each planet with its description.
Look at the drawing of the planets to help with your answers.

Sun a. b. c. d. e. f. g. h. i.

___ 1. the planet between Venus and Mars — A. Mercury
___ 2. the smallest planet — E. Jupiter
___ 3. the planet with sixteen moons — C. Earth
___ 4. the planet closest to the sun — I. Pluto

Short Essay Use complete sentences to answer each question.

1. Why does the moon appear to have different shapes?

2. Name two things scientists know about Saturn.

108

* Answers to masters on pages 246E–246H

TEACHING PLAN

Purpose

This feature gives students background information on several careers that involve earth science. Students might not have heard of some of the careers that are introduced. Geology, oceanography, and hydrology are other areas with occupations related to earth science.

Teaching Tips

- Discuss early surveying efforts, such as the Lewis and Clark expedition, or have students research and report their findings to the class.
- If possible, bring in some examples of aerial photography. Ask students to *explain* why it is difficult to show a large part of the earth on one photograph. (The earth is round.)
- Have students *list* various kinds of severe storms. Ask students who have experienced one of these storms to *describe* it to the class.
- Discuss the importance of careful spacecraft design and assembly. Point out that people in the aerospace industry are working hard to correct design flaws of past spacecraft.

Careers

Can you imagine trying to measure the height of a mountain? How would you measure the distance around a very large lake? If you were a **surveyor,** you would know how to solve these problems. Surveyors measure different distances on the earth. They might figure out the boundary lines between houses. They might measure the length and width of rivers. People who make maps use their measurements. Surveyors spend a great deal of time outdoors. Sometimes they must walk long distances carrying heavy equipment. Surveyors learn their skills by taking special classes after high school. They also learn by working with other surveyors.

Aerial photographers also help map makers. These photographers use special cameras to take pictures from above the earth. They might take pictures while riding in an airplane or a helicopter. A map maker might use these pictures to check the exact location of a lake. Then he or she can place the lake correctly on a map. Aerial photographers take courses in photography. They improve their skills by practicing on the job.

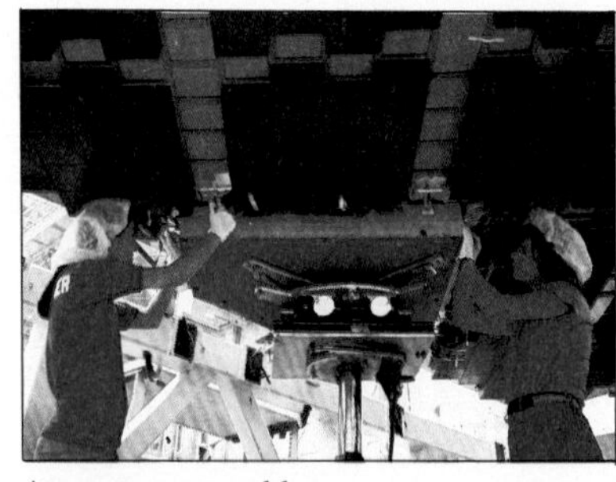

Aerospace assemblers

Many other careers involve earth science. **Volcanologists** study volcanoes. They collect lava samples from volcanoes. One of their most important jobs is to try to predict when volcanoes will erupt. You need to go to college for at least four years to become a volcanologist.

Other careers involve watching for severe storms. Hurricanes are dangerous storms with strong winds, rain, and thunder and lightning. They are like tornadoes at sea. When hurricanes come near land, they can cause great damage to communities. **Hurricane hunters** search for these storms. They try to find hurricanes in time to warn people in the path of a hurricane.

Many other careers involve the world beyond earth. For example, **aerospacecraft assemblers** build spacecraft. The space shuttles are built by teams of assemblers. Aerospacecraft assemblers must be very careful to put parts together in exactly the right way. They learn their skills during two years of college.

Teaching Options

Enrichment

Invite a local surveyor to demonstrate the use of surveying equipment to the class. Arrange a field trip to a nearby museum that has an exhibit of aerial photography, volcanoes, or hurricanes and other storms. Show films or pictures of some historic moments in space exploration.

Where To Write

Surveyor: American Congress on Surveying, 210 Little Falls Street, Falls Church, VA 22046

Aerial Photographer: Professional Photographers of America, Inc., 1090 Executive Way, Des Plaines, IL 60018.

Volcanologist: American Geological Institute, 4220 King Street, Alexandria, VA 22302.

Hurricane Hunter: National Weather Service, Gramax Building, 8060 13th Street, Silver Springs, MD 20910.

Aerospacecraft Assembler: NASA, Education Services Branch, LCG-9, 400 Maryland Avenue SW, Washington DC 20546.

How It Works

Space Suits

This astronaut is ready to travel in space outside her spacecraft. Living in outer space will be very different from living on earth. For one thing, she will find no oxygen in space. She must have oxygen to breathe and stay alive. Also, space will be very cold. She will feel no gravity in space. Gravity is the force that holds you to the earth. Without gravity, a person just floats around.

Also, space contains many kinds of harmful rays like the ones that cause sunburn. In space, these powerful rays can kill a person easily. This diagram shows parts of the astronaut's space suit that protect her.

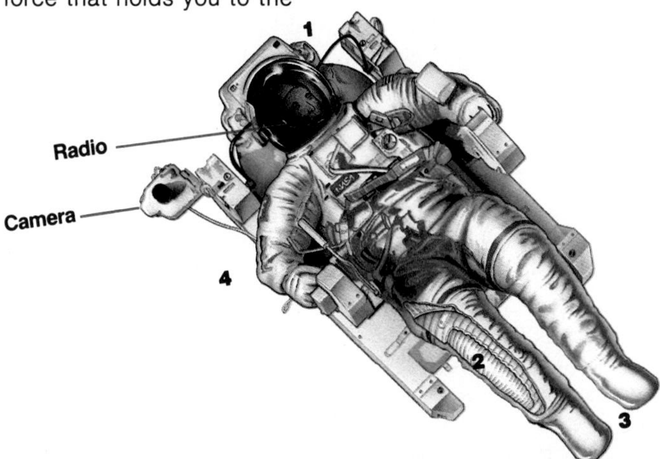

1 First, the suit contains a tank filled with oxygen. The tank is carried in her backpack. It has enough oxygen for her to work in space for many hours.

2 The suit is made of special materials that keep heat inside. These materials are like the ones in a winter coat, but they are much better at holding in heat.

3 The astronaut's shoes are special too. Sometimes they have a sticky material on the bottom. The material helps the astronaut stick to the spacecraft. Sometimes the shoes contain heavy weights. The weights would help the astronaut walk on the moon, where gravity is low.

4 The outside coating on the suit contains a thin layer of metal. The metal reflects harmful rays and keeps them out of the suit.

5 The suit also contains other special items. Inside the helmet is a radio set. The radio lets the astronaut talk and listen to other astronauts and to people on the earth.

How It Works

Teaching Tips

- You might consider asking students to design a space suit ***before*** they read the feature. Have them ***predict*** the kinds of problems astronauts might confront in space travel and then suggest ways to deal with those problems.
- Review with students the settings in which a space suit might be needed, such as in a spacecraft, during EVA, and exploring the moon or another planet. Then discuss how the demands for a space suit might differ in each of these settings.
- Ask students to consider any special properties space suit materials need to have. (For example, the metal coating on the suit must be thin, reflective, and flexible.)
- From books on the Apollo space program, you can find photographs and descriptions of space suits used in those missions. Have students try to *identify* differences between those space suits and the newer shuttle models.
- Have students read about pressure suits used by deep-sea divers. Modern space suits are direct descendents of those devices.

Science Background

The space suit described in this feature was designed for extra-vehicular activity (EVA) in the U.S. shuttle program. Astronauts wear no space suits inside the shuttle, but don the suit for work outside the shuttle. This suit differs from earlier models, such as those worn in the Apollo program. Earlier models were custom designed for individual astronauts and could not be reused by another astronaut. Shuttle suits are modular products. Each suit consists of an upper torso, lower torso, arm assembly, glove, and helmet component. All components are made in many sizes, allowing any one astronaut to be fitted with components "off the shelf."

Complete the Sentence

1. decay	**7.** tornado
2. craters	**8.** sedimentary rock
3. precipitation	**9.** core
4. axis	**10.** phase
5. ores	**11.** weather
6. magma	**12.** dew

Short Answer

1. It will contain materials of different colors. *(Restating ideas)*

2. Minerals and soil are two kinds of natural resources. *(Recalling facts)*

3. Both are processes in which parts of the earth's surface are worn away. *(Identifying similarities)*

4. Water runs faster down a hill, thereby washing away greater amounts of soil. *(Cause and effect)*

5. Magma, melted rock under the earth's surface, becomes lava when it erupts to the surface. Since both are the same material, they must contain the same minerals. *(Restating ideas)*

6. Both are condensed water vapor, but fog is close to the ground, while a cloud is higher in the air. *(Recognizing differences)*

7. Hurricanes get their energy from warm ocean waters. They die out over land. Therefore, places far from the oceans cannot be hit by the full strength of a hurricane. *(Drawing conclusions)*

8. once *(Comprehending meaning)*

9. No, erosion cannot occur without wind or water. *(Drawing conclusions)*

Essay

1. Water falls to earth as precipitation and collects in rivers and the ocean. The sun's energy causes the water to evaporate, forming water vapor which condenses into droplets that form clouds. When the drops become heavy, they fall to the earth. *(Summarizing)*

2. They all revolve around the sun. Earth and some other planets are made of rock, but other planets are made of gases. Earth is warmer and closer to the sun than some planets, but colder and farther from the sun than others. Earth has one moon; some planets have no moons, while others have many moons. Some planets are smaller than Earth, while others are larger. *(Identifying similarities and differences)*

Unit 3 Review

Complete the Sentence

Fill in the blank with the correct word or words from the list.

axis	ores
core	precipitation
craters	phase
decay	sedimentary rock
dew	tornado
magma	weather

1. Plants and animals ___ when they die.
2. Objects from space create ___ when they crash into the moon.
3. Snow and rain are two kinds of ___.
4. The earth rotates around its ___.
5. Useful minerals are found in some rocks called ___.
6. A volcano erupts when ___ comes up through the crust.
7. A ___ has the most powerful winds on earth.
8. The layers in ___ are formed when material is pressed together.
9. The ___ is the center part of the earth and is its hottest layer.
10. A complete circle can be seen during the ___ called the full moon.
11. Water and gases in the air ___ rocks slowly.
12. During the night, ___ often forms on cold surfaces.

Short Answer

Write a short answer for each question or statement.

1. How can you tell that a rock has many different minerals by looking at it?
2. What are two kinds of natural resources?
3. How are weathering and erosion alike?
4. Why does erosion happen faster on hilly land than on flat land?
5. How do you know magma and lava are made of the same minerals?
6. What is the main difference between fog and a cloud?
7. Why are places far away from oceans safe from hurricanes?
8. How many times does the earth orbit the sun in one year?
9. The moon has no water and almost no wind. Would you see erosion on the moon? Explain your answer.

Essay

Write a paragraph for each question or statement.

1. Describe the steps in the water cycle.
2. Tell how the earth and other planets are alike and how they are different.

270

Test Book page 113 *

Name ______________________

Unit Test 3

Multiple Choice Choose the best answer.

1. Rocks are made of one or more
   a. carbons.
   b. minerals.
   c. nutrients.
2. Minerals and soil that people use are
   a. natural resources.
   b. igneous rocks.
   c. nutrients.
3. Materials from the soil that plants need are
   a. clays.
   b. sediments.
   c. nutrients.
4. The wearing down or breaking apart of rocks is called
   a. an eruption.
   b. mining.
   c. weathering.
5. Which is a mineral?
   a. copper
   b. wood
   c. cotton

Use the picture to answer questions 6 and 7.

6. In this picture of a volcano, magma is
   a. 1. b. 2. c. 3.
7. Which number shows lava from the volcano?
   a. 1 b. 2 c. 3
8. Plants can help change the earth's
   a. mantle.
   b. center.
   c. crust.
9. The layer of the earth with soil is called the
   a. crust.
   b. mantle.
   c. core.

1. ⓐⓑⓒ 4. ⓐⓑⓒ 7. ⓐⓑⓒ
2. ⓐⓑⓒ 5. ⓐⓑⓒ 8. ⓐⓑⓒ
3. ⓐⓑⓒ 6. ⓐⓑⓒ 9. ⓐⓑⓒ

113

Test Book page 114 *

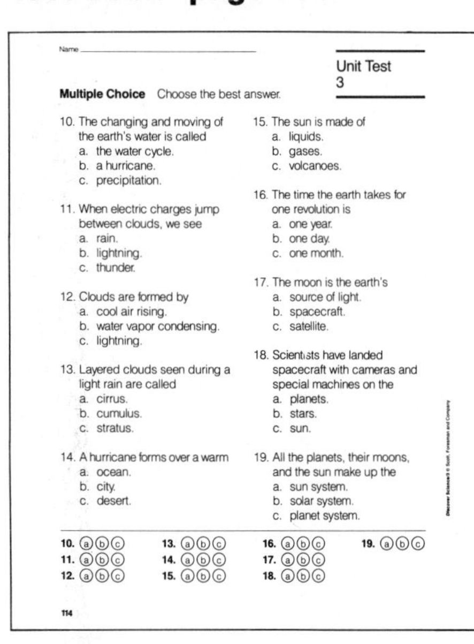

Name ______________________

Unit Test 3

Multiple Choice Choose the best answer.

10. The changing and moving of the earth's water is called
    a. the water cycle.
    b. a hurricane.
    c. precipitation.
11. When electric charges jump between clouds, we see
    a. rain.
    b. lightning.
    c. thunder.
12. Clouds are formed by
    a. cool air rising.
    b. water vapor condensing.
    c. lightning.
13. Layered clouds seen during a light rain are called
    a. cirrus.
    b. cumulus.
    c. stratus.
14. A hurricane forms over a warm
    a. ocean.
    b. city.
    c. desert.
15. The sun is made of
    a. liquids.
    b. gases.
    c. volcanoes.
16. The time the earth takes for one revolution is
    a. one year.
    b. one day.
    c. one month.
17. The moon is the earth's
    a. source of light.
    b. spacecraft.
    c. satellite.
18. Scientists have landed spacecraft with cameras and special machines on the
    a. planets.
    b. stars.
    c. sun.
19. All the planets, their moons, and the sun make up the
    a. sun system.
    b. solar system.
    c. planet system.

10. ⓐⓑⓒ 13. ⓐⓑⓒ 16. ⓐⓑⓒ 19. ⓐⓑⓒ
11. ⓐⓑⓒ 14. ⓐⓑⓒ 17. ⓐⓑⓒ
12. ⓐⓑⓒ 15. ⓐⓑⓒ 18. ⓐⓑⓒ

114

Unit Projects and Books

Science Projects

1. When animals of long ago died, some fell into soft mud or clay. This made a hole. The animals rotted, and the hole, or mold, was filled in with sand, mud, and other materials. The materials hardened and formed a cast. Molds and casts are two kinds of fossils. You can make a mold and a cast. Obtain some clay. Press it out into a rectangle. Place the clay in a shallow box. Press objects, such as a shell and a twig into the clay. Remove the objects from the clay. Now mix some plaster of Paris with water so that it looks like thick soup. Pour the plaster into the molds. Allow the plaster to dry. Remove the plaster cast from the clay mold.
2. Use a clear plastic container with straight sides as a simple rain gauge. Set your rain gauge outside. After rain has fallen, place a centimeter ruler up against the outside of the container. How many centimeters of rain did you collect?

Books About Science

Flash, Crash, Rumble, and Roll by Franklyn M. Branley. Crowell, 1985. Find out what happens during a thunderstorm.

To Space and Back by Sally Ride, with Susan Okie. Lothrop, 1986. Learn about the space shuttle flight with Sally Ride, the first woman astronaut. Gr. 3-6

Science and Society

Land Use About fifty people live in the forests and meadows of Bayonne Ridge. The town is close to a large city. Some people want a new housing development. Here is what some people of Bayonne Ridge are saying about this development: "We need better schools and new stores." "We like the Ridge just the way it is. We love the trees, the animals, and space." "But if we build new houses, city people would move here and create new businesses. So what if we have to tear down a few trees?" "Maybe, but clearing the land might cause erosion and water pollution." What are some good reasons to build a new development on Bayonne Ridge? What are some good reasons not to build the development?

271

Unit Projects and Books

TEACHING PLAN

Science Projects

1. Have students bring their casts to class to display.

2. Compare students' readings to see if the amount of rain that fell was the same throughout the city.

Science and Society

Encourage students to explore both sides of the issue in class or on their own. In class, you may wish to set up a debate on the issue. On their own, students might do further research and choose one side of the issue to support. Throughout this exercise, emphasize the importance of respecting the opinions of others. **Thinking Skill:** *Judging and evaluating*

Books in Spanish

De la tierra al cosmos. Bilingual Publications, 1984. Introduction to astronomy for children. (grades 2-5)

Lewellen, John. *La luna, el sol, y las estrellas.* Childrens Press, 1984. Introduction to the sun, moon, and stars. (grades 3-4)

Teaching Options

Science Background

Science and Society Questions of land use involve many issues, including political, economic, social, and aesthetic. In less developed areas, the major argument may be one of aesthetics. People in areas like Maine, Idaho, and Montana sometimes argue that they moved there to escape urban sprawl. They do not want to see the same thing happen all over again in their new homes. Others respond that they are ignoring some economic facts. People need to make a living and commercial and/or industrial developments are sometimes the only way to keep an area from economic ruin. However, land development may also raise ecological and environmental issues.

*Answers to masters on pages 246E-246H

UNIT 4/The Human Body

Planning Guide

Unit Components		Pages
Unit Overview/*SCIENCE IN THE NEWS*		272–27
Chapter 13 The Body's Support	**Lesson 1** What Makes Up Your Body? **Lesson 2** How Are Bones Important? **Lesson 3** How Are Muscles Important?	274–29
Chapter 14 Your Body's Health Needs	**Lesson 1** How Can You Stay Healthy? **Lesson 2** What Causes Disease? **Lesson 3** How Can Alcohol, Tobacco, and Drugs Affect the Body?	292–31
Careers	sports medicine physician, physical therapist, registered nurse, medical laboratory technician, physical education teacher	312
How It Works	Canning of Food	313
Unit Review		314
Unit Projects and Books	Science Projects Science and Society Books About Science	315

Science Process Skills Book

The Science Process Skills Book contains worksheets that can be used to teach and then assess student mastery of the basic science process skills. In addition, other worksheets in this book teach students the manipulative skills they will need to use basic science equipment. You might use some or all of these worksheets

- at the beginning of the year to teach and assess skills.
- directly before students encounter a particular skill in an activity.
- as a remedial tool when students begin to have trouble with specific skills.

cience Resources for the Unit

Resource Books

hr, Amy C. *It's OK to Say No: A Book for Parents and Children to Read Together*. Grosset & Dunlap, 1986. Teaches children mportant self-protection skills.

ing, Mary. *The Macmillan Book of the Human Body*. Macmillan, 1986. Examines each major body system, focusing on the organs, he purpose of the system, and ow it works.

tts, Eve. *Understanding Your mmune System*. Avon, 1986. Discusses how the immune ystem works; common ubstances that can weaken it; utrition's role in a strong mmune system; the prevention nd treatment of AIDS; and the ffects of aging, lifestyle, and stress.

ller Skating Rink Operators Association. *Your Muscles in Action*. A poster presenting diagrams showing some of the muscle groups that go into action when you roller skate. Classroom quantities available.

Community Resources

spitals, medical schools, and armacies often provide models d diagrams of the human body. Visit a local hospital to learn how rays are used. Invite a representative from a al law enforcement agency, a al health agency, MADD or AA speak to the class about the ngers of drug abuse.

Audio-Visual Resources

Bones and Muscles. International Film Bureau. Film or video, 15 minutes. Shows how the human skeletal, muscular, and nervous systems move the body.

Drugs! A Trick—A Trap. Britannica. Sound filmstrip series, 70 frames/9 minutes each. Five filmstrips reveal how drugs can trick the unwary and how serious the consequences can be.

Eating Right with Harv and Marv. Higgins. Film, 11 minutes. Uses animation to introduce the concept of a balanced diet.

Health: Communicable Diseases. AIMS. Film or video, 10 minutes. An animated germ tells children how people get sick, where germs live, and how germs spread.

Health: Exercise, Rest, and Sleep. AIMS. Film or video, 10 minutes. Shows how exercise, good posture, balanced meals, and plenty of rest and sleep are necessary for growth

Health: Personal Cleanliness. AIMS. Film or video, 8 minutes. Teaches how to clean the skin, hair, fingernails, and teeth.

The Healthy Way in Wonderland. Walt Disney. Filmstrip series. Six filmstrips teach lessons in nutrition, physical fitness, and the basics of daily hygiene routines.

How to Catch a Cold. Walt Disney. Filmstrip. Shows the symptoms of a cold; explains transmission of cold viruses; and explains the importance of a balanced diet, rest, and exercise in keeping the immune system strong.

Computer Software

Body Awareness. Learning Well. Apple IIe+, IIe, IIc, 48K. Students locate and match words with body parts and match body parts with seasonal clothing.

The Body Transparent. Design Ware. Apple IIe+, IIe, IIc, 48K. Uses a game format to teach and drill on the locations and functions of bones and organs.

Bones and Muscles: A Team to Depend On. Marshware. Apple IIe+, IIe, IIc, 48K. Presents the relationship between bone and muscle, their composition, their function, and how to keep them healthy.

Food Group Puzzles. Marshware. Apple II+, IIe, IIc, 48K. Students place pictures of individual foods in their food groups and create a balanced meal.

Health Watch. Learning Well. Apple II+, IIe, IIc, 48K. Three arcade-type games drill students on the concepts of proper diet, exercise, and dental care.

TEACHING PLAN

Unit Overview
This unit describes the human body's support system—bones and muscles. It explains how the skeletal and muscular systems function. The unit also describes the health needs of the body, such as nutrition, exercise, and sleep. It introduces the concepts of communicable and noncommunicable diseases, how drugs can be harmful to the body, and stresses ways to keep healthy.

About the Photograph
The three children in the photograph are involved in active play. Exercise is only one aspect of how to keep the human body healthy. Eating a balanced diet, getting enough sleep, and avoiding drugs are also necessary to maintain good health. Exercise like these children are getting is especially helpful in reducing stress by helping them relax.

272

Teaching Options

Chapter 13 Preview

The Body's Support
This chapter presents how cells are grouped into tissues, organs, and systems. The students will learn to distinguish between voluntary and involuntary muscles. The functions of bones and joints are discussed and students will learn how joints help the boy move.

Chapter 14 Preview

Your Body's Health Needs
In this chapter students will learn the importance of proper nutrition, exercise, and sleep. They will also learn how some diseases spread from person to person, and that some diseases do not spread. The chapter also stresses the harmful effects of alcohol, tobacco, and other drugs.

Unit 4

Human Body

These children know that active play can help them get exercise. Getting enough exercise helps keep your bones and muscles strong. It also helps you feel your best.

Following good health habits can help keep all the parts of the body healthy. In this unit, you will discover how your bones and muscles work. You also will learn about ways you can reduce your chances of getting some illnesses.

SCIENCE IN THE NEWS During the next few weeks, look in newspapers or magazines for stories about bones and muscles. Also, look for news about eating habits that can help people stay healthy. Share the news with your class.

Teaching Tips

- Direct the students' attention to the photograph on page 272. Questions: **What are the children in the picture doing?** (one child is jumping rope, the other two are turning the rope) **What body parts help the children move?** (Students should *infer* that bones and muscles help the body move.)
- Ask students to *describe* where they can feel bones and muscles in their bodies. Direct them to feel the bones and muscles in the upper and lower arm as they move their arms.
- Discuss the importance of regular exercise. Ask students to tell what activities they do regularly to get enough exercise. List the activities on the chalkboard.

13 Chapter 13 The Body's Support Planning Guide

TEACHING PLAN

Chapter Components	Skills	Materials
Chapter Opener/*DISCOVER:* Observing Body Parts pp. 274–275	*DISCOVER* p. 275 Science Process Skill *Observing*	*DISCOVER* p. 275 (groups of 2) no materials needed
Lesson 1 What Makes Up Your Body? pp. 276–278	Thinking Skills Challenge!: *Inferring* Find Out On Your Own: *Making models*	Demonstration p. 276 puzzle with 10–20 pieces
Science and Technology Using Computers to Make Joints p. 279	Thinking Skills *Observing, Communicating*	
Lesson 2 How Are Bones Important? pp. 280–282	Thinking Skills Challenge!: *Making analogies* Find Out On Your Own: *Comparing information, Communicating*	Demonstration p. 280 picture of steel or wood framework of a building, model of a skeleton
Activity Observing Cells p. 283	Science Process Skills *Observing, Collecting and interpreting data*	(groups of 2) 15 micrsocope slides, 15 cover slips, 1 large onion, 15 hand lenses, 15 paper clips, 15 sheets black construction paper
Lesson 3 How Are Muscles Important? pp. 284–286	Thinking Skills Challenge!: *Inferring* Find Out On Your Own: *Restating or explaining ideas*	Demonstration p. 284 no materials needed
Activity Making a Model of Arm Muscles p. 287	Science Process Skills *Observing, Communicating, Inferring, Making models*	(groups of 2) 15 large sheets lightweight cardboard, 15 pairs scissors, 15 paper punches, 15 paper fasteners, 1 ball string, 1 roll masking tape
Skills for Solving Problems Using Metric Scales and Line Graphs pp. 288–289	Problem Solving Skills *Making decisions/Identifying and solving problems, Interpreting charts, maps, and graphs*	
Chapter Review pp. 290–291	Thinking Skills *Sequencing, Restating or explaining ideas, Contrasting, Recognizing cause and effect, Inferring*	

Teaching Options

Strategies	Extensions		Resource Masters
Cooperative Learning p. 274 (Also see p. T23.) Applying Whole Language p. 275 (Also see p. T30.)			Family Letter: *Resource Book* p. 147
Reading Strategies p. 276 (Also see pp. T26–T29.)	Reinforcement p. 277 Special Education p. 277	Science and Art p. 278 Reteaching Suggestion p. 278	Vocabulary Preview: *Workbook* p. 73
Reading Strategies p. 280 (Also see pp. T26–T29.)	Special Education p. 281 Reinforcement p. 281	Reteaching Suggestion p. 282	Science Activity: *Workbook* p. 74 Science Skills: *Workbook* p. 75
			Activity Worksheet: *Resource Book* p. 151
Reading Strategies p. 284 (Also see pp. T26–T29.)	Special Education p. 285 Reinforcement p. 285	Reteaching Suggestion p. 286	Science and Language Arts: *Workbook* p. 76 Vocabulary Puzzle: *Workbook* p. 77
			Activity Worksheet: *Resource Book* p. 153
	Reteaching Suggestion p. 289		Science and Reading: *Resource Book* p. 155
Cooperative Learning p. 290 (Also see p. T23.)			Chapter Tests: Forms A and B *Test Book* pp. 117–120

Classroom Management

Advance Preparation

Demonstration, page 276
Bring a small jigsaw puzzle with 10–20 pieces to class.

Demonstration, page 280
Find a photograph of the framework of a building and a model of a human skeleton for this demonstration.

Activity, page 283
Make sure each group has a piece of onion skin, a microscope slide, a cover slip, and a hand lens.

Activity, page 287
Collect two pieces of thin shirt-cardboard for each group. Also provide each group with one paper fastener and two 18-cm (12-in.) pieces of string.

Vocabulary Review

Use the following sentences with your students to review the meanings of the italicized words.

1. Your body is made of billions of *cells*.
2. An ant, a flower, a person, and a horse are four kinds of *organisms*.
3. The *X ray* of the patient's arm showed the doctor where the bone was broken.
4. Your *skeleton* includes all the bones in your body.
5. A *nerve cell* sends a message to the brain to signal the muscles to move.

High-Potential Students

Suggest that high-potential students construct a model skeleton that shows how muscles move bones. Provide cardboard, rubber bands, and paper fasteners. Students could cut bones from cardboard and connect them at the joints with paper fasteners. They can use string or rubber bands for muscles. Ask them to demonstrate certain motions, such as walking, bending, or throwing, with their skeleton.

Mainstreamed Students

Visually Impaired
Let students with visual impairments form pieces of clay into different shapes to reinforce that cells have different shapes. Have them put the pieces together to make a tissue.

Hearing Impaired
When discussing the bones, muscles, and tendons have students with hearing impairments locate these body parts on a model of a human skeleton. Make dramatic movements with your arms and legs to demonstrate how muscles move bones.

Science Fair Projects

The Experiment Skills on p. 272 and the Investigate feature in the chapter can be used for science fair projects. You might also encourage interested students to do one of the following projects:

1. Compare animal and human skeletons by preparing diagrams of animal skeletons and of a human skeleton. Use different colors to illustrate the skull bones, ribcage, leg bones, and joints on each skeleton.
2. Research the types of movable joints such as hinge joints and pivot joints. Construct a model of each type of joint. Make a diagram of the human body showing where these joints are located and what kind of movement they allow.

lassroom Resources

Bulletin Board

Challenge students to think of other examples that are similar to the "cells to organs to systems" organization. Examples could include the different stages of construction on a house before it becomes a home, the different parts that go together to form an entire car, and so on.

CELLS TO SYSTEMS

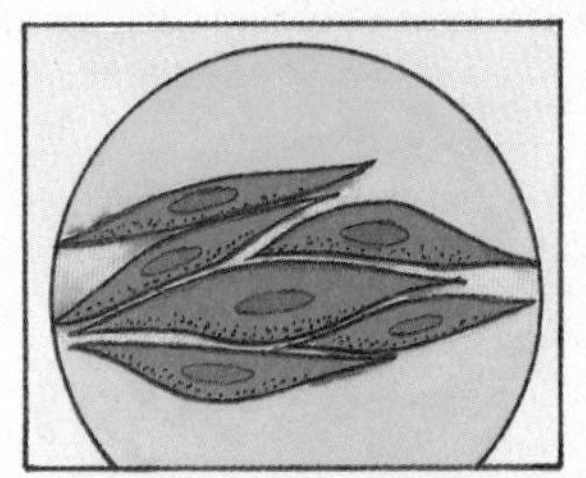

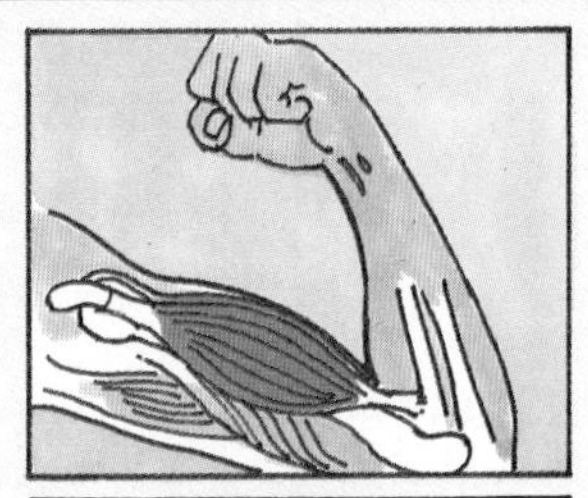

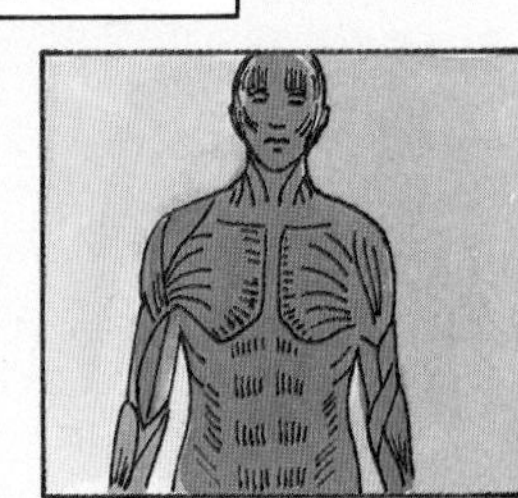

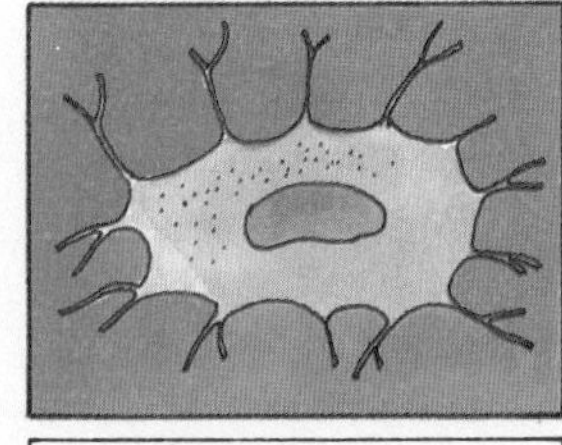

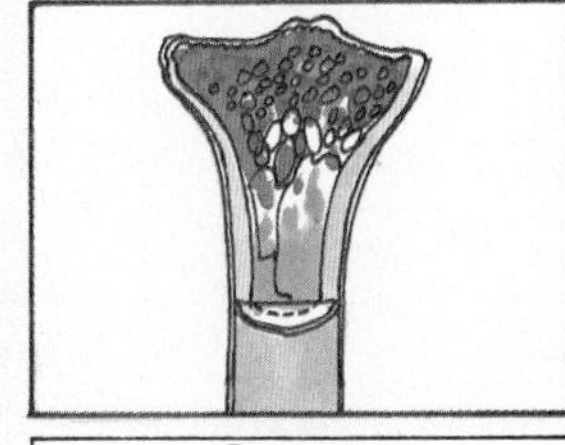

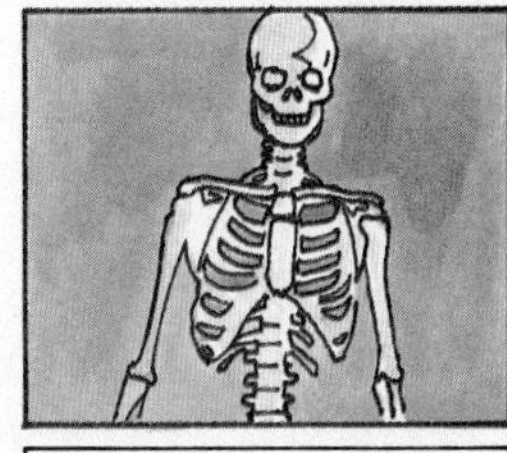

Chapter 13 Poster

Science Discovery Center

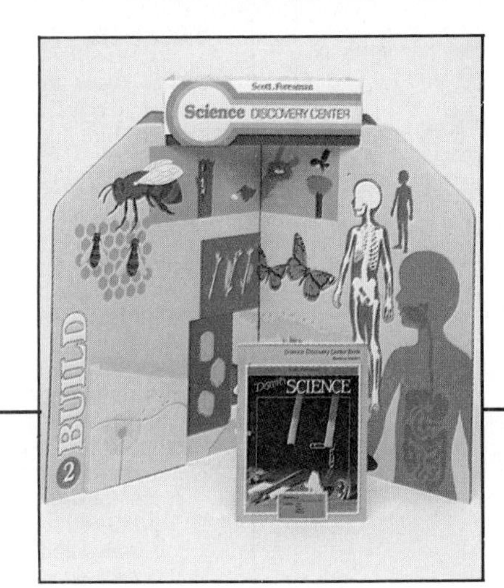

Use pages 135–140 from the *Science Discovery Center Book*. Place these worksheets in the appropriate pockets in the Science Discovery Center.

Overhead Transparencies

Use Transparencies 22 and 23 from the package of color overhead transparencies.

CHAPTER 13 COPY MASTERS

Teacher's Resource Book

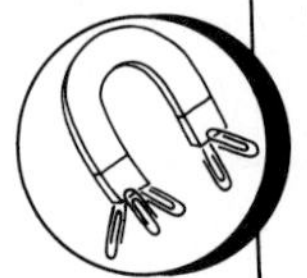

Dear Family,
Your student will be reading **Chapter 13: The Body's Support** in *Discover Science,* published by Scott, Foresman. We will learn that the bones provide structure and the muscles help the body move.
You and your student can learn about the muscles of the face by doing this activity together. Most animal faces cannot move, but human faces have many small muscles that enable them to create expressions that communicate. These exercises will help you learn about your face muscles.

Face Exercises

1. Stand before a mirror.
2. Try each exercise. Ask, "Can you see your muscles move?"
3. Gently touch your face as you do each exercise. Ask, "Can you feel your muscles move?"
4. On the diagram, find the muscles that you used.

Pull scalp back.

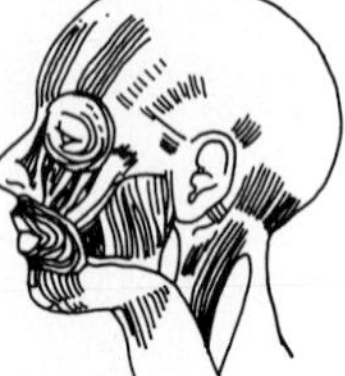
Muscles of the Face

Wiggle ears.

Open wide.

Frown.

Wink each eye.

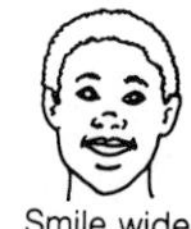
Smile wide.

Workbook

Name ____________
Use with Lesson 1: pages 276-278

Chapter 13
Vocabulary Preview

The Body's Support

Chapter Vocabulary			
contract	joint	system	tissue
involuntary muscle	organ	tendon	voluntary muscle

Vocabulary Cards

1. Write each word on a card.
2. Find each word in the glossary. Copy the pronunciation under the word on the card.
3. Practice saying the words with a partner.

Word Meanings

The word *voluntary* means "done on purpose." The word *involuntary* means "not done on purpose." Read the following list of body actions. On the given line, mark each action as being voluntary or involuntary.

raising your arm **voluntary**

walking across the room **voluntary**

sneezing **involuntary**

your heart pumping blood **involuntary**

throwing a ball **voluntary**

food moving through your stomach **involuntary**

scratching your nose **voluntary**

blood flowing through your body **involuntary**

At Home

1. Practice saying the words. Learn their meanings.
2. Look for the words and pictures of the words in newspapers and magazines. You can make a poster.

Teacher's Notes: Encourage the students to write the definitions on the word cards as they come across the words in the chapter.

Teacher's Resource Book

Name ____________
Use with Lesson 1: page 283

Chapter 13
Activity Worksheet

Observing Cells

Record Your Results

How the onion skin looks

State Your Conclusion

1. Can you see the pieces that make up an onion skin without a hand lens?
 ____
2. Do you see the tiny pieces that make up the onion skin more clearly in the light or dark places of the onion skin?
 ____

Use What You Learned

If the onion skin is made of thousands of pieces, what can you infer about the whole onion?
____
____

Workbook

Name ____________
Use with Lesson 2: pages 280-282

Chapter 13
Science Activity

What Do Your Bones Do?

Gather These Materials

- paper • scissors • 3 soda straws
- clay • tape

Follow This Procedure

1. On a piece of paper, draw a picture of yourself standing up. Make the picture 25 cm long. Cut out the picture. What happens when you try to stand it up?
2. Make a support system for your picture. Cut one soda straw 17 cm long. Then cut four pieces of straw 10 cm long.
3. Use clay to put together the pieces of straw as shown.
4. Tape the top of your picture to the straw support system. What happens when you try to stand it up?

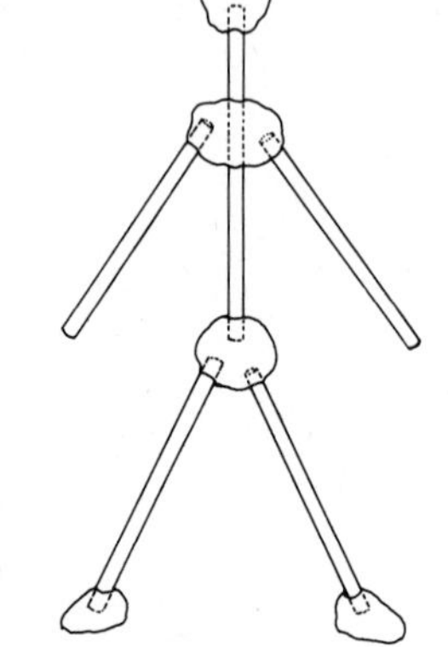

Record Your Results

1. What happens when your picture stands alone?
 It does not stand up.
2. What happens when your picture stands with a straw support system?
 It stands up.

State Your Conclusion

1. What part of your body is the picture of you like?
 skin or muscles
2. What part of your body is the straw support system like?
 bones or skeleton
3. What would happen if you did not have any bones?
 Without bones, I could not stand up.

Teacher's Notes: Encourage students to build a variety of straw-and-clay structures and to test the strength of the structures.

Workbook

Name ______________________

Use with Lesson 2: pages 280-282

Chapter 13

Science Skills

Classifying Joints

Read about the different kinds of joints in your body. Color the joints as directed.

1. Most of the joints in your body help you move. But most joints in your head do not move. These are called fixed joints. Color the skeleton's fixed joints red.
2. Some of your joints are like the hinge of a door. These hinge joints move back and forth. Your knees, elbows, fingers, and toes are hinge joints. Color the hinge joints green.
3. Shake your head no. You just moved pivot joints in your neck. Your elbows also are pivot joints. They can move from side to side. Put an *X* on each of these pivot joints.
4. The joints with the most movement are ball and socket. The joints in your hips and shoulders are ball and socket. Color these joints purple.

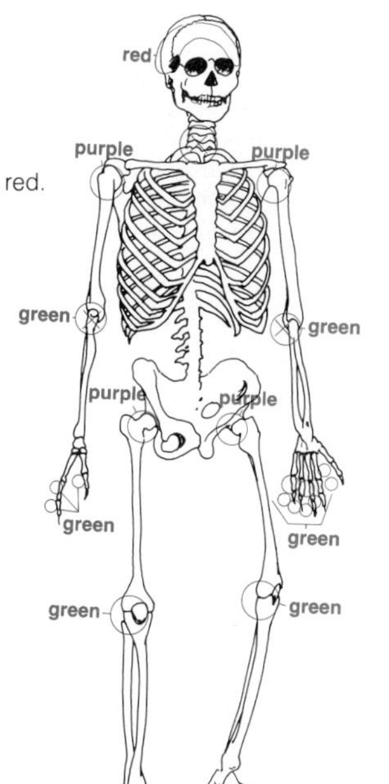

Teacher's Notes: Have students move several of their own joints to see and feel how different joints move.

Workbook

Name ______________________

Use with Lesson 3: pages 284-286

Chapter 13

Science and Language Arts

Choosing the Correct Book

Mike is writing a report. The list on the left shows topics he might write about. Match each topic with the book that would help Mike. Write the book title on the line. A book can be used more than once.

1. How involuntary muscles work
   The Muscle System
2. The parts of a heart
   The Heart
3. The ways joints can move
   The Book of Bones
4. How skin cells protect your body
   Know Your Cells
5. Ways to keep your muscles fit
   Exercise for Health
6. How the eye works
   Sight
7. What your lungs do for you
   All About Breathing

Teacher's Notes: Ask your librarian to show students how your library is organized and how to find books on specific topics.

Workbook

Name ______________________

Use with Lesson 3: pages 284-286

Chapter 13

Vocabulary Puzzle

Wordsearch

Unscramble the words from the word bank.
Then find and circle each word in the puzzle below.

Word Bank			
contract	joint	system	tissue
involuntary muscle	organ	tendon	voluntary muscle

TONIJ **JOINT** NEDTON **TENDON**

RANGO **ORGAN** CRANTOCT **CONTRACT**

STEMYS **SYSTEM** SUESIT **TISSUE**

NATOLVURY SEMCLU **VOLUNTARY MUSCLE**

TANOLVUNIRY CLUSME **INVOLUNTARY MUSCLE**

T I S S U E J E Y I T J O I N T G U
E K Y B K W G O R G A N L E V R C D
N T S C O N T R A C T F S R P L C B
D B T C P B Q L M H T I Y E J Q M C
O Q E V O L U N T A R Y M U S C L E
N M M L D V A F K T W J B K W G L V
I N V O L U N T A R Y M U S C L E D

Teacher's Notes: Ask students to write a definition for each of the words from the puzzle.

Teacher's Resource Book

Name ______________________

Use with Lesson 3: page 287

Making a Model of Arm Muscles

Chapter 13

Activity Worksheet

Record Your Results

String pulled	How arm model looks
Back edge	
Front edge	

State Your Conclusion

1. What happens when you raise each of the two strings?
2. Compare your arm muscles to the model and explain how muscles work in pairs.

Use What You Learned

How do your arm muscles change in length when you raise and lower your lower arm?

Name ____________________

Comprehension: drawing conclusions

Chapter 13

Science and Reading

Tell Me Why

Read the story. Then write a sentence to answer each question.

Mike is very interested in how the bones and muscles work in his body. Someday he wants to be a doctor. He is excited because today his class is going on a tour of a hospital operating room.

When the class gets to the hospital, they must put on special uniforms to cover their clothes. Even their shoes must be covered. They also have to wear hats.

Inside the operating room, everything looks shiny and clean. The temperature is kept very cold because the big hot lights over the table make the room very warm. The operating room table moves up and down. It is important for the doctors to be comfortable while they stand and operate.

After visiting the hospital, Mike is even more sure that he wants to be a doctor.

1. Why did the class have to put on special uniforms?

   The special uniforms cover up any germs or dirt that might be brought into the operating room.

2. Why is the operating room temperature kept cold?

   The temperature is kept cold because the lights are very hot.

3. Why does the table move up and down?

   The table moves up and down so that both short and tall doctors will be comfortable working on it.

Name ____________________

Chapter 13 Test A

Multiple Choice Choose the best answer.

1. Every body part is made up of (1-1)
   a. bones.
   [b.] cells.
   c. muscles.

2. Each kind of cell in the body has its own (1-2)
   a. muscle.
   b. organ.
   [c.] job.

3. A group of cells that look alike and do the same job is (1-3)
   [a.] a tissue.
   b. an organ.
   c. a system.

4. A group of organs working together to do a job in the body is (1-3)
   [a.] a system.
   b. a tissue.
   c. an organism.

5. Bones help a person (2-1)
   a. see.
   [b.] move.
   c. think.

6. The body parts that allow bones to move in many different ways are called (2-3)
   [a.] joints.
   b. ribs.
   c. tendons.

7. A muscle that works without a person controlling it is called (3-3)
   a. a joint.
   [b.] an involuntary muscle.
   c. a voluntary muscle.

8. A person can control head and arm movements with (3-3)
   a. tendons.
   b. involuntary muscles.
   [c.] voluntary muscles.

9. Muscles move bones by (3-3)
   a. relaxing.
   b. connecting.
   [c.] contracting.

10. Muscles are connected to bones with a tissue called a (3-3)
    a. joint.
    b. skeleton.
    [c.] tendon.

Numbers in parentheses after each question refer to the lesson number and the objective of that lesson.

Name ____________________

Chapter 13 Test A

Short Answer Write the name of each body part in the blank below the picture. Use these words: joint, skeleton, tendon, muscle.

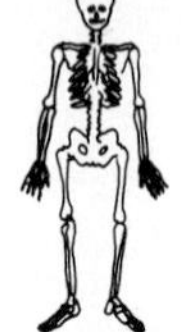

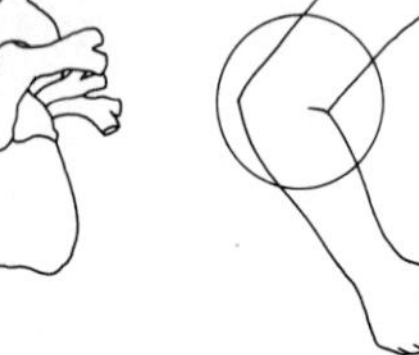

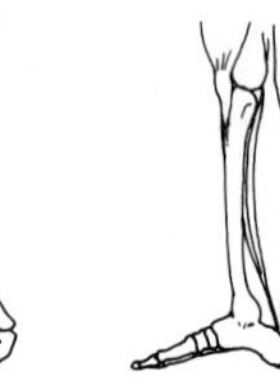

1. **skeleton** 2. **muscle** 3. **joint** 4. **tendon**

Short Essay Use complete sentences to answer each question.

1. Name three ways bones help the body. (2-1)

   Bones help the body by protecting organs and holding a person up. Bones also work with muscles to move the body.

2. Name three joints and tell why joints are important. (2-2, 2-3)

   Three joints are the shoulder, elbow, and finger. Joints are important because they allow bones to move in different ways.

Numbers in parentheses after each question refer to the lesson number and the objective of that lesson.

Name ____________________

Chapter 13 Test B

Multiple Choice Choose the best answer.

1. All parts of the body are made up of (1-1)
   a. bones.
   [b.] cells.
   c. muscles.

2. Each kind of cell in the body has its own (1-2)
   a. muscle.
   b. organ.
   [c.] job.

3. A group of the same kind of tissues forms (1-3)
   [a.] an organ.
   b. an organism.
   c. a system.

4. The body parts that help a person stand and move about are the (2-1)
   a. ribs.
   [b.] bones.
   c. lungs.

5. The places where bones are joined together are called (2-2)
   [a.] joints.
   b. ribs.
   c. tendons.

6. The tissues that move bones in a body are (3-1)
   a. organs.
   b. skeletons.
   [c.] muscles.

7. Muscles that a person can control are called (3-2)
   a. joints.
   b. involuntary muscles.
   [c.] voluntary muscles.

8. Muscles that help a person breath are (3-2)
   a. joints.
   [b.] involuntary muscles.
   c. voluntary muscles.

9. The tissues that connect muscles to bones are the (3-3)
   [a.] tendons.
   b. organs.
   c. muscles.

10. When a muscle contracts, it becomes (3-3)
    a. thinner.
    b. longer.
    [c.] shorter.

Numbers in parentheses after each question refer to the lesson number and the objective of that lesson.

Name ____________________

Chapter 13
Test B

Short Answer Each picture shows the use of a muscle. Write whether the muscle being used is a voluntary or involuntary muscle. (3-2, 3-3)

1. voluntary 2. involuntary 3. voluntary 4. voluntary

Short Essay Use complete sentences to answer each question.

1. Name three joints. Why are joints important? (2-2, 2-3)

Three joints are shoulder, elbow, and finger. Joints are important because they allow bones to move in different ways.

2. List three ways that bones help the body. (2-1)

Bones help the body by protecting organs and holding a person up. Bones also work with muscles to move a person about.

 Numbers in parentheses after each question refer to the lesson number and the objective of that lesson.

TEACHING PLAN

Major Concepts

Lesson 1 The body is made up of cells that are grouped together to form tissues, organs, and systems.
Lesson 2 The body has different kinds of bones that give the body support, protection, and allow for movement.
Lesson 3 Muscles give the body shape, keep body organs working, and work with the bones to help the body move.

Chapter Vocabulary

contract, involuntary muscle, joint, organ, system, tendon, tissue, voluntary muscle

Getting Started

Ask students to *explain* what the person in the photograph is doing. (a handstand on a balance beam, probably for a gymnastics competition) Ask whether or not a handstand is difficult to do, and why. (It is difficult because it requires arm strength, balance, and flexibility.) Ask what parts of the body must be trained to do gymnastics. (The muscles and tendons must be made stronger and more flexible; the nerves must be trained to coordinate fine muscle movements required for balance and grace.) Allow students to talk about other physical activities they enjoy.

Chapter 13

The Body's Support

How is this person using her body? Notice how she twists and turns her body in different ways. She is using many different parts of her body.

274

Teaching Options

Cooperative Learning ♦

Jigsaw Format **(See page T23.)**
Assign the following topics at random to your cooperative learning teams.

Topic A: Explain how cells are organized into the tissues and systems that give the body support and describe the support structures.
Topic B: Explain the roles of bones and joints in the body, and describe how they make a selected body part move.
Topic C: Define *involuntary muscle* and *voluntary muscle*, and explain the roles of each in the body, providing at least two examples for each type.
Topic D: Select a movable body part and explain how the structures that give support to the body work together to allow its movement.

Have students search for information on their topic as they read the chapter. Then let all students with the same topic meet in an expert group to discuss the information. When students return to their teams, they may take turns presenting their topics to the team. Then give students a test covering all topics to complete individually (Chapter 13 Test A or B in the *Test Book)*. Award Superteam certificates to teams whose average test scores exceed 90%, and Greatteam certificates to teams whose average test scores exceed 80%.

♦ ***Suitable as a language development activity***

Introducing the Chapter

In this chapter you will learn about what makes up your body. You will also learn how bones and muscles are important to your body. The activity below will help you learn how one part of your body works.

Observing Body Parts

Bend your right elbow. Have your partner put a hand on your upper arm as shown. Squeeze your right hand into a fist and then let go. Notice how your upper arm changes. Repeat several times. Ask your partner to feel different parts of your arm moving. Exchange places with your partner and repeat the activity.

Talk About It

1. How did your upper arm change when you made a fist?
2. What parts of your arm would move if you lifted a book?

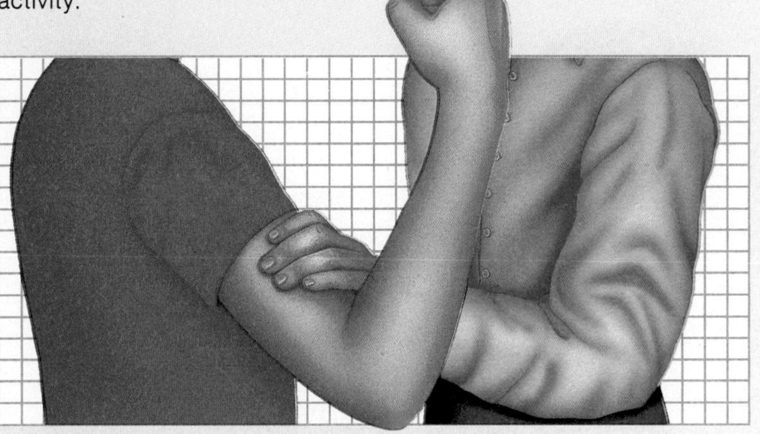

DISCOVER

Objective ♦

This optional *DISCOVER* activity will help students explore and build background information about the body's support. Later in the chapter, students will be able to draw on this experience to help them assimilate the new content.

Science Process Skill

Observing

Safety Tip (See page T24.)

- Caution students not to upset their desks as they push with their arms.

Teaching Tips

- Question: **What other movements make your arm muscles change?** (picking up a book, pulling on pants, exercising, and so on)
- **Helpful Hint:** Have students shake their arms after the activity to relax their muscles.

Answers

Talk About It

1. Students should respond that the muscle became bigger or harder.

2. Students should indicate that muscles all along their arm would move if they lifted a heavy book.

Applying Whole Language ♦

Discuss the whole language framework with each Teaching Option you select. Here is an example applied to the Reinforcement option on p. 281.

1. Purpose: To infer the positions of skeletons on pictures of animals

2. Context: Students look at photographs or drawings of animals and use tracing paper to draw the positions of skeletons

3. Decisions: What is the function of the different body parts? Which parts need support? How can we check our drawings for accuracy?

4. Evaluation: How accurate were our drawings? (See p. T30.)

Resource Book page 147 *

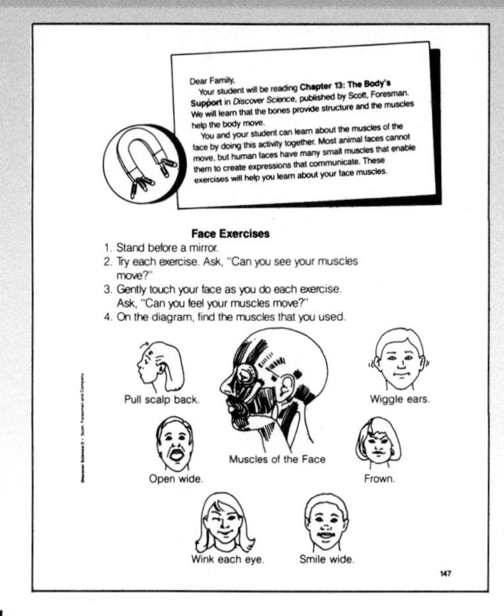

Dear Family,

Your student will be reading **Chapter 13: The Body's Support** in *Discover Science*, published by Scott, Foresman. We will learn that the bones provide structure and the muscles help the body move.

You and your student can learn about the muscles of the face by doing this activity together. Most animal faces cannot move, but human faces have many small muscles that enable them to create expressions that communicate. These exercises will help you learn about your face muscles.

Face Exercises

1. Stand before a mirror.
2. Try each exercise. Ask, "Can you see your muscles move?"
3. Gently touch your face as you do each exercise. Ask, "Can you feel your muscles move?"
4. On the diagram, find the muscles that you used.

Pull scalp back.
Wiggle ears.
Muscles of the Face
Open wide.
Frown.
Wink each eye.
Smile wide.

147

Science Background

Muscles work in pairs to move body parts. They alternately contract and relax to move the different parts of the body. Extended periods of muscle work can cause a muscle to tire. If this activity were repeated over and over again, the students would have the opportunity to experience muscle fatigue.

*** *Answers to masters on pages 274E—274H***

TEACHING PLAN

Lesson Objectives
- *Define* *cell* and *name* kinds of cells that are found in the body.
- *Explain* how cells form tissues, organs, and systems.

Lesson Vocabulary
organ, system, tissue

1. Motivate

Demonstration Activity ♦
Obtain a puzzle with 10–20 pieces. Begin to put the puzzle together, but leave the puzzle unfinished.

Discussion
Question: **Is this puzzle complete?** (No.) Ask a student to finish the puzzle. Ask students to *identify* what the puzzle is made of. (pieces of wood or cardboard) Point out that our bodies are also made up of many tiny pieces known as cells. Explain that just as puzzle pieces form a puzzle, groups of cells form the tissues, organs, and systems of our bodies.

1 What Makes Up Your Body?

LESSON GOALS

You will learn
- that different kinds of cells are found in the body.
- how cells form tissues, organs, and systems in the body.

Notice how the students in the picture use small blocks to make a house with different parts. The house has windows, walls, and a roof. Your body has many different parts too. You have bones and muscles. You also have a brain, a stomach, and a heart. You need all these parts for your body to work properly.

Cells

[1]different shapes and colors
[2]both made of blocks

Cells are the building blocks of the body.

How is the roof in the picture different from the walls?[1] How are the roof and the walls alike?[2] The parts of your body also are different from each other. Yet all these parts are alike in one way. You learned in Chapter 3 that a cell is the basic unit of an organism. All the parts of your body are made of cells.

Teaching Options

Science Background

Each cell consists of a cell membrane, cytoplasm, and a nucleus. The nucleus is the center of all cell activity. Chromosomes in the nucleus contain the instructions for the inherited traits of each individual. Since chromosomes are duplicated each time a body cell divides, each new cell receives a set of chromosomes that is identical to the original set.

Reading Strategies ♦

1. Guide students' pre-reading by asking: What ideas do you think you would find in the lesson to answer the question-title?
2. Assign these strategies: Writing a Memory Sentence and Visualizing Information for the subheadings (See pages T26–T29.)
3. Pair students to share what information is clear and unclear and initiate discussion using students' unanswered questions.

♦ *Suitable as a language development activity*

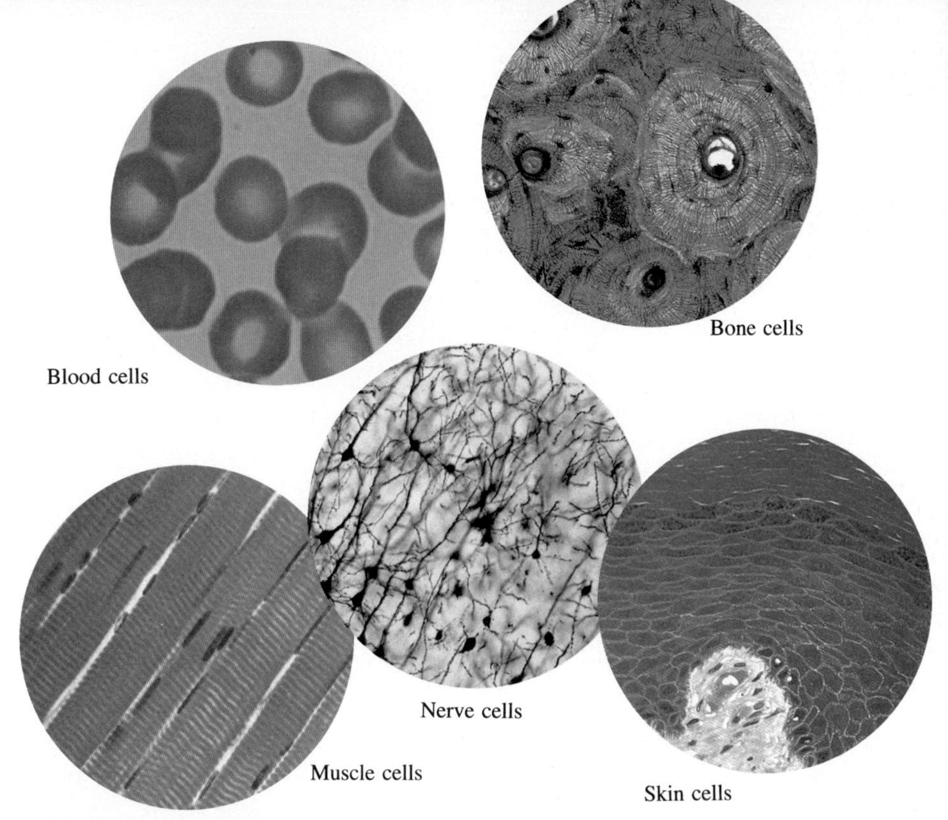

Blood cells

Bone cells

Nerve cells

Muscle cells

Skin cells

How Cells Make Up the Body

Billions of cells make up your body. Notice how the cells in the picture look different from each other. Each kind of cell does a different job in the body. For example, skin cells cover and protect your body. Notice how the nerve cells look different from the muscle cells. Find the blood cells in the picture.

Groups of cells form the parts of your body. A group of the same kind of cells forms a **tissue.** A group of bone cells forms bone tissue. A group of muscle cells forms muscle tissue. What does a group of nerve cells form?[1]

[1]nerve tissue

SCIENCE IN YOUR LIFE

Scientists use microscopes to study cells. Most body cells are colorless. Cells can be stained with special dyes. Then the cells can be seen under a microscope.

tissue (tish′ü), a group of cells that look alike and do the same job.

2. Teach

Teaching Tips

- Tell students to look at the palms of their hands. Ask them to ***identify*** the kind of tissue that they see. (skin tissue) Point out that the skin is the body's first defense against disease.
- Question: **What do you think your body uses to grow new skin or bone tissue if you fall and scrape a knee or break a bone?** (Guide students to *infer* that the body uses new skin and bone cells to repair the skin and bone tissue.)

Workbook page 73 *

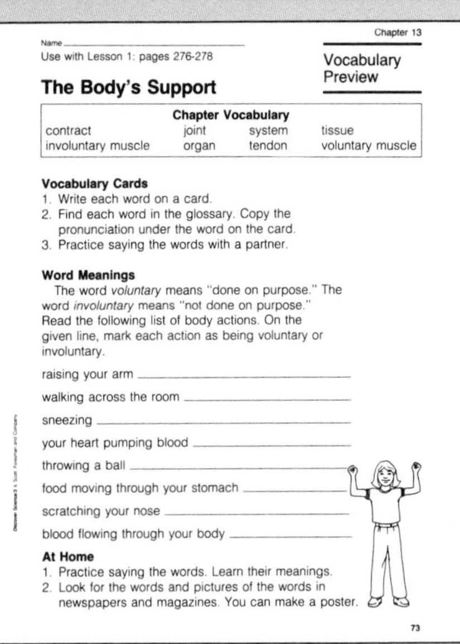

Chapter 13

Name ____________

Use with Lesson 1: pages 276-278

Vocabulary Preview

The Body's Support

Chapter Vocabulary			
contract	joint	system	tissue
involuntary muscle	organ	tendon	voluntary muscle

Vocabulary Cards
1. Write each word on a card.
2. Find each word in the glossary. Copy the pronunciation under the word on the card.
3. Practice saying the words with a partner.

Word Meanings

The word *voluntary* means "done on purpose." The word *involuntary* means "not done on purpose." Read the following list of body actions. On the given line, mark each action as being voluntary or involuntary.

raising your arm ____________

walking across the room ____________

sneezing ____________

your heart pumping blood ____________

throwing a ball ____________

food moving through your stomach ____________

scratching your nose ____________

blood flowing through your body ____________

At Home
1. Practice saying the words. Learn their meanings.
2. Look for the words and pictures of the words in newspapers and magazines. You can make a poster.

73

Special Education

Have students with learning disabilities work with modeling clay to form models of cells and tissues. Direct them to make cells of different colored clay, then to mold the like cells together to form tissues.

Reinforcement

With students, make a list of cells on the chalkboard. (It may include skin, bone, muscle, and nerve cells.) Ask students to *explain* the function of each type of cell. (Skin cells form the skin that protects the body, bone cells form bones that support the body, and so on.) Remind students that each cell is able to grow and reproduce and, when joined with cells of the same kind, to form tissues and organs that serve special functions in the body.

TEACHING PLAN

Teaching Tip

• **Possible Misconception:** Some students might think that blood is part of the veins and arteries. Review that blood consists of blood cells and fluid.

3. Assess

Lesson Review

1. cells
2. different organs working together
3. Challenge! Responses should include bones, muscles, and skin. **Thinking Skill:** *Inferring*

Find Out On Your Own

Student pictures can include any of the body systems with a brief description of how that system functions in the body. **Thinking Skill:** *Making models*

organ (ôr′gən), a group of tissues that work together to do a job in the body.

system (sis′təm), a group of organs that work together to do a job in the body.

A group of different tissues forms an **organ.** The tissues in an organ work together to keep you alive. Your heart, stomach, eyes, and brain are some of your body organs.

Different organs in your body work together. The picture shows your heart and other organs that move blood through your body. Different organs working together are called a **system.** Your body has many systems. Each body system works to keep you healthy.

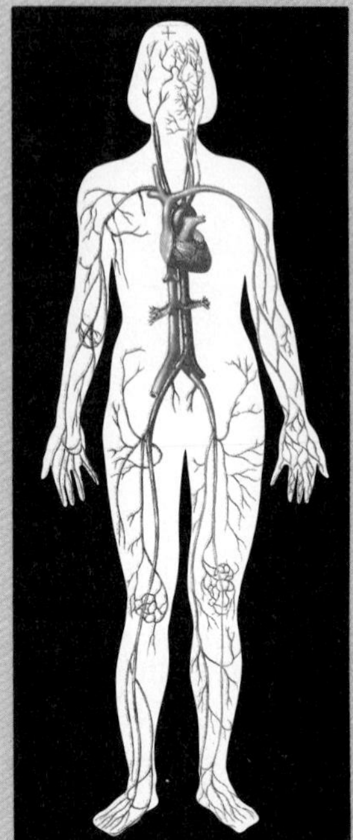

System that moves blood through the body

Lesson Review

1. What makes up all of the parts of the body?
2. What makes up a body system?
3. **Challenge!** Look at your hand. Feel the different parts. What different kinds of tissues do you have in your hand?

Study on your own, pages 340–341.

HUMAN BODY
FIND OUT ON YOUR OWN

Look in library books to find out about a body system. Draw a picture that shows the different parts of the system. Write a few sentences explaining how this system helps keep the body alive and healthy.

278

Teaching Options

Science Anecdote

The heart pumps over seven thousand liters (about 7,400 quarts) of blood each day. The pathway of blood from the heart to the lungs and back is the *pulmonary circulation*. The pathway of blood from the heart to other parts of the body and back is the *systemic circulation*.

Reteaching Suggestion ♦

Have students refer to the micrographs on page 277 to draw pictures of skin cells, bone cells, muscle cells, and blood cells. Next to each picture, have them draw a corresponding part of the body that each type of cell makes up.

Science and Art

Have students make collages of living things that contain cells. Students may use magazine pictures, photographs, drawings, and specimens from nature in their collages.

♦ ***Suitable as a language development activity***

Using Computers to Make Joints

The Problem More than 31 million people in this country have arthritis—a disease of the joints. People with arthritis feel pain, stiffness, or swelling in their joints. Arthritis cripples many people. Often, a joint of an elderly person will simply wear out. This form of joint disease cannot be cured. Doctors can only try to relieve the pain.

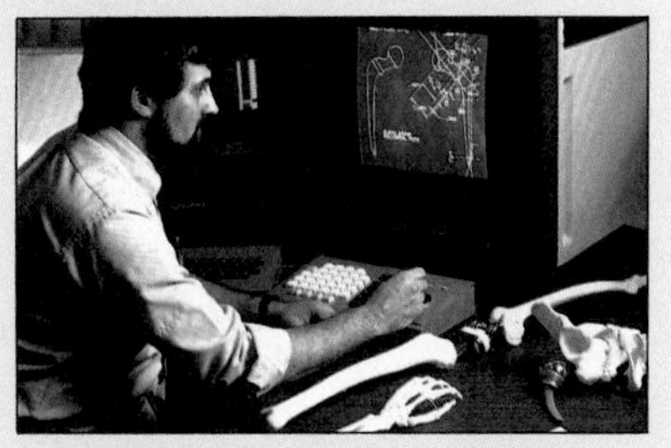

Designing a joint with a computer

The Breakthrough For many years, doctors have tried to replace badly damaged joints. In 1890, a German doctor made the first artificial hip. He replaced the natural joint of the hip with a joint made of ivory. Soon, doctors in America and Europe began using other materials to repair joints. It was not easy to make a new joint. First, a designer looked at the patient's X ray. Then a model of the new joint was drawn. A technician would try to shape a block of metal to fit the patient. The doctor sometimes would have to cut away bone and tissue to make the new joint fit.

New Technology Today, with the help of computers, people make artificial joints that fit the patient better. These new joints take less time to make and are cheaper. The computer gets some information about the patient from the X rays. Then the doctor tells the computer how much the patient weighs and how active the patient is. The computer uses this and other information to choose a design. The design appears on the screen, as you can see in the picture. The doctor checks the design to make sure it will fit. The doctor tells the computer when everything looks fine. The computer then sends instructions to the computer-run equipment. The equipment cuts out the new joint. During surgery, the doctor replaces the patient's joint with the new one.

What Do You Think?

1. Why are computers better at making new joints than people?
2. How can computers be used to design and make cars?

Science and Technology

Discussion

Questions: **What is a joint?** (a place where bones join together, which allows for movement) **Where are the joints located in your body?** (Students may name knees, elbows, ankles, or wrists.) Have students *describe* how each joint they name allows the body to move.

Teaching Tip

- Point out that more than 100,000 joint replacements are performed each year. The two joints most commonly replaced are the hip and the knee. Most, but not all, of these operations are performed on people over the age of 65.

Answers

What Do You Think?

1. The computer is more precise, and works with more information. **Thinking Skill:** *Observing*

2. Computer-aided design and manufacture works in the same way as for artificial joints. Engineers enter information about the car into the computer, and the computer makes a design. Engineers then view the design and make changes on the screen. The information is sent to computerized equipment that makes parts exactly as designed. **Thinking Skill:** *Communicating*

Science Background

The two bones that come together at the hip joint are the hipbone and the thighbone. The hipbone contains a socket. The thighbone fits into the socket like a ball. Artificial hips and other joints may be fixed to the bone with special cement. A new type of artificial joint is coated with tiny metal beads. These beads allow the surrounding tissues to grow into the prosthesis. In this way, the prosthesis becomes part of the body.

The technology described in the feature is called CAD/CAM. The term is an acronym for computer-aided design and computer-aided manufacture. CAD/CAM is making more precise parts for cars, airplanes, nuclear generators, and many other products formerly designed by hand.

TEACHING PLAN

Lesson Objectives
- *List* the functions of bones.
- *Describe* joints and *explain* how joints help the body move.

Lesson Vocabulary
joint

1. Motivate

Demonstration Activity ♦
Obtain a photograph or drawing of the steel or wood framework of a building and a model of a skeleton. Point out what each item is. Allow students to examine the pictures.

Discussion
Questions: **What is the function of the framework?** (to support the building and give it shape) **What is the function of the skeleton?** (to support the body and give it shape) Ask students to *infer* which parts of the skeleton are like the steel or wood beams. (bones) Point out that students will learn about some of the parts of the skeleton in this lesson.

LESSON GOALS

You will learn
- that bones protect certain body parts and give the body shape and support.
- that joints allow bones to move in different ways.

2 How Are Bones Important?

What can make this puppet move? What might happen if a person let go of the puppet? The hand inside the puppet holds up the puppet. The hand also makes the puppet move.

Bones

Think about how your body is like the puppet. Your body needs help to stay up and to move. The bones of your body make up your skeleton. Your skeleton helps hold you up. Your bones also work with your muscles to help you move.

Your bones help protect organs in your body. Look at this picture of the skeleton. Find the ribs. These bones protect your heart and lungs. The bones of the head protect your brain.

Look at the many different kinds of bones in this skeleton. Notice how the long, narrow parts of your body have long, narrow bones. All the different kinds of bones help give your body shape.

Teaching Options

Science Background

The human skeleton is made up of 206 bones. The functions of these bones are many: they allow the body to move; they give the body a definite shape; they protect the internal organs; they serve as a storage site for minerals (such as calcium) and fats; and tissues in certain bones manufacture blood cells. Because bones have different functions, scientists divide the skeleton into two parts. The *axial* skeleton consists of bones that protect the body's organs. These bones include the skull, the vertebrae, and the sternum. The *appendicular* skeleton is made up of the bones that allow body movement. These include the bones of the arms and legs.

Reading Strategies ♦

1. Guide students' pre-reading by asking: What words/ideas would you expect to find under each subheading?
2. Assign these strategies: Writing a Memory Sentence for each Subheading and Visualizing Information (See pages T26–T29.)
3. Pair students to share what information is clear and unclear and initiate discussion using students' unanswered questions.

♦ ***Suitable as a language development activity***

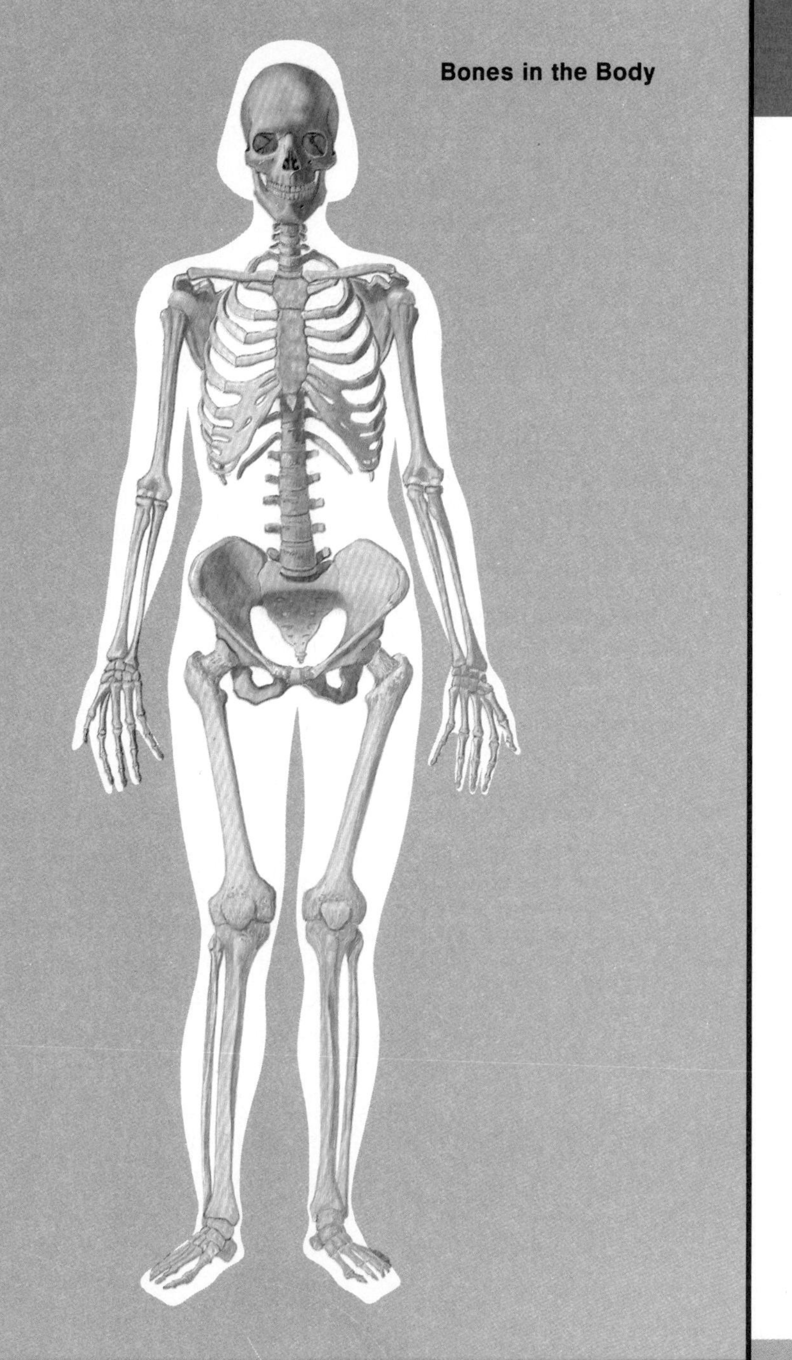

2. Teach

Teaching Tips

- Ask students to imagine what their bodies would be like without bones. (shapeless)
- Darken the room and turn on an overhead projector. Ask a volunteer to place his/her hand on the projector. Have students ***compare*** what they see on the screen to the hand bones on the skeleton illustration. Ask students to ***describe*** the shape of the bones in the fingers. (long, thin, straight)
- **Possible Misconception:** Some students might think that all bones are long and straight. Point out that the bones of the pelvic area, ribs, and shoulder blades are shaped differently.
- Ask students to take off their shoes and wiggle their feet. Questions: **What shape do you think these bones are?** (thin, straight, short) Check students' responses with the skeleton illustration. **Why do you think the bones of the feet are shaped like this?** (Guide students to *infer* that these thin short bones allow the foot to move easily.)
- If possible, obtain some X-ray pictures to show to the class using an overhead projector. Ask students to *identify* the joints and bones in each X ray.

Workbook page 74 *

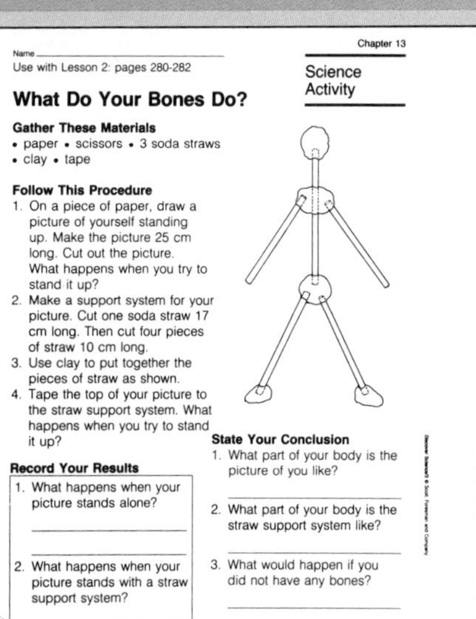
Name ________
Use with Lesson 2: pages 280-282

Chapter 13
Science Activity

What Do Your Bones Do?

Gather These Materials
- paper • scissors • 3 soda straws
- clay • tape

Follow This Procedure
1. On a piece of paper, draw a picture of yourself standing up. Make the picture 25 cm long. Cut out the picture. What happens when you try to stand it up?
2. Make a support system for your picture. Cut one soda straw 17 cm long. Then cut four pieces of straw 10 cm long.
3. Use clay to put together the pieces of straw as shown.
4. Tape the top of your picture to the straw support system. What happens when you try to stand it up?

Record Your Results
1. What happens when your picture stands alone?
2. What happens when your picture stands with a straw support system?

State Your Conclusion
1. What part of your body is the picture of you like?
2. What part of your body is the straw support system like?
3. What would happen if you did not have any bones?

74

**** Answers to masters on pages 274E-274H***

Special Education

If possible, bring in some clean chicken bones for students with learning disabilities to examine. Keep the bones refrigerated until they are used. Allow students to examine the texture and shape of each bone. Encourage them to *infer* the function of each bone. Cut open a bone to expose the marrow. Discuss how some bone marrow produces blood cells. *CAUTION:* Direct students to wash their hands after handling bones.

Reinforcement

Use the skeleton illustration on this page to discuss the different shape and function of bones. Point to different bones such as the rib cage, collarbones, femur, and so on. Have students *identify* the bones used in each action on the illustration and discuss their functions.

TEACHING PLAN

Investigate!

Accept any testable hypothesis. Students' data will support various hypotheses. One possible hypothesis is: *A chicken can move its leg back and forth and around and around.* Students should find out that the joints in animals move in similar ways to those in humans. For example, a chicken's thigh bone is relatively short and meets the pelvic girdle in a ball-and-socket joint.

3. Assess

Lesson Review

1. Bones give the body shape and support, protect organs, and work with muscles to help the body move.
2. Bones come together at joints. They allow the body to move.
3. Challenge! The knee joint, elbow joint, and joints in the fingers and toes allow back and forth movement. **Thinking Skill:** *Making analogies*

Find Out On Your Own

Elbows and knees are fulcrums. The arm and leg bones attached to these fulcrums act as levers. The levers move back and forth to lift objects. **Thinking Skills:** *Communicating, Comparing*

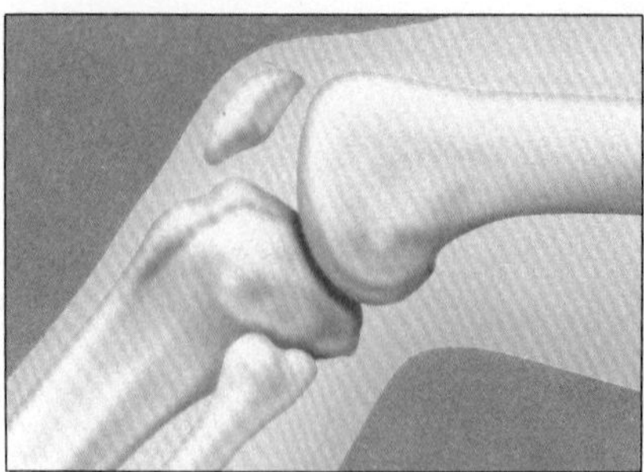
Knee joint

Shoulder joint

INVESTIGATE!

Animals with backbones have joints. Find out how the different joints help them move in different ways. Write a hypothesis and test it with an experiment. You might compare the joints in a chicken's skeleton to those in your body.

joint, a place where bones join together; different joints allow different movement.

Joints

The pictures show places in the body where two bones come together. These places are called **joints.** Move your arm at your shoulder joint. Now move your arm at your elbow joint. Move your fingers. Notice how you can move different parts of your body. Different kinds of joints allow bones to move in different ways.

Lesson Review

1. What are three ways bones help the body?
2. How do joints help the body?
3. **Challenge!** Hinges on a door allow the door to move back and forth. Where is one joint that allows parts of your body to move back and forth?

Study on your own, pages 340–341.

PHYSICAL SCIENCE FIND OUT ON YOUR OWN CONNECTION

Some parts of your body are simple machines. What parts of your arms and legs act like levers? How do the levers move?

282

Teaching Options

Science Anecdote

Damage to joints is the most common type of injury in sports. Diseases, such as arthritis, bursitis, and rheumatism, can also damage joints.

Reteaching Suggestion ♦

Point to different bones on your body such as the clavicle, femur, cervical vertebrae, and so on. Ask students to move each bone identified. Then ask students to *describe* the function of each bone. Finally, have students *infer* how movement would be limited if that bone were broken.

Workbook page 75 *

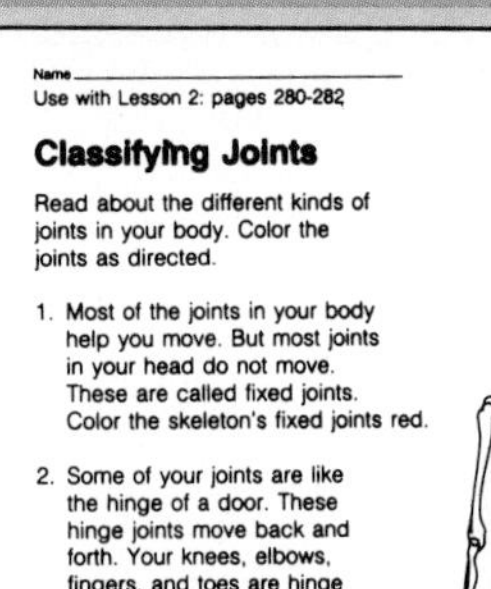

Name ____________________

Use with Lesson 2: pages 280-282

Chapter 13

Science Skills

Classifying Joints

Read about the different kinds of joints in your body. Color the joints as directed.

1. Most of the joints in your body help you move. But most joints in your head do not move. These are called fixed joints. Color the skeleton's fixed joints red.
2. Some of your joints are like the hinge of a door. These hinge joints move back and forth. Your knees, elbows, fingers, and toes are hinge joints. Color the hinge joints green.
3. Shake your head no. You just moved pivot joints in your neck. Your elbows also are pivot joints. They can move from side to side. Put an X on each of these pivot joints.
4. The joints with the most movement are ball and socket. The joints in your hips and shoulders are ball and socket. Color these joints purple.

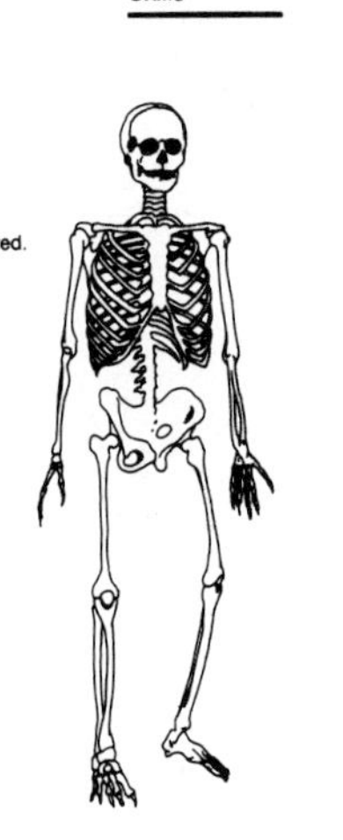

75

♦ ***Suitable as a language development activity***

Observing Cells

Suggested grouping: 2-3 students

Purpose
Observe the cells in an onion skin.

Gather These Materials
• microscope slide • hand lens • onion skin • paper clip • cover slip • black construction paper

Follow This Procedure
1. Use a chart like the one shown to record your observations.
2. Wet the microscope slide.
3. Get a very thin slice of onion skin from your teacher. Lay the onion skin flat on the wet slide.
4. Use the paper clip to flatten out the onion skin.
5. Place the cover slip on the slide over the onion skin. *CAUTION: Be careful handling the glass slide and the cover slip.*
6. Stand near a bright light or window. Hold your hand lens and your slide above a black piece of paper.
7. Look at the onion skin through the hand lens as shown in the picture. Does it look like the onion skin is made of little boxes?

Record Your Results

How the onion skin looks
has many tiny pieces

State Your Conclusion
1. Can you see the pieces that make up an onion skin without a hand lens?
2. Do you see the tiny pieces that make up the onion skin more clearly in the light or dark places of the onion skin?

Use What You Learned
If the onion skin is made of thousands of pieces, what can you infer about the whole onion?

283

Activity

Concept
All living things are composed of cells.

Objectives/Process Skills
- *Observe* cells in onion skin.
- *Record* observations.
- *Describe* the appearance of the onion cells.

Time Allotment
Allow 20 minutes.

Safety Tips (See page T24.)
- Remind students to immediately wipe up any water spilled on the floor to prevent slipping accidents.
- For young students, it is best to use plastic slides and cover slips. This will help prevent breakage and cuts.
- Remind students not to eat any plant parts being studied unless specifically directed to do so.
- Instruct students to use artificial light as the light source for the microscope; directly reflected sunlight could cause eye injuries.
- Direct students to wash their hands thoroughly when finished with this activity, as onion juice may be irritating.

Teaching Tip
- *Elodea* is an aquatic plant, which most tropical fish stores stock. The cells in the leaves are large and can be seen with a good hand lens.

Answers
State Your Conclusion
1. No.
2. light places

Use What You Learned
A whole onion must be made of many more tiny pieces than the onion skin.
Thinking Skill: *Inferring*

Resource Book page 151

Name
Use with Lesson 1: page 283
Observing Cells

Chapter 13
Activity Worksheet

Record Your Results

How the onion skin looks

State Your Conclusion
1. Can you see the pieces that make up an onion skin without a hand lens?
2. Do you see the tiny pieces that make up the onion skin more clearly in the light or dark places of the onion skin?

Use What You Learned
If the onion skin is made of thousands of pieces, what can you infer about the whole onion?

151

Activity Results
The students should be able to see and draw the cells in the onion skin.

*** Answers to masters on pages 274E—274H**

TEACHING PLAN

Lesson Objectives

- *Define* voluntary and involuntary muscles.
- *Explain* how voluntary muscles help the body move.

Lesson Vocabulary

contract, involuntary muscle, tendon, voluntary muscle

1. Motivate

Demonstration Activity ♦

Ask several volunteers to stand in front of the class. Direct each student to demonstrate a different movement such as winking, bending, kicking, sitting down, and reaching. *CAUTION:* Make certain that volunteers do not have physical limitations that would rule out certain body movements.

Discussion

Question: **What part of the body helps these students move?** (muscles, though some students might say bones) Ask the volunteers to repeat their movements. Have the class ***hypothesize*** about the location of the muscles used in each movement.

3 How Are Muscles Important?

LESSON GOALS

You will learn
- that the body has different kinds of muscles.
- how muscles help the body move.

voluntary (vol′ən ter′ē) **muscle,** the kind of muscle a person can control.

involuntary (in vol′ən ter′ē) **muscle,** the kind of muscle that works without a person's control.

These children are playing soccer. They use their muscles to run and kick the ball. Your muscles help you move your body.

Kinds of Muscles

Muscles also help you in another way. Feel the back of your leg and the top of your arm. Muscles give your body its shape.

Look at the picture of the muscles on the next page. You can control many of these muscles. Muscles you can control are called **voluntary muscles.** You use voluntary muscles to turn your head and move your arms.

Other muscles in your body work without you thinking about them. Muscles that work without your control are called **involuntary muscles.** Some kinds of involuntary muscles help you breathe and keep your heart working.

Muscles help people move.

284

Teaching Options

Science Background

There are three types of muscle in the body: voluntary, involuntary, and cardiac. Skeletal muscles, which are voluntary, allow bones to move. They are sometimes called striated muscle because the muscle fiber consists of tiny threads of protein filaments arranged in light and dark stripes. Smooth muscles make up the walls of internal organs such as the stomach and blood vessels. Smooth muscles, which are involuntary, help move food through the digestive tract and allow blood vessels to constrict. The heart is made of cardiac muscle, which allows the heart to beat to its own rhythm. Cardiac muscle consists of protein filaments similar to skeletal muscle. Like smooth muscle cells, it is also involuntary.

Reading Strategies ♦

1. Guide students' pre-reading by asking: What ideas do you think you would find in the lesson to answer the question-title?

2. Assign these strategies: Visualizing Information and Writing a Memory Sentence (See pages T26–T29.)

3. Pair students to share what is clear and unclear and initiate discussion using students' unanswered questions.

♦ ***Suitable as a language development activity***

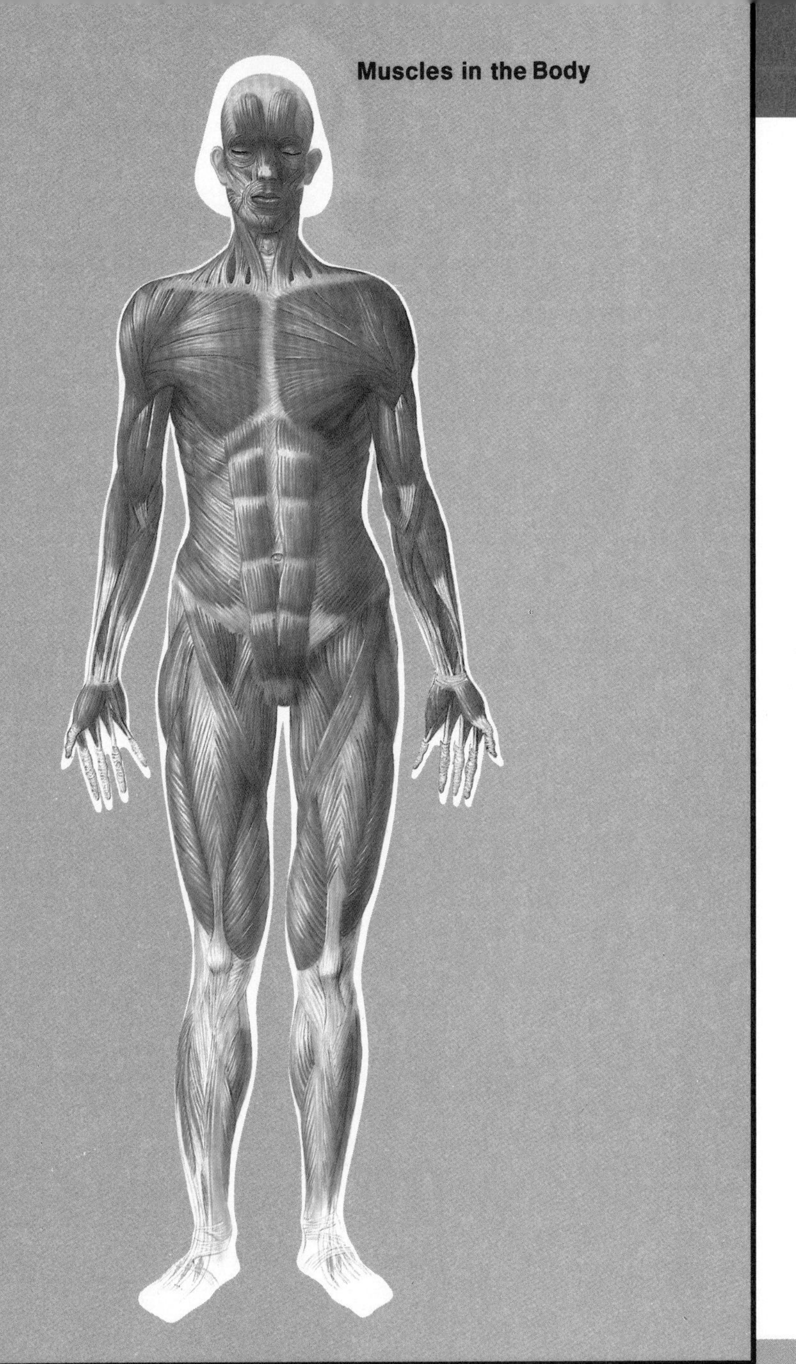

2. Teach

Teaching Tips

- Point out that muscles, like bones, have many shapes and sizes.
- Remind students of the *DISCOVER* on page 275, where they observed how arm muscles help the arm to move. Ask students to ***name*** some muscle movements that they can control. (chewing, walking, smiling, and so on) Have students ***describe*** the voluntary muscles involved in each movement.
- Question: **What type of muscles move food through the digestive tract?** (involuntary muscles) Have students ***name*** other movements controlled by involuntary muscles. (blinking, movement of blood, heart beating, and so on)
- Guide students to ***infer*** that involuntary muscles are important because they keep the body functioning during sleep or other activities and do not require a person's control.
- Point out various muscles shown on the illustration and have students locate and move their corresponding muscles.

Workbook page 76 *

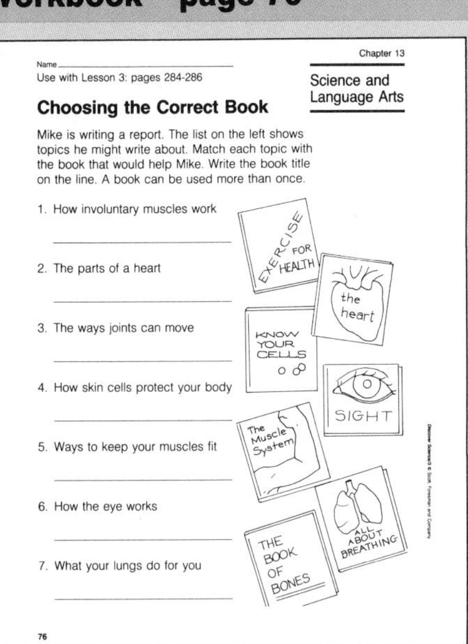
Name ____________

Use with Lesson 3: pages 284-286

Chapter 13

Science and Language Arts

Choosing the Correct Book

Mike is writing a report. The list on the left shows topics he might write about. Match each topic with the book that would help Mike. Write the book title on the line. A book can be used more than once.

1. How involuntary muscles work ____________
2. The parts of a heart ____________
3. The ways joints can move ____________
4. How skin cells protect your body ____________
5. Ways to keep your muscles fit ____________
6. How the eye works ____________
7. What your lungs do for you ____________

76

* Answers to masters on pages 274E-274H

Special Education

Have students with learning disabilities move their arm or leg muscles to feel them contract. Then have them ***observe*** as you stretch and release a rubber band. Have students ***compare*** the rubber band's movement with the contraction and relaxation of their muscles.

Reinforcement

Help students ***identify*** which muscles contract when they raise and lower their upper and lower legs. Instruct students to feel the muscles tighten as they contract. Next, have students feel which leg muscles contract when they raise and lower their feet. Have students ***identify*** these muscles, referring to the illustration on page 285.

TEACHING PLAN

Teaching Tips

• Refer students to the illustration on page 285. Have them *identify* places where tendon tissue is connected to bones. Then ask students to move their corresponding muscles and bones.

• Instruct students to spread their fingers apart. Question: **What are the long cords just under the skin?** (Guide students to *infer* that these are tendons.) Have students open and close one hand while they feel the tendons with their other hand. Ask students to *describe* how the tendons feel. (springlike)

3. Assess

Lesson Review

1. Muscles give the body its rounded shape, keep body organs working, and work with bones to move parts of the body.

2. voluntary, involuntary

3. Challenge! Involuntary, because a person cannot control how these muscles move. **Thinking Skill:** *Inferring*

Find Out On Your Own

Student answers will vary, but should *describe* a slow gentle stretching as opposed to bouncing. **Thinking Skill:** *Restating or explaining ideas*

Muscles change shape.

SCIENCE IN YOUR LIFE

A muscle is made of many muscle cells. When a person exercises the muscle cells get larger.

contract (kən trakt′), the action that occurs when a muscle becomes shorter.

tendon (ten′dən), a tissue that holds a muscle to a bone.

[1] long and thin

How Muscles Work

Muscles change shape when you move. The picture shows how one muscle becomes shorter and thicker, or **contracts,** to lift the hand. How does this muscle look when the hand is down?[1]

Bones and muscles work together. Muscles are connected to bones by tissues called **tendons.** When a muscle contracts, the tendon causes a bone to move.

Lesson Review

1. What do muscles do?
2. What are two kinds of muscles?
3. **Challenge!** What kind of muscle helps move food around in your stomach? Explain your answer.

Study on your own, pages 340–341.

HUMAN BODY
FIND OUT ON YOUR OWN

Proper stretching helps keep muscles healthy. Find a book about exercise for children. Learn a stretching exercise. Write the directions for the exercise on a piece of paper. You might want to teach this exercise to others in your classroom.

286

Teaching Options

Science Anecdote

The average heart muscle of a child contracts and relaxes about 70–80 times per minute, or about 100,800–115,200 times per day.

Reteaching Suggestion ♦

Have students make a mural for the class that illustrates activities controlled by voluntary and involuntary muscles. Label one side of the mural *Voluntary Muscles* and the other side *Involuntary Muscles.* Have students draw or find pictures in magazines that illustrate various activities controlled by each type of muscle. Instruct students to place pictures under the correct label.

Workbook page 77 *

Name ____________
Use with Lesson 3: pages 284-286

Chapter 13
Vocabulary Puzzle

Wordsearch

Unscramble the words from the word bank.
Then find and circle each word in the puzzle below.

Word Bank
contract, joint, system, tissue, involuntary muscle, organ, tendon, voluntary muscle

TONIJ ____________ NEDTON ____________
RANGO ____________ CRANTOCT ____________
STEMYS ____________ SUESIT ____________
NATOLVURY SEMCLU ____________
TANOLVUNIRY CLUSME ____________

T I S S U E J E Y I T J O I N T G U
E K Y B K W G O R G A N L E V R C D
N T S C O N T R A C T F S R P L C B
D B T C P B Q L M H T I Y E J Q M C
O Q E V O L U N T A R Y M U S C L E
N M M L D V A F K T W J B K W G L V
I N V O L U N T A R Y M U S C L E D

77

♦ ***Suitable as a language development activity***

Making a Model of Arm Muscles

Suggested grouping: pairs

Purpose
Observe how muscles work in pairs by *making a model* of upper arm muscles.

Gather These Materials
• cardboard • scissors • paper punch • paper fastener • string • tape

Follow This Procedure
1. Use a chart like the one shown to record your observations.
2. Ask a partner to trace your lower arm and hand on a piece of cardboard. Draw a rectangle about the same size as your upper arm on another piece of cardboard. The rectangle will be used for the upper arm shape. Cut out the shapes.
3. The picture shows how the pieces of cardboard should be joined. Punch holes in the upper and lower arm shapes. Join the two elbow holes with the paper fastener.
4. Thread the strings through the holes. Tie the lower end of each string in a loop. Place a piece of tape over each string.
5. Hold the back edge of the upper arm. Pull up on the string on the inside of the elbow, as shown in the picture.
6. Hold the front edge of the upper arm. Pull on the string on the back side of the elbow.

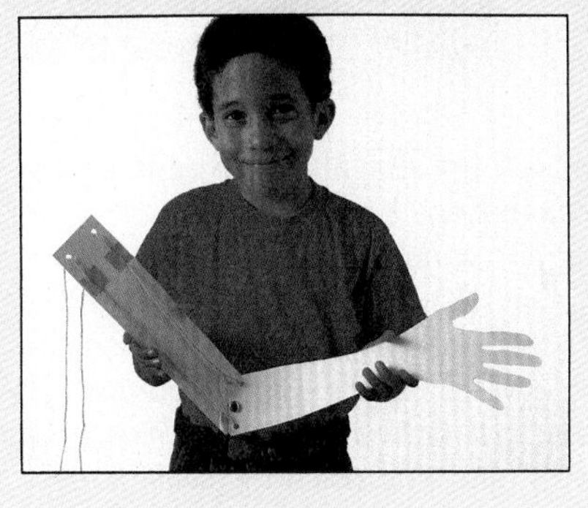

Record Your Results

String pulled	How arm model looks
Back edge	arm straight
Front edge	arm bent

State Your Conclusion
1. What happens when you raise each of the two strings?
2. Compare your arm muscles to the model and explain how muscles work in pairs.

Use What You Learned
How do your arm muscles change in length when you raise and lower your lower arm?

287

Activity

Concept
Arm muscles work in pairs to help move the arm.

Objectives/Process Skills
- *Make a model* of the arm muscles.
- *Observe* how the arm model moves when the strings are pulled.
- *Record* observations.
- *Infer* that muscles work in pairs.

Time Allotment
Allow 30 minutes.

Safety Tips (See page T24.)
- Have students use only round-tip scissors for this activity.
- Remind students to be careful when punching holes in cardboard so as not to injure themselves or others.
- Remind students that paper cuts can result from sliding their hands or fingers along the edge of the cardboard.
- Instruct students not to wave their models about in the air, as this could result in injuries to others.

Teaching Tips
- Make a similar model of the leg and foot to demonstrate how the foot flexes and points.
- Make a string puppet and have students *observe* how pulling the strings makes the parts of the puppet move.

Answers
State Your Conclusion
1. Pulling the string at the inside of the elbow raises the lower arm. Pulling the string at the back side of the elbow lowers the lower arm.
2. The string in front of the model is like the front upper arm muscle. The string at the back of the model is like the back upper arm muscle. The muscles work in pairs to raise and lower the lower arm.

Use What You Learned
When you raise your lower arm, the front upper arm muscle becomes shorter and the back upper arm muscle becomes longer. When you lower the lower arm, the front upper arm muscle becomes longer and the back upper arm muscle becomes shorter. **Thinking Skill:** ***Applying information to new situations***

Resource Book page 153

Name
Use with Lesson 3: page 287
Making a Model of Arm Muscles

Chapter 13
Activity Worksheet

Record Your Results

String pulled	How arm model looks
Back edge	
Front edge	

State Your Conclusion
1. What happens when you raise each of the two strings?

2. Compare your arm muscles to the model and explain how muscles work in pairs.

Use What You Learned
How do your arm muscles change in length when you raise and lower your lower arm?

153

Activity Results

***** ***Answers to masters on pages 274E–274H***

TEACHING PLAN

Purpose
To develop the skills of collecting and organizing information using metric measures and line graphs to solve problems.

1. Motivate

Discussion
Display containers that measure volume in metric units, such as graduated cylinders and measuring cups. Point out the scales on the containers. Question: **What might you measure using this kind of container?** (liquid such as drinks, soap, detergent, and so on)

2. Teach

Teaching Tips
- Point out that the scale on a container runs from the smaller value at the bottom to the larger value at the top.
- **Helpful Hint:** Point out that in this activity, the containers are inverted. The scales are read from the smaller value at the top to the larger value at the bottom.

Using Metric Scales and Line Graphs

Problem: How does the amount of air boys can breathe out change with age?

Part A. Using Metric Scales to Collect Information

1. An involuntary muscle below your lungs helps you breathe. You can meaure how many cc of air you can let out. What does cc stand for?
2. Look at the picture. People breathed into the tubes that go to the upside down jars. Air rises up into the top part of the jar. The metric scale on each jar shows how many cc of air entered the jar. Look at the level of air in the first jar. The scale reads 1100 cc. How much air did the 7-year-old boy breathe out? the 8-year-old?
3. How is the amount of air a 6-year-old breathed out different from that of a 9-year-old?

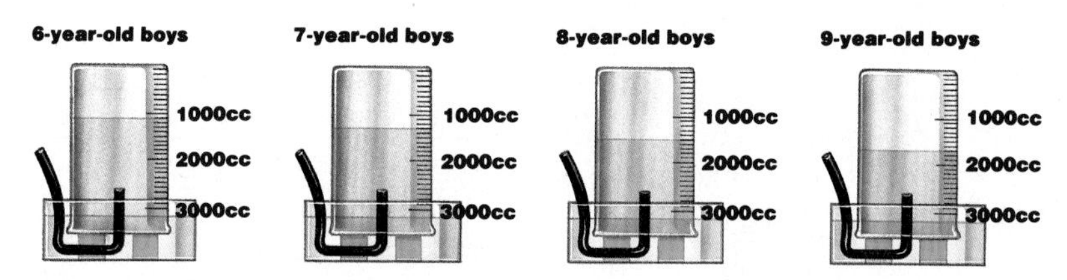

Part B. Using a Line Graph to Organize and Interpret Information

4. The line graph contains the information you collected about how much air boys of different ages can breathe out. What does the scale on the left stand for? the scale at the bottom?
5. Look at the first dot on the graph. Move your finger down until you reach the scale at the

Teaching Options

Sample Line Graph for Part C

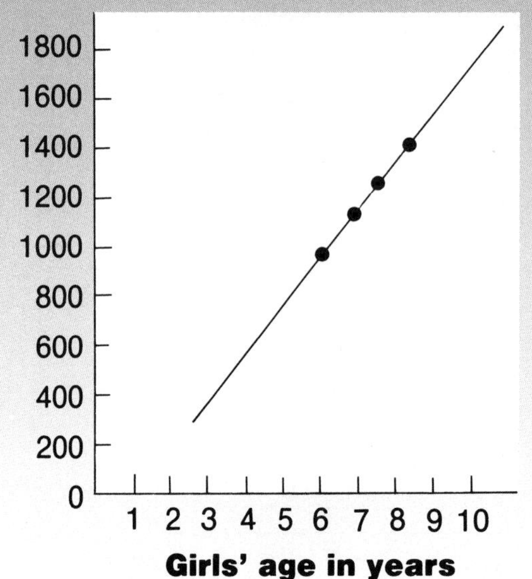

♦ ***Suitable as a language development activity***

bottom of the graph. What does the dot show? Move your finger to the scale at the left. What else does the dot show?

6. Look at the next dot. How much air can 7-year-olds breathe out? 8-year-olds? 9-year-olds?
7. Does the amount of air a boy can breathe out change as he gets older? How much air do you think a 10-year-old boy could breathe out?

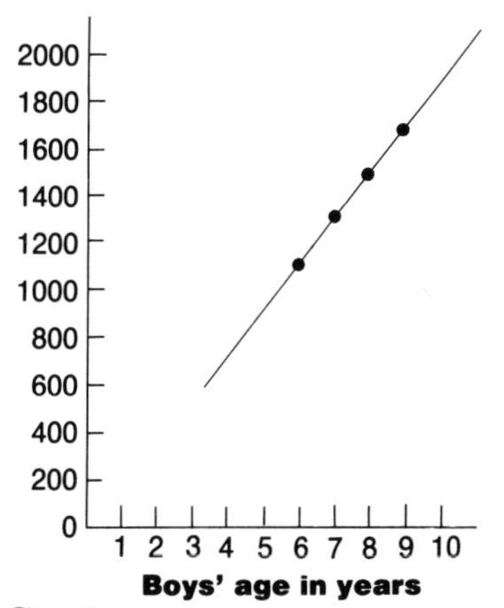

Part C. Using Metric Scales and Line Graphs to Solve a Problem

Problem: How does the amount of air girls can breathe out change as they get older?

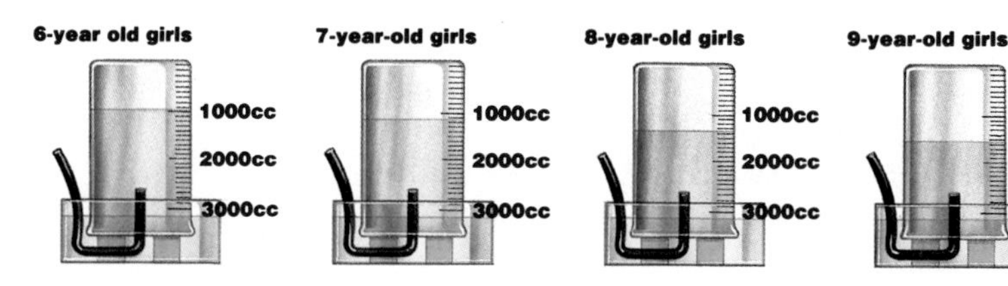

8. Use the picture to collect the information you need to solve the problem. Make a line graph similar to the one shown in Part B to organize your information.
9. Look at your line graph. Does the amount of air girls can breathe out change as they get older? How is the amount of air 6-, 7-, 8-, and 9-year-old girls can breathe out different?
10. Compare your graph with the line graph in Part B. Is the amount of air girls can breathe out different from the amount boys can breathe out? How?
11. You might want to do this experiment and use your own results to make a line graph.

289

3. Assess

Part A

1. cubic centimeters
2. 1,300 cc; 1,500 cc
3. It is 600 cc lower.

Part B

4. the number of cc of air; the boys' age in years
5. information about 6-year-old boys; the number of cc of air they breathed out
6. 1,300 cc; 1,500 cc; 1,700 cc
7. Yes, it increases. A 10-year-old would probably breathe out about 1,900 cc of air.

Part C

8. See Sample Line Graph for Part C in Teaching Options.
9. Yes, it increases. The amount of air breathed out is higher as the girls are older.
10. Yes, boys have greater lung capacities.
11. If students do the activity, have them use 3,000 cc plastic containers. For easier measurement, mark the scale on the side of the jars with a nontoxic waterproof marker. Place the containers in pans large enough to hold the displaced water. *CAUTION:* Immediately wipe up any water that spills on the floor. Place the containers on blocks of wood so that the tubes are not flattened. *CAUTION*: Warn students not to share tubes.

Reteaching Suggestion ♦

Present information on how much air women breathe out by drawing jars with metric scales similar to the ones on page 268. Mark them as follows: 20-year-old women—3100 cc; 25-year-old women: 3500 cc; 33-year-old women: 3300 cc; 35-year-old women: 3200 cc; 40-year-old women: 3000 cc. Have students draw line graphs using this information. Lead students to *conclude* that women between the ages of 20 and 40 expel between 3000 and 3500 cc of air.

Resource Book page 155 *

Name ____________________ Chapter 13

Comprehension: drawing conclusions

Science and Reading

Tell Me Why

Read the story. Then write a sentence to answer each question.

Mike is very interested in how the bones and muscles work in his body. Someday he wants to be a doctor. He is excited because today his class is going on a tour of a hospital operating room.

When the class gets to the hospital, they must put on special uniforms to cover their clothes. Even their shoes must be covered. They also have to wear hats.

Inside the operating room, everything looks shiny and clean. The temperature is kept very cold because the big hot lights over the table make the room very warm. The operating room table moves up and down. It is important for the doctors to be comfortable while they stand and operate.

After visiting the hospital, Mike is even more sure that he wants to be a doctor.

1. Why did the class have to put on special uniforms?

2. Why is the operating room temperature kept cold?

3. Why does the table move up and down?

155

* *Answers to masters on pages 274E–274H*

REVIEW PLAN

Reviewing Science Words
1. involuntary muscles
2. joints
3. organ
4. tissue
5. system
6. voluntary muscles
7. contracts
8. tendon

Reviewing What You Learned
1. d
2. a
3. c
4. b
5. c
6. c
7. a
8. d

Chapter 13 Review

Chapter Main Ideas

Lesson 1 • The body is made up of billions of cells. Many different kinds of cells make up the body. • Groups of cells work together, forming tissues, organs, and systems.

Lesson 2 • Bones support the body and help a person move. Some bones protect organs in the body. • The body has different kinds of bones that come together at places called joints. Different kinds of joints allow the body to move in different ways.

Lesson 3 • The body has muscles that work with bones to help a person move. • A person can control the use of voluntary muscles. Involuntary muscles work without a person's control.

Reviewing Science Words

contracts	organ	tissue
involuntary muscles	system	voluntary muscles
joints	tendon	

Copy each sentence. Fill in the blank with the correct word or words from the list.

1. Muscles that a person cannot control are ___.
2. ___ allow the bones in the knee and the shoulder to move in different ways.
3. A group of tissues that work together form an ___.
4. A group of the same kind of cells form a ___.
5. The stomach works with other organs to form a ___.
6. Muscles a person can control are called ___.
7. A muscle becomes shorter and thicker when it ___.
8. A ___ is a tissue that connects a muscle to a bone.

Reviewing What You Learned

Write the letter of the best answer.

1. The building blocks of the body are
   (a) tissues. (b) systems. (c) organs. (d) cells.

290

Review Options

Cooperative Learning ♦

STAD Format (See page T23.)
Assign students to work in four- to five-member teams to study Chapter 13 Review. Students should work together to make sure that they and their teammates know the material in the chapter. After students have had enough time to study together, give them a test to complete individually (Chapter 13 Test A or B in the *Test Book*). Award Superteam certificates to teams whose average test scores exceed 90%, and Greatteam certificates to teams whose average test scores exceed 80%.

Test Book page 117 *

Name ______________

Chapter 13 Test A

Multiple Choice Choose the best answer.

1. Every body part is made up of
   a. bones.
   b. cells.
   c. muscles.
2. Each kind of cell in the body has its own
   a. muscle.
   b. organ.
   c. job.
3. A group of cells that look alike and do the same job is
   a. a tissue.
   b. an organ.
   c. a system.
4. A group of organs working together to do a job in the body is
   a. a system.
   b. a tissue.
   c. an organism.
5. Bones help a person
   a. see.
   b. move.
   c. think.
6. The body parts that allow bones to move in many different ways are called
   a. joints.
   b. ribs.
   c. tendons.
7. A muscle that works without a person controlling it is called
   a. a joint.
   b. an involuntary muscle.
   c. a voluntary muscle.
8. A person can control head and arm movements with
   a. tendons.
   b. involuntary muscles.
   c. voluntary muscles.
9. Muscles move bones by
   a. relaxing.
   b. connecting.
   c. contracting.
10. Muscles are connected to bones with a tissue called a
    a. joint.
    b. skeleton.
    c. tendon.

117

♦ ***Suitable as a language development activity***

2. A group of bone cells form bone
   (a) tissue. (b) organ. (c) organisms. (d) system.
3. A heart is an example of
   (a) a cell. (b) a tissue.
   (c) an organ. (d) a system.
4. Different organs that work together are called a
   (a) tissue. (b) system. (c) cell. (d) brain.
5. Bones help a person
   (a) think. (b) see. (c) move. (d) feel.
6. An elbow is a
   (a) nerve. (b) muscle. (c) joint. (d) bone.
7. The muscles of the upper arm are
   (a) voluntary muscles. (b) involuntary muscles.
   (c) systems. (d) tendons.
8. When a muscle contracts it becomes
   (a) longer. (b) thinner. (c) softer. (d) shorter.

Interpreting What You Learned

Write a short answer for each question or statement.

1. Put the following groups in the correct order: tissue, systems, cells, organs.
2. What is one way bones protect inside parts of the body?
3. What are two joints in your body?
4. How do voluntary muscles and involuntary muscles differ?
5. Tell what happens to the muscle on the top of the upper arm when a person bends the elbow.

Extending Your Thinking

Write a paragraph to answer each question or statement.

1. What do different sized and shaped bones do to help the body? Give examples.
2. How could a bone injury change the way a person uses his or her muscles?

To explore scientific methods, see Experiment Skills on pages 372–373.

Interpreting What You Learned

1. cells, tissues, organs, systems. **Thinking Skill:** *Sequencing*
2. The rib bones form a cage that protects the heart and lungs; bones of the head protect the brain. **Thinking Skill:** *Restating or explaining ideas*
3. Accept answers that mention any of the body's joints. Students might mention shoulder, elbow, wrist, finger joint, hip, knee, ankle, toe joint. **Thinking Skill:** *Restating or explaining ideas*
4. A person can control the action of a voluntary muscle and a person cannot control the action of an involuntary muscle. **Thinking Skill:** *Contrasting*
5. The upper arm muscle contracts when a person bends his or her elbow. **Thinking Skill:** *Recognizing cause and effect*

Extending Your Thinking

1. Different size and shape of bones are needed to do the different jobs and to move in different ways. **Thinking Skill:** *Inferring*
2. Bones and muscles work together. If a bone cannot be moved or used, a person might not be able to move or use the muscles attached to that bone. **Thinking Skill:** *Inferring*

Test Book page 118 *

Name ____________

Chapter 13
Test A

Short Answer Write the name of each body part in the blank below the picture. Use these words: joint, skeleton, tendon, muscle.

1. ________ 2. ________ 3. ________ 4. ________

Short Essay Use complete sentences to answer each question.

1. Name three ways bones help the body.

2. Name three joints and tell why joints are important.

118

Test Book page 119 *

Name ____________

Chapter 13
Test B

Multiple Choice Choose the best answer.

1. All parts of the body are made up of
   a. bones.
   b. cells.
   c. muscles.
2. Each kind of cell in the body has its own
   a. muscle.
   b. organ.
   c. job.
3. A group of the same kind of tissues forms
   a. an organ.
   b. an organism.
   c. a system.
4. The body parts that help a person stand and move about are the
   a. ribs.
   b. bones.
   c. lungs.
5. The places where bones are joined together are called
   a. joints.
   b. ribs.
   c. tendons.
6. The tissues that move bones in a body are
   a. organs.
   b. skeletons.
   c. muscles.
7. Muscles that a person can control are called
   a. joints.
   b. involuntary muscles.
   c. voluntary muscles.
8. Muscles that help a person breath are
   a. joints.
   b. involuntary muscles.
   c. voluntary muscles.
9. The tissues that connect muscles to bones are the
   a. tendons.
   b. organs.
   c. muscles.
10. When a muscle contracts, it becomes
    a. thinner.
    b. longer.
    c. shorter.

119

Test Book page 120 *

Name ____________

Chapter 13
Test B

Short Answer Each picture shows the use of a muscle. Write whether the muscle being used is a voluntary or involuntary muscle.

1. ________ 2. ________ 3. ________ 4. ________

Short Essay Use complete sentences to answer each question.

1. Name three joints. Why are joints important?

2. List three ways that bones help the body.

120

*** Answers to masters on pages 274E—274H**

Chapter 14 Your Body's Health Needs Planning Guide

TEACHING PLAN

Chapter Components	Skills	Materials
Chapter Opener/*DISCOVER:* Observing How Exercise Changes Breathing pp. 292–293	*DISCOVER* p. 293 Science Process Skills *Observing, Collecting information*	*DISCOVER* p. 293 (groups of 2) 1 large clock with a second hand
Lesson 1 How Can You Stay Healthy? pp. 294–296	Thinking Skills Challenge!: *Suggesting alternatives* Find Out On Your Own: *Organizing information*	Demonstration p. 294 no materials needed
Activity Testing Foods for Sugar p. 297	Science Process Skills *Observing, Classifying, Collecting and interpreting data*	(groups of 2) 6 bottles of sugar test paper, assorted food samples
Lesson 2 What Causes Disease? pp. 298–301	Thinking Skills Challenge!: *Drawing conclusions* Find Out On Your Own: *Collecting information, Communicating*	Demonstration p. 298 rubber cement or glue
Activity Making a Model of Bacteria Growth p. 302	Science Process Skills Making models, Measuring, Collecting and interpreting data, Predicting	(groups of 2) 5 bags popcorn kernels, 15 bottles white glue, 120 file cards
Science and People Becoming a Doctor p. 303	Thinking Skills *Recognizing cause and effect, Ranking ideas and information according to criteria*	
Lesson 3 How Can Alcohol, Tobacco, and Drugs Affect the Body? pp. 304–307	Thinking Skills Challenge!: *Identifying and suggesting alternatives* Find Out On Your Own: *Collecting information, Communicating, Listing*	Demonstration p. 304 adult-sized clothes, ball, waste basket
Skills for Solving Problems Using Diagrams and Bar Graphs pp. 308–309	Problem Solving Skills *Making decisions/Identifying and solving problems, Interpreting charts, maps, and graphs*	
Chapter Review pp. 310–311	Thinking Skills *Comprehending meaning, Summarizing, Restating or explaining ideas, Drawing conclusions, Judging and evaluating, Recognizing relevant information and data, Recognizing cause and effect*	

Teaching Options

Strategies	Extensions		Resource Masters
Cooperative Learning p. 292 (Also see p. T23.) Applying Whole Language p. 293 (Also see p. T30.)			Family Letter: *Resource Book* p. 159
Reading Strategies p. 294 (Also see pp. T26–T29.)	Special Education p. 295 Reteaching Suggestion p. 296		Vocabulary Preview: *Workbook* p. 79 Science Activity: *Workbook* p. 80 Science Skills: *Workbook* p. 81
			Activity Worksheet: *Resource Book* p. 163
Reading Strategies p. 298 (Also see pp. T26–T29.)	Reinforcement pp. 299, 300 Special Education p. 299 Science and Social Studies p. 299	Enrichment p. 300 Game Suggestion p. 301 Reteaching Suggestion p. 301	Science and Social Studies: *Workbook* p. 82
			Activity Worksheet: *Resource Book* p. 165
Reading Strategies p. 304 (Also see pp. T26–T29.)	Reinforcement pp. 305 306 Special Education p. 305 Science and Social Studies p. 305	Enrichment p. 306 Game Suggestion p. 307 Reteaching Suggestion p. 307	Vocabulary Puzzle: *Workbook* p. 83
	Reteaching Suggestion p. 309		Science and Reading: *Resource Book* p. 167
Cooperative Learning p. 310 (Also see p. T23.)			Chapter Tests: Forms A and B *Test Book* pp. 125–128

Classroom Management

Advance Preparation

DISCOVER, page 293
Make sure that a clock or watch with a second hand is visible to all students.

Activity, page 297
Sugar test paper is a commercial product used by diabetics to detect sugar in their urine. It can be purchased from a pharmacy and used to test foods for the presence of sugar. Provide 5–10 pieces of sugar test paper for each group. Bring in samples of different foods for students to test. Suggestions include meats, poultry, vegetables, breakfast cereals, fruits, potatoes, and prepared foods.

Activity, page 302
Put about 300 kernels of popcorn in a sandwich bag for each group.

Demonstration, page 304
Bring adult-sized clothes to class for this demonstration.

Vocabulary Review

Use the following sentences with your students to review the meanings of the italicized words.

1. Bicycle riding, jogging, and running are all forms of *exercise.*
2. Doing something every day, such as brushing your teeth, is a *habit.*
3. Food contains *nutrients*, which are substances your body needs to grow.
4. Colds are caused by *germs*, which are living things too small for you to see.
5. You have an *infection* when you are ill with a disease caused by germs.
6. When you are ill, *medicines* can sometimes help you get better.
7. Medicines are a type of *drug* that people take to help get rid of an illness.
8. Leaves from the *tobacco* plant are used in making cigarettes.

High-Potential Students

Have students design a hand puppet that represents a germ that causes a common disease, such as influenza, colds, chicken pox, measles, or mumps. Provide suitable materials for students to use and ask them to give their puppets a gruesome face and an appropriate name. Have students research the germ their puppet represents and write a story or play about how the germ acts, the disease it causes, and the symptoms the disease produces. Students' stories should include how the disease can be treated, and how it can be prevented. Have students use their puppets to tell their stories to the class.

Mainstreamed Students

Emotionally Handicapped
Have students who are emotionally handicapped keep track of what they eat and drink for 3 days. Help the students decide which of the foods were nutritious. Help students recognize other foods which are high in sugar, fats, or preservatives. Ask these students to help show interested classmates how to keep daily records of what they eat and how to analyze their diet.

Science Fair Projects

The Experiment Skills on p. 374 and the Investigate feature in the chapter can be used for science fair projects. You might also encourage interested students to do one of the following projects:

1. Research different types of communicable diseases and draw pictures of them, labeling whether they are caused by viruses or by bacteria.
2. Find out about diseases that can be caused by use of alcohol or tobacco. Make a display of information about one or more of these diseases. You might want to direct students to research diseases such as lung cancer, cirrhosis of the liver emphysema, and heart disease.

Classroom Resources

Bulletin Board

Encourage students to bring in other pictures that illustrate good disease prevention practices.

DISEASE PREVENTION PRACTICES

DON'T DRINK ALCOHOL

GET PLENTY OF REST

EAT NUTRITIOUS FOODS

DON'T SHARE GERMS

GET PLENTY OF EXERCISE

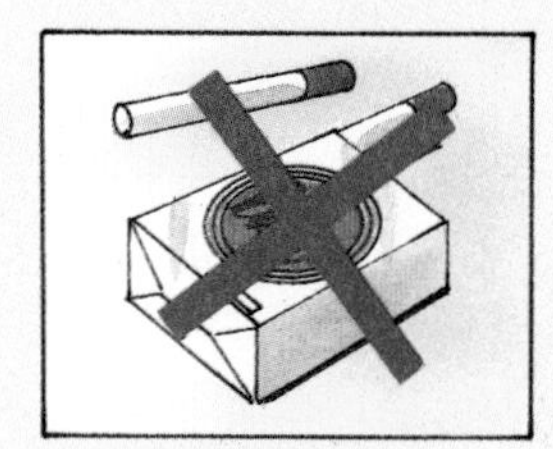

DON'T USE TOBACCO

Chapter 14 Poster

How do you say no to drugs?

Science Discovery Center

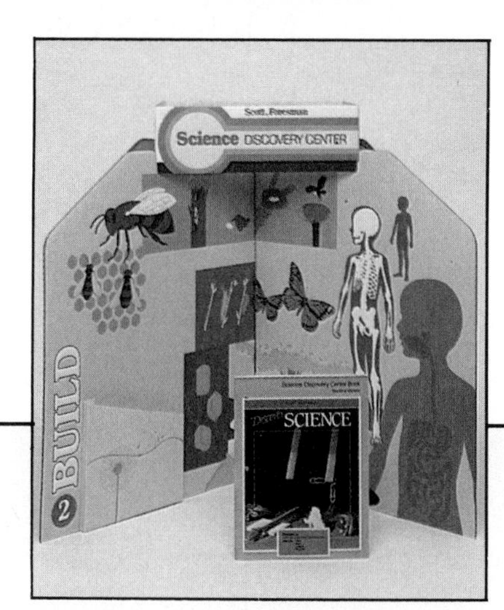

Use pages 145–150 from the *Science Discovery Center Book*. Place these worksheets in the appropriate pockets in the Science Discovery Center.

Overhead Transparencies

Use Transparency 24 from the package of color overhead transparencies.

Teacher's Resource Book

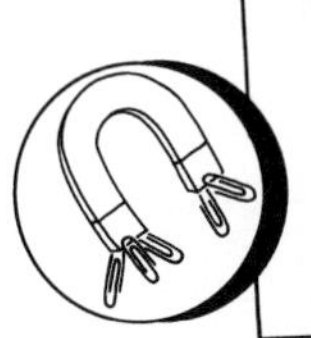

Dear Family,

Your student will be reading **Chapter 14: Your Body's Health Needs** in *Discover Science*, published by Scott, Foresman. We will learn that eating properly, exercising, and getting enough sleep promote good health. Diseases can be caused by bacteria and viruses. Good health habits help protect against disease. We will also discuss the dangers of alcohol, nicotine and smoking, and illegal drugs.

You and your student can do this activity together to learn more about proper diet.

Do You Eat a Good Diet?

1. Help your student keep track of the foods he or she eats for 3 days. Use the chart below. Each day, write in the proper row the name of each food eaten.
2. After three days, discuss these questions:
   - Are you eating enough foods from each group? See the textbook for the proper number of servings.
   - What changes could you make to improve your diet?

Food Groups	Day 1	Day 2	Day 3
Milk			
Meat&Fish&Beans			
Vegetables&Fruit			
Bread & Cereal			
Foods That Did Not Fit in Above Groups			

Workbook

Name ______________________

Use with Lesson 1: pages 294-296

Chapter 14

Vocabulary Preview

Your Body's Health Needs

Chapter Vocabulary

alcohol	diabetes	illegal drug	physical fitness
cancer	disease	nicotine	virus

Vocabulary Cards

1. Write each word on a card.
2. Find each word in the glossary. Copy the pronunication under the word on the card.
3. Practice saying the words with a partner.

Vocabulary Riddles

Use the words from the vocabulary list to answer the riddles. Write the answer on the line.

1. It can help you work and play. Walking, running, or playing sports helps you build it.

   What is it? **physical fitness**

2. It is found in beer, wine, and liquor. It can harm the body.

   What is it? **alcohol**

3. It is a very small germ that can make you ill. It can cause the flu and the chicken pox.

   What is it? **virus**

4. It is in tobacco and can harm a person's heart. People take it in when they smoke cigarettes.

   What is it? **nicotine**

At Home

1. Practice saying the words. Learn their meanings.
2. Look for the words and pictures of the words in newspapers and magazines. You can make a poster.

Teacher's Notes: Have students find the hidden words in the term *illegal drug*. Examples: ill, legal, leg, rug, drill, and real.

Workbook

Name ______________________

Use with Lesson 1: pages 294-296

Chapter 14

Science Activity

What Do Food Labels Tell You?

Gather These Materials

- four food labels from cans, jars, or boxes of food

Follow This Procedure

1. Study the food labels at the right. One label shows a food that has a lot of salt. That food also has few vitamins and little protein.
2. Examine your food labels. On the chart, record some of the information from the labels.

Chips
Serving Size 1 Ounce
Sodium .... 250 mg

PERCENTAGE OF U.S. RECOMMENDED DAILY ALLOWANCES (U.S. RDA)

Protein	2
Vitamin A	2
Vitamin C	0
Iron	2

Peas
Serving Size 3 1/2 Ounces
Sodium .... 90 mg

PERCENTAGE OF U.S. RECOMMENDED DAILY ALLOWANCES (U.S. RDA)

Protein	8
Vitamin A	15
Vitamin C	30
Iron	8

Record Your Results

	Label One	Label Two	Label Three	Label Four
How much salt, or sodium?				
How much protein?				
How much Vitamin A?				
How much Vitamin C?				
How much iron?				

When students answer the questions below, make sure that they compare the same units, for example, grams to grams or percentages to percentages.

State Your Conclusions **Answers will vary depending on labels chosen.**

1. Which foods have much salt? **chips, pretzels**
2. Which foods have many vitamins? **vegetables, fruits**
3. Which foods have much iron? **raisins, spinach**
4. Which foods have much protein? **peanut butter, tuna fish**

Teacher's Notes: Explain that salt is also referred to as *sodium*.

Workbook

Name ______________________

Use with Lesson 1: pages 294-296

Chapter 14

Science Skills

Classifying Foods

These are the foods one student ate in a day. Write the name of each food in its food group below.

Vegetable and Fruit Group	Bread and Cereal Group	Milk and Cheese Group	Meat, Poultry, Fish, and Bean Group
apple	cereal	milk	fish
carrots	toast	yogurt	pork chop
peas	bread	cheese	

Teacher's Notes: Invite students to chart and classify the foods they eat for one day.

Name

Use with Lesson 1: page 297

Chapter 14

Activity Worksheet

Testing Foods for Sugar

Record Your Results

Food	Sugar (yes/no)
1.	
2.	
3.	

State Your Conclusion

1. Which of your foods have sugar? How do you know?

2. Which of your foods do not contain sugar?

Use What You Learned

If you want to eat only snacks that do not have sugar, what foods can you eat?

Name

Use with Lesson 2: pages 298-301

Chapter 14

Science and Social Studies

Dr. Lister Kills Germs

Read the story. Then answer the questions.

Joseph Lister was an English doctor. He lived during the 1800s. During this time, many people who had surgery died. They died because they got infections.

Dr. Lister realized that the infections were caused by germs. He knew that the way to prevent infections was to make the operating room free of germs. So he experimented with many different liquids. He discovered a liquid that killed the germs. Dr. Lister used this liquid to wash down the operating rooms. The liquid was also used to wash the instruments used during surgery. The surgeons washed their hands with the liquid, too.

The use of this germ-killing liquid made a big difference. Few people died of infections after surgery when this liquid was used. Dr. Lister's methods became known as antiseptic surgery. The term antiseptic means "preventing the growth of germs." The liquids used to kill germs are called antiseptics.

1. Why did so many people die from surgery in the 1800s?
   They died from infections.
2. Why did Dr. Lister want to kill germs in operating rooms?
   Germs caused the infections.
3. What did Dr. Lister's methods become known as?
   Antiseptic surgery
4. What is the term for liquids that kill germs?
   Antiseptics

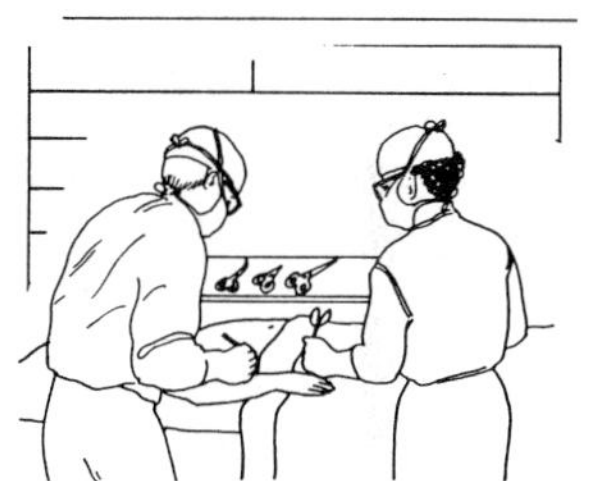

Teacher's Notes: Today, all surgery done in hospitals and doctors' offices is antiseptic surgery.

Name

Use with Lesson 2: page 302

Chapter 14

Activity Worksheet

Making a Model of Bacteria Growth

Record Your Results

Time	Bacteria
12:00	
12:30	
1:00	
1:30	
2:00	
2:30	
3:00	
3:30	

State Your Conclusion

1. How many bacteria did you have at 3:30?

2. How many bacteria would you have at 4:00?

Use What You Learned

How many hours do you think it would take to go from one bacteria cell to a thousand cells?

Name

Use with Lesson 3: pages 304-307

Chapter 14

Vocabulary Puzzle

Crossword

Write the words in the puzzle. Use the word bank and the sentences below.

Word Bank			
alcohol	diabetes	illegal drugs	physical fitness
cancer	disease	nicotine	virus

1 diabetes; 2 illegal drugs; 3 alcohol; 4 physical fitness; 5 disease; 6 nicotine; 7 cancer; 8 virus

Across

1. ______ is a disease in which the body cannot use sugar properly.
2. A person should always say *no* to ______.
3. ______ is a drug found in beer, wine, and liquor.
4. Exercise can help build your ______.
5. A cold is a kind of ______.

Down

6. Tobacco contains a drug called ______.
7. ______ is a disease in which cells that are not normal destroy healthy cells.
8. A ____ is a kind of germ that can cause the flu.

Teacher's Notes: Make a reverse crossword puzzle by copying the puzzle with the answers. Have students write clues.

Name __________________________

Chapter 14

Science and Reading

Comprehension: steps in a process

Making a Tossed Salad

Read the paragraph. Then circle the letter of the sentence that best answers each question.

A tossed salad is a healthy food that is fun to make. There are several steps to follow when making a tossed salad. First, get out a large bowl, the lettuce, and any other vegetables that you want to put in your salad. Then, thoroughly wash all the vegetables. Next, tear the lettuce into small pieces and put it in the bowl. Then, carefully cut up the rest of the vegetables and spread them on top of the lettuce. Finally, right before serving the salad, gently toss it with two spoons.

1. What is the first thing to do when making a salad?
   a. Gently toss the salad with two spoons.
   (b) Get out a large bowl and the vegetables.
   c. Tear the lettuce into small pieces.

2. When do you put the lettuce in the bowl?
   a. after putting in all the rest of the vegetables
   b. before washing the vegetables
   (c) before putting in the rest of the vegetables

3. When do you toss the salad?
   a. right before you cut up the vegetables
   b. right before you put the lettuce in the bowl
   (c) right before the salad is served

Name __________________________

Chapter 14 Test A

Multiple Choice Choose the best answer.

1. In order to have a healthy body, a person should **(1-3)**
   (a) get enough sleep.
   b. eat more food.
   c. never exercise.

2. Eating the right foods from each food group helps the body get enough **(1-1)**
   a. nicotine.
   b. exercise.
   (c) nutrients.

3. Walking, swimming, and riding a bicycle are three **(1-2)**
   a. unhealthy activities.
   b. illegal drugs.
   (c) healthy exercises.

4. Illnesses are often caused by **(2-1)**
   a. exercises.
   b. cold weather.
   (c) germs.

5. Germs that are smaller in size that bacteria are **(2-1)**
   a. diseases.
   (b) viruses.
   c. colds.

6. Two diseases that are not passed from one person to another are **(2-2)**
   (a) cancer and diabetes.
   b. colds and measles.
   c. flu and chicken pox.

7. People who do not smoke have less chance of getting **(2-2)**
   a. skin cancer.
   b. diabetes.
   (c) lung cancer.

8. The heart gets stronger when a person **(1-2)**
   a. reads.
   (b) exercises.
   c. watches television.

9. Tobacco contains a harmful drug called **(3-2)**
   a. nitrate.
   b. nitrogen.
   (c) nicotine.

10. Cocaine is an example of **(3-3)**
    a. a legal drug.
    (b) an illegal drug.
    c. a helpful drug.

Numbers in parentheses after each question refer to the lesson number and the objective of that lesson.

Name __________________________

Chapter 14 Test A

Short Answer Look at the pictures. Draw a square around each food in the bread and cereal group. Draw a line under fruits and vegetables. Draw a circle around meats, and draw a triangle around milk products. Put an X through foods that do not belong to any of the four good groups. **(1-1)**

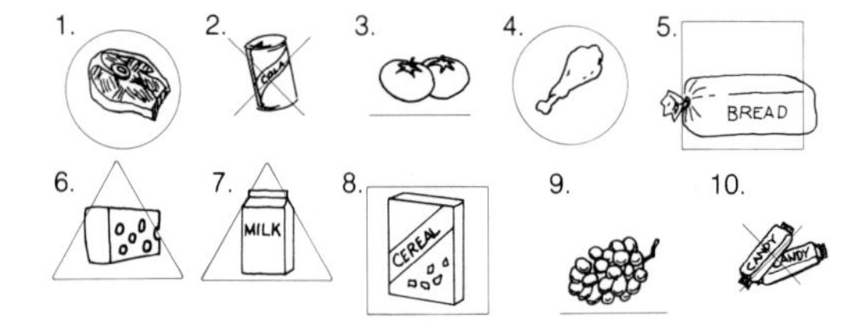

Short Essay Use complete sentences to answer each question.

1. Name three good health habits. **(1-1)**

   Eat foods from each of the food groups.

   Get plenty of rest.

   Exercise regularly.

2. Name two harmful drugs and tell why they are harmful. **(3-1)**

   Tobacco can be harmful because its use can cause lung cancer.

   Too much alcohol can cause a person to have trouble thinking, walking, and talking.

Numbers in parentheses after each question refer to the lesson number and the objective of that lesson.

Name __________________________

Chapter 14 Test B

Multiple Choice Choose the best answer.

1. One way to help your body stay healthy is to **(1-3)**
   a. eat candy.
   b. stay in the house.
   (c) get enough sleep.

2. The ability to work, play, and exercise without getting tired or injuring yourself too easily is called **(1-2)**
   (a) physical fitness.
   b. a virus.
   c. sports.

3. In order to stay healthy, a person should not eat too many **(1-1)**
   (a) sweets.
   b. vegetables.
   c. breads.

4. Colds and some other diseases are caused by **(2-1)**
   a. people.
   (b) germs.
   c. measles.

5. Flu, chicken pox, and measles are some of the diseases that are caused by **(2-1)**
   a. coughs.
   (b) viruses.
   c. diseases.

6. Diseases such as cancer and diabetes are caused by **(2-2)**
   a. germs.
   (b) changes inside the body.
   c. viruses.

7. One way to help reduce the chance of getting lung cancer is to **(2-2)**
   (a) not smoke.
   b. not eat candy.
   c. not exercise.

8. One disease that is not spread from one person to another person is **(2-2)**
   (a) heart disease.
   b. measles.
   c. chicken pox.

9. Alcohol and tobacco are two types of **(3-1)**
   a. medicines.
   (b) drugs.
   c. nutrients.

10. One example of an illegal drug is **(3-3)**
    (a) cocaine.
    b. aspirin.
    c. tobacco.

Numbers in parentheses after each question refer to the lesson number and the objective of that lesson.

Test Book

Name ______________________

Chapter 14 Test B

Short Answer Label the basic four food groups.

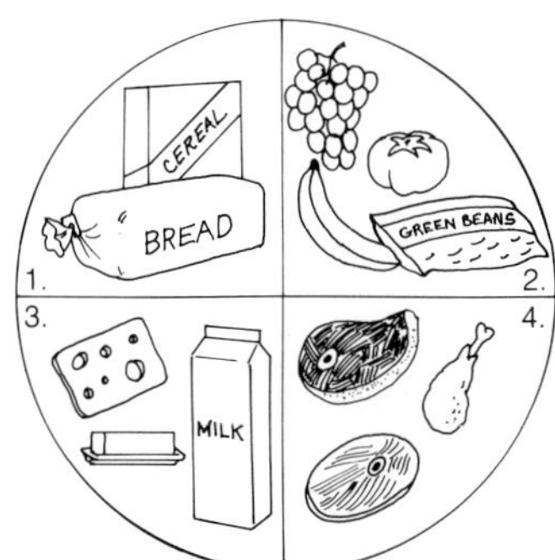

1. **breads and cereals**
2. **fruits and vegetables**
3. **milk products**
4. **meats**

Short Essay Use complete sentences to answer each question.

1. Why should a person never drink alcohol and drive? **(3-1)**

   People who drink alcohol do not think or act quickly and may go to sleep. They can have an accident if they drive.

2. List three good health habits. **(1-1)**

   Eat foods from each of the four food groups each day.

   Get plenty of rest.

   Exercise regularly.

Numbers in parentheses after each question refer to the lesson number and the objective of that lesson.

Test Book

Name ______________________

Unit Test 4

Multiple Choice Choose the best answer.

1. Every body part is made up of **(13-1-1)**
   a. bones.
   ⓑ cells.
   c. muscles.

2. A group of organs working together to do a job is **(13-1-3)**
   ⓐ a system.
   b. a tissue.
   c. an organism.

3. The places where bones join together are called **(13-2-2)**
   ⓐ joints.
   b. ribs.
   c. tendons.

4. Bones help a person **(13-2-1)**
   a. see.
   ⓑ move.
   c. think.

5. Muscles are connected to bones with a tissue called a
   a. joint.
   b. skeleton.
   ⓒ tendon. **(13-3-3)**

6. All the bones of the body make up the **(13-2-1)**
   a. system.
   ⓑ skeleton.
   c. ribs.

7. A muscle that works without a person controlling it is
   a. a joint.
   ⓑ an involuntary muscle.
   c. a voluntary muscle. **(13-3-3)**

8. Muscles that a person can control are called **(13-3-2)**
   a. joints.
   b. involuntary muscles.
   ⓒ voluntary muscles.

9. Muscles move bones by
   a. relaxing.
   b. connecting.
   ⓒ contracting. **(13-3-3)**

10. The tissues that move bones in a body are **(13-3-1)**
    a. organs.
    b. skeletons.
    ⓒ muscles.

1. ⓐ ● ⓒ **4.** ⓐ ● ⓒ **7.** ⓐ ● ⓒ **10.** ⓐ ⓑ ●
2. ● ⓑ ⓒ **5.** ⓐ ⓑ ● **8.** ⓐ ⓑ ●
3. ● ⓑ ⓒ **6.** ⓐ ● ⓒ **9.** ⓐ ⓑ ●

Numbers in parentheses after each question refer to the chapter and lesson numbers and the objective of that lesson.

Test Book

Name ______________________

Unit Test 4

Multiple Choice Choose the best answer.

Use the picture to answer questions 11 and 12.

11. Which food group includes vegetables? **(14-1-1)**
    ⓐ 2 b. 3 c. 4

12. Milk products are shown in
    a. 1. b. 2. ⓒ 3. **(14-1-1)**

13. The heart gets stronger when a person **(14-1-2)**
    a. reads.
    ⓑ exercises.
    c. watches television.

14. Tobacco contains a harmful drug called **(14-3-2)**
    a. nitrate.
    b. nitrogen.
    ⓒ nicotine.

15. Germs that are smaller in size than bacteria are **(14-2-1)**
    a. diseases.
    ⓑ viruses.
    c. colds.

16. One disease that is not spread from person to person is
    ⓐ heart disease.
    b. measles.
    c. chicken pox. **(14-2-3)**

17. People who do not smoke have less chance of getting **(14-2-2)**
    a. skin cancer.
    b. diabetes.
    ⓒ lung cancer.

18. Cocaine is an example of
    a. a legal drug.
    ⓑ an illegal drug.
    c. a helpful drug. **(14-3-3)**

19. One way to help your body grow stronger is to **(14-1-2)**
    a. eat candy.
    b. sleep late each morning.
    ⓒ get enough exercise.

11. ● ⓑ ⓒ **14.** ⓐ ⓑ ● **17.** ⓐ ⓑ ●
12. ⓐ ⓑ ● **15.** ⓐ ● ⓒ **18.** ⓐ ● ⓒ
13. ⓐ ● ⓒ **16.** ● ⓑ ⓒ **19.** ⓐ ⓑ ●

Numbers in parentheses after each question refer to the chapter and lesson numbers and the objective of that lesson.

TEACHING PLAN

Major Concepts
Lesson 1 Good nutrition, exercise, and adequate sleep can help the body stay healthy.
Lesson 2 Communicable and noncommunicable diseases harm the body; many of these diseases can be prevented, treated, or cured.
Lesson 3 Drugs which are misused and abused can have harmful effects on the body.

Chapter Vocabulary
alcohol, cancer, diabetes, disease, illegal drug, nicotine, physical fitness, virus

Getting Started
Ask students why bicycle riding helps the family in the picture enjoy good health. (because exercise strengthens the muscles, reduces stress, and helps the heart work more efficiently) Ask students to name activities that help keep people healthy (sports, dancing, walking). Ask students what kinds of exercise they themselves enjoy.

Chapter 14

Your Body's Health Needs

All of the members of this family enjoy good health. They know that good health helps them work and play better.

292

Teaching Options

Cooperative Learning ♦

Jigsaw Format (See page T23.)
Assign the following topics at random to your cooperative learning teams.
Topic A: What can a person do to help him/herself stay healthy?
Topic B: What effects does exercise have on the body, and how do these effects promote health?
Topic C: What factors can cause disease, and how can illness be prevented?
Topic D: What substances may harm the body, and what are their effects?
Have students search for information on their topic as they read the chapter. Then let all students with the same topic meet in an expert group to discuss the information. When students return to their teams, they may take turns presenting their topics to the team. Then give students a test covering all topics to complete individually (Chapter 14 Test A or B in the *Test Book*). Award Superteam certificates to teams whose average test scores exceed 90%, and Greatteam certificates to teams whose average test scores exceed 80%.

♦ ***Suitable as a language development activity***

Introducing the Chapter

Exercise is a good health habit that can help all the parts of your body. The activity below will help you learn about exercising. In this chapter, you will learn how you can help protect yourself from certain diseases. Following good health habits and saying *no* to drugs also can help keep you healthy.

Observing How Exercise Changes Breathing

Find out what happens to your breathing when you exercise. First, sit quietly in a chair. Use a clock or watch with a second hand to count how many times you breathe in one minute. Write this number on a sheet of paper.

Now run in place about twenty-five times. When you stop running, count how many times you breathe in one minute. Write this number on the sheet of paper. Compare the two numbers on your paper.

Talk About It

1. Did you breathe faster before or after you exercised?
2. What other way did your breathing change when you exercised?

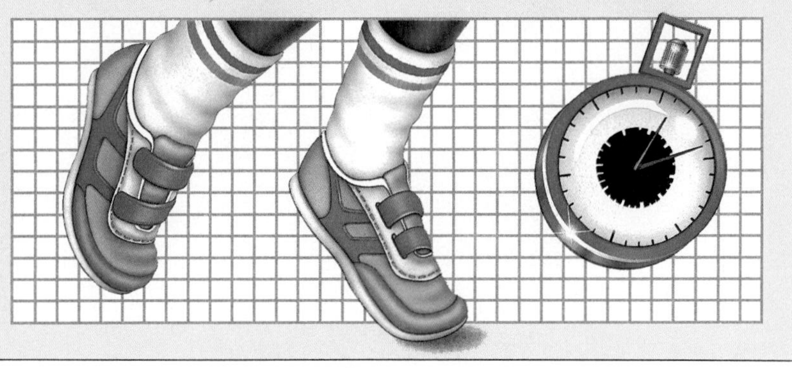

293

DISCOVER

Objective ♦

This optional *DISCOVER* activity will help students explore and build background information about the concept that exercise affects respiration. Later in the chapter, students will be able to draw on this experience to help them assimilate the new content.

Science Process Skills

Observing, Collecting information

Materials

For the class: 1 large clock with a second hand

Safety Tips (See page T24.)

- Divide class into pairs for this activity.
- Prior to this activity, make certain that no student has physical restrictions that would require a doctor's approval for running in place.

Teaching Tip

- Question: **When you're playing hard, what changes happen in your body that let you know it is time to rest?** (being short of breath, becoming tired, feeling hot, and so on)

Answers

Talk About It

1. after exercise
2. breathed more deeply

Applying Whole Language ♦

Discuss the whole language framework with each Teaching Option you select. Here is an example applied to the Reinforcement option on p. 306.

1. Purpose: To make posters reinforcing the theme, "Say No to Drugs"
2. Context: Variety of situations, reasons, and dialog to show different coping strategies. Small group project; whole group sharing. School display.
3. Decisions: Poster ideas. Situations. Characters. Dialog.
4. Evaluation: How did these posters help us? others? (See p. T30.)

Resource Book page 159

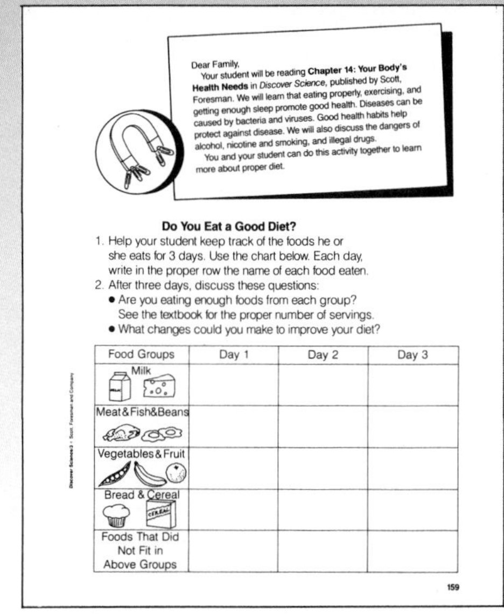

Dear Family,
Your student will be reading **Chapter 14: Your Body's Health Needs** in *Discover Science*, published by Scott, Foresman. We will learn that eating properly, exercising, and getting enough sleep promote good health. Diseases can be caused by bacteria and viruses. Good health habits help protect against disease. We will also discuss the dangers of alcohol, nicotine and smoking, and illegal drugs.
You and your student can do this activity together to learn more about proper diet.

Do You Eat a Good Diet?

1. Help your student keep track of the foods he or she eats for 3 days. Use the chart below. Each day, write in the proper row the name of each food eaten.
2. After three days, discuss these questions:
   - Are you eating enough foods from each group? See the textbook for the proper number of servings.
   - What changes could you make to improve your diet?

Food Groups	Day 1	Day 2	Day 3
Milk			
Meat & Fish & Beans			
Vegetables & Fruit			
Bread & Cereal			
Foods That Did Not Fit in Above Groups			

159

Science Background

Exercise helps the respiratory system become more efficient. People who exercise regularly breathe more slowly and deeply, taking in more oxygen and giving off more carbon dioxide with each breath. Exercising to build cardiovascular fitness can help a person have a healthy heart, and healthy lungs and blood vessels.

LESSON 1 pages 294–296

TEACHING PLAN

Lesson Objectives

- *Describe* how proper nutrition helps the body stay healthy.
- *Describe* how exercise helps maintain good health.
- *Explain* why adequate sleep is important in maintaining good health.

Lesson Vocabulary

physical fitness

1. Motivate

Demonstration Activity ♦

List the following foods on the chalkboard: taco (ground beef, lettuce, tomato, cheese, corn shell); pizza (tomato sauce, cheese, green peppers, whole wheat crust); and chili (beef, tomato sauce, beans, green pepper, onions).

Discussion

Questions: **Which of the foods are nutritious? Which are not nutritious?** (Allow students to respond.) Guide students to understand that all of the foods are nutritious because they contain nutrients that help keep the body healthy.

LESSON GOALS

You will learn
- that eating properly can help you stay healthy.
- that exercising can help keep your body healthy.
- that getting enough sleep is important for your health.

1 How Can You Stay Healthy?

Eating, exercising, and getting enough sleep are good health habits. Following these habits can help keep your body working well.

Eating Properly

How can you plan healthy meals? You can follow the four food group plan shown in the picture. You need to eat foods from each of these food groups. This plan tells you how much of each kind of food to eat each day.

You learned in Chapter 9 that plants and animals need nutrients. Foods contain nutrients that your body uses to grow and stay healthy. Eating the right amounts of foods from each food group helps you get all the nutrients you need.

Vegetable-fruit group
Four servings

Bread-cereal group
Four servings

294

Teaching Options

Science Background

The body uses six different types of nutrients—carbohydrates, proteins, fats, vitamins, minerals, and water—to maintain good health. The first three of these nutrients provide the body with energy. They also are used, along with some minerals, for the growth and repair of body tissues. Proteins, vitamins, fats, and some minerals help regulate body processes.

Exercise also helps the body maintain good health. It improves cardiovascular fitness, reduces stress, increases strength, and increases stamina. It also helps the digestive system perform more efficiently by increasing food digestion and absorption.

Reading Strategies ♦

1. Guide students' pre-reading by asking: What ideas do you think you would find in the lesson to answer the question-title?
2. Assign these strategies: Writing a Memory Sentence for each subheading and Mapping Examples of each subheading (See pages T26–T29.)
3. Pair students to share what information is clear and unclear and initiate discussion using students' unanswered questions.

♦ *Suitable as a language development activity*

Exercising Properly

What kind of activities do you enjoy? Exercising can help build **physical fitness.** Physical fitness helps you work and play without getting tired or hurt easily. Playing sports and active games can help build physical fitness. Walking, dancing, or riding a bicycle also can help you keep fit.

Your heart is a muscle. Your heart beats faster during exercise. When you exercise properly, you help your heart get stronger. When your heart is strong, you can work and play for a long time without getting tired.

Exercise helps all the parts of your body work well. Exercise also helps you sleep well. Getting enough exercise helps you look and feel your best. How can playing every day give you the exercise you need?[1]

INVESTIGATE!

Find out if different kinds of exercises affect how fast your heart beats. Write a hypothesis and test it with an experiment. You might count your heartbeats for one minute after walking. Then run in place and count your heartbeats again for a minute.

physical (fiz′ə kəl) **fitness,** the ability to work, play, and exercise without getting tired or injured easily.

[1]by playing active games that exercise the muscles.

Meat-poultry-fish-bean group
Two servings

Milk-cheese group
Three servings

2. Teach

Teaching Tips

- Ask students to *list* foods they eat for snacks, and to *classify* the foods according to the four food groups. If the snacks cannot be placed into a food group, ask students to *list* some healthy snack alternatives.
- When discussing the benefits of exercise, refer students to the *DISCOVER* on page 293, in which they experienced the effects of exercise on breathing, and have them ***describe*** what they learned from the activity.
- **Possible Misconception:** Students might think that if a little of something (exercise, a certain food, sleep) is good for them, then more of that thing is even better for them. Help students understand that excessive amounts can be as unhealthy as inadequate amounts.

Investigate!

Accept any testable hypothesis. Students' data will support various hypotheses. One possible hypothesis is: *Some exercises make the heart beat faster.* Students should find out that some exercises cause a considerable increase in heart rate. For example, running in place for one minute causes the heart to beat faster than walking for one minute.

Workbook page 79 *

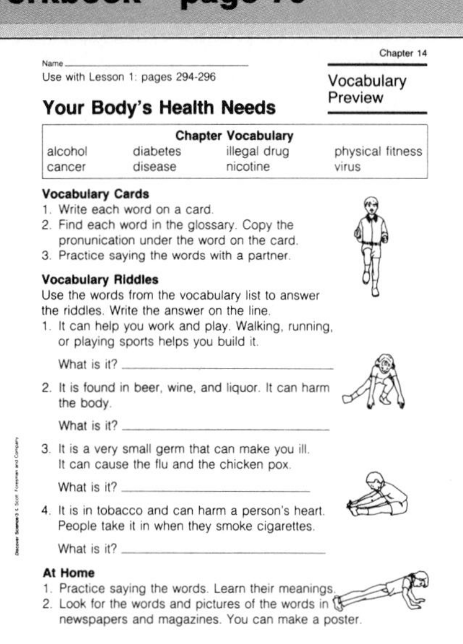

Name ____________
Use with Lesson 1: pages 294-296

Chapter 14

Vocabulary Preview

Your Body's Health Needs

Chapter Vocabulary

alcohol	diabetes	illegal drug	physical fitness
cancer	disease	nicotine	virus

Vocabulary Cards

1. Write each word on a card.
2. Find each word in the glossary. Copy the pronunciation under the word on the card.
3. Practice saying the words with a partner.

Vocabulary Riddles

Use the words from the vocabulary list to answer the riddles. Write the answer on the line.

1. It can help you work and play. Walking, running, or playing sports helps you build it.
   What is it? ____________
2. It is found in beer, wine, and liquor. It can harm the body.
   What is it? ____________
3. It is a very small germ that can make you ill. It can cause the flu and the chicken pox.
   What is it? ____________
4. It is in tobacco and can harm a person's heart. People take it in when they smoke cigarettes.
   What is it? ____________

At Home

1. Practice saying the words. Learn their meanings.
2. Look for the words and pictures of the words in newspapers and magazines. You can make a poster.

79

Workbook page 80 *

Name ____________
Use with Lesson 1: pages 294-296

Chapter 14

Science Activity

What Do Food Labels Tell You?

Gather These Materials

- four food labels from cans, jars, or boxes of food

Follow This Procedure

1. Study the food labels at the right. One label shows a food that has a lot of salt. That food also has few vitamins and little protein.
2. Examine your food labels. On the chart, record some of the information from the labels.

Chips
Serving Size 1 Ounce
Sodium .... 250 mg
PERCENTAGE OF U.S. RECOMMENDED DAILY ALLOWANCES (U.S. RDA)
Protein 2
Vitamin A 2
Vitamin C 0
Iron 2

Peas
Serving Size 3 1/2 Ounces
Sodium .... 90 mg
PERCENTAGE OF U.S. RECOMMENDED DAILY ALLOWANCES (U.S. RDA)
Protein 8
Vitamin A 15
Vitamin C 30
Iron 8

Record Your Results

	Label One	Label Two	Label Three	Label Four
How much salt, or sodium?				
How much protein?				
How much Vitamin A?				
How much Vitamin C?				
How much iron?				

State Your Conclusions

1. Which foods have much salt? ____________
2. Which foods have many vitamins? ____________
3. Which foods have much iron? ____________
4. Which foods have much protein? ____________

80

Special Education

Have students with learning disabilities draw pictures of their favorite games or sports. Ask the students to ***explain*** to the class how the activity can help people get exercise. Display students' pictures or make a class book of the different activities students like to participate in on a regular basis.

* Answers to masters on pages 292E-292H

TEACHING PLAN

3. Assess

Lesson Review

1. You can plan healthy meals by including the proper number of servings from each of the four food groups.
2. Exercise helps build a strong heart and strong muscles, helps the body work better, and helps a person look and feel better.
3. Sleep gives the body a chance to rest. It also helps the body grow and make new cells.
4. Challenge! Students might suggest playing a type of sports or an active game. **Thinking Skill:** *Suggesting alternatives*

Find Out On Your Own

Student posters and explanations will vary. **Thinking Skill:** *Organizing information*

This girl works better when she gets enough sleep.

Getting Enough Sleep

Your body needs time to rest after a busy day. You need sleep to grow properly. Your body makes and repairs cells while you sleep. Sleep also helps give you energy for the next day. This girl knows she can work, play, and learn better when she gets enough sleep.

Lesson Review

1. How can you plan healthy meals for a day?
2. How does exercise help a person stay healthy?
3. How does sleep help the body?
4. **Challenge!** What are some ways to make exercise fun?

Study on your own, pages 342–343.

HUMAN BODY
FIND OUT ON YOUR OWN

Cut out pictures from magazines to make a poster about healthy meals. Show healthy meals and snacks for a day. The pictures of the four food groups will give you some ideas for foods to choose. Write a few sentences telling about your poster.

296

Teaching Options

Science Anecdote

Muscles that are not used atrophy, or waste away. The word *atrophy* is from a Greek word that means "ill fed." Exercise does not produce more muscles, it increases the size of muscle cells.

Reteaching Suggestion ♦

Have students pretend that they have a friend who is out of shape, feeling tired and weak, and acting irritably. Ask the students to write a letter to the friend, suggesting some possible causes of the friend's problems. Instruct the students to *explain* types of changes in lifestyle that might help the friend feel and look better, have more energy, and gain strength.

Workbook page 81 *

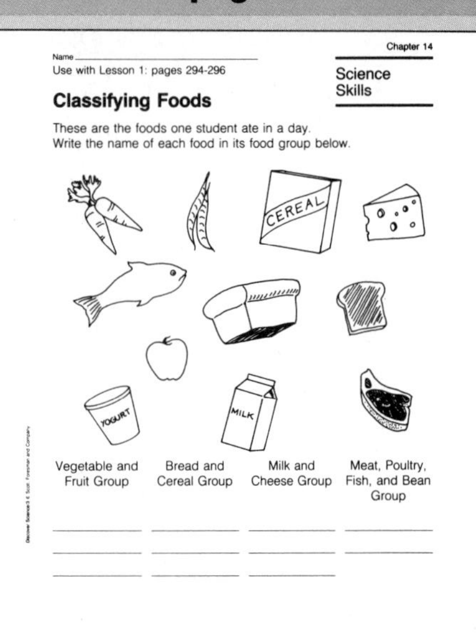

Name ______

Use with Lesson 1: pages 294-296

Chapter 14

Science Skills

Classifying Foods

These are the foods one student ate in a day.
Write the name of each food in its food group below.

Vegetable and Fruit Group	Bread and Cereal Group	Milk and Cheese Group	Meat, Poultry, Fish, and Bean Group

81

♦ *Suitable as a language development activity*

Testing Foods for Sugar

Suggested grouping: 2-4 students

Purpose
Classify different types of foods according to the presence or absence of sugar.

Gather These Materials
• sugar test paper • several kinds of food

Follow This Procedure
1. Use a chart like the one shown to record your observations.
2. Touch the sugar test paper against each of your foods as shown in the picture. The paper will change color when it touches foods that have sugar. Record what you see.

Record Your Results

Food	Sugar (yes/no)
1.	
2. Answers will vary according to food.	
3.	

State Your Conclusion
1. Which of your foods have sugar? How do you know?
2. Which of your foods do not contain sugar?

Use What You Learned
If you want to eat only snacks that do not have sugar, what foods can you eat?

297

Activity

Concept
A variety of foods contain sugar.

Objectives/Process Skills
- *Test* various foods for sugar content.
- *Observe* changes in the test paper.
- *Record* observations.
- *Classify* foods according to sugar content.

Time Allotment
Allow 30 minutes.

Safety Tips (See page T24.)
- Instruct students not to put the test paper into their mouths. Remind them never to place any materials used in science activities into their mouths unless specifically directed to do so by the teacher.
- Remind students to immediately wipe up any liquids or foods that spill on the floor to prevent slipping accidents.

Teaching Tip
- All packaged foods have their ingredients listed. Have students look at the ingredients found in breakfast cereals.

Answers
State Your Conclusion
1. Answers will vary. Answers might include fruits or cereals. Foods with sugar cause the test tape to change color.
2. Answers will vary. Answers might include meat and poultry.

Use What You Learned
Answers might include meat, poultry, and some dairy products. **Thinking Skill:** *Inferring*

Resource Book page 163

Name ______
Use with Lesson 1: page 297
Chapter 14
Activity Worksheet

Testing Foods for Sugar

Record Your Results

Food	Sugar (yes/no)
1.	
2.	
3.	

State Your Conclusion
1. Which of your foods have sugar? How do you know?

2. Which of your foods do not contain sugar?

Use What You Learned
If you want to eat only snacks that do not have sugar, what foods can you eat?

163

Activity Results

See chart overprint above.

* *Answers to masters on pages 292E-292H*

LESSON 2 pages 298–301

TEACHING PLAN

Lesson Objectives
- *Explain* how communicable diseases are spread.
- *Name* some noncommunicable diseases.

Lesson Vocabulary
cancer, diabetes, disease, virus

1. Motivate

Demonstration Activity ♦
Place a sticky substance (rubber cement or glue) on your hands. Touch several washable objects in the room. Explain that the glue on your hands represents germs that cause disease.

Discussion
Ask a student to retrace your steps and touch the things you touched. Question: **What did you notice?** (The student should notice that the things you touched are sticky.) Explain that some germs are transferred from people's hands to objects in the same way that the glue was transferred from your hands.

2 What Causes Disease?

LESSON GOALS

You will learn
- how some diseases can spread from person to person.
- about diseases that do not spread from person to person.

disease (də zēz′), an illness.

Think about a time when you felt sick. Most people become sick from time to time. In many cases, you can help keep yourself well by following good health habits.

Diseases That Spread

The girl in the picture has a cold. Her nose is runny and her eyes are watery. A cold is a **disease,** or illness, that can spread from person to person.

Germs cause colds and can cause other diseases that spread. Germs are too small for you to see. They are found everywhere. They are in air and water. They are in food and on other objects. You have germs inside you and on your skin all the time.

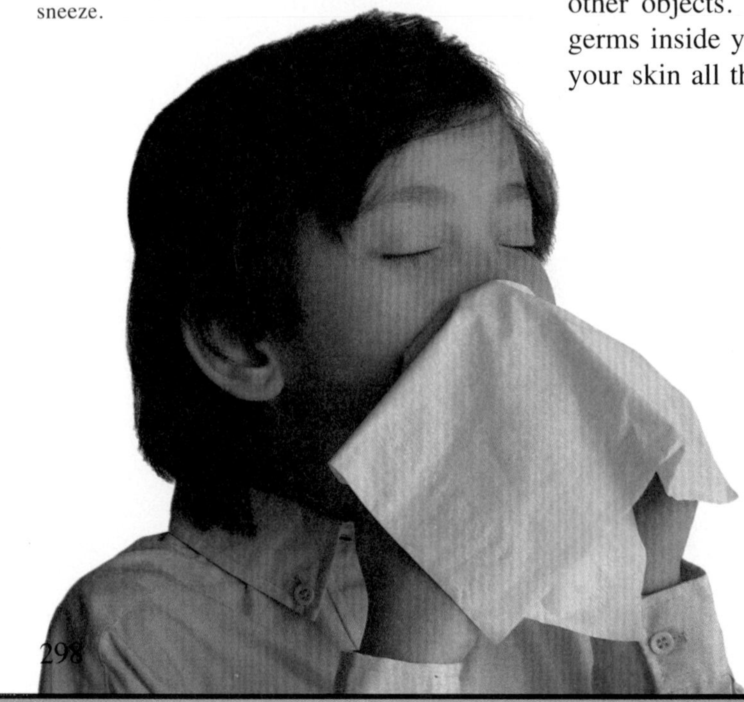

Cover your nose and mouth when you sneeze.

298

Teaching Options

Science Background

Most contagious diseases can be prevented, treated or cured. Disease prevention includes getting vaccinations to protect against specific diseases, and following good health practices. Treatment often includes giving medicines to relieve symptoms or cure diseases.

Some noncommunicable diseases might be prevented by following good health practices. Eating a proper diet, exercising, and not smoking can help in the prevention and treatment of heart disease and can reduce the chances of getting cancer. Diabetes is either inherited or precipitated by diet, and is treated with insulin and/or by controlling sugar intake.

Reading Strategies ♦

1. Guide students' pre-reading by asking: Which parts of the lesson are familiar, and which parts are new?
2. Assign these strategies: Writing a Memory Sentence for each subheading and Mapping Examples of each subheading (See pages T26–T29.)
3. Pair students to share what information is clear and unclear and initiate discussion using students' unanswered questions.

♦ ***Suitable as a language development activity***

Germs can get into your body through a cut on your skin. You might breathe germs from the air. You might take in germs on food or touch an object that has germs. Sometimes disease germs grow inside your body. Then you feel sick.

You get different diseases from different kinds of disease germs. You learned in Chapter 3 that some bacteria cause diseases. For example, bacteria germs can cause strep throat and other infections. **Viruses** are germs that are smaller in size than bacteria. Flu, chicken pox, and measles are caused by viruses.

The pictures show some ways you can help keep germs from spreading. Keeping your skin clean is important. Washing with soap and water can get rid of many germs on your skin. Proper food, exercise, and sleep can help keep your body strong enough to fight off many diseases.

SCIENCE IN YOUR LIFE

Medicines called antibiotics can kill many kinds of bacteria that cause diseases. Taking antibiotics for a disease caused by certain bacteria can help a sick person get well.

virus (vī′rəs), a kind of disease germ that causes diseases such as colds and flu.

Use your own glass.

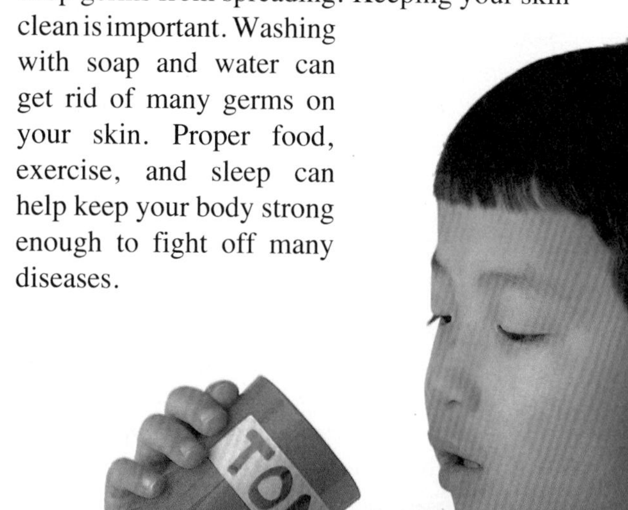

2. Teach

Teaching Tips

• **Possible Misconception:** Some students might think that only objects such as eating utensils, or dirty objects such as trash cans have germs on them. Make sure students understand that germs are everywhere.

• Direct students to read pages 298 and 299. Question: **Where are some specific places that germs can be found?** (in air and water, inside the body, in food and on objects, on the skin) Question: **Why are people not sick all of the time, since germs are everywhere?** (because some germs are not harmful; proper food, exercise, and sleep help the body stay strong enough to fight many diseases; and following good health practices helps people avoid becoming ill)

• Ask students to look at the girl in the picture. Question: **How could this girl avoid spreading her cold to others?** (Guide students to *infer* that the girl could cover her sneezes and coughs, and avoid sharing objects with others.)

Reinforcement

Ask students to draw a flow chart that shows how a person who has a communicable disease can spread it to another person, and that person can spread the disease to a third person. (For example, girl 1 sneezes, boy 1 breathes germs and gets sick, boy 1 covers his sneeze with his hand and hands pencil to girl 2, girl 2 gets sick.)

Special Education

Have students with learning disabilities *list* 5 ways to help prevent the spread of communicable diseases. Help students implement a plan to lessen the spread of germs (Place box of tissues in classroom. Have students make a sign to place over the sink or in the restroom that reminds people to wash their hands.)

Science and Social Studies

Have students use a world map to mark the locations of the places where some of the world's well-known scientists made discoveries in preventing or treating diseases. (for example Louis Pasteur—France, Jonas Salk—United States, Christian Barnard—South Africa) Help the students find out what these people discovered or contributed to the prevention or treatment of disease.

LESSON 2 pages 298–301

TEACHING PLAN

Teaching Tips

- Tell students that the types of diseases that do not spread are more common in today's society than they were in past generations. Stress that people can help prevent some of these diseases. Questions: **How can a person help avoid getting skin cancer?** (Students may be aware that exposure to the sun, which can be avoided, can cause skin cancer.) **What are some other types of cancer?** (If students do not mention lung cancer, make a point of telling students about this type of cancer. Stress that lung cancer is usually caused by smoking.)
- Ask students to *describe* how diabetes can be controlled. (by diet, exercise, and medicine) Explain that people sometimes develop diabetes from being overweight and consuming too much sugar. Mention that this type of diabetes can usually be controlled or reversed by losing weight and limiting sweets.
- Remind students that the heart is a muscle, and that exercise helps make the heart stronger. Question: **What can weaken the heart and lead to heart disease?** (smoking, eating fatty foods and foods that are high in salt)

diabetes (dī′ə bē′tis), a disease in which the body cannot use sugar properly.

cancer, a disease in which cells that are not normal destroy healthy body cells.

Diseases That Do Not Spread

Some diseases are not passed from person to person. These diseases are caused by changes that happen inside a person's body.

Diabetes is one kind of disease that you cannot catch from another person. Diabetes is a disease in which the body cannot use sugar properly. Children with this disease need to take medicine every day. People with diabetes also must eat carefully. They can help control their disease by eating fewer sweets and fats, and exercising regularly.

Cancer is a disease that cannot spread from person to person. In this disease, cells that are not normal destroy healthy body cells. Operations, medicines, and other treatments can control some kinds of cancers. Doctors are trying to find better ways to treat cancer.

People can help reduce their chances of getting some kinds of cancer. Not smoking can help keep people from getting lung cancer. Not spending long periods of time in the sun can help keep people from getting skin cancer.

300

Teaching Options

Science Anecdote

There are over 600 different viruses that can cause cold symptoms. There are over 100 different types of cancer.

Reinforcement

Divide the class into several groups. Ask each group to generate a list of all of the communicable and noncommunicable diseases they can think of. Have students *list* the diseases under an appropriate heading. Allow students to use reference books.

Enrichment

Help students develop a game or obstacle course that will provide them with exercise and enjoyment. Ask the students to *name* the types of noncommunicable diseases that they can help prevent by exercising.

♦ ***Suitable as a language development activity***

Heart disease can keep the heart from working properly. Heart disease does not spread from person to person. Following good health habits can help keep a person's heart healthy. What ways to keep your body strong and healthy are shown on these posters?[1]

[1]avoiding too much sun, not smoking, eating right, exercising

Lesson Review

1. How can diseases spread from person to person?
2. What are some diseases that do not spread from person to person?
3. **Challenge!** Why should you cover your nose when you sneeze?

Study on your own, pages 342–343.

Some people have allergies to certain plants. Find out what an allergy is. What are some plants that people have allergies toward?

LIFE SCIENCE
FIND OUT ON YOUR OWN
CONNECTION

3. Assess

Lesson Review

1. Diseases can spread by touching a person or an object that has germs, by breathing in germs from the air, by eating food that has germs, or through a cut in the skin.
2. Cancer, diabetes, and heart disease do not spread from person to person.
3. Challenge! You should cover sneezes to prevent germs from getting into the air and spreading to another person. **Thinking Skill:** *Drawing conclusions*

Find Out On Your Own

An allergy is a harmful reaction in some people's bodies to certain substances. Some plants people have allergies toward are ragweed, Bermuda grass, sagebrush, and English plantain. **Thinking Skills:** *Collecting information, Communicating*

Workbook page 82 *

Name ____________
Use with Lesson 2: pages 298-301

Chapter 14
Science and Social Studies

Dr. Lister Kills Germs

Read the story. Then answer the questions.

Joseph Lister was an English doctor. He lived during the 1800s. During this time, many people who had surgery died. They died because they got infections.

Dr. Lister realized that the infections were caused by germs. He knew that the way to prevent infections was to make the operating room free of germs. So he experimented with many different liquids. He discovered a liquid that killed the germs. Dr. Lister used this liquid to wash down the operating rooms. The liquid was also used to wash the instruments used during surgery. The surgeons washed their hands with the liquid, too.

The use of this germ-killing liquid made a big difference. Few people died of infections after surgery when this liquid was used. Dr. Lister's methods became known as antiseptic surgery. The term antiseptic means "preventing the growth of germs." The liquids used to kill germs are called antiseptics.

1. Why did so many people die from surgery in the 1800s?
2. Why did Dr. Lister want to kill germs in operating rooms?
3. What did Dr. Lister's methods become known as?
4. What is the term for liquids that kill germs?

32

Game Suggestion ♦

Have students play a game of tag. Tell the person who is "it" that he or she is a germ that can cause disease. Whoever is tagged gets sick. Designate a student to be "medicine" and to retag the sick people to make them well. If a person who has been made well is tagged again by the germ, he or she is out of the game. Let students take turns being the germ and the medicine. *CAUTION:* Before this activity, make certain that no student has physical restrictions requiring a doctor's approval to run.

Reteaching Suggestion ♦

Play "Disease Roll Call" with students. Name a disease discussed in the chapter and have the students *identify* it as communicable or noncommunicable, state what causes or prevents the disease, and discuss what treatment is available.

* ***Answers to masters on pages 292E-292H***

Concept
Bacteria can multiply very rapidly over a short period of time.

Objectives/Process Skills
- ***Make a model*** of bacteria growth.
- ***Measure*** the number of bacteria each half hour for three hours.
- ***Record*** data.
- ***Predict*** the number of bacteria in another half hour.

Time Allotment
Allow 20 minutes.

Safety Tips (See page T24.)
- Use only nontoxic glue for activities involving students.
- Remind students not to eat foods used in science activities unless specifically directed to do so by the teacher.

Teaching Tip
- Have students use wood glue to place the corn kernels on the cards.

Answers
State Your Conclusion

1. 128

2. 256

Use What You Learned

In 5 hours, the bacteria would have grown from 1 cell to 1024 cells. **Thinking Skill:** *Drawing conclusions*

Making A Model of Bacteria Growth

Suggested grouping: pairs

Purpose

Observe how bacteria increase in their number.

Gather These Materials
- package of popcorn kernels
- glue • 8 file cards

Follow This Procedure
1. Use a chart like the one shown to record your observations.
2. Label your first file card 12:00. Label the second file card 12:30. Continue labeling the file cards with the times of every half hour. Label the last file card 3:30.
3. Pretend the popcorn kernels are bacteria. On the 12:00 card, glue 1 popcorn kernel as shown in the picture.
4. At 12:30, your bacteria cell divides to form 2 cells. Glue 2 kernels on the 12:30 card.
5. At 1:00, the two bacteria divide to make 4 bacteria cells. Glue 4 kernels on the 1:00 card.
6. At 1:30, the 4 bacteria divide again. Glue the right number of kernels on the 1:30 card. Keep going until you have finished the 3:30 card.
7. Count the number of popcorn kernels on each card.

Record Your Results

Time	Number of Bacteria
12:00	1
12:30	2
1:00	4
1:30	8
2:00	16
2:30	32
3:00	64
3:30	128

State Your Conclusion
1. How many bacteria did you have at 3:30?
2. How many bacteria would you have at 4:00?

Use What You Learned

How many hours do you think it would take to go from one bacteria cell to a thousand cells?

302

Activity Results

See chart overprint above.

Resource Book page 165

Chapter 14

Activity Worksheet

Name ____________

Use with Lesson 2: page 302

Making a Model of Bacteria Growth

Record Your Results

Time	Bacteria
12:00	
12:30	
1:00	
1:30	
2:00	
2:30	
3:00	
3:30	

State Your Conclusion
1. How many bacteria did you have at 3:30?
2. How many bacteria would you have at 4:00?

Use What You Learned

How many hours do you think it would take to go from one bacteria cell to a thousand cells?

165

Becoming a Doctor

When José Sandoval was your age, he was already a farm worker. He traveled with his family, and they found jobs wherever they could. They picked cotton, chilies, and other crops. José was so busy working, he could not go to school regularly.

When José was twelve years old, his father made an important decision. He decided that his family would not travel and be farm workers anymore. His children would go to school full-time. José studied science because he was interested in spiders and bugs. He became a very good student even though he spoke very little English until he went to school.

José went to college in California. Then, he got a job as a United States Peace Corps volunteer in a faraway country. There, he became interested in medicine. He wanted to help people get well and stay healthy. He returned to California to go to medical school and became a doctor.

Dr. Sandoval especially likes taking care of people who cannot easily afford care. He teaches his patients how important it is to take care of themselves. He reminds them to see a doctor when they show early signs of illness.

Dr. José Sandoval

Dr. Sandoval also works with other doctors to help encourage young people to become doctors. He feels that if young people see that he worked hard and became a doctor, they will feel like they could do it too.

What Do You Think?

1. Why do you think Dr. Sandoval tells people to see a doctor when they first feel ill?
2. If you were a doctor, what would you tell your patients about how to stay healthy?

TEACHING PLAN

Discussion

Have a discussion with students about setting and reaching goals. Ask students if they have ever set goals that seemed very difficult to reach. Discuss how they felt when they reached these goals. Point out that some goals might be fairly easy to achieve, while others might take years to reach.

Teaching Tip

- Point out that Dr. Sandoval encourages young people to accomplish what they set out to do. Ask students to think of people whom they admire and would like to model themselves after.

Answers

What Do You Think?

1. so that their illnesses will not get worse and become difficult or impossible to cure **Thinking Skill:** *Recognizing cause and effect*

2. Answers might include to eat right; to exercise; to get enough rest; to try not to spread diseases; and to see a doctor regularly. **Thinking Skill:** *Ranking ideas and information according to criteria*

Teaching Options

Science Background

José Sandoval grew up in Brawley, California, a small farming community about 200 miles from Los Angeles. After graduating from UCLA Medical School, he returned to Brawley to treat the farm community. He has worked at the San Bernadino County Medical Center, overseeing two clinics that serve mostly Hispanic patients. In addition, he was a founder and the first president of the La Raza Medical Association, an organization whose purpose is to encourage Latinos to enter the field of medicine. He is currently Assistant Professor at the University of California, Irvine, where he sees patients, trains medical students and residents, and works with students interested in becoming doctors.

LESSON 3 pages 304–307

TEACHING PLAN

Lesson Objectives
- *Describe* some immediate and long-term effects of alcohol use.
- *Describe* some effects of tobacco use.
- *Describe* some effects of drug abuse and misuse.

Lesson Vocabulary
alcohol, nicotine, illegal drug

1. Motivate

Demonstration Activity ♦
Bring adult-sized clothes to school. Have a volunteer take a ball, run to a waste basket, put the ball in the basket, take it out, return to the starting place, and repeat the activity for two minutes. Then have the volunteer put on the clothes and try the activity again.

Discussion
Explain that drugs hamper a person's abilities somewhat like the clothes did. Question: **How were the volunteer's speed, accuracy, and coordination affected by the clothing?** (Let students respond.) Point out that drug users are often unaware that they are impaired.

3 How Can Alcohol, Tobacco, and Drugs Affect the Body?

LESSON GOALS

You will learn
- how alcohol can harm the body.
- how tobacco is harmful to a person's health.
- how using harmful drugs can be dangerous to a person's health.

alcohol (al′kə hôl), a drug that can be harmful and is found in beer, wine, and liquor.

This boy is taking medicine because he is ill. Medicines can help sick people get better. Some medicines can keep people from getting ill. You need to be careful with medicines. They can be dangerous if taken in the wrong way. Medicines are a kind of drug. All drugs cause changes in the body. Many drugs can harm a person's health. Some drugs can change the way a person thinks, feels, or acts.

Alcohol
Some drinks have a drug called **alcohol.** This drug is found in beer, wine, and liquor. Drinking too much alcohol can harm the body. Alcohol changes the way a person's brain works. A person who drinks too much alcohol might have trouble thinking clearly.

Medicines can help sick people.

304

Teaching Options

Science Background
Alcohol is the most abused drug in the country. (About 10 million Americans are alcoholics.) Long-term drug abuse of any type can harm the body and even lead to death. Children of parents who abuse drugs often feel responsible for their parents' behavior. They are also more likely to repeat the pattern of drug abuse. Children who abuse drugs might not grow mentally or physically as they should. Breaking the pattern of drug abuse is a complex problem, which requires professional help over an extended period of time.

Reading Strategies ♦
1. Guide students' pre-reading by asking: What ideas do you think you would find in the lesson to answer the question-title?
2. Assign these strategies: Visualizing Information and Writing a Memory Sentence (See pages T26–T29.)
3. Pair students to share what information is clear and unclear and initiate discussion using students' unanswered questions.

♦ ***Suitable as a language development activity***

People who drink too much alcohol can become quarrelsome. Alcohol can cause a person to get a stomachache or a bad headache. Alcohol also affects the way a person moves. A person who drinks too much alcohol can have trouble walking. Drinking too much alcohol can change the way a person talks. The person might forget the correct words or not be able to speak clearly.

Alcohol can keep a person from acting quickly. A person who drinks too much alcohol might feel dizzy and have trouble seeing clearly. Alcohol also can make a person feel tired and go to sleep. A person who has been drinking alcohol should not ride a bicycle or drive a car.

SCIENCE IN YOUR LIFE

Some drugs that doctors use to help people are placed in patches that a person wears like a bandage. These patches can be placed on the skin so that the medicine goes into the body through the skin.

305

2. Teach

Teaching Tips

- **Possible Misconception:** Students might think that drugs prescribed by a doctor cannot be abused. Explain that any drug that is taken too much or in the wrong manner is being abused and can be dangerous.
- Discuss with students things that people do that can be harmful, such as crossing a street without looking, riding a bike without a helmet, and so on. If desired, display pictures of people engaging in such activities. Question: **Do these people seem intelligent and mature while doing this?** (Allow students to respond.) Explain that people who abuse drugs and alcohol harm themselves every day.

Reinforcement

Cut out alcohol advertisements and paste on a poster board. Have students ***describe*** the kinds of people and behavior depicted by the advertisements. Discuss what the advertisements do not reveal about alcohol use.

Special Education

To reinforce the concept that even prescription drugs may be harmful, ask students to ***explain*** how a drug can be both good and bad. (When taken as directed, medicine can treat and cure an illness; when taken improperly, it can cause illness.) Review safety rules associated with prescription medicines (take only as prescribed, do not take anyone else's prescription, report side effects to an adult, and so on)

Science and Social Studies

Help students find out how alcohol and drug use affect young people's growth and development. If possible, you might want to invite the school nurse or other health professional to speak to the class about the special problems associated with drug and alcohol use in youth.

LESSON 3 pages 304–307

TEACHING PLAN

Teaching Tips

- Tell students that cigarettes are addictive because of the nicotine they contain. Explain that many people start smoking with the idea that they can stop later if they have health problems. Question: **Why is that a bad idea?** (Students should realize that people who start smoking might not be able to stop, even though they develop health problems.) Tell students that people who do stop smoking have to go through withdrawal. Explain that withdrawal is somewhat like being hungry but not being allowed to eat.
- Explain that marijuana is a drug that is smoked. Mention that it is thought to be more damaging to the lungs than cigarette smoke, and affects a person's coordination and thinking in much the same way as alcohol does. Point out that cocaine can stop a person's heart and/or breathing. Ask students to ***suggest alternatives*** to drug and alcohol use. (Accept reasonable responses.)

Tobacco

nicotine (nik′ə tēn′), a drug in tobacco that can harm a person's body.

Cigarettes and cigars are made from the leaves of the tobacco plant. Tobacco contains a drug called **nicotine.** This drug can be harmful to the health. Nicotine makes the heart beat faster than normal. A person who smokes is more likely to get heart disease than a person who never smokes.

Cigarette smoke also can be harmful. Some of the materials in cigarette smoke can damage the lungs. A person who smokes is more likely to get lung cancer and other lung diseases than a person who never smokes. Cigarette smoke can harm the health of people who are around smokers. Many buildings have signs like the one in the picture. How can this sign help protect people's health?[1]

[1] keep smokers from smoking; protect nonsmokers from cigarette smoke.

Find the sign that warns people not to smoke.

306

Teaching Options

Reinforcement

Ask students to make a list of the reasons to say "No" to drugs. Have students share their lists with the class. You might want to encourage students to use the lists to make a classroom poster to display in a school hallway.

Enrichment

Help students plan and carry out an anti-drug rally for their grade. Invite the school principal and administration to attend.

Science Anecdote

Caffeine is a stimulant drug. It is found in coffee, tea, colas, and chocolate. Caffeine stimulates the nervous system. In moderate quantities, caffeine is not harmful; larger amounts may cause sleeplessness, irritability, and loss of appetite. There is also evidence that caffeine may be habit-forming.

◆ ***Suitable as a language development activity***

Other Harmful Drugs

Sometimes people use drugs that are against the law. These **illegal drugs** include marijuana and cocaine. These drugs can damage body organs and change the way the brain works. They can change the way a person thinks or acts.

A person should always say *no* to illegal drugs. The boy in the picture enjoys doing activities that make him feel good. He knows that staying away from harmful drugs will help keep him healthy.

illegal (i lē′gəl) **drug,** a drug that is against the law to use or have.

Exercising can help a person feel good.

Lesson Review

1. How can alcohol harm a person?
2. How can tobacco harm the body?
3. How are illegal drugs harmful?
4. **Challenge!** What are some ways a person can say no to illegal drugs?

Study on your own, pages 342–343.

HUMAN BODY

FIND OUT ON YOUR OWN

Caffeine is a drug found in many drinks. Caffeine speeds up the heartbeat and can make some people very nervous. Use library books to find out about caffeine. Make a list of drinks with caffeine. Also make a list of healthy drinks people can choose.

3. Assess

Lesson Review

1. Alcohol can change the way a person thinks, feels, acts, and moves, and can damage the body.
2. Tobacco can cause lung disease, heart disease, and cancer.
3. Illegal drugs change the way the brain works, and can damage other body organs.
4. Challenge! Accept all reasonable ways of saying no to drugs. **Thinking Skill:** *Identifying and suggesting alternatives*

Find Out On Your Own

Drinks that contain caffeine include tea, coffee, many soft drinks, and cocoa. Healthy alternatives to caffeinated drinks include water, orange juice, other fruit juices, and milk. **Thinking Skills:** *Collecting information, Communicating, Listing*

Workbook page 83 *

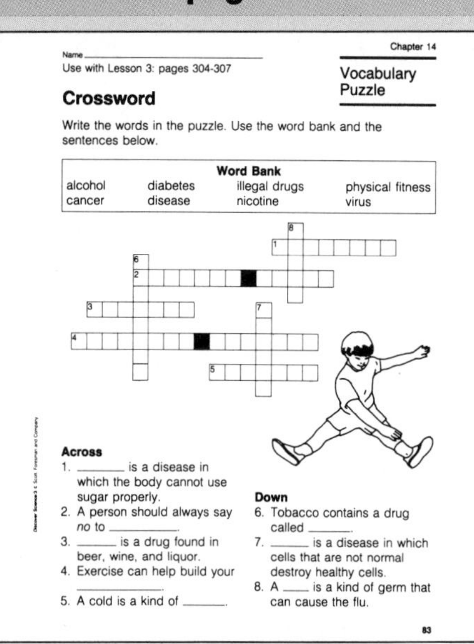

Name ____________

Use with Lesson 3: pages 304-307

Chapter 14

Vocabulary Puzzle

Crossword

Write the words in the puzzle. Use the word bank and the sentences below.

Word Bank

alcohol	diabetes	illegal drugs	physical fitness
cancer	disease	nicotine	virus

Across
1. ______ is a disease in which the body cannot use sugar properly.
2. A person should always say *no* to ______.
3. ______ is a drug found in beer, wine, and liquor.
4. Exercise can help build your ______.
5. A cold is a kind of ______.

Down
6. Tobacco contains a drug called ______.
7. ______ is a disease in which cells that are not normal destroy healthy cells.
8. A ____ is a kind of germ that can cause the flu.

83

* ***Answers to masters on pages 292E—292H***

Game Suggestion ♦

Have a game day when students can bring in games from home to share with the class. You could call the day *Fun Without Drugs Day.* Let students vote on what their favorite game was during that day.

Reteaching Suggestion ♦

Divide the class into two teams. Let Team 1 tell some reasons people might take drugs. Let Team 2 answer or counter with a reason not to take drugs. After one round, have the teams switch sides of the debate.

TEACHING PLAN

Purpose
To develop the skills of collecting and organizing information using diagrams and bar graphs to solve problems.

1. Motivate

Discussion
Provide or ask students to bring in pictures of a variety of foods. Encourage students to discuss which foods they would select for a single meal, and help them identify the food groups to which the foods belong. Question: **How well is your chosen meal balanced?** (Answers will vary.)

2. Teach

Teaching Tips
- Remind students of the importance of good nutrition and the need for a balanced diet. Review with students serving sizes and suggestions for using the four food groups.
- **Helpful Hint:** Explain that all the information needed for Part C is found in the diagram shown in Part A.

Skills for Solving Problems

Using Diagrams and Bar Graphs

Problem: How are the servings a child needs from the four food groups different from the servings an adolescent needs?

Part A. Using Diagrams to Collect Information

1. The diagram shows the four basic food groups. Eating the right amounts of foods from each group helps you get all the nutrients you need. What are the four food groups?
2. The diagram shows how many servings from each group children, teenagers, and adults need each day. A child needs four servings a day from the bread-cereal group. How many servings does a teenager need? an adult?
3. How many servings from the fruit-vegetable group does a child need? a teenager? an adult?
4. How many servings from the meat-poultry-fish-bean group does a child need? a teenager? an adult?
5. How many servings from the milk group does a child need? a teenager? an adult?

Food group	Total number of servings		
	Child	Teenager	Adult
Milk group	3	4	2
Meat poultry fish bean group	2	2	2
Fruit vegetable group	4	4	4
Bread cereal group	4	4	4

Teaching Options

Sample Bar Graph for Part C

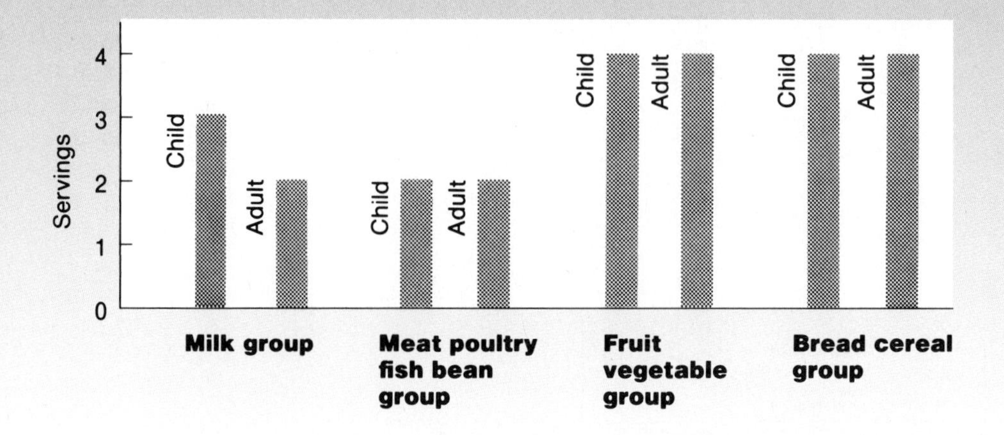

♦ *Suitable as a language development activity*

Part B. Using a Bar Graph to Organize and Interpret Information

6. The bar graph contains information you collected about the servings needed of the four food groups. The bars show how many daily servings of the four food groups a child and a teenager need. Compare the two bars shown for each food group. For which groups do a child and a teenager need the same number of servings? a different number of servings?
7. Why might a teenager need a different number of servings of a food group than a child does?

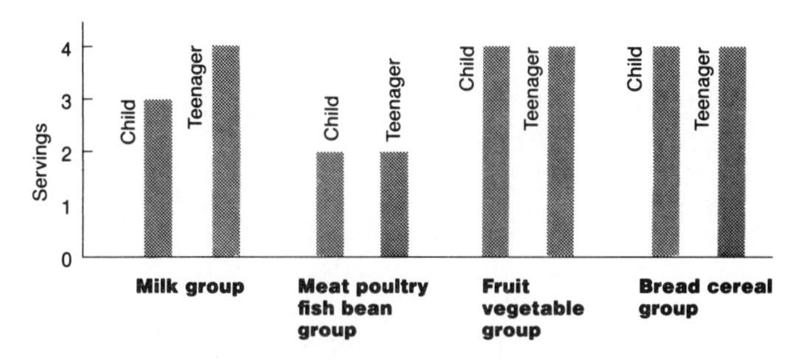

Part C. Using Diagrams and Bar Graphs to Solve a Problem

Problem: How are the numbers of servings a child needs from the four food groups different from the numbers of servings an adult needs?

8. Use the diagram in Part A to collect the information you need to solve the problem. Make a bar graph similar to the one shown in Part B to organize your information.
9. Look at your bar graph. For which food groups does an adult need a different number of servings than a child does? Why might an adult need a different number of servings of a food group than a child does?

3. Assess

Part A

1. bread-cereal group, fruit-vegetable group, milk group, meat-poultry-fish-bean group
2. four; four
3. four; four; four
4. two; two; two
5. three; four; two

Part B

6. A child and a teenager need the same number of servings from the bread-cereal group, the fruit-vegetable group, and the meat-poultry-fish-bean group. They need a different number of servings from the milk group.
7. A teenager might need more servings of milk than a child because the teenager is in a period of more rapid growth and milk is an important food source for growth.

Part C

8. See Sample Bar Graph for Part C in Teaching Options.
9. An adult needs fewer servings from the milk group than a child does. This might be because an adult is in a period of slower growth than a child is.

Reteaching Suggestion ♦

Have students interview three children, 3 teenagers, and 3 adults to find out the approximate number of hours they exercise and sleep each day. Help students find the average for each activity in each group. (total number of hours for each activity divided by 3) Have them create a diagram and a bar graph similar to the ones on page 308 to show their findings. Encourage students to *compare* and *contrast* the time spent exercising and sleeping for each age group.

Resource Book page 167 *

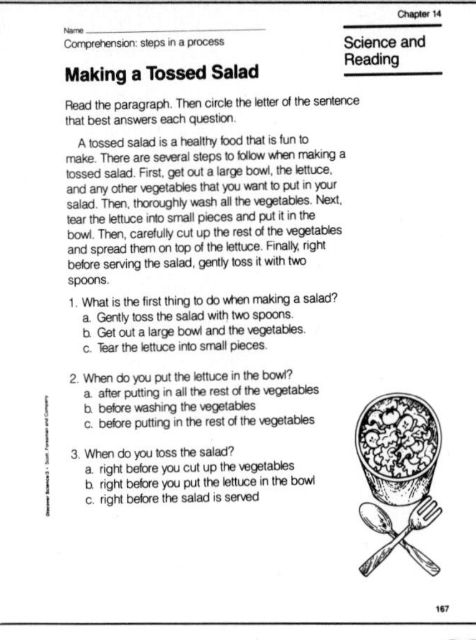
Chapter 14

Name ____________

Comprehension: steps in a process

Science and Reading

Making a Tossed Salad

Read the paragraph. Then circle the letter of the sentence that best answers each question.

A tossed salad is a healthy food that is fun to make. There are several steps to follow when making a tossed salad. First, get out a large bowl, the lettuce, and any other vegetables that you want to put in your salad. Then, thoroughly wash all the vegetables. Next, tear the lettuce into small pieces and put it in the bowl. Then, carefully cut up the rest of the vegetables and spread them on top of the lettuce. Finally, right before serving the salad, gently toss it with two spoons.

1. What is the first thing to do when making a salad?
   a. Gently toss the salad with two spoons.
   b. Get out a large bowl and the vegetables.
   c. Tear the lettuce into small pieces.
2. When do you put the lettuce in the bowl?
   a. after putting in all the rest of the vegetables
   b. before washing the vegetables
   c. before putting in the rest of the vegetables
3. When do you toss the salad?
   a. right before you cut up the vegetables
   b. right before you put the lettuce in the bowl
   c. right before the salad is served

167

* *Answers to masters on pages 292E—292H*

REVIEW PLAN

Reviewing Science Words

1. physical fitness
2. diabetes
3. alcohol
4. disease
5. illegal drugs
6. nicotine
7. cancer
8. virus

Reviewing What You Learned

1. b **3.** c **5.** b **7.** b
2. d **4.** a **6.** c

Interpreting What You Learned

1. Getting the right amount of sleep and exercise, and eating the right foods can help a person stay healthy. **Thinking Skill:** *Recognizing cause and effect*

Chapter 14 Review

Chapter Main Ideas

Lesson 1 • Eating the right amounts of foods from each of the four food groups helps a person get all the needed nutrients. • Exercising properly helps all the parts of the body work well. • Getting the right amount of sleep also helps a person stay healthy.

Lesson 2 • People can help keep themselves well by following good health habits. Germs such as bacteria and viruses can cause diseases that spread from person to person. • Some diseases do not spread from person to person.

Lesson 3 • Drinking too much alcohol can harm the body. • Smoking is harmful to the health. • Illegal drugs can be harmful to the body.

Reviewing Science Words

alcohol, cancer, diabetes, disease, illegal drugs, nicotine, physical fitness, virus

Copy each sentence. Fill in the blank with the correct word from the list.

1. The ability to exercise, work, and play without getting tired easily is ___.
2. ___ is a disease in which the body cannot use sugar properly.
3. ___ is a drug that is contained in beer, wine, and liquor.
4. An illness is called a ___.
5. Marijuana and cocaine are two kinds of ___ that can harm a person's health.
6. Tobacco contains a drug called ___ that makes the heart beat faster and can be harmful to health.
7. ___ is a disease in which cells that are not normal destroy healthy body cells.
8. A ___ is a kind of germ that can cause flu, chicken pox, and measles.

310

Review Options

Cooperative Learning ♦

STAD Format (See page T23.) Assign students to work in four- to five-member teams to study Chapter 14 Review. Students should work together to make sure that they and their teammates know the material in the chapter. After students have had enough time to study together, give them a test to complete individually (Chapter 14 Test A or B in the *Test Book*). Award Superteam certificates to teams whose average test scores exceed 90%, and Greatteam certificates to teams whose average test scores exceed 80%.

Test Book page 125 *

Name ____________

Chapter 14
Test A

Multiple Choice Choose the best answer.

1. In order to have a healthy body, a person should
   a. get enough sleep.
   b. eat more food.
   c. never exercise.
2. Eating the right foods from each food group helps the body get enough
   a. nicotine.
   b. exercise.
   c. nutrients.
3. Walking, swimming, and riding a bicycle are three
   a. unhealthy activities.
   b. illegal drugs.
   c. healthy exercises.
4. Illnesses are often caused by
   a. exercises.
   b. cold weather.
   c. germs.
5. Germs that are smaller in size that bacteria are
   a. diseases.
   b. viruses.
   c. colds.
6. Two diseases that are not passed from one person to another are
   a. cancer and diabetes.
   b. colds and measles.
   c. flu and chicken pox.
7. People who do not smoke have less chance of getting
   a. skin cancer.
   b. diabetes.
   c. lung cancer.
8. The heart gets stronger when a person
   a. reads.
   b. exercises.
   c. watches television.
9. Tobacco contains a harmful drug called
   a. nitrate.
   b. nitrogen.
   c. nicotine.
10. Cocaine is an example of
   a. a legal drug.
   b. an illegal drug.
   c. a helpful drug.

125

♦ ***Suitable as a language development activity***

Reviewing What You Learned

Write the letter of the best answer.

1. Materials in food that help the body grow and stay healthy are
   (a) bread. (b) nutrients. (c) viruses. (d) fruit.
2. One way to help the heart get stronger is to
   (a) read. (c) watch television.
   (b) sleep. (d) play active games.
3. How many servings do you need each day from the milk-cheese group?
   (a) one (b) none (c) three (d) six
4. Which disease can be passed from person to person?
   (a) flu (b) cancer (c) diabetes (d) heart disease
5. Cocaine is a drug that is
   (a) legal. (b) illegal. (c) not harmful. (d) weak.
6. A disease that cannot spread from person to person is
   (a) flu. (b) measles. (c) cancer. (d) a cold.
7. How many servings do you need each day from the meat-poultry-fish-bean group?
   (a) one (b) two (c) six (d) five

Interpreting What You Learned

Write a short answer for each question or statement.

1. What are some ways a person can stay healthy?
2. Explain how a person can get the nutrients he or she needs for good health.
3. Why should a person never drink alcohol and drive?
4. Explain how a person with diabetes can help control his or her disease.

Extending Your Thinking

Write a paragraph to answer each question or statement.

1. Why is it important for people to wash their hands before they cook food or eat a meal?
2. What advice would you give to someone who is thinking of smoking a cigarette?

To explore scientific methods, see Experiment Skills on pages 374–375.

2. Eating the right amount of foods from the four food groups can provide needed nutrients. **Thinking Skill:** ***Comprehending meaning***

3. Drinking alcohol can keep a person from thinking and seeing clearly, and from acting quickly. **Thinking Skill:** *Summarizing*

4. Diabetes can be controlled by diet, exercise, and medicine. **Thinking Skill:** *Restating or explaining ideas*

Extending Your Thinking

1. Students' answers should include an explanation of how germs can be passed from a person's hands, to food, to other people. **Thinking Skill:** *Drawing conclusions*

2. Students should include facts about how cigarette smoking can harm a person by causing lung diseases, cancer, and heart disease. They might also mention that smoking harms nonsmokers' health. **Thinking Skills:** *Judging and evaluating, Recognizing relevant information and data*

Test Book page 126 *

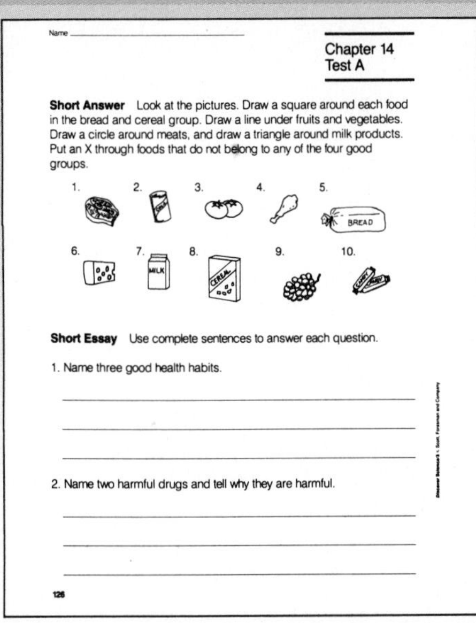

Name ______

Chapter 14
Test A

Short Answer Look at the pictures. Draw a square around each food in the bread and cereal group. Draw a line under fruits and vegetables. Draw a circle around meats, and draw a triangle around milk products. Put an X through foods that do not belong to any of the four good groups.

1. 2. 3. 4. 5.
6. 7. 8. 9. 10.

Short Essay Use complete sentences to answer each question.

1. Name three good health habits.
2. Name two harmful drugs and tell why they are harmful.

126

Test Book page 127 *

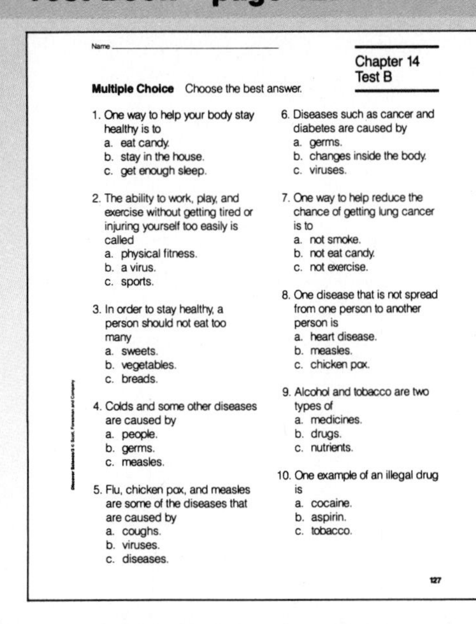

Name ______

Chapter 14
Test B

Multiple Choice Choose the best answer.

1. One way to help your body stay healthy is to
   a. eat candy.
   b. stay in the house.
   c. get enough sleep.
2. The ability to work, play, and exercise without getting tired or injuring yourself too easily is called
   a. physical fitness.
   b. a virus.
   c. sports.
3. In order to stay healthy, a person should not eat too many
   a. sweets.
   b. vegetables.
   c. breads.
4. Colds and some other diseases are caused by
   a. people.
   b. germs.
   c. measles.
5. Flu, chicken pox, and measles are some of the diseases that are caused by
   a. coughs.
   b. viruses.
   c. diseases.
6. Diseases such as cancer and diabetes are caused by
   a. germs.
   b. changes inside the body.
   c. viruses.
7. One way to help reduce the chance of getting lung cancer is to
   a. not smoke.
   b. not eat candy.
   c. not exercise.
8. One disease that is not spread from one person to another person is
   a. heart disease.
   b. measles.
   c. chicken pox.
9. Alcohol and tobacco are two types of
   a. medicines.
   b. drugs.
   c. nutrients.
10. One example of an illegal drug is
    a. cocaine.
    b. aspirin.
    c. tobacco.

127

Test Book page 128 *

Name ______

Chapter 14
Test B

Short Answer Label the basic four food groups.

BREAD
CEREAL
GREEN BEANS
MILK

1. ______
2. ______
3. ______
4. ______

Short Essay Use complete sentences to answer each question.

1. Why should a person never drink alcohol and drive?
2. List three good health habits.

128

***** *Answers to masters on pages 292E—292H*

TEACHING PLAN

Purpose

Several careers that involve health are presented in this feature. Background information is given on some familiar careers and on some that the students might not have heard of.

Teaching Tips

- Students might want to discuss injuries they have had while playing a sport. Emphasize the importance of being careful to avoid injuries during sports.
- Ask students to *list* as many things as possible that a nurse does for patients. The list might include checking blood pressure and temperature, recording weight, giving innoculations, advising patients on maintaining health, and recording information on the patient's chart.
- Discuss the importance of laboratory tests on such samples as blood or throat cultures. Changes in the numbers of different kinds of cells, changes in the chemical composition of blood, or the presence of bacteria give doctors information that help them diagnose diseases.

Careers

How do you help your body stay healthy? You can eat healthy foods and exercise regularly. Many people work to keep other people healthy.

When you think of a doctor, perhaps your family doctor comes to mind. Family doctors treat all kinds of health problems. Some doctors only treat certain health problems. **Sports medicine physicians** help athletes get back into top physical condition after injuries. They also help athletes keep in good physical shape. To become a sports medicine doctor, you must go to medical school after college. Then you must work with experienced doctors for several years.

Some athletes and others might lose the use of one or more body parts from injuries. **Physical therapists** help these people recover the use of their body parts. A physical therapist works out a treatment plan for each patient. This plan might include exercise, massage, and whirlpool baths. If a patient will remain partly disabled, the therapist will show him or her how to do everyday tasks. To become a physical therapist, you must go to college for at least four years.

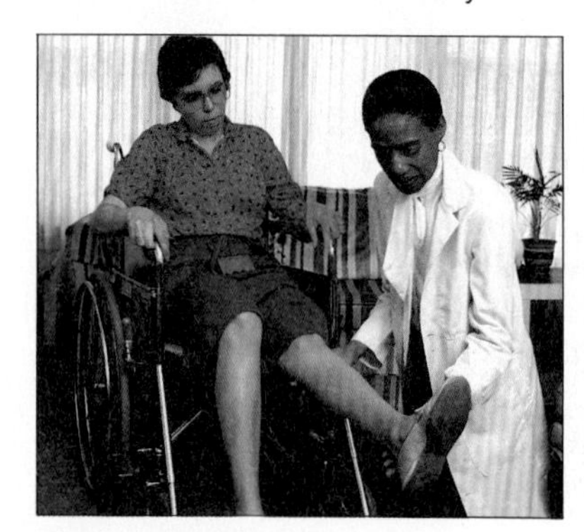

Physical therapist

Registered nurses care for hospital patients in many ways. They observe and record a patient's progress in getting well. They teach patients how to care for themselves when they get home. Many registered nurses work in hospitals. Registered nurses also work in doctors' offices, nursing homes, schools, homes, and industry. Nurses who work in hospitals or nursing homes might have to work nights and weekends. Registered nurses go to school for two to five years after high school.

Sometimes a nurse takes a sample of your blood. A **medical laboratory technician** then looks at the sample under a microscope. He or she can find out more about your illness by seeing this sample. Medical laboratory technicians might also look at samples from other body tissues. Medical laboratory technicians take classes for two years after high school.

Teaching Options

Enrichment

A medical laboratory technician might be willing to talk to the class about what happens to the various samples and specimens a patient is asked to provide. Invite a physical therapist to talk with the students. Ask him or her to bring some assistive devices used by people who are undergoing physical therapy. Ask your school nurse or another nurse in your area to discuss the demands of that profession with the class.

Where To Write

For more information about these careers contact:

Sports Medicine Physician: American Medical Association, 535 North Dearborn Street, Chicago, IL 60610.

Physical Therapist: American Physical Therapy Association, 1111 North Fairfax Street, Alexandria, VA 22314.

Registered Nurse: American Nurses' Association, 2420 Pershing Road, Kansas City, MO 64108.

Medical Laboratory Technician: International Society for Clinical Laboratory Technology, 818 Olive Street, St. Louis, MO 63101.

Canning of Food

Food spoils when germs grow in it. Germs grow best when food is warm and moist. Air also helps most germs to grow.

Food can be stored for a long time if germs and air are kept away from it. Some foods can be dried or frozen to keep them safe from germs. Canning is another way to keep food from spoiling. The picture shows how food is packed in cans.

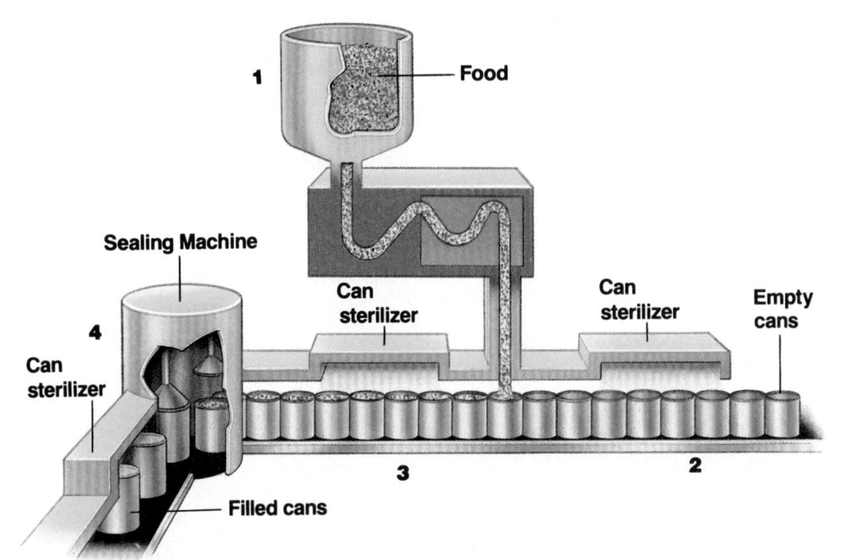

1 First, the food is washed and prepared. Fruits and vegetables might be peeled or cut up. Some foods, such as soup, are cooked just like you cook soup at home. Cooking helps kill germs in the food.

2 Next, the food is sent to the canning machine. The machine contains empty cans with no tops. The cans are heated to a very high temperature, or sterilized, to kill germs in them. Then the machine fills the cans with food.

3 The filled cans are then sterilized. This kills germs that might be left in the cans. Also, heating pushes air out of the cans. The hot food expands and fills the can to the top.

4 Finally, a machine puts a top tightly on each can. Then, the can is sterilized again to make sure all the germs have been killed.

Teaching Tips

- Have students report on another method of food preservation, such as freezing, drying, salting, smoking, or irradiation. They might want to make a bulletin board display that shows how the process works.
- Have a class discussion on the advantages and disadvantages of buying canned foods versus fresh foods.
- Have a representative from your local public health department explain the kinds of problems that may arise from using improper methods for canning foods.
- Ask a local grocery store manager or food wholesaler to discuss how a consumer can tell if food in a can or jar has spoiled.

Science Background

The exact steps in canning vary somewhat depending on the kind of food involved. Pre-cooking may or may not be necessary. Two methods are used to remove air from a filled can. In one, the filled can is heated, causing foods to expand. This forces air out of the can. When the can cools, the foods contract, producing a vacuum in the upper part of the can when the lid is put into place. In this process, filling may be done in the presence of an inert gas, such as nitrogen or carbon dioxide. In the second process, air is removed mechanically by means of a vacuum pump. The processing temperature varies from 100°C (212°F) for high-acid foods to about 120°C (248°F) for foods that are less acidic, neutral, or basic.

Unit 4 Review

Complete the Sentence

1. diabetes
2. organ
3. contracts
4. virus
5. cancer
6. alcohol
7. joint
8. tendon
9. nicotine
10. illegal
11. physical fitness

Short Answer

1. because each type of cell performs a different job in the body *(Drawing conclusions)*

2. A shoulder joint allows a person to move his or her arm in all directions. A knee joint allows a person to bend the lower leg in only one direction. *(Identifying differences)*

3. One muscle contracts and the other stretches. *(Restating ideas)*

4. Exercise, eat right, and get plenty of sleep. *(Summarizing)*

5. playing sports and active games, walking, dancing, or riding a bicycle *(Restating ideas)*

6. Germs can enter cuts, or enter through mouth or nose. *(Making inferences)*

7. Diabetes is a disease that people do not catch from others. *(Identifying differences)*

8. not smoking and protecting themselves from the sun *(Summarizing)*

9. A person can get a stomachache, headache, feel dizzy, not see clearly, or feel tired. *(Comparing similarities)*

10. Nonsmokers might breathe in smoke from other people's cigarettes and be affected by the nicotine and other harmful materials. *(Visualizing)*

Essay

1. Answers should include that muscles move by contracting, muscles move bones, bones move at joints, and arm and hand muscles are voluntary. *(Visualizing, Organizing information)*

2. Smoking harms the heart and makes breathing more difficult. *(Comprehending meaning)*

Unit 4 Review

Complete the Sentence

Fill in the blank with the correct word or words from the list.

alcohol, nicotine, cancer, organ, contracts, physical fitness, diabetes, tendon, illegal, virus, joints

1. When a person has the disease of ___, the body cannot use sugar properly.
2. Tissues in an ___ work together to do a job in your body.
3. A muscle ___ when it becomes shorter.
4. A kind of germ that causes diseases such as colds and flu is a ___.
5. A disease in which cells that are not normal destroy healthy body cells is ___.
6. The drug ___ is found in beer and wine.
7. A place where two bones join together is a ___.
8. A ___ holds a muscle to a bone.
9. Tobacco has a drug called ___ that can harm a person's heart.
10. Cocaine and marijuana are ___ drugs.
11. Being able to exercise, work, and play without getting tired easily is ___.

Short Answer

Write a short answer for each question or statement.

1. Why does your body need many different kinds of cells?
2. Compare the ways a shoulder joint and a knee joint help a person move.
3. What happens to upper arm muscles when you bend your arm at the elbow?
4. Name three ways to keep your body healthy.
5. In what ways can a person build physical fitness?
6. How can germs on the skin get inside the body and make a person sick?
7. How is diabetes different from measles, flu, or a cold?
8. Name two ways that people can help reduce their chances of getting cancer.
9. In what ways can drinking alcohol make people feel sick?
10. How can cigarette smoke harm the health of people who don't smoke?

Essay

Write a paragraph for each question or statement.

1. Describe what the muscles and bones do when a person throws a ball.
2. How does smoking cigarettes harm a person's health?

314

Test Book page 133 *

Name ______

Unit Test 4

Multiple Choice Choose the best answer.

1. Every body part is made up of
   a. bones.
   b. cells.
   c. muscles.
2. A group of organs working together to do a job is
   a. a system.
   b. a tissue.
   c. an organism.
3. The places where bones join together are called
   a. joints.
   b. ribs.
   c. tendons.
4. Bones help a person
   a. see.
   b. move.
   c. think.
5. Muscles are connected to bones with a tissue called a
   a. joint.
   b. skeleton.
   c. tendon.
6. All the bones of the body make up the
   a. system.
   b. skeleton.
   c. ribs.
7. A muscle that works without a person controlling it is
   a. a joint.
   b. an involuntary muscle.
   c. a voluntary muscle.
8. Muscles that a person can control are called
   a. joints.
   b. involuntary muscles.
   c. voluntary muscles.
9. Muscles move bones by
   a. relaxing.
   b. connecting.
   c. contracting.
10. The tissues that move bones in a body are
    a. organs.
    b. skeletons.
    c. muscles.

1. ⓐⓑⓒ 2. ⓐⓑⓒ 3. ⓐⓑⓒ 4. ⓐⓑⓒ 5. ⓐⓑⓒ 6. ⓐⓑⓒ 7. ⓐⓑⓒ 8. ⓐⓑⓒ 9. ⓐⓑⓒ 10. ⓐⓑⓒ

133

Test Book page 134 *

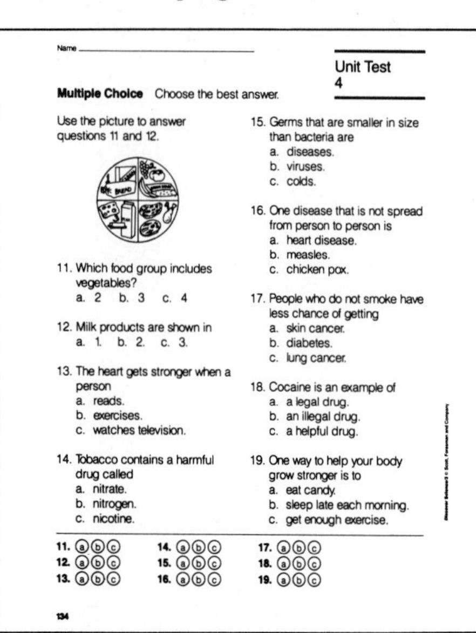

Name ______

Unit Test 4

Multiple Choice Choose the best answer.

Use the picture to answer questions 11 and 12.

11. Which food group includes vegetables?
    a. 2 b. 3 c. 4
12. Milk products are shown in
    a. 1. b. 2. c. 3.
13. The heart gets stronger when a person
    a. reads.
    b. exercises.
    c. watches television.
14. Tobacco contains a harmful drug called
    a. nitrate.
    b. nitrogen.
    c. nicotine.
15. Germs that are smaller in size than bacteria are
    a. diseases.
    b. viruses.
    c. colds.
16. One disease that is not spread from person to person is
    a. heart disease.
    b. measles.
    c. chicken pox.
17. People who do not smoke have less chance of getting
    a. skin cancer.
    b. diabetes.
    c. lung cancer.
18. Cocaine is an example of
    a. a legal drug.
    b. an illegal drug.
    c. a helpful drug.
19. One way to help your body grow stronger is to
    a. eat candy.
    b. sleep late each morning.
    c. get enough exercise.

11. ⓐⓑⓒ 12. ⓐⓑⓒ 13. ⓐⓑⓒ 14. ⓐⓑⓒ 15. ⓐⓑⓒ 16. ⓐⓑⓒ 17. ⓐⓑⓒ 18. ⓐⓑⓒ 19. ⓐⓑⓒ

134

Unit Projects and Books

Science Projects

1. Some bones are not covered by very many muscles. These bones can be felt easily. Feel your face, your sides, your arms, and your legs. Make a list of the bones you can feel.
2. One kind of joint allows you to bring one part of your body closer to another part of your body. For example, if you bend your elbow, you bring your lower arm closer to your upper arm. What happens when you bend your fingers and your knees. How does the movement of these joints differ from the movement of your hip joint?
3. Make a poster with a slogan that says no to drinking, smoking, or using harmful drugs.

Books About Science

The Skeleton Inside You by Philip Balestrino. Crowell, 1989. Explore your skeleton including what happens when a bone breaks.

Muscles and Movement by Gwynne Vevers. Lothrop, 1984. Find out about different kinds of muscles and what they help you do. Gr. 1-4

Germs Make Me Sick! by Melvin Berger. T.Y. Crowell, 1985. Learn about germs and how your body fights them. Gr. ps-3

Science and Society

Smokeless Flights People on a flight from San Francisco to Los Angeles are talking about smoking. A recent law banned smoking on all airline flights within California. Some people like the ban on smoking. ''This new policy has really cleared the air,'' they say. ''Smokers do not have the right to harm our health.'' The smokers are not happy with the law. ''We have rights too,'' they say. ''We do not tell others what to eat or drink. How can they tell us not to smoke?'' The nonsmokers reply, ''We have the right to clean air. In a crowded airplane, we get sick when other people smoke.'' The smokers answer that they can smoke on flights in other parts of the country. What are reasons for banning smoking on airplane flights? What are reasons for allowing smoking in one section of an airplane?

Unit Projects and Books

TEACHING PLAN

Science Projects

1. Students' answers will vary. Answers might include jaw bones, cheek bones, nose bone, skull, ribs, hip bones, arm bones, leg bones, finger and toe bones, heel, and ankle bones.

2. The hinge joints in the fingers and knees allow for back-and-forth movement. The ball-and-socket joint in the hip allows for circular movement.

3. Display completed posters in the classroom or, with your principal's permission, display posters throughout the school.

Science and Society

Encourage students to explore both sides of the issue in class or on their own. In class, you may wish to set up a debate on the issue. On their own, students might do further research and choose one side of the issue to support. Throughout this exercise, emphasize the importance of respecting the opinions of others. **Thinking Skill:** *Judging and evaluating*

Teaching Options

Science Background

Science and Society The main issue here is a conflict of rights: the right of nonsmokers to clean air versus the right of smokers to practice a personal habit they enjoy. In such conflicts, decisions must be made on a case-by-case basis, and may involve a compromise between rights. The evidence that smoking harms health is overwhelming. Do people have the right to take such action? As long as it affects only the individual, the answer may be yes. But nonsmokers as well as smokers pay the cost of treating smoking-related illnesses through higher insurance and medical costs for everyone. Also, there is some indication that the health of nonsmokers can be damaged by breathing secondhand smoke.

*Answers to masters on pages 292E-292H

Lesson 1
1. Weed roots are long and thick. Grass roots are thin and spread out.
2. hold it in the soil, and take in water and minerals
3. minerals
4. stems
5. through small tubes
6. leaves
7. through tiny openings in the leaves
8. sugar

Lesson 2
1. seeds
2. (a) the center of the flower; (b) the cone

Independent Study Guide

Use the *Independent Study Guide* to review the lessons in each chapter.

Chapter 1 Study Guide

On a separate sheet of paper, write the word or words that best complete the sentence or answer the question.

LESSON 1
pages 10–13

1. How are the roots of weeds and grasses different?
2. What two things do roots do for a plant?
3. Materials in the soil that were never alive are ___.
4. The ___ of most plants hold up the leaves and other plant parts that grow above the ground.
5. How do water and minerals get from the roots of a plant to its leaves?
6. The ___ of green plants make most of the food for the plant.
7. How does a plant take in carbon dioxide?
8. Most plants store some ___ in their roots and stems.

LESSON 2
pages 16–19

1. The ___ from flowers grow into new plants.
2. What part of the plant in each of the pictures below makes seeds?

316

3. What must happen before a flower can form seeds?
4. When the center part of a flower changes into a fruit, the flower ___ dry up and fall off.
5. What does the fruit of a plant do for the seeds?
6. How do bees move pollen between flowers?
7. Name two other animals that help pollinate flowers.
8. Animals and ___ can scatter pollen.
9. How are corn and other grasses pollinated?
10. The ___ on some trees have pollen inside.
11. Most trees with cones have needle-shaped ___.
12. What is one kind of tree that has cones?

LESSON 3
pages 20–22

1. What does a seed coat do for the seed?
2. Water, wind, and ___ help scatter seeds.
3. A seed needs enough air and water and the proper ___ to germinate.
4. A young plant that has just pushed through the ground is called a ___.
5. The time during which a plant grows from a seed until it makes its own seeds is the plant's ___.
6. How would you put the following pictures in the order that they happen?

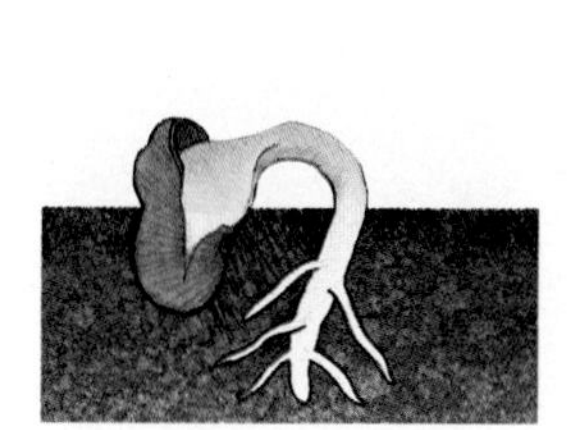

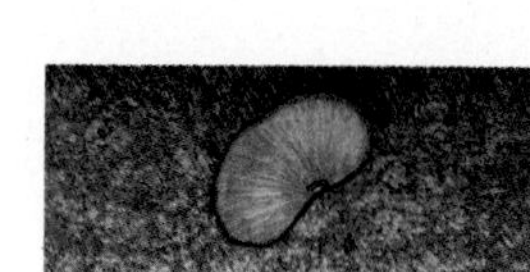

317

3. Pollen must move to the center of the flower.
4. petals
5. It protects them as they grow.
6. Pollen sticks to bees' bodies and falls off onto other flowers.
7. butterflies, moths, and humming-birds
8. wind
9. by the wind carrying pollen
10. cones
11. leaves
12. pine trees, spruce trees, and fir trees

Lesson 3
1. protects it
2. animals
3. temperature
4. seedling
5. life cycle
6. B, A, C

Lesson 1
1. backbone
2. to stand straight, to bend and move
3. Bees do not have backbones, and birds do.

Lesson 2
1. with lungs
2. Parent birds bring them food in their beaks.
3. cold-blooded
4. eggs
5. fins
6. through gills
7. They have scales, are cold-blooded, and hatch from eggs.
8. Young fish care for themselves as soon as they hatch.
9. The snake and fish are cold-blooded; the bird is warm-blooded.

Chapter 2 Study Guide

On a separate sheet of paper, write the word or words that best complete the sentence or answer the question.

LESSON 1
pages 30–31

1. One way that scientists classify animals is by whether or not they have a ___.
2. What does your backbone help you do?
3. Why would bees and birds not be classified into the same group?

LESSON 2
pages 32–36

1. How do birds breathe?
2. How do baby birds get food?
3. Snakes are ___ because their body temperatures change with the temperature of the air around them.
4. Young reptiles come from ___.
5. A fish's ___ help it swim.
6. How do fish breathe?
7. How are fish and reptiles the same?
8. What happens to young fish after they hatch?
9. Which of the animals in the pictures is warm-blooded and which are cold-blooded?

10. Animals with backbones and hair or fur are ___.
11. What are two kinds of mammals that live on land?
12. What is one kind of mammal that lives in water?
13. Mammals breathe with ___.
14. How are the ways mammals and birds are born different from one another?
15. People belong in the group of ___.
16. What food do young mammals eat?

LESSON 3
pages 38–41

1. A ___ has a thin, soft body and no legs.
2. Earthworms make ___ in the soil.
3. A ___ has a soft body and a hard shell.
4. A spider has two main body parts and ___ legs.
5. A spider spins a ___ from silk that it makes inside its body.
6. The largest group of animals without backbones are the ___.
7. How are insects different from spiders?
8. The butterfly has ___ stages in its life cycle.
9. Describe the larva of the butterfly.
10. These pictures show the way a butterfly changes. Describe how you would put the pictures in the correct order.

10. mammals
11. dogs and horses
12. whales and dolphins
13. lungs
14. Mammals grow inside the mother, and birds grow inside eggs.
15. mammals
16. their mothers' milk

Lesson 3
1. worm
2. tunnels
3. snail
4. eight
5. web
6. insects
7. They have three body parts, six legs, and some have wings.
8. four
9. The caterpillar is the larva of the butterfly. The larva looks different from the adult.
10. B, A, C, D

Chapter 3 Study Guide

Lesson 1

1. organisms
2. cells
3. through a microscope
4. five
5. protists
6. roots
7. A—bacteria; B—fungus; C—plant; D—mammal

Lesson 2

1. population
2. size
3. It grows larger.
4. community

Chapter 3 Study Guide

On a separate sheet of paper, write the word or words that best complete the sentence or answer the question.

LESSON 1
pages 50–52

1. Living things are called ▬.
2. Organisms are made of one or more ▬.
3. How can you see cells?
4. Scientists have divided organisms into ▬ different groups.
5. Some ▬ move by whipping their tails back and forth.
6. A plant's ▬ keep it attached to the soil.
7. Name the kind of organism in each picture.

LESSON 2
pages 54–56

1. A group of sheep living in one place together form a ▬.
2. The ▬ of a population often changes because of the number of births and deaths in the group.
3. What happens to the size of a population when it has plenty of food?
4. The organisms in a ▬ depend on each other for food and shelter.

320

5. Why do the squirrels in a forest community need the trees?
6. What do insects do for a forest community?
7. What organisms live in the community you see in the picture below?

8. An organism gets everything it needs from its ▒.
9. How is a habitat different from a community?

LESSON 3
pages 58-60

1. Why are green plants producers?
2. Mice, deer, and rabbits are ▒ because they eat other organisms for food.
3. Snakes and lions are ▒ because they hunt for their food.
4. Lions hunt for antelope, so antelope are the lion's ▒.
5. Grass growing in a meadow and then a mouse eating the grass is an example of a part of a ▒.
6. Where are food chains found?

5. for nests
6. pollinate plants
7. trees, grass, ducks
8. habitat
9. Only one organism lives in a habitat; many live in a community.

Lesson 3
1. because they make their food
2. consumers
3. predators
4. prey
5. food chain
6. in soil, on land, and in water

Lesson 1
1. food, water, shelter
2. pollution
3. by burning fuels
4. by dumping chemical wastes into lakes and streams
5. habitats
6. endangered
7. A, extinct; B, endangered

Lesson 2
1. parks
2. by removing chemical wastes from water before returning it to lakes and streams

Chapter 4 Study Guide

On a separate sheet of paper, write the word or words that best complete the sentence or answer the question.

LESSON 1
pages 68–71

1. What are three things organisms get from their habitats?
2. Air, water, and land habitats are often destroyed by ___.
3. How do factories cause air pollution?
4. How do factories cause water pollution?
5. Litter can change the ___ of plants and animals.
6. When many organisms of the same kind die, that kind of organism can become ___.
7. Look at the pictures below. Tell whether each organism is endangered or extinct.

LESSON 2
pages 72–74

1. People cannot hunt animals or collect plants in ___.
2. What is one way factories can keep water from becoming polluted?

322

3. What is one way factories can keep the air from becoming polluted?
4. How do scientists help protect endangered plants?
5. Some endangered animals are kept in ____ to keep them from becoming extinct.

LESSON 3
pages 76–79

1. What are two things you get from plants and animals?
2. Fruits, grains, and ____ come from plants.
3. Name three different kinds of grains.
4. What part of the celery plant do people eat?
5. What part of a plant is an apple?
6. Cheese, butter, yogurt, and ice cream come from ____.

7. Look at the pictures above. Tell whether each food comes from a plant or an animal.
8. People use ____ from trees to make paper.
9. Where does linen come from?
10. Leather comes from the ____ of animals.
11. Name a plant that can cause some people to get an itchy rash.

3. by using fuels that cause less pollution
4. by saving their seeds
5. zoos

Lesson 3

1. food and clothing
2. vegetables
3. wheat, oats, and corn
4. the stem
5. the fruit
6. milk
7. plant—tomato, bread, muffin; animal—cheese, egg, meat
8. wood
9. flax plants
10. skins
11. poison ivy

Lesson 1
1. space
2. volume
3. larger
4. by using a balance
5. more
6. They are made of matter.
7. properties
8. size, shape, color, smell, and taste
9. solid
10. a gas
11. Solids keep a certain shape, but liquids take the shape of their containers.
12. It spreads out into the room.
13. Answers will vary, but might indicate size, shape, color, or weight of objects.

Chapter 5 Study Guide

On a separate sheet of paper, write the word or words that best complete the sentence or answer the question.

LESSON 1
pages 94–98

1. The objects on your desk are alike because they all take up ___.
2. The amount of space an object fills is its ___.
3. A car has a ___ volume than a bicycle does.
4. How can you measure how much mass an object has?
5. An apple has ___ mass than an eraser.
6. How are people like objects in a room?
7. When you describe an object, you tell about its ___.
8. What are five properties you can use to describe an object?
9. What state of matter is a rock?
10. What state of matter is the air you breathe?
11. How are solid objects and liquid objects different from one another?
12. What happens to the air in a balloon when the balloon breaks?
13. Using a chart like the one below, list some of the properties of three objects you have at your desk.

Object	Properties

324

LESSON 2

pages 100–102

1. Two or more ___ join to form larger particles of matter.
2. A solid has particles that are ___ than the particles in liquids.
3. How do liquids change their shape?
4. In which of the states of matter do the particles have the weakest pull?
5. A ___ can spread to fill any space.

LESSON 3

pages 104–108

1. Changes in the shape of an object made of clay are ___ changes.
2. What are three kinds of physical changes?
3. How can liquid water be changed to a solid?
4. 0° Celsius is the ___ of water.
5. What does heat do to the particles in ice?
6. Over time, water in a puddle changes into a ___.
7. Liquid water ___ to form water vapor.
8. What happens to liquid water when it is heated to a temperature of 100° Celsius?
9. Water vapor ___ when it cools.
10. Burning wood causes a ___ to take place.
11. How does rust form on a can?
12. Rusting is a ___ change.
13. A change in the color of silver is a ___ change.
14. What kind of change has taken place below?

325

Lesson 2

1. atoms
2. closer
3. The pull between the particles weakens and they can move around each other.
4. gas
5. gas

Lesson 3

1. physical
2. changes in size, shape, and state
3. by cooling it to the freezing point
4. freezing point
5. It speeds their movement, making them move apart.
6. gas
7. evaporates
8. It changes into water vapor.
9. condenses
10. chemical change
11. The material of the can mixes with the air.
12. chemical
13. chemical
14. chemical change

Lesson 1
1. force
2. more to move a chair
3. gravity
4. mass
5. 85
6. There is friction between the ball and the floor.
7. work
8. energy

Lesson 2
1. a machine with few or no moving parts
2. lever and wedge, screw, inclined plane

Chapter 6 Study Guide

On a separate sheet of paper, write the word or words that best complete the sentence or answer the question.

LESSON 1
pages 116–120

1. A ▬ changes the way an object moves.
2. Would you need more force to move a chair or a pencil?
3. A ball thrown in the air is pulled down by the force of ▬.
4. The more ▬ an object has, the more gravity pulls on it.
5. If you weigh 85 pounds the pull of the earth's gravity on you is ▬ pounds.
6. What is one reason a ball stops when it is rolled across the floor?
7. You do ▬ when you move a book across the room.
8. You are able to do work because you have ▬.

LESSON 2
pages 122–127

1. What is a simple machine?
2. What kind of simple machines are each of the objects in the pictures?

326

3. If you use a board to lift a heavy object, you use the board as a ▒.
4. Pushing down on a lever makes it move back and forth on its ▒.
5. What kind of simple machine is a seesaw?
6. An ▒ is a simple machine with a flat surface that is higher at one end.
7. How do inclined planes help a person do work?
8. What are two examples of inclined planes?
9. How does a wedge help do work?
10. What simple machine is like an inclined plane wrapped around a rod?
11. The axle turns when you put force on the ▒.
12. A ▒ can be used to move a load up, down, or sideways.
13. What kind of simple machine would you use to move an object to a hard-to-reach place?

LESSON 3

pages 130–131

1. What makes a paddle boat a compound machine?
2. The ▒ in a pencil sharpener sharpen the pencil.
3. Following ▒ can help keep you safe when using machines.
4. How is the person in this picture using a machine safely?

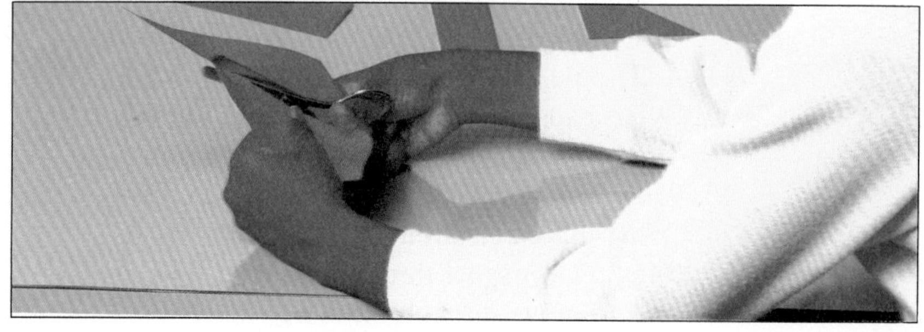

3. lever
4. fulcrum
5. a lever
6. inclined plane
7. by helping move objects to a higher or to a lower place
8. a slanted road, a ramp, a path up a hill
9. It can be used to push objects apart.
10. a screw
11. wheel
12. pulley
13. a pulley

Lesson 3

1. It is made of two or more simple machines put together.
2. wedges
3. safety rules
4. pointing scissors away from oneself

Chapter 7 Study Guide

Lesson 1

1. energy
2. the kind of energy moving objects have
3. It changes to light energy and heat energy.
4. sun

Lesson 2

1. Their electric charges move from one object to another.
2. attract
3. repel
4. battery
5. the moving of electric charges from one place to another
6. circuit
7. rubber, plastic, and glass
8. It keeps the electric charges from moving outside the cord.
9. poles
10. attract
11. repel
12. The nail becomes an electromagnet.
13. by making more turns of wire in the loops
14. that they can be turned on and off
15. electric current

Chapter 7 Study Guide

On a separate sheet of paper, write the word or words that best complete the sentence or answer the question.

LESSON 1
pages 138–139

1. You use ___ when you move objects from one place to another.
2. What is energy of motion?
3. What happens to the energy from electricity when you turn on an electric lamp?
4. Plants get energy from the ___.

LESSON 2
pages 140–145

1. What happens when you rub two objects together?
2. Unlike charges ___ each other.
3. Like charges ___ each other.
4. A ___ pushes electric charges from place to place.
5. What is electric current?
6. A bulb lights when the ___ is complete.
7. Name two kinds of materials that do not carry electric current well.
8. What does the rubber on an electric cord do?
9. Magnetism is strongest at a magnet's ___.
10. The unlike poles of two magnets near each other will ___ each other.
11. Like poles of magnets will ___ each other.
12. What happens when electric current moves through loops of wire wrapped around a nail?
13. How can you make an electromagnet stronger?
14. What makes electromagnets useful?
15. Moving a magnet through loops of wire causes ___ to move through the wire.

328

LESSON 3

pages 148–151

1. What is a source?
2. ■ is the flow of energy from warmer places and objects to cooler ones.
3. Tell why the ice cube in the picture below is melting.

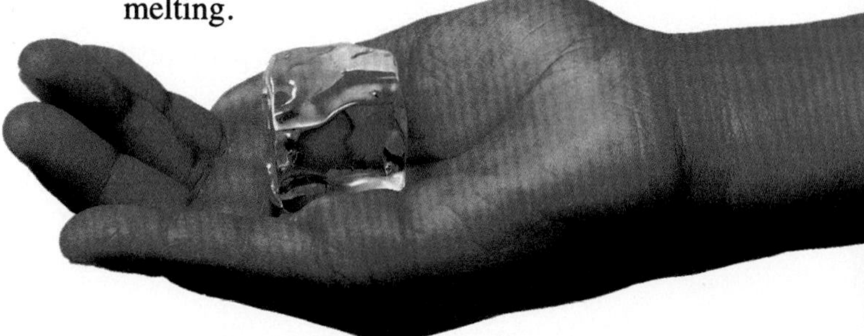

4. Why do many cooking pans have wooden handles?
5. Most liquids and gases are ■ energy conductors.
6. ■ is a measure of how fast particles of matter are moving.
7. The faster the particles of matter move, the ■ the temperature of the matter.
8. Some thermometers measure temperature in degrees ■.
9. What are thermostats used for?
10. Lining a building's roof and walls with an ■ helps keeps heat from moving into the outside air.

LESSON 4

pages 152–156

1. What happens when light strikes an object?
2. Most of the light you see is ■ light.
3. Light energy travels in ■.
4. Light waves ■ when they pass through a lens.
5. What are two uses of lenses?
6. Muscles in the iris of the eye control the size of the ■.

Lesson 3

1. a place from which everything comes
2. heat
3. Heat moves from the person's hand to the ice cube.
4. because wood is an insulator
5. poor
6. temperature
7. higher
8. Celsius
9. to control heating and cooling systems
10. insulator

Lesson 4

1. It bounces off, or reflects.
2. white
3. waves
4. bend
5. Answers will vary, but might include: eyeglasses, magnifying glasses, cameras, telescopes, and microscopes.
6. pupil

Chapter 8 Study Guide

Lesson 1

1. vibrate
2. the loudness or softness of a sound
3. louder
4. less
5. how high or low a sound is
6. quickly
7. slowly
8. by shortening it
9. the rubber band around the red box

Lesson 2

1. in waves
2. air
3. weaker
4. matter

Chapter 8 Study Guide

On a separate sheet of paper, write the word or words that best complete the sentence or answer the question.

LESSON 1
pages 164–166

1. When you hit a drum, parts of the drum ___.
2. What is volume?
3. The more an object vibrates, the ___ the sound.
4. A soft sound has ___ volume than a loud sound.
5. What is pitch?
6. When an object vibrates ___, it makes a sound with a high pitch.
7. An object that vibrates ___ makes a sound with a low pitch.
8. How can you make a string vibrate more quickly?
9. Look at the rubber bands in the picture below. Tell which rubber band would have the sound with the highest pitch.

LESSON 2
pages 168–170

1. How does sound travel?
2. When a bell rings, it causes the ___ around it to vibrate.
3. Sound waves get ___ as they move away from the object that made them.
4. You can only hear sounds when they travel through ___.

5. Sound moves more ▩ through liquids than it does through solids.
6. When sound bounces back from an object, you hear an ▩.
7. What kind of surfaces produce echoes best?

LESSON 3

pages 172–174

1. What happens to your vocal cords when you speak?
2. Vocal cords are thin flaps at the top of your ▩.
3. When you talk, air comes from your ▩ and passes between your vocal cords.
4. What does the outer part of the ear do?
5. What do sound waves do to the eardrum?
6. The eardrum makes the ▩ in the middle of your ear vibrate.
7. What part inside the ear tells your brain about sounds you hear?
8. Name the parts of the ear shown in the picture below.

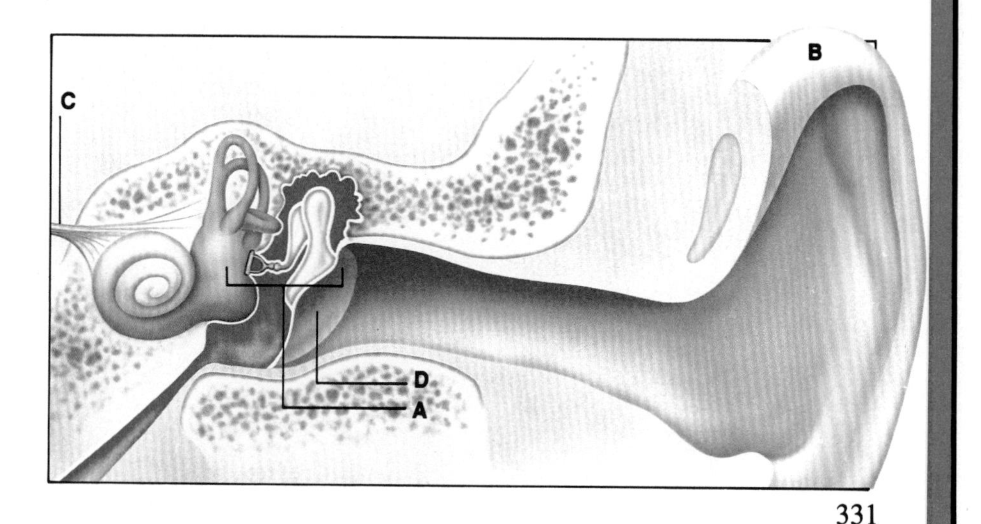

5. slowly
6. echo
7. smooth, hard surfaces

Lesson 3

1. They move back and forth, or vibrate.
2. windpipe
3. lungs
4. It moves sound waves to the part of the ear inside your head.
5. make it vibrate
6. bones
7. the nerve
8. A—bones; B—outer ear; C—nerve; D—eardrum

Lesson 1
1. properties
2. different colors, shiny, dull, soft, or hard
3. minerals
4. because it is very hot
5. They cool and harden.
6. They form sedimentary rock.
7. metamorphic rock
8. sedimentary
9. It becomes marble.
10. harder

Lesson 2
1. through holes and cracks
2. The plant's roots can break the rock apart.

Chapter 9 Study Guide

On a separate sheet of paper, write the word or words that best complete the sentence or answer the question.

LESSON 1
pages 188–191

1. Each mineral has its own special ▒.
2. What are some properties a mineral might have?
3. What makes the different colors you might see in a rock?
4. Why do minerals melt when they are deep inside the earth?
5. What happens when melted rocks reach the earth's surface?
6. What happens as layers of rocks and shells harden at the bottom of lakes and oceans?
7. Pressure can change igneous or sedimentary rock to ▒.
8. Limestone is a ▒ rock.
9. What happens to limestone when it is heated and squeezed inside the earth?
10. Marble is much ▒ than limestone.

LESSON 2
pages 194–196

1. How does water get inside a rock?
2. How can this plant break apart the rock?

3. How can water break apart a rock?
4. Over many years, rock is broken down and forms ▒.
5. What materials make up soil?
6. What gives soil its dark color?
7. Soil can have pieces of different sizes and ▒.
8. Tiny grains in ▒ make it feel smooth.
9. Why might plants be unable to grow in clay soil?
10. ▒ is loose and easy to dig.
11. Why might plants not grow well in sandy soil?
12. Loam is a mixture of clay, sand, and ▒.
13. Which soil type would be best for growing plants?

LESSON 3

pages 198–201

1. The glass you drink from is made of ▒ that come from rock.
2. People need soil to get ▒ to make clothing.
3. What object do you use each day in school that was made from trees?
4. People use the mineral ▒ to make coins, electric wires, and pots and pans.
5. A rock with large amounts of copper would be copper ▒.
6. People who dig deep tunnels to search for minerals are ▒.
7. A ▒ is something people use that comes from the earth.
8. What is one way people can prevent wasting natural resources?
9. Why should farmers change the kinds of plants they grow from year to year?

333

3. by freezing and melting
4. soil
5. rock, air, water, and humus
6. humus
7. color
8. clay soil
9. The grains are too close.
10. sandy soil
11. It does not hold water, and has few nutrients.
12. humus
13. loam

Lesson 3
1. minerals
2. cotton
3. a book, a pencil
4. copper
5. ore
6. miners
7. natural resource
8. by recycling them
9. to build up lost nutrients

Lesson 1
1. crust, mantle, core
2. crust
3. melted
4. The earth's core is made of mostly iron.
5. outside
6. inside
7. movements

Chapter 10 Study Guide

On a separate sheet of paper, write the word or words that best complete the sentence or answer the question.

LESSON 1
pages 208–210

1. Name the three layers of Earth shown below.

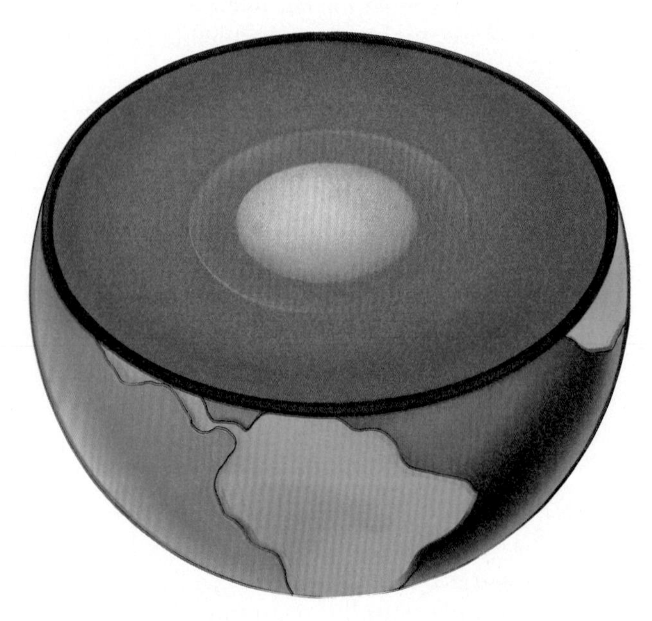

2. The land under the oceans makes up part of the earth's ___.
3. Some of the rock inside the earth's mantle is solid and some of it is ___.
4. What is the earth's core made of?
5. The ___ part of the core has liquid iron.
6. The ___ part of the core has solid iron.
7. ___ of the earth's crust give scientists information about the inner layers of the earth.

334

LESSON 2
pages 212–215

1. What happens when rocks weather?
2. Water that ___ and melts weathers rocks.
3. What parts of plants help weather rocks?
4. A material made from the ___ in the air and water helps weather rocks.
5. Underground ___ often form in weathered rock.
6. What has happened in the picture below?

LESSON 3
pages 218–221

1. What makes earthquakes dangerous?
2. Scientists try to ___ earthquakes to help save lives.
3. What should you do if you are indoors during an earthquake?
4. What should you do if you are outdoors during an earthquake?
5. What is magma?
6. How does lava form new crust?
7. Why are volcano eruptions dangerous?

LESSON 4
pages 222–223

1. How does building roads and buildings change the earth's crust?
2. What can happen to the land when people cut down plants and trees?
3. How do people change the earth's crust when they search for ores?
4. How do earthworms change the earth's crust?
5. How do plants protect the land from erosion?

335

Lesson 2
1. They crumble and wear away.
2. freezes
3. roots
4. carbon dioxide
5. caverns
6. Wind erosion caused piles of sand to form.

Lesson 3
1. Many people can be hurt or killed by falling objects.
2. predict
3. Stay away from windows and furniture that could fall on you.
4. Stay away from tall buildings and objects that could fall.
5. melted rock deep inside the earth
6. It comes out of an erupting volcano, and hardens into rock.
7. Hot lava can burn homes and injure people.

Lesson 4
1. People must dig up soil and rocks to build them.
2. Wind and water can erode the soil.
3. They dig up the earth to make mines.
4. by digging through the soil
5. by holding down the soil

Lesson 1

1. evaporates
2. condenses
3. Water vapor condenses in the air.
4. cools
5. cloud
6. Fog is close to the ground.
7. cirrus and cumulus
8. tiny pieces of ice
9. stormy
10. stratus
11. Cool air has caused water vapor to condense to fog.

Chapter 11 Study Guide

On a separate sheet of paper, write the word or words that best complete the sentence or answer the question.

LESSON 1

pages 230–232

1. After you swim, water on your skin ▒ and becomes water vapor.
2. Water vapor in the air at night ▒ into drops of dew.
3. Why can you see your breath on a cold day?
4. As warm air rises, it ▒.
5. When air condenses into tiny droplets of water high in the sky, a ▒ forms.
6. How is fog different from a cloud?
7. What kinds of clouds might you see in good weather?
8. What are cirrus clouds made of?
9. The weather can become ▒ when cumulus clouds become large.
10. You might see ▒ clouds during a light rain.
11. What has happened to the air in this picture?

336

LESSON 2

pages 234–236

1. Why can the tiny drops of water in clouds float in the air?
2. When drops of water become larger they might fall as ▒.
3. Snow is one kind of ▒.
4. Explain how the pictures below show the steps in the water cycle.

LESSON 3

pages 238–240

1. Thunderstorms occur when ▒ air moves in to replace warm air that is quickly rising.
2. You see ▒ when electric charges jump between clouds.
3. What do you often hear after lightning flashes?
4. Where do hurricanes form?
5. What two things make hurricanes dangerous?
6. What do tornados look like?
7. Why might a tornado be dangerous when it touches ground?
8. Where should people go when there is a hurricane or a tornado?

Lesson 2

1. They are very light.
2. rain
3. precipitation
4. The photograph shows evaporation of water from a lake, condensation (clouds), and precipitation.

Lesson 3

1. cool
2. lightning
3. thunder
4. over warm oceans
5. powerful winds and heavy rains
6. funnel-shaped clouds
7. Its powerful winds can destroy objects in its path.
8. inside, away from windows

Chapter 12 Study Guide

Lesson 1

1. rotates
2. the imaginary line through the earth's center that the earth rotates around
3. 24 hours
4. because as the earth rotates, a different part of the earth faces the sun at different times of the day
5. the imaginary path on which it moves around the sun
6. circle
7. 365 days, or one year
8. The moon does not have air or water.
9. revolves
10. one month
11. sun's
12. different amounts of the lighted part of the moon facing the earth as the moon revolves

Lesson 2

1. gases
2. larger
3. because it is closer to Earth
4. cooler
5. a large body that revolves around the sun

Chapter 12 Study Guide

On a separate sheet of paper, write the word or words that best complete the sentence or answer the question.

LESSON 1

pages 248–252

1. The earth is like a spinning top in the way that it ___.
2. What is the earth's axis?
3. How long does it take the earth to complete one rotation?
4. Why does the earth have daylight and darkness each day?
5. What is the earth's orbit?
6. The earth's orbit is shaped like a ___.
7. How long does it take the earth to complete one revolution around the sun?
8. What is one way in which the earth and the moon are different?
9. The moon is a satellite of the earth because it ___ around the earth.
10. How long does it take for the moon to complete one revolution around the earth?
11. The moon's light comes from the ___ light reflecting off the moon.
12. What are phases of the moon?

LESSON 2

pages 254–257

1. Stars are made of hot, glowing ___.
2. The closer an object is to you, the ___ it looks.
3. Why does the sun look larger than other stars?
4. The surface temperature of the sun is ___ than the temperature in the center of the sun.
5. What is a planet?

338

6. Name the nine planets that revolve around the sun.
7. Why are the temperatures on Mercury and Venus very hot?
8. The temperatures on Earth are close to those on ▒.
9. Saturn is ▒ in size than Earth.
10. Jupiter is made mostly of ▒.
11. The rings around Saturn are made of ▒.
12. Look at the picture below. Which planet has the coolest temperatures.

LESSON 3

pages 260–263

1. What did the telescope help scientists learn about the earth?
2. What are astronauts?
3. What do scientists think caused the craters on the moon?
4. Venus is always covered by ▒.

6. Mercury, Venus, Earth, Mars, Jupiter, Saturn, Uranus, Neptune, and Pluto
7. because they are closest to the sun
8. Mars
9. larger
10. gases
11. ice
12. Jupiter

Lesson 3

1. that it revolves around the sun
2. scientists who travel in space
3. rocks from space that crashed into the moon
4. clouds

Lesson 1
1. a cell
2. billions
3. skin
4. tissue
5. heart, stomach, eyes, and brain
6. system

Lesson 2
1. muscles
2. protect
3. long and narrow
4. shoulder joint
5. allows them to move different ways

Chapter 13 Study Guide

On a separate sheet of paper, write the word or words that best complete the sentence or answer the question.

LESSON 1
pages 276–278

1. What is the basic unit of an organism?
2. About how many cells make up your body?
3. What kind of cells cover and protect your body?
4. A group of nerve cells forms a nerve ■.
5. Name four body organs.
6. Your heart and the other organs that work together to move blood through your body form a ■.

LESSON 2
pages 280–282

1. Your bones and ■ work together to help you move.
2. Your bones help ■ organs in your body.
3. What kind of bones would the long parts of your body, like your legs, have?
4. Your shoulder and arm come together at the ■.
5. What do the joints below do for the body?

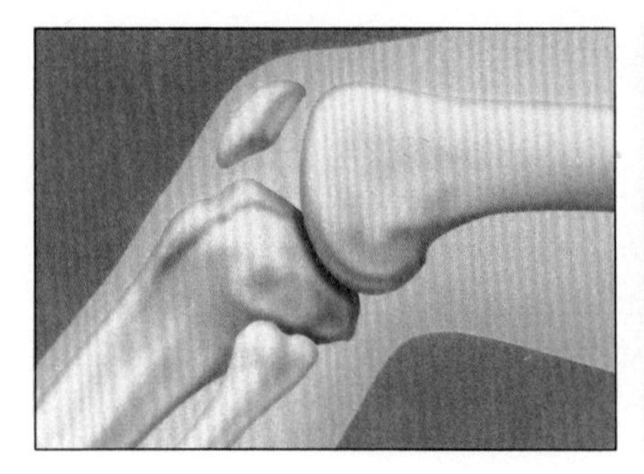

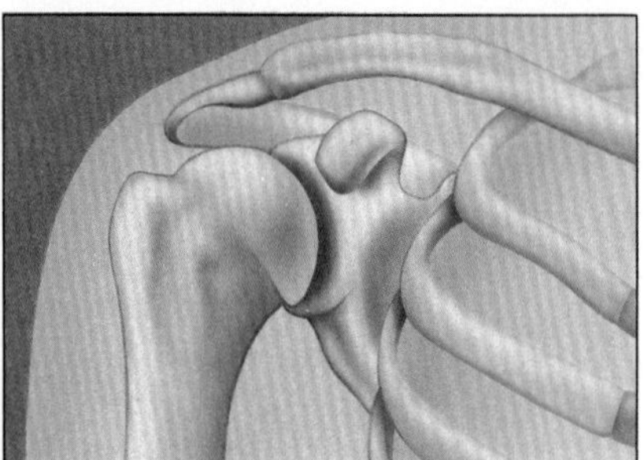

340

LESSON 3

pages 284–286

1. What helps give your body its shape?
2. When you move your arms or your legs you are using ▒ muscles.
3. The ▒ muscles of your body help you breathe and keep your heart working.
4. What kind of muscles work without you thinking about them?
5. When you move your body, your muscles ▒ or become shorter and thicker.
6. When a muscle contracts, the ▒ causes a bone to move.
7. Look at the boy in the picture. Tell what kind of muscles he is using to exercise.

Lesson 3

1. muscles
2. voluntary
3. involuntary
4. involuntary muscles
5. contract
6. tendon
7. voluntary muscles

Lesson 1

1. nutrients
2. vegetable-fruit group: four servings; bread-cereal group: four servings; meat-poultry-fish-bean group: two servings; milk-cheese group: three servings
3. It includes the servings from each food group.
4. Physical fitness helps people work or play without getting tired or hurt easily.
5. walking, dancing, or riding a bike
6. muscle
7. It helps the heart grow stronger.
8. Your body makes new cells while you sleep.

Chapter 14 Study Guide

On a separate sheet of paper, write the word or words that best complete the sentence or answer the question.

LESSON 1
pages 294–296

1. What do you get from foods that help you grow and stay healthy?
2. How many servings from each food group are needed each day?
3. Look at the meal in the picture below. Tell why this is a healthy meal.

4. Why is it important to build physical fitness?
5. What are three types of exercise you can do to build physical fitness?
6. Your heart is a ___.
7. How does exercising help your heart?
8. How does sleeping help your body grow properly?

342

LESSON 2

pages 298–301

1. What is a disease that can spread from person to person?
2. Germs that might be on your skin can get into your body through a ▒.
3. What makes you feel sick?
4. What are two diseases caused by viruses?
5. Why is it important to keep your skin clean?
6. People who have diabetes cannot use ▒ properly.
7. What are two ways in which people with diabetes can control their disease.
8. How does cancer hurt the body?
9. What are two ways doctors can treat some kinds of cancer?
10. What is one way people can help keep from getting lung cancer?
11. People protect themselves from ▒ cancer by not spending long periods of time in the sun.
12. Cancer and ▒ disease do not spread from person to person.

LESSON 3

pages 304–307

1. What is one way medicines help people?
2. Medicines are a kind of ▒.
3. Alcohol changes the way a person's ▒ works.
4. What are two skills people might have trouble doing after they have drunk too much alcohol?
5. What drug is contained in tobacco?
6. What disease are people who smoke more likely to get than those who do not smoke?
7. Drugs that are against the law to use are ▒ drugs.
8. What are two drugs that are against the law?
9. How can people protect themselves from the harmful effects of drugs?

343

Lesson 2

1. a cold
2. cut
3. disease germs growing inside your body
4. flu, chicken pox, and measles
5. to get rid of germs on your skin
6. sugar
7. by eating fewer sweets and fats, and by exercising regularly
8. It destroys normal body cells.
9. through medicines or operations
10. by not smoking
11. skin
12. heart

Lesson 3

1. They help the sick get better, and keep people from getting sick.
2. drug
3. brain
4. walking and writing
5. nicotine
6. lung cancer
7. illegal
8. cocaine and marijuana
9. by saying no to illegal drugs

TEACHING PLAN

Teaching Tips

- Students can use these two pages to remind them of the basic steps of scientific methods. A more complete explanation of the methods, with examples, appears on pages xii–5.
- As a review of scientific methods, gather a list of questions about the natural world from students. Write appropriate questions on the board. Then have the class brainstorm how scientific methods might be used to answer each question.

Using Scientific Methods

Scientists ask many questions. No one may know the answers. Then scientists use scientific methods to find answers. Scientific methods include steps like the ones on the next page. Sometimes scientists use the steps in different order. You can use these steps to do the experiments in this section.

Identify Problem
The problem is usually a question such as, "Does sound travel faster through water than through air?"

Make Observations
Notice many things about an object such as its size, color, or shape.

State Hypothesis
Try to answer the problem.

Test Hypothesis
If possible, do an experiment to see if your hypothesis is correct. Then you should do the experiment again to be sure.

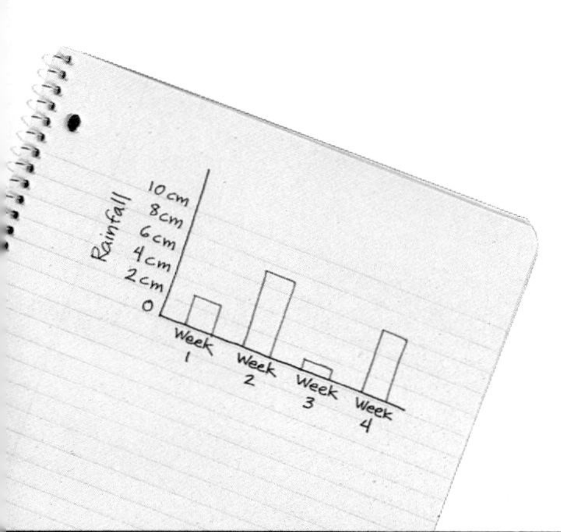

Collect Data
Your observations from the experiment are your data.

Study Data
You can understand your data better if you put it in charts and graphs.

Make Conclusions
Decide if your hypothesis is correct.

Teaching Tip

- Ask students why the steps listed here are not numbered. (The order of the steps often changes depending on the problem.) However, the order in which the steps appear, from *Identify Problem* to *Make Conclusions,* shows an accepted general sequence.

TEACHING PLAN

Teaching Tip

- Ask a group of students to make a safety poster listing safety tips for experiments or cardboard banners for each safety tip. Display the poster or banners in appropriate places throughout the room, such as over a sink or at work stations.

Safety in Science

Scientists are careful when they do experiments. You need to be careful too. The next page shows some rules to remember.

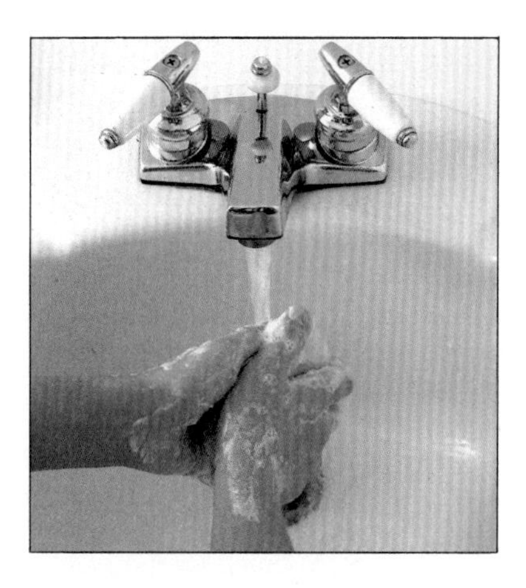

- Never taste or smell unknown things.
- Handle thermometers carefully.
- Read each experiment carefully.
- Wear cover goggles when needed.
- Clean up spills right away.

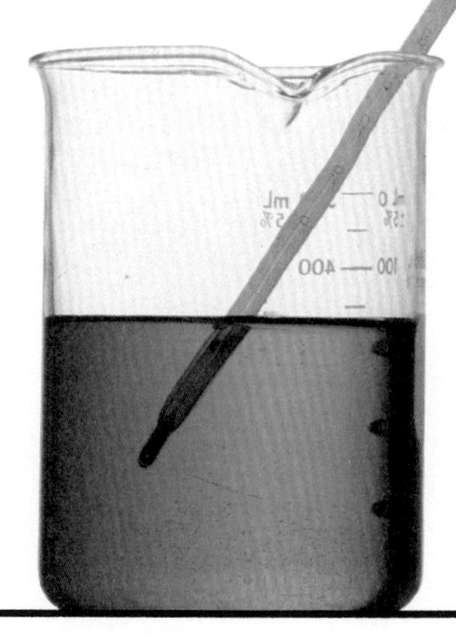

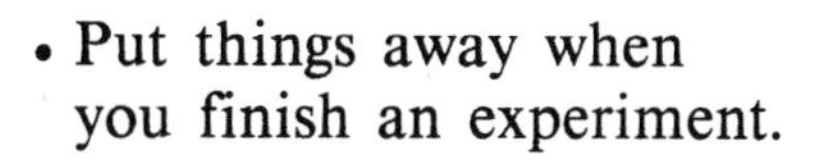

- Put things away when you finish an experiment.
- Wash your hands after each experiment.

Teaching Tips

- Underscore the importance of each safety tip by asking students what could happen if each tip is not followed.
- Invite a scientist or high-school science teacher to your class. Ask him/her to discuss the importance of safety procedures in the classroom, in the laboratory, and in the field.

TEACHING PLAN

Purpose
To develop the skill of setting up a control in an experiment that tests the effect of colored light on plant growth

Objectives/Process Skills
- *Identify variables* in an experiment about how colored light affects plant growth.
- *Set up a control* in the experiment.

1. Motivate

Discussion
Use a prism to show students that white light is made up of light of different colors. Encourage them to list the different colors of light they see. Ask students if they think plants would grow better in white light, which contains all colors of light, or in light of one color. Explain to students that their answers are hypotheses, which can be tested in an experiment.

2. Teach

Safety Tips (See page T24.)
- If students do this experiment, they should wear cover goggles when stapling the cellophane cones.
- In order to prevent falls, have students wipe up spills immediately.

Teaching Tips
- Let students discuss what they think is involved in setting up a scientific experiment. Guide students to realize that the first step in doing an experiment is forming a hypothesis to test in the experiment. Point out the hypothesis on page 349. Then direct students to the other steps on the page. Elicit the idea that drawing conclusions from the data is the final step of an experiment.

Chapter 1 Experiment Skills

Setting Up a Control

Marianne wanted to have colored glass put in her bedroom window. She had several plants in her room. She wondered if colored light coming in through the window would change how the plants grew. She decided to do an experiment with plants and light to find out. In her experiment, Marianne found out that plants grow better in regular light than in colored light.

Read Marianne's experiment on the next page. Then answer the questions on this page.

Thinking About the Experiment

1. How many cups did Marianne use for her experiment?

Anything in an experiment that could change is called a variable. Only one variable should change during an experiment.

2. What variable did Marianne change between the two cups using cellophane?
3. Why was it important for the two cups to have all the same things except for the kind of cellophane?

The cup with a clear cellophane cone was the control in Marianne's experiment. It let in all light, just as clear glass would. Marianne could compare the plant from the control cup with the plant from the other cup.

4. Suppose Marianne did not have a control cup. Could she tell if the plant with red cellophane was growing any differently? Why or why not?

See pages 344-347 to review scientific methods and safety.

Experimenting with Plants

Problem
How does red light affect bean seed growth?

Hypothesis
A bean plant will not grow as well in red light as in regular sunlight.

Materials

2 bean seeds	potting soil
cellophane, clear and red squares	stapler
metric ruler	tape
2 plastic cups	water

Procedure
1. Fill one of the cups with water. Soak 2 bean seeds in it overnight.
2. Empty the water from the cup. Fill both cups more than half full with potting soil.
3. Put a seed in each. Add a little soil on top of the seeds.
4. Water both cups well.
5. Make cones out of the two squares of cellophane. Use a staple to keep their shapes.
6. Tape a cone onto each cup.
7. Place the cups in a sunny spot.
8. After a week, measure the height of the plants. Count the number of leaves on each.

Data and Observations

	Red wrap	Clear wrap
Plant height		
Number of leaves		

Conclusions
You might do Marianne's experiment to find your own data and make observations. Draw conclusions based on your data.

Practice Setting Up a Control

Plants need minerals to grow. Suppose you want to find out if plants grow better when minerals are added to the soil.
1. How would you change Marianne's experiment to find out how well plants grow with added minerals?
2. What would the control in your experiment be?

349

Experiment Results

Plants will grow better under the clear cellophane cone than under the red cellophane cone.

Teaching Tips
- Write the words *variable* and *control* on the chalkboard. Help students define both terms.
- After students have read the experiment, ask them to list the variables in the experiment. (soil, seeds, containers for seeds, water, amount of light, color of light) Encourage them to note that the color of light is the variable that changes because the two plants are covered by a different color of cellophane. Since color of light is the only variable that changes, any differences in the growth of the two seeds is probably due to this variable.
- Ask students which setup is the control. Students should realize that the control has clear cellophane, which admits all colors of light.
- Note that most types of plants use blue and red light most efficiently during photosynthesis. However, plants can use all colors of the spectrum. Therefore, a plant grown under a few colors of light will not grow as well as one grown in white light, which contains all colors.

3. Assess

Answers
Thinking About the Experiment
1. two
2. the color of the cellophane
3. In this way, Marianne could conclude that differences in plant growth were due to differences in the color of light.
4. No, she would have nothing to compare her results with.

Practice Setting Up a Control
1. Answers will vary. Students might suggest planting bean plants in two different cups, giving the same amount of soil, water, and light, but adding extra minerals to only one plant.
2. The plant without the extra minerals would be the control.
You might encourage students to do this experiment.

TEACHING PLAN

Purpose
To develop the skill of making observations in an experiment with mealworms

Objectives/Process Skills
- *Set up an experiment* that tests a hypothesis about mealworms.
- *Make observations* about how mealworms behave.
- *Make conclusions* based on observations in an experiment.

1. Motivate

Discussion
Bring in some mealworms for students to observe. These insects are available from scientific supply companies or they can be collected from moist soil. Let students talk about some of the things they notice about the mealworms. Point out to students that the things they notice are *observations.* Tell them that they will be reading about an experiment with mealworms that involves making careful observations.

2. Teach

Safety Tips (See page T24.)
- In order to prevent falls, have students wipe up spills immediately.
- Make sure students wash their hands after handling the mealworms.
- Dispose of the mealworms promptly and properly after completing the experiment.

Making Observations

Tom liked visiting his uncle's farm. One day he noticed some tiny animals called mealworms. They were living in a damp sack of grain that an animal had spilled. Tom wondered if the mealworms like to live in a damp place or a dry place. He decided to do an experiment with ten mealworms to find out.

At the end of his experiment, Tom observed that only two of the ten mealworms crawled to a dry place. All the rest crawled to a wet place.

Read Tom's experiment on the next page. Then answer the questions on this page.

Thinking About the Experiment

1. Where did Tom first observe the mealworms?
2. What was Tom trying to find out in his experiment?
3. What materials did Tom use in his experiment?

Making observations means watching closely. Tom used his observations to make a conclusion.

4. How many mealworms did Tom observe at the dry sponge?
5. How many mealworms did Tom observe at the wet sponge?
6. What conclusion do you think Tom made?

See pages 344-347 to review scientific methods and safety.

Experimenting with Mealworms

Problem
Do mealworms prefer to live in a damp place or a dry place?

Hypothesis
Mealworms prefer to live in a damp place.

Materials

small cake pan	2 sponges
10 mealworms	water
cardboard	sand

Procedure
1. Fill the bottom of the pan with sand.
2. Wet 1 sponge. Then squeeze it.
3. Place the wet sponge in the sand on one side of the pan.
4. Place the dry sponge opposite the wet sponge.
5. Use a piece of cardboard to put the mealworms on the sand in the middle of the pan.
6. Observe the mealworms after 30 minutes. Write down how many are at each sponge.

Data and Observations

	Number of mealworms
Wet sponge	8
Dry sponge	2

Conclusions
You might do Tom's experiment to find your own data and make observations. Draw conclusions based on your data.

Practice Making Observations

1. Suppose you observed the mealworms for five days. Set up the data table that you would use.
2. Suppose you wanted to see if the mealworms grew. How could you make observations about growth?
3. What would you use to measure how much they grew?

351

Experiment Results

Although the numbers of mealworms that go to the damp sponge may vary, more mealworms will go to the damp sponge than to the dry sponge.

Teaching Tips

- Make sure students handle the mealworms carefully. Mealworms can be injured or crushed with rough handling.
- Tell students that mealworms are larvae of beetles.

3. Assess

Answers

Thinking About the Experiment

1. in a damp sack of grain.
2. whether mealworms prefer damp or dry places
3. cardboard, small cake pan, 10 mealworms, sand, 2 sponges, water
4. 2
5. 8
6. Mealworms prefer damp places

Practice Making Observations

1. Data tables should show additional columns for each day of observation. One way is shown below.

Number of Mealworms

	Day 1	Day 2	Day 3	Day 4	Day 5
Wet sponge					
Dry sponge					

2. measure lengths
3. a metric ruler
You might encourage students to do this experiment.

TEACHING PLAN

Purpose

To develop the skill of stating a hypothesis in an experiment about the effect of moisture on mold growth

Objectives/Process Skills

- *State a hypothesis* about mold growth.
- *Set up an experiment* that tests the hypothesis.
- *Make observations* in the experiment.

1. Motivate

Discussion

Ask students to describe any molds or other fungi they have seen. Some will have seen mold on food or mildew in damp places. Explain that some fungi, such as mushrooms and the molds in some cheeses, are helpful. Question: **Where do molds grow?** (Accept any reasonable answers.) **Will mold grow faster in a moist place than in a dry place?** (Accept any answer.) Explain to students that their answers are hypotheses which can be tested by an experiment.

2. Teach

Safety Tips (See page T24.)

- Molds can aggravate allergies. Do not open the plastic bags or allow students to open them. Dispose of the sealed bags after the experiment is completed.
- In order to prevent falls, have students wipe up spills immediately.
- Warn students not to eat any food used in the experiment.
- Direct students to wash their hands after doing the experiment.

Chapter 3 Experiment Skills

Stating a Hypothesis

When Robert started to make a peanut butter sandwich, he saw that mold was growing on the bread. He also noticed water on the bread wrapper. Robert wondered if water helped the mold to grow. He thought mold would grow faster on wet bread than on dry bread. He did an experiment with dry and wet bread to see if he was right. In his experiment, Robert observed that more mold grew on the wet bread than on the dry bread.

Read Robert's experiment on the next page. Then answer the questions on this page.

Thinking About the Experiment

1. What question did Robert have about how fast bread mold grows?
2. How would Robert finish this sentence? "Mold grows faster on . . ."

Robert's sentence is his hypothesis. A good hypothesis is one that can be tested with an experiment.

3. What materials did Robert use to test the hypothesis he made?
4. What procedure did Robert follow to test his hypothesis?

Suppose Robert wanted to show that mold grow better in the dark than in light.

5. What would be his new hypothesis?

See pages 344-347 to review scientific methods and safety.

Experimenting with Mold

Problem
Will mold grow faster on wet bread than on dry bread?

Hypothesis
Give your own hypothesis for this experiment.

Materials
2 slices bread
marker
masking tape
2 plastic sandwich bags
water

Procedure
1. Leave 2 slices of bread out overnight. This will allow mold to start growing on the bread.
2. The next day, place 1 of the slices in a plastic sandwich bag.
3. Wet the second slice. Let several drops of water fall from your finger onto the bread in several places. Place the slice in a plastic sandwich bag.
4. Tape both bags shut. Label them by writing on the tape.
5. Place the bags in a warm, dark place.
6. Observe the bread slices through the plastic every other day for 1 week. Write your observations in a data table.
7. After the experiment, throw the closed bags away.

Data and Observations

	Day			
	1	3	5	7
Wet bread				
Dry bread				

Conclusions
You might do Robert's experiment to find your own data and make observations. Draw conclusions based on your data.

Practice Stating a Hypothesis

Some breads have chemicals called preservatives that keep the bread from getting moldy. Other breads do not have preservatives. Do preservatives keep mold from growing on bread?
1. Write a hypothesis to answer this question.
2. How could you test this hypothesis?

353

Experiment Results

Mold should appear on the wet bread after 3–4 days. Mold may not appear on the dry bread for 10 or more days.

Teaching Tips
- Use white bread for the experiment. Mold will be easier to see on white bread than on dark bread.
- After students have read page 353, question: **What observations did Robert make before he set up the experiment?** (Answers will vary, but students should mention that he saw mold and moisture in a package of bread.)
- You may wish to explain to students that molds that grow on food feed on the food as they grow on it. Water helps the molds to grow. The antibiotic called penicillin is obtained from a mold that grows on bread and fruits.

3. Assess

Answers
Thinking About the Experiment
1. Answers may vary. Robert's question might have been: Does moisture help mold grow? or Does mold grow faster on wet bread than on dry bread?
2. wet bread than on dry bread.
3. 2 slices of bread, a marker, masking tape, 2 plastic bags, and water
4. He observed wet and dry bread for one week.
5. Mold grows better in darkness than in light.

Practice Stating a Hypothesis
1. Answers will vary, but students should suggest that preservatives prevent mold from growing on bread or that mold grows more slowly on bread with preservatives.
2. Set up an experiment using bread with and without preservatives. Keep other conditions the same, including equal amounts of water, darkness, and temperature.
You might encourage students to do this experiment.

TEACHING PLAN

Purpose
To develop the skill of setting up an experiment about how fertilizer affects plant growth

Objectives/Process Skills
- *Set up an experiment* that tests the effect of fertilizer on plant growth.
- *Identify variables* in the experiment.

1. Motivate

Discussion
Write the word *fertilizer* on the chalkboard. Help students understand that fertilizers are chemicals that help plants grow. Discuss with students how people use fertilizers. Farmers use fertilizers to help crops grow; homeowners use fertilizers to make their lawn greener; florists use fertilizers to keep houseplants healthy. Tell students they will be reading about an experiment that deals with fertilizers and plant growth on page 355.

2. Teach

Safety Tips (See page T24.)
- If students do this experiment, warn them that the fertilizer is a strong chemical that can hurt their skin. Make sure that students wear rubber gloves and cover goggles for protection while they set up and do this experiment.
- In order to prevent falls, have students wipe up spills immediately.

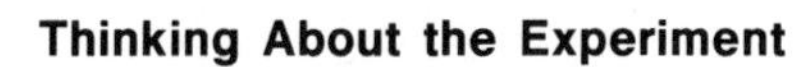
Chapter 4 Experiment Skills

Setting Up an Experiment

Lori's grandmother lived on a farm with a big pond. A garden was near the pond. Lori noticed that the water nearest the garden looked greener than the rest of the pond. This water was covered with many tiny floating duckweed plants. Her grandmother said she used fertilizer in the garden. Lori wondered if the fertilizer had washed into the pond and made the duckweed near the garden grow quickly.

She set up an experiment using pond water, duckweed, and fertilizer to find out. In her experiment, Lori found out that duckweed with fertilizer grows faster than duckweed without fertilizer.

Read Lori's experiment on the next page. Then answer the questions on this page.

Thinking About the Experiment

1. What was Lori's hypothesis?
2. How many cups did Lori use to set up her experiment?

When Lori set up her experiment, she made sure the cups were the same size. She put the same amount of pond water with duckweed in each cup. Then she put the cups in the sun. Lori changed one variable. This was the variable she was testing.

3. What variable did Lori change?

The cup without fertilizer was Lori's control.

4. How did the control help Lori with her observations?

See pages 344-347 to review scientific methods and safety.

Experimenting with Fertilizer on Plants

Problem
Does fertilizer change how fast duckweed grows?

Hypothesis
Fertilizer helps duckweed grow quickly in water.

Materials

gloves	phosphate fertilizer
2 large plastic cups	water with duckweed
marker	medicine dropper

Procedure
1. Fill each cup half full with pond water.
2. Put on the gloves. Add 10 drops of fertilizer to one of the cups. Stir the fertilizer in the water.
3. Place 20 duckweed plants in each cup.
4. Record the number of plants in your data table beside "At start."
5. Label the cups. Put them in a sunny place.
6. Observe the cups once a week for two weeks. Count the duckweed plants in each cup each time. Record the numbers on your data table.

Data and Observations
Number of Duckweed

	With fertilizer	Without fertilizer
At start		
After 1 week		
After 2 weeks		

Conclusions
You might do Lori's experiment to find your own data and make observations. Draw conclusions based on your data.

Practice Setting Up an Experiment

Does the temperature of water affect how fast duckweed plants grow? Suppose you wanted to set up an experiment to find out. You could use two cups of pond water with duckweed.
1. What would be the same about each cup?
2. What would be the variable?
3. What would be the control?

Experiment Results

There should be more duckweed in the cup with fertilizer than in the cup without fertilizer.

Teaching Tips

- Read the top part of page 354 aloud with the students. Help students realize that Lori wondered if fertilizer washing into the pond made duckweed grow faster at the grassy edge of the pond.
- Define *hypothesis* for the students. Point out that a hypothesis answers a question posed by the experimenter. A hypothesis is given in the form of a statement and is tested in the experiment. Ask students what they think Lori's hypothesis was. Write their responses on the chalkboard. Have students check their ideas against the hypothesis given on page 355.

3. Assess

Answers

Thinking About the Experiment
1. Answers may vary, but students' responses should include the statement that fertilizer helps duckweed grow.
2. two
3. Lori added fertilizer to only one cup.
4. She could compare how fast the duckweed grew with and without fertilizer.

Practice Setting Up an Experiment
1. The number of duckweed, the amount of water, the amount of food, and the amount of light should all be the same.
2. The temperature of the water would differ.
3. A cup with room temperature water would be the control.
You might encourage students to do this experiment.

TEACHING PLAN

Purpose
To develop the skill of identifying variables in an experiment in which sugar is mixed in water of different temperatures

Objectives/Process Skills
• *Identify variables* in an experiment about mixing sugar and water.
• *Study data* collected in an experiment.

1. Motivate

Discussion
Ask students to share their experiences with mixing substances in water. Students may have mixed bouillon cubes in hot water, fruit-drink powders in cold water, or frozen concentrate in cold water. You may wish to demonstrate how well these or other substances mix in water. Let students discuss whether most substances dissolve faster in hot water, warm water, or cold water. Help students form hypotheses based on their experiences. Tell them they will be reading about an experiment that tests how water temperature affects the mixing of sugar and water.

2. Teach

Safety Tips (See page T24.)
• Tell students not to taste any substance used in the experiment.
• In order to prevent falls, have students wipe up spills immediately.

Chapter 5 Experiment Skills

Identifying Variables

Mario was making a snack. His father had given him a cup of warm water. Mario was mixing a packet of instant tomato soup in the water. He noticed that the instant soup mixed quickly in the warm water. He wondered if the temperature of water changes how fast something will mix. Mario decided to do an experiment to find out. In his experiment, he mixed sugar in water of different temperatures.

Read Mario's experiment on the next page. Then answer the questions on this page.

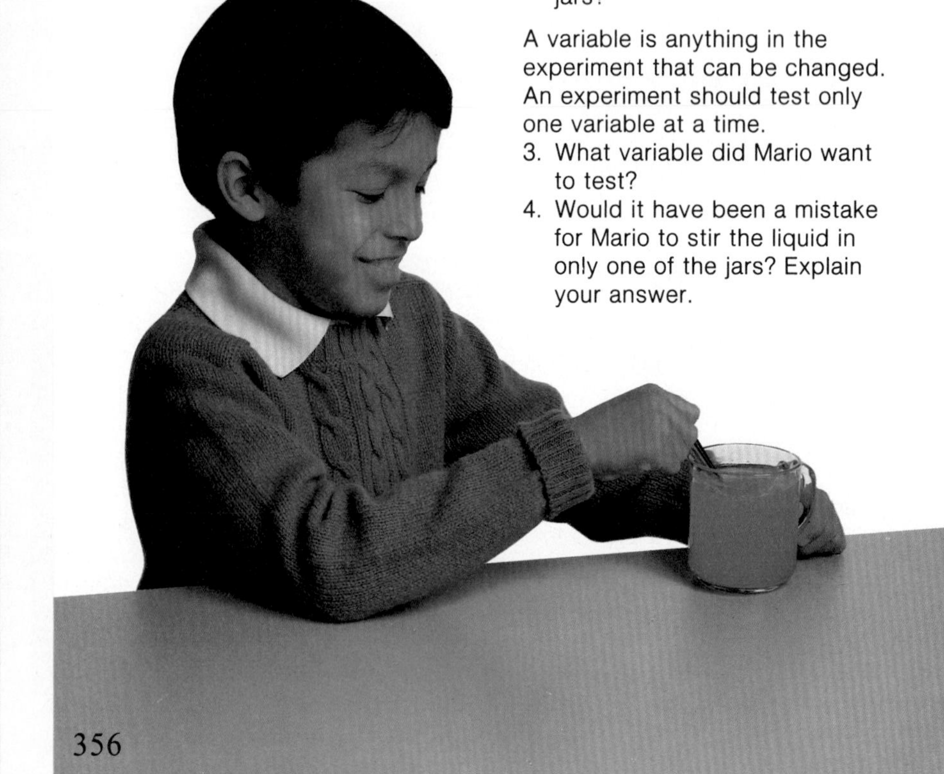

Thinking About the Experiment

In his experiment, Mario used jars that were the same size. He put the same amount of water and of sugar in each jar.

1. What else was the same for all the jars?
2. What was different in the three jars?

A variable is anything in the experiment that can be changed. An experiment should test only one variable at a time.

3. What variable did Mario want to test?
4. Would it have been a mistake for Mario to stir the liquid in only one of the jars? Explain your answer.

See pages 344-347 to review scientific methods and safety.

Experimenting with Mixtures

Problem
Does the temperature of water change how fast something will mix in it?

Hypothesis
A substance will mix faster in warm water than in cold water.

Materials

warm water	sugar
room temperature water	spoon
cold water	clock
3 jars of the same size	tape
graduated cylinder	marker

Procedure
1. Label the jars *Warm, Cold,* and *Room temperature.*
2. Put 25 mL of sugar in each jar.
3. Measure 100 mL of warm water. Pour it into the jar marked *Warm.* Observe the time.
4. Stir the sugar and water until the sugar crystals disappear.
5. Observe how much time the sugar took to disappear. Write the time on a chart like this one.
6. Follow the same procedure again. Add cold water to the jar marked *Cold.* Add room temperature water to the jar marked *Room temperature.*

Data and Observations

Water temperature	Time for sugar to disappear
Warm	
Room temperature	
Cold	

Conclusions
You might do Mario's experiment to find your own data and make observations. Draw conclusions based on your data.

Practice Identifying Variables

Suppose you wanted to do an experiment to find out if the amount of water changes how fast sugar will dissolve.
1. How would you set up the experiment?
2. What would have to be the same in your experiment?
3. What variable changes?

357

Experiment Results

The sugar will dissolve faster in warm water than in cold water.

Teaching Tips

- After students have read the experiment on page 357, ask them to name some variables in the experiment. List these variables on the chalkboard. Help students realize that the amount of water, the temperature of the water, the substance being mixed, the amount of the substance being mixed, and how the substance is mixed are all variables in the experiment.
- Ask students to identify the variable that changes in this experiment. Point out that only one variable changes. The other variables remain the same throughout the experiment. Students should realize that the variable that changes is the temperature of the water.

3. Assess

Answers

Thinking About the Experiment
1. He stirred the sugar and water in each jar.
2. the temperature of the water
3. water temperature
4. Yes, it would have added another variable being tested.

Practice Identifying Variables
1. Set up the experiment similarly to Mario's, only use different amounts of water.
2. temperature of water, amount of sugar, size of the jars, stirring
3. amount of water
You might encourage students to do this experiment.

TEACHING PLAN

Purpose
To develop the skill of setting up an experiment that tests the performance of different kinds of wedges

Objectives/Process Skills
- *Set up an experiment* that tests the performance of wedges of different shapes.
- *Make a model* of a wedge using chalk.
- *Make observations* concerning the amount of force needed to use wedges of different shapes.

1. Motivate

Discussion
Ask students if they have watched someone split wood. Let students discuss the types of tools people might use to split wood. They might suggest using an ax or a heavy hammer with a wedge-shaped tool. You may want to demonstrate how to use a wedge and hammer to split wood. If you do this demonstration, make sure to wear safety goggles and have the students stand back to prevent injury from flying wood fragments. Discuss the amount of force needed to split the wood and the importance of the wedge's shape.

Setting Up an Experiment

Jeff's father used a wedge to split wood. The hammer pushed the wedge into the wood and the wood split or cracked. Jeff wondered if the sharpness of a wedge makes a difference in how well it works. He decided to set up an experiment to find out. Jeff knew it was not safe to use his father's sharp wedge. He used wedges made out of chalk.

Read Jeff's experiment on the next page. Then answer the questions on this page.

Thinking About the Experiment

Jeff set up his experiment carefully. He kept every part of the setup the same except for one variable. That was the variable he wanted to test.

1. What variable is Jeff testing?
2. What variables did Jeff keep the same in the experiment?
3. What kind of data did Jeff get from the experiment?

See pages 344-347 to review scientific methods and safety.

Experimenting with Wedges

Problem
Does the sharpness of a wedge make a difference in how easily it can be pushed into something?

Hypothesis
A wedge is most useful if it has a sharp edge.

Materials
3 pieces of chalk
large ball of modeling clay
rough surface

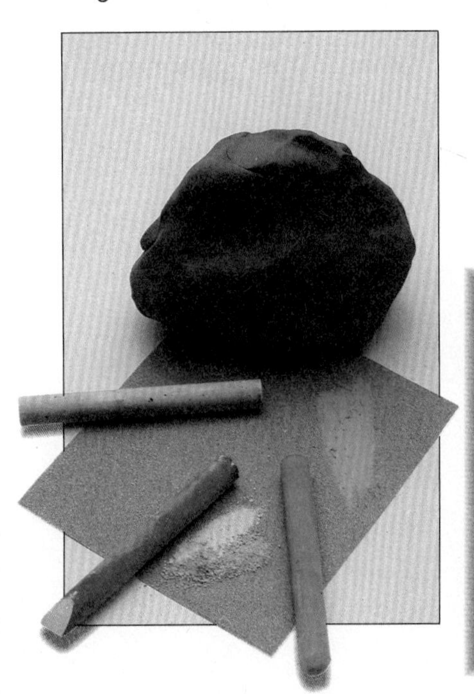

Procedure
1. Rub one end of each piece of chalk on the rough surface to make a wedge shape. Make the ends of the wedges sharp, rounded, and flat.
2. Push each chalk wedge into the ball of clay.
3. Record how hard or easy it is to push each wedge.

Data and Observations

Wedge shape	Force needed
Rounded	
Flat-ended	
Sharp	

Conclusions
You might do Jeff's experiment to find your own data and make observations. Draw conclusions based on your data.

Practice Setting Up an Experiment

Suppose you wanted to know if the material a wedge is made of affects how well it works.
1. What might be your hypothesis?
2. What variable would you test?
3. How would you set up your experiment?

Experiment Results

Sharp wedges can be pushed into the clay more easily than round or flat-ended wedges.

2. Teach

Teaching Tips

- Point out that in the experiment on page 359, the chalk is a model of a wedge and the clay represents wood. Discuss with the students why Jeff decided to use models instead of a real wedge and wood.
- **Helpful Hint:** Pass out small pieces of coarse sandpaper for students to use in shaping their chalk wedges. Grade 80 works well. If sandpaper is not available, have them rub the edges of the chalk wedges against dark construction paper to shape them.

3. Assess

Answers

Thinking About the Experiment
1. the shape of the wedges
2. material the wedges are made of; material the wedges are pushed into; who is pushing the wedges into the material
3. information on the force needed to push each wedge into the clay

Practice Setting Up an Experiment
1. A wedge made of hard material works better than a wedge made of soft material.
2. the material wedges are made of
3. Make the wedges out of several different materials, but make them all the same shape. Push the wedges into the same ball of clay and observe the amount of force needed to push each one.
You might encourage students to do this experiment.

TEACHING PLAN

Purpose
To develop the skill of making conclusions in an experiment with magnets and paper clips.

Objectives/Process Skills
- *Collect data* about making magnets.
- *Make conclusions* about the best way to magnetize a paper clip.

1. Motivate

Discussion
Ask students to share their knowledge about magnets. Discuss the idea that magnets have two poles. A bar magnet has a pole at either end. Students should realize that magnets attract objects with iron in them. Encourage students to suggest ways they could use a bar magnet to turn iron objects into magnets. Help students state their ideas in the form of a hypothesis.

2. Teach

Safety Tips (See page T24.)
- Direct students to handle their straightened paper clips carefully to avoid hurting themselves and others with the sharp, pointed ends.
- You may wish to provide cover goggles for students doing this experiment.

Teaching Tip
- After students have read the experiment on page 361, compare the ways students suggested making magnets from paper clips with those given in the procedure of the experiment. Let students predict which of the ways listed will work the best.

Making Conclusions

Kelly was studying magnets at school. Her teacher said there were ways you could make your own magnet. He said that to do this you would need only a paper clip and a bar magnet. These were the only directions her teacher would give.

Kelly decided she would do an experiment to find out how to make a magnet. She tried to make a magnet in three different ways.

Read Kelly's experiment on the next page. Then answer the questions on this page.

Thinking About the Experiment

Kelly tested the paper clip that the laid next to the magnet. It did not pick up any small paper clips.

1. What is Kelly's conclusion about making a magnet this way?

Kelly tested the paper clip that she rubbed back and forth with a magnet. The paper clip picked up another paper clip.

2. What is Kelly's conclusion?

Next, Kelly tested the paper clip that she rubbed in one direction with the magnet. The paper clip picked up four other paper clips.

3. What is Kelly's conclusion about the best way to make a magnet?

See pages 344-347 to review scientific methods and safety.

Experimenting with Magnets

Problem
How can you make a magnet?

Hypothesis
Rubbing a paper clip with one end of a bar magnet can turn the paper clip into a magnet.

Materials
three large paper clips
bar magnet
several small paper clips

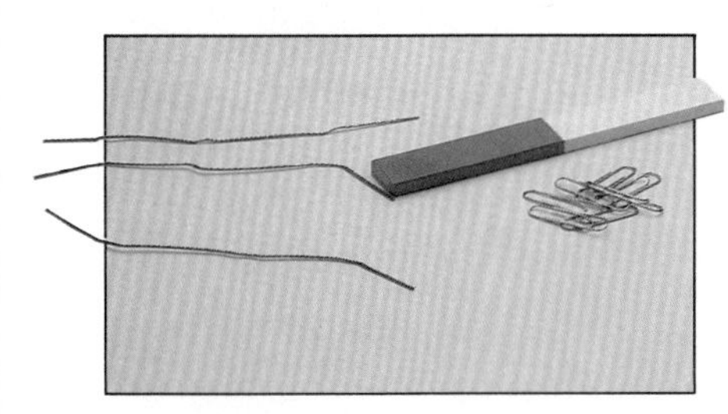

Procedure
1. Straighten out the large clips.
2. Set one end of the magnet next to one end of a straightened clip for 30 minutes.
3. Rub one end of the magnet back and forth over the end of another paper clip. Do this at least 20 times.
4. Rub one end of the magnet over the end of the third paper clip. But this time rub the clip in only one direction. Do this at least 20 times.
5. Test each of your straightened clips to see how well they are magnetized. Write down how many small paper clips are attracted to each of your homemade magnets.

Data and Observations

Straightened paper clips	Number of paper clips picked up
Laid next to magnet	0
Rubbed back and forth	1
Rubbed in only one direction	4

Conclusions
You might do Kelly's experiment to find your own data and make observations. Draw conclusions based on your data.

Practice Making Conclusions

1. Suppose the paper clip laid next to the magnet was able to pick up the most paper clips. What would be your conclusion to this experiment?
2. Suppose a conclusion does not agree with a hypothesis. Explain why the experiment is still useful.

Experiment Results

By trial and error, students should find that stroking the paper clip in one direction is the best way to make a magnet. Use one end of a bar magnet to stroke the paper clip.

Teaching Tips
- Direct students' attention to the data table on page 361. Let students compare their data for this experiment. If students do not do the experiment, give them a sample value for each entry. Question: **Which of the magnetized paper clips picked up the most small paper clips?** (paper clip in step 4) **Which picked up the fewest paper clips?** (paper clip in step 2) Help students *draw conclusions* from the data.
- **Helpful Hint:** You might want to straighten the paper clips for the students before they do the experiment.

3. Assess

Answers
Thinking About the Experiment
1. Setting a magnet next to a paper clip is not a good way to make a magnet.
2. Rubbing a paper clip back and forth with a magnet will turn the paper clip into a magnet.
3. The best way to magnetize a paper clip is by rubbing a magnet in one direction along the paper clip.

Practice Making Conclusions
1. Laying a paper clip near a magnet is the best way to turn the paper clip into a magnet.
2. The experiment still provides useful data. A conclusion that does not agree with the hypothesis indicates that another hypothesis has to be considered. *You might encourage students to do this experiment.*

TEACHING PLAN

Purpose
To develop the skill of collecting data in an experiment about sound

Objectives/Process Skills
- *Identify a problem* that is the basis for an experiment about sound.
- *Identify variables* in an experiment.
- *Collect data* to compare the pitch of sounds.

1. Motivate

Discussion
Ask students to share what they know about how different types of instruments produce sounds. Students might describe how hammers hitting stretched wires produce sounds in a piano; blowing through a hole produces sound in wind instruments; and tapping on objects of different shapes produces sounds in percussion instruments. Tell students that, in this experiment, they will collect data about how different sounds are produced by tapping on bottles filled with different amounts of water.

2. Teach

Safety Tip (See page T24.)
- In order to prevent falls, have students wipe up spills immediately.

Collecting Data

Ann tapped gently with a spoon on her glass of fruit juice. She was keeping time with music on the radio. Then she drank most of the juice. When she tapped the glass again she noticed that the sound was different. The sound was higher with less juice in the glass.

Ann wondered if this would be true if she tapped bottles. She thought of this hypothesis to answer her question. *The more liquid in a bottle, the lower the sound made when you tap on the bottle.* She decided to set up an experiment to test her hypothesis.

Read Ann's experiment on the next page. Then answer the questions on this page.

Thinking About the Experiment

In an experiment, you collect data about the problem you want to solve.

1. What is the problem in Ann's experiment?
2. What did Ann collect data about?
3. How did Ann organize the data she collected?
4. After looking at her data, what do you think Ann's conclusions were?

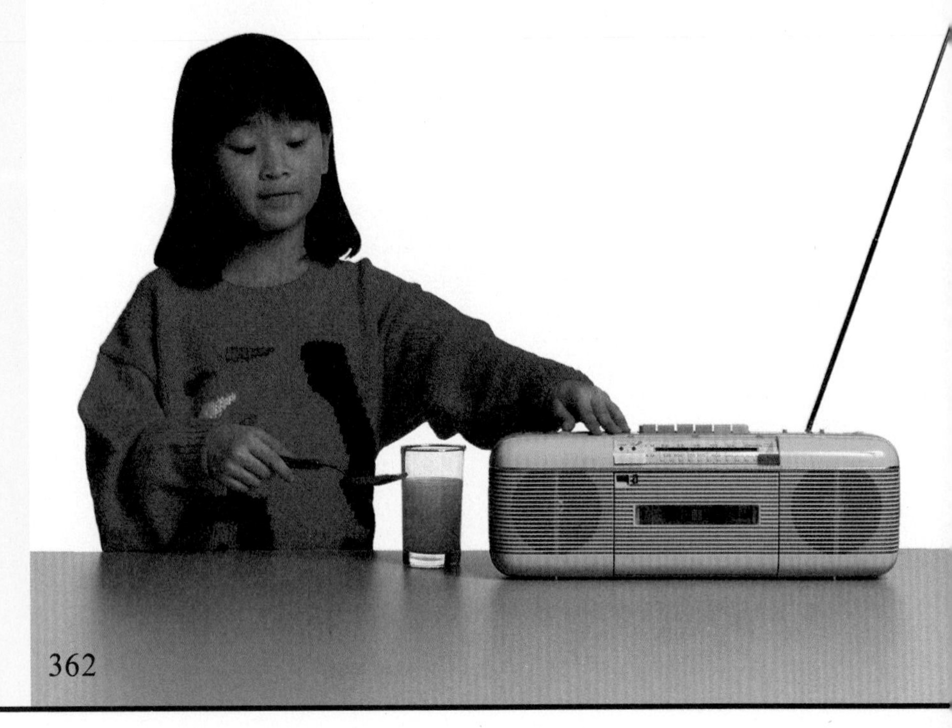

See pages 344-347 to review scientific methods and safety.

Experimenting with Sound

Problem
Does the amount of liquid in a bottle make a difference in the sound made by tapping on the bottle?

Hypothesis
The more liquid in a bottle, the lower the sound made when you tap on the bottle.

Materials
4 bottles of the same size
water
spoon

Procedure
1. Put four bottles on a desk or table.
2. Leave the first bottle empty. Put a little water in the second bottle. Fill the third bottle about halfway. Fill the fourth bottle almost full.
3. Tap the first and second bottles gently with a spoon. Compare the sounds. Record which is higher.
4. Tap the third bottle. Compare the sound with the sounds of the first and second bottles. Record your observations.
5. Tap the fourth bottle and compare the sounds. Record your observations.

Data and Observations

Amount of water in bottle	Sound from tapping
Empty	highest
A little water	lower
Half full	still lower
Full	lowest

Conclusions
You might do Ann's experiment to collect your own data and make observations. Make conclusions based on your data.

Practice Collecting Data

Suppose you wanted to do an experiment to find out if the size of the bottle makes a difference in the sound produced.
1. What might your hypothesis be?
2. How could you set up an experiment to test the hypothesis?
3. What data is important to collect in your experiment?

Experiment Results

The more liquid in the bottle, the lower the sound produced when the bottle is tapped.

Teaching Tips

- Play a scale or a simple tune for the students on a piano or other musical instrument. Ask them to identify which notes are high and which are low. Make sure students can distinguish high notes from low notes. If a musical instrument is not available, you may wish to use several bottles filled with different levels of water. Wrap the bottles with paper so that students are forced to determine the relative pitch of the sounds without seeing the water levels.
- If you have hearing-impaired students in your class, you may wish to suggest that these students touch the rim or side of the bottle after they tap it, to observe how the glass vibrates. Fast vibrations indicate a high sound; slower vibrations indicate a lower sound.

3. Assess

Answers

Thinking About the Experiment

1. Does the amount of liquid in a bottle make a difference in the sound made by tapping on the bottle?
2. how high or low a sound is with each level of water
3. She put the information in a table.
4. The more water in the bottle, the lower the pitch of the sound.

Practice Collecting Data

1. Answers will vary. Students might say larger bottles, when tapped, make a higher (or lower) sound than smaller bottles.
2. Tap on similar bottles of several sizes. Bottles should all be empty or full. Observe and compare sounds produced.
3. sounds produced when various sized bottles are tapped
You might encourage students to do this experiment.

TEACHING PLAN

Purpose

To develop the skill of making observations in an experiment that tests the effect of soil type on plant growth

Objectives/Process Skills

- *Identify a problem* concerning how well plants grow in different types of soil.
- *Make observations* about the growth of plants in various types of soil.

1. Motivate

Discussion

Bring in a plant that students can observe. For example, you may wish to bring in a flower bud that is just opening. Ask students to look at the flower at the beginning of the day. Put the flower in a warm, sunny place. Ask students to look at the flower in the afternoon. Encourage students to discuss any changes they notice in the flower. Point out that the things the students noticed about the flower and how it changed are observations.

2. Teach

Safety Tips (See page T24.)

- If students do this experiment, remind them not to play with the beans or put anything that they use in an experiment in their mouth.
- In order to prevent falls, have students wipe up spills immediately.

Chapter 9 Experiment Skills

Making Observations

Alice helped her father get the soil in the garden ready for new plants. They mixed decaying leaves and grass clippings into the soil. Alice wondered if plants grow better in mixed soil than in sandy soil or in gravel. Alice did an experiment to find out.

She planted bean seeds in different kinds of soil and in gravel. She observed and measured the growing plants. In the experiment, Alice found out that plants grow best in mixed soil.

Read Alice's experiment on the next page. Then answer the questions on this page.

Thinking About the Experiment

The first step in solving a problem is to make observations.

1. What observations did Alice make about the plants?
2. What units of measurement did Alice use?
3. Where did Alice record her observations and measurements?

Alice treated the bean seeds alike except for one thing.

4. What variable did Alice change?
5. What was Alice trying to find out about the plants?

See pages 344-347 to review scientific methods and safety.

Experimenting with Soil

Problem
Do plants grow better in some substances than in others?

Hypothesis
Plants grow better in some substances than in others.

Materials
9 bean seeds soaked in water overnight
3 paper or plastic cups
loam, sand, and fine gravel
large plastic or metal tray
measuring cup
water
metric ruler
pencil
masking tape
marker

Procedure
1. Label the cups *1*, *2*, and *3*.
2. Poke a hole in the bottom of each cup with a pencil.
3. Put soil in cup *1*, gravel in cup *2*, and sand in cup *3*.
4. Plant 3 seeds in each cup. Plant the seeds about 2 cm deep.
5. Put the tray of cups on a windowsill.
6. Every 3 days, add 15 mL of water to each cup.
7. On days you water the cups, observe each plant. Measure the height of the plants in centimeters. Record your observations and measurements on a chart like the one shown.

Data and Observations

Cup	Substance	Height and appearance
1	loam	Answers will
2	gravel	vary for the measurements
3	sand	in the chart.

Conclusions
You might do Alice's experiment to find your own data and make observations. Draw conclusions based on your data.

Practice Making Observations

Suppose you wanted to find out if the amount of water makes a difference in how a plant grows.
1. What hypothesis could you make?
2. How would you set up an experiment to test your hypothesis?
3. What observations would you make?

365

Experiment Results

Seeds planted in soil grow better than those planted in sand or in gravel.

Teaching Tips

- Direct students to read the top of page 364. Question: **What kind of soil do we have in our area?** (Accept all answers.) Ask students what they have observed that makes them think they have a particular type of soil in their area.
- Discuss some of the factors students think might affect plant growth. Students might mention sunlight, water, temperature, and soil type.
- Let students evaluate the idea that soil type affects plant growth. Ask them if they think plants will grow better in sand, gravel, or soil. Help students state their answers as hypotheses. Then direct students to read the experiment on page 365.
- **Helpful Hints:** If possible, use potting soil for the soil in this experiment, since soil gathered from different locations will vary in quality.
- You may wish to add other types of soil to the experiment, including clay soil and peat moss.

3. Assess

Answers

Thinking About the Experiment
1. She observed height and appearance.
2. centimeters
3. on a chart
4. substance plants grew in
5. their change in height and appearance when grown in different substances.

Practice Making Observations
1. Plants grow best when they get enough water.
2. Grow the same kind of plants in the same kind of soil but add different amounts of water to each plant.
3. height and appearance of the plants
You might encourage students to do this experiment.

TEACHING PLAN

Purpose
To develop the skill of identifying variables in an experiment about erosion

Objectives/Process Skills
- *Identify variables* in an experiment.
- *Use a model* of rain eroding a hill of dirt.
- *Make observations* about erosion.

1. Motivate

Discussion
Ask students if they have ever seen rain water rushing down a hill of sand or dirt. Encourage students to share their experiences. Question: **What is the name of the process in which water washes away soil and dirt?** (erosion) Write this word on the chalkboard. Inform students that they will be making models of soil erosion. Using their models in an experiment they will identify some of the factors that affect the rate at which rain washes away soil.

2. Teach

Safety Tips (See page T24.)
- In order to prevent falls, have students wipe up spills immediately.
- Direct students to wash their hands after doing this experiment.

Identifying Variables

Bob was digging in his garden when it started to rain. He ran inside and left a pile of dirt out in the rain. Then he watched the pouring rain wash away the dirt. He wondered if a lot of rain washed away more dirt than only a little bit of rain could. He decided to do an experiment using sand and water to find out.

In his experiment Bob found out that a lot of water washes away more sand than a small amount of water. Read Bob's experiment on the next page. Then answer the questions on this page.

Thinking About the Experiment

Bob made three mounds of sand all the same size. He poured a different amount of water over each mound.

1. Why was it important for Bob to make all three mounds of sand the same size?
2. What did Bob use to shape the mounds?

Only one variable should change in an experiment. That is the variable you want to test.

3. What variable did Bob want to test?
4. How did he test this variable?

See pages 344-347 to review scientific methods and safety.

Experimenting with Sand and Water

Problem
Does a large amount of water wash away more sand than a small amount of water does?

Hypothesis
A large amount of water washes away more sand than a small amount does.

Materials

1 plastic cup
3 pans of the same size
measuring cup
3 plastic straws
sand
marker
water

Procedure

1. Fill the plastic cup with sand. Pack sand firmly in the cup.
2. Turn the cup over into a pan. Remove the cup so that the sand forms a mound.
3. Repeat step 2 to form the same size sand mounds in each of the other pans.
4. Stand a straw up in the center of each mound. Mark how high the sand comes on each straw.
5. Pour 50 mL of water over the first mound. Observe what happens to the sand. Record how the height of the sand changes.
6. Pour 250 mL of water over the second mound. Pour 500 mL of water over the third mound. Record your observations.

Data and Observations

Effect of Water on Mounds

	How mounds look after using water
Mound 1	
Mound 2	
Mound 3	

Conclusions
You might do Bob's experiment to find your own data and make observations. Make conclusions based on your data.

Practice Identifying Variables

Suppose you wanted to do an experiment to find out how much sand is washed away when the same amount of water is poured onto different amounts of sand.

1. What should be the variable in this experiment?
2. How would this experiment be different from Bob's experiment?

Experiment Results

A large amount of water washes away more soil than a small amount does.

Teaching Tips

- Ask students to identify some variables in the experiment. (the amount of soil, the type of soil, how firmly the soil is packed, the shape of the hill, the amount of water, how the water is poured). Help the class identify the variable that changes in this experiment. (the amount of water)
- Make sure students understand the purpose of the straw in this experiment. Help them realize that the straw will be used to measure the amount of soil that washes away in the experiment.
- You may wish to demonstrate to the class how to set up a sample soil mound with a straw marker.
- **Helpful Hint:** You may wish to suggest students do this experiment with different types of soil, including loam, sand, and clay.

3. Assess

Answers

Thinking About the Experiment

1. so that he could compare the effects of different amounts of water, not different amounts of dirt
2. a cup
3. amount of water
4. by pouring different amounts of water over mounds of the same size

Practice Identifying Variables

1. amount of dirt
2. the amount of dirt would be different in each mound; the amount of water would stay the same for each mound. (In the experiment on page 367 the amount of dirt was the same and the amount of water was different for each.)
You might encourage students to do this experiment.

TEACHING PLAN

Purpose
To develop the skill of testing a hypothesis in an experiment about air temperature

Objectives/Process Skills
- *Test a hypothesis* about how covering a box affects the temperature of air in it.
- *Set up a control* that tests changes in air temperature.
- *Study data* about air temperatures.

1. Motivate

Discussion
Review different types of clouds with the students. Question: **Which type of clouds form in layers that are often spread across the sky?** (stratus) Ask students to cut out pictures of these clouds from magazines. Explain that clouds not only keep us from seeing sunlight but they also have an effect on air temperature as well.

2. Teach

Safety Tips (See page T24.)
- Caution students to handle thermometers with care.
- Tell students to keep plastic wrap away from nose and mouth.
- Warn students not to touch the hot light bulb or allow it to touch the plastic wrap or the box. Hot light bulbs can cause burns or fire.

Chapter 11 Experiment Skills

Testing a Hypothesis

Nathan was keeping a weather chart as part of his science project. He noticed that the air seemed to get colder on clear nights than on cloudy nights. He wondered if clouds really affected the temperature of the air below them. He thought of this hypothesis as a possible answer to his question: *Covering air in some way helps keep it warm.*

Nathan did an experiment to test his hypothesis. He placed thermometers in two boxes. Then he covered both boxes and placed them in the sun for twenty minutes.

Read Nathan's experiment on the next page. Decide if Nathan's hypothesis was correct. Then answer the questions on this page.

Thinking About the Experiment

Usually in an experiment, parts of the experiment take the place of things in nature. For example, the bottom of Nathan's shoe boxes act like the ground.

1. What does the plastic cover on the boxes take the place of?
2. Why did Nathan remove the cover from one of his boxes?
3. To test his hypothesis, did Nathan really need to remove the cover from one box? Why?
4. What do you think happened to the temperature of the air in each of the boxes?

See pages 344-347 to review scientific methods and safety.

Experimenting with Heating Air

Problem
Do clouds help keep air warm?

Hypothesis
Covering air in some way helps keep it warm.

Materials

2 shoe boxes the same size	plastic wrap
2 thermometers	masking tape
	60-watt bulb

Procedure
1. Place a thermometer in the bottom of 2 shoe boxes. Cover both with a double thickness of clear plastic wrap.
3. Set the boxes under a lighted bulb, about 10 cm from the bulb. Be careful, light bulbs can cause burns!
4. After 20 minutes, remove the plastic wrap from one of the boxes.
5. Record the temperature in each box every minute for the next 5 minutes. Do not remove the thermometers.

Data and Observations
Temperature of Air in Each Box

Time	Uncovered Box	Covered Box
0	43°C	43°C
1	30°C	37.5°C
2	25°C	34°C
3	25°C	33.5°C
4	25°C	32°C
5	25°C	31.5°C

Conclusions
You might do Nathan's experiment to find your own data and make observations. Draw conclusions based on your data.

Practice Testing a Hypothesis

1. Suppose you wanted to do an experiment to find out if the color of the ground affects how fast air heats up. What would be your hypothesis?
2. How could you change Nathan's experiment to test your hypothesis?
3. What variable would change in testing this hypothesis?

Experiment Results

Results will vary, but students should observe that the air in the covered box will stay warmer longer than the air in the uncovered box.

Teaching Tips
- Point out that removing the cover from one box allows the warm air to mix with room temperature air in a short time. The uncovered box is the control. Temperatures measured in this box give data to compare with data taken from the covered box.
- Ask students how Nathan organized his data so he could study it. (He recorded it in a table.)
- **Helpful Hint:** Baby shoe boxes work best for this experiment. Boxes 11 cm by 17 cm were used to collect the data shown in the table.

3. Assess

Answers
Thinking About the Experiment
1. the cloud cover
2. to show what happens to air temperature without cloud cover
3. Yes, this serves as a control. It shows what happens when no cloud cover is present and provides a basis for comparison.
4. Answers will vary. Students might suggest that the box with the cover will stay warm longer.

Practice Testing a Hypothesis
1. Answers will vary, but students might suggest that the darker (or lighter) the color of the ground, the more the air is heated up.
2. Students might suggest placing white paper in the bottom of one box and black paper in the bottom of another, and covering both boxes with plastic wrap.
3. the color of the paper
You might encourage students to do this experiment.

TEACHING PLAN

Purpose
To develop the skill of using models in an experiment about eclipses

Objectives/Process Skills
• *Use a model* to test a hypothesis about eclipses.
• *Test a hypothesis* about what happens during a lunar eclipse.
• *Study data* that show the positions of the sun, moon, and earth during a lunar eclipse.

1. Motivate

Discussion
Review with the students how the earth and moon move relative to each other and to the sun. Ask volunteers to represent earth, sun, and moon and to demonstrate how these bodies move. Explain that the volunteers are making a model.

2. Teach

Safety Tips (See page T24.)
• Make sure the students do not shine the flashlights into the eyes of their classmates.
• Caution students to be careful with the dowel rods.

Chapter 12 Experiment Skills

Using Models

Lupe observed an eclipse of the moon one evening. She watched the shape of the full moon change as though something was covering it. She knew that the shadow of the earth is important in an eclipse. She wondered what the positions of the moon, sun, and earth are during an eclipse. She made this hypothesis to answer her question. *The earth blocks the sun's light from shining on the moon during an eclipse of the moon.* Lupe decided to experiment with a model to understand how an eclipse takes place.

Read Lupe's experiment on the next page. Then answer the questions on this page.

Thinking About the Experiment

Models stand for the real thing.

1. In Lupe's model what do the flashlight, the globe, and the plastic ball stand for?
2. How did Lupe's model help her test her hypothesis?
3. After reading Lupe's experiment, draw sketches of where you think the earth, moon, and sun would be during an eclipse of the moon.

Experimenting with Models

Problem
What causes an eclipse of the moon?

Hypothesis
The earth blocks the sun's light from shining on the moon during an eclipse of the moon.

Materials

globe	dowel rod
flashlight	modeling clay
plastic foam ball	meter stick

Procedure
1. Push one end of the dowel rod halfway into the ball.
2. Push the other end of the rod into the clay. Use the clay to hold the rod and ball upright.
3. Place the globe on the floor or on a table. Place the rod and ball 50 cm from the globe.
4. Darken the room. Shine the light on the globe from a distance of 1 m.
5. Move the light around the globe until the shadow of the globe covers the ball. Observe the positions of the light, globe, and ball.
6. Draw three circles in the data table to show the positions in Step 5. Your table shows a model of an eclipse.

Data and Observations

Sun	Earth	Moon

Conclusions
You might do Lupe's experiment to collect your own data and make observations. Draw conclusions based on your data.

Practice Using Models
Suppose you wanted to find the positions of the earth, moon, and sun during the moon's phases.
1. What would you use to make models to experiment with the moon's phases?
2. How could you find the moon's positions at different phases?

Experiment Results

During an eclipse of the moon, the earth keeps the sun's light from shining on the moon. Therefore, the moon's surface darkens during an eclipse.

Teaching Tips
• Place a basketball on a large piece of white paper. Dim the lights and place a flashlight or desk light so that it shines on the object. Point out the shadow behind the basketball. Move a baseball into the shadow. Note that the baseball changes in appearance. In the light, it appears light; in the shadow, it darkens. Tell students that the earth also casts a shadow. Point out that the moon can move into the shadow of the earth. An eclipse of the moon occurs when the moon moves through the earth's shadow.
• **Helpful Hint:** If a globe is not available, use a basketball for the earth and a baseball for the moon. You could use an overhead projector or a slide projector for the sun.

3. Assess

Answers
Thinking About the Experiment
1. The flashlight stands for the sun, the globe for the earth, and the plastic ball for the moon.
2. By observing what happened in his model, he could find the positions of the objects during an eclipse.
3. The data table shows the positions of the sun, earth, and moon during a lunar eclipse.

Practice Using Models
1. Answers might vary, but students could use the same objects that were used in the experiment.
2. Answers might vary, but students could move the moon model around the earth and observe how its lighted surface appears from earth.
You might encourage students to do this experiment.

TEACHING PLAN

Purpose
To develop the skill of testing a hypothesis in an experiment about muscles

Objectives/Process Skills
- *Test a hypothesis* comparing voluntary and involuntary muscles.
- *Identify variables* in an experiment.
- *Study data* about how long voluntary and involuntary muscles can work.

1. Motivate

Discussion
Encourage students to share their experiences of being tired from strenuous activity. Point out that they probably felt tired because their voluntary muscles were overworked. Note that voluntary muscles move hands, arms, legs, and feet, and involuntary muscles make internal organs, such as the stomach and heart, work.

2. Teach

Safety Tip (See page T24.)
- Check with the school nurse to make sure no students have medical conditions that might make it unwise for them to participate in this activity.

Chapter 13 Experiment Skills

Testing a Hypothesis

Andy helped his father rake leaves. The muscles in his arms got tired and sore. He had learned at school that he had two kinds of muscles, voluntary and involuntary. He knew that involuntary muscles work all the time. He wondered if they got tired and sore like voluntary muscles. He thought of this hypothesis to answer his question. *Involuntary muscles can work longer without getting sore than voluntary muscles can.*

Andy set up an experiment to test his hypothesis. He compared how voluntary muscles and involuntary muscles felt after they each worked for a minute.

Read Andy's experiment on the next page. Then answer the questions on this page.

Thinking About the Experiment
1. What muscles did Andy test in his experiment?
2. How long did Andy test each muscle?

A variable is anything in the experiment that changes. An experiment should have only one variable that changes. The variable is the one you are testing.

3. What variable is being tested in this experiment?
4. What things stayed the same in the experiment?

See pages 344-347 to review scientific methods and safety.

Experimenting with Muscles

Problem
Do involuntary muscles get tired and sore as easily as voluntary muscles?

Hypothesis
Involuntary muscles can work longer without getting sore than voluntary muscles can.

Materials
clock or watch with a second hand

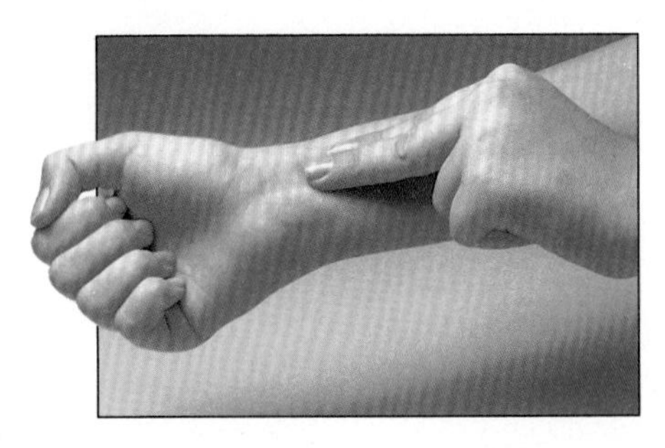

Procedure
1. Make a fist. Squeeze and open your fist for 1 minute. Try to squeeze 90 times in a minute.
2. Record how many times you squeezed your hand. Record how your hand muscles feel.
3. Count the number of times your heart beats in a minute. Feel your heartbeat by putting your hand on your chest. You can count the beats by feeling the pulse on your wrist.
4. Record the number of heartbeats in a minute. Record how your heart muscle feels.

Data and Observations

Kind of muscle muscle	Number of beats	How the muscles feel
Hand (voluntary)	Answers	will vary
Heart (involuntary)	for each	student.

Conclusions
You might do Andy's experiment to find your own data and make observations. Draw conclusions based on your data.

Practice Testing a Hypothesis

Suppose you wonder if one kind of voluntary muscle can work faster than another kind.
1. What hypothesis might you make?
2. How could you set up an experiment to test your hypothesis?
3. What variable would you change in your experiment?

373

Experiment Results

Involuntary muscles can work longer without getting sore than voluntary muscles can.

Teaching Tips
• Make sure students can locate and count their pulses. Direct students to find their pulses using their index and middle fingers, rather than their thumbs.
• Note that the average pulse for a student this age is 90 beats per minute. However, some students' pulses may be slightly higher and others may be slightly lower. Point out that the heart muscle works continuously for many years without stopping. Question: **Could voluntary muscles do this job without tiring?** (No.)
• **Helpful Hint:** As an alternative, students could squeeze tennis balls or other rubber balls, rather than repeatedly opening and closing their hands. Be aware that squeezing a tennis ball will be much more tiring than opening and closing the hand.

3. Assess

Answers
Thinking About the Experiment
1. involuntary (heart) and voluntary muscles (hand)
2. one minute
3. voluntary muscles
4. how long the muscles worked

Practice Testing a Hypothesis
1. Answers will vary, but students might suggest that one kind of voluntary muscle can work faster (or more slowly) than another kind.
2. Answers will vary, but students might suggest exercising as fast as possible and comparing the number of times the different kinds of muscles contracted.
3. the different kinds of muscles being observed
You might encourage students to do this experiment.

TEACHING PLAN

Purpose
To develop the skill of making conclusions in an experiment about foods

Objectives/Process Skills
- *State a hypothesis* about fat in foods.
- *Collect data* about foods containing fats.
- *Make conclusions* using data collected in an experiment.

1. Motivate

Discussion
Discuss with students what they have learned about foods that are high in fat. Question: **What are some characteristics of fatty foods?** (Students might note that foods high in fat often leave grease or oil on their fingers, make their lips feel oily, or leave grease stains on cloth and paper.)

2. Teach

Safety Tip (See page T24.)
- Remind students not to eat any of the foods they use in the experiment.

Teaching Tip
- Direct students to read pages 374 and 375. You may wish to write the hypothesis for this experiment as a class. Ask students for suggestions and have the class vote on the best hypothesis.

Chapter 14 Experiment Skills

Making Conclusions

Sarah had some pizza in a box. She noticed a shiny grease spot on the box. It came from the fat in the pizza. She knew that eating too much fat was not good for a person's health. Sarah wondered what foods she ate had fat in them. She decided to do an experiment to find out.

In Sarah's experiment, she put several kinds of foods into paper bags. The next day, she looked at the bags. The foods that made grease spots had fat in them.

Read Sarah's experiment on the next page. Then answer the questions on this page.

Thinking About the Experiment

1. What was Sarah trying to find out in her experiment?

Sarah tested an apple, a potato, some oatmeal, and peanut butter.

2. Which of these foods do you think have fat? Write your answer in a sentence. This sentence is your hypothesis.

When Sarah looked at the bags the next day, she saw only one grease spot.

3. Which of the foods had fat in it? Write your answer in a sentence.

Your answer to question 3 is a conclusion. You can also make another conclusion from this experiment.

4. Which of the foods did not have fat in them? Write your answer in a sentence.

374

See pages 344-347 to review scientific methods and safety.

Experimenting with Fat

Problem
What are some foods that have fat in them?

Hypothesis
Write your own hypothesis for this experiment.

Materials
1 slice of apple
dry oatmeal
peanut butter
newspaper
1 slice of raw potato
4 small paper bags

Procedure
1. Place the potato slice and apple slice into different paper bags.
2. Put a tablespoon of oatmeal and tablespoon of peanut butter into different bags.
3. Fold the bags over so the food does not spill out. Place the bags on the newspaper.
4. Let the bags sit overnight.
5. The next day, look on the outside of each bag. A fatty food will leave a grease spot.

Data and Observations

Food	Grease spot	No grease spot
Apple		X
Oatmeal		X
Potato		X
Peanut butter	X	

Conclusions
You might do Sarah's experiment to find your own data and make observations. Draw conclusions based on your data.

Practice Making Conclusions

Matt put peanuts, almonds, cashews, walnuts, and pecans in five different paper bags. All the nuts made greasy spots in the bags.
1. What conclusion did he make about the nuts he tested?
2. Should he make the same conclusion about all nuts? Explain.

Experiment Results

The peanut butter contains a great deal of fat, but the other foods do not contain enough fat to leave grease marks on the bags.

Teaching Tip
• **Helpful Hint:** Students could work in pairs. Each pair of students could test a different set of foods. Some foods you could test are bread, an orange slice, a tablespoon of jelly, uncooked rice, a banana slice, lettuce leaves, cheese, green beans, crackers, plain popcorn, grapes, lunch meat, a tablespoon of butter, dried cereal, onion slices, and yams. Put foods that spoil easily in the refrigerator overnight. Make a class chart of all the foods tested so students can compare their results.

3. Assess

Answers
Thinking About the Experiment
1. which of the foods she eats contain fat
2. Accept all statements that include predictions about which of the tested foods have fat.
3. Peanut butter contains fat.
4. Apple, oatmeal, and potato do not contain fat.

Practice Making Conclusions
1. He probably concluded that nuts contain fat.
2. No, although he might suspect that all nuts contain fat, he should not draw this conclusion until he has tested many different kinds of nuts.
You might encourage students to do this experiment.

Glossary

Pronunciation Key

The pronunciation of each word is shown just after the word, in this way: **ab bre vi ate** (ə brē′vē āt). The letters and signs used are pronounced as in the words to the right. The mark ′ is placed after a syllable with primary or heavy accent, as in the example above. The mark ′ after a syllable shows a secondary or lighter accent, as in **ab bre vi a tion** (ə brē′vē ā′shən).

a	hat, cap
ā	age, face
ä	father, far
b	bad, rob
ch	child, much
d	did, red
e	let, best
ē	equal, be
ėr	term, learn
f	fat, if
g	go, bag
h	he, how
i	it, pin
ī	ice, five
j	jam, enjoy
k	kind, seek
l	land, coal
m	me, am
n	no, in
ng	long, bring
o	hot, rock
ō	open, go
ȯ	order, all
oi	oil, voice
ou	house, out
p	paper, cup
r	run, try
s	say, yes
sh	she, rush
t	tell, it
th	thin, both
ᴛʜ	then, smooth
u	cup, butter
u̇	full, put
ü	rule, move
v	very, save
w	will, woman
y	young, yet
z	zero, breeze
zh	measure, seizure
ə	represents: **a** in about, **e** in taken, **i** in pencil, **o** in lemon, **u** in circus

A

alcohol (al′kə hôl), a drug that can be harmful and is found in beer, wine, and liquor.

amphibian (am fib′ē ən), an animal that has a backbone and lives some of the time in water and some of the time on land.

astronaut (as′trə nôt), a person who travels in space.

atom (at′əm), a small particle that makes up matter.

axis (ak′sis), imaginary line through the center of the earth around which the earth rotates.

B

bacteria (bak tir′ē ə), organisms made of one cell that can be seen through a microscope.

C

cancer, a disease in which cells that are not normal destroy healthy body cells.

carbon dioxide (kär′bən dī ok′sīd), a gas in the air that plants use to make food.

cell (sel), the basic unit of an organism.

chemical (kem′ə kəl) **change,** a change that causes matter to become a new kind of matter.

376

circuit (sėr′kit), the path along which electric current moves.
cirrus (sir′əs) **clouds,** feathery white clouds made of ice.
clay soil, tightly packed soil with tiny grains.
cold-blooded (kōld′ blud′id), an animal with a body temperature that changes with the temperature of the air or water around it.
community (kə myū′nə tē), all organisms that live in a place.
compound machine (kom′pound mə shēn′), a machine made of two or more simple machines.
condense (kən dens′), to change from a gas to a liquid.
conductor (kən duk′tər), a material that carries energy.
consumer (kən sü′mər), an organism that eats food.
contract (kən trakt′), the action of a muscle becoming shorter.
control (kən trōl′), the part of an experiment that does not change.
core (kôr), Earth's center part.
crater (krā′tər), a large hole in the ground shaped like a bowl.
crust (krust), the outer layer of the earth.
cumulus (kyü′myə ləs) **clouds**, clouds that look like cotton.

D

data (dā′tə), information gathered from an experiment.
decay (di kā′), to slowly rot.
degrees Celsius (di grēz′ sel′sē əs), metric unit for measuring temperature.
dew (dü), water vapor that condenses on cool surfaces, usually during the night.
diabetes (dī′ə bē′tis), a disease in which the body cannot use sugar properly.
dinosaur (dī′nə sôr), one of a group of extinct reptiles that lived millions of years ago.
disease (də zēz′), an illness.

E

eardrum, thin skin that covers the middle ear.
earthquake (ėrth′kwāk′), a shaking of the earth's crust.
echo (ek′ō), a sound that bounces back from an object.
electric charges (i lek′tric chärjəz), tiny bits of electricity in all matter.
electric current (kėr′ənt), the smooth flow of electric charges from one place to another.
electromagnet (i lek′trō mag′nit), a wire coil that is a magnet when electric current moves through it.

377

endangered (en dān′jərd) **organisms,** kinds of organisms that are very few in number and might someday no longer be found on the earth.

energy (en′ər jē), the ability to do work.

energy of motion (mō′shən), energy of moving objects.

erosion (i rō′zhən), the movement of soil or rocks by wind or water.

erupt (i rupt′), to burst out.

evaporate (i vap′ə rāt′), to change from a liquid to a gas.

extinct (ek stingkt′) **organisms,** organisms that no longer are found on the earth.

F

fibers (fī′bərz), strong thin threads in plants that can be used to make cloth.

fog, a cloud that forms just above the surface of the earth.

food chain, the way food passes from one organism to another organism in a community.

force (fôrs), a push or pull.

fossil (fos′əl), the hardened remains or traces of an animal or plant of a former age.

friction (frik′shən), force caused by objects rubbing together that slows moving objects.

fuel (fyü′əl), coal, wood, oil, or any other material that can be burned to produce useful heat or power.

fulcrum (ful′krəm), point on which a lever is supported and turns.

fungus (fung′gəs), an organism, such as a mold or mushroom, that gets food from dead material or by growing on food or a living thing. [Plural: **fungi** (fun′jī)]

G

gear (gir), a wheel with jagged edges like teeth.

germinate (jėr′mə nāt), begin to grow and develop.

gills (gils), the parts of fish that are used to take in oxygen from the water.

gravity (grav′ə tē), the force that makes objects pull toward each other.

H

habitat (hab′ə tat), the place where an organism lives.

humus (hyü′məs), the decayed matter in soil.

hurricane (hėr′ə kān), a huge storm that forms over a warm ocean and has strong winds and heavy rains.

hypothesis (hī poth′ə sis), a possible answer to a problem.

I

igneous (ig′nē əs) **rock,** a rock formed from melted minerals.

illegal (i lē′gəl) **drug,** a drug that is against the law to use or have.

image (im′ij), a copy.

inclined (in klīnd′) **plane,** a simple machine that is a flat surface with one end higher than the other.

insect (in′sekt), an animal with three main body parts, six legs, and no backbone.

insulator (in′sə lā′tər), a material through which energy cannot easily flow.

involuntary (in vol′ən ter′ē) **muscle,** a muscle that works without a person's control.

iris (ī′ris), colored part of the eye.

J

joint, a place where bones join together; different joints allow different movement.

L

larva (lär′və), the young of an animal that is different from the adult. [Plural: **larvae** (lär′vē)]

lava (lä′və), hot, melted rock that flows from a volcano.

lens (lenz), a piece of clear material that bends light waves that pass through it.

lever (lev′ər), a simple machine made of a bar that is supported underneath at some point.

life cycle (sī′kəl), the stages in the life of a plant or animal.

load (lōd), an object that is being moved.

loam (lōm), soil that is a mixture of clay, sand, and humus.

M

machine (mə shēn), a tool that makes work easier.

magma (mag′ma), hot, melted rock deep inside the earth.

magnet (mag′nit), object that attracts objects with iron in them.

magnetism (mag′nə tiz′əm), the force around a magnet.

mammal (mam′əl), an animal with a backbone and hair.

mantle (man′tl), the middle layer of the earth.

mass (mas), how much matter an object contains.

matter (mat′ər), a substance of which all objects are made.

medicine (med′ə sən), a drug that can help protect people from sickness, or treat or cure a disease.

metamorphic (met′ə môr′fik) **rock,** an igneous or sedimentary rock that was changed by heat or pressure.

379

mineral (min′ər əl), material that was never alive and that can be found in soil.

N

natural resource (nach′ər əl ri′sôrs), something people use that comes from the earth.
nerve (nėrv), a body part that carries messages to the brain.
nicotine (nik′ə tēn′), a drug in tobacco that can harm the body.
nutrient (nü′trē ənt), a material that plants and animals need to live and grow.

O

orbit (ôr′bit), a closed, curved path an object follows as it moves around another object.
ore (ôr), rock with useful minerals.
organ (ôr′gən), a group of tissues that work together to do a job in the body.
organism (ôr′gə niz′əm), a living thing.
oxygen (ok′sə jən), a gas in air that living things need to stay alive.

P

particle (pär′tə kəl), a little bit.
petal (pet′l), the outside, colored part of a flower.
phase (fāz), the shape of the lighted part of the moon as it is seen from the earth.
physical (fiz′ə kel) **change,** a change in the size, shape, state, or appearance of matter.
physical (fiz′ə kəl) **fitness,** the ability to work, play, and exercise without getting tired or injured easily.
pitch (pich), how high or low a sound is.
planet (plan′it), a large body of matter revolving around the sun.
pollen (pol′ən), a fine yellowish powder in a flower.
pollinate (pol′ə nāt), to carry pollen to the center part of a flower.
pollution (pə lü′shən), anything harmful added to the air, water, or land.
population (pop′yə lā′shən), organisms of the same kind that live in the same place.
precipitation (pri sip′ə tā′shən), moisture that falls to the ground.
predator (pred′ə tər), organism that captures and eats other organisms.
prey (prā), organism that is captured and eaten by another organism.

prism (priz′əm), a clear piece of glass or plastic that is used for separating white light.
producer (prə dü′sər), an organism that makes food.
property (prop′ər tē), something about an object that can be observed, such as size or shape.
protist (prō′tist), an organism that lives in a wet place and has one or more cells.
pulley (pul′ē), a simple machine made of a wheel and a rope.
pupa (pyü′pə), stage in the insect life cycle between larva and adult. [Plural: **pupae** (pyü′pē)]
pupil (pyü′pəl), the opening in the eye that lets in light.

R

recycle (rē sī′kəl), to change something so it can be reused.
reflect (ri flekt′), to turn back.
reptile (rep′təl), a cold-blooded animal with a backbone, scales, and lungs.
revolution (rev′ə lü′shən), movement of an object in an orbit around another object.
rotation (rō tā′shən), the act of spinning on an axis.

S

sandy soil, loose soil with large grains.
satellite (sat′l īt), object that revolves around another object.
scale, one of the thin, hard plates that cover fish and reptiles.
scientific methods, organized ways of solving problems.
screw (skrü), a simple machine used to hold objects together.
sedimentary (sed′ə men′tər ē) **rock,** rock that forms when layers of material are pressed together.
seed coat (sēd kōt), the outside covering of a seed.
seed leaf (sēd lēf), a part that looks like a leaf and is inside each seed.
seedling (sēd′ling), a young plant that grows from a seed.
simple machine (sim′pəl mə shēn′), one of six kinds of tools with few or no moving parts that makes work easier.
skeleton (skel′ə tən), the bones of a body, fitted together in their natural places.
solar system (sō′lər sis′təm), the sun, the planets and their moons, and other objects that revolve around the sun.
source (sôrs), a place from which something comes.
star, a ball of hot, glowing gases.
states of matter, the three forms of matter—solid, liquid, and gas.
stratus (strā′təs) **clouds,** clouds that form in sheets or layers.

381

system (sis′təm), a group of organs that work together to do a job in the body.

T

tadpole (tad′pōl′), young frog.
temperature (tem′pər ə chər), a measurement of the speed at which particles of matter are moving.
tendon (ten′dən), a tissue that holds a muscle to a bone.
thermometer (thər mom′ə tər), tool for measuring temperature.
thermostat (thėr′mə stat), a tool that controls temperature in a home or building.
tissue (tish′ü), a group of cells that look alike and do the same job.
tornado (tôr nā′dō), a funnel cloud that has very strong winds and moves along a narrow path.

V

variable (ver′ē ə bəl), anything in an experiment that can be changed.
vibrate (vī′brāt), move quickly back and forth.
virus (vī′rəs), a kind of disease germ that causes diseases such as colds and flu.
visible spectrum (viz′ə bəl spek′trəm), the band of colors formed when a wave of white light is bent.

vocal cords (vō′kəl kôrdz), thin flaps at the top of the windpipe.
volcano (vol kā′nō), a mountain with an opening through which lava, ashes, rocks, and other materials come out.
volume (vol′yəm), (1) amount of space an object takes up; (2) loudness or softness of a sound.
voluntary (vol′ən ter′ē) **muscle,** a muscle a person can control.

W

warm-blooded (wôrm′ blud′id), animals that usually keep about the same body temperature.
water cycle (sī′kəl), the movement of water by evaporation, condensation, and precipitation.
water vapor (wô′tər vā′pər), water in the form of gas.
weather (weᴛʜ′ər), to wear down or break apart rocks.
wedge (wej), a simple machine used to cut or split an object.
wheel and axle (hwēl ənd ak′səl), a simple machine with a center rod attached to a wheel.
work (wėrk), something done when a force moves an object through a distance.

Index

A **bold-faced** number indicates a page with a picture about the topic

385

Acknowledgments

Unless otherwise acknowledged, all photos are the property of Scott, Foresman and Company. Page positions are as follows: (T)top, (C)center, (B)bottom, (L)left, (R)right, (INS)inset.

iv: Biophoto Associates/Photo Researchers **viii(1):** Stuart Cohen **viii(c):** FPG **x(tl):** Ed Reschke **x(tc):** Centre National de Recherches Iconographiques **x(tr):** Eric V. Grave/Phototake **x(bl):** Manfred Kage/Peter Arnold, Inc. **x(br):** Ed Reschke **xii:** Jim Steers/Chicago Symphony Orchestra **4T:** David Burnett/Woodfin Camp & Associates **4B:** Courtesy Moog Music, Inc. **5B:** Roger Ressmeyer/Starlight **6:** Mitch Reardon/Tony Stone Worldwide/Masterfile **8:** Lynn M. Stone **13ALL:** Walter Chandoha **15:** Anitra Thorhaug **16:** Robert E. Lyons/Color Advantage **17T:** Dan Suzio **18(BOTH):** Dwight R. Kuhn **19T:** Ruth Dixon **19BL:** William E. Ferguson Photography **19BR:** Ed Cooper **21L:** Robert E. Lyons/Color Advantage **21R:** Lynn M. Stone **28:** Wolfgang Bayer Productions **30L:** Jim Brandenburg **30BR:** Jim Brandenburg **30TR:** Gwen Fidler **31TL:** D. Wilder **31R:** Marty Snyderman **31BL:** Dwight R. Kuhn **32:** Don and Pat Valenti **33L:** Lynn M. Stone **33R:** Carl Roessler **34L:** G.I.Bernard/ANIMALS ANIMALS **34R:** Zig Leszczynski/ANIMALS ANIMALS **35TL:** Dr. Merlin D. Tuttle **35BL:** C.Allan Morgan/Peter Arnold, Inc. **35R:** Loren McIntyre **36:** G.Ziesler/Peter Arnold, Inc. **38:** Dwight R. Kuhn **39L:** Doug Wechsler **39R:** Dwight R. Kuhn **41:** Bill Ivy **43B:** Transactions of the Academy of Science of St. Louis, Vol.XXIV, No. 9, December 1923/Academy of Science of St. Louis **43T:** Don and Pat Valenti **48:** Bob & Clara Calhoun/Bruce Coleman Inc. **50T:** Manfred Kage/Peter Arnold, Inc. **50BL:** CNRI/Science Photo Library/Photo Researchers **50BR:** Biophoto Assoc./Photo Researchers **51L:** Leonard LaRue III/Bruce Coleman Inc. **56T:** Johnny Johnson/DRK Photo **56BL:** Dr. M.P.Kahl/DRK Photo **56BC:** Alex Kerstitch/Sea of Cortez Enterprises **56BR:** Don and Pat Valenti **59:** Stephen J. Krasemann/DRK Photo **61:** WWF/Timm Rautert/Bruce Coleman Ltd. **66:** Bill Ivy **68:** C.C.Lockwood/Cactus Clyde Productions **69:** Milt & Joan Mann/Cameramann International, Ltd. **70L:** Larry West **70R:** Fred Bavendam/Peter Arnold, Inc. **71T:** Jane Burton/Bruce Coleman Inc. **72:** Lynn M. Stone **74:** Jose Azel/Contact Press Images **79L:** J. Serrao **79R:** Lynn M. Stone **81T:** Sierra Club/William E. Colby Memorial Library **81B:** Harald Sund **90:** Michael Melford/The Image Bank **92:** Harald Sund **97:** Jeffrey L. Rotman **103:** Fermilab Photo Dept. **118:** (c) Mickey Plefeger 1987 **123R:** Jeanne Trombly **124L:** Tom Bean/DRK Photo **124BR:** Yoav/Phototake **126:** Greg Pease **129:** Milt & Joan Mann/Cameramann International, Ltd. **130R:** Milt & Joan Mann/Cameramann International, Ltd. **136:** Greg Pease **144T:** Harald Sund **147T:** UPI/Acme/Bettmann Newsphotos **147B:** Peter Menzel **151:** Courtesy Owens-Corning **153:** David R. Frazier Photolibrary **162:** Howard Hall **175:** Brown Brothers **180B:** Bob Daemmrich **184:** Luis Padilla/The Image Bank **186:** Tom Algire **188ALL:** Stuart Cohen **189L:** Harald Sund **189R:** David R. Frazier Photolibrary **190:** Gary Braasch **191L:** FPG **191R:** Stuart Cohen **193:** G. Marche/FPG **194-195:** Harald Sund **195INS:** Lawrence Hudetz **196(ALL):** Barry L. Runk/Grant Heilman Photography **199:** Milt & Joan Mann/Cameramann International, Ltd. **200:** L.L.T.Rhodes/Taurus Photos, Inc. **200(ALL):** L.L.T.Rhodes/Taurus Photos, Inc. **201:** Charlton Photos **206:** Tom Algire **210:** U.S. Geological Survey **212:** Harald Sund **213:** Chip Clark **214:** Steven C. Wilson/Entheos **215:** Lawrence Hudetz **217:** Rob Lewine Photography **218:** James Balog/Black Star **219:** Guillermo Aldana **220:** Greg Vaughn/Black Star **222:** Don and Pat Valenti **228:** John Foster/Masterfile **230:** Ewing Galloway **232L:** Gary Braasch **232C:** David R. Frazier Photolibrary **232R:** David R. Frazier Photolibrary **236L:** Tom Branch/Photo Researchers **236TR:** Gary Braasch **236BR:** Tom Algire **239:** Howie Bluestern/Science Source/Photo Researchers **241:** Dan McCoy/Rainbow **246:** NASA **251L:** NASA **251R:** Dennis Milon **252:** NASA **255:** NASA **259:** NASA **261ALL:** NASA **262:** James Sugar/Black Star **263:** NASA **268:** NASA **274:** David Black **277BC:** Centre National de Recherches Iconographiques **277BR:** Ed Reschke **277TL:** Ed Reschke **277TR:** Manfred Kage/Peter Arnold, Inc. **277BL:** Eric V. Grave/Phototake **279:** Medichrome/Tsiaras/The Stock Shop **303:** Courtesy Dr.Jose Sandoval **307:** (c) Mickey Pfleger 1987 **312:** Brent Jones **316L:** Lynn M. Stone **316R:** Ruth Dixon **318L:** Carl Roessler **318C:** Bob & Clara Calhoun/Bruce Coleman Inc. **318R:** Lynn M. Stone **320CL:** Leonard LaRue III/Bruce Coleman Inc. **320L:** CNRI/Science Photo Library/Photo Researchers **322L:** The Bettmann Archive **322R:** Fred Bavendam/Peter Arnold, Inc. **335: ALL** Lawrence Hudetz **337L:** Tom Branch/Photo Researchers **337TR:** Gary Braasch **337BR:** Tom Algire

387

Using Metric

Metric Conversion Tables

Measurement

When You Know	Multiply by	To Find	Multiply by	To Find
		When You Know		
inches	2.54	centimeters	0.394	inches
feet	0.3048	meters	3.281	feet
yards	0.914	meters	1.0936	yards
miles	1.609	kilometers	0.62	miles
square inches	6.45	square centimeters	0.155	square inches
square feet	0.093	square meters	10.76	square feet
square yards	0.836	square meters	1.196	square yards
acres	0.405	hectares	2.471	acres
square miles	2.59	square kilometers	0.386	square miles
cubic inches	16.387	cubic centimeters	0.061	cubic inches
cubic feet	0.028	cubic meters	35.315	cubic feet
cubic yards	0.765	cubic meters	1.31	cubic yards
fluid ounces	29.57	milliliters	0.0338	fluid ounces
quarts	0.946	liters	1.057	quarts
gallons	3.785	liters	0.264	gallons
ounces	28.35	grams	0.0353	ounces
pounds	0.4536	kilograms	2.2046	pounds
tons	0.907	metric tons	1.102	tons

Temperature

When You Know		
Fahrenheit	subtract 32; then *divide* by 1.8	to find Celsius
Celsius	multiply by 1.8; then *add* 32	to find Fahrenheit

Kindergarten Scope and Sequence

Life Science

Plants

Chapter 4
Seeds and Plants

Lesson 1 Kinds of seeds
Lesson 2 Growth from seeds to plants
Lesson 3 Kinds of plants
Lesson 4 Plant parts
Lesson 5 Needs of plants
Lesson 6 Uses of plants

Animals

Chapter 5
Animals

Lesson 1 Kinds of animals
Lesson 2 Animal growth from baby to adult
Lesson 3 Animal movements
Lesson 4 Needs of animals
Lesson 5 Animal habitats
Lesson 6 Uses of animals

Ecology

Chapter 3
Living and Nonliving

Lesson 1 Living things
Lesson 2 Nonliving things
Lesson 3 Using living and nonliving things

Physical Science

Matter

Chapter 6
Comparing and Grouping Matter

★Lesson 1 Color
★Lesson 2 Shape
Lesson 3 Mass
★Lesson 4 Number
★Lesson 5 Position

Energy

Chapter 7
Heat and Cold

★Lesson 1 Sources of heat
Lesson 2 Uses of heat
★Lesson 3 Heat and cold change things

Chapter 8
Sound

★Lesson 1 Loud sounds
★Lesson 2 Soft sounds

Chapter 9
Movement

Lesson 1 How things move
★Lesson 2 Magnets
★Lesson 3 Sinking and floating

Earth Science

Earth

Chapter 12
Seasons

★Lesson 1 Winter
★Lesson 2 Spring
★Lesson 3 Summer
★Lesson 4 Fall

Chapter 13
Helping Our World

Lesson 1 Protecting resources
★Lesson 2 Saving electricity
★Lesson 3 Saving water
Lesson 4 Saving paper
Lesson 5 Picking up litter

Weather

Chapter 11
Weather

★Lesson 1 Kinds of weather
★Lesson 2 Air and wind
★Lesson 3 Temperature
Lesson 4 Effects of weather changes

Space

Chapter 10
Earth and Sky

Lesson 1 Land and water
★Lesson 2 Day and night
Lesson 3 Sun and moon

Human Body

Chapter 1
Your Senses

★Lesson 1 Seeing
★Lesson 2 Hearing
★Lesson 3 Smelling
★Lesson 4 Touching
★Lesson 5 Tasting

Chapter 2
Growing and Changing

Lesson 1 Body parts
Lesson 2 Growth from baby to adult
★Lesson 3 Getting new teeth
★Lesson 4 Eating and exercising

Note: If time does not permit you to cover all of the material in this book, it is recommended that the lessons with a star be presented as a core program. This core program represents a balanced treatment of topics throughout the series.

Grade 1 Scope and Sequence

Life Science

Plants

Chapter 4
Learning About Plants

- ★Lesson 1 How plants are alike and different
- ★Lesson 2 How plants grow
- ★Lesson 3 What plants need to grow
- Lesson 4 Why people need plants

Animals

Chapter 5
Learning About Animals

- ★Lesson 1 Ways animals are different
- ★Lesson 2 How animals grow
- ★Lesson 3 Why people need animals
- Lesson 4 How to care for a pet

Ecology

Chapter 3
Living and Nonliving

- ★Lesson 1 What a living thing is
- ★Lesson 2 What living things need
- ★Lesson 3 What nonliving things are like

Physical Science

Matter

Chapter 6
Grouping Things

- ★Lesson 1 Ways to group things
- Lesson 2 Matter takes up space
- ★Lesson 3 What solids and liquids are like
- ★Lesson 4 What gases are like

Energy

Chapter 7
Light, Sound, and Heat

- ★Lesson 1 How light can change
- ★Lesson 2 How sound can change
- Lesson 3 What can be learned about heat

Chapter 8
Moving and Working

- Lesson 1 Ways objects move
- Lesson 2 What can move objects (magnets)
- ★Lesson 3 Work machines can do

Earth Science

Earth

Chapter 9
The Earth

- ★Lesson 1 Mountains, valleys, soil, rocks, and resources
- ★Lesson 2 Location of water on earth
- ★Lesson 3 How air is useful
- Lesson 4 How people use land and water

Weather

Chapter 10
Weather and Seasons

- ★Lesson 1 Different kinds of weather
- Lesson 2 How weather can change in seasons
- ★Lesson 3 How weather is important to people

Space

Chapter 11
The Sky

- ★Lesson 1 What is seen in the sky
- Lesson 2 What the sun is like
- ★Lesson 3 What the moon is like
- ★Lesson 4 What the stars are like

Human Body

Chapter 1
Your Senses

- Lesson 1 Learning
- ★Lesson 2 Using the senses

Chapter 2
Growing and Changing

- ★Lesson 1 How people change
- ★Lesson 2 What helps people grow

Note: If time does not permit you to cover all of the material in this book, it is recommended that the lessons with a star be presented as a core program. This core program represents a balanced treatment of topics throughout the series.

Grade 2 Scope and Sequence

Life Science

Plants

Chapter 1
How Plants Are Different

- ★Lesson 1 How plant parts can differ
- ★Lesson 2 What plants use to make food
- Lesson 3 Different ways plants grow
- Lesson 4 Where plants grow

Animals

Chapter 2
How Animals are Different

- ★Lesson 1 Some groups of animals
- ★Lesson 2 Where animals live
- ★Lesson 3 What animals need to stay alive
- Lesson 4 How animals change as they grow

Ecology

Chapter 3
Life on Earth Long Ago

- ★Lesson 1 Evidence of life long ago
- ★Lesson 2 What dinosaurs were like
- ★Lesson 3 Other animals that lived long ago
- Lesson 4 What can change plants and animals
- ★Lesson 5 Coal, oil, and gas made from plants and animals that lived long ago

Physical Science

Matter

Chapter 4
Matter Around You

- ★Lesson 1 What matter is
- Lesson 2 How matter can be grouped
- ★Lesson 3 How matter can change

Energy

Chapter 5
Heat, Light, and Sound

- ★Lesson 1 What heat is
- ★Lesson 2 What light is
- Lesson 3 How heat and light can be used safely
- ★Lesson 4 What sound is
- Lesson 5 How sound is helpful

Chapter 6
Machines and Electricity

- ★Lesson 1 What machines can do
- ★Lesson 2 What magnets can do
- ★Lesson 3 What electricity can do
- Lesson 4 How machines and electricity can be used safely

Earth Science

Earth

Chapter 7
Water and Air

- ★Lesson 1 Where to find fresh water
- ★Lesson 2 What oceans are like
- ★Lesson 3 Why clean water is important
- ★Lesson 4 What air is like
- ★Lesson 5 Why clean air is important

Weather

Chapter 8
Changes in Weather

- ★Lesson 1 How weather can change
- ★Lesson 2 What makes clouds, rain, and snow
- Lesson 3 What weather is like around the world

Space

Chapter 9
The Sun and Other Stars

- ★Lesson 1 What the sun is
- ★Lesson 2 Planets move around the sun
- Lesson 3 Why most stars look small

Human Body

Chapter 10
How Your Body Works

- Lesson 1 How bones and muscles work
- Lesson 2 What the brain can do
- ★Lesson 3 How the heart and lungs work
- ★Lesson 4 What happens to food that is eaten

Chapter 11
Keeping Healthy

- ★Lesson 1 What is needed for good health
- ★Lesson 2 Protection from sickness
- ★Lesson 3 How to stay safe

Note: If time does not permit you to cover all of the material in this book, it is recommended that the lessons with a star be presented as a core program. This core program represents a balanced treatment of topics throughout the series.

Grade 3 Scope and Sequence

Life Science

Plants

Chapter 1
Plant Growth

- ★Lesson 1 Importance of roots, stems, and leaves
- ★Lesson 2 Importance of flowers and cones
- Lesson 3 How plants grow from seeds

Animals

Chapter 2
How Animals Grow

- Lesson 1 How animals can be grouped
- ★Lesson 2 How some animals with backbones grow and change
- ★Lesson 3 How some animals without backbones grow and change

Ecology

Chapter 3
Living Things Need Each Other

- Lesson 1 The five groups of living things
- ★Lesson 2 How organisms live together
- ★Lesson 3 How organisms get food

Chapter 4
How People Affect Plants and Animals

- ★Lesson 1 How people change the lives of plants and animals
- Lesson 2 How people protect plants and animals
- Lesson 3 How people use plants and animals

Physical Science

Matter

Chapter 5
Properties of Matter

- Lesson 1 What matter is
- ★Lesson 2 The make-up of matter
- ★Lesson 3 How matter can change

Energy

Chapter 6
Work and Machines

- Lesson 1 What work is
- ★Lesson 2 What simple machines are
- ★Lesson 3 What compound machines are

Chapter 7
Forms of Energy

- ★Lesson 1 Some kinds of energy
- ★Lesson 2 What electricity is
- ★Lesson 3 How matter is heated
- ★Lesson 4 How light travels

Chapter 8
Sound

- ★Lesson 1 What sound is
- ★Lesson 2 How sound travels
- Lesson 3 How people make sound

Earth Science

Earth

Chapter 9
Rocks and Soil

- Lesson 1 How rocks are formed
- ★Lesson 2 What soil is made of
- ★Lesson 3 How people use rocks and soil

Chapter 10
Changes in the Earth

- Lesson 1 What the inside of the earth is like
- ★Lesson 2 How water and wind change the earth's crust
- ★Lesson 3 How earthquakes and volcanoes change the earth's crust
- Lesson 4 How living things change the earth's crust

Weather

Chapter 11
Clouds and Storms

- ★Lesson 1 How clouds form
- ★Lesson 2 Water in clouds
- ★Lesson 3 Causes of storms

Space

Chapter 12
The Sun, Moon, and Planets

- ★Lesson 1 How the earth and the moon move
- Lesson 2 What the sun and the planets are like
- ★Lesson 3 What scientists learn from space travel

Human Body

Chapter 13
The Body's Support

- ★Lesson 1 The make-up of the body
- ★Lesson 2 The importance of bones
- ★Lesson 3 The importance of muscles

Chapter 14
Your Body's Health Needs

- ★Lesson 1 How to stay healthy
- Lesson 2 Causes of disease
- ★Lesson 3 How alcohol, tobacco, and drugs affect the body

Note: If time does not permit you to cover all of the material in this book, it is recommended that the lessons with a star be presented as a core program. This core program represents a balanced treatment of topics throughout the series.

S C O T T , F O R E S M A N D I S C O V E R S C I E N C E

Grade 4 Scope and Sequence

Life Science

Plants

Chapter 1
Flowering Plants

- Lesson 1 How plants are classified
- Lesson 2 What the parts of a flower do
- Lesson 3 How seeds and fruits develop
- Lesson 4 How seeds scatter and grow

Animals

Chapter 2
Animal Behavior

- Lesson 1 How animals live in groups
- Lesson 2 How animals care for their young
- Lesson 3 How animals behave

Ecology

Chapter 3
Food Chains and Food Webs

- Lesson 1 Where green plants get energy
- Lesson 2 Where animals get food
- Lesson 3 What a food chain is
- Lesson 4 What a food web is

Chapter 4
Animal and Plant Adaptations

- Lesson 1 How adaptations help animals and plants survive
- Lesson 2 How animals are adapted to their environments
- Lesson 3 How plants are adapted to their environments

Physical Science

Matter

Chapter 5
Measuring Matter

- Lesson 1 What matter is
- Lesson 2 The composition of matter
- ★Lesson 3 How length and volume are measured
- ★Lesson 4 How mass and density are measured

Energy

Chapter 6
Work and Energy

- Lesson 1 What makes things move
- Lesson 2 How work and energy are related
- ★Lesson 3 How machines use energy to do work

Chapter 7
Electricity and Magnetism

- ★Lesson 1 What electricity is
- ★Lesson 2 What magnetism is
- ★Lesson 3 How electricity and magnetism are related

Chapter 8
Light and Sound

- ★Lesson 1 What light is
- ★Lesson 2 How light travels
- Lesson 3 How light and sound are similar and different

Earth Science

Earth

Chapter 10
Changes in Landforms

- ★Lesson 1 The composition of the earth
- ★Lesson 2 How volcanoes and earthquakes change landforms
- Lesson 3 How weathering changes landforms
- ★Lesson 4 How rocks are made

Chapter 11
Oceans

- Lesson 1 What oceans are and how they are important
- ★Lesson 2 How ocean water moves
- ★Lesson 3 The ocean bottom

Weather

Chapter 9
Measuring Weather Conditions

- Lesson 1 Causes of different air temperatures
- ★Lesson 2 How temperature affects air pressure and wind
- ★Lesson 3 How low temperature causes clouds and precipitation
- ★Lesson 4 How weather is predicted

Space

Chapter 12
Movement in the Solar System

- ★Lesson 1 How the earth moves
- ★Lesson 2 Phases of the moon
- ★Lesson 3 How planets move
- Lesson 4 Other objects in the solar system that move

Human Body

Chapter 13
Digestion and Circulation

- ★Lesson 1 How the digestive system works
- ★Lesson 2 How the circulatory system works
- Lesson 3 How to keep body systems healthy

Chapter 14
Your Brain and Your Sense Organs

- Lesson 1 How the brain gets information
- ★Lesson 2 How the eyes work
- ★Lesson 3 How the ears work
- ★Lesson 4 How the tongue and nose work
- Lesson 5 How the skin gathers information

Note: If time does not permit you to cover all of the material in this book, it is recommended that the lessons with a star be presented as a core program. This core program represents a balanced treatment of topics throughout the series.

Grade 5 Scope and Sequence

Life Science

Plants

Chapter 2
Plant Processes

- ★Lesson 1 Plant cells
- Lesson 2 How plants take in materials
- ★Lesson 3 How plants make and use food
- ★Lesson 4 How plants produce seeds
- Lesson 5 How plants reproduce without seeds

Animals

Chapter 3
Invertebrates and Vertebrates

- ★Lesson 1 Organisms in the animal kingdom
- ★Lesson 2 How invertebrates are classified
- ★Lesson 3 How arthropods are classified
- ★Lesson 4 How vertebrates are classified

Ecology

Chapter 1
Classifying Living Things

- ★Lesson 1 Ways all living things are alike
- ★Lesson 2 Classifying living things into kingdoms
- ★Lesson 3 Dividing kingdoms into smaller groups

Chapter 4
Populations and Communities

- ★Lesson 1 What populations and communities are
- Lesson 2 How populations share environments
- ★Lesson 3 How populations affect each other
- Lesson 4 How populations change
- Lesson 5 How communities change naturally

Physical Science

Matter

Chapter 5
Investigating Matter

- ★Lesson 1 Properties of matter
- ★Lesson 2 What atoms are
- ★Lesson 3 What elements are
- ★Lesson 4 Combining atoms
- Lesson 5 Solutions and suspensions

Energy

Chapter 6
Heat and Matter

- ★Lesson 1 Heat and temperature
- Lesson 2 How matter is heated
- ★Lesson 3 How heat affects matter
- Lesson 4 How temperature is measured

Chapter 7
Changing Forms of Energy

- ★Lesson 1 How energy changes form
- ★Lesson 2 How electric energy is produced
- Lesson 3 How electric energy is changed and used
- ★Lesson 4 Measuring work and force

Chapter 8
Energy Resources

- ★Lesson 1 Fossil fuels as sources of energy
- ★Lesson 2 Atoms as a source of energy
- Lesson 3 Moving water as a source of energy
- ★Lesson 4 How people use sunlight and wind
- Lesson 5 Using heat inside the earth as a source of energy

Earth Science

Earth

Chapter 9
Earth's Changing Crust

- ★Lesson 1 How weathering changes the earth's crust
- ★Lesson 2 How moving water causes erosion
- ★Lesson 3 How glaciers and wind cause erosion
- Lesson 4 How rocks change
- Lesson 5 How fossils form

Chapter 10
Protecting the Environment

- ★Lesson 1 Protecting water
- ★Lesson 2 Protecting air
- ★Lesson 3 Protecting the land
- Lesson 4 Solving local pollution problems

Weather

Chapter 11
Climate

- ★Lesson 1 What climate is
- ★Lesson 2 How sunlight affects climate
- ★Lesson 3 How land and water affect climate
- Lesson 4 How the shape of the land affects climate
- Lesson 5 Factors that cause climates to change

Space

Chapter 12
Mapping the Stars

- ★Lesson 1 How scientists learn about the universe
- ★Lesson 2 Measuring distance in space
- ★Lesson 3 How people map the stars
- ★Lesson 4 Brightness of the stars

Human Body

Chapter 13
Body Support, Movement, and Growth

- ★Lesson 1 How the bones help the body
- ★Lesson 2 How muscles help the body work
- ★Lesson 3 How to keep bones and muscles healthy
- Lesson 4 How the body grows

Chapter 14
Respiration and Excretion

- ★Lesson 1 How the body takes oxygen from the air
- Lesson 2 How the body uses oxygen
- ★Lesson 3 How the body gets rid of cell wastes
- Lesson 4 How to care for the respiratory and excretory systems

Note: If time does not permit you to cover all of the material in this book, it is recommended that the lessons with a star be presented as a core program. This core program represents a balanced treatment of topics throughout the series.

SCOTT, FORESMAN DISCOVER SCIENCE

Grade 6 Scope and Sequence

Life Science

Plants

Chapter 4
Plant Response
★Lesson 1 How plants respond to environments
Lesson 2 Stimuli
Lesson 3 Plants flower at different times
Lesson 4 Light and temperature affect leaves

Animals

Chapter 1
Cells and Heredity
★Lesson 1 How living things are organized
★Lesson 2 Cell reproduction
★Lesson 3 How many-celled organisms reproduce
★Lesson 4 How traits are inherited
★Lesson 5 Dominant and recessive genes

Ecology

Chapter 2
The Fossil Record
★Lesson 1 What scientists learn from fossils
★Lesson 2 Patterns in the fossil record
Lesson 3 Divisions of geologic time
Lesson 4 Inferring about dinosaurs

Chapter 3
Change Through Time
★Lesson 1 Adaptations
★Lesson 2 How species change
Lesson 3 How species become extinct

Chapter 5
Ecosystems and Biomes
★Lesson 1 Energy flow in a community
★Lesson 2 How materials cycle through ecosystems
Lesson 3 Land biomes
Lesson 4 Water biomes

Physical Science

Matter

Chapter 6
Structure of Matter
★Lesson 1 Describing matter
★Lesson 2 The building blocks of matter
★Lesson 3 Classifying matter

Chapter 7
Investigating Compounds
★Lesson 1 Compounds
★Lesson 2 Chemical reactions
Lesson 3 Acids and bases

Chapter 8
Gravity and Motion
★Lesson 1 What gravity is
★Lesson 2 What motion is
★Lesson 3 First law of motion
★Lesson 4 Second law of motion
★Lesson 5 Third law of motion

Energy

Chapter 9
Electrical Energy
★Lesson 1 What electricity is
★Lesson 2 How electromagnets work
★Lesson 3 How electricity is made and used
Lesson 4 How electronic devices work

Chapter 10
Investigating Light and Sound
★Lesson 1 What light is
★Lesson 2 How light behaves
★Lesson 3 How color is seen
★Lesson 4 What sound is
Lesson 5 How people communicate using light and sound

Earth Science

Earth

Chapter 12
Earth's Moving Plates
Lesson 1 How continents have changed positions
★Lesson 2 How plate tectonics explains movement of continents
★Lesson 3 How plate tectonics explains earthquakes and volcanoes

Chapter 13
Resources and Conservation
Lesson 1 How replacement rates of earth's resources vary
★Lesson 2 Resources provided by air and land
★Lesson 3 Resources provided by water
Lesson 4 How people affect the earth's resources

Weather

Chapter 11
Forecasting Weather
Lesson 1 How forecasts benefit people
Lesson 2 Data needed to make a forecast
★Lesson 3 How data are used to forecast weather
★Lesson 4 How severe weather is predicted

Space

Chapter 14
Exploring Space
★Lesson 1 About the sun
★Lesson 2 Composition of the universe
Lesson 3 How the universe formed
Lesson 4 Exploring space

Human Body

Chapter 15
The Body's Control Systems
Lesson 1 How the body controls life processes
Lesson 2 How the endocrine system is involved in control
★Lesson 3 How messages travel through the body
★Lesson 4 How the body responds to the environment

Chapter 16
Growing Up Healthy
Lesson 1 Stages in the human life cycle
★Lesson 2 Causes of disease
★Lesson 3 Protections from disease
★Lesson 4 How drug abuse can affect health

Note: If time does not permit you to cover all of the material in this book, it is recommended that the lessons with a star be presented as a core program. This core program represents a balanced treatment of topics throughout the series.

Audio-Visual Sources

AIMS Media
6901 Woodley Avenue
Van Nuys, CA 91406

Benchmark Films, Inc.
145 Scarborough Road
Briarcliff Manor, NY 10510

BFA Educational Media
Phoenix/BFA Film and Video, Inc.
468 Park Avenue South
New York, NY 10016

Bullfrog Films, Inc.
Oley, PA 19547

Churchill Films
662 N. Robertson Boulevard
Los Angeles, CA 90069

Coronet Films and Video
108 Wilmot Road
Deerfield, IL 60015

Encyclopaedia Britannica Education Corporation
425 North Michigan Avenue
Chicago, IL 60611

General Educational Media, Inc.
701 Beaver Valley Road
Wilmington, DE 19803

Great Plains Films
P.O. Box 80669
Lincoln, NE 68501

Higgins Productions, Inc.
9100 Sunset Blvd.
Los Angeles, CA 90069

Hubbard Scientific
1946 Raymond
Northbrook, IL 60062

International Film Bureau, Inc.
332 S. Michigan Avenue
Chicago, IL 60604

Jam Handy
School Service, Inc.
2781 E. Grand Boulevard
Detroit, MI 48211

Knowledge Unlimited
P.O. Box 52
Madison, WI 53701

Macmillan
34 MacQuesten Parkway, South
Mt. Vernon, NY 10550

Marshfilm, Inc.
P.O. Box 8082
Shawnee Mission, KS 66208

McGraw-Hill Training Systems
Box 641
674 Via De Lavalle
Delmar, CA 92014

Media, Inc.
P.O. Box 496
Media, PA 19063

NASA Film Library
National Aeronautics and Space Administration
National Audiovisual Center
General Services Administration
Washington, DC 20409

National Educational Media, Inc.
21601 Devonshire
Chatsworth, CA 91311

National Geographic Society
17th and M Streets NW
Washington, D.C. 20036

National Oceanic and Atmospheric Administration
Motion Picture Service
12231 Wilkins Avenue
Rockville, MD 20852

Palisades Wildlife Film Library
1205 S. Ogden Drive
Los Angeles, CA 90019

Roller Skating Rink Operators Association
P.O. Box 81846
Lincoln, NE 68501

Sigma Electronics, Inc.
1184 Enterprise Road
East Petersburg, PA 17520

Stanton Films
2417 Artesia Boulevard
Redondo Beach, CA 90278

Sterling Educational Films
241 East 34th Street
New York, NY 10016

United States Department of Agriculture
Soil Conservation Service
Room 0054-S
P.O. Box 2890
Washington, D.C. 20013

Visual Instruction Productions
112 Brentwood Drive
Stamford, CT 06903

Walt Disney Educational Media Co.
500 S. Buena Vista Street
Burbank, CA 91521

Computer Software Sources

Aquarius People Materials
P.O. Box 128
Indian Rocks Beach, FL 33535

Astronomical Society of the Pacific
1290 24th Avenue
San Francisco, CA 94122

Bohle Company
1901 Avenue of the Stars
Los Angeles, CA 90067

Brain Bank, Inc.
200 Fifth Avenue
New York, NY 10001

CBS Interactive Learning
One Fawcett Place
Greenwich, CT 06836

CC Publications
P.O. Box 23699
Tigard, OR 97005

COMPress
P.O. Box 102
Wentworth, NH 03282

Concept Educational Software
P.O. Box 6184
Allentown, PA 18001

Diversified Educational Enterprises, Inc.
725 Main Street
Lafayette, IN 47901

Educational Activities, Inc.
P.O. Box 392
Freeport, NY 11520

Educational Computing Systems, Inc.
136 Fairbanks Road
Oak Ridge, TN 27830

Educational Materials and Equipment Co.
P.O. Box 2805
Danbury, CT 06813

Focus Media, Inc.
839 Stewart Avenue
Garden City, NY 11530

D. C. Heath
125 Spring Street
Lexington, MA 02173

Holt, Rinehart, & Winston
383 Madison Avenue
New York, NY 10017

HRM
175 Tompkins Avenue
Pleasantville, NY 10570

Ideatech, Inc.
P.O. Box 62451
Sunnyvale, CA 94088

J & S Software
140 Reid Avenue
Port Washington, NY 11050

Micrograms, Inc.
P.O. Box 2146
Loves Park, IL 61130

Mindscape Inc.
3444 Dundee Road
Northbrook, Illinois 60062

Minnesota Educational Computing Consortium
2520 Broadway Drive
St. Paul, MN 55113-5199

Orange Cherry Media
7 Delano Drive
Bedford, NY 10507

Queue
5 Chapel Hill Drive
Fairfield, CT 06432

Right-On Programs
P.O. Box 977
Huntington, NY 11743

Scholastic Software
730 Broadway
New York, NY 10003

Science Research Associates
155 Wacker Drive
Chicago, IL 60606

Scott, Foresman and Co.
1900 East Lake Avenue
Glenview, Illinois 60025

Society for Visual Education
1345 Diversey Parkway
Chicago, IL 60614

Spectrum Software
75 Todd Pond Road
Lincoln, MA 01773

Sunburst Communications, Inc.
39 Washington Avenue
Pleasantville, NY 10570

Science Equipment Suppliers

Catalogs from these suppliers will provide you with information about availability, costs, and alternative equipment or supplies.

Carolina Biological Supply Co.
2700 York Rd.
Burlington, NC 27215

Central Scientific Co. (CENCO)
11222 Melrose Ave.
Franklin Park, IL 60131

Fisher Scientific Co.
Educational Materials Division
4901 W. LeMoyne St.
Chicago, IL 60651

Frey Scientific Co.
905 Hickory Ln.
Mansfield, OH 44905

Nasco
901 Janesville Ave.
Fort Atkinson, WI 53538

Sargent-Welch Scientific Co.
7300 N. Linder Ave.
Skokie, IL 60077

Science Kit and Boreal Labs
777 E. Park Dr.
Tonawanda, NY 14150

Ward's Natural Science Establishment, Inc.
5100 West Henrietta Rd.
P.O. Box 92912
Rochester, NY 14692-9012